FRAME

The Back Issue

THE ESSENTIAL GUIDE TO FRAME'S FIRST 50 ISSUES

Frame Publishers
Amsterdam
—
Birkhäuser —
Publishers for
Architecture
Basel·Boston·Berlin
—

Find Back

CONTENTS

Our Issue

HOW TO FRAME A MAGAZINE?

Having published 50 issues of Frame, we thought it would be a nice idea to celebrate this special jubilee with a compilation of the best articles featured in nearly a decade of devotion to design. It wasn't easy to make a selection. Which articles have stood the test of time and deserve to be commemorated? And which can better remain hidden between the covers of a magazine?

The idea persisted, and – pretentious though it may be – you're holding the result of our celebratory efforts in your hands. We abandoned the notion of a 50-kilogram tome early on, opting instead for a chic handbook small enough to carry around and interesting enough to keep, cherish and use. Frame enthusiasts curious about the magazine's origins will find information at the back of the book on how it all got started and how we turned our ambitions into reality.

We hope you enjoy Frame: The Back Issue and that you continue to buy and enjoy Frame magazine. One hundred issues – here we come!

Peter Huiberts,
publisher

—

Robert Thiemann,
editor

—

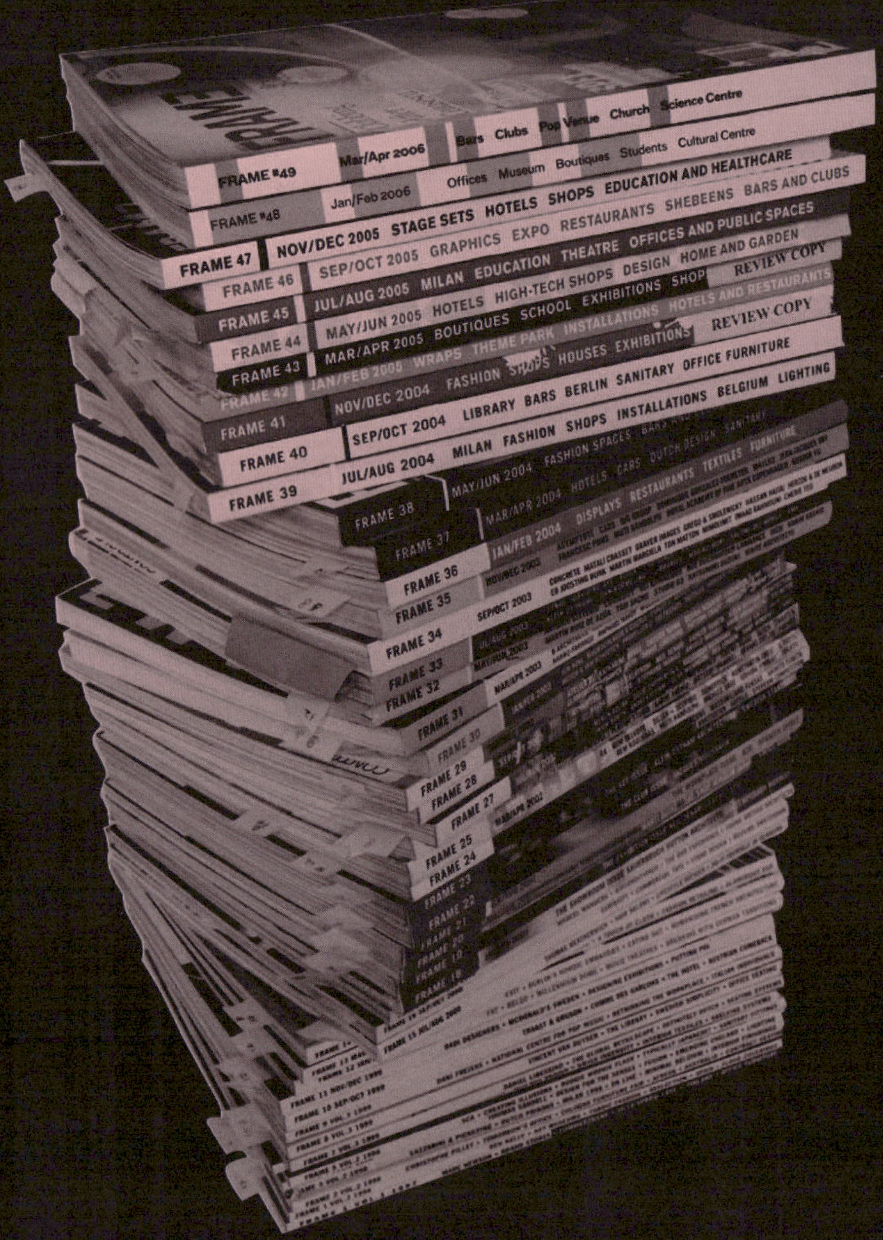

FRAME #49 Mar/Apr 2006 Bars Clubs Pop Venue Church Science Centre
FRAME #48 Jan/Feb 2006 Offices Museum Boutiques Students Cultural Centre
FRAME 47 NOV/DEC 2005 STAGE SETS HOTELS SHOPS EDUCATION AND HEALTHCARE
FRAME 46 SEP/OCT 2005 GRAPHICS EXPO RESTAURANTS SHEBEENS BARS AND CLUBS
FRAME 45 JUL/AUG 2005 MILAN EDUCATION THEATRE OFFICES AND PUBLIC SPACES
FRAME 44 MAY/JUN 2005 HOTELS HIGH-TECH SHOPS DESIGN HOME AND GARDEN
REVIEW COPY
FRAME 43 MAR/APR 2005 BOUTIQUES SCHOOL EXHIBITIONS SHOP
REVIEW COPY
FRAME 42 NOV/DEC 2004 FASHION SHOPS HOUSES EXHIBITIONS
FRAME 41 SEP/OCT 2004 LIBRARY BARS BERLIN SANITARY OFFICE FURNITURE
FRAME 40 JUL/AUG 2004 MILAN FASHION SHOPS INSTALLATIONS BELGIUM LIGHTING
FRAME 39
FRAME 38 MAY/JUN 2004
FRAME 37 MAR/APR 2004 DISPLAYS
FRAME 36 JAN/FEB 2004
FRAME 35
FRAME 34 SEP/OCT 2003
FRAME 33
FRAME 32
FRAME 31
FRAME 30
FRAME 29
FRAME 28
FRAME 27
FRAME 25
FRAME 24
FRAME 23
FRAME 22

FRONT PAGES
009-536

Read Back

COVERS & FEATURES

Welcome to the Hall of Frame: 50 issues, 50 covers, 50 features. Learn all about the key articles that helped shape a magazine. Do the featured projects still exist? What has become of the people who designed them? Look, read and enjoy.

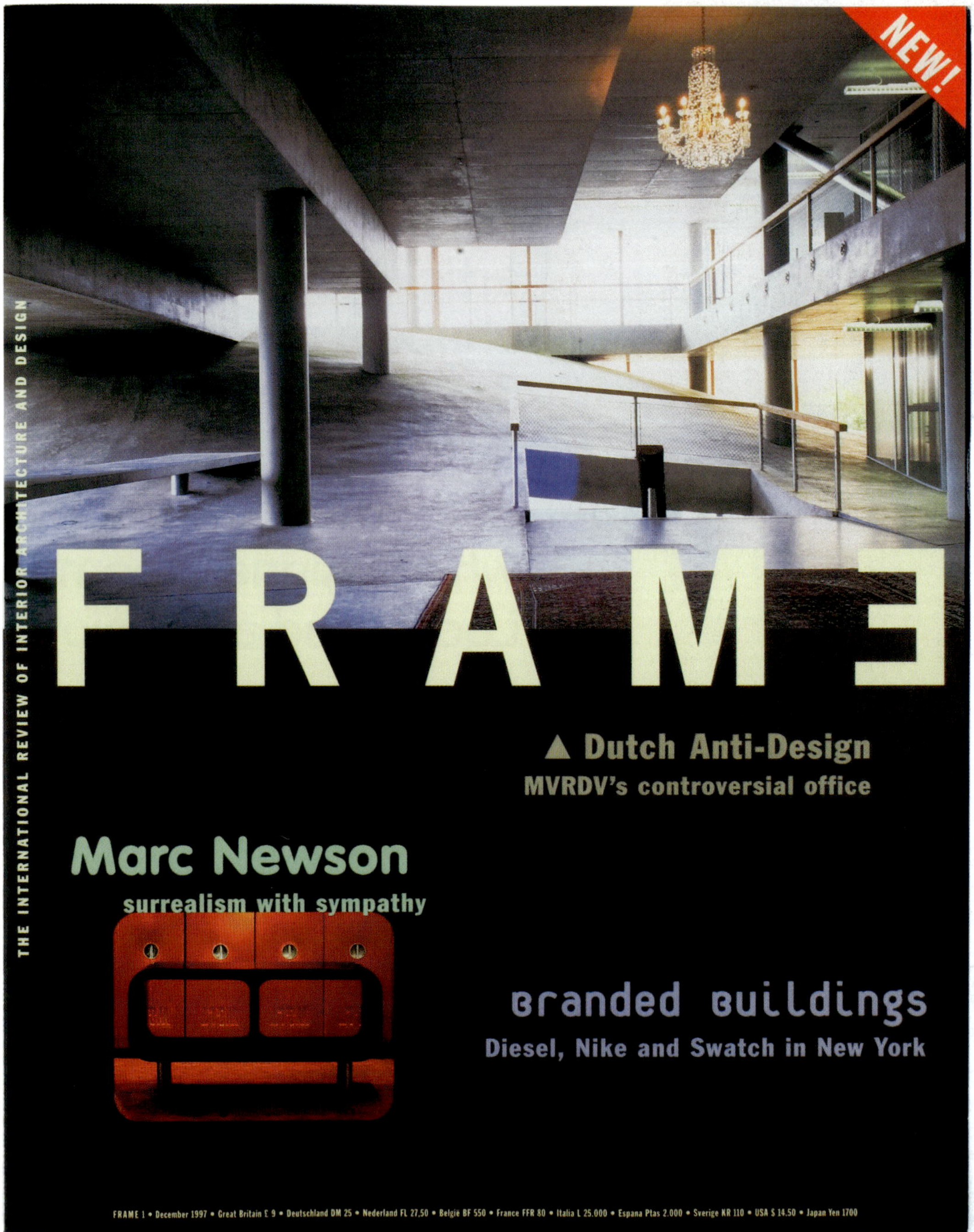

ART DIRECTOR
Ton Homburg,
Opera Designers

00

Brand New World

Designers Daniel Weil and James Biber about the project

Timely. Late 20th century. Still proud of it.

Comments on the published article

Very nice to see drawings. As I recall, one of the first attempts to bring together the new brand-store phenomenon of the early '90s.

Did the article have an effect?

It made us feel great! And very proud to be part of your courageous first issue.

DESIGNERS
Gordon Thompson III (Nike)
www.nike.com
Daniel Weil and James Biber (Pentagram)
www.pentagram.com
Wieneke van Geemeren (Diesel)
www.diesel.com

SUBJECT
Niketown
Swatch Timeship
Diesel megastore

LOCATION
New York, USA

PHOTOGRAPHER
—

WRITER
Janet L. Rumble

Shop

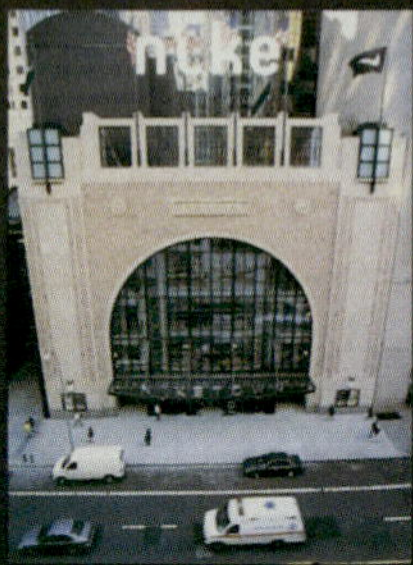

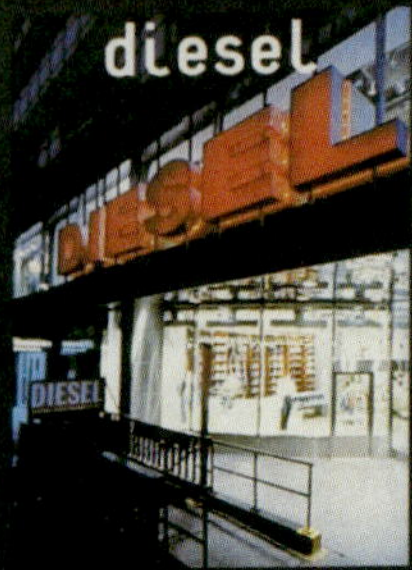

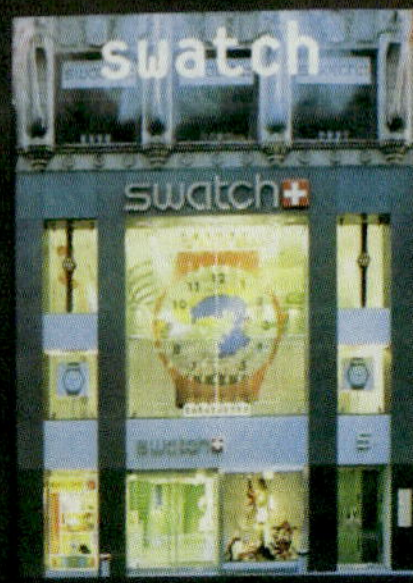

brand new world

Niketown, the Swatch Timeship, and Diesel's megastore are among the most visible landmarks in New York City's fertile retail 'brandscape'.

by Janet L. Rumble

Make no mistake: Niketown New York is a temple. But to what exactly? With nearly a thousand styles of sneakers for sale, scores of interactive displays, and a series of inspirational films, fans see a highly diverting tribute to the noble values of good sportsmanship. Opponents, however, think this looming, five-story retail megalopolis on posh 57th Street is little more than a powerful multinational adrift on an expensive ego trip. But whether you like it or you don't, you must admit the place has presence, and this, more than anything, is what successful brand retailing has become. Niketown is by no means the only megastore to have elbowed its way into the marketing mecca of Midtown Manhattan; the new Swatch Timeship, just across street, and Diesel's hip headquarters around the corner are two other able players in this hot corporate game known as raising the flagship.

Time was, you could build a better mousetrap and then sit back and wait for the world to line up at your doorstep. But no more. Even the most ingenious and dependable product is not a guaranteed success because in today's glutted marketplace, function alone no longer stands as an adequate differential. Now the makers of everything from potato chips to computer chips are flexing their marketing muscle through branding in order to distinguish themselves from competitors. And the marketplace isn't the only thing that's changing - so are the public's attitudes toward the things they buy. Today's consumer wants to be brand loyal. Faced with so many choices, shoppers are anxious to feel secure in their purchases, and branding, by conveying a company's commitment to its wares, plays a key role in providing that security.

And yet, there's a lot more to it than coming up with a fetching logo - branding is a promise of quality that customers can rely on, and ultimately, the only way to make good on this promise is by delivering sound merchandise. Advertising, no matter how clever, only goes so far. However, couple an innovative and visible marketing campaign with a winning product line and you've got solid gold. Just ask Nike, Swatch, and Diesel; although based in the U.S., Switzerland, and Italy, respectively, these brands are advertised and distributed worldwide. Within a short time, Nike and Swatch have become global icons; Diesel, a newer company, is well on its way to ultimate fame. That said, why the need at this point for an actual physical existence in the form of a retail outlet?

Futuristic armature Building Niketown was a way to have more control over the company's image, says Gordon Thompson III, vice president of design for the Portland, Oregon-based sportswear manufacturer. "The point was to expose the brand beyond a 30-second TV spot or a page in a magazine to explain what we're all about." For the most part, Thompson means having the room to show off Nike's many inventive products. But he also admits that a 66,500-square-foot billboard like Niketown New York enables the company to better manage - and soften - a reputation made shaky by reports of abused workers overseas. The message that Nike prefers to disseminate is one of an exacting design standard, and according to Thompson, this is what determines the architectural concept behind Niketown New York. Conceived as a building within a building, the outer shell takes its cues from an old school gymnasium a few blocks away, complete with the words "Courage, Commitment, Teamwork, and Honor" inscribed into the stone pediment above the Palladian window. Inside, a gleaming, futuristic armature showcasing five floors of merchandise huddles around a soaring central atrium. "The idea is that it's a ship in a bottle," Thompson explains, a metaphor for "how we represent innovation in the world of sports."

Hopefully this notion resonates with Nike's target consumers down to their competitive core, fitness enthusiasts "of all walks of life,"

Thompson describes, "from the most elite to the weekend warrior." Such a mind-set transcends age, culture, and gender. And yet, most of those browsing around on a recent summer afternoon looked more apathetic than athletic. Sure, plenty of sweatsuits were in evidence, but they tended to be sported by the middle-aged from the Midwest, something to keep them comfortable as they methodically wend their way from the Warner Bros. Studio store across the street to the Harley Davidson Café around the corner.

Participatory exhibits The truth is, people generally venture to Niketown just to do it. Along with places like the Virgin Records and Diesel megastores, the Disney, Levi's, and Swatch outposts, the Hard Rock and Fashion cafés, Niketown is competing with MoMA, the Met, and other high cultural temples for the public's attention - as it happens, Niketown Chicago is already that city's leading tourist destination. (Niketowns can also be found in Portland, Costa Mesa, Atlanta, Los Angeles, Seattle, San Francisco, Boston, and Honolulu; coming soon, London and Berlin.) Shopping, while it does occur, is not the point. "The amount of space geared toward merchandise is minimal compared to the exhibits," one man tells me; I found him leaning against the turnstiles near the entrance with his wife. Up from Miami on holiday, the couple is waiting for a friend who'd stopped to purchase a t-shirt somewhere up above. "A t-shirt?," I ask, incredulous. As if he couldn't come by one of those at home? Of course he could, but it wouldn't be a Niketown t-shirt. There's a difference. One is roughly two yards of cotton fabric; the other, proof of a pilgrimage.

But before I can delve into all of this with my Floridian friends, the atrium darkens, a hush falls, and everyone gathers around us. Twice an hour, a 36-by-22-foot screen descends, showing one of five three-minute sports documentaries to the captive audience. This one's about a Brazilian soccer team, but I'm not really paying attention. I'm thinking about the whole notion of interactivity. Thompson and his colleagues went to great trouble to devise a number of "participatory" exhibits intended to keep customers entertained, such as the one highlighting Nike Air technology on the third floor, or the regulation punching bag set up on the fourth. But can you really call denting a plastic air sac with your finger or taking a jab at the bag "interactive" experiences? There's far more substance to NGage, Nike's "digital footwear sizing system" - sort of a full-body Brannock Device that issues a computer print-out - but every single one I hunted out was broken.

The film has ended, the lights are back up, and the couple's friend resurfaces. "Look," he says to them excitedly, digging into his shopping bag, "the receipt is in its own little envelope!" It may seem like a small detail, but when experienced with all the rest - the New York Marathon route spelled out in terrazzo, the impeccable stainless swooshes on the railings, the curious in-house soundtrack with its echoes of sneakers screeching on gym floors and tennis balls connecting with catgut - Niketown New York successfully emerges as the multimedia, three-dimensional advertisement that it is. And if some merchandise happens to move during the course of a day, well that's fine too.

Space-age Across the street at the Timeship, the Swiss watchmaker Swatch has set out to build an environment evocative of its singular mode of timekeeping: "Equal parts emotion and accuracy," explains architect and industrial designer Daniel Weil, head of the Pentagram's London office. Along with architect James Biber of Pentagram New York, Weil managed to turn the narrow, three-story space, once a swank ▶

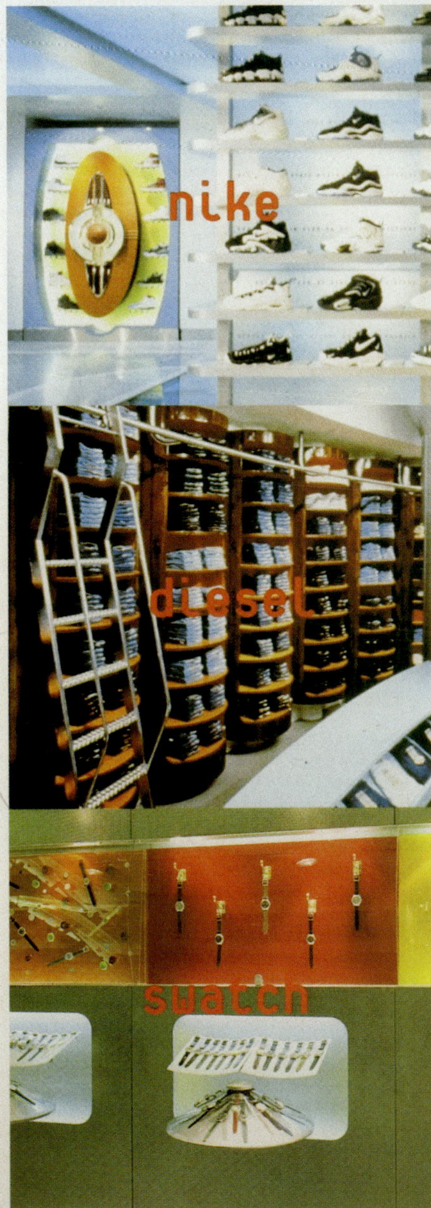

SUBJECT Niketown, Swatch Timeship, Diesel megastore
LOCATION New York, USA
DESIGNERS Various, see page 013
PHOTOGRAPHER —
WRITER Janet L. Rumble

Chanel boutique, into a playful and exceedingly people-friendly shop. "Nothing renders time more accurately than movement," Weil declares. So true to the analog model of seconds, minutes, and hours, "each floor of the Timeship is an expression of a certain speed," he says, to appeal to a broad range of consumers. Those well-versed in all things Swatch, for example, typically enjoy the very casually paced top floor, which features a gallery of memorable styles and monitors showing ad campaigns from the good ol' days. There's also a bar where you can order not just beverages, but actual products - you can sidle up to the counter, slam your fist down, and say to the barkeep, "I'd like the red one from the Irony series and a coffee chaser, please." While you sit back and sip away, your new watch is delivered from the stock room to the bar via a space-age pneumatic tube.

Spinning saucers Things on the mezzanine are a little more edgy, but no less cliquey. There are lots of references to something known as the Swatch Club (members receive notice of special events and product editions), and a place to purchase new batteries, straps, and other accessories. But the ground floor is the definitive second-hand, "geared toward a fast-paced retail experience," explains Weil. It features an innovative series of waist-level, rotating pedestals onto which dozens of Swatch watches are attached like spokes. It's an ingenious solution to the crux of mass selling, because it enables visitors to hold the watches directly up to their own wrists, but not pocket them. In most places where watches are sold, you have to wait around for a salesperson to patiently transport each and every timepiece that catches your eye from display case to countertop for closer examination. The pedestals were "inspired by vertical postcard towers and twirling costume jewellery displays," Weil says, and like these efficient devices, they make shopping more fun.

It's over one of these spinning saucers that I become acquainted with Laura Hoyer, a 26-year-old graphic designer who, although clutching a natty little Swatch shopping bag in one hand, is already contemplating her next purchase. As it turns out, the one she just bought brings her own personal Swatch collection, begun "way back in 1984," to 50. She says she prefers the Timeship - which, by the way, is the only store of its kind - to the small parcels of real estate Swatch stakes out in department stores because the selection here is vast, and the "fun and futuristic design better suits the product."

While much of the Timeship's festive flavor is meant to reach out to loyal fans like Hoyer, its design really serves to remind all consumers that a Swatch watch is far more than just a pretty clock face - it is a technological wonder that manages to produce precise quartz movement with about half the components of most watches, a clever simplification that enables the company to manufacture them inexpensively. Unfortunately, it also spawned a host of two-bit imitators from which Swatch would very much like to distance itself, and what better way to spread the word than with 5,000 square feet of mega-retail space in the heart of New York City? You couldn't reach more people if you shouted from the top of the Empire State Building.

The Manhattan megastores of Nike, Swatch and Diesel: the future of retail

Fashion stylists The strategy at Diesel seems to be more along the lines of saturation bombing, as its Manhattan flagship is one of ten such superstores in five countries. Like the sundry Niketowns, however, each one has a distinctive personality, says Wieneke van Gemeren, the Italian clothing company's director of interior design. "There is a basic concept," she explains, but like the timely medium of fashion itself, "it's constantly developing. Essentially, all the stores present the elements of the 'Diesel lifestyle' in an atmosphere that's new, refreshing, but sometimes perhaps confusing."

Diesel specializes in sporty clothes on the cutting edge of cool. Think jeans, flash nylon jackets festooned with reflective strips, and wraparound sunglasses - the kind of thing favored by the skate/snowboard set. To be honest, I was bracing myself for an obnoxious experience fraught with sensory overload and pushy poseurs. What I found was a laid back, unfussy environment. It's kind of a curious space, because at 10,000 square feet, the second floor is four times the size of the first. There's really not much to do on the ground floor except hop on the funky escalator that jettisons you to the mezzanine. There, you can chat with the full-time deejay as he spins some groovy tunes, play a game or two of Ms. Pac-Man, sit and listen to new CDs via headphones, or like me, laze around on a comfy couch in the café/lounge area reading fashion magazines. Oh, and you can shop, too. Interestingly, neither Van Gemeren nor any of her colleagues has a background in architecture or interior design; they regard themselves as fashion stylists, and the Diesel store interiors are where they both explore and create their own visual language. In New York, this consists of an inspired mix of the mechanical and the whimsical. Pre-cast concrete and galvanized metal elements - like the rotating washing machine drums at street level - are interspersed with such Sixties Modern icons as Saarinen chairs and framed Marimekko-type fabric panels. These, along with the video games, give the place a relaxed rec-room feel that's meant to appeal to those who, like myself, grew up in the late Sixties and early Seventies.

Puzzle Although quite different in size, layout, and tenor, all three superstores embody the future of retail. And while there does seem to be a tried-and-true formula for success, one that combines state-of-the-art design, up-to-the-minute marketing tactics, and an extra helping of real estate, clearly there is no pat recipe. What emerges as the essential ingredient in this hodgepodge is rather abstract - the experience of place. Apparently people are hungry for it. They flock to these stores to obtain the last and largest piece of a puzzle they've held in mind for some time now, an incomplete portrait cobbled together from actual products and images of products. It's like finally visiting someone's home for the first time: here, at last, you get to see how they really live, what they read, what they keep in the medicine cabinet. Yes, yes, I know - these are huge corporations, not friends. But what makes Niketown New York, the Swatch Timeship, and the Diesel megastore compelling destinations is that which sets them - and for that matter, people - apart: their uniqueness, their character, and their integrity. ▶

*Every inch of the 66,500-square-foot Niketown New York beckons consumers to **Just Buy It***

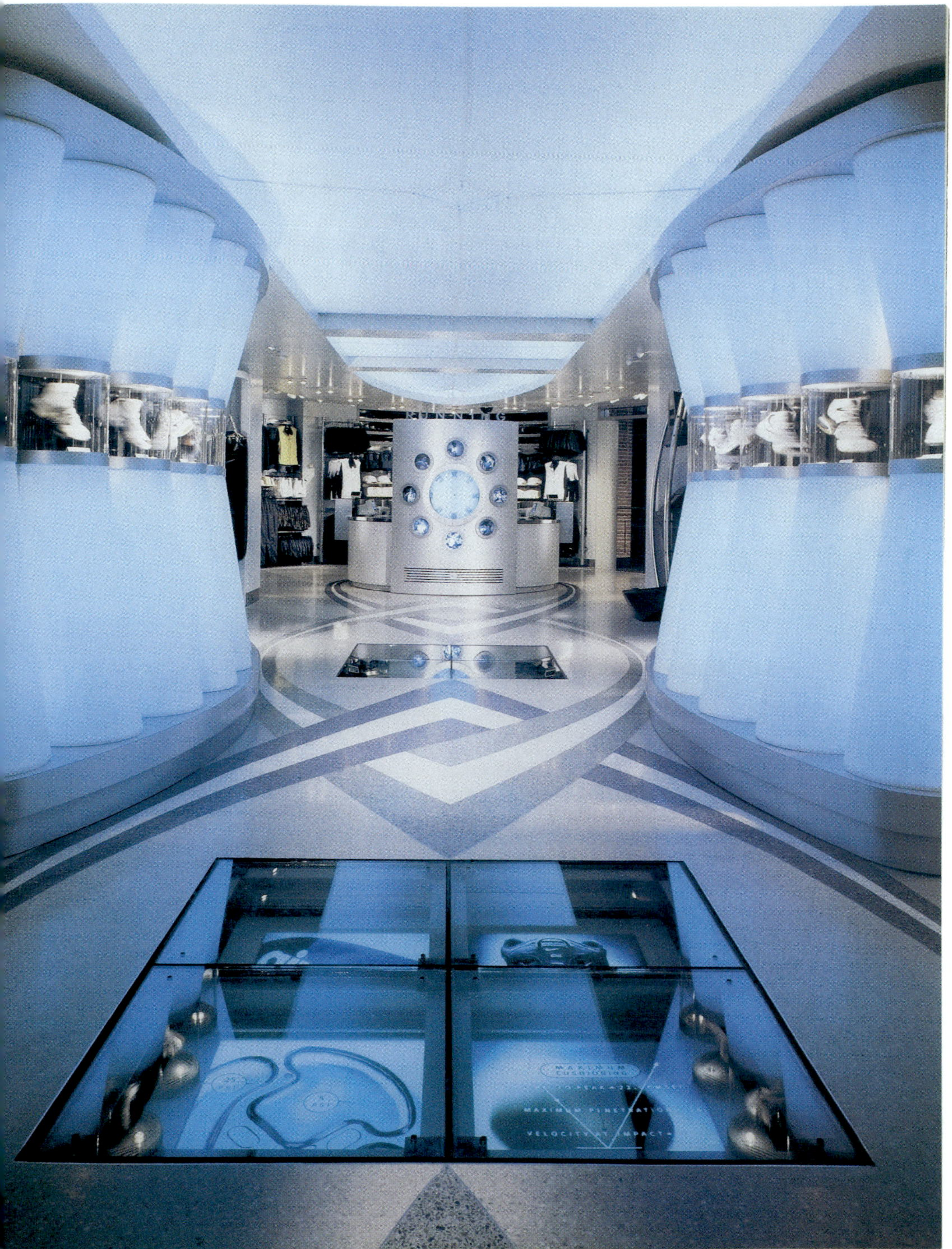

SUBJECT
Niketown, Swatch Timeship, Diesel megastore

LOCATION
New York, USA

DESIGNERS
Various, see page 013

PHOTOGRAPHER
—

WRITER
Janet L. Rumble

Niketown New York succesfully emerges as a multimedia, three-dimensional advertisement

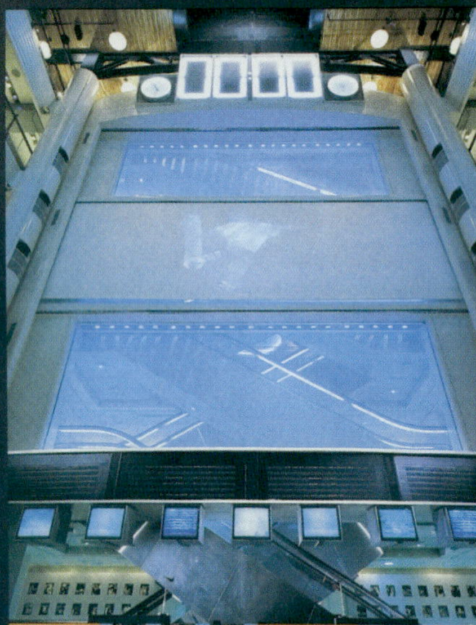

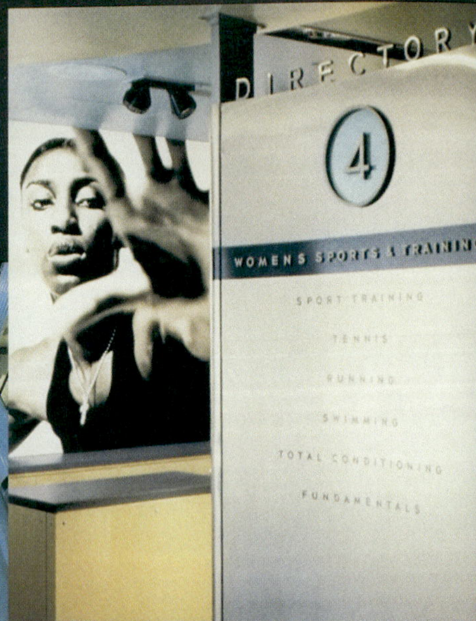

Interactive displays meant to keep shoppers busy can be found on each of the five floors. Every half hour, a huge, luminous screen drops into the main space and onlookers are treated to a mini sports documentary film

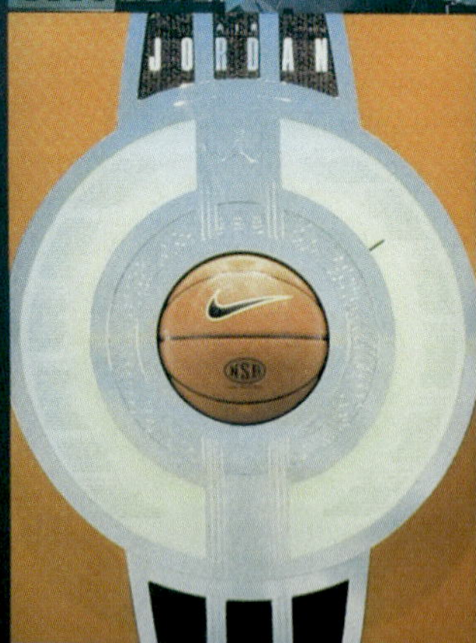

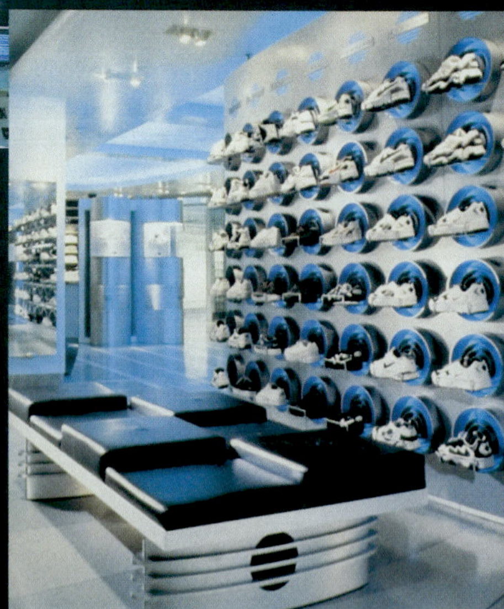

First floor display plan

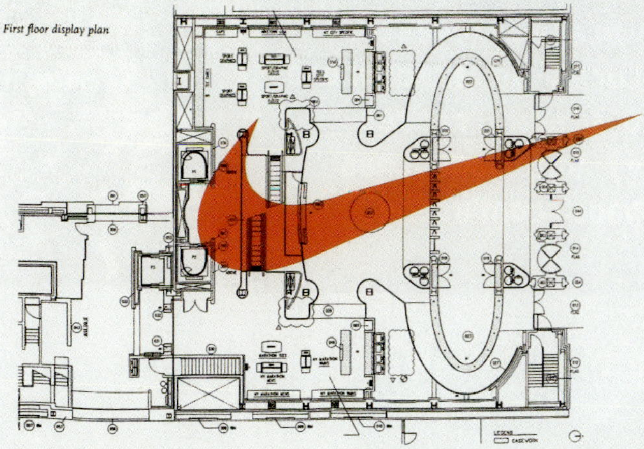

Fifth floor display plan

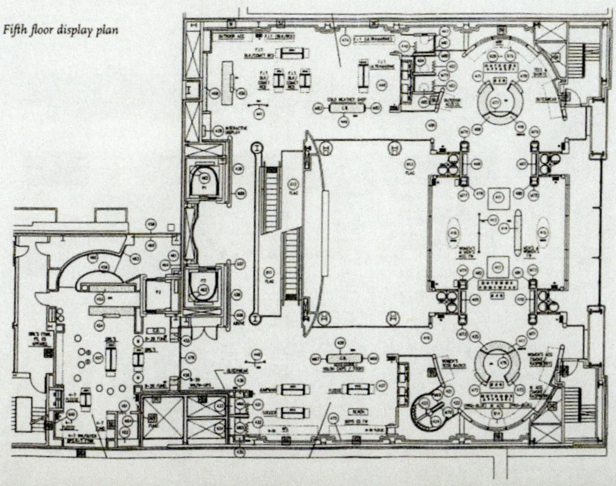

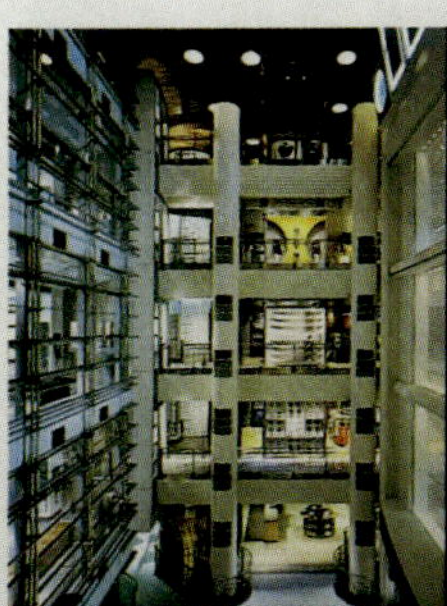

From the outside, Niketown New York resembles an old school gymnasium, but the interior features futuristic displays of merchandise huddled around a soaring central atrium

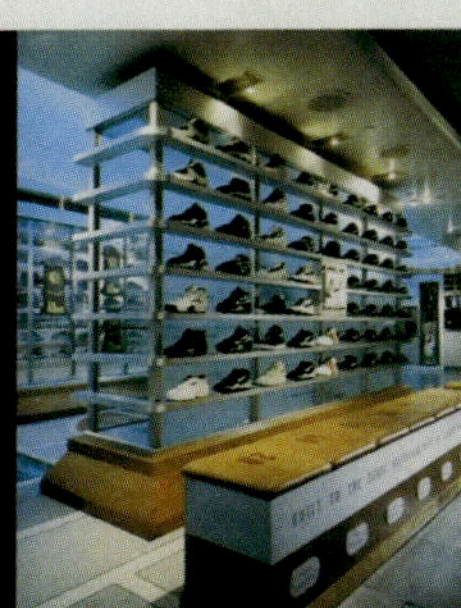

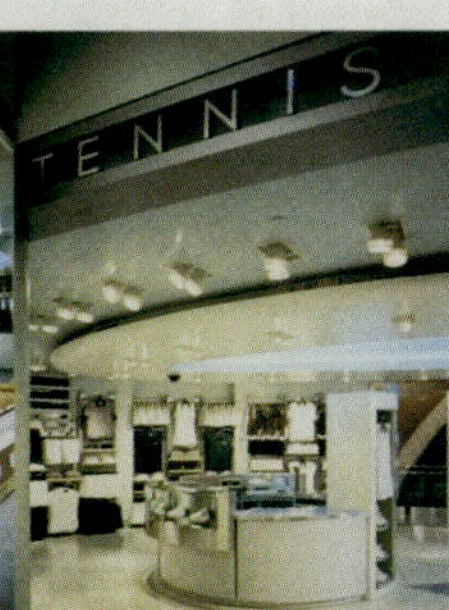

SUBJECT
Niketown, Swatch Timeship, Diesel megastore

LOCATION
New York, USA

DESIGNERS
Various, see page 013

PHOTOGRAPHER
—

WRITER
Janet L. Rumble

The Diesel megastore features an atmosphere that's new, refreshing, but sometimes perhaps confusing

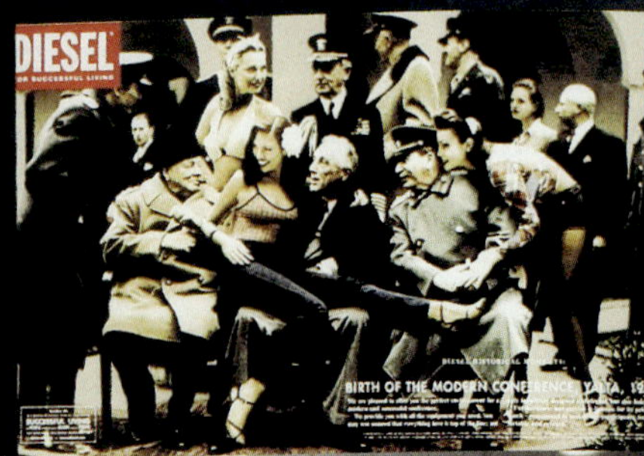

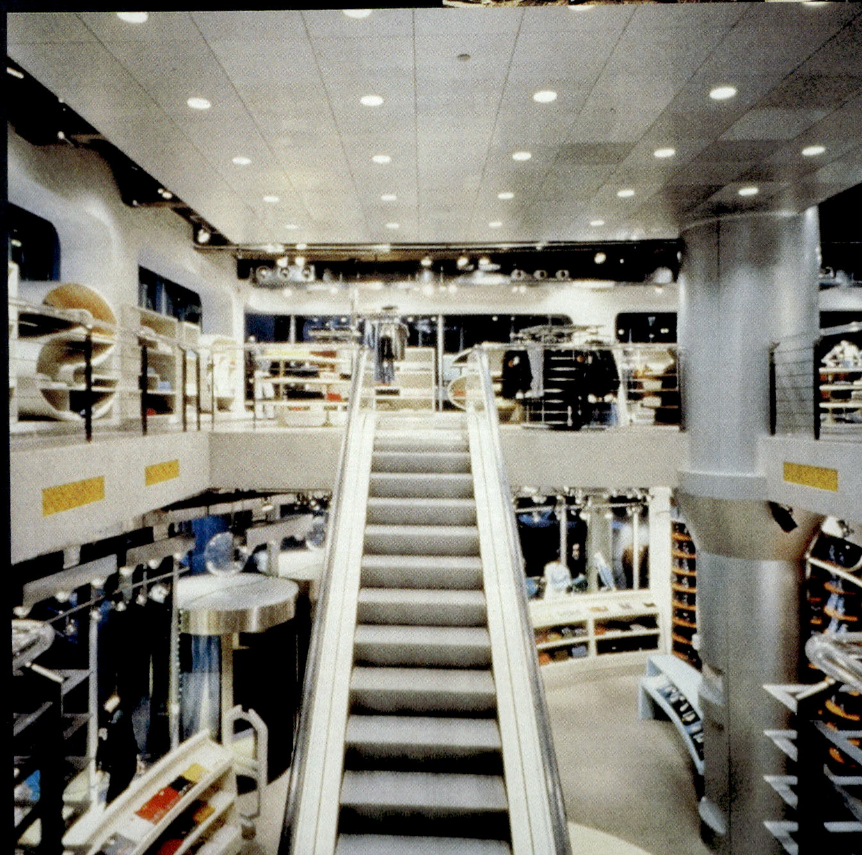

The 'cool' quotient we've come to recognize in the Diesel clothing line as well as in its advertising campaigns has been transformed into a three-dimensional tableau at the Manhattan megastore

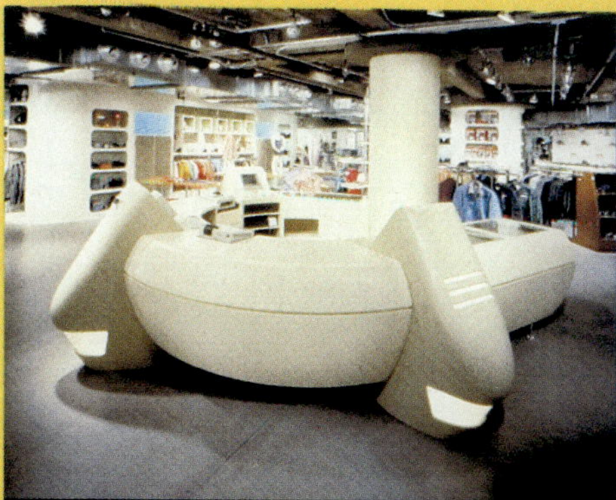

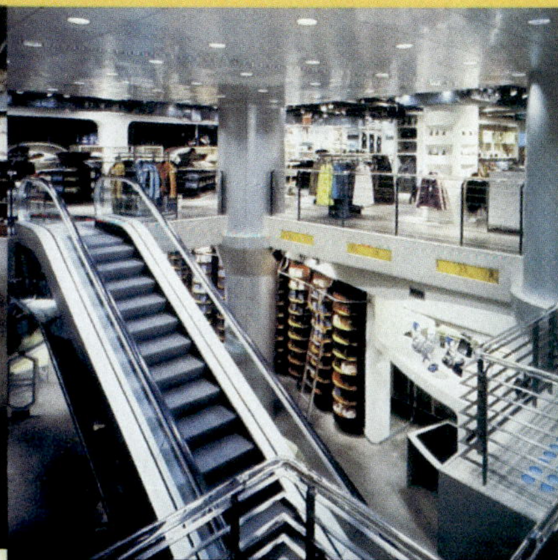

Throughout the 10,000-square-foot, two-level space, pre-cast concrete and galvanized metal elements are mixed with Sixties Modern furnishings and Seventies video games. This relaxed rec-room feel seems to appeal to style-savvy Generation-Xers

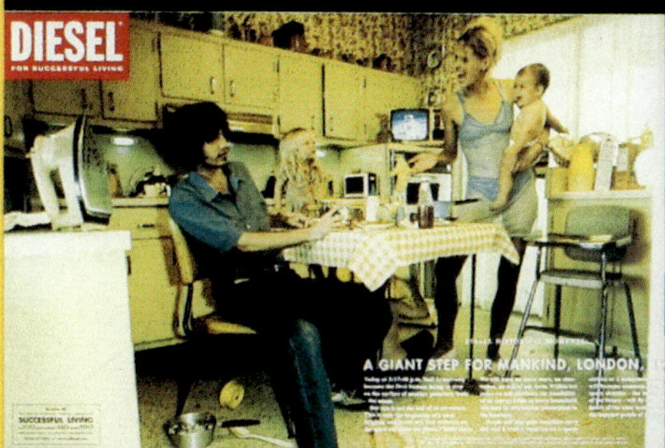

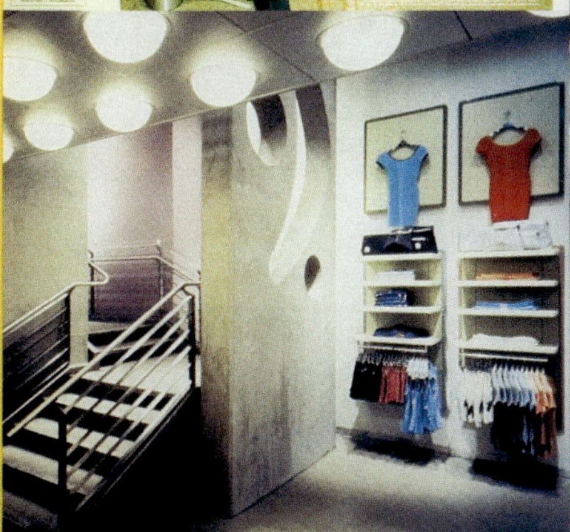

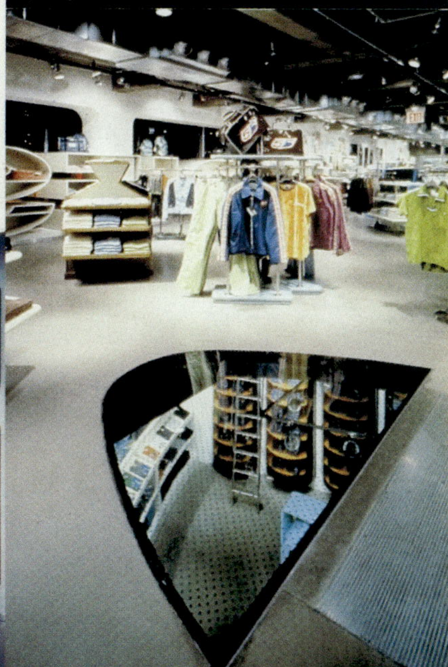

SUBJECT
Niketown, Swatch Timeship, Diesel megastore

LOCATION
New York, USA

DESIGNERS
Various, see page 013

PHOTOGRAPHER
—

WRITER
Janet L. Rumble

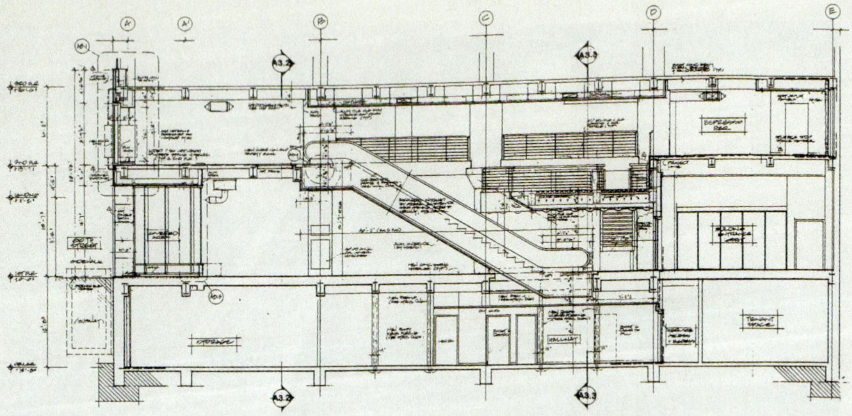

Section

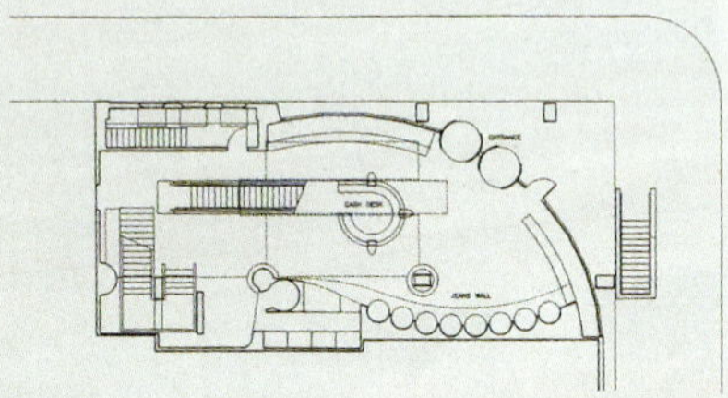

Ground floor plan

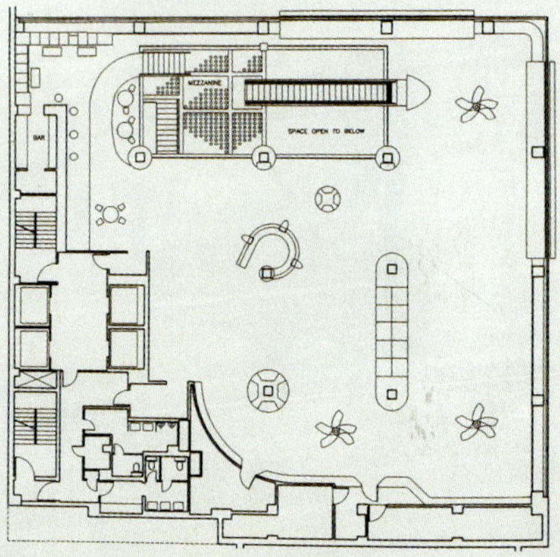

First floor plan

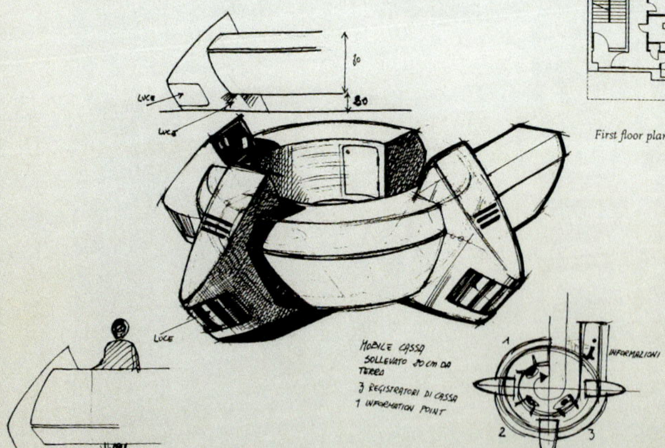

Cash desk

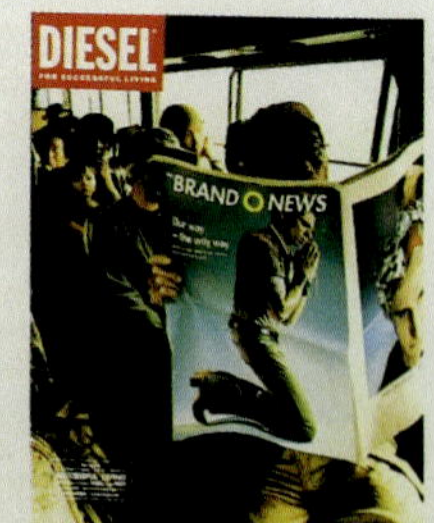

The Timeship reminds consumers that a Swatch watch is far more than just a pretty clock face

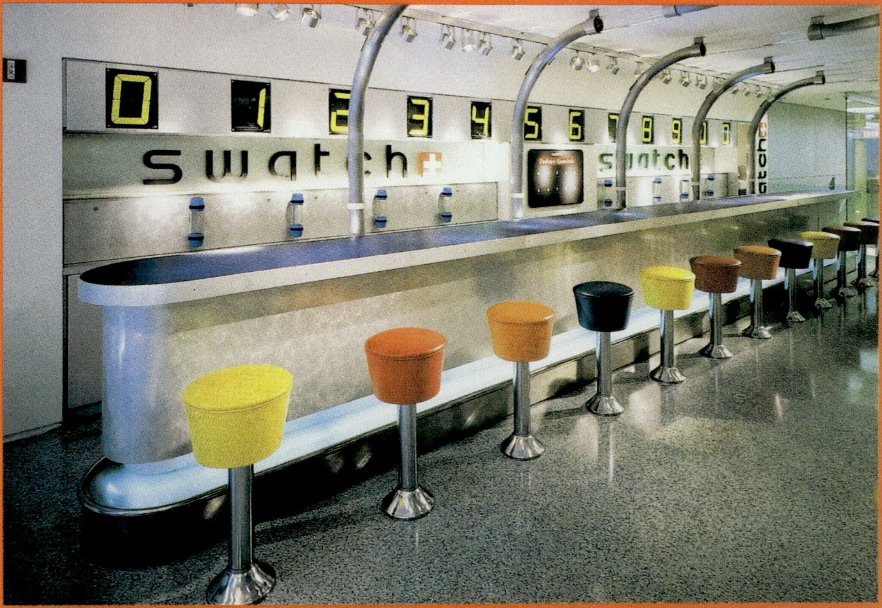

Swatch's forward-thinking fantasy land features such space-age innovations as a 'product' bar that delivers goods via pneumatic tubes and a series of rotating display units reminiscent of flying saucers

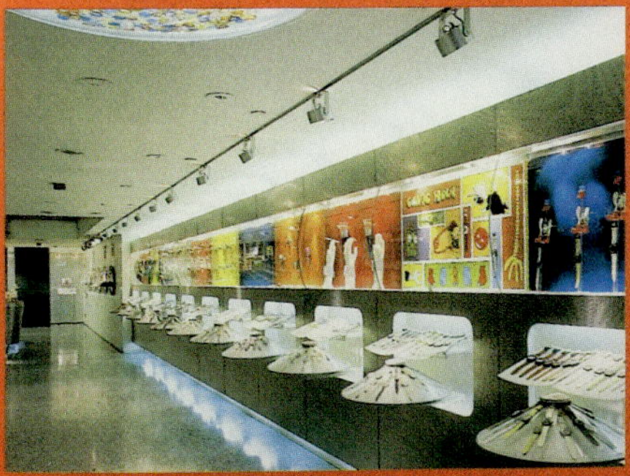

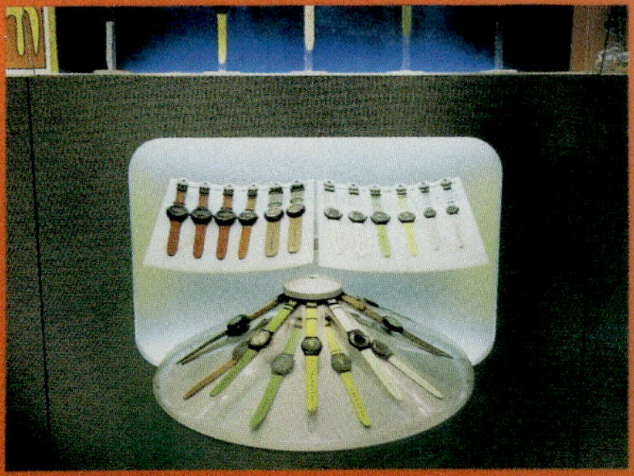

SUBJECT
Niketown, Swatch Timeship, Diesel megastore

LOCATION
New York, USA

DESIGNERS
Various, see page 013

PHOTOGRAPHER
—

WRITER
Janet L. Rumble

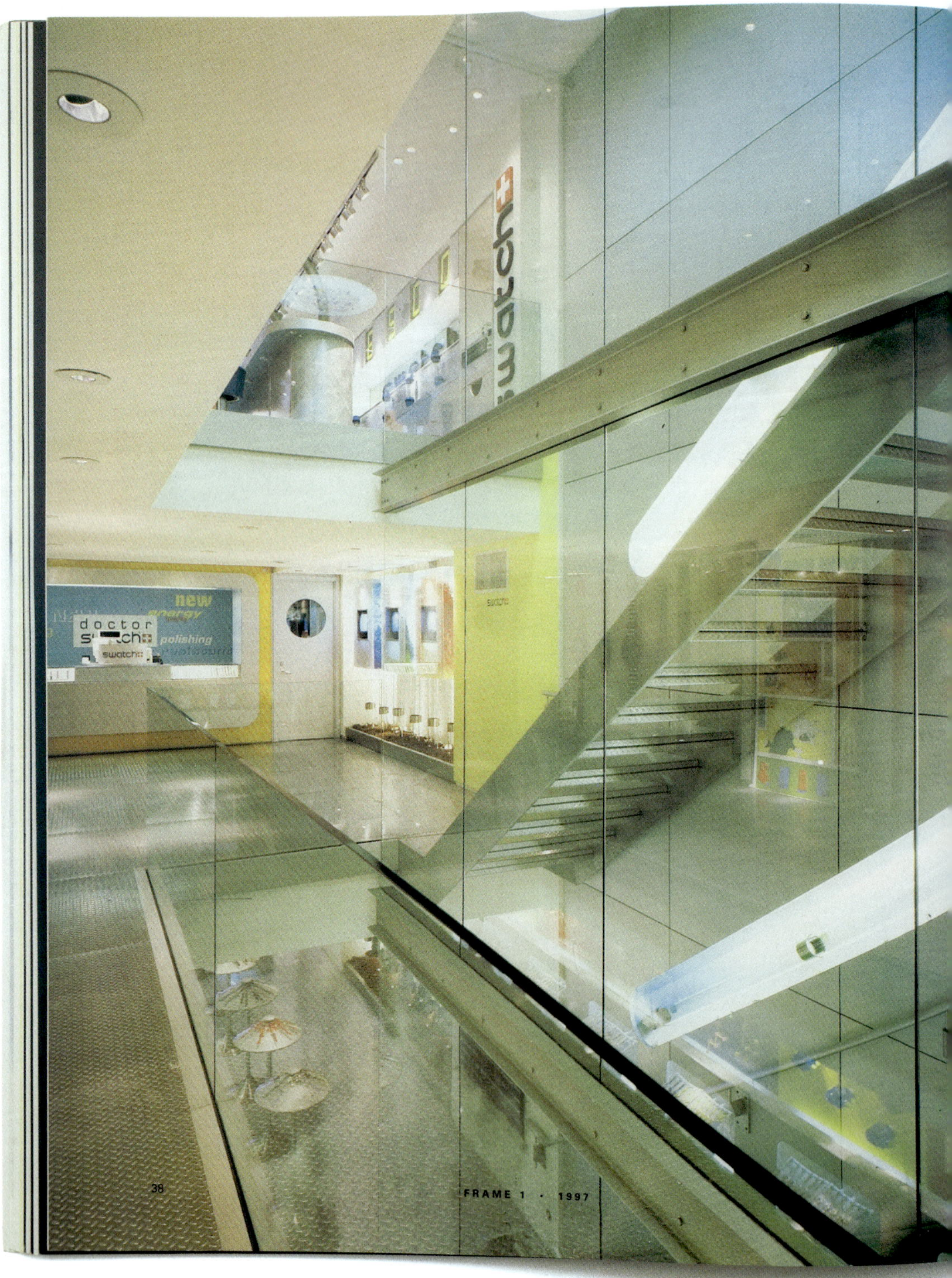
swatch
doctor
new
energy
polishing
38
FRAME 1 · 1997

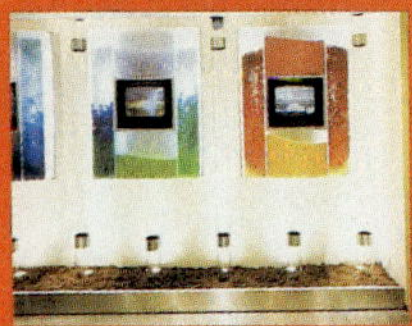

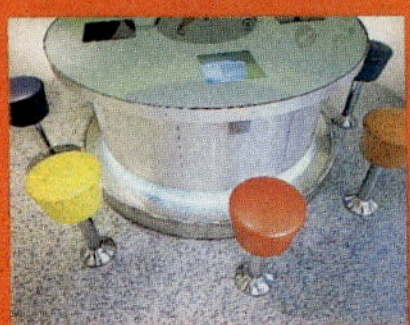

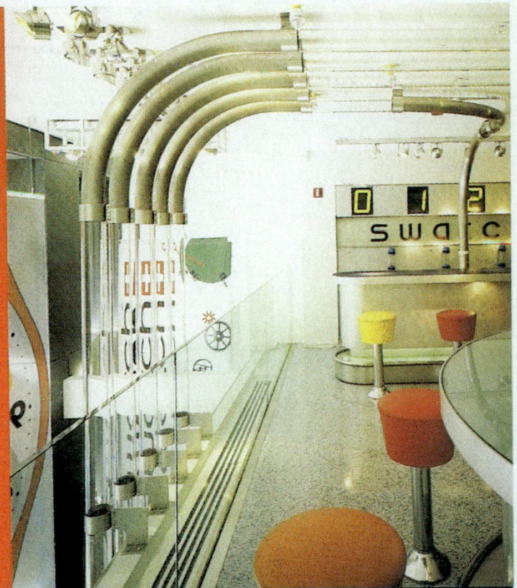

Using the analog timekeeping model of seconds, minutes, and hours, each floor of the Timeship was designed to be an expression of a different speed

Orthogonal projection

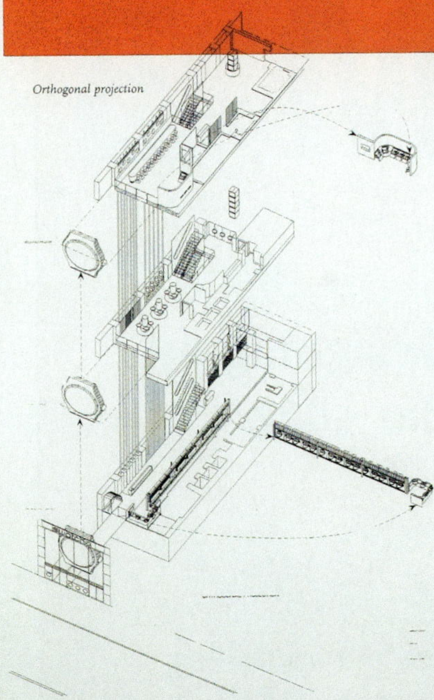

The narrow, three-story space, once an exclusive Chanel boutique, is now a playful and exceedingly people-friendly place

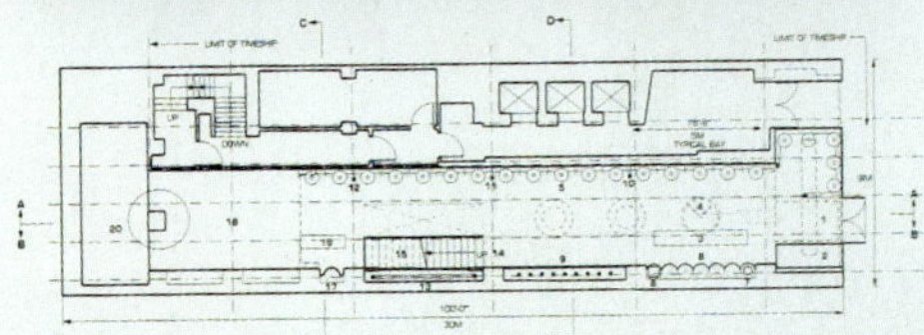

Ground floor plan

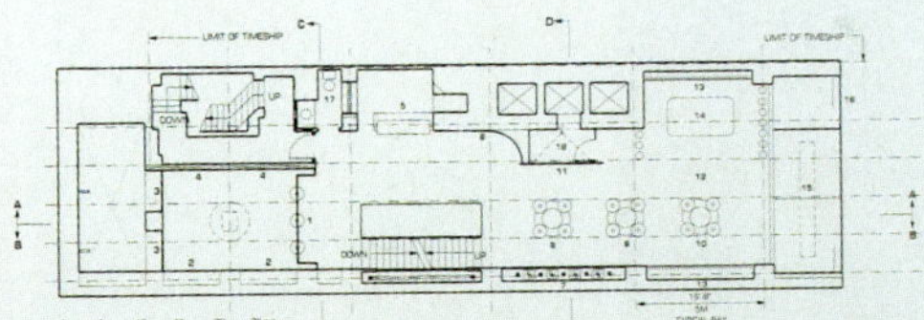

First floor plan

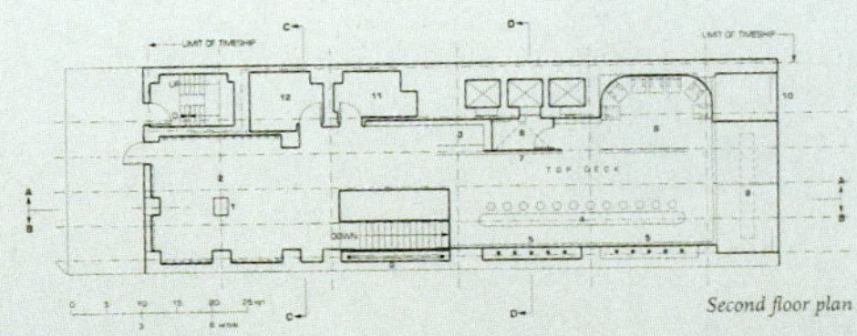

Second floor plan

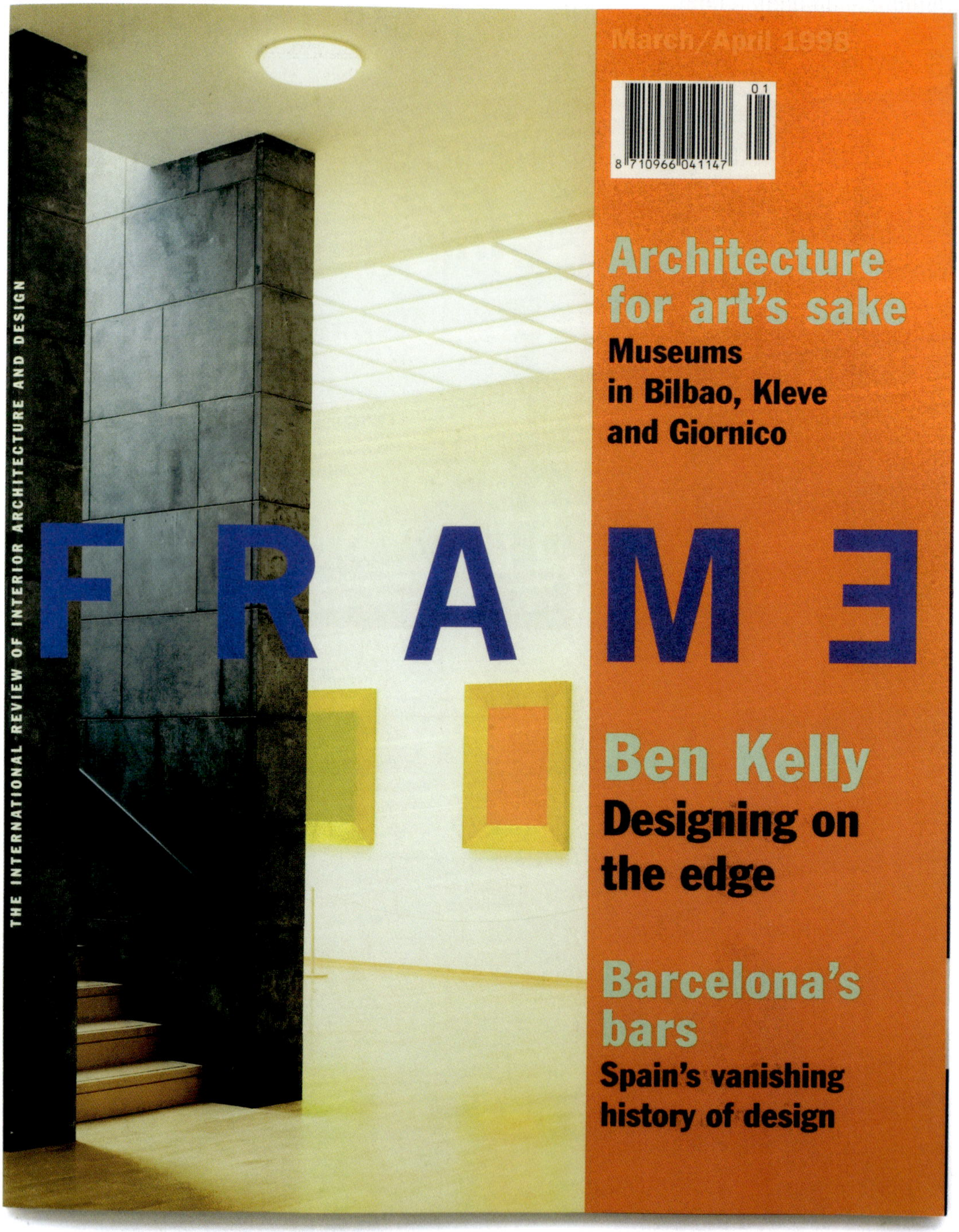

ART DIRECTOR
Ton Homburg,
Opera Designers

01

Not Looking for Visitors

Frame editor Robert Thiemann about the project

We published the Hans Josephson Museum, also known as La Congiunta, as a counterbalance to the then brand-new Guggenheim Museum in Bilbao. Whereas the Gehry building caused quite a stir because of its voluptuous architecture, Swiss architect Peter Märkli's museum is nothing more than a concrete box in the Alps. And whereas the Guggenheim is a market-driven exhibition machine boasting 19 galleries, an auditorium, a restaurant, a café, a library and a shop, La Congiunta has no parking, no reception and no posted information. Peter Märkli, a great architect who shuns publicity (we couldn't even get him on the phone for this book), created an austere museum featuring beautifully proportioned exhibition spaces, rough concrete walls and skylights. It's always open – just ask for the key at the neighbouring 'Gasthof'. Our approach – then and now – involved more than simply featuring two museums. We like to show our readers opposites.

DESIGNER
Peter Märkli

SUBJECT
Museum Hans Josephson

LOCATION
Giornico, Switzerland

PHOTOGRAPHER
Friedrich Busam/
Architekturphoto
www.architekturphoto.de

WRITER
Verena Huber

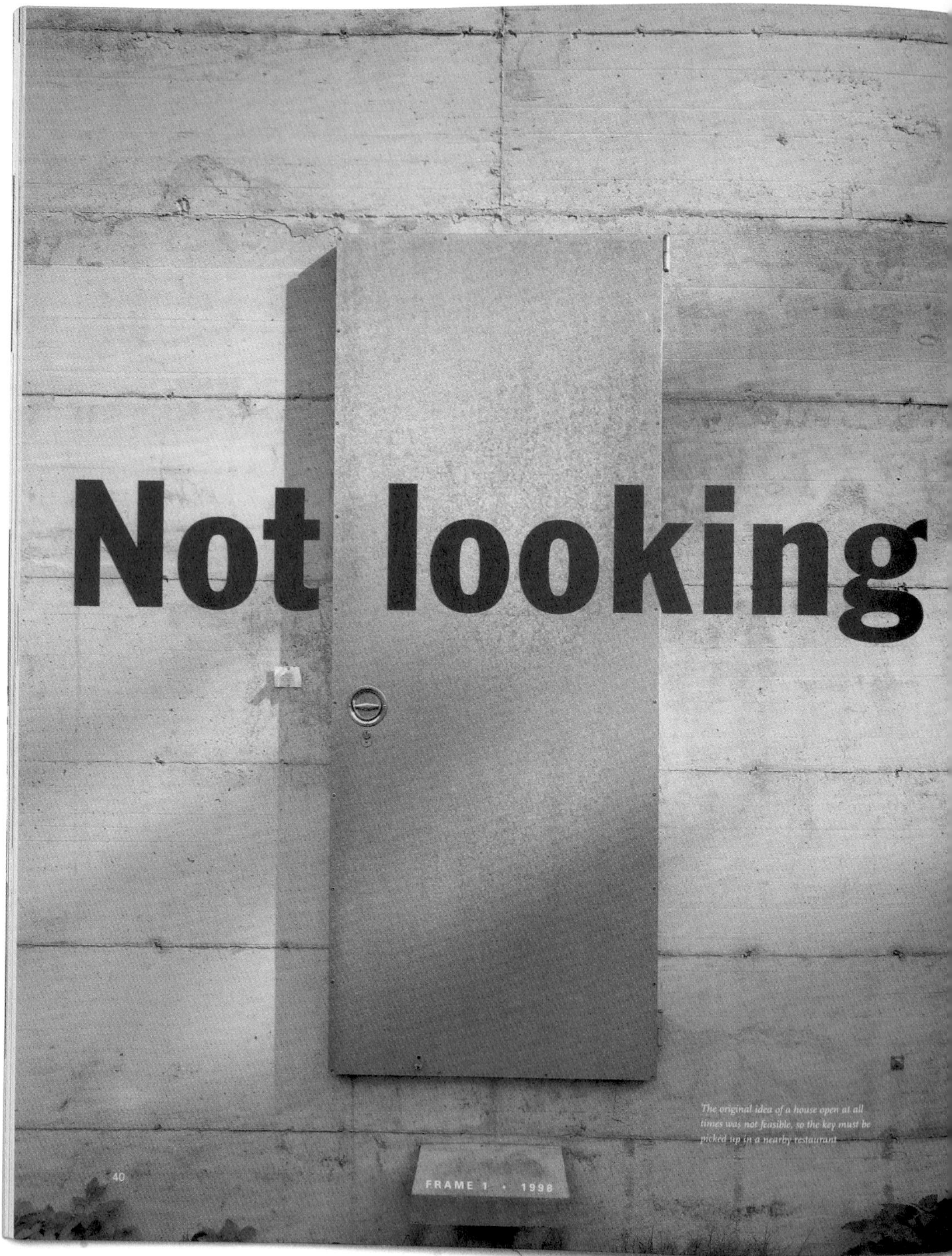

Not looking

The original idea of a house open at all times was not feasible, so the key must be picked up in a nearby restaurant

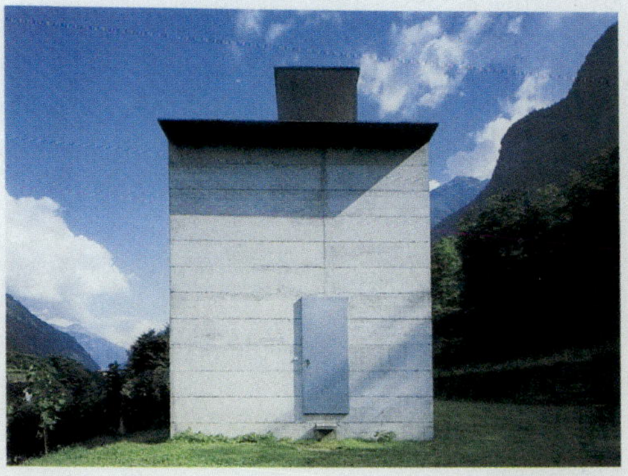

Visitors approach the building from the back and have to go around it in order to find the modest entrance

for visitors

High in the Swiss Alps stands the Museum Hans Josephson. Those who can find this uninviting concrete box, a form reduced to the utterly essential, will be rewarded for their efforts. They will themselves become part of the exceptional museum's exhibition.

by Verena Huber photography F. Busam/Architekturphoto

Can we compare the museums of today with the churches of the past? Perhaps this is best possible in the Swiss village of Giornico, on a terrace in the south of St. Gotthard pass, a village famous for its 12th century church San Nicolao. Giornico was found to be an ideal place for housing the works of the master sculptor Hans Josephson. Peter Märkli, who completed his professional education with Josephson, was considered the ideal architect.

In 1992, Märkli developed an architectural structure comparable to an essay. The architecture of Museum Hans Josephson exists for itself, reduced to the essential, searching the archetypical. The architect and the sculptor express this theme in corresponding ways. The museum is not looking for visitors - no parking, no reception, no explanations. But people who are really interested can find the museum: their efforts are richly subsequently rewarded.

The original idea of a house open at all times was not feasible, so you must pick up the key in a nearby restaurant. You approach the building from the back and have to go around it in order to find the modest entrance. This enables you to see the proportions of the interior spaces from the outside. A longitudinal sequence of three rooms, defined by different geometrically based orders, gives needed space to Josephson's collection of bronze reliefs and sculptures, arranged in chronological order.

The architecture is in the service of the art. The varying room heights allow you to experience different effects of spaces and natural lightening in conjunction with the reliefs and sculptures. Moving through the calming 'monolythic' concrete rooms, you become part of the exhibition. The three different rooms end in four lateral cells, similar to the lateral chapels in a church. Conceived for the art of Hans Josephson, the museum still allows for architectural perception.

The trains of the St. Gotthard line do not stop in Giornico anymore. It is necessary to take the bus in order to reach the village. But the train still passes close to the Medieval churches of Giornico and to this exceptional house of art and architecture, an archaic monument in a dramatic landscape of stepped canyons on the south side of the St. Gotthard pass. ▶

WRITER Verena Huber

PHOTOGRAPHER Friedrich Busam/ Architekturphoto

DESIGNER Peter Märkli

LOCATION Giornico, Switzerland

SUBJECT Museum Hans Josephson

Museum Hans Josephson Giornico

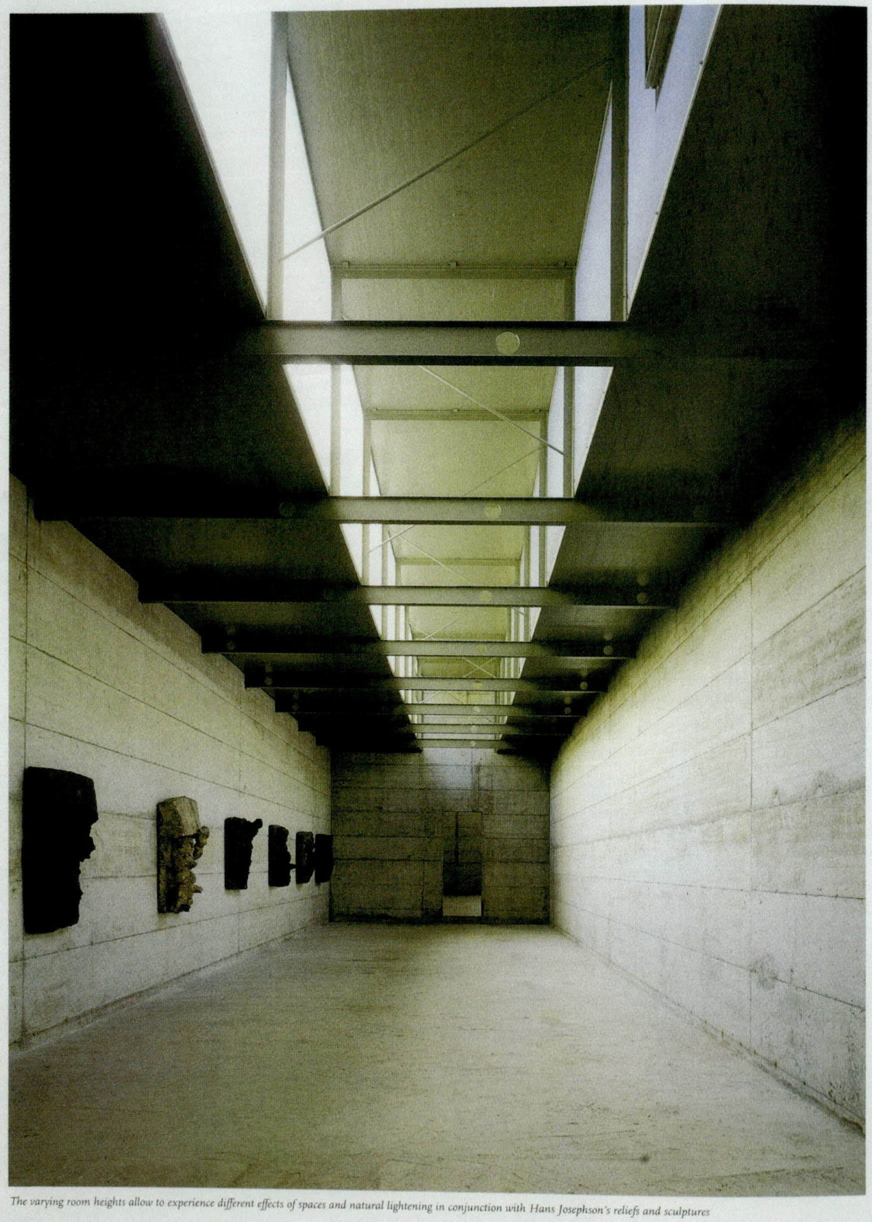

The varying room heights allow to experience different effects of spaces and natural lightening in conjunction with Hans Josephson's reliefs and sculptures

Moving through the calming concrete rooms, visitors become part of the exhibition

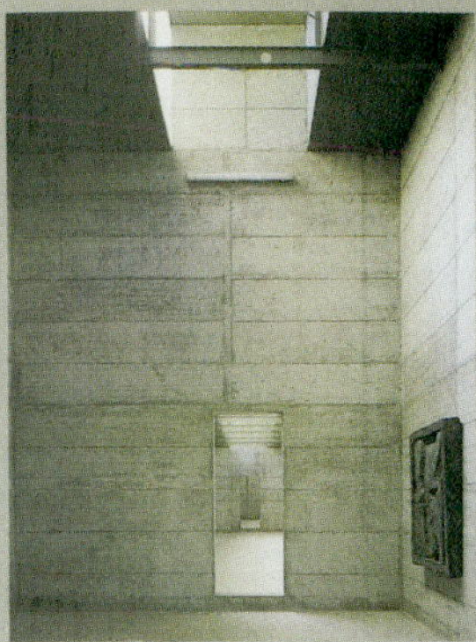

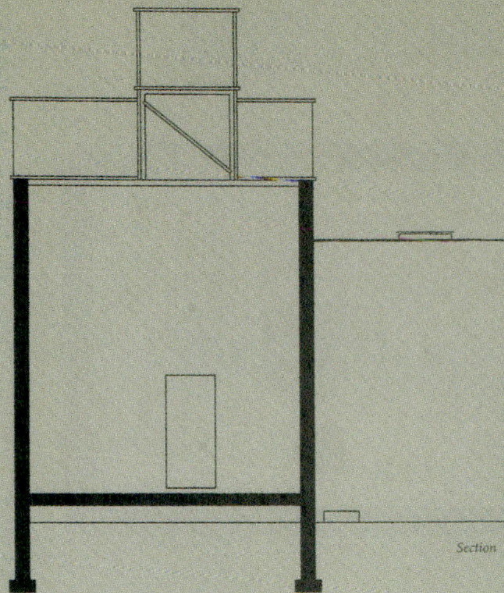

Section

The architecture of Museum Hans Josephson exists for itself, reduced to the essential, searching the archetypical

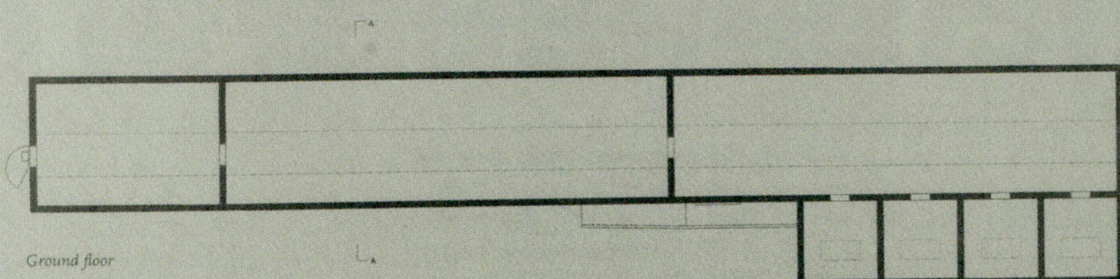

Ground floor

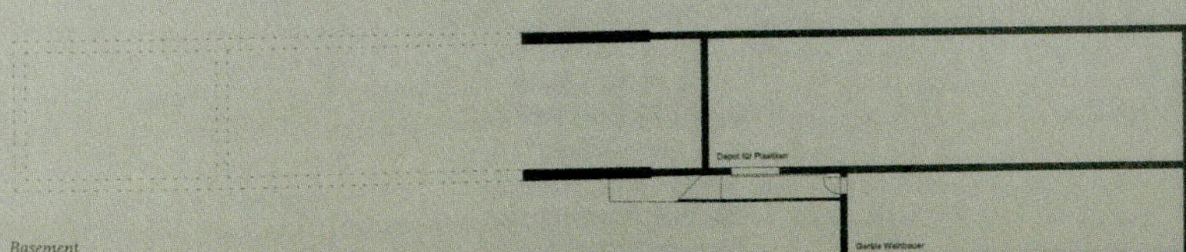

Basement

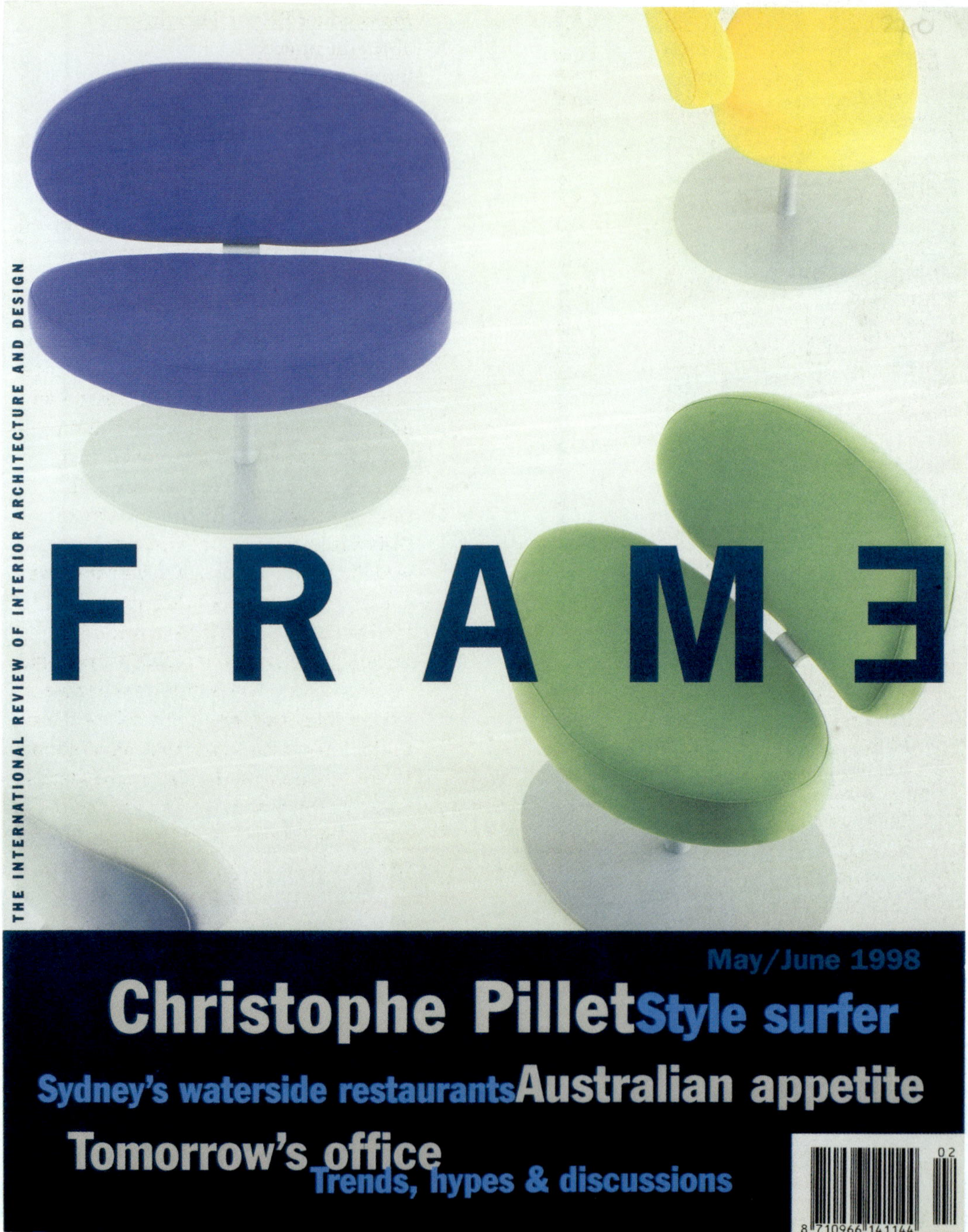

ART DIRECTOR
Ton Homburg,
Opera Designers

02

When Art Meets Architecture

Frame editor Robert Thiemann about the project

Right from the beginning, Frame has featured both world-famous and completely unknown architects and designers. Everyone admires (or hates) the stars but respects the anonymous, hard-working people who are still hoping to achieve success. Johannes Peter Hölzinger is not exactly a household name. This German architect first stepped into the limelight in 1978, when he completed a private house with office in Bad Nauheim, a design that made an impact in Germany and subsequently became the subject of a monograph. Twenty years later, we published his restaurant building for the Ministry of Defence in Bonn. What interested us was the strong relationship between the sculptural works of art that dominated the interior and Hölzinger's pyramidal building. The art and the building not only reinforced each other but were almost impossible to separate. We believed the project would be a source of inspiration for interior designers. Now, eight years later, the building still exists, although part of the Ministry of Defence has moved to Berlin. Mr Hölzinger, currently listed as a member of a German architects' association, will celebrate his 70th birthday this year.

DESIGNER
Johannes Peter Hölzinger

SUBJECT
Restaurant Kasino Bonn

LOCATION
Bonn, Germany

PHOTOGRAPHER
Studio Ivan Nemec

WRITER
Robert Thiemann

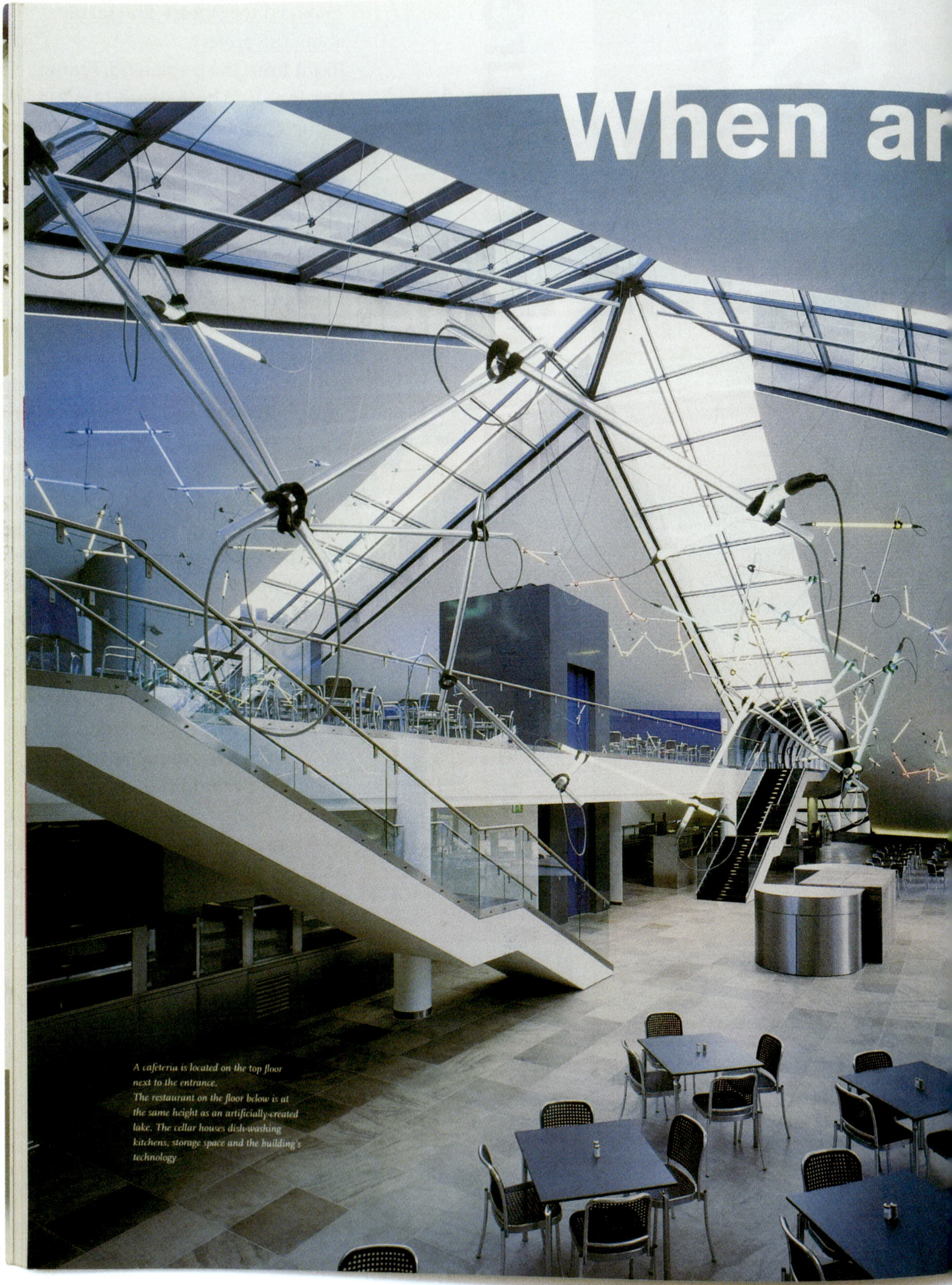
When ar
A cafeteria is located on the top floor next to the entrance.
The restaurant on the floor below is at the same height as an artificially-created lake. The cellar houses dish-washing kitchens, storage space and the building's technology

meets architecture

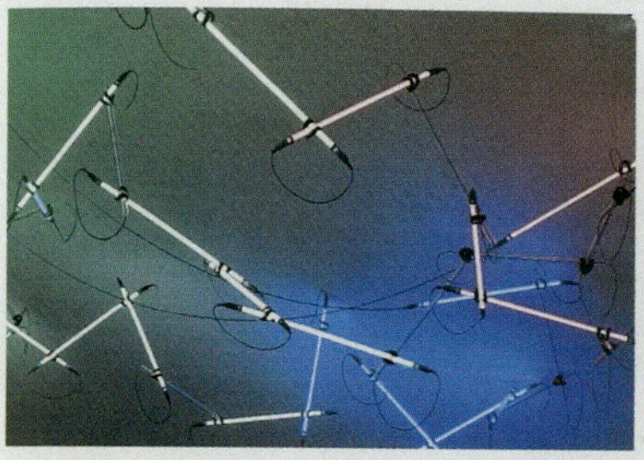

Kasino Bonn, a restaurant built for the German ministry of defence, demonstrates what happens when independently-created art and architecture acquire a common form.

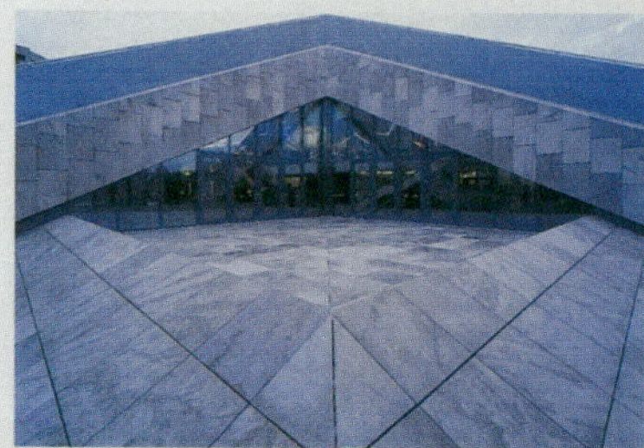

by Robert Thiemann photography Studio - Ivan Nemec

Many town halls include a sculpture in the main entrance, paintings counter the sterility of hospital corridors and striking works (of varying quality) often hang above the director's desk in an attempt to add a little lustre to his position. Yet, no matter how honourable the original intentions, the general result is that the art becomes merely decorative. These works have not been made for the space that they end up in. Nor was the space designed with these works in mind. Consequently there is no chance of achieving a harmonious unity that is more than simply the sum total of the separate parts.

Johannes Peter Hölzinger (1936) is an architect who specializes in the interaction between art and architecture. He feels that art must play an essential role in public spaces. "To create a really fruitful relation with society, the art must not only be the result of the artist's boundless imagination, it must also react to the environment", argues Hölzinger. In his opinion, this kind of art is diametrically opposed to the autonomous art that relates exclusively to itself. "Autonomous art derives its principle of experimentation, and even the breaks with its tradition, exclusively from its own history." For similar reasons, Hölzinger believes that architecture must be imbedded in the surrounding urban and natural context.

In Kasino Bonn, a restaurant intended for use by the German ministry of defence, Hölzinger has been able to take his ideas to an extreme. The building consists of three levels and is located between an artificial lake and a number of no-nonsense office buildings that date from the 1970s. Hölzinger designed the Kasino by tracing a cross in the ground and raising the four triangles that nestle between its arms. Additional incisions then created the necessary entrances and windows. With the passing of time, the roof will become completely overgrown with vegetation and will merge into the surrounding nature. Conversely, the building's diagonal structure refers to the rectangular construction of the adjoining offices. Hence, the Kasino mediates between nature and the built environment.

Yet in its interior, it's the art and architecture that are inextricably connected. Norbert Müller-Everling has covered the approach to the Kasino - a pedestrian bridge over the lake - with a spiral-shaped object. The spiral does not simply reflect the dynamism of the constant stream of visitors, it also functions as a protection against the wind.

Colourful sparks The art of Leonardo Mosso is unashamedly spectacular. His contribution to the Kasino consists of a network of colourful neon tubes that fan out dramatically and have been positioned beneath the gigantic roof. Because these floating strips of light have extricated themselves from the strictly geometric architecture, they imbue the building with a softness and a more human scale. Moreover, Mosso's artwork illuminates the building at night.

During the day, the neon tubes are reflected in the windows that Hölzinger has included in the roof. They look like colourful sparks that have been ejected by the sun only to vanish once more into the heavens. In Hölzinger's Kasino, it is almost impossible to separate the architecture from the art; they are interwoven and emphasize each other. ◀

SUBJECT Restaurant Kasino Bonn

LOCATION Bonn, Germany

DESIGNER Johannes Peter Hölzinger

PHOTOGRAPHER Studio Ivan Nemec

WRITER Robert Thiemann

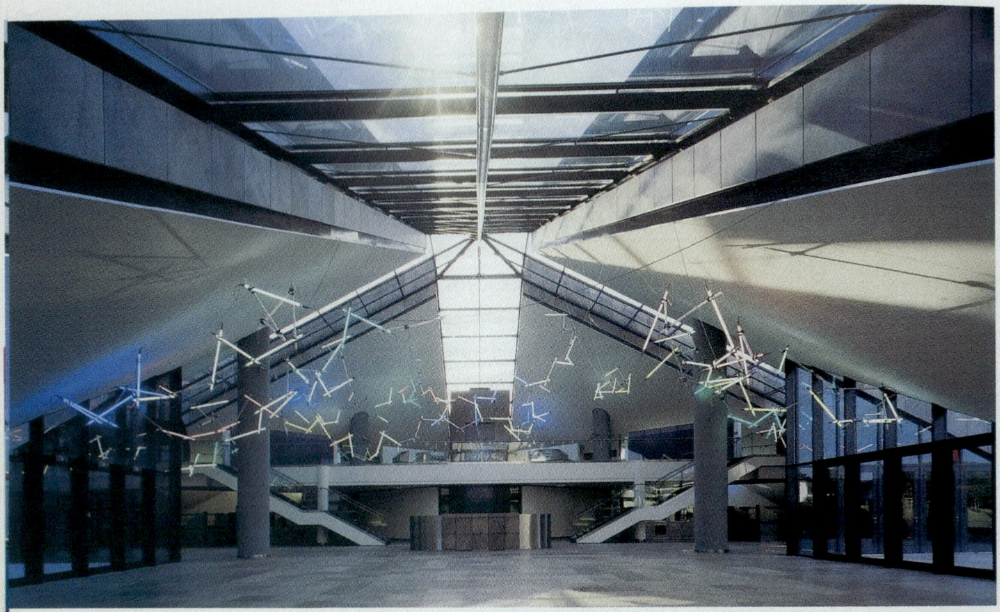

Leonardo Mosso's network of neon tubes soften the impact of Kasino Bonn's strictly geometric interior. As lights, they also create an atmospheric effect at night

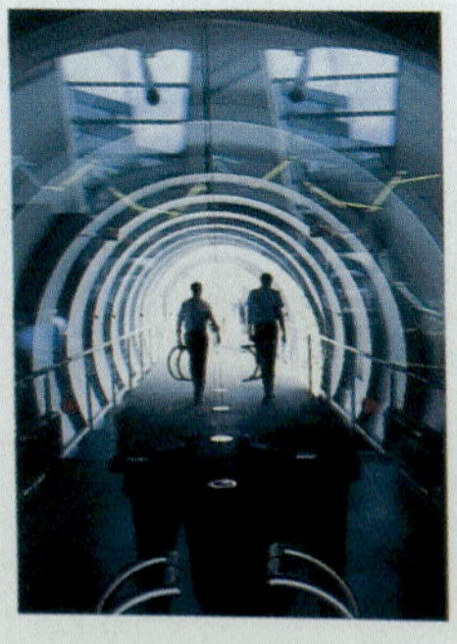

The entrance, a pedestrian bridge across the water, has been covered with a spiral-shaped artwork by Norbert Müller-Everling. Photo: Johannes Peter Hölzinger

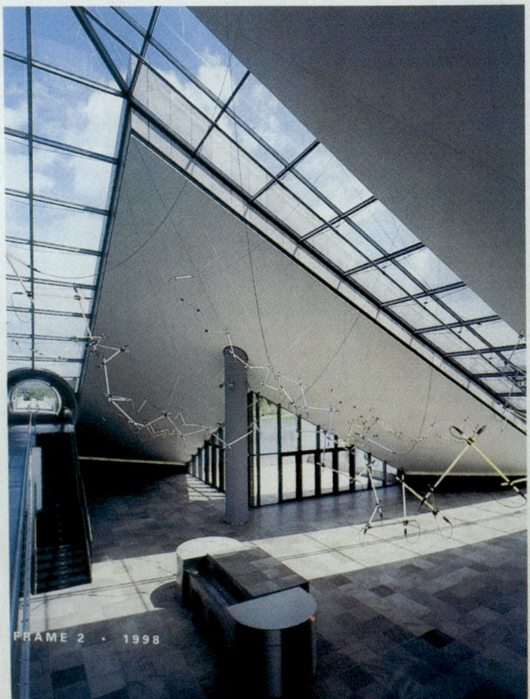

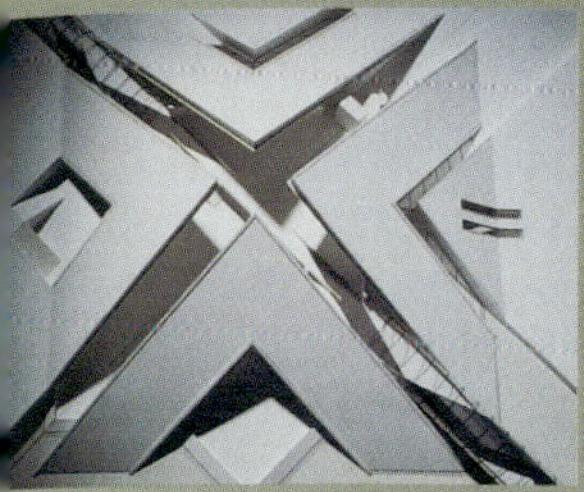

A maquette of the building shows how its form has been determined by tracing a cross in the landscape and raising the resulting triangles. Photo: Johannes Peter Hölzinger

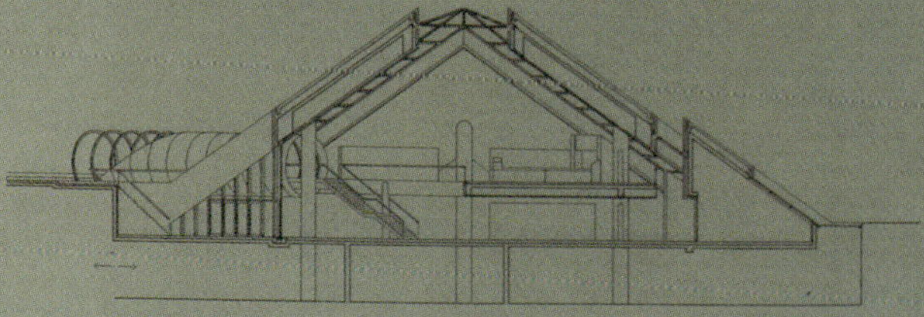

Section

"To create a really fruitful relation with society, art must react to the environment"

Ground floor

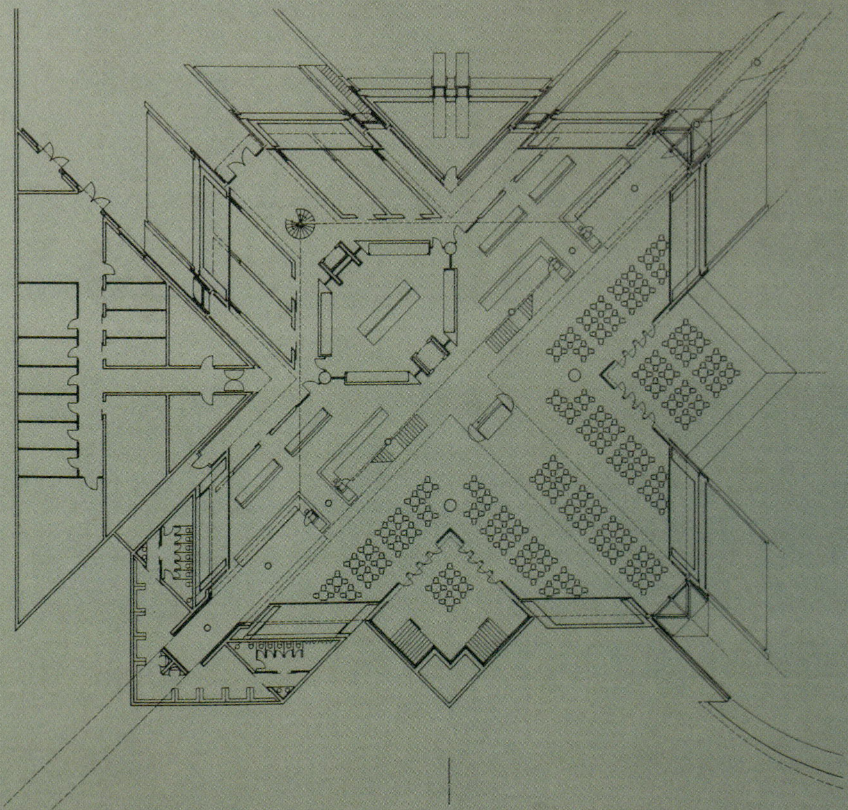

Issue 03 July/August 1998 038-039

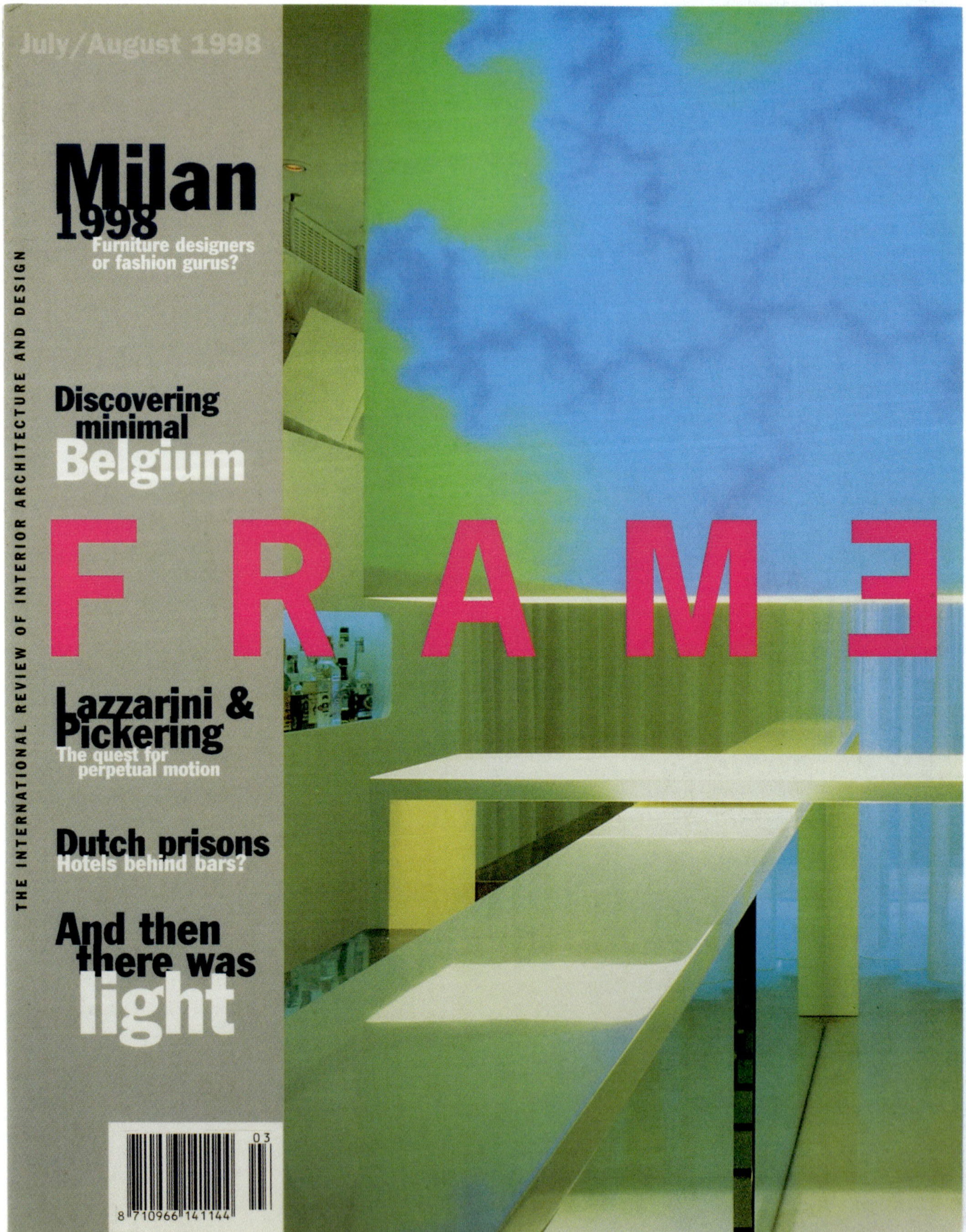

ART DIRECTOR
Ton Homburg,
Opera Designers

03

25 Frames per Second

Architects Stanley Felderman and Nancy Keatinge of Felderman Keatinge + Associates about the project

The MTV Networks project was instrumental in helping to invade an alternative architecture, one that was more lyrical, iconic and playful. It was a leader in the change of culture. It reflected the irreverent quality of MTV. Often design has a linear aspect, but this project precipitated a major shift. It was at the cusp of the internet/dotcom era and helped pave the way for what was to come in design.

Comments on the published article

The article in its description and graphics portrayed the essence of the project. The overlay of the graphics and the fluid nature of the article picked up on the vibrancy that was evoked by the project.

Recent news or information about the project featured in *Frame*

The project reinforces and still evokes a sense of humour and lyrical quality. It is as fresh today as when we designed it. Our firm still gets calls and positive response to the project.

Current projects

Investment company of Pacific Capital Group // Law firms of Pillsbury Winthrop Shaw Pittman LLP and Sheppard Mullin // Richter & Hampton // BMG Music // City Front Place condominiums // Escada.

DESIGNERS
Felderman Keatinge
+ Associates
www.fkadesign.com

SUBJECT
MTV Networks
West Coast
headquarters offices

LOCATION
Santa Monica, USA

PHOTOGRAPHER
—

WRITER
Robert Thiemann

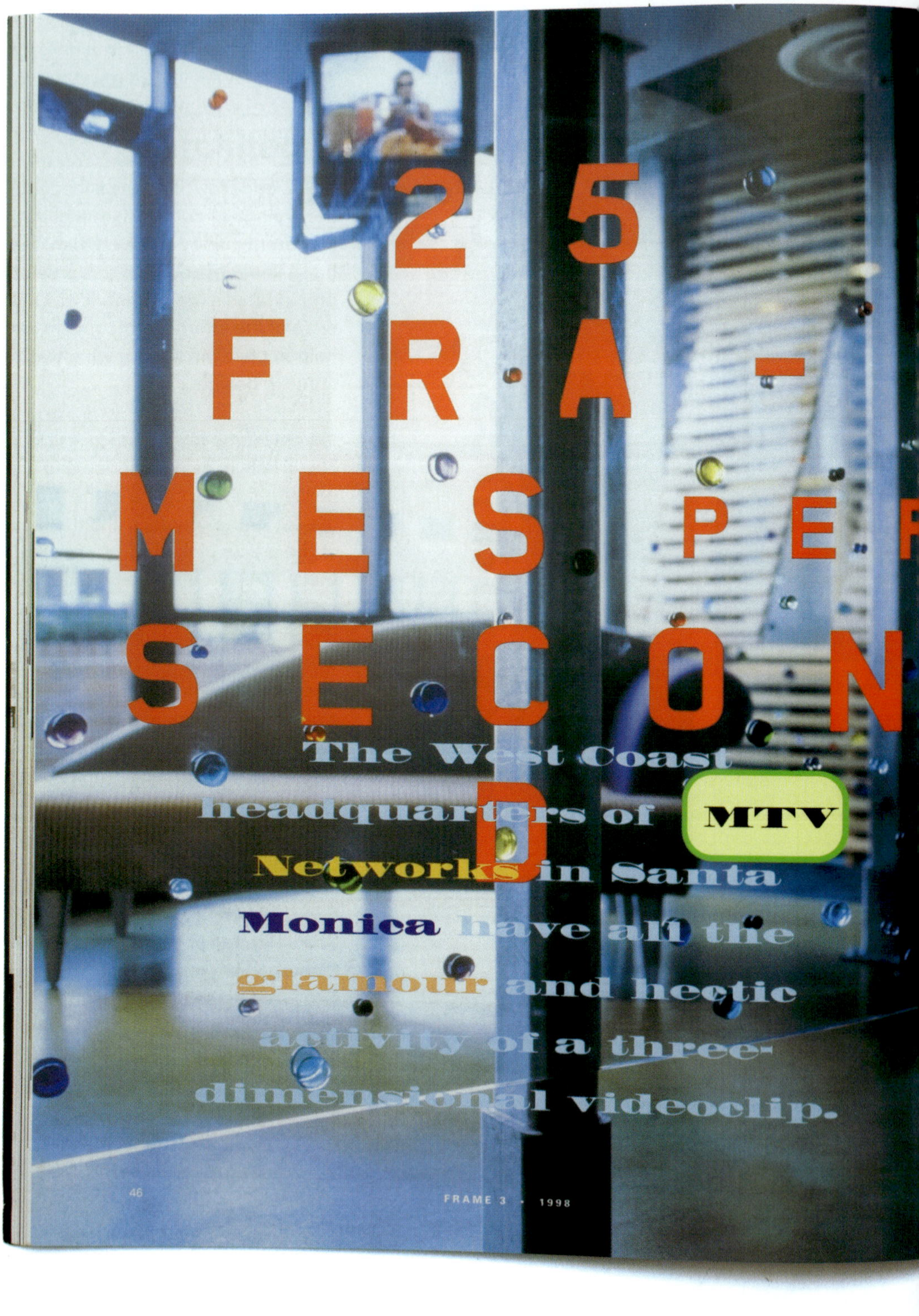
25 FRA-
MES PER
SECOND
The West Coast headquarters of MTV Networks in Santa Monica have all the glamour and hectic activity of a three-dimensional videoclip.
46
FRAME 3 • 1998

Turn on MTV and you will be immediately overwhelmed by a deluge of sounds and images. Trendy videoclips and high-speed commercials are responsible for most of the hurtling succession of visual impulses. In fact, MTV only comes into its own in the relatively brief gaps between the clips and the ads. And then some hip VJ will simply announce which clips are coming up or the station will launch into one of its own pulsating jingles that looks more like video art than a regular commercial.

It was through this swirling combination of pop music and up-tempo visual stimuli that MTV established itself in the 1980s as a trendsetter for young people. The station determined who and what was hip. Throughout the world it influenced not only teenagers and twenty-somethings, but also those who sought to ingratiate themselves with this group and their generous spending habits: a queue headed by the ad agencies and the purveyors of soft drinks and jeans.

But the way of the trendy world is that at a certain point the new replaces the old or there is a upsurge of competition. How do you keep ahead of the rest? In MTV's case, this has meant gradually concentrating on other fields including a TV station for children called Nickelodeon.

It is for this expanded group, MTV Networks, that the Santa Monica team of Felderman + Keatinge Associates were asked to design the West Coast headquarters. This involved transforming a dull office block in the middle of the busy media city of Santa Monica into a home away from home for MTV Networks' 450 employees. The design was to include something of the sun, sand and surf ethos of this city on the Pacific Rim, a sensibility that is closely linked with MTV's basic philosophy. Nonetheless an environment was required that was not only fun but also functional - for after all MTV means big business. However, the fly in the ointment was that while each of the five floors was to acquire its own identity reflecting the character of the various subsidiaries housed in this building, as a whole MTV was still to maintain a uniform corporate look.

Reference Felderman + Keatinge's designs were strongly influenced by Santa Monica's painterly environment: the pier, the beach with its palm trees and bungalows, the artists' district and the local avant-garde architecture with its industrial materials. Consequently MTV's presence is heralded by a bright-red boat-like form at the building's entrance. The name MTV Networks is sculpted on it in rough-hewn carbon steel.

Inside the oval lobby, visitors are greeted by a large, shiny aluminium-coated Airstream trailer which dates from 1957. It stands on green Astroturf alongside a blue Astroturf pond. This somewhat kitschy image is reinforced by the presence of pink flamingos and vintage aluminium garden chairs. Inside the renovated Airstream, which serves both as a waiting-room and a conference space, you can while away the time with ▶

MTVs OFFICES PRIMARILY CREATE A SENSE OF ABUNDANCE

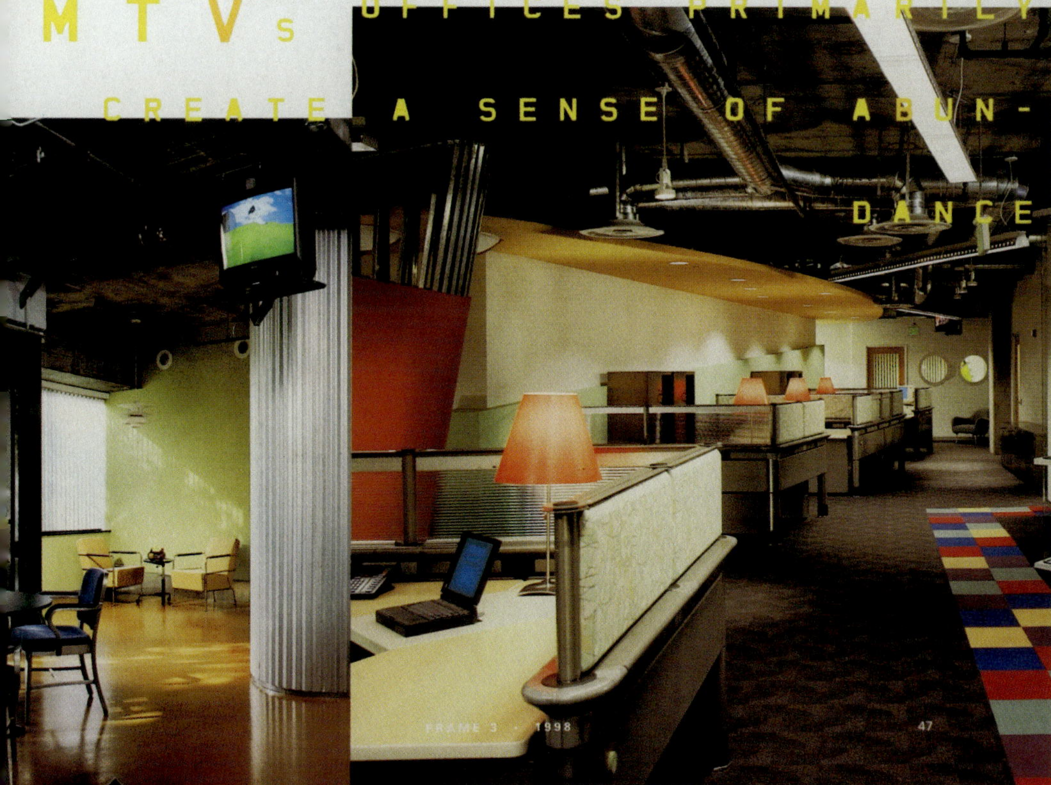

SUBJECT
MTV Networks West Coast headquarters offices

LOCATION
Santa Monica, USA

DESIGNERS
Felderman Keatinge + Associates

PHOTOGRAPHER
—

WRITER
Robert Thiemann

a trip down memory lane. There are fabrics with 1950s prints of boomerangs and flying saucers, pink carpeting and black-and-white linoleum, a Formica-topped kitchen table and all the trinkets from a bygone age which imbue this interior with a look that hovers somewhere between cult and kitsch.

Right next to the trailer is another image which is at least as strong: two enormous talking heads with eyes and mouths that consist of interactive television monitors. On the opposite side is a boat covered with brushed aluminium that has been transformed into a reception area. The final element in the lobby is a conference room that is dominated by a copper-topped table.

In the offices, Felderman + Keatinge again have constantly referred to the elements that are so typical of Santa Monica. The outer walls of these offices have been transformed into facades that can also be found in the city. Some are fronted with wood-lathe siding and casement windows while other divisions are a metaphor for low, green hedges that imbues

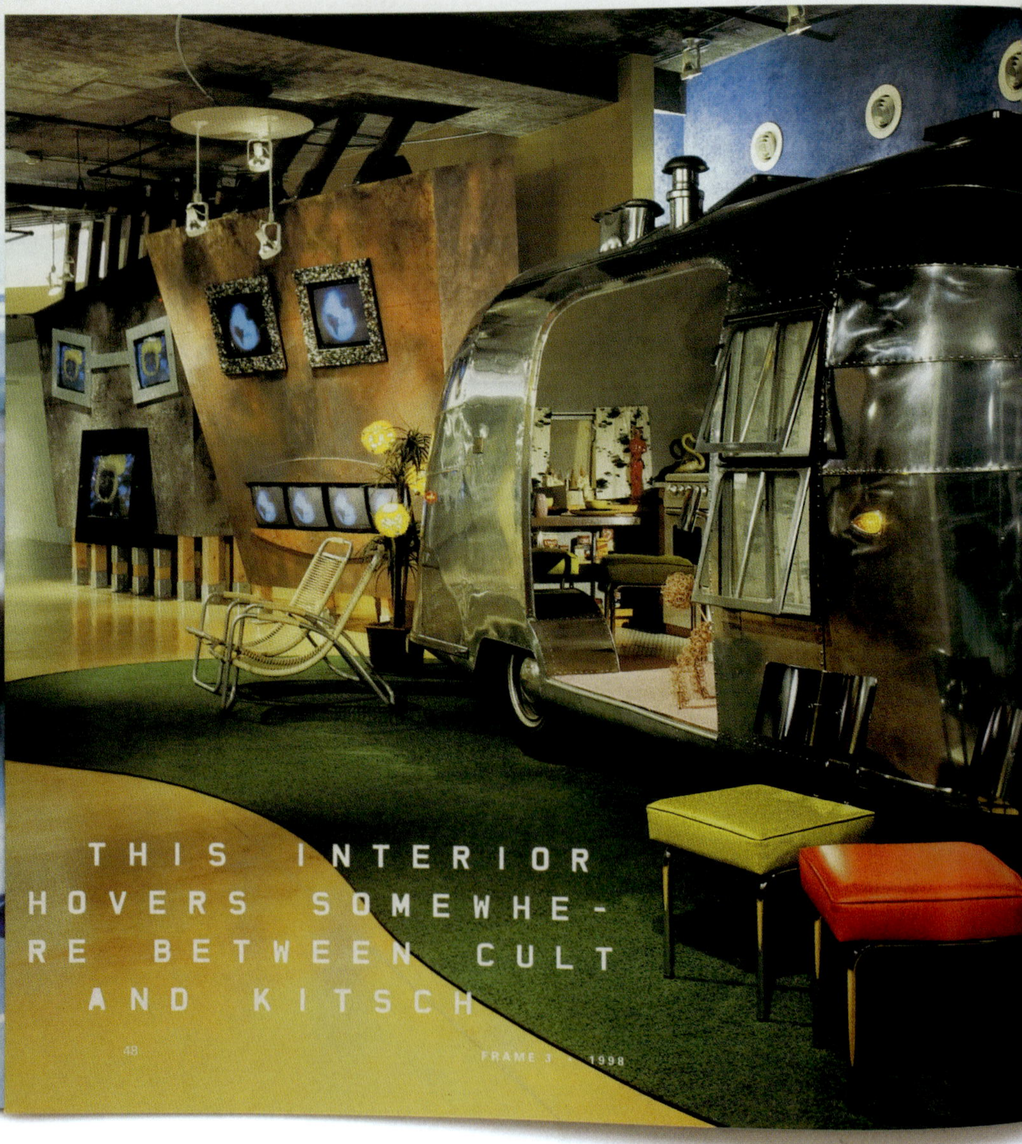

this working environment with a friendly quality. The ceilings of the office floors have been left unfinished: the air conditioning and wiring is clearly visible and is a reference to the industrial character of the surrounding architecture. F + K has responded to the demand for informal conference spaces for meetings and brainstorm sessions by providing each floor with two spacious areas complete with large rugs, lounge furniture and (inevitably) television monitors.

F + K's MTV headquarters primarily create a sense of abundance. Wherever you work and wherever you relax, your senses are being constantly bombarded by forms, colours and materials. Whether or not this is practicable, it is at least sometimes hip. However, there are other places where Felderman + Keatinge have been carried away by their own enthusiasm. In short, the designers have created the perfect, three-dimensional equivalent of MTV's antics on the box. ◀

RT

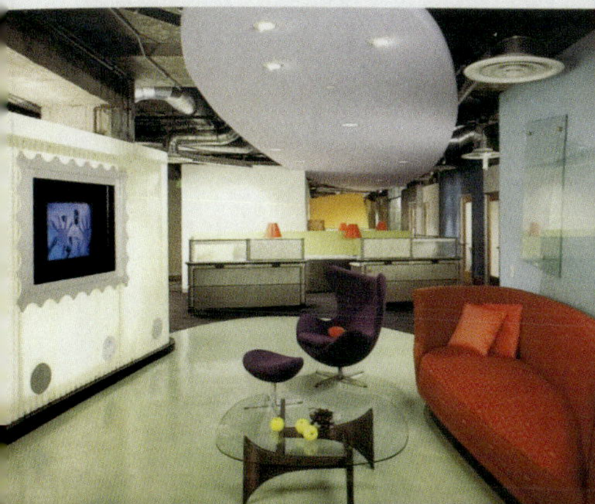

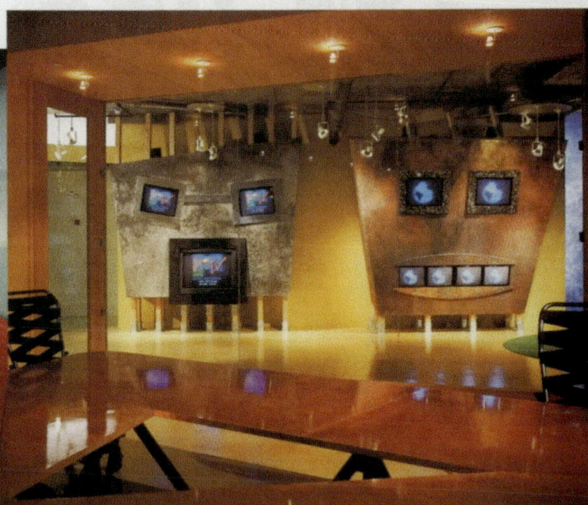

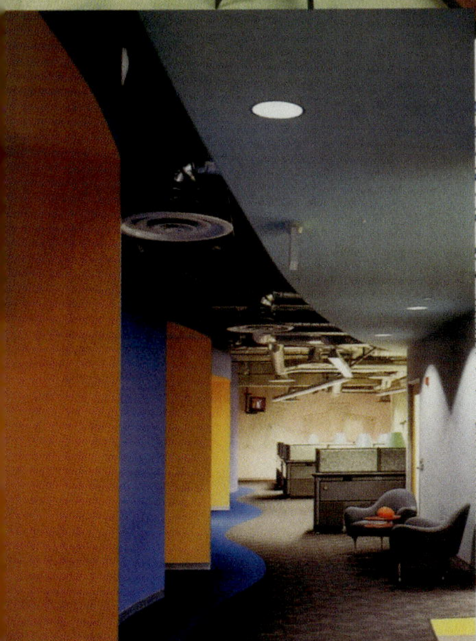

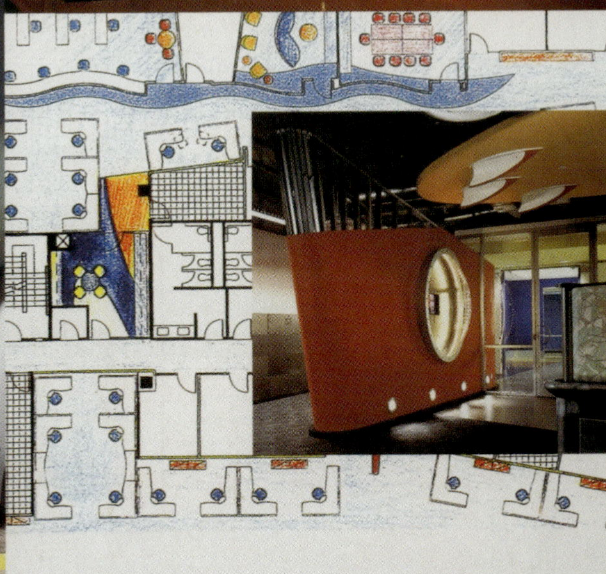

SUBJECT
MTV Networks West Coast headquarters offices

LOCATION
Santa Monica, USA

DESIGNERS
Felderman Keatinge + Associates

PHOTOGRAPHER
—

WRITER
Robert Thiemann

September/October 1998

Issue 04

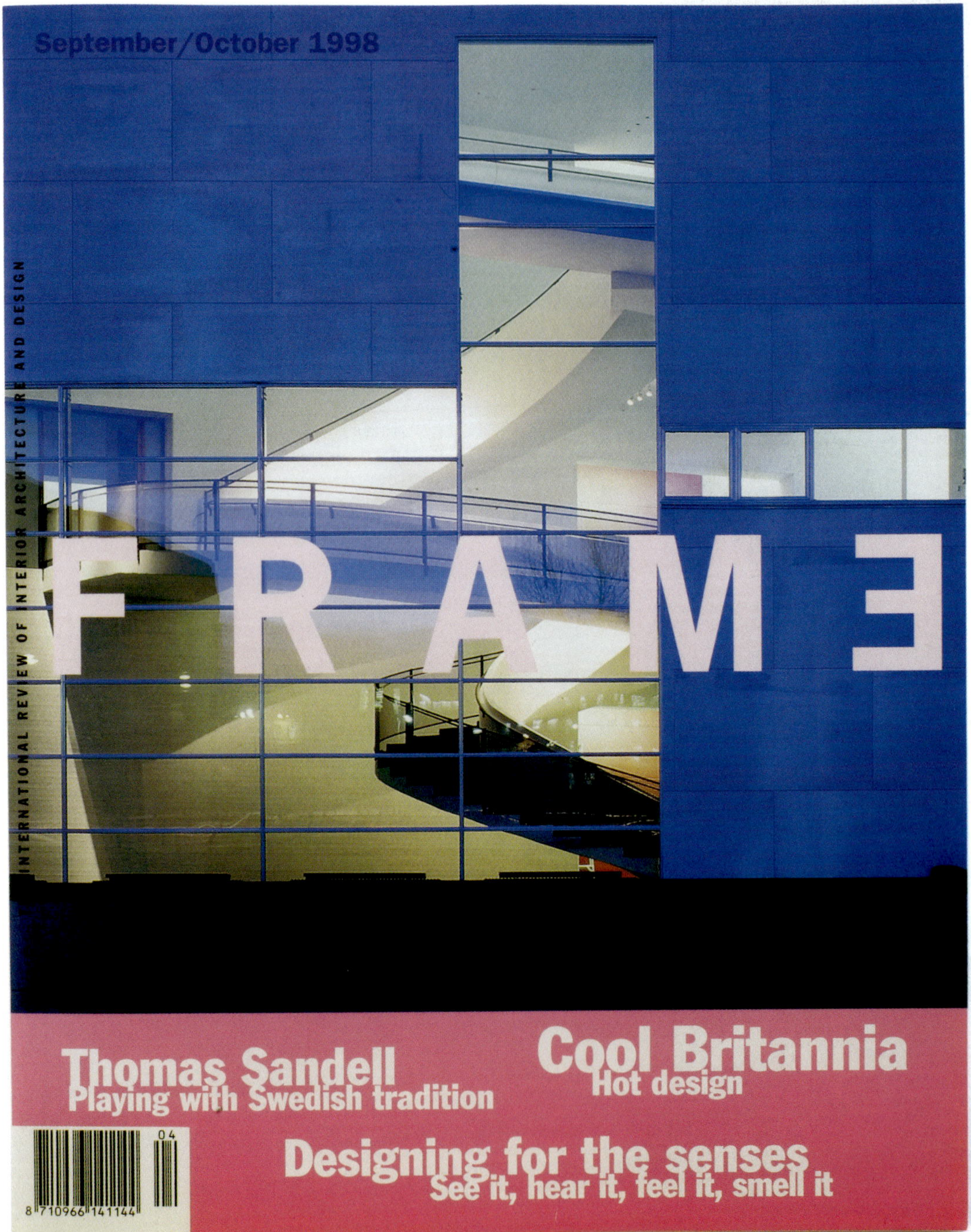

ART DIRECTOR
Ton Homburg,
Opera Designers

04

Frozen Waves

Architect Lars Spuybroek (NOX Architects) about the project

Totally great.

Did the article have an effect?

No article ever has an effect.

Recent news or information about the project featured in *Frame*

We inserted a horizontal floor two years ago, and now it is not just beautiful but functions much better.

Current projects

Whispering Garden, a public art work // myHouse, a series of private houses

DESIGNER
NOX Architects
www.noxarch.com

SUBJECT
Media centre V2_Lab

LOCATION
Rotterdam, Netherlands

PHOTOGRAPHER
NOX

WRITER
Robert Thiemann

Frozen waves

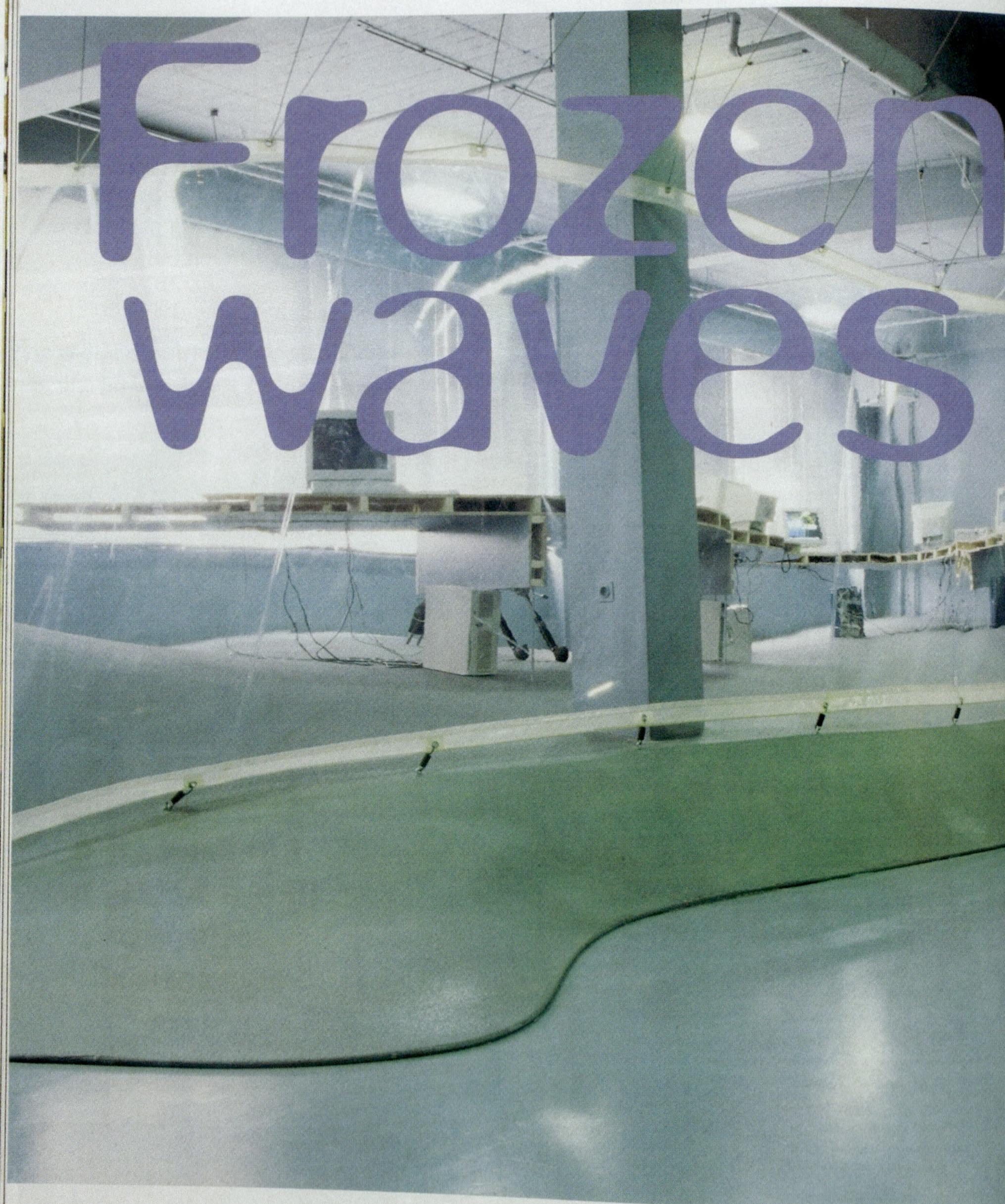

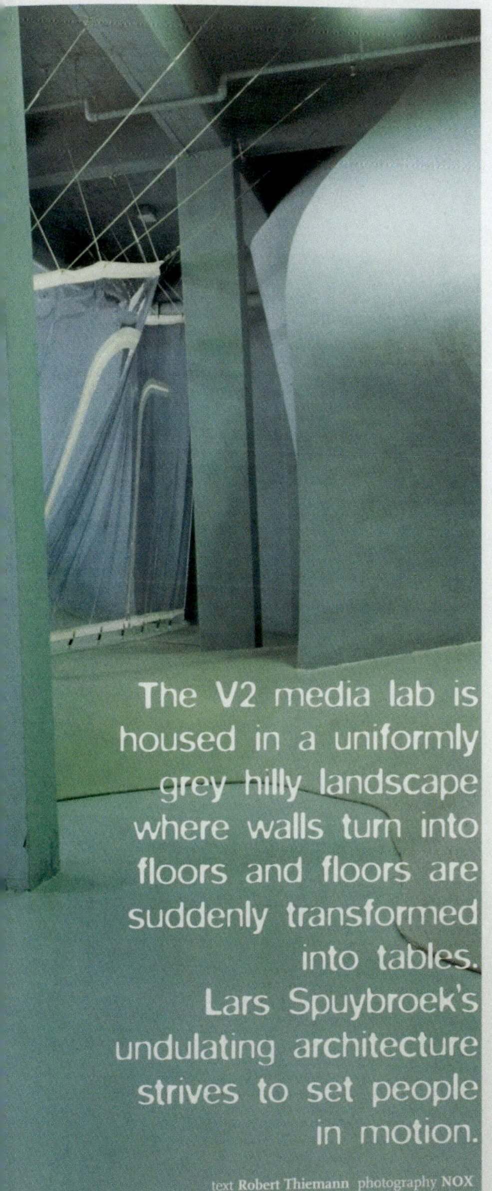

The V2 media lab is housed in a uniformly grey hilly landscape where walls turn into floors and floors are suddenly transformed into tables. Lars Spuybroek's undulating architecture strives to set people in motion.

text Robert Thiemann photography NOX

Enter an ugly door in an anonymous building in Rotterdam city centre and on the first floor you will find the V2_Lab, an international workplace for the new media. Here, up to twelve artists, scientists and technicians will work together on digital artworks, software and computer interfaces which examine the interaction between people and machines and the boundaries between real and virtual spaces.

Their activities take place in an extraordinary space that is as volatile as the media themselves. Most of the floor's surface looks like a hilly landscape with a summit measuring 1.35 metres. An endless table has been attached to this wavy floor that snakes its way through the space. Office chairs have just two wheels and are supported by two legs with gas suspension to achieve the desired stability. Elsewhere two computers have been placed on the floor that appears to be functioning here as a table. At this point the floor becomes a rocky wall. Everything in this fluid space has been painted grey: the floor, the walls, the ceiling and the work-tables.

Once you've negotiated the level entrance to this hilly landscape, you find yourself constantly having to adjust your balance. Suddenly it's not as easy as it seems to walk from one place to the next. Sometimes you have to struggle to walk up hill and then you must pull up sharp to avoid tumbling down the other side. Here too, the easiest route between two points is not necessarily the shortest.

Vibrations This interior has been designed by Lars Spuybroek of NOX Architects. His aim is to get people moving. "I'm trying to create an architecture of waves that influences every step you take. For me, topology is a machine that charges your body like a spring." This means that he is trying to achieve the complete opposite of most of his colleagues. "Generally architects try to solve a program of demands with simple solutions. You have to go from A to B through Corridor One and from B to C via Staircase Two. This means that people don't have to think for themselves; the architect has already done that for them. Here, the aim is to immobilize their bodies. But what I offer them is a variety of ways of going from A to B. Using the body as my basis, I want to dispense with all that unambiguousness." He has certainly succeeded in this although the question is whether this working environment provides the media artists with much by way of comfort.

Spuybroek is fascinated by what happens when people can use a space in a number of different ways. For instance, what happens when the floor is also a table. "I guarantee that if an artist here has just heard that his subsidy has been axed, he will immediately stand on this table." Another example: the long, curving work-table has been suspended in a way that leads to inevitable vibrations. "If you're angry and you strike the table, people will sit up and take notice on the other side of the space because the table will vibrate. The space reacts to human behaviour and vice versa."

This is what distinguishes Spuybroek from his fellow architects who prefer to describe their work as baroque, fluid or organic. By contrast this Rotterdam architect is seeking haptical effects rather than optical ones. That is also why it is only the floor, the furniture and some of the walls that undulate in the V2_Lab and not the ceiling or the columns – because this would create an exclusively optical effect. Spuybroek's design procedure also differs radically from architects such as Erick van Egeraat and even Peter Eisenmann. "Their basis is a modernist box or set of boxes which they then model from the outside. I generate a space ▶

WRITER Robert Thiemann

PHOTOGRAPHER NOX

DESIGNERS NOX Architects

LOCATION Rotterdam, Netherlands

SUBJECT Media centre V2_Lab

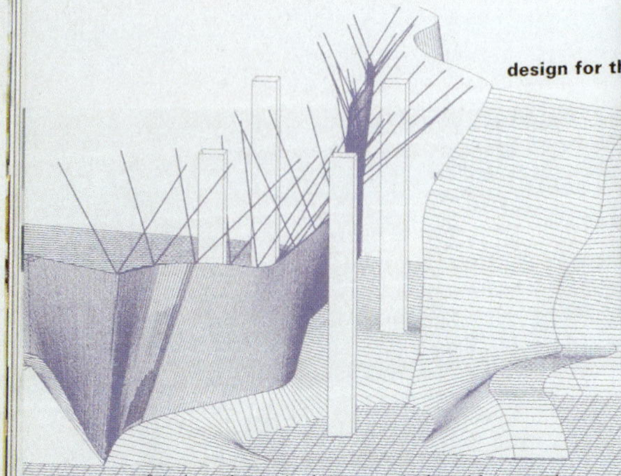

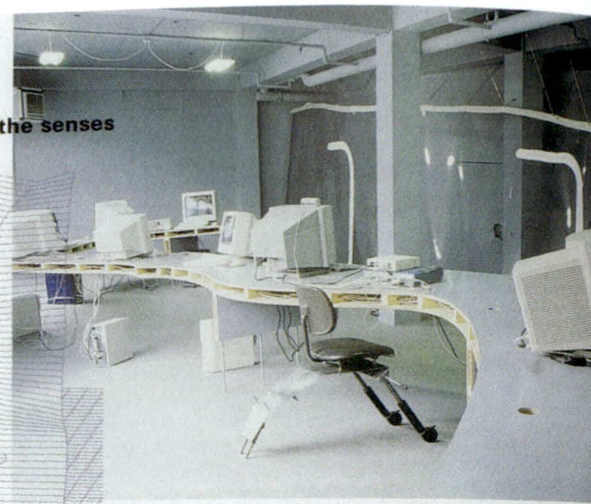

The working table snakes its way through the grey space. Office chairs have ju two wheels and are supported by two legs with gas suspension

from the inside." Spuybroek uses extremely advanced design procedures. They provide him with the potential of designing moving geometries which he describes as 'time-spaces'. The basis of this is a modified system of co-ordinates. The X, Y and Z axes, which combine to describe a three-dimensional space, normally converge to create a single point. Spuybroek's intervention consists of axes that converge in a hole that can contract and expand, which he describes as a spring. By releasing all kinds of modulations on that spring, the system of co-ordinates also begins to vibrate as if it were a heart muscle that contracts and expands to create waves throughout the space. Spuybroek defines the geometry of a space by periodically freezing time and recording the co-ordinates. The final result is the sum of geometries that have been frozen in time.

Not aesthetic The space is a uniformly grey colour because Spuybroek wants to provide a "neutral neurological background" for his experiential architecture. This focuses attention on human behaviour. Just as important for him is that there is no aesthetic finish: "I've noticed that the effect of this architecture is diminished when you neatly finish off all the details." That's why the undulating floor abruptly turns into a normal one and an anti-slip polyurethane coating has only been applied to the hilly landscape. The side of the long work-table has also been left open and the thick transparent screen that curves through the space, and which the architect prefers to describe as a fluid window, has been crudely attached to the floor with springs and steel cables.

To some extent, the screen - which consists of 4 mm-thick plastic that is normally used in cold storage - divides the long table where the artists and technicians work from the V2 staff table. Here and there, the plastic falls in folds which in turn create reflections that vaguely distort the image. This is one optical effect that Spuybroek is proud of.

This lunar landscape is as exciting to look at as it is to walk around in. Soon it will become even more fascinating when people actually begin to work here. But this is also the fly in the ointment. Somehow it's difficult to imagine serious work taking place in this amusement park. There's something about it that's too contrived. Spuybroek's ideas are fascinating and the computer models look fantastic but the question is whether this virtual architecture is really suited to everyday reality. Or will we learn to love it once we've recovered from the shock of the new? ◀

The V2_Lab reacts to human behaviour and vice versa

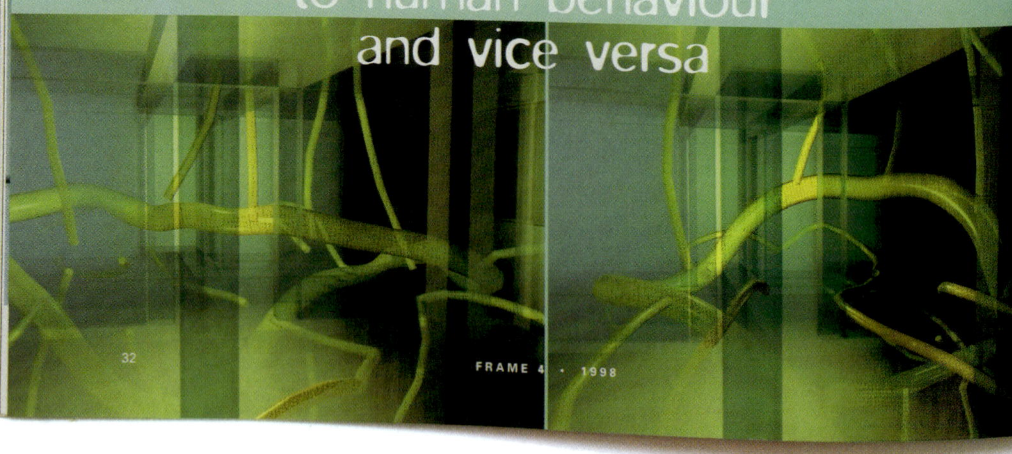

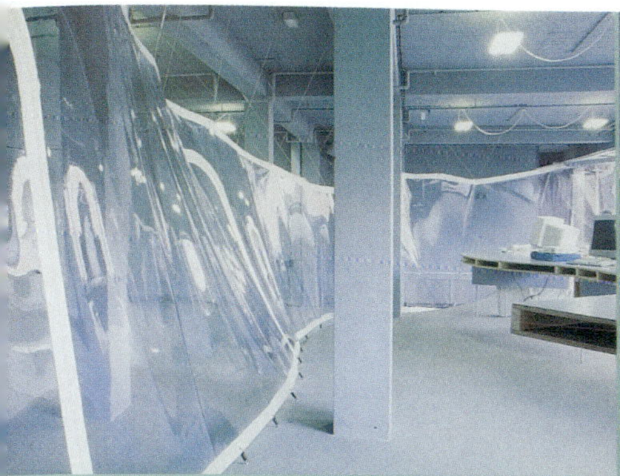

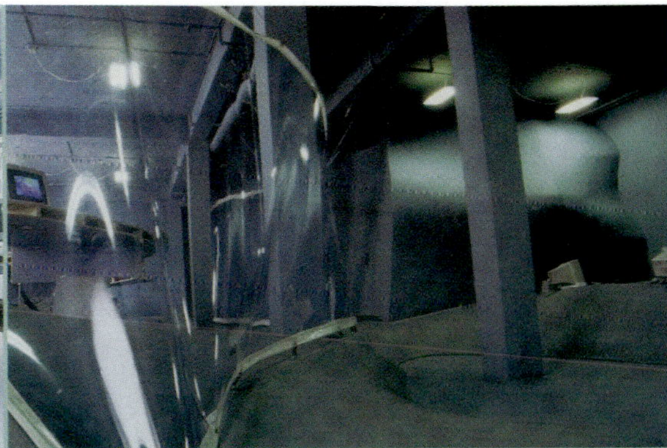

The 4 mm-thick transparent screen that separates the working area from the V2 staff creates reflections that vaguely distort the image

Left, the working area and, right, the staff area. Here, the floor transforms into a working table

The effect of this architecture is diminished when you neatly finish off all the details

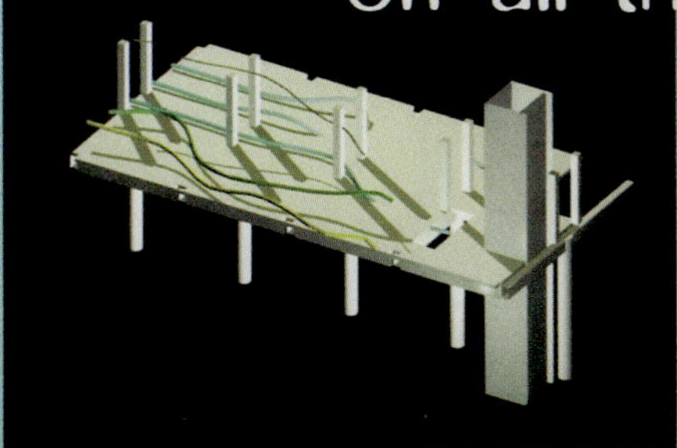

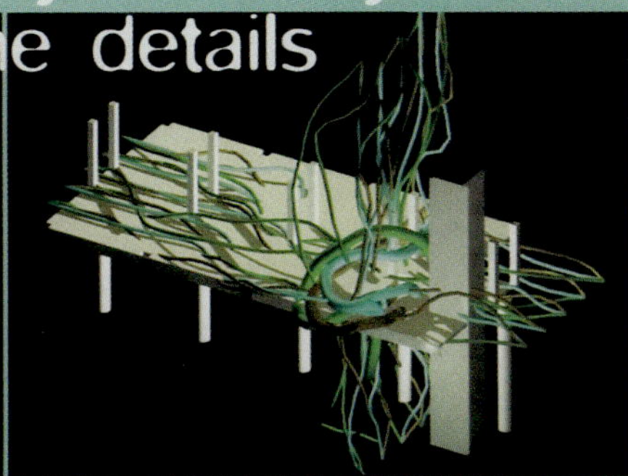

The basis of Lars Spuybroek's architecture is a modified system of co-ordinates. The X, Y and Z axes converge in a hole instead of a single point. Like a spring, this hole can contract and expand, creating vibrating geometries

Using extremely advanced software, Lars Spuybroek designs moving geometries which he describes as 'time-spaces'. The architect defines the definite geometry by periodically freezing time and recording the co-ordinates

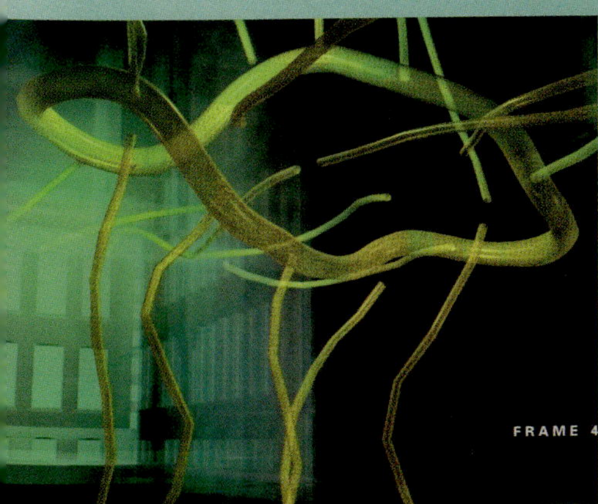

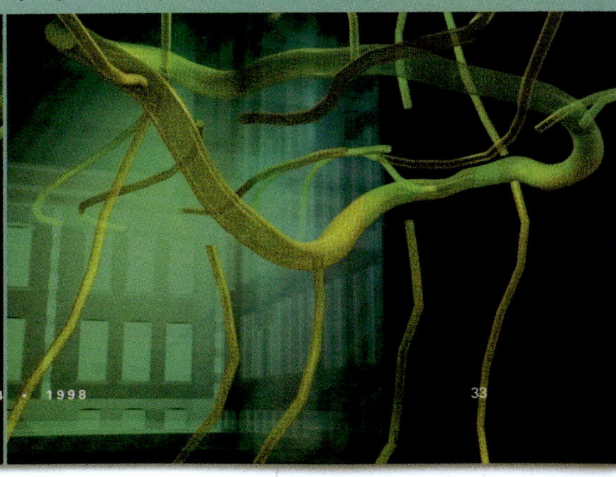

SUBJECT Media centre V2_Lab
LOCATION Rotterdam, Netherlands
DESIGNERS NOX Architects
PHOTOGRAPHER NOX
WRITER Robert Thiemann

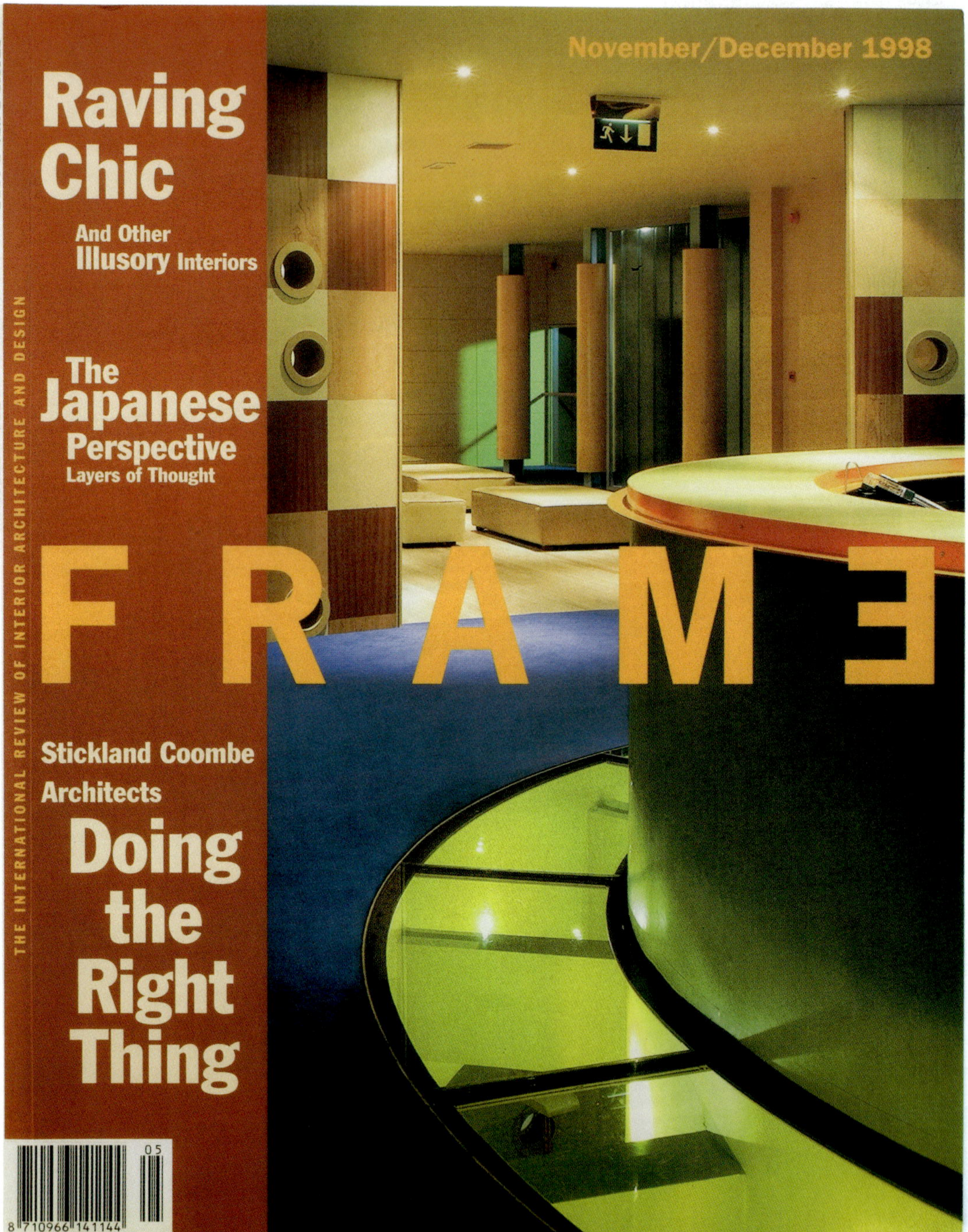

ART DIRECTOR
Ton Homburg,
Opera Designers

05

The Arabian Nights

Frame editor Robert Thiemann about the project

In Frame 05 we featured a thematic series of articles on the architecture of illusion. We talked about how Disney parks, casino hotels and Planet Hollywood immersed visitors in a dreamland. Then photographer and writer Eleanor Curtis – with whom we hadn't worked before – called out of the blue to say that she had discovered a fairy-tale five-star hotel designed by Michael Graves on the coast of Egypt's Red Sea. This 284-room Miramar Hotel covers an area of no less than 200,000 square metres, a sixth of which is filled with an artificial lagoon and another quarter with palm trees. The architecture of the village-like resort hovers somewhere between post-modernism à la Graves and Egyptian vernacular. The whole place is geared towards luring guests into tales from 1001 nights. Now, if that isn't about creating illusions... Eight years later, we're no less enthusiastic about the super-sized italics of the title that introduced the article and the photos of the gorgeous hotel bar featured in the piece. By the way, the hotel still exists – the Arabian nights, however, may have become less fairy-like.

DESIGNER
Michael Graves
www.michaelgraves.com

SUBJECT
Miramar Hotel

LOCATION
El Gouna, Egypt

PHOTOGRAPHER
Eleanor Curtis
www.eleanorcurtis.com

WRITER
Eleanor Curtis

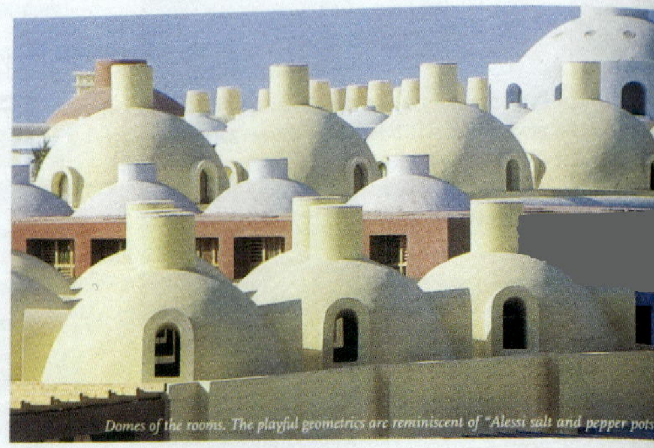

Domes of the rooms. The playful geometrics are reminiscent of "Alessi salt and pepper pots

The Arabian

Michael Graves' Miramar Hotel, on the coast of Egypt's Red Sea, combines his preference for post-modern classicism with the local vocabulary. The architect hopes that hotel guests will identify with the shapes, colours and details of their surroundings, rather than to be left with nothing more than another international hotel experience.

text and photography **Eleanor Curtis**

Coming from the main road that divides the mountainous desert from the glorious Red Sea on Egypt's west coast is a surprise party of strikingly colourful geometrics. Canary yellows, rustic reds and sky blues in all shapes and sizes litter this amalgam of small islands, surrounded by water at every angle and fronting the ocean – a bright and positive hotel architecture of colour and form amongst the dusty desert sands and heated haze.

The Miramar Hotel is a new project designed by American architect Michael Graves, internationally renowned more for his monumental post-modern classicism in the US and Japan than for playful geometrics in Egypt's sands. Commissioned by Orascom, one of Egypt's largest engineering and construction companies, for their new stand-alone tourist village of El-Gouna, the ITT Sheraton Miramar Hotel is the first completed project of three designed by Graves for this site.

This 284-room, five-star hotel covers an area of 200,000 square metres, a sixth of which is filled with artificial turquoise lagoons and a quarter terraced with local palm trees. A building characterised by width rather than by Graves' usual monumental heights, the Miramar is a series of shaped islands and themed clusters that are connected by wooden bridges across the water.

Circular lagoons enclose pools of water that bear their reflections on the still surface; great yellow drums encircle staircases; pointy, red 'pigeon-houses' pierce arcades; Arabic numerals – 7s and 8s – hide in brick lattice; and tall towers afford panoramic scenes across a landscape of ▶

Nights

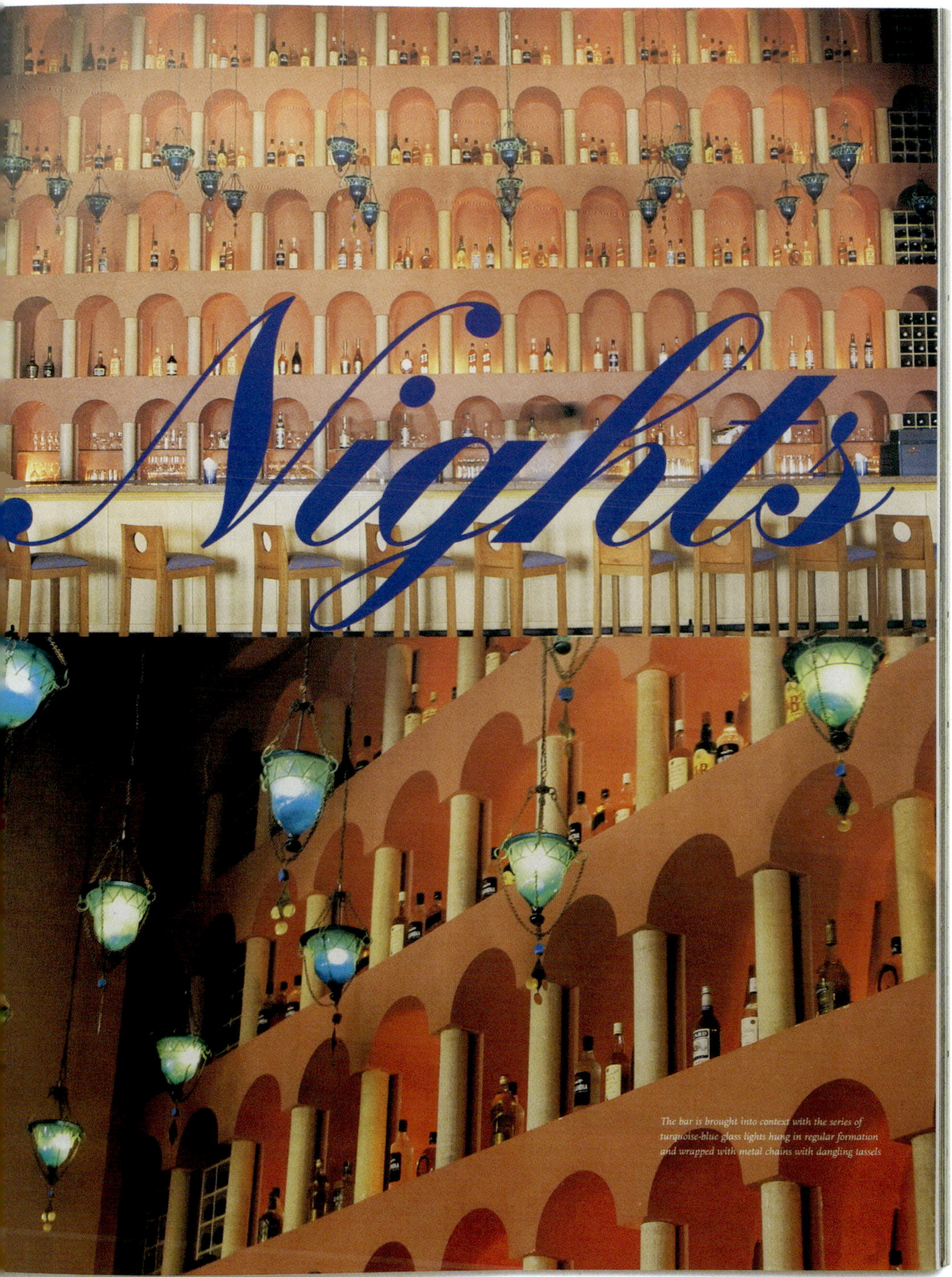

The bar is brought into context with the series of turquoise-blue glass lights hung in regular formation and wrapped with metal chains with dangling tassels

Graves took the dome and vault forms – the design of which typif

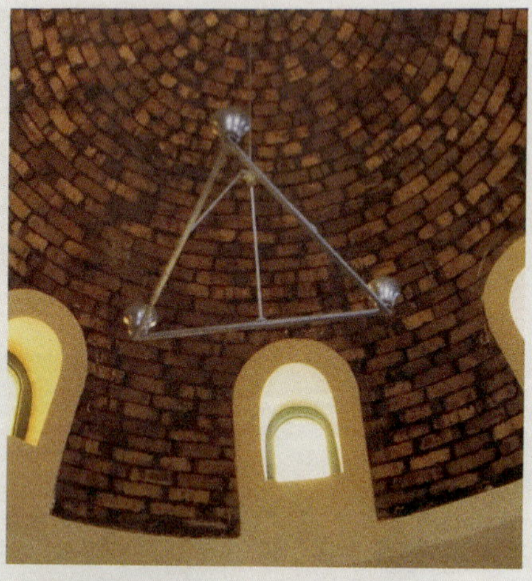

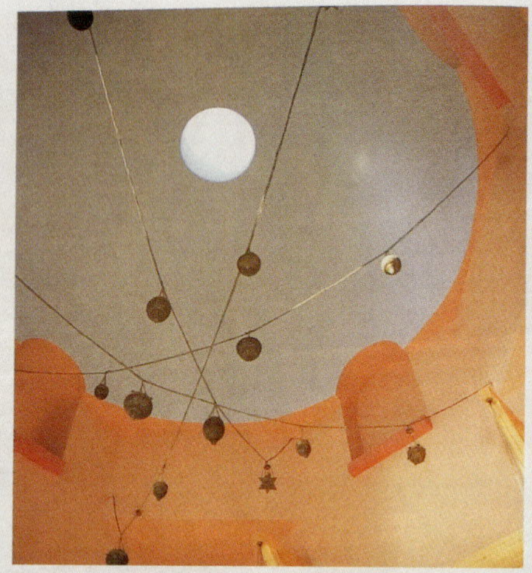

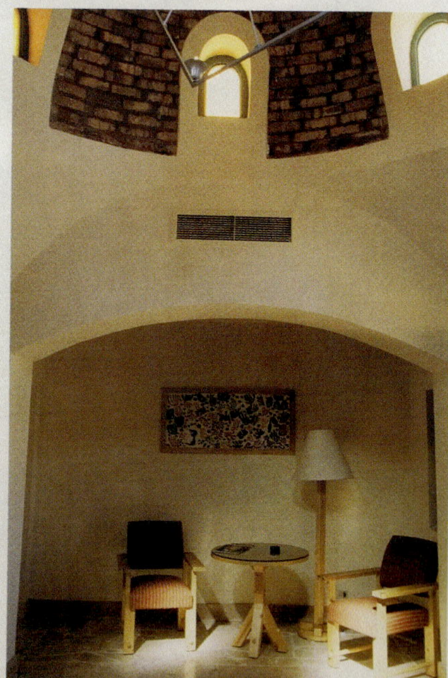

Guestroom interiors remain simplistic, exhibiting an almost ivory-coloured stucco that serves to enhance the raw, red brick of the dome

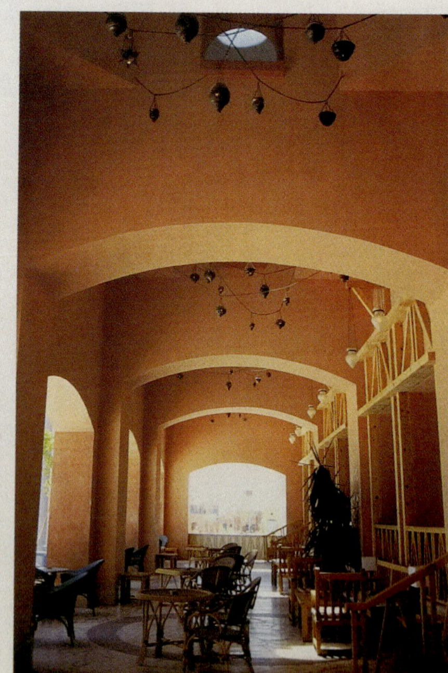

Bar terrace, coloured with biscuit browns, blue domes and chains of small, metal Islamic lights

Egyptian rural vernacular architecture – as his point of departure

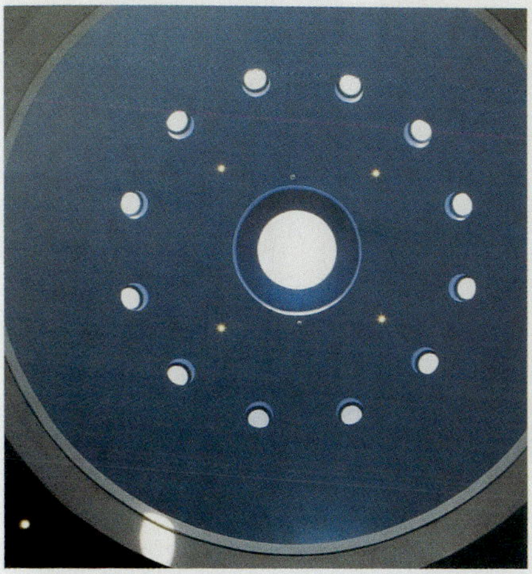

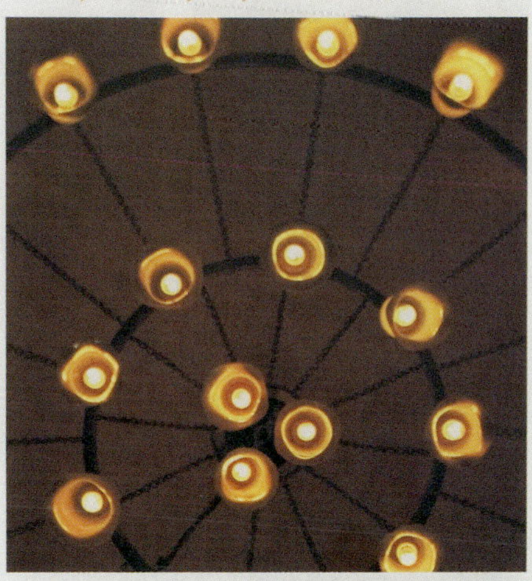

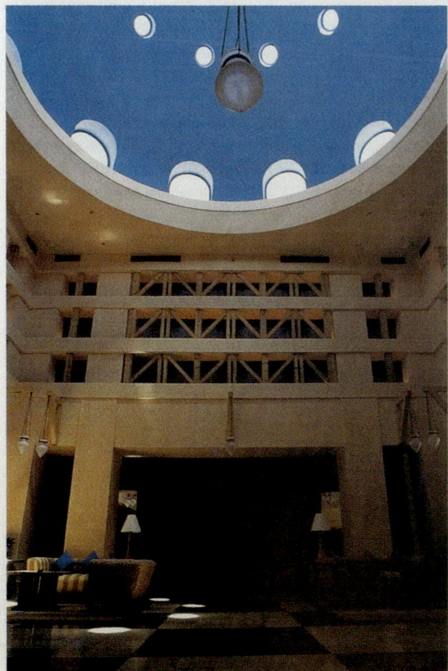

Entering the lobby there is an expansive sense of light and space gathering focus under the luminescence of the great blue dome

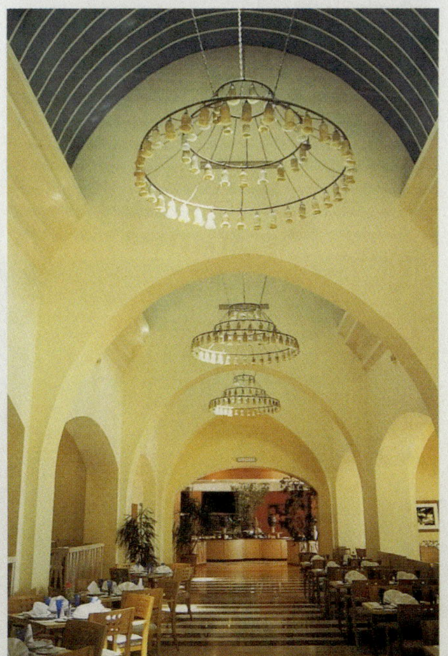

Throughout the public spaces are great chandeliers suspended from domes and hanging sequences of colourful glass or decorated metal lanterns

SUBJECT
Miramar Hotel
LOCATION
El Gouna, Egypt
DESIGNER
Michael Graves
PHOTOGRAPHER
Eleanor Curtis
WRITER
Eleanor Curtis

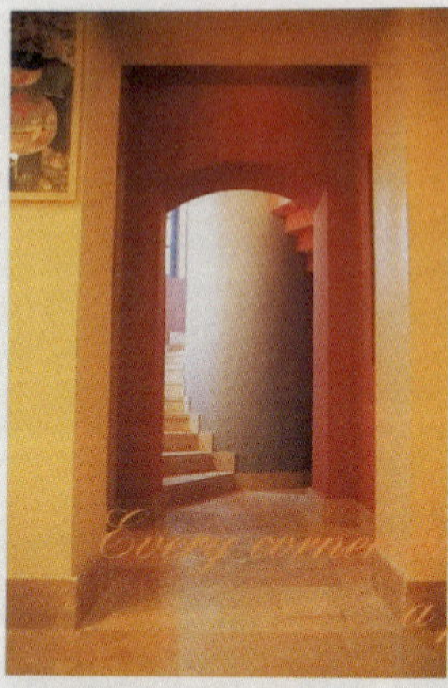
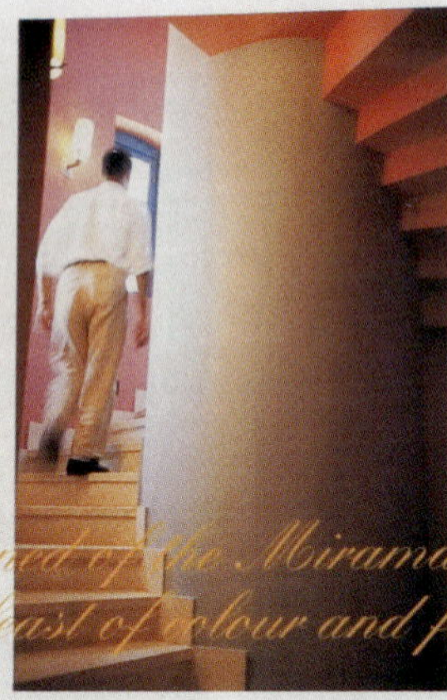
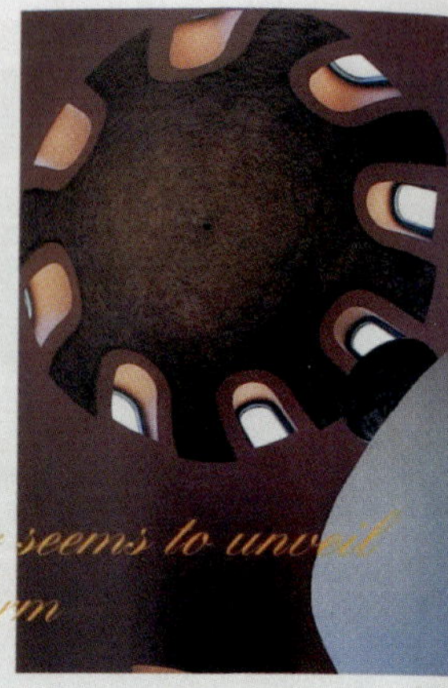

Every corner turned of the Miramar seems to unveil a feast of colour and form

Executive Suite staircase. Pinks wrap blues, yellows frames pinks, and ivories comfort yellows

domes, cones and curves. Viewing the structure from a distance, one visiting English journalist likened the playful geometrics that perforate the skyline to "Alessi salt and pepper pots".

Rustic furniture At the client's request, Graves took the dome and vault forms found in the Kafr el-Gouna (the village centre) – the design of which typifies Egyptian rural vernacular architecture – as his point of departure.

The clearest example of Graves' homage to the vernacular is to be found in guestroom domes, which remain in raw brick with the bed positioned directly underneath, affording a 'raw' awakening for guests. As Graves explains: "In terms of stuccoing versus leaving bare certain brick elements, we attempted to persuade the client to leave many of the domes and vaults bare brick. We feel that the construction of these architectural elements is so wonderfully made that it would lessen the whole if some of these roofs were not left exposed."

In his derivations from the local vocabulary Graves has successfully combined the indigenous with his geometrical elevations and decorations, resulting in a busy and lively exterior and a rich interior architecture full of surprises. "While the exteriors have much of my personality in them, we have tried to be sympathetic to the constructional possibilities and vernacular methods which has led to interesting tangency's to the local character and context. Although the furniture we have designed is not explicitly Egyptian, I do feel that its 'rustic' character leads to both an archaic and vernacular quality."

This 'rustic' assortment of tables, chairs and lights found throughout the hotel make reference not only to local form but also to local manufacturing possibilities. As the Graves team discovered during the production phase, the outcome includes more adaptation than precision, giving a '-cruder' line of furniture than originally designed. A long-held tradition, as Graves points out: "I have always been interested in early Egyptian furniture developing mostly from the anamorphic character of many of the basic elements such as chairs and tables. I am also fascinated with the accommodation to the construction of the architecture relative to how furniture might adapt to it. For example, a long rectangular table might have only three legs, two at one end and one at another, because floors were often rough and uneven and three points determined the plane."

Though none of the furniture actually goes to these extremes, the uneven floor undulations of the long gallery that stretches from the bar to the pool cluster plays with this idea on a larger scale. With nose to the ground, one can see a ripple of solid waves that give an almost water-like motion to the long space.

Brick dome Every corner turned of the Miramar seems to unveil a feast of colour and form. Entering the lobby there is an expansive sense of light and space gathering focus under the luminescence of the great blue dome. With light streaming in from all sides the soft blues and yellows subdue the tone, and, combined with the peaceful trickling of the water of the Roman fountain, it is altogether a very relaxing initiation to the hotel.

Directly in front is a small *sheesha* terrace (otherwise known as the 'hubbly-bubbly pipes' smoked throughout the Arab world) with small water fountains between each table, and on either side the terraces of the bar and restaurant – coloured with biscuit browns, blue domes and chains of small, metal Islamic lights.

To the right is the galleria, where the sun beams into the arcades, playing on the reds and turning them into dusty pinks. Graves explains: "We chose colours that from the day they were painted would, from the intense exposure to the sun, fade from the original bright colour to faded colours. We think that the stucco surfaces when rendered in this way take on a patina that links them more closely with their climate."

Interiors of rooms remain more simplistic, exhibiting an almost ivory-coloured stucco that serves to enhance the raw, red brick of the dome. In contrast, the 'Japanese' bathrooms are tiled in a deep blue. Behind the doors of the more upmarket Executive and Presidential Suites, however, pinks can be found wrapping blues, yellows framing pinks, and ivories comforting yellows. "People have been attracted to the colours and always comment on the brick dome in the room. They especially love the Japanese bathrooms with the blue tiles," comments Yasser Hamed, who has worked at the Miramar since its opening. Themes of colour run throughout the Miramar at all levels of detail: from the cool-blue blown glass on the restaurant tables to every dome, vault and roof. In creating bold, varied designs, the Graves team were hoping that guests would be able to identify with the shapes, colours and details, rather than to be left with nothing more than another 'international' hotel experience.

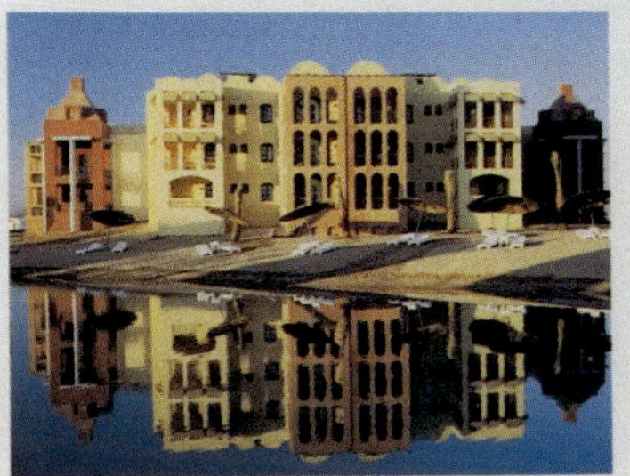

Room cluster. Michael Graves created a bright hotel architecture of colour and form amongst the dusty desert sands and heated haze

Squiggly Throughout the public spaces are great chandeliers suspended from domes and hanging sequences of colourful glass or decorated metal lanterns. True to the traditions of Egyptian lighting found in the old and great palaces are the loose strings of metal chains that connect the otherwise invisible planes of domes. Fascinated by the traditional lighting used in other site projects, Graves felt that their character was appropriate for some of the public spaces and so invited the local lighting designer to lend shape to these spaces. The bar, designed in a style that one would expect to find in the trendy quarters of Barcelona, is brought into context with the series of turquoise-blue glass lights hung in regular formation and wrapped with metal chains with dangling tassels. Another small but delightful detail is found in the bar's armchairs, covered with textiles that weave golden Pharaonic stars into cloth.

In contrast, lighting for the bare domes and vaults of the bedrooms takes on a deliberately modern style, clearly distinguishable from the traditional. "I thought it was important to be explicit about the lights that were seen as our design and therefore modern. The lights for the guestrooms were designed to illuminate the vaults and domes that were minimal in character." The result juxtaposes the old with the new.

Remaining ever playful, the design of the main pool follows irregular, 'squiggly' lines that reach their final point directly under the balcony of a room. The children's pools in the adjacent court take on the shapes of fish, waves and a circle, at depths of 50cm, 30cm and 5cm. The wave theme also washes into the pool bar furnishings, into the canopy of the main and Oriental restaurant terraces, interweaving coarse hessian through blue, and into the rippling handrails of guest-room balconies. The momentum continues as pathways undulate their way through the site, crossing bridges and running under pavilions, joining galleries and arcades.

Huge chandeliers The restaurant is a grand spectacle of striped vaults leading to the 'buffet-in-the-round' at the far end, with its decorative floor. "It reminds me of the cathedrals that we have back home in Spain. It is so monumental that we now refer to it as 'the cathedral'," comments Assistant Manager Pilar Monzon, who hails from Seville. Blue-striped vaults make explicit reference to tents, a structure used throughout Egypt's deserts by the Bedouin tribes; says Graves: "I have always loved the kind of light that comes through the canvas of tents, especially while dining." Though the light in the restaurant is not quite as subtle as that through canvas, the huge chandeliers do create the illuminative effect that Graves was seeking.

While taking in the panoramic view afforded by the "tower" over the Oriental restaurant on the beach, Reservations Manager Mohammed Hafez summed up, in plain terms, Graves' play on the old in creating the new: "The entire site seems like an amalgam of different palaces, but with a touch of the modern. The old and new sit very well together in this landscape." ◀

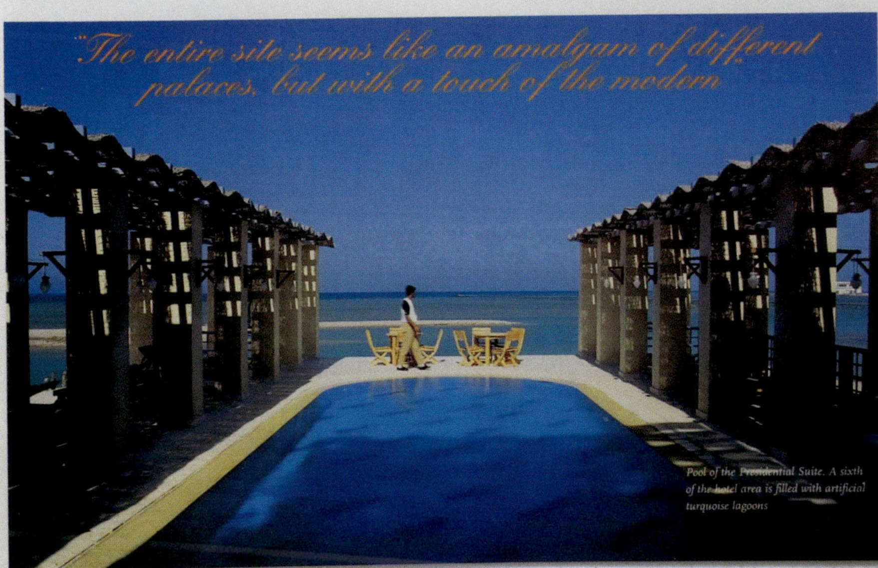

Pool of the Presidential Suite. A sixth of the hotel area is filled with artificial turquoise lagoons

Issue 06

January/February 1999

058-059

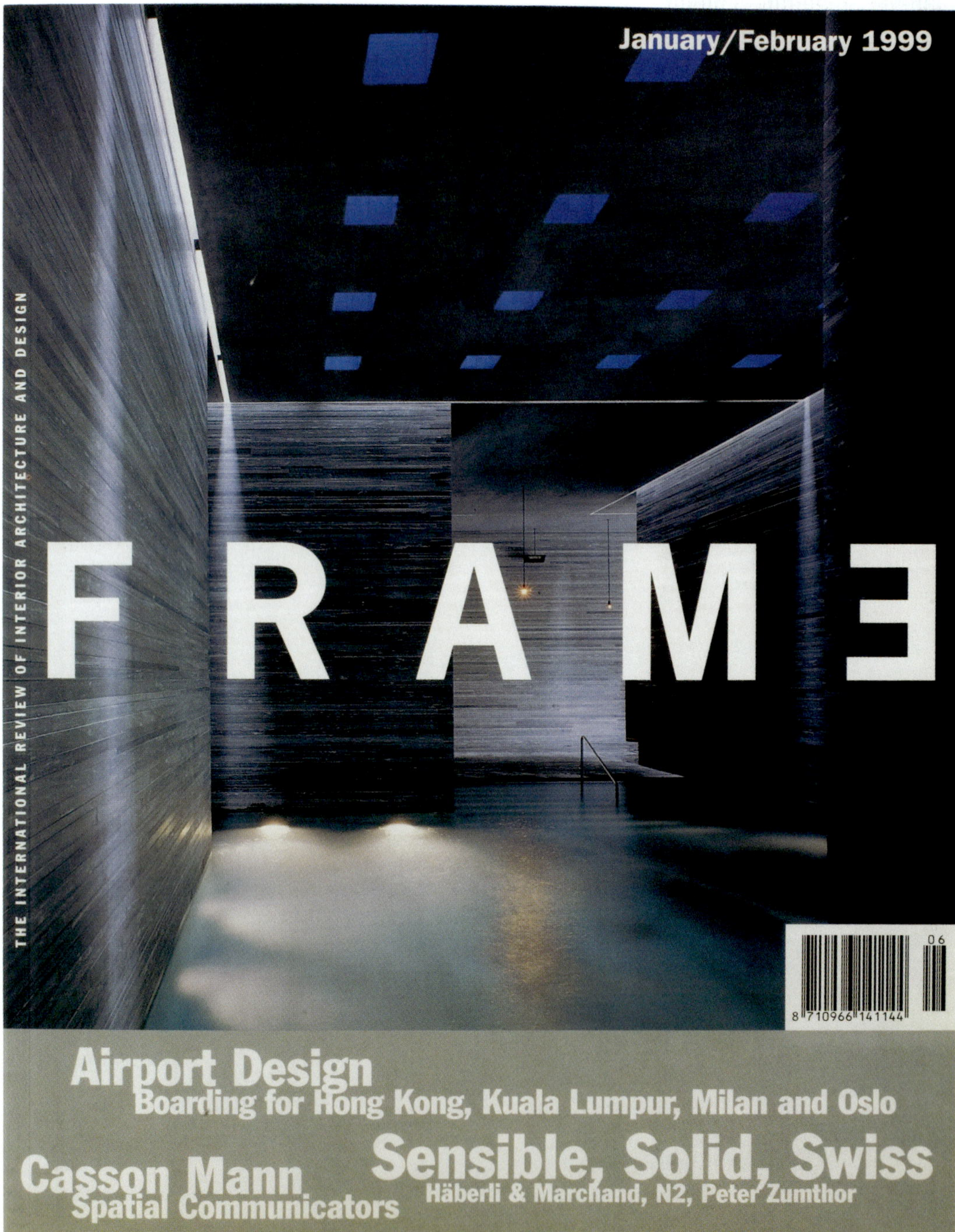

ART DIRECTOR
Ton Homburg,
Opera Designers

06

Sneaky-Clean Spaciousness

Frame editor Robert Thiemann
about the project

From the very beginning – in 1997 – we wanted to make a truly international magazine. But at that time it wasn't as easy as it is today to feature high-quality, innovative projects. So we were really happy when our first Japanese contributor, Chiharu Watabe, came up with images of Kuala Lumpur International Airport. Malaysia represented a territory totally unknown to us. The airport, designed by Kisho Kurokawa, proved to be spacious and spectacular, though perhaps less unfamiliar than we'd hoped. Main features were large glazed curtain walls, huge load-bearing columns, a small rain forest and squeaky-clean floors. Apart from the palm trees and the bamboo, however, we were looking at an airport that could have been in Paris or London. We've not yet featured another Malaysian project.

DESIGNER
Kisho Kurokawa
www.kisho.co.jp

SUBJECT
Kuala Lumpur
International Airport

LOCATION
Kuala Lumpur, Malaysia

PHOTOGRAPHERS
Tomio Ohashi
Chiharu Watabe

WRITER
Chiharu Watabe

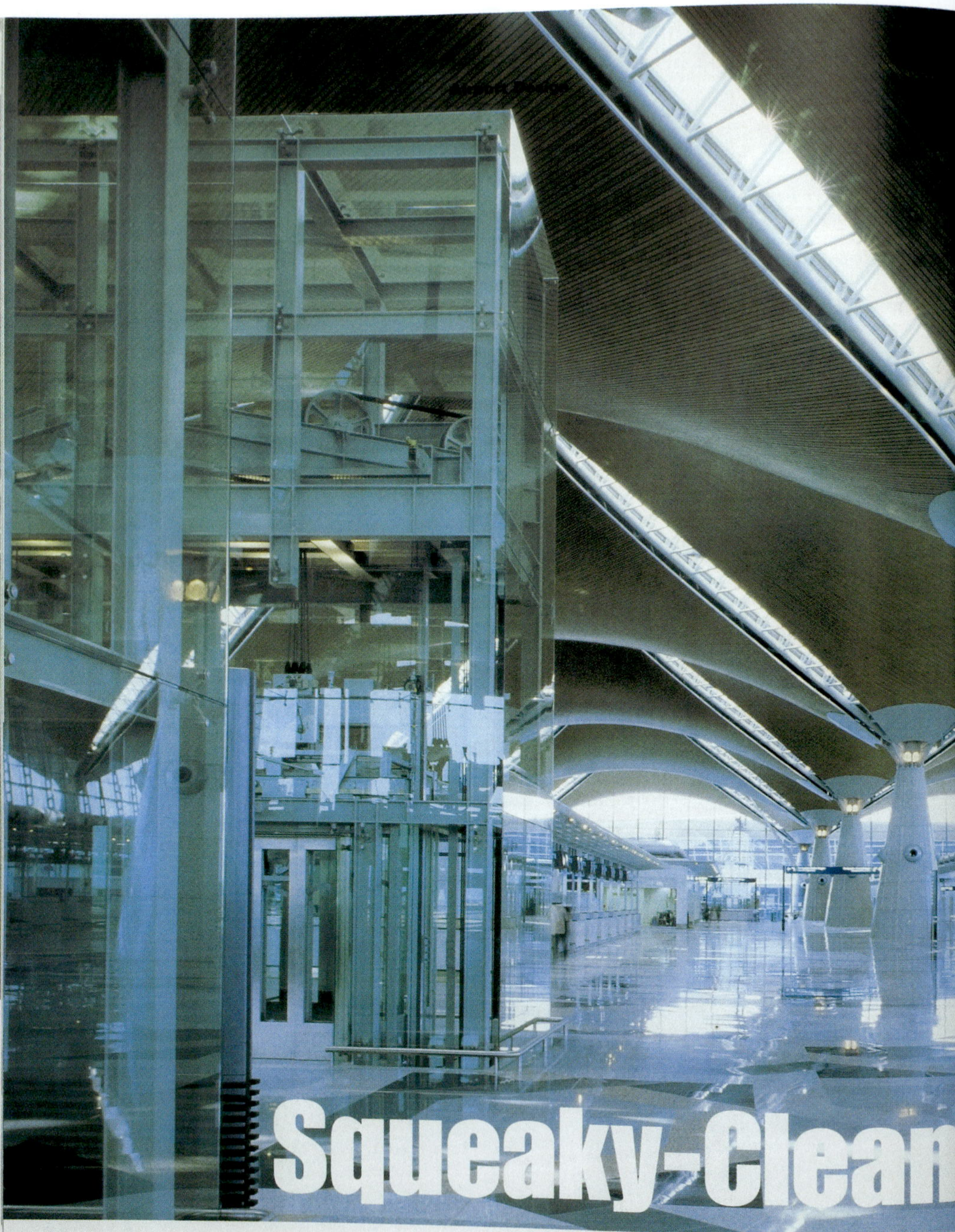

Squeaky-Clean

The departures hall of Kuala Lumpur International Airport: literally stainless

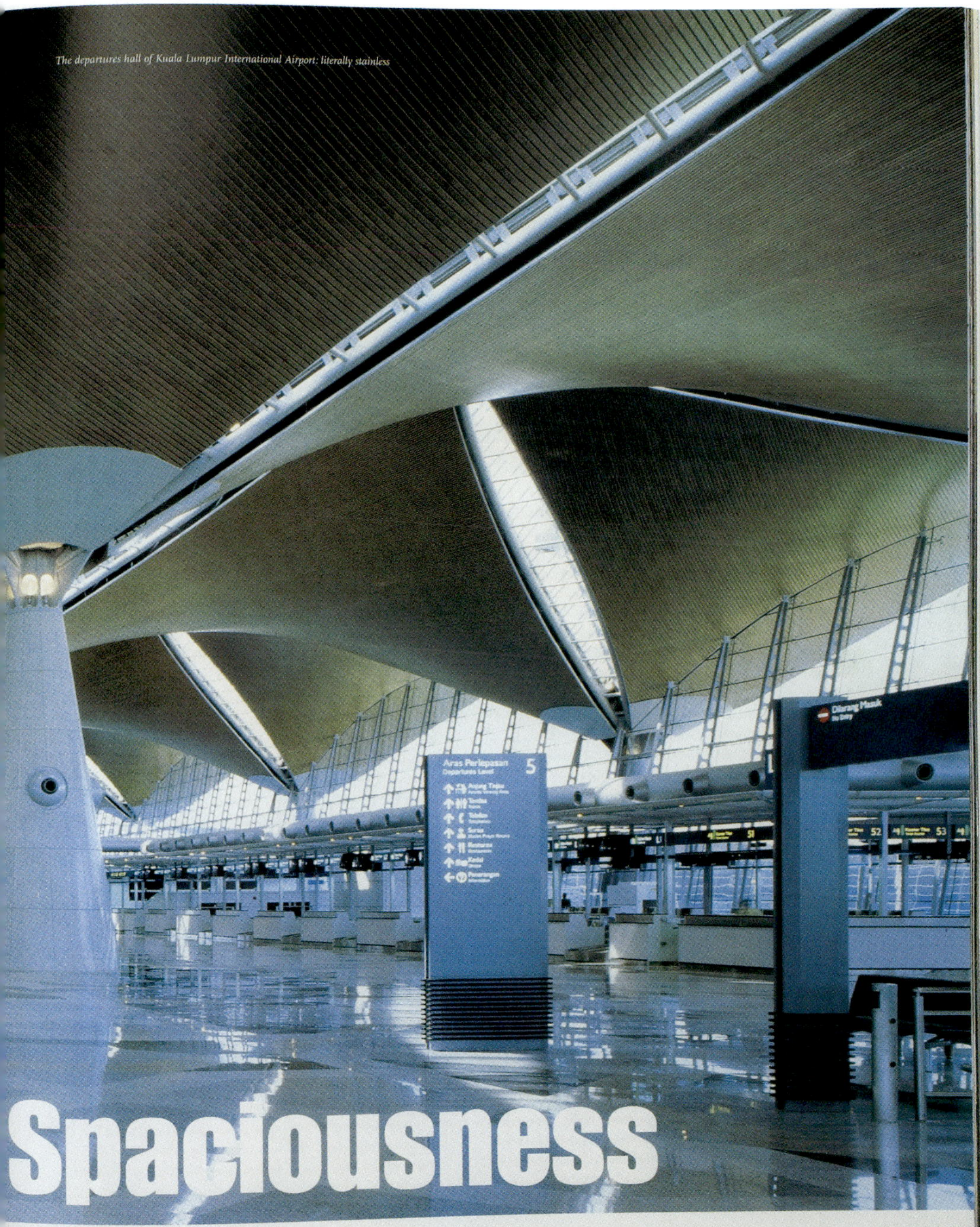

Spaciousness

SUBJECT
Kuala Lumpur
International Airport

LOCATION
Kuala Lumpur, Malaysia

DESIGNER
Kisho Kurokawa

PHOTOGRAPHERS
Tomio Ohashi
Chiharu Watabe

WRITER
Chiharu Watabe

The first impression of Kuala Lumpur International Airport is one of extreme purity. There is not a spot or stain, and the abundant use of stainless steel and glass increases the impression of hygiene. In this sea of glass and steel, green implants provide a link to the surroundings.

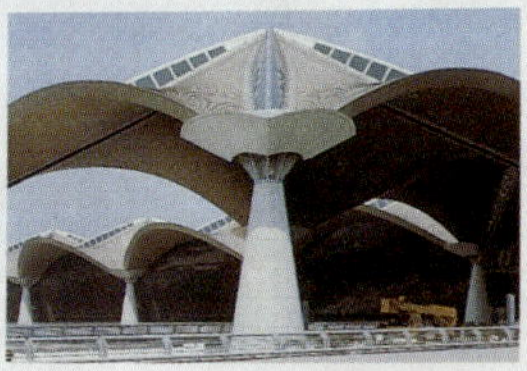

Roof of the main terminal

text **Chiharu Watabe** photography **Tomio Ohashi**

The reason I decided to travel from Tokyo to Kuala Lumpur, the capital of Malaysia, was a practical one. I was heading for England and, by chance, the cheapest flight was with Malaysian Airlines. I didn't know much about Malaysia or its new airport – opened this year – apart from the TV reports of the problem on the opening day (rather than about the opening ceremony). It's quite possible that the Japanese media wouldn't even have mentioned the airport if there hadn't been a problem, despite the fact that the Malaysian government expects it to become the busiest and most important transit airport in Asia.

Kuala Lumpur is commonly known as KL. The new airport is also known by its initials: KLIA – I wonder if the name is a kind of play on the word 'clean'! My first impression of Kuala Lumpur airport is its extreme cleanliness. Although all of the airports I have ever frequented are decently clean, this one is literally stainless; there is no rubbish or dirt anywhere. The abundant use of stainless steel and the combination of stainless steel and glass give the impression of a professional kitchen facility or medical laboratory. I imagine there are more cleaners than other staff, perhaps even more than passengers. Presumably this has something to do with the Malaysian national fondness for hygiene. Cleanliness and newness are considered to be signs of civilisation...

The other significant impression KLIA makes on most passengers is its spaciousness. Here are some statistics: total floor area, 479,404 square metres; highest elevation, 21 metres; lowest elevation, 10 metres; aircraft gates, 46; aircraft stands, 106; passengers per hour at peak times, 7,130. There are already plans to expand the airport.

The impression of spaciousness is somewhat exaggerated by the fact that passenger numbers have not yet reached their maximum. Many retail units in the shopping area are still not open and queues are rarely seen. Accordingly, there is no need for a large number of information personnel and the information counter is sparsely manned. Instead, there are computer information kiosks at every corner, which is not very helpful if you just want to find the nearest bathroom. But even if the airport were running to capacity, it would not seem as full as Narita (Tokyo) – and most other airports. The leaflet states that there are at least 55 square metres per passenger at peak hour.

Great refreshment It is not only a lack of people that makes KLIA pleasant, but also the huge glass curtain walls and a 'forest' just outside the window. The architect responsible for the airport's design concept, Kisho Kurokawa, describes it as 'an airport in the forest; a forest in the airport'. Between buildings and inside the buildings, there is greenery. The green area at the core of the cross-shaped satellite building is especially remarkable. An abundance of plants – rain-forest trees, palm-trees, bamboo, shrubs – and a powerful waterfall have been placed within an inverted conical space 50 metres in diameter and surrounded by a glazed curtain wall.

Since all four concourses meet at this core, every passenger comes across this miniature forest. It is enormously refreshing, especially for passengers like me arriving after a long flight (7 hours from Tokyo, 13 from London). Situated around the glass wall are lounge spaces and two cafes. I really regret not having a cup of tea in the cafe. It has luxury sofas and cakes on trays in the English high-tea manner – it must be one of the most beautiful cafes in the airport. The roof of the main terminal is reminiscent of the undulating oil palm plantations of Malaysia. The conical columns inside reinforce the tree image. Owing to local fire authority concerns, however, Kurokawa's proposal to use timber for the ceiling of the main area was not realised.

Other facilities will appeal to transit passengers. In addition to shopping malls there is a gym, an 80-room hotel, and some areas of worship for Muslims. Or if you only have an hour or so to kill, you can simply take a shower. Even if you are not boarding a plane, you are still free to use most of the facilities in the main terminal. The check-in area is accessible to the public, since there is no baggage inspection before check-in. There is so much to do there that you could easily spend a half a day browsing around. Have a delicious (and inexpensive) local meal in the viewing point or decent Italian espresso in the coffee bar – both reason enough for a visit. ◀

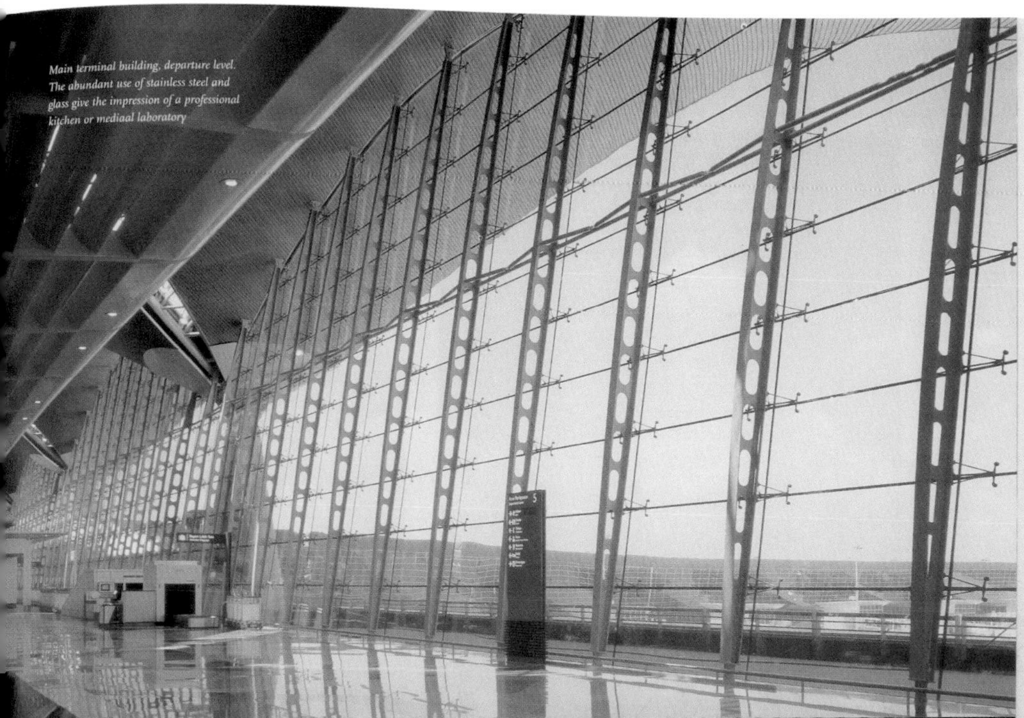

Main terminal building, departure level. The abundant use of stainless steel and glass give the impression of a professional kitchen or mediaal laboratory

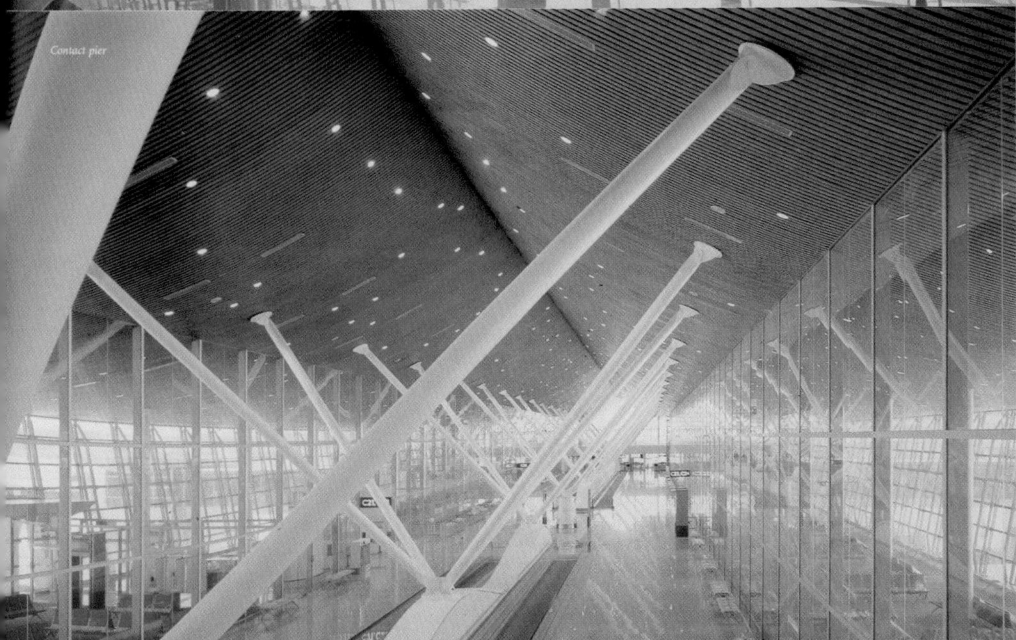

Contact pier

SUBJECT
Kuala Lumpur International Airport

LOCATION
Kuala Lumpur, Malaysia

DESIGNER
Kisho Kurokawa

PHOTOGRAPHERS
Tomio Ohashi
Chiharu Watabe

WRITER
Chiharu Watabe

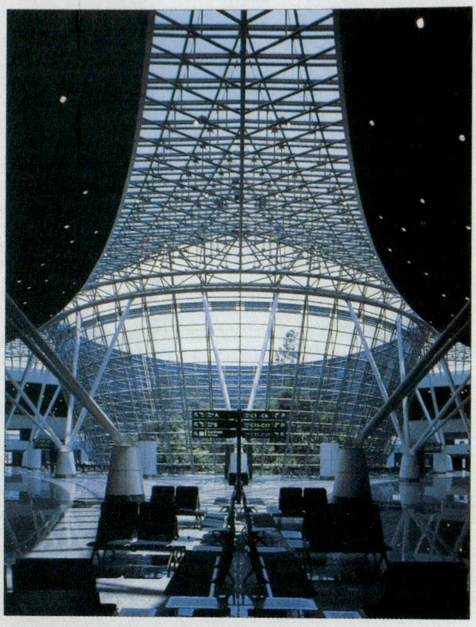

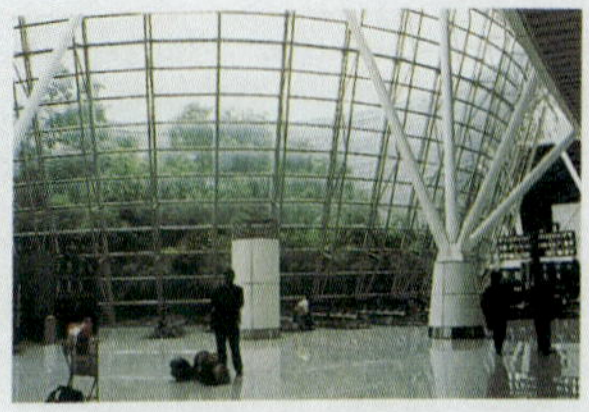

Photo: Chiharu Watabe

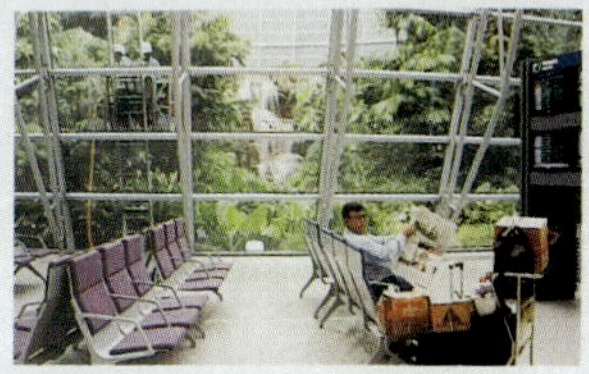

Photo: Chiharu Watabe

The cross-shaped satellite building features a green area in the core. An abundance of plants and a powerful waterfall have been placed within an inverted conical space surrounded by a glazed curtain wall

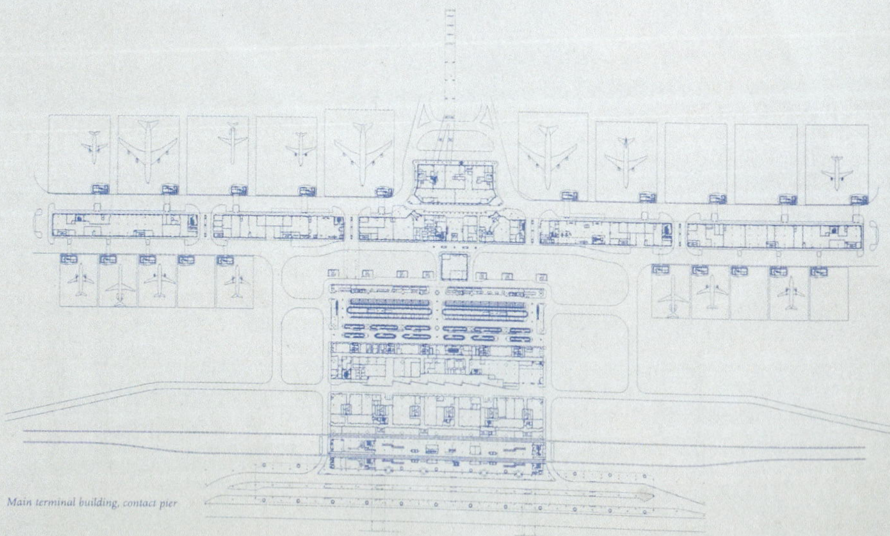

Main terminal building, contact pier

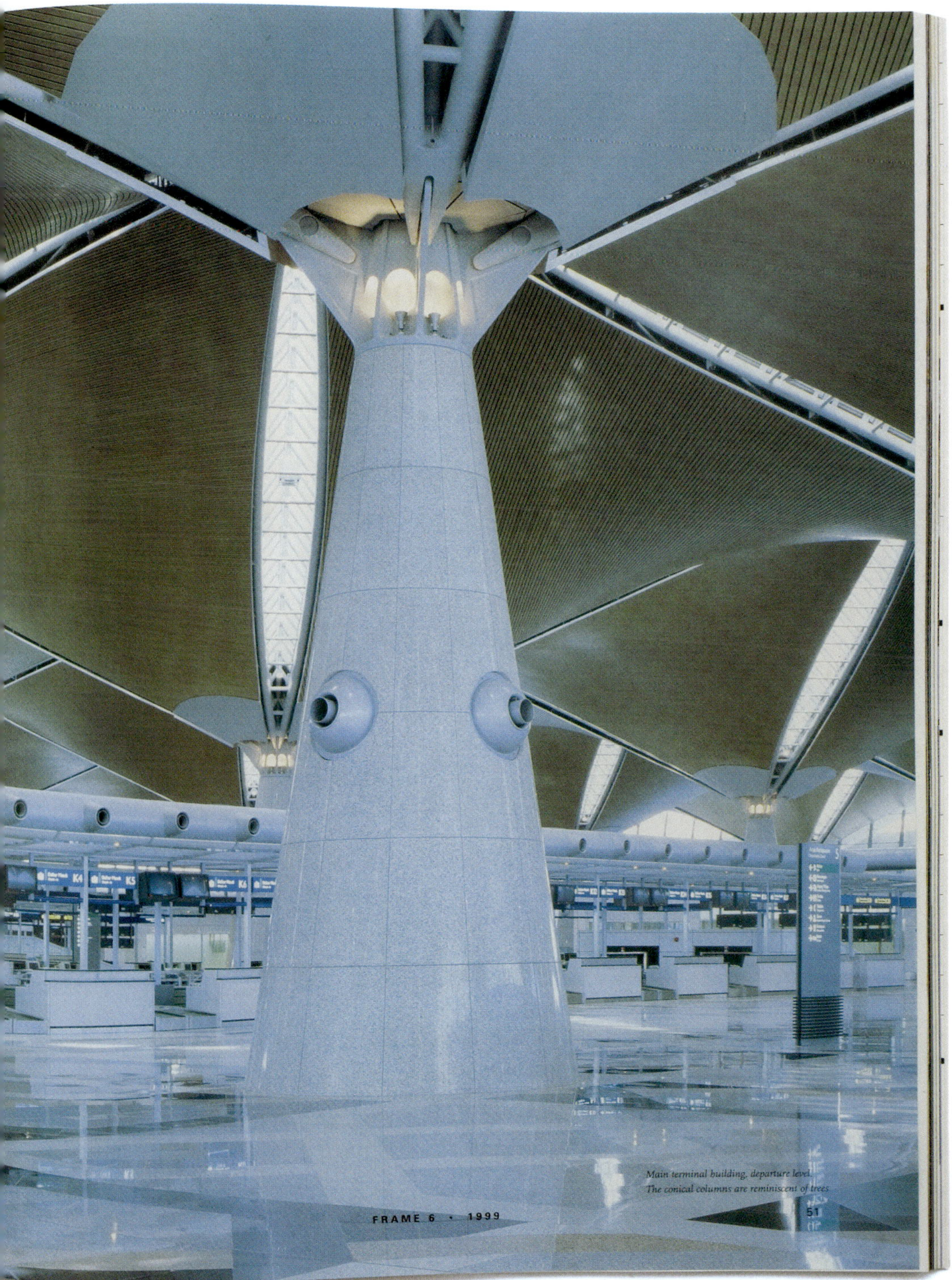

SUBJECT
Kuala Lumpur
International Airport

LOCATION
Kuala Lumpur, Malaysia

DESIGNER
Kisho Kurokawa

PHOTOGRAPHERS
Tomio Ohashi
Chiharu Watabe

WRITER
Chiharu Watabe

066-067

March/April 1999

Issue 07

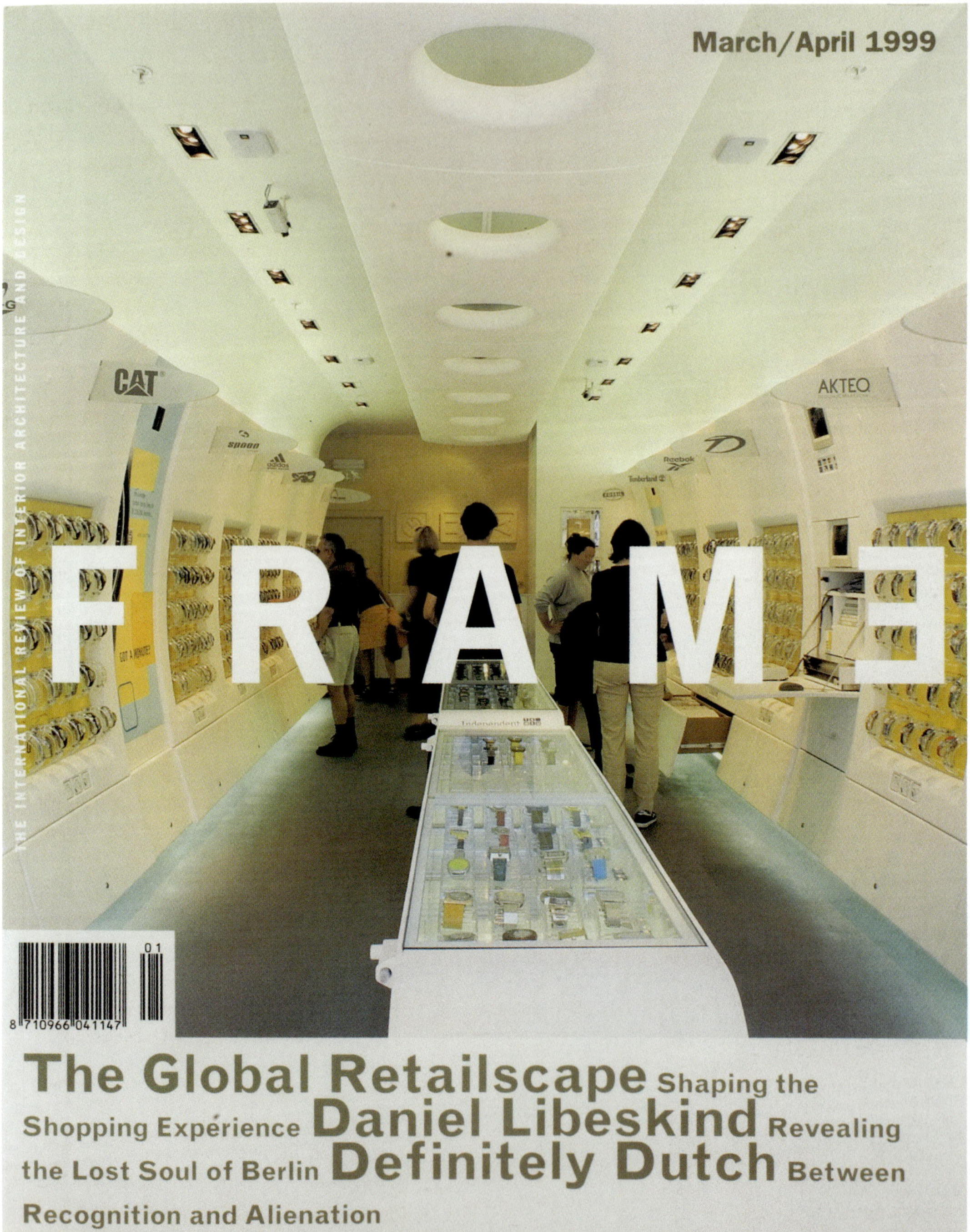

ART DIRECTOR
Ton Homburg,
Opera Designers

07

Competing With the Glamour / Building a Better Beehive

DESIGNERS
John Herbert Partnership Design (JHP Design)
www.jhp-design.co.uk
Greig + Stephenson Architects
www.gands.co.uk
Virgile and Stone
www.virgileandstone.com
Merkx+Girod Architects
www.merkx-girod.nl

SUBJECT
Selfridges versus De Bijenkorf

LOCATIONS
Manchester, England
Amstelveen, Netherlands

PHOTOGRAPHERS
Roos Aldershoff
www.roosaldershoff.com
Alexander van Berge
Chris Gascoigne

WRITERS
Robert Preece
Jules Marshall

Designer Raj Wilkinson (JHP Design) about the project

A hugely successful extension of the Selfridges' brand outside its home town of London and into the UK's second city: Manchester.

Did the article have an effect?

It helped to celebrate the success of Selfridges' vision of being a destination store and to communicate JHP's excitement at having been asked to be involved with the project.

Recent news or information about the project featured in *Frame*

The project was looked at as a stepping stone for Selfridges' further expansion, which included a second store in Manchester city centre and its radical new store in Birmingham.

—

Architect Nigel Stephenson (Greig + Stephenson Architects) about the project

The De Bijenkorf department stores – Holland's equivalent of New York's Bloomingdale's – have a strong reputation in Europe. Each De Bijenkorf store has a clear sense of identity. The largest challenge architecturally was to create a well-known, large-format department store in an area half the normal trading size. The brief was to incorporate the five 'worlds' of the store within a space that had many constraints. We believe we achieved this successfully.

Comments on the published article

The article was comprehensively written and the format of the article well considered. The comparison between Selfridges and De Bijenkorf worked well. It was particularly interesting to read the comments from such a wide range of contributors to the respective schemes.

of similar items in different colours and designs. This is the conformity of the unique: the same products available in a variety of styles to a wide range of customers.

The topography of this type of shopping is not the heart of town or a traditional street lined with shops, but a destination known perhaps as the Plaza, the Hive or the Tower: the shopping street transformed into three levels under one roof, unaffected by the elements. Here, where it's always warm and light, where water murmurs like a soft melody, the wicked world outside seems far away. As long as your visit lasts, you bathe in luxury and extravagance. Those who don't buy, look.

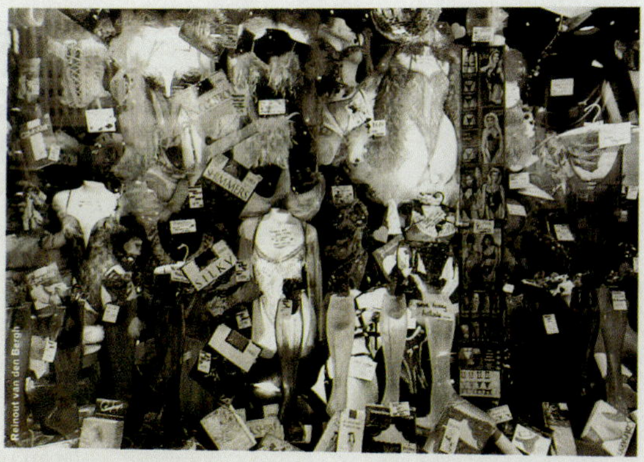

Reinout van den Bergh

Those weary of shopping take a break; every department store has a bar, a café, a restaurant – both upstairs and down. 'Fun shopping' means spending money on every level.

This new shopping style has an international flavour as well. Things once available only in faraway lands are now at your fingertips. The tea set on display is not just any tea set, but the *Méditerranée* ensemble 'straight from the Provence': one distinctive model available in a wide range of colours and motifs. This is the world of the 'look', a personal style to match every taste. The Biblical Paradise was a vegetable garden shimmering in shades of green and full of trees 'pleasant to the sight and good for food'. A river flowing through this garden separated into four streams, one of which surrounded a land 'where there is gold' as well as 'bdellium and the onyx stone'. A pastoral idyll for the agrarian soul. For metropolitan residen Paradise is a palace of consumption, a horn of plenty. In today's land of milk and honey trees have become natural wood, water has solidifie into cool mirrors, gold sparkles in the form of brass, and bdelliur and onyx are granite. The contemporary Garden of Eden is in a palace old or new, located where shopping streets are flanked by the void of an atrium on one side and the fullness of 'shops in a shop' on the other. It is a re-creatio of the old heart of town, a labyrinth of streets surrounding empty 'squares' topped by a glass roof, a world without thresholds and doors in which you're led, unwittingly, from one enviro ment to another. Each shop has a different '-theme', which is geared to label and interior, and your shopping experience is segmented according to time and subculture. There's something here for everybody, but all under th same roof, like a house for the ideal, old-fashioned family. The shopping search is on for something that no longer exists at home: a shared experience, unity in variety. ◂

Competing with the Clamour

Manchester's new **Selfridges** department store is housed in a giant, Las-Vegas-like shopping mall. The big issue was taming this over-the-top, post-modern beast. *By Robert Preece.* ▸

Building a Better Beehive

The newest addition to Holland's most prestigious department-store chain, **De Bijenkorf**, is also the smallest. Despite its size, however, customers are treated to a sense of spaciousness. *By Jules Marshall.* ▸

SUBJECT
Selfridges versus De Bijenkorf

LOCATIONS
Manchester, England
Amstelveen, Netherlands

DESIGNERS
Various, see page 067

PHOTOGRAPHERS
Various, see page 067

WRITERS
Robert Preece
Jules Marshall

Make no mistake; the Trafford Centre is fierce visual competition for any department store. From a distance you might confuse the panorama of the mall for the domes of Rome – for a split second, that is. Learning from Las Vegas, the £600-million Trafford Centre – a retail and entertainment complex in suburban Manchester – hits you with a semiologic punch that packs the power and precision of a semi-automatic weapon.

Climaxing at the 'town square', or two-level food court, the celebration of appropriation becomes overwhelming. Public toilets and Pizza Hut emerge behind an ancient Egyptianesque façade of hieroglyphics and the columns of kings; and McDonalds' golden arches rise beneath their Moroccan equivalents. This place has everything: a French Quarter à la New Orleans, China's Imperial Palace and a tropical rainforest with talking birds. It even boasts a triumphal arch, but despite the claim of a spokesperson that the structure is a reference to Venice, we put our money on Rome and, specifically, the Arch of Constantine. To top it all off, our symbolic voyage around the world takes place on an enormous ship. Amidst this power, can design efforts at the mall's new anchor store – Selfridges & Co. – possibly compete?

Publicised as 'the £20-million store focusing on design', this 200,000-square-foot Selfridges is the company's first branch outside London. Founded in 1909, the store sells a variety of wares and clothes at different price levels, 'from a £5 T-shirt to a £5,000 designer dress'. According to General Manager Kathy McGowan, Selfridges aims for difference. 'Shopping should be about fun, energy, and excitement,' she tells us. 'Selfridges Manchester offers all these things, plus a friendly staff, great departments and a stunning environment in which to shop.' Selfridges goes for its goals by integrating shopping with eating and entertainment, a scheme that affects the design and its usage.

Positioned at the centre of the mall, Selfridges stands at the intersection of an Egyptianised shortcut to the food court and a two-level arcade that combines plastic obelisk-shaped shrubbery, Roman-inspired sculptural forms and murals with a Renaissance look set below a barrel-shaped glass roof. If this isn't enough, circular vanity portraits of the new Trafford Centre Medici stare out and over passing shoppers. At this mega-confluence of historical fetishism, Rome's grandest dome – adorned with more murals and accommodating a bronze-lined, glass lift next to side-by-side escalators – extends upwards, creating the effect of Viagra-induced verticality. Framing Selfridges, rose-coloured marbleised columns sport gold-painted bases and capitals. Called in to compete with this 'busi-ness', John Herbert Partnership (JHP) – the lead designers – had their work cut out for them. 'For the mall front, we used Rosso Alicante marble, which had to be approved by the Trafford Centre,' says a spokesperson. 'We needed to make a strong enough statement to not get lost in the clamour.'

Meanwhile, the street-side exterior refers to Selfridges' flagship store in London, with its manorial façade and, to maximise the symbolism, a lawn, shrubbery, and fountain.

Unfortunately, however, this entrance appears under-utilised and awkward as a result of the store's dominant mall orientation. The situation here – an entrance that takes customers directly into the men's department, where they're confronted with the back of a grand staircase – has a negative effect on the design as a whole. ▶

Opposite page: The mall entrance. The transition [...] dark-coloured marble co[...] trasts strongly with the b[...] lights and white gleam o[...] cosmetics department

The affluent Amsterdam suburb of Amstelveen has been around since the 12th century – though you'd never guess so today. It's a bland, rather bourgeois place, distinguished from newer Amsterdam satellites only by the presence of the CoBrA Museum. To reconfirm Amstelveen's municipal autonomy, earlier in the 1990s its leaders drew up plans to extend the cultural heart of the centre. Part of this project involved the creation of a new town square and an accompanying shopping mall: the Rembrandthof.

Similar to the commotion that hit Leeds in 1997 when Harvey Nichols opened a shop there, it was a newsworthy event – a feather in the cap of the town as a whole – when De Bijenkorf, Holland's most prestigious chain of department stores, signed on as a magnet for the mall. De Bijenkorf ('The Beehive') has branches in most of the country's major cities and, starting with its first establishment (opened in 1914 in Amsterdam), has built a reputation for erecting landmark buildings. The store in The Hague, for example – designed in 1926 by Piet Kramer – is a listed national monument. As the company's first newly built store in fifteen years, the Amstelveen project would be both an important philosophical statement on where the organisation stands, in retail terms, at the end of the century and, as such, a benchmark for a new generation of stores.

In 1995 the UK firm of Greig + Stephenson (which had recently extended the store in The Hague) was chosen as project and interior architect. They were joined on the project by fellow Brits Virgile & Stone and Amsterdam-based Merkx + Girod, two organisations jointly responsible for interior design. 'It was a very difficult project,' says Ria Baauw, product manager of the Bijenkorf's in-store design department and co-ordinator of the project from their end. 'For a start, it was not a building owned by the Bijenkorf, but had to fit in with a pre-existing "master plan" of another architectural firm.' In fact the design of the Rembrandthof, by Atelier PRO Architekten, was already well advanced at the time the Bijenkorf team was commissioned. It's true that Greig + Stephenson found themselves constrained by an existing design, but in a sense, having a clear idea of the context of the new Bijenkorf was a liberating introduction to their job.

At 7,500 square metres, spread over three storeys, the branch in Amstelveen is De Bijenkorf's smallest store. Nonetheless, the priority in the brief was to give an impression of spaciousness. This was made harder by the space allotted them by the mall designers: a triangular atrium on two floors plus a basement, all underneath a residential development. 'It's very hard to fit everything around that,' says Neil Tomlinson, project architect for Greig + Stephenson. 'The tendency is to end up with unusable corners. The tolerances are minimal, making it a hard space to use.'

The solution was to open up one end of the triangle, blurring the boundary and allowing the store to diffuse into the mall space. 'We chose to rub away at the boundaries of surfaces, to blur the edges,' says Tomlinson.

And since the Bijenkorf was invited in as a main anchor store, they were given relative freedom to define these edges. This is achieved primarily through a ceiling of triangular beech tiles that ▶

Opposite page: Along with the monumental doors, the ceiling of triangular beech tiles succeeds in creating a feeling of being in the Bijenkorf before actually entering

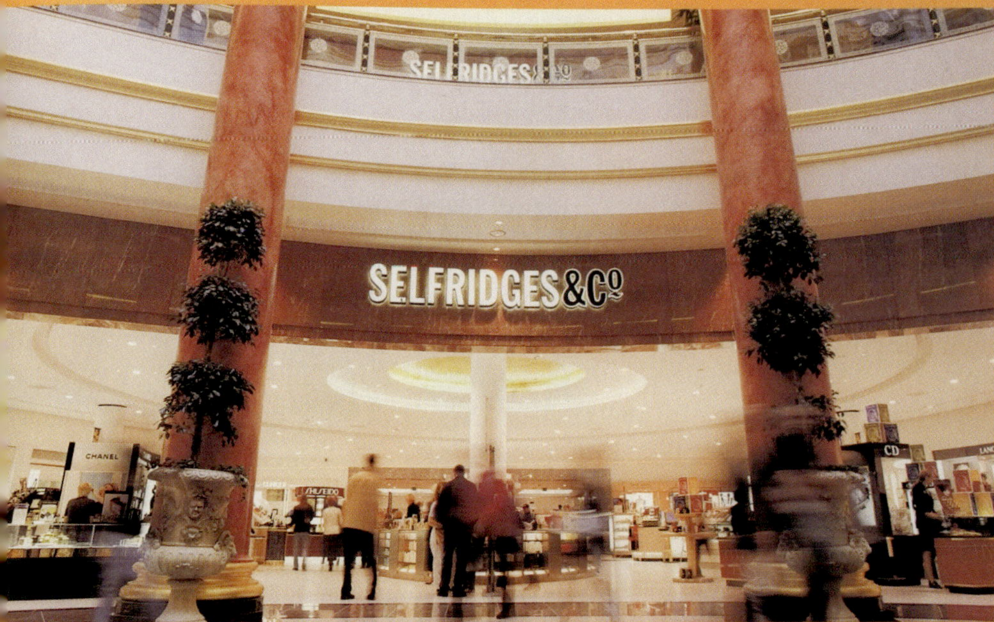

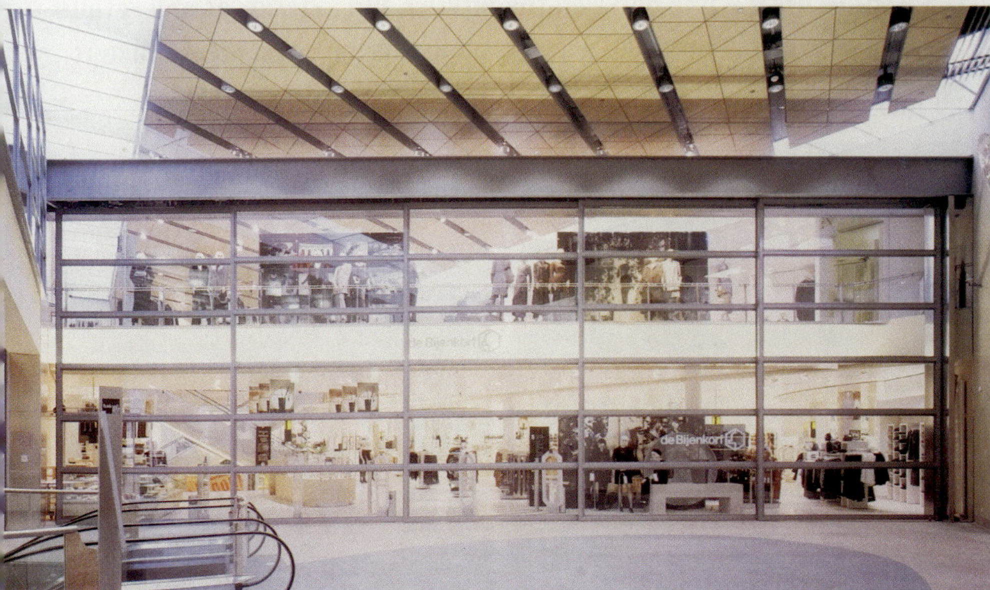

SUBJECT
Selfridges versus
De Bijenkorf

LOCATIONS
Manchester, England
Amstelveen, Netherlands

DESIGNERS
Various, see page 067

PHOTOGRAPHERS
Various, see page 067

WRITERS
Robert Preece
Jules Marshall

At the mall entrance the transition of dark-coloured marble, which relates to the interior of the dome, contrasts strongly with the bright lights and white gleam of cosmetics flanked by Ralph Lauren and Tommy Hilfiger. While JHP designed the space with a sleek, open, radial layout, the vernacular Christmas usage that greeted us last December transformed the space, making the original design almost unrecognisable. Redesigned for confrontation, promotional displays created an obstacle course of logos and samplers, emphasising a distinction between JHP's beautiful drawings and the store's seasonal get-up.

Beyond cosmetics, two dominant focal points emerge amidst the ground-floor fashion sections, also the responsibility of JHP. At the store's centre a giant X of criss-crossing escalators catches the eye, owing to its central location and vertical dominance. At the time of our visit, a Jamaican band played at the foot of the escalator, infusing the atmosphere with a sense of 'leisure'.

Unfortunately, this first focal point, marked by simplicity, seems to have received less attention than the second: the Moleanos limestone staircase. This grand gesture instils a note of 'architec- ▶

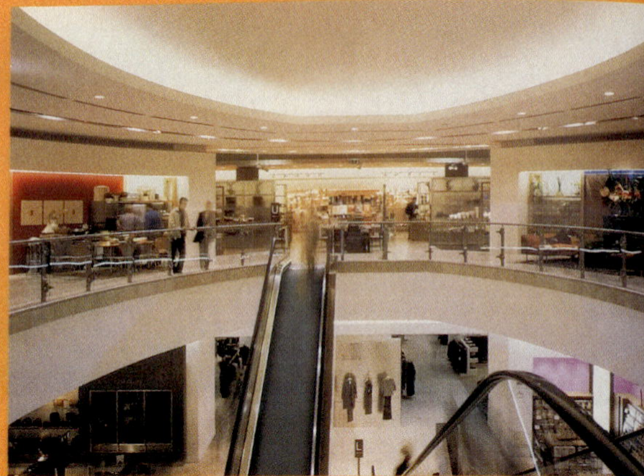

Above: Repeated throughou the store, the ceiling's uniqu exposure delineates space, creates and accentuates planes, reinforces directiona ity and zoning, and results i different strategies geared t different areas. Photo: Chri Gascoigne/View

Opposite page: Fashion department

Cool and restrained, the store's design balances retail commerce and new concepts, while making an occasional, selective gesture towards the mall and its apparent goals

It's hard to draw a line between the interior and exterior of any retail project, but in this case blurred edges and multiple designers make it even more so

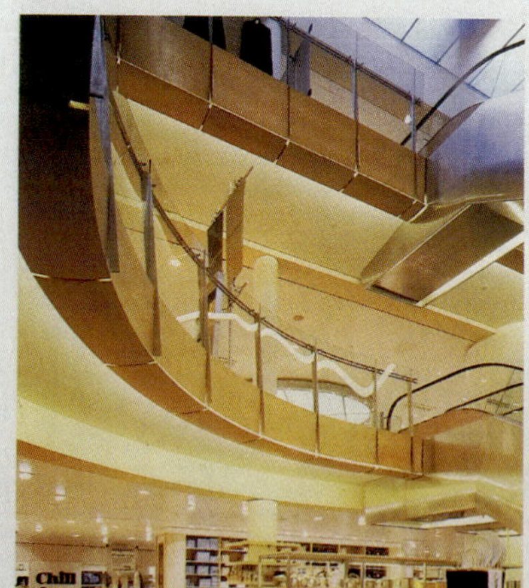

form a subtle reference to the Bijenkorf's famous hexagonal honeycomb logo. The ceiling extends out through five monumental, two-storey (8-metre-high) glass doors running on a steel girder, and on into the mall space.

Surprisingly in this litigious age, Baauw says that nothing more than a 'gentleman's agreement' was needed to enable the identity of the Bijenkorf to be realised from the exterior and the interior atrium space. Even so, it took an 18-month struggle to finalise the look of the ceiling and to reach an accord on how far into the atrium it could project. But, along with the monumental doors, the ceiling succeeds in creating a feeling of being in the store before actually entering.

At the exit leading out into the public square, a stainless steel ridge extending out over the heads of customers also generates a welcoming sense of being inside before one actually is. It's hard to draw a line between the interior and exterior of any retail project, but in this case blurred edges and multiple designers make it even more so.

Providing the store with its own identity was an equally difficult and complex process, and one that required the three design ▶

This and opposite page: Onl a handful of materials have been used: white beech, lam nated glass combined with satin stainless steel, and Capri limestone coupled wit a darker, contrasting Cumbrian stone

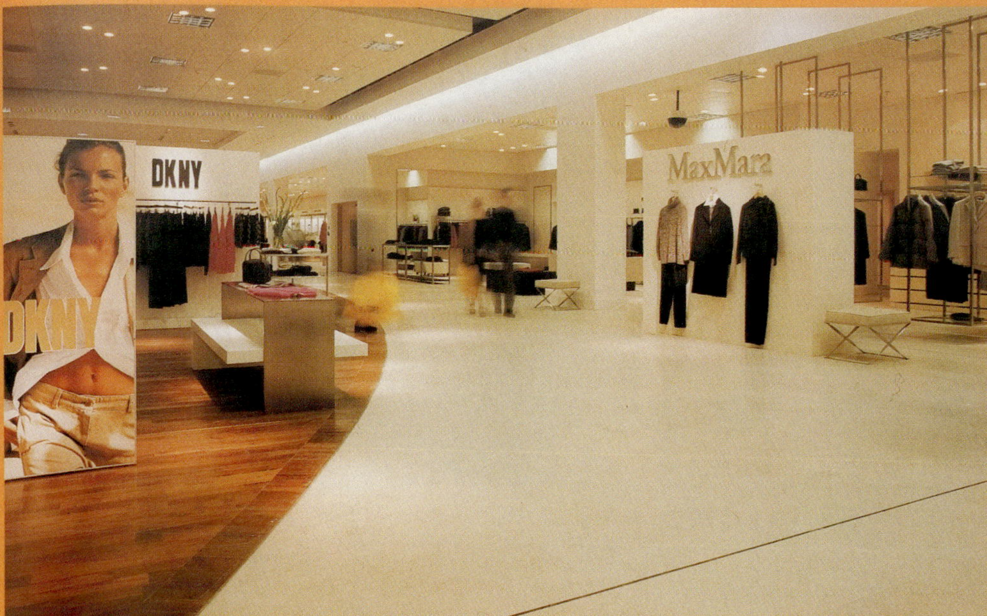

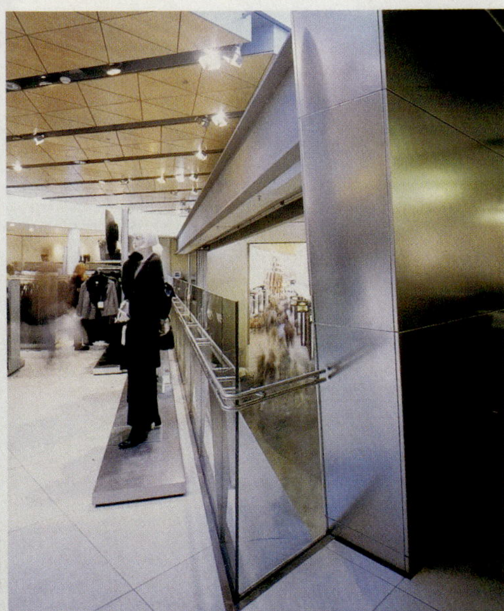

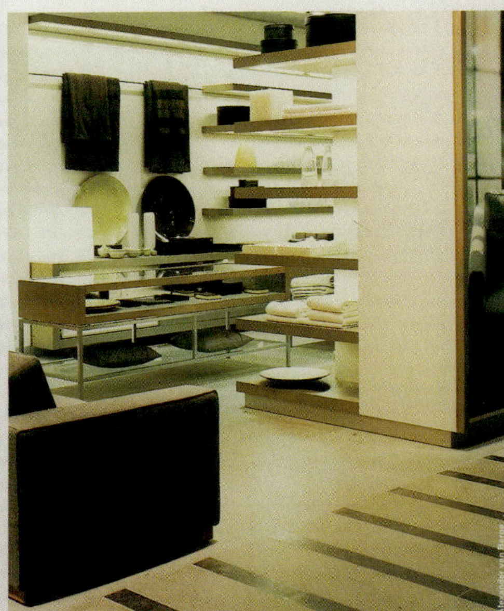

Alexander van Berge

SUBJECT
Selfridges versus
De Bijenkorf

LOCATIONS
Manchester, England
Amstelveen, Netherlands

DESIGNERS
Various, see page 067

PHOTOGRAPHERS
Various, see page 067

WRITERS
Robert Preece
Jules Marshall

tural drama', refers to the exterior and becomes a visual curiosity encouraging travel deeper into the space. From either left or right, the view from below is dynamic – a sculptural form that requires a second look, and then a third.

At the mid-landing of a staircase set beneath an elliptical skylight, sixteen TV monitors – a video wall – multiply imagery and encourage those reluctant to walk up a large flight of stairs. During our visit, standard music videos played in a predictable MTV-like way. A welcome alternative would be dynamic video or CD-ROM art; or perhaps the wall could be used to pull viewers into a discourse on music, fine art, technology, fashion and products from around the world. At the top of the staircase, Selfridges Spirit provides a club-like aura for its 'urbanwear' section, in which the video wall is complemented by metallic track lighting. Occasionally, a live DJ pumps realia into an abstracted atmosphere geared towards a subcultural market.

Two particularly pleasing cafe-restaurants – designed by Conran CD Partnership – can be found at both ends of the staircase, to the rear. Above, Selfridges Spirit Cafe features a combination of retro Havana pendant lighting and Bertoia metal chairs, which are echoed in the material of the barstools. For the exterior window covering, pale blue and white sheer cotton fabric is used to block out the unappealing exterior view while allowing light to filter in, maintaining the cafe's aura. A glass wall, which divides the cafe from Spirit, liberates the space and keeps the focus on shopping.

For Conran CD Partnership, the upstairs cafe, characterised by openness, signifies 'light' and 'sky', whereas the Sienna ▶

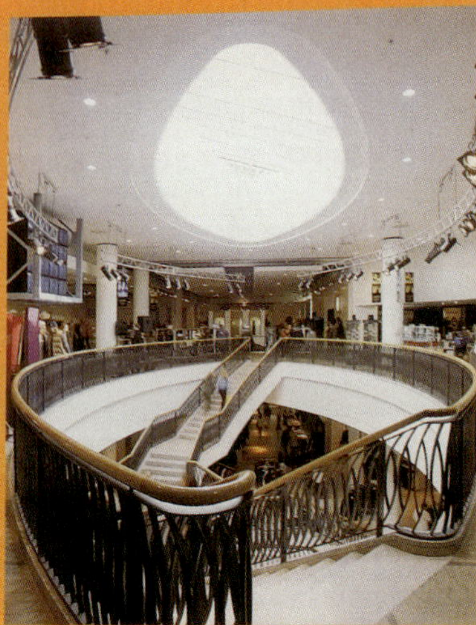

This page: Selfridges Spiri provides a club-like aura fo its 'urbanwear' section, fea turing sixteen TV monitors and metallic track lighting

Opposite page: Fashion department. At the store's centre a giant X of criss-crossing escalators catches the eye, owing to its central location and vertical dominance

firms to approach their collaboration with a great deal of tact. The Bijenkorf, wishing to offer its customers a highly varied shopping experience, liked the idea of using more than one designer. Essentially, Merkx + Girod did the hardware and Virgile & Stone the fashion-related items. At the beginning of the project ideas flew back and forth from one designer to another, according to Mark van der Geest of Merkx + Girod, a swapping process that ultimately led to a number of general principles that became the basis of a joint presentation to the Bijenkorf.

As for the interior, Greig + Stephenson pared it all down to a John Pawson-type way of retailing. 'Let the merchandise speak for itself, and keep the materials simple and natural,' as Tominson puts it, belying the effort that went into agreeing precisely how this was to be done. The pallet reveals no embellishment and only a handful of materials: white beech, laminated glass combined with satin stainless steel, and Capri limestone coupled with a darker, contrasting Cumbrian stone. Very natural-'90s.

Natural light has been deliberately invited into the store, something retail designers are generally advised to avoid, since the presence of daylight is linked with a loss of luminance control. This store, though, has shafts of diffuse light from above, as well as from the area around the entrance, and vertical elements from the central circulation void that accommodates the lifts. 'The guiding ethos from our point of view was that visibility is what sells the scheme,' says Tomlinson. And Baauw interjects that the Bijenkorf reacted positively to designers' ideas of minimalism, simplicity, and 'letting high quality materials create the subtle feeling of luxury without having to shout it out.'

Yet another layer of design is added to the store with the concepts of in-house designers responsible for the actual merchandise display. Such a small store requires them to be more flexible, to play more. Adds Baauw, 'We can use fewer brand-name shops – only Hugo Boss and Sony. The key is to be flexible but not flat. Allow the products to speak for themselves.'

The interior design solution to the space problem was to avoid cramped areas by creating 'combination departments'. For example, Merkx + Girod gathered together products under an overall concept of 'furnishings' rather than having one section for towels, another for bed linen, and so on.

The designers could have gone further, says Van der Geest. 'We suggested adding food and drink to the furnishings department by integrating the restaurant or pastry department, for example.' He does concede, however, that this was a case of 'getting more or less what you want by aiming higher than the client would accept': a strategy aimed at a solution satisfactory to all.

After the discussions and general presentations, the design groups concentrated on their individual tasks for the most part, checking with one another once in a while to ensure they were still on parallel tracks. This allowed different design practices, each with its own skills and methods of working, to do their best work. 'I'd say Virgile aims for big, broad statements, while Merkx are more about paying attention to the details of their designs,' says Van der Geest. In the end, Merkx designed the basement ▶

Opposite page: The booksho (left) and the cookery depart ment in the basement, designed by Merkx + Girod. To prevent further division o the relatively little space available, products are presented not individually but i coherent 'worlds'. Tablecloth are in the cookery departme – along with everything else related to dining – rather tha next to the linens. Photography: Alexander van Berge

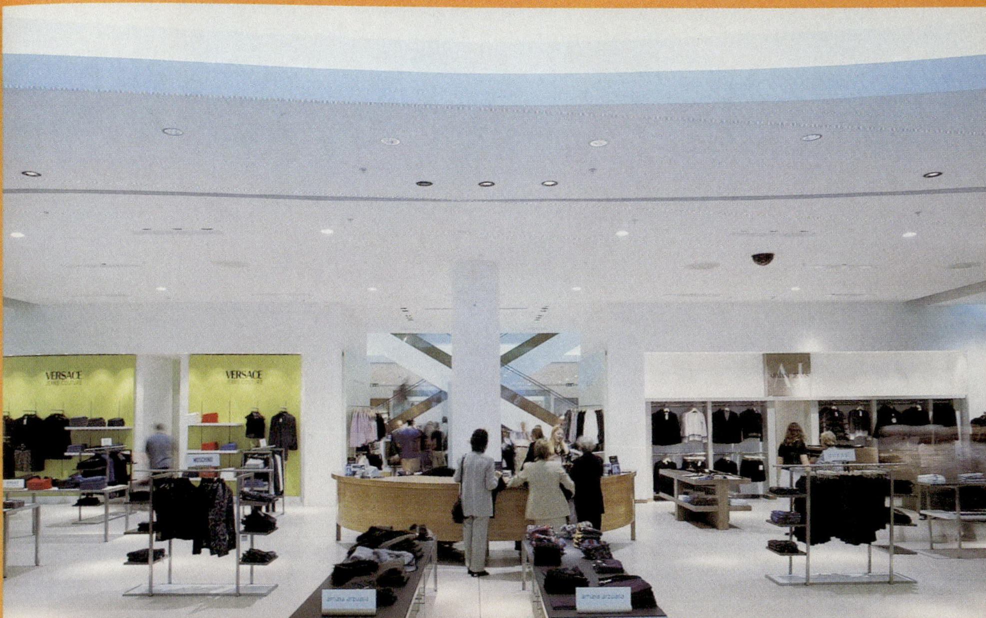

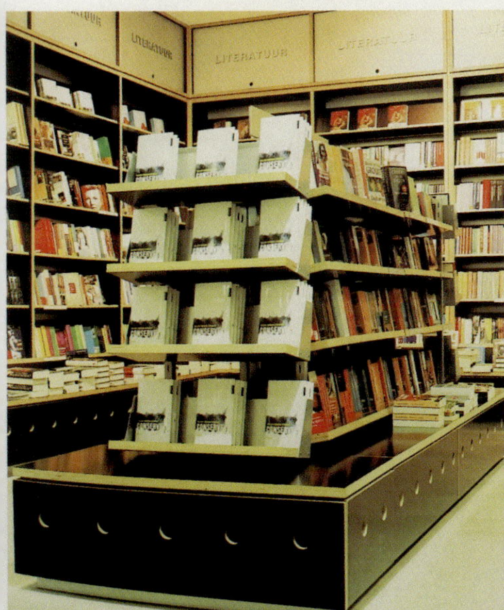

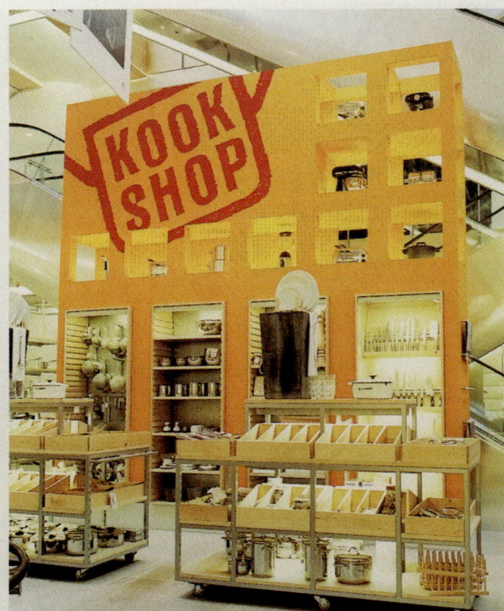

SUBJECT
Selfridges versus
De Bijenkorf

LOCATIONS
Manchester, England
Amstelveen, Netherlands

DESIGNERS
Various, see page 067

PHOTOGRAPHERS
Various, see page 067

WRITERS
Robert Preece
Jules Marshall

Restaurant on the ground floor conveys 'earth' and 'warmth'. Unfortunately, this restaurant, located in a somewhat awkward position – in the back and off to the side – is in danger of being overlooked entirely. The designers, given a small space with a low ceiling, had to counter the image of a storeroom. A narrow entrance provides a feeling of exclusivity, while glass and bright colours help to maximise the space. The cool greys of an exposed kitchen, separated from the dining area only by glass, contrast with a bright yellow wall and an accentuated horizontal red stripe for food passage. The play on depth produced by this design and its colours is striking in this intimate space and adds an element of drama. An impressive result, it leaves us savouring the discovery of a rare find.

Moving from Spirit to housewares means moving into a milieu designed by Gerard Taylor/Aldo Cibic, who highlighted the flooring in this section. Under our feet an array of textures – beech and fumigated oak, handmade tile, carpeting – offers a scale of different sounds and, with that, the various psychological effects of flooring, while also relating to the products found here. Yet it is in this department that the ceiling work dominates. Repeated throughout the store, its unique exposure delineates space, creates and accentuates planes, reinforces directionality and zoning, and results in different strategies geared to different areas.

On the ground floor, exposure to the subdued silver colours of air-circulation vents and power and data sources becomes apparent as you pass the cosmetics area. It comes into fruition in a diamond-shaped, cut-ceiling void underneath the staircase, which is intend- ▶

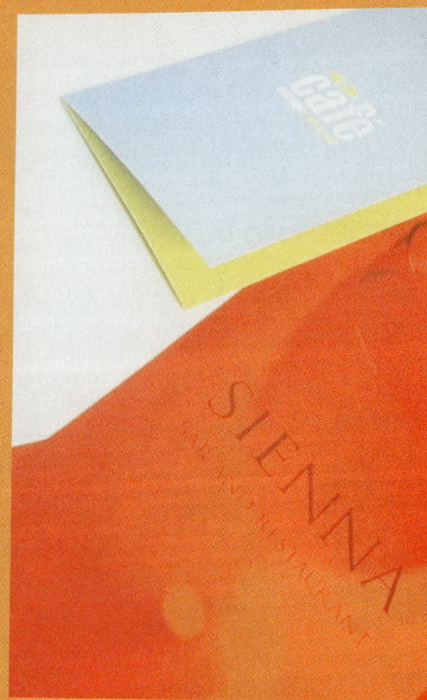

SIENN
BAR AND RESTAU

café
Selfridges spirit caf

Logos of the Selfridges Spiri Cafe and the Sienna Restaurant, both designed by CD Partnership. The upstairs cafe, characterised by openness, signifies 'light' and 'sky', whereas the Sienna Restaurant on the ground floor conveys 'earth' and '-warmth'

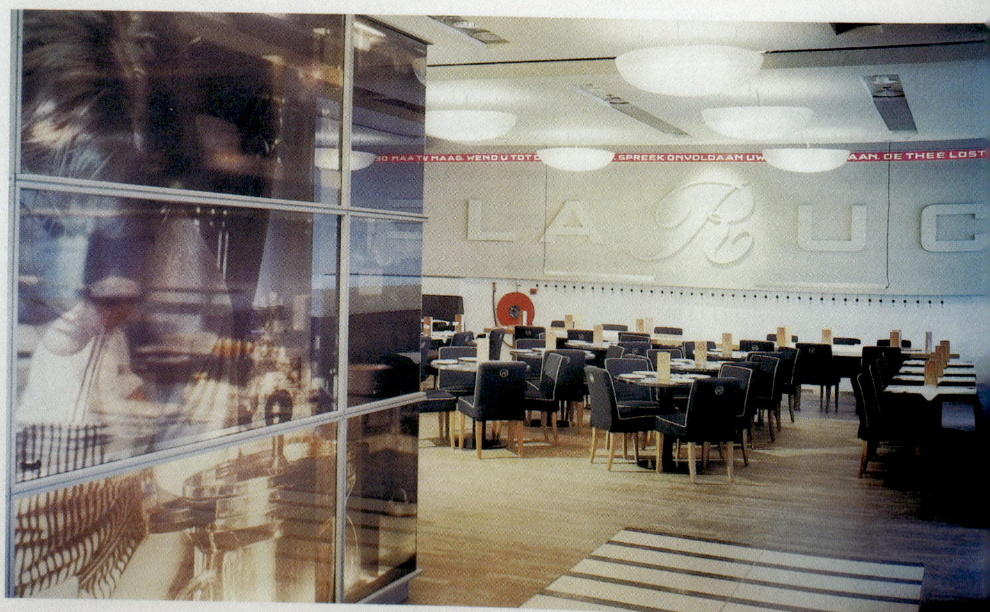

The food hall takes on a high-end New York deli concept, separates and abstracts it, and offers a spatial combination of opportunities for purchasing comestibles, as well as for eating at counters. Photo: Chris Gascoigne/ View

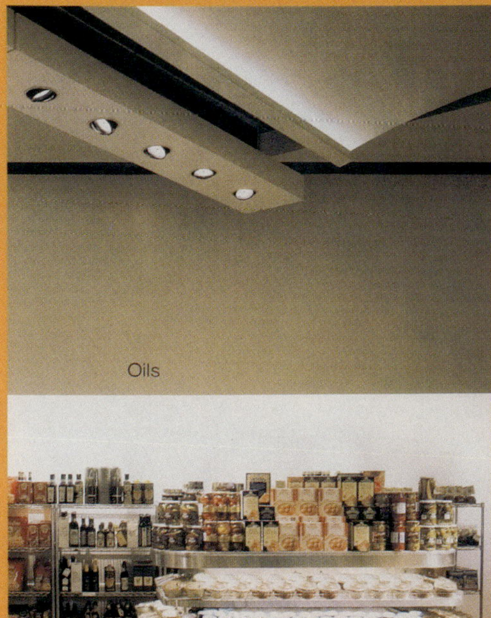

The food hall, energised by a more interactive location and purpose, is a space buzzing with action and powered by a life of its own

Previous and this page: Restaurant La Ruche in Amstelveen's new Bijenkorf was designed, from menu to furniture, by Merkx + Girod. The most striking feature is the large transparent box that accommodates the kitchen. The box stands at the entrance to the restaurant, where it functions as a partition between restaurant and retail activities. The use of laminated glass makes certain parts more transparent than others. Photography: Roos Aldershoff

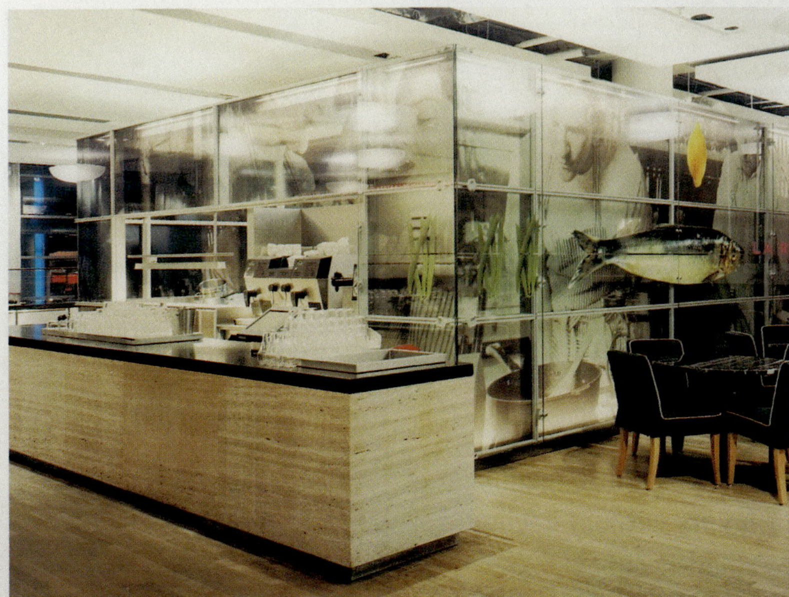

SUBJECT Selfridges versus De Bijenkorf
LOCATIONS Manchester, England Amstelveen, Netherlands
DESIGNERS Various, see page 067
PHOTOGRAPHERS Various, see page 067
WRITERS Robert Preece Jules Marshall

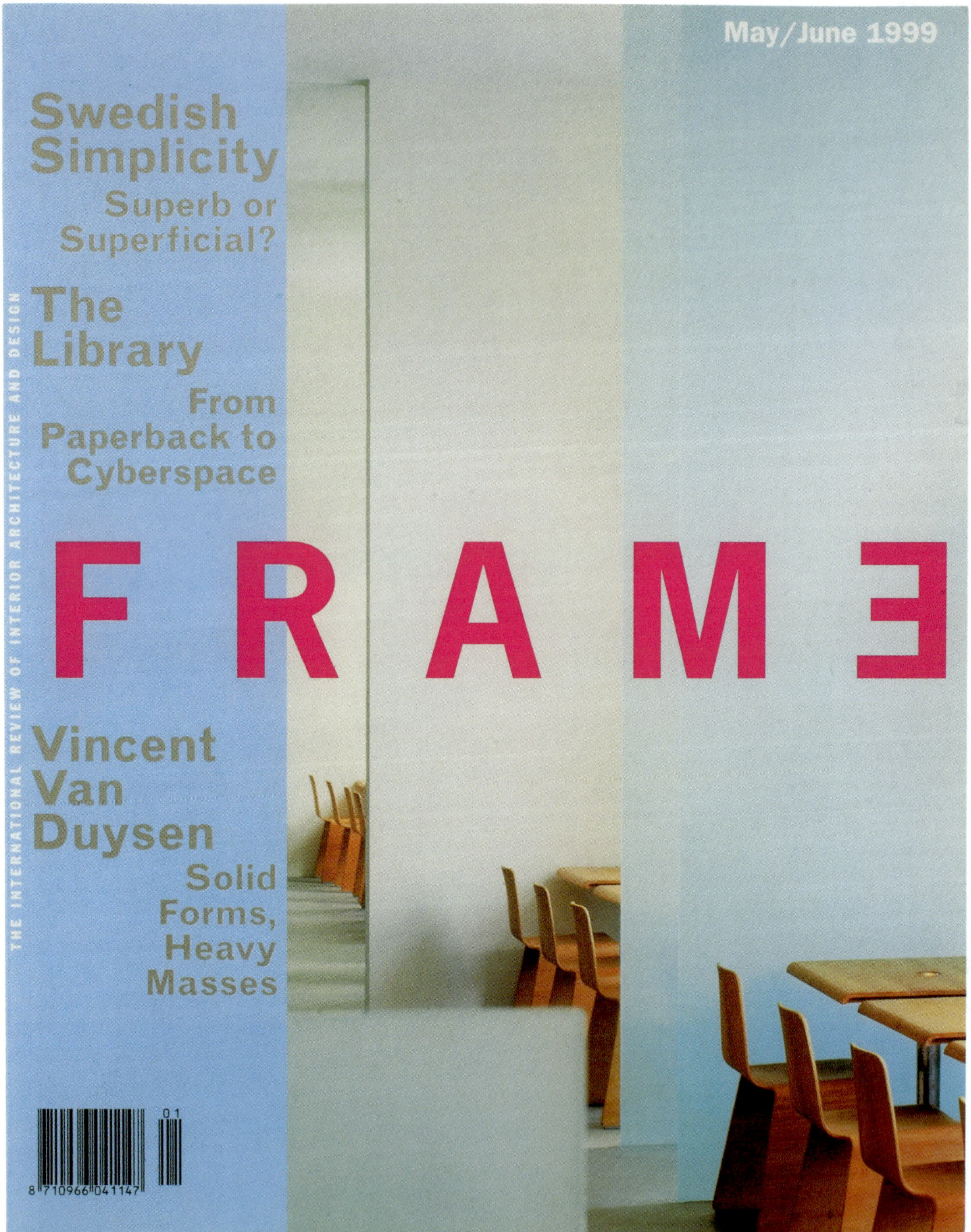

ART DIRECTOR
Ton Homburg,
Opera Designers

08

Grass, Glass and Cyberspace

DESIGNERS
Mecanoo Architects
www.mecanoo.com

SUBJECT
Library of the University of Technology

LOCATION
Delft, Netherlands

PHOTOGRAPHER
Fas Keuzenkamp

WRITER
Eleonoor Jap Sam

Architect Francine Houben (Mecanoo Architects) about the project
This building showed the world that a building can be like a landscape. Comments on the published article

Comments on the published article
I like the title, 'Grass, Glass and Cyberspace', very much. The article gives a very good description of the experience of the building.

Recent news or information about the project featured in *Frame*
The library is still very successful. The building symbolizes the new identity of Delft University of Technology. Mecanoo is working on the master plan of the campus of this university.

Current projects
Montevideo, Rotterdam, Netherlands // Philips Business Innovation Centre/ FiftyTwoDegrees, Nijmegen, Netherlands // La Llotja Theatre and Congress Centre, Lleida, Spain // Marnix Sports Centre, Amsterdam, Netherlands.

GRASS, GLASS AND CYBER-SPACE

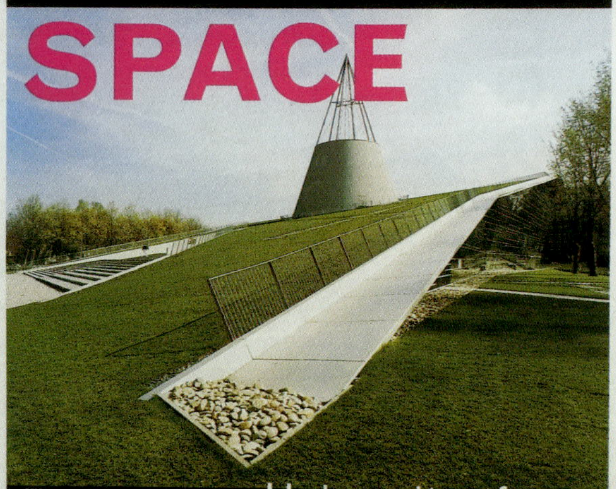

The library of the University of Technology at Delft attracts attention thanks to its sloping grass roof marked by a tall white cone. Inside, the visitor is surrounded by the efficiency of an airport. The presence of books and magazines is a virtually intangible feature. *By Eleonoor Jap Sam. Photography by Fas Keuzenkamp.*

The library of the University of Technology, located in the Dutch town of Delft and designed by Mecanoo Architects, is intriguing even at first glance. In the heart of this otherwise introverted university neighbourhood, the building lies concealed beneath a sloping field of crisp, green grass. Where the greensward is elevated, the architects have positioned exterior glass wall at a slant, leaving the building exceptionally transparent and revealing to those outside what's going on behind the facade. But the most striking feature is the white cone that pierces the field of grass. The architects see it as a symbol of the technical knowledge amassed within the building. The cone contains reading rooms and fulfils an important role in the illumination of the building as well. A glass structure – the 'pencil point' of the cone – allows daylight to penetrate right down into the heart of the building. Towering nearly 40 metres above ground level, the cone functions as a beacon, day and night. Electric light illuminates the structure after dark.

A monumental, concrete, deep-stepped staircase leads to the first-floor entrance of the building. Once inside, you are confronted with an enormous sense of space, particularly in the immense central hall, which stretches before you immediately upon stepping inside. This hall is encircled by the various components of the library: an area of workstations on the north side, and offices along the walls facing east and south.

A library is, first and foremost, a collection point for the storage of information. Today, of course, more and more information providers are electronic in nature, although we have been assured that books are not willing to be written off anytime soon. It's interesting to see how Mecanoo has interpreted this phenomenon in the architecture of the interior. ▶

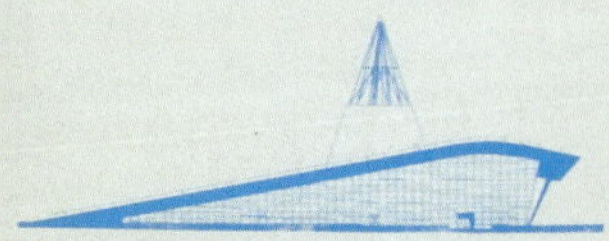

South facade

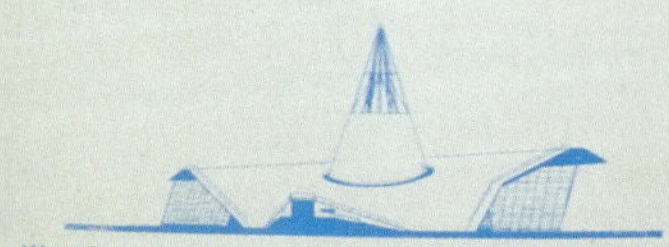

West facade

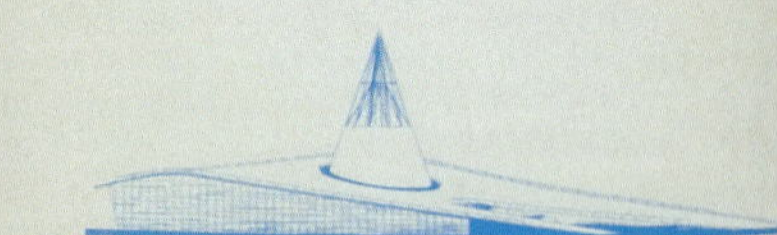

North facade

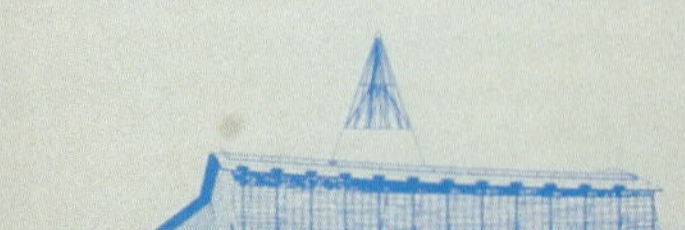

East facade

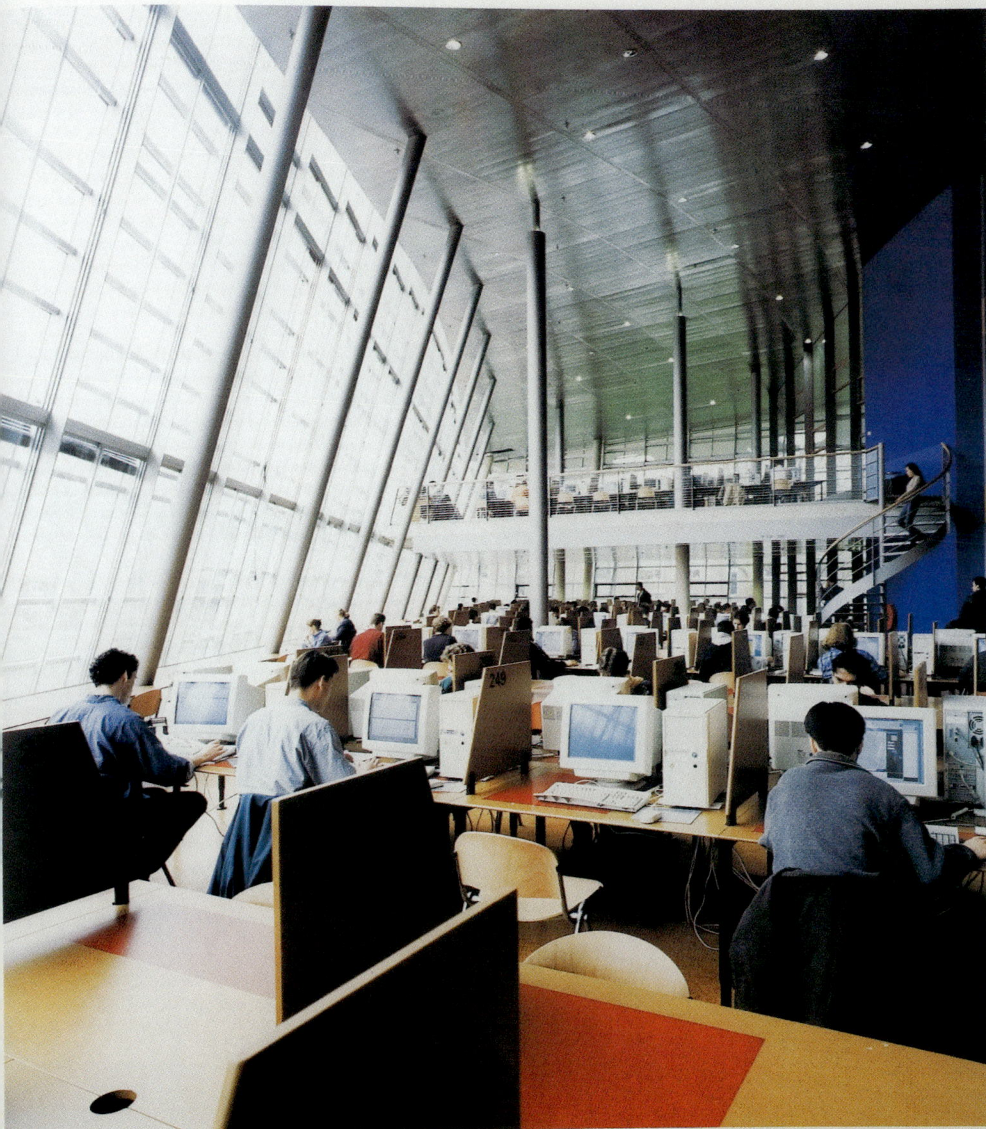

Of the 1,000 study cubicles available, 300 are equipped with computer terminals. Partitions separating the workstations provide privacy and lend a sense of visual organisation to the displays as a whole

Previous page: The library lies beneath a sloping field of grass pierced by a 40-metre-high cone

SUBJECT Library of the University of Technology

LOCATION Delft, Netherlands

DESIGNERS Mecanoo Architects

PHOTOGRAPHER Fas Keuzenkamp

WRITER Eleonoor Jap Sam

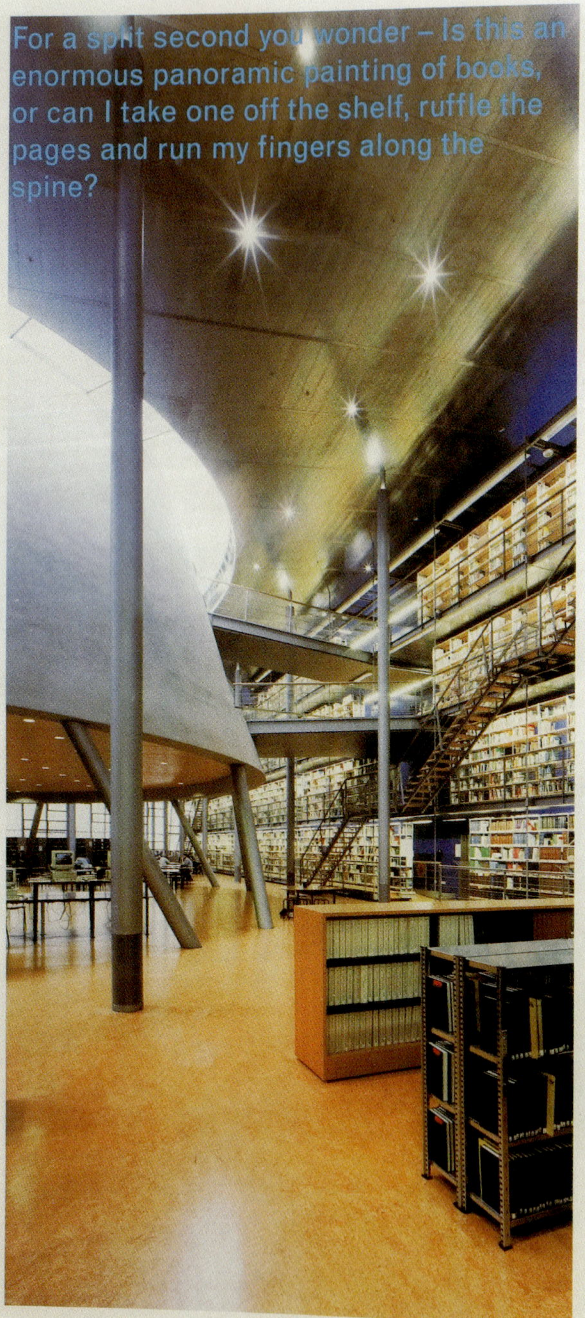

Most of the books here – especially volumes unavailable to the general public – are stored in a repository in the basement, out of sight. The stairway leading to the entrance is above this reposito-ry, so that those entering the building are literally walking over books. A lift made of glass connects the hall to the underground repository. Books are processed and stored according to a conven-tional system. Advanced search systems, which work with auto-mated probes and barcodes, have not been installed yet.
Books available to the public are found aboveground, on book-shelves towering four levels high and backed by a monumental, deep-blue wall. Visitors free of vertigo can reach any of the 80,000 or so volumes via steel stairways and catwalks.
Transparent displays consisting of glass panels that continue to the floor – a design both simple and surprising – accommodate 2,500 periodicals.

Although the gigantic bookcase gives the space the old, familiar appearance of a library, something else is going on here as well. The conspicuously blue wall behind the shelves, together with the shelves and the books themselves, creates a 'backdrop'. For a split second you wonder – Is this an enormous panoramic painting of books, or can I take one off the shelf, ruffle the pages and run my fingers along the spine? This sensation of unreality produces a somewhat cool and technical atmosphere, unlike the feeling you get in a cosy, old-fashioned library.
The sensation is reinforced by the relatively large number of com-puters in the building. If you're looking for digital information, no less than 300 of the library's 1,000 study cubicles are equipped with computer terminals. This many computers and t[he] subsequent heat emitted required special measures. Partitions separating the workstations provide privacy and lend a sense of visual organisation to the displays as a whole. Tabletops boast re[d] desk protectors with a Mondrian-like design, a nostalgic referen[ce] to the leather pads still found in certain older libraries. To provi[de] the computer area with its own climate-regulation system, the architects placed a glass wall between this area and the central hall. The ceiling, however, appears to be one continuous expans[e], which reinforces the sense of spaciousness.
Together with the lighting, this spacious feeling is one of the building's strong points. The hard, metal ceiling is softened by light radiating from supporting columns. Grids at the bottom of the columns allow warm air to be distributed throughout the building: these columns literally support, illuminate and heat th[e] building. Natural light enters the building through glass walls (outside and inside), through a glass ring encircling the cone at its junction with the grass roof, and through a glass construction that composes the tip of the cone.
Interior furnishings are somewhat disappointing. The powerful impression given by the building itself is not matched by the choice of furniture, fixtures and materials used to complete the interior. The rather 'clinical' atmosphere found inside is ▶

This and opposite page: The central hall with the cone is enormously spacious. One wall accommodates all book[s] that may be handled withou[t] special request. The blue wa[ll] behind the shelves creates the illusion of a panoramic 'backdrop', a huge painted mural

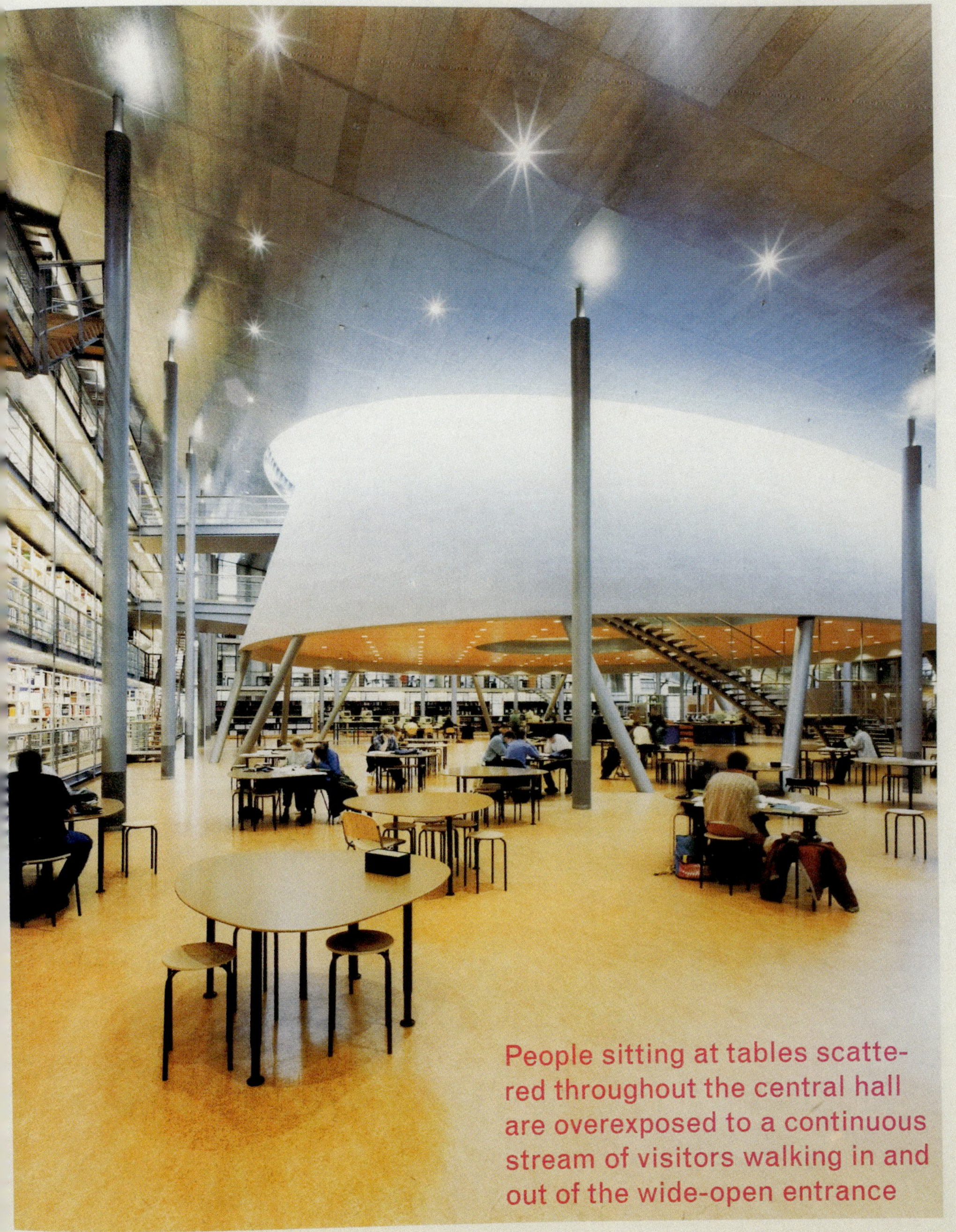

People sitting at tables scattered throughout the central hall are overexposed to a continuous stream of visitors walking in and out of the wide-open entrance

SUBJECT
Library of the University of Technology

LOCATION
Delft, Netherlands

DESIGNERS
Mecanoo Architects

PHOTOGRAPHER
Fas Keuzenkamp

WRITER
Eleonoor Jap Sam

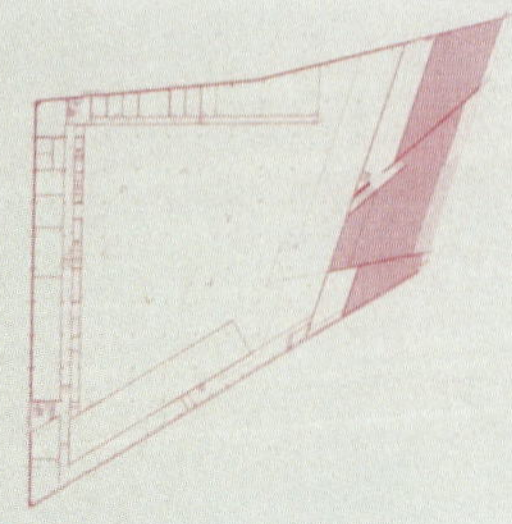
Ground floor

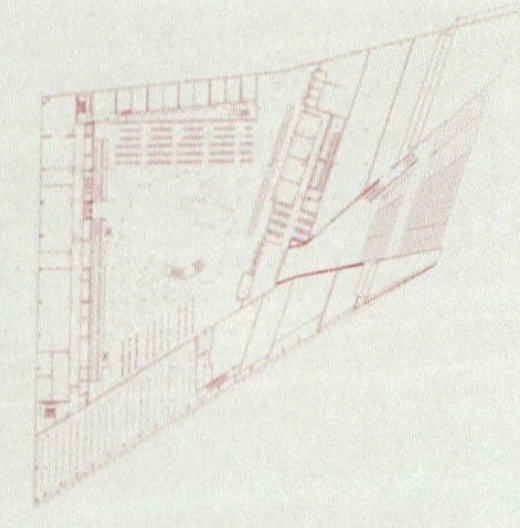
First floor

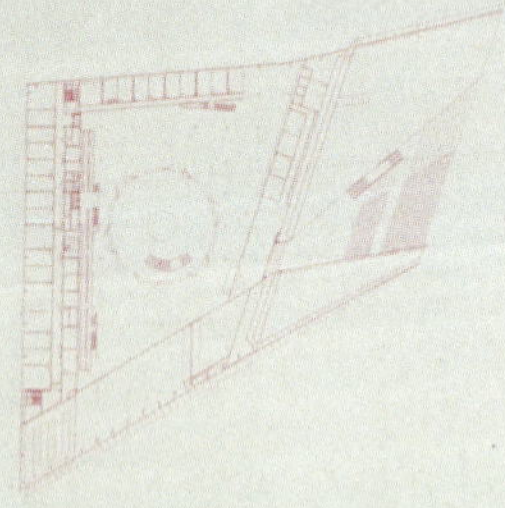
Second floor

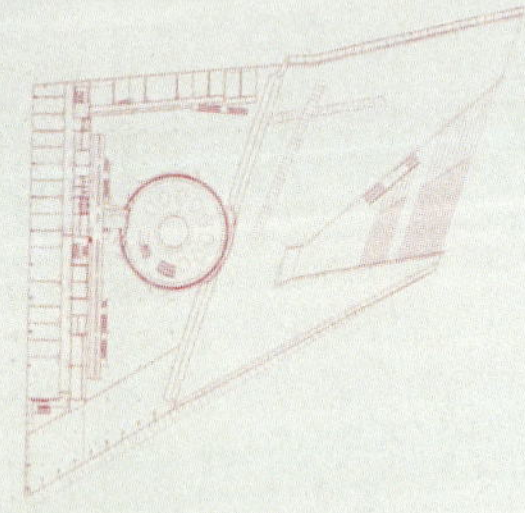
Third floor

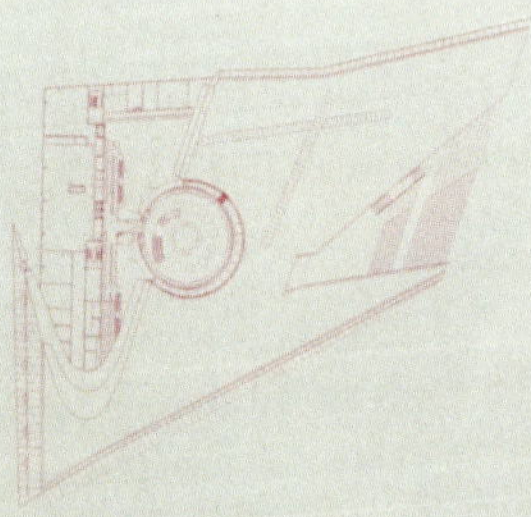
Fourth floor

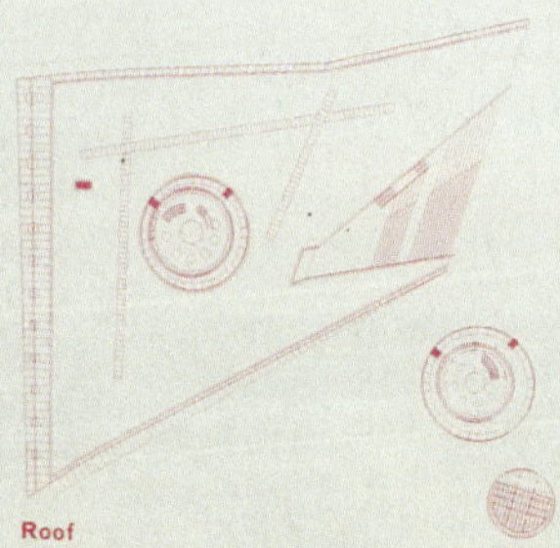
Roof

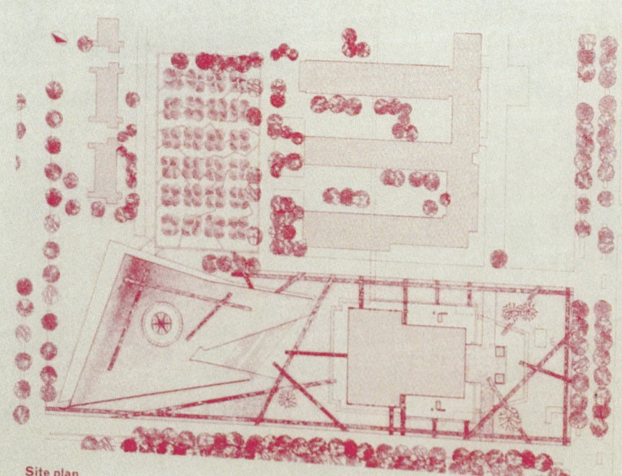
Site plan

determined, to a great extent, by the materials selected. Why d Mecanoo opt for a Sahara-coloured linoleum floor for such a l surface area instead of a seamless resilient floor, with or withc terrazzo finish? Furnishings in the main hall, composed main of MDF, consist of a counter whose organic shape is derived f the Caucasian Zelkova, round tables and low stools. Neither th organic shapes nor the stools are really user-friendly. Visitors a unlikely to sit on stools, with no support for their backs, for or second longer than necessary. Furthermore, the counter is a somewhat directionless element.

Another problem lies in the transparency of the central hall, which allows the movement of other visitors and a multitude of images to disturb the concentration of those tryin to study. People sitting at tables scattered throughout this area overexposed to a continuous stream of visitors walking in and of the wide-open entrance. A greater degree of privacy exists in the cone, but once inside, a new problem raises its head: poor acoustics. What might have worked better, both acoustically an visually, was a system of bookshelves against the concrete walls the cone.

Mecanoo opted for an ingenious architectonic design that includes an interior organisation in which books and magazine are not immediately tangible or even perceptible. The result is anything but an archetypal library with an atmosphere conduci to hours of contented browsing. On the other hand, the compu area, which draws a crowd of students every day, certainly satis the desire of the client for optimal access to the digital highway Here it can be said that the architects provided the university w what it wanted: a functionally regulated flow of information accommodated by a building with the efficiency of an airport. Above ground level, it's a completely different story. A vibrant, green landscape opens before you, inviting you to stretch out ar contemplate the heavens. It might not be a bad idea to hand ou laptops on sunny days and send visitors outdoors to surf the Internet on a green hill, surrounded by the reality of nature. ◄

Mecanoo opted for an ingenious architectonic design that includes an interior organisation in which books and magazines are not immediately tangible or even perceptible

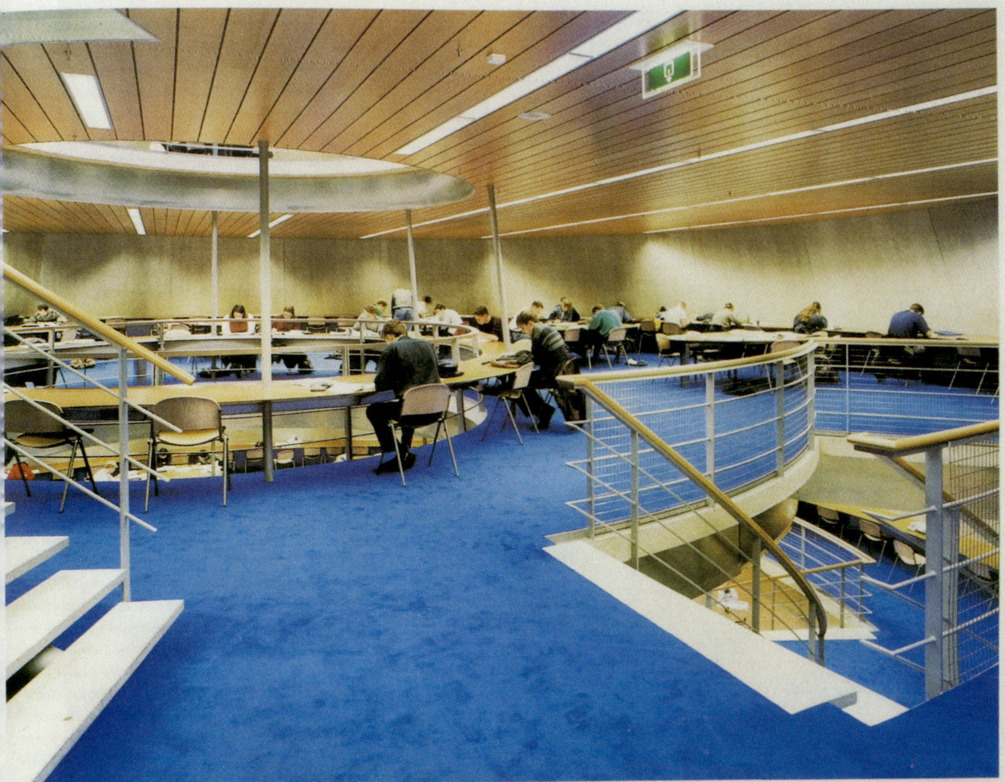

Above: Reading rooms in the cone, illuminated by both artificial and natural light from above, offer more privacy

Right: Entrance to the cone. The severe technical effect of the metal ceiling is softened somewhat by sophisticated lighting

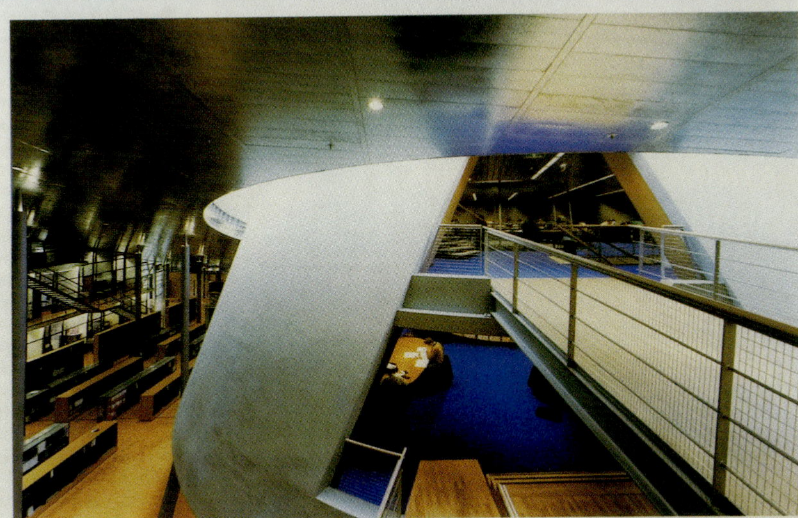

SUBJECT Library of the University of Technology

LOCATION Delft, Netherlands

DESIGNERS Mecanoo Architects

PHOTOGRAPHER Fas Keuzenkamp

WRITER Eleonoor Jap Sam

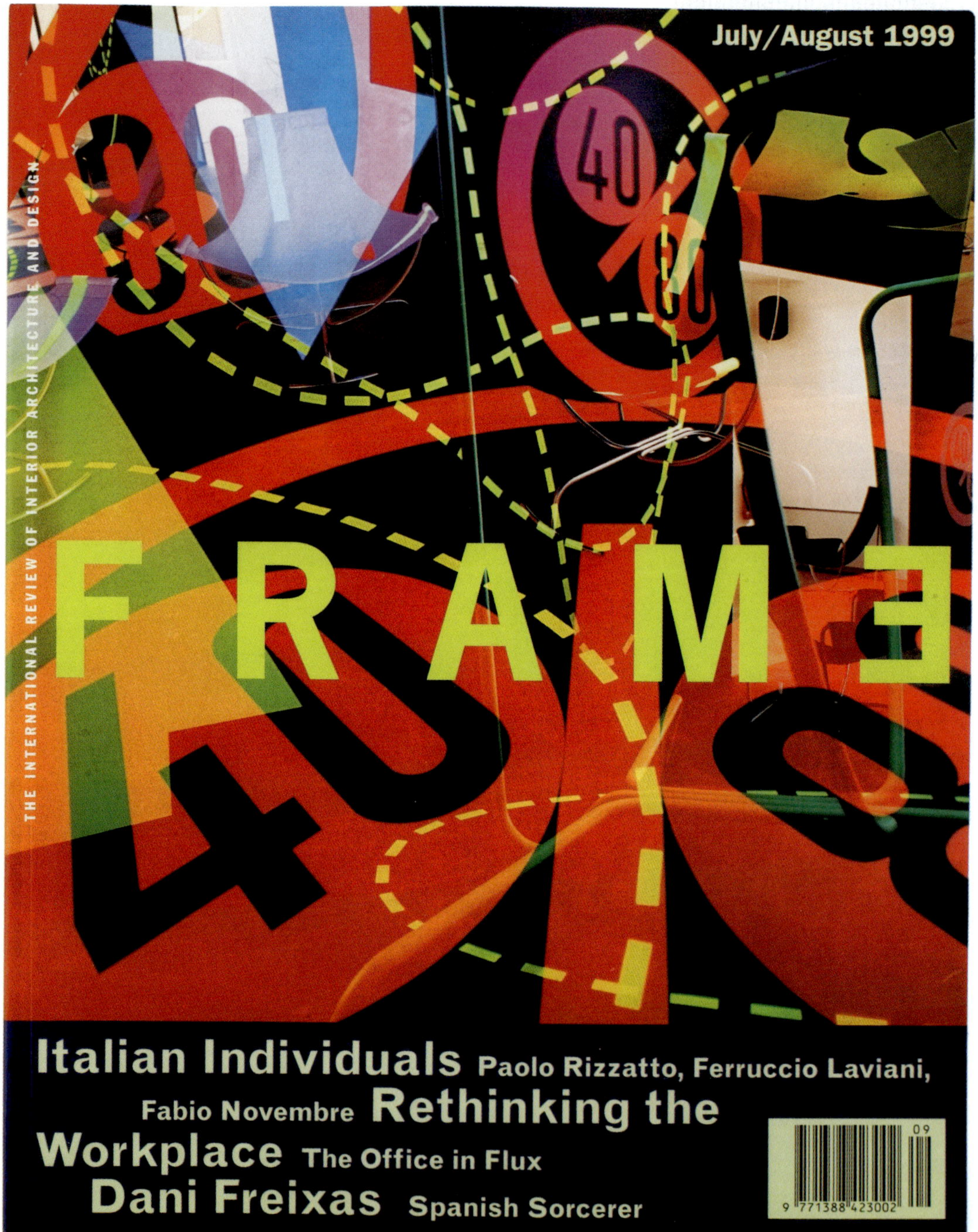

ART DIRECTOR
Ton Homburg,
Opera Designers

09

Be Your Own Messiah

Architect Fabio Novembre about the project. Did the article have an effect?

The article had a big effect, especially on me, for two main reasons. The first had to do with the realization that my work was appreciated and that it made sense to someone besides me. The second was learning to know the people who were showing interest in my work: a team of young, cool guys trying to make with a magazine what I was trying to do with my spaces. Since then, I can say I have good friends in Amsterdam.

Current project

I'm currently working on my first real architectural experience. It's going to be the headquarters of a fashion brand, Meltin' Pot. The 7000-sq-m interior is located in the south of Italy, very close to the place where I was born. Back to the origin!

DESIGNER
Fabio Novembre
www.novembre.it

SUBJECT
Portrait of
Fabio Novembre

LOCATION
Milan, Italy

PHOTOGRAPHER
Alberto Ferrero
www.albertoferrero.it

WRITER
Leo Gullbring
www.calimero.se

Be Your Own Messiah

He can't draw, but he's a good writer. He doesn't care about rationalists or the modern movement. Manifestos leave him cold. And he calls architecture arrogant. Meet Fabio Novembre, an injection of fresh blood from southern Italy who's upsetting the laid-back impostors of design. 'Architecture is like a beautiful woman – you want her nude.' *By Leo Gullbring. Photography by Alberto Ferrero.*

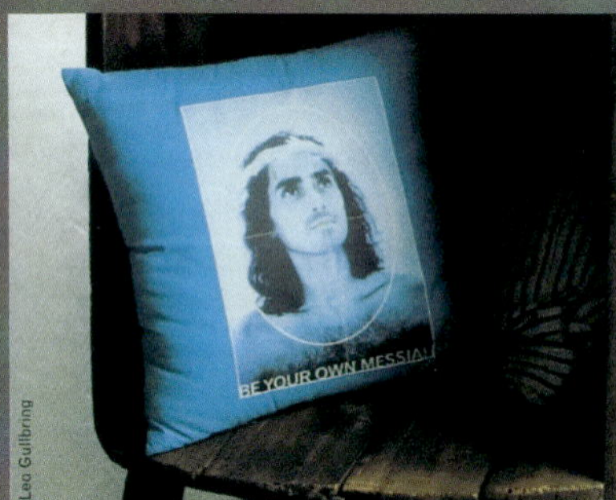

Leo Gullbring

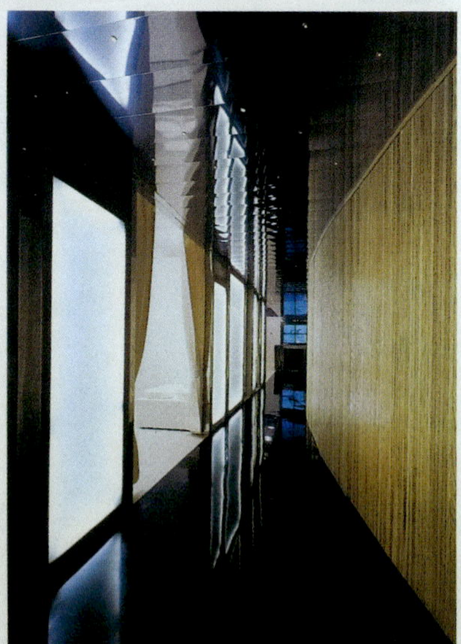

Right: ON Centro Benessere Naturale, Milan. This establishment, which offers a range of natural treatments for good health and well-being, features the practice of Chinese medicine, massage therapy and instruction in modern relaxation techniques

Opposite page: B2, a Hong Kong shop with a range of young, funky fashion labels. B2's design is based on the concept of an acrylic shell around a core of beauty. Polished steel panels and colourful acrylic surfaces, both transparent and opaque, are the cladding for an interior in which new visual effects are easy to create by moving elements from one place to another. Photo: Charlie Chee

My destination is a place on the outskirts of Milan, somewhere along the line of an old, dilapidated tramway. 'Fabio?' says the shopkeeper. 'He's right here, at the back of the building – please, make your way between the shelves. Take that door over there.' I'm directed to a flight of red stairs running along the exterior wall of a small factory. They lead to a spacious studio, where sunshine pouring through large windows splashes a pale grey concrete floor. A king-size bed hides reluctantly behind dividers, and a couple of drawing desks discreetly face the windowsill. The scene is completed by incense and freak-out furniture, including several moulded items and sofas scattered throughout the room. My host is Fabio Novembre, an injection of fresh blood from southern Italy who's here to enliven and upset the laid-back impostors of design. I put the question to him without hesitation: 'What does it take to make a good designer?'

'Have you heard of Carmelo Bene? I love the man. When I first heard him talk about himself, my life changed. He said that he's not *making* art – he *is* art! Each person is a piece of art, each individual, each life. If you live your life like a work of art, you have no choice but to be generous. In other words, the world is me, and the logical consequence of that thought is encapsulated in the Italian expression *lasciarti mangiare.*' I admit that the term appeals to me: 'Let yourself be eaten.' It stands for a quality that I encounter everywhere I go in Italy – a sense of inviting, almost naive, generosity. Be welcoming. Be hospitable. Be yourself. Be ►

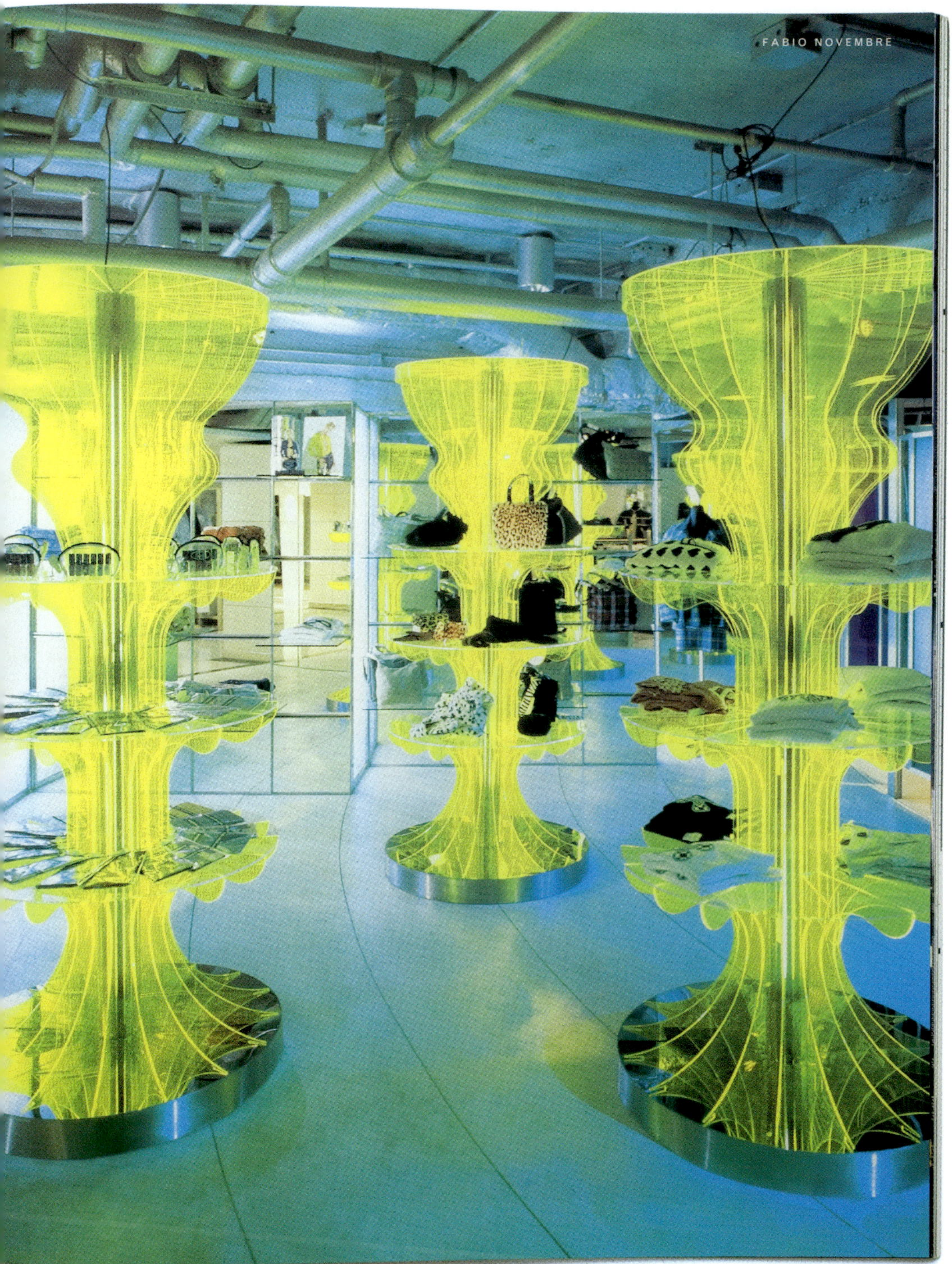

SUBJECT
Portrait of
Fabio Novembre

LOCATION
Milan, Italy

DESIGNER
Fabio Novembre

PHOTOGRAPHER
Alberto Ferrero

WRITER
Leo Gullbring

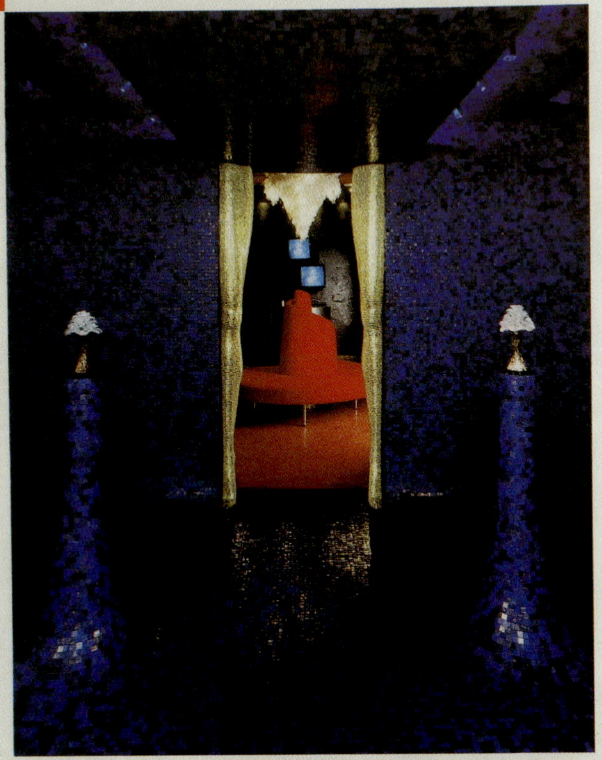

easy-going. Be good.
The adjectives are a perfect description of Fabio himself. Not satisfied with the idea of an interview over a cup of coffee, he says that his sister will arrive shortly to cook a delicious Italian meal for us. Turning on the amp, he picks up his guitar, all the while expounding on subjects that arouse his existential curiosity: the meaning of life, the universe, everything.

'I feel that I've been born in an era marked by a total lack of understanding. Where are the political theories? The ideologies? I have no desire to join societies and movements that have no real value. The only thing that counts is living. If you reduce the whole ball of wax down to the bare essentials, you get the act of living – *Vivere*, with a capital V. I don't care about rationalists or the modern movement. Manifestos leave me cold. I don't believe any of it. Each of us has a unique story, and I'm here to tell mine. Every 20th-century manifesto is simply a testimony to life. The big challenge translates into one's own, personal challenge. Everybody has to carry his own backpack of responsibilities.'
Sifting through those thoughts, I'm not surprised to hear that Fabio is a big fan of Timothy Leary. I'm also reminded of an Italian acquaintance, a semiotician named Omar Calabrese, who once argued that all this talk about post-modernism is nonsense; it fails to get to the heart of the matter. Drawing on Sarduy, Calabrese referred to the contemporary era – with its aesthetic of repetition as an apparent substitute for new styles and ideologies – as neo-baroque. Admittedly, an analysis of the present is always risky. It's not easy to decipher the code to history in the making. Or, as Fabio puts it, to sort the good stuff from the bad at a place like the furniture fair.
'But in the end,' he continues, 'the important thing is not some professional label you wear, but what you say. You can express yourself in a million different ways – by writing a book, painting a picture, working in a bank . . . The important thing is to say something with your life. Along the way, your work automatically transforms you into an expressive medium.'
Novembre contemplates the strings of his guitar. 'Our profession is damaged by people who don't think big. The ones who do are my role models. Take Marc Newson, for example. He's part of a unique race, along with Joe Colombo, Verner Panton, Carlo Mollino – guys who opened up a lot of new roads, who enlightened our universe. I like that, the idea of stars to look up to, to use as references. Gives me the strength to go on.'
Fabio's black hair curls like the snakes of Medusa. His dark eyes sparkle. And his words have a certain affinity with the radical movement of the '60s and '70s, with a longing for something different – a better future or an altered present. That desire failed to become the collective surge that many had anticipated, however, and proliferated only at the individual level. Can we accept the idea that what it all boils down to is one life? A society of the personal? Is that the context we need to appreciate visions introduced by Archizoom, Superstudio, Gaetano Pesce, Vico Magistretti and all the others vying for our attention?

I'm also reminded of Ettore Sottsass, who argues in the preface to Novembre's book, *A Sud di Memphis*, that existence is sensory, not mental. He writes: 'Objects are not read by your mind but by your cock, your stomach, your tongue, your eyes, your ears, your senses.' This sensory existence is not a primitive phenomenon. But apart from what the senses tell us, surely something spiritual makes its claim on life as well. The failure of ▶

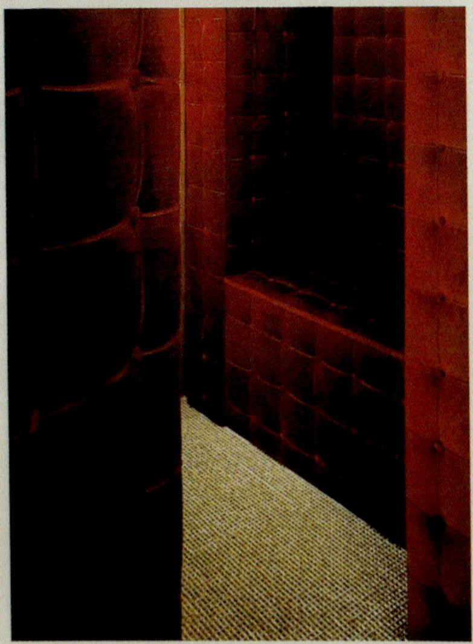

This and opposite page: Anna Molinari – Blumarine, fashion shop, London. Colours used in this design refer to the Union Jack. The upholstered space with the metallic look (opposite page) is a changing room, as is the one done in red velvet (right). The mosaic-finished blue room featuring female legs and shoes is used for the purpose of display (above)

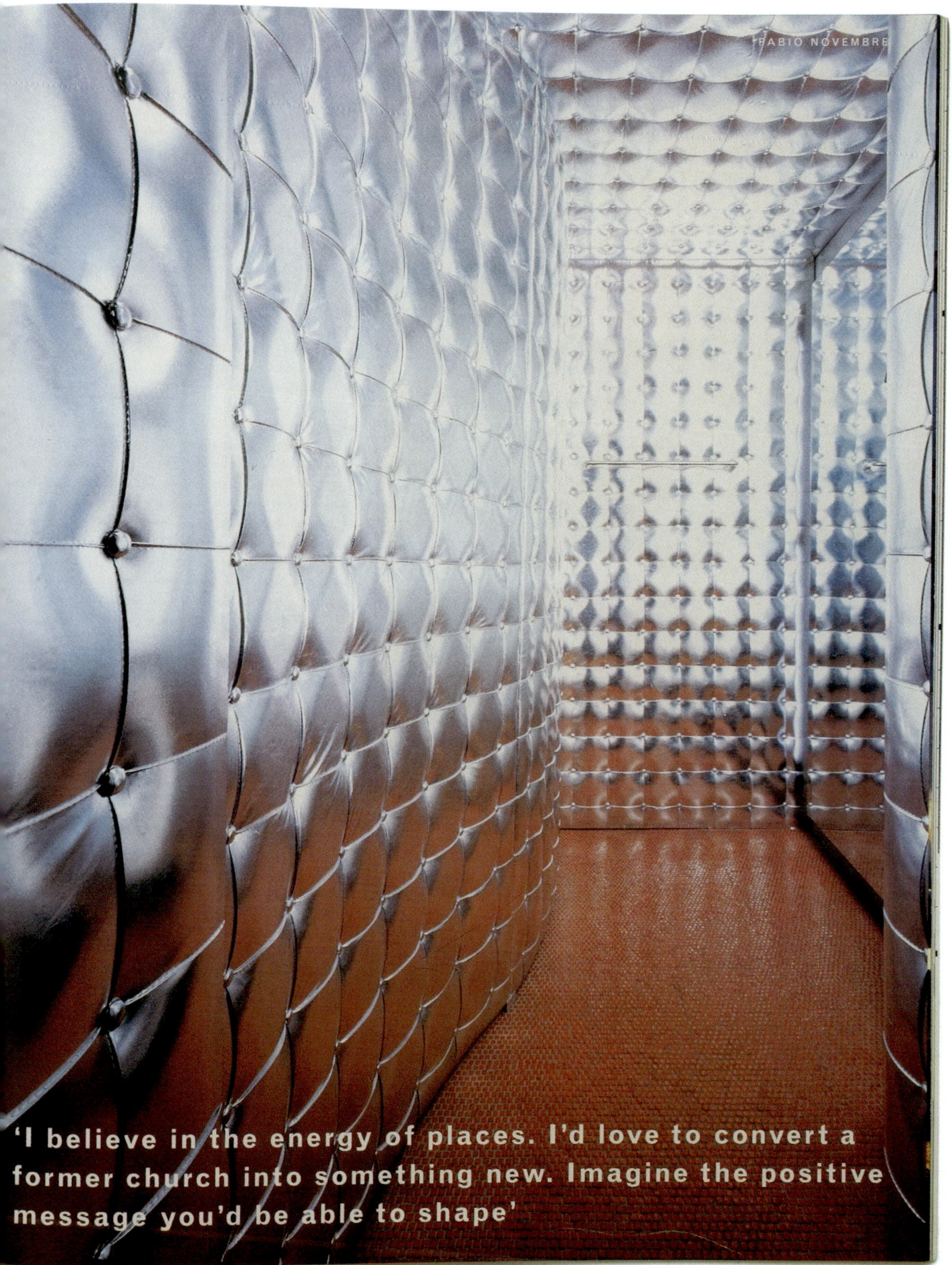

'I believe in the energy of places. I'd love to convert a former church into something new. Imagine the positive message you'd be able to shape'

SUBJECT
Portrait of Fabio Novembre

LOCATION
Milan, Italy

DESIGNER
Fabio Novembre

PHOTOGRAPHER
Alberto Ferrero

WRITER
Leo Gullbring

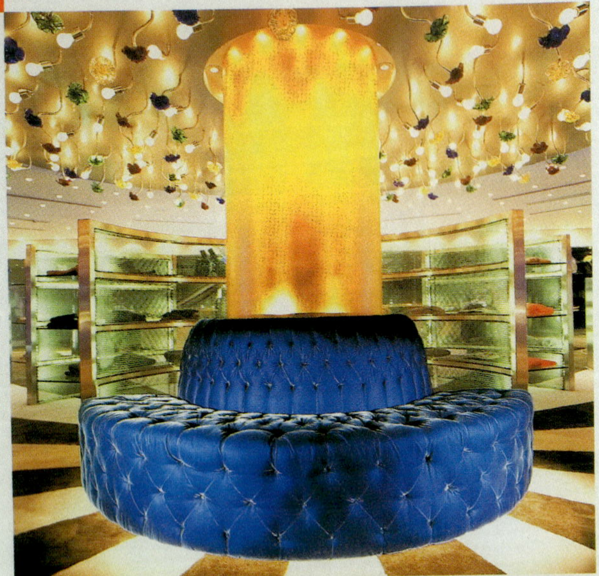

This page: Anna Molinari, fashion shop, Hong Kong. Displays are made of metal wire framed in polished stainless steel. Photo: Guy Bertrand

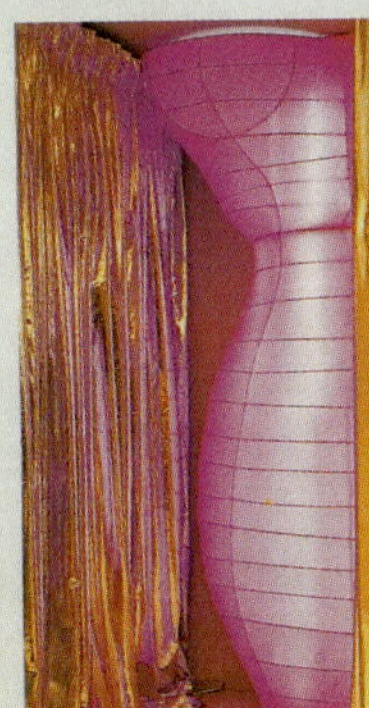

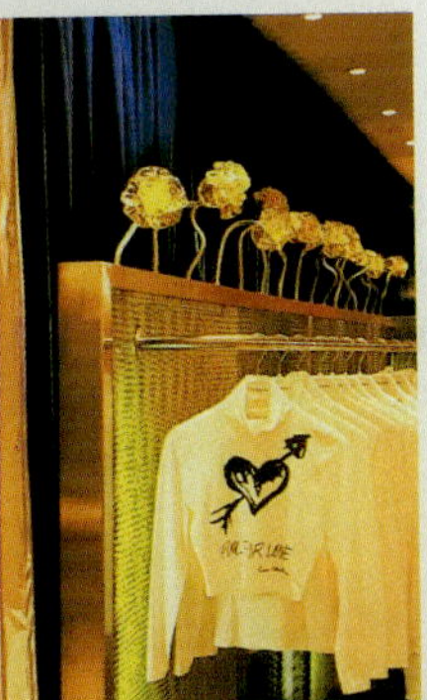

Opposite page: Bar-restaurant-discotheque l'Atlantique, Milan. The brief contained no description of the establishment's various functions. Novembre came up with flower-like chairs, a huge chandelier of fibre-optics and crap-metal androids

'Architecture is so arrogant. Why don't we put an expiry date on architecture? Why isn't the death of architecture part of the design? Why not put explosives into the concrete? We can't ignore death'

the Pope to get in sync with the '90s doesn't nullify a religious tendency – a desire to believe – that I find among Italian designers, young and old.

When I couple the lingering scent of incense to the need for religious experience, Fabio corrects me. He says it's not about religion. The Oriental influence of sandalwood may evoke thoughts of Buddhism, but Buddhists are not waiting for a prophet. He's convinced that the secret to everything is life, plain and simple. He talks about concentration, about relaxation, urging me to be myself and not buy into some functionalist negation of life. 'Just because your blood stops flowing and your heart stops pumping doesn't mean that you're dead.' Pointing to a photograph of Marcello Mastroianni, he continues: 'See his expression, his face, his leap? Where is that energy? It's still here. It will be here forever.' My eye falls on another photograph, this time an image of Fabio himself on a postcard with the text, 'Be Your Own Messiah'. His dark hair wears a crown of fibre-optic thorns. Blasphemous? I think not.

Having propped his guitar against a chair, he moves to the CD player. His philosophical musings gradually shift to his line of work. 'I used to find interior design quite limiting. I thought photography or another branch of the arts would be much more satisfying. But when I reassessed architecture, I saw it as the ultimate challenge presented by three-dimensional space. It's the last medium to acknowledge that we're made of flesh and blood, that in the end there's only us, that we're surrounded by the tangible, by people we can touch, embrace, take to bed. Architecture is the medium that defends three-dimensional space.'

True or false or somewhere in-between, I can imagine Fabio searching for love at l'Atlantique, a perfect setting of his own making, a bar-restaurant-discotheque packed with people sharing emotions, burning with passion. An attraction that appealed to the fair's party crowd, l'Atlantique – a place too elegant and upper-crust for my taste – is decorated with a bunch of crap-metal androids, which infect the atmosphere with a sort of bourgeois, Mad-Max-inspired virus. The brief for l'Atlantique contained no description of the establishment's various functions. Waxing lyrical, Fabio wrote to the client: 'Architecture is like a beautiful woman – you want her nude, sculptural, showing her most intimate, most structural forms. You want her as perfectly true as a Helmut Newton picture, authentic in her fierce, screaming sexuality.' Fabio turns up the volume, and Skunk Anansie blasts the message into my brain: his work is about sex; work and sex share the same energy. He wouldn't mind experiencing a global epidemic – more sex would make the world a better place. 'Why no drawings?' I ask. Fabio willingly admits that he can't draw but claims that, in his case, the lack of one talent is compensated by another. He's a good writer and can spin a tale that evolves into an idea for a project. 'I've got this capacity to imagine *in absentia*. When I think of a place, I picture how it's going to look, down to the smallest detail. A good imagination is one thing, however. Conveying my vision to others – the client, the carpenters, anyone else involved – is my main problem. There I stand with my spiel – "Here's this column that I want attached to the ceiling, which will be gold over here, and this part should be blue, like the sky, and this lamp has to be a shower of fibre optics that cast sprinkles of light all over the bar and, and, and . . ." Organising and orchestrating on-site work is definitely the hard part of the job.' The word *imagineer* springs to mind, a term that ▶

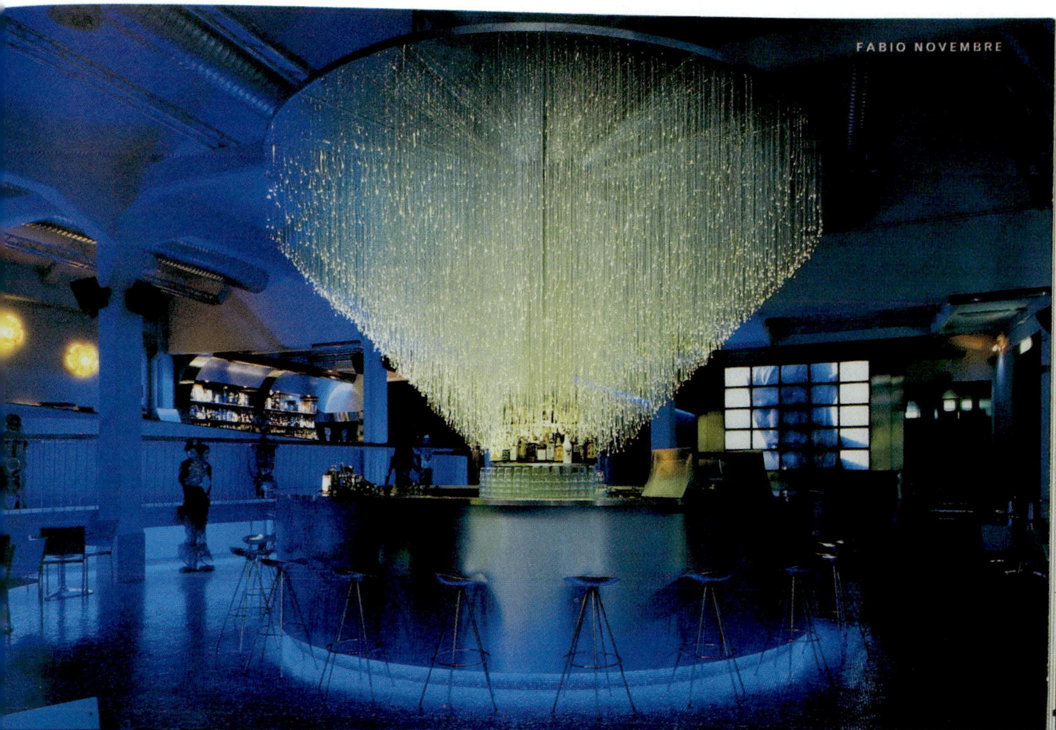

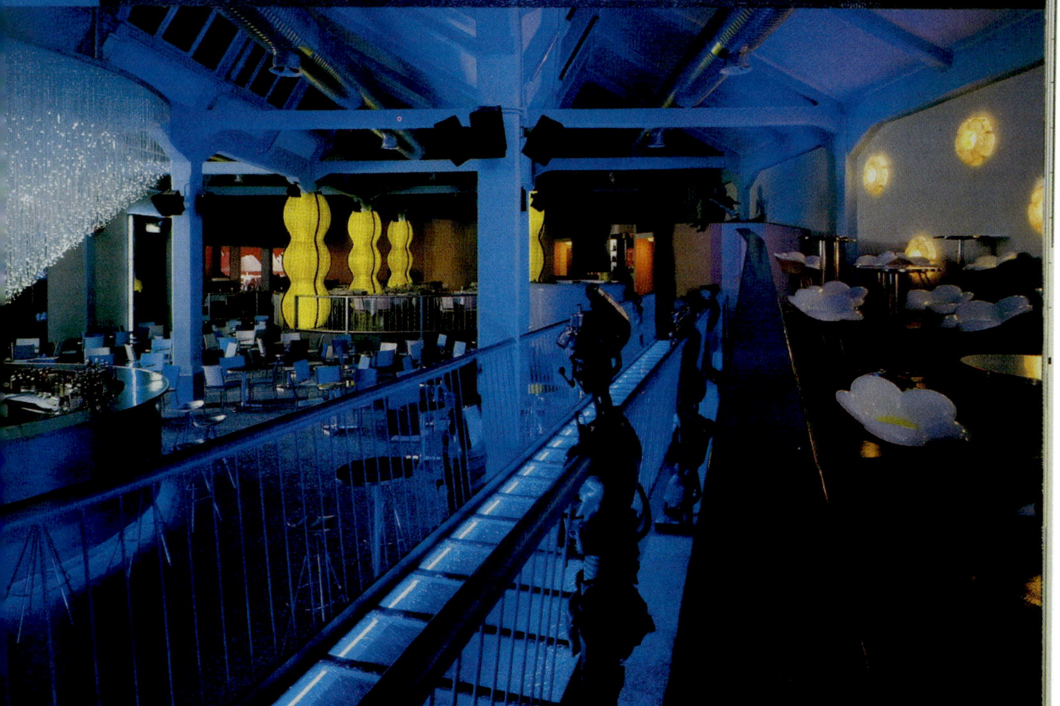

SUBJECT
Portrait of Fabio Novembre

LOCATION
Milan, Italy

DESIGNER
Fabio Novembre

PHOTOGRAPHER
Alberto Ferrero

WRITER
Leo Gullbring

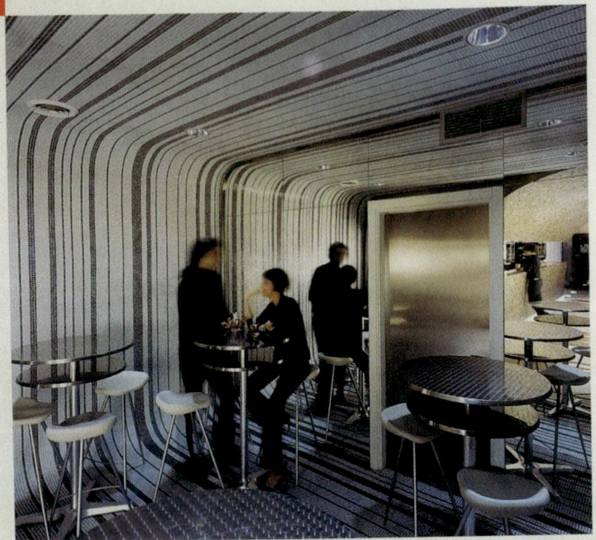

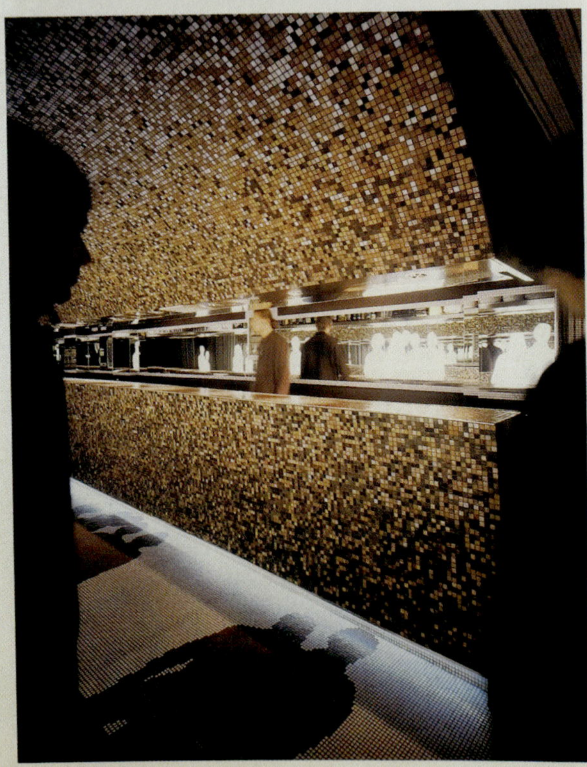

I'd use to describe many a famous Italian designer. They all want to break down barriers, rise above the level of the architect, be a *progettista*, a visionary.

Design may be about communication, but image isn't everything. Fabio wants to peel away the superficial and get down to reality, to bring his work into the circle of life. But how does one go about transforming abandoned shops into something completely different? Where does he make the connection? 'I believe in the energy of places. You have to listen to the ambience and feel the flow of energy. Some places carry a negative charge and some a positive charge. I'd love to convert a former church into something new. Imagine the positive message you'd be able to shape.' He ponders aloud the journey from one dimension to another, calling a thought a 'zero-dimensional vision', a drawing a 'two-dimensional concept', the implementation the 'three-dimensional reality', and then retracing his steps: two-dimensional photograph of three-dimensional object becomes zero-dimensional thought.

I'm not wrong in speculating that Fabio has plans to make a film someday. Diploma in hand, he didn't make a beeline to the nearest architecture firm to share the good news, but took off for New York City to study filmmaking. While furthering his education and moonlighting in an art gallery, he got a call from Italian stylist Anna Molinari, who offered him his first commission: the interior design of a shop in Hong Kong. 'I'm fascinated by film. It's about giving away dreams – a *grandioso* way of condensing art, music, words and images into a hundred or so minutes of magic. It's fantastic.' His all-time favourite is *Blade Runner*, undeniably an incarnation of the contemporary spirit. Films like *Blade Runner* are proof that design transcends the drawing board. Real designers are conceptual geniuses – people like William Gibson, Walter Benjamin and Italo Calvino. Big minds full of big ideas.

'I'm ready to get into outdoor architecture as well. But compared to film, architecture is so arrogant. Look at this building in front of mine, which blocks a beautiful panoramic view. Why don't we put an expiry date on architecture? Why isn't the death of architecture part of the design? Why not put explosives into the concrete? We can't ignore death. It's the only thing that makes us human. The fact that we die is the absolute democratic principle. It's horrendous that we don't apply this principle to architecture, to everything.'

Food is about to be served. Fabio's girlfriend, Norwegian model Melissa Johansen, is laying the table, and I'm still scrambling for an answer to my question: 'What does it take to make a good designer?' With Ledoux in mind, I mention to Fabio that his obsession with sex might be the source of a design for a brothel. 'Dead on target,' he laughs. 'The ultimate dream.' ◄

Fabio Novembre's website: www.novembre.it

This and opposite page: Bar Lodi, located in Lodi, Italy. The entrance, with its rounded corners, is clad in a mosaic design that mimics the barcode on Novembre's book: A Sud di Memphis. A large mirror to the right of the bar is a major space-enhancer. A mosaic-covered 'tunnel' conceals all mechanical systems. Fluorescent lighting that passes through a number of silhouettes made by removing part of the mirror's silver coating gives an impression of 'customers', even when the bar is empty. These silhouettes are repeated in the mosaic on the floor

'Architecture is the last medium to acknowledge that we're made of flesh and blood, that in the end there's only us, that we're surrounded by the tangible, by people we can touch, embrace, take to bed'

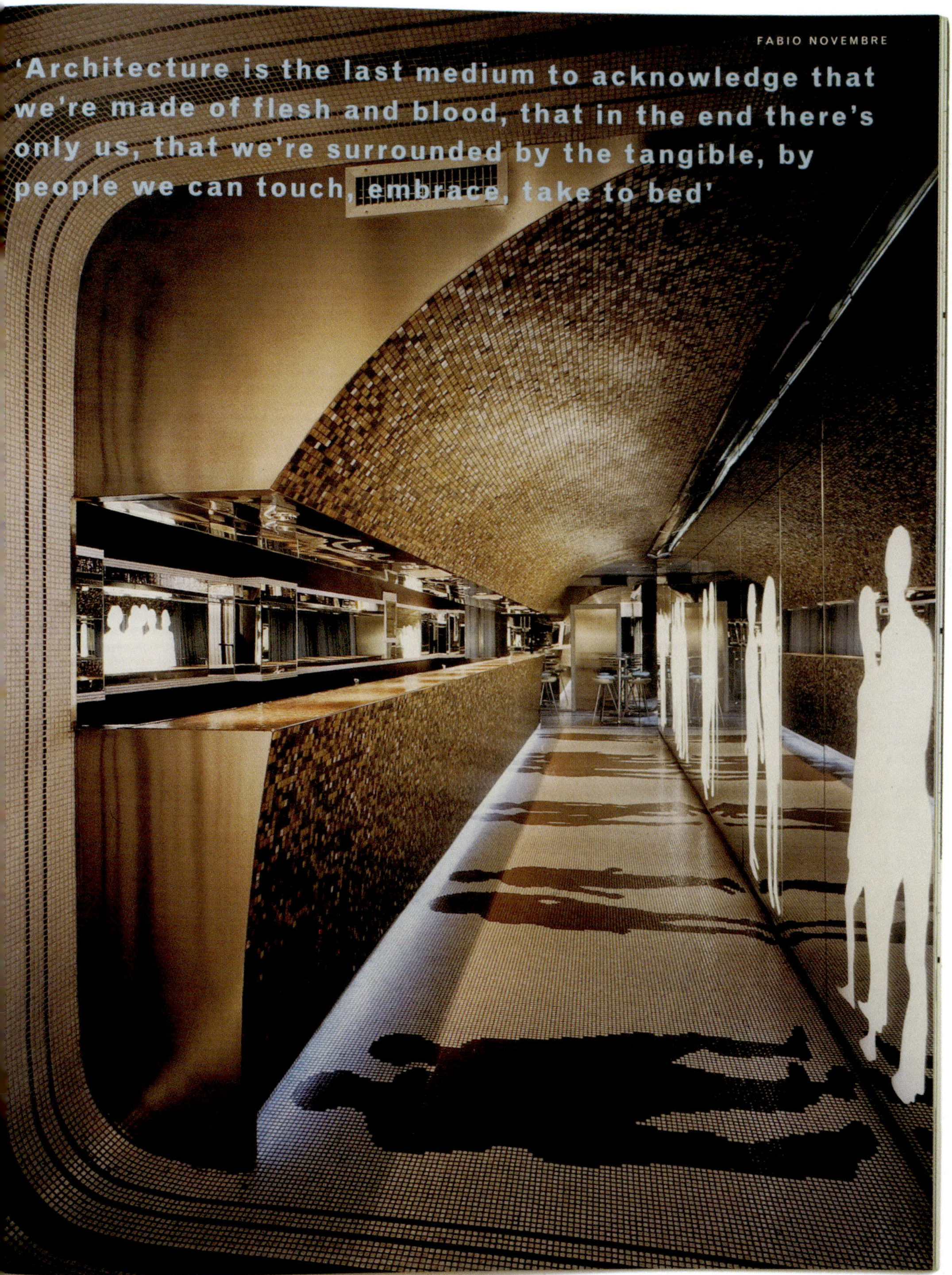

SUBJECT
Portrait of
Fabio Novembre

LOCATION
Milan, Italy

DESIGNER
Fabio Novembre

PHOTOGRAPHER
Alberto Ferrero

WRITER
Leo Gullbring

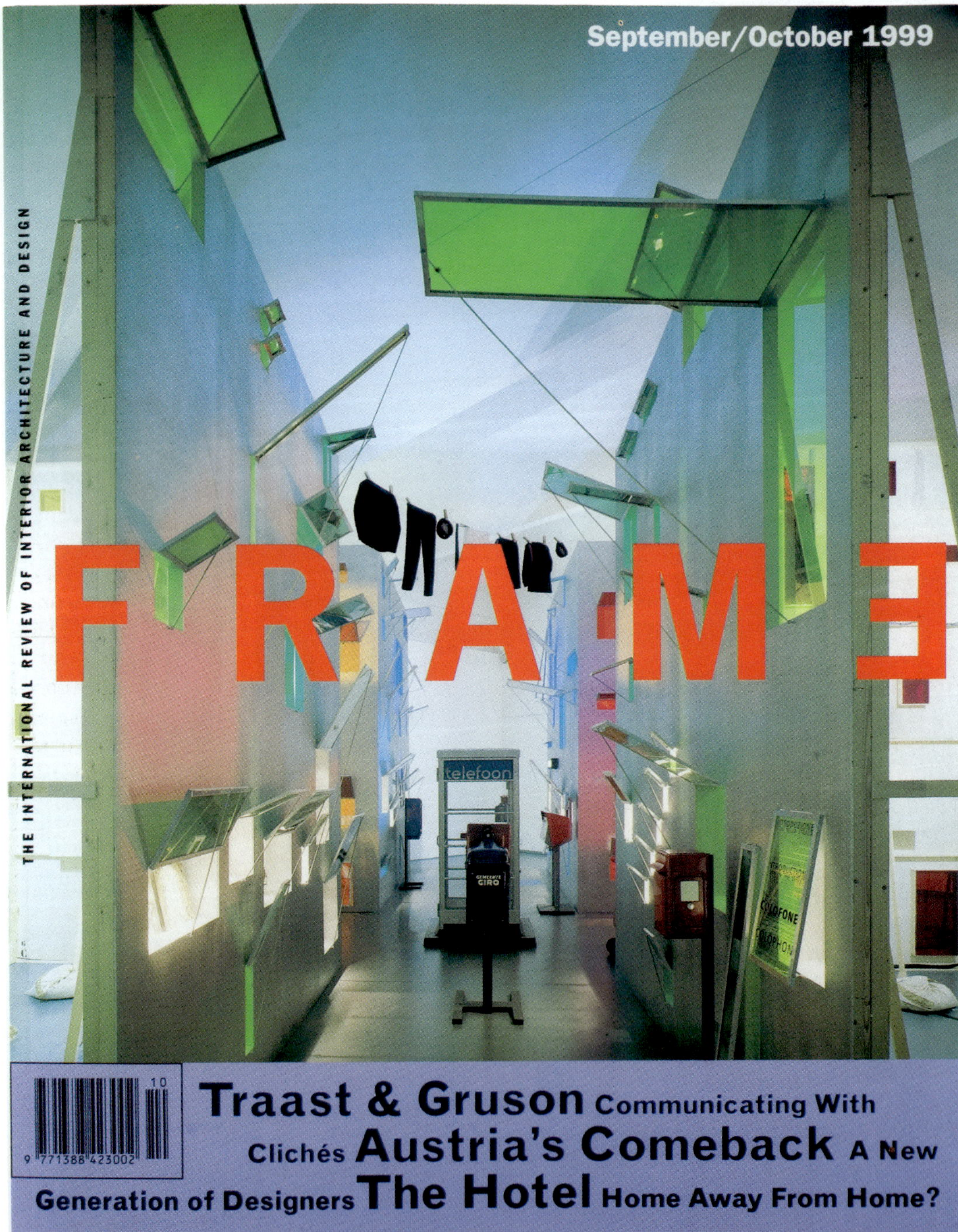

ART DIRECTOR
Ton Homburg,
Opera Designers

10

Comme Kawakubo

Frame editor Robert Thiemann about the project

Rei Kawakubo first appeared on our pages in Frame 10. This Japanese fashion maven manages time and again to turn the boutique phenomenon upside down, as exemplified in this article, which focuses on her New York shop and two Comme boutiques in Tokyo. Any normal retailer would give her eyeteeth for both an A location – a good spot that attracts shoppers – and the biggest shop window possible for displaying merchandise and tempting passers-by. Kawakubo doesn't follow the rules, however. In New York she set up shop in Chelsea, a neighbourhood that in 1999 was unexplored territory for retailers and had little more going for it than a smattering of art galleries. 'Those in the know will discover it soon enough' was and is Kawakubo's theme song. And why add a display window? She asked the same question in Tokyo, where a bright orange door (and nothing else) announced one of her two boutiques, and the display window of the other was covered in blue dots that largely concealed everything behind it. Kawakubo has proved to be trendsetter: 21st-century Chelsea is hot. And the absence of shop windows has hindered the growth of Rei's empire not a jot.

DESIGNERS
Rei Kawakubo
(Comme des Garçons)
Future Systems
www.future-systems.com
Takao Kawasaki
Studio Morsa

SUBJECT
Comme des Garçons shops

LOCATIONS
Tokyo, Japan
New York, USA

PHOTOGRAPHERS
Todd Eberle
www.toddeberle.com
Masayuki Hayashi
J. Francois Jose

WRITERS
Asko Ahokas
Carolien van Tilburg

Comme

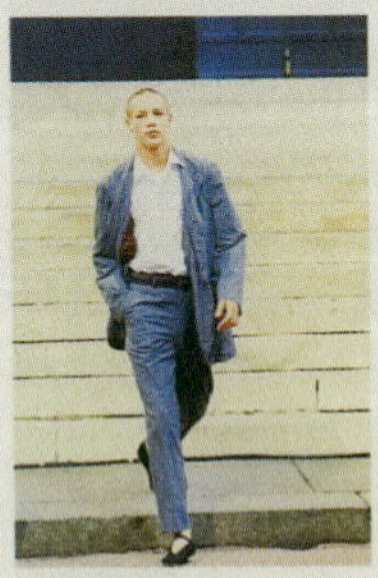

Comme des Garcons, men's fashion, summer 1999. Photo: J. Francois Jose

With its latest stores in New York and Tokyo, Comme des Garçons once again sets itself apart from the mainstream. Sacral spaces literally turn their back on the serious fashion aficionados they're aiming to seduce. *By Asko Ahokas and Carolien van Tilburg.*

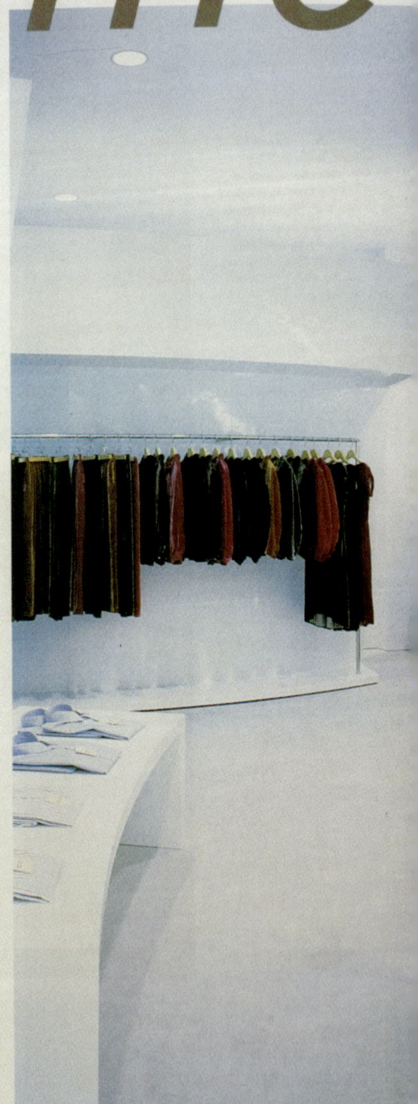

Comme des Garçons was the first establishment of its kind to open in New York's Soho district more than fifteen years ago. Soho's artistic cache has since attracted many more fashion retailers to the once bohemian neighbourhood – names that range from Prada and Louis Vuitton to those targeting a more mainstream clientele, such as Victoria's Secret. A decade and a half later, the time had come for Rei Kawakubo to relocate her store to a new and unexplored territory.

Kawakubo has always maintained strict control of the interiors in which her clothes are sold. Comme des Garçons' first stores mirrored the *Zeitgeist* with their spare, box-like design, which matched the clothes and created a strong, minimal environment marked by a faint aura of inaccessibility. Following in Kawakubo's steps, many other designer labels went on to create flagship stores that were colossal statements defying minimalism. In the late '90s, however, Kawakubo's designs are once more featuring sculptural elements, and it seems that a new look for the body has necessitated a new look for the shop interior.

The same deconstructivist instincts present in Kawakubo's clothes are reflected in her retail premises. The new Comme des Garçons shop in New York City's Chelsea area is another pioneering establishment. Located at 22nd Street and 10th Avenue, it is the only retail store in a neighbourhood that's home to a growing colony ▶

Kawakubo

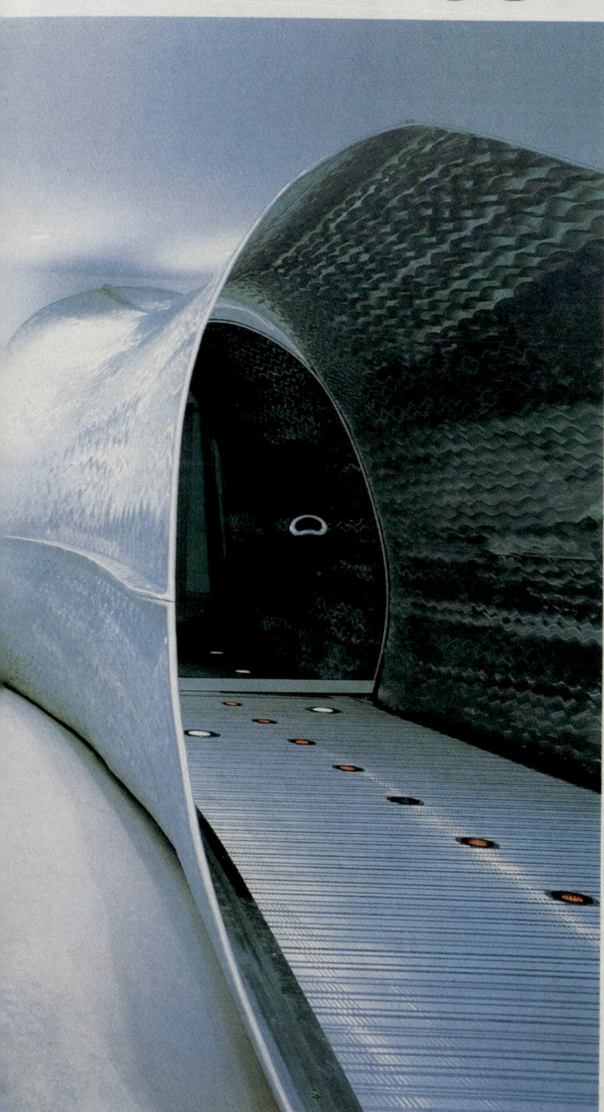

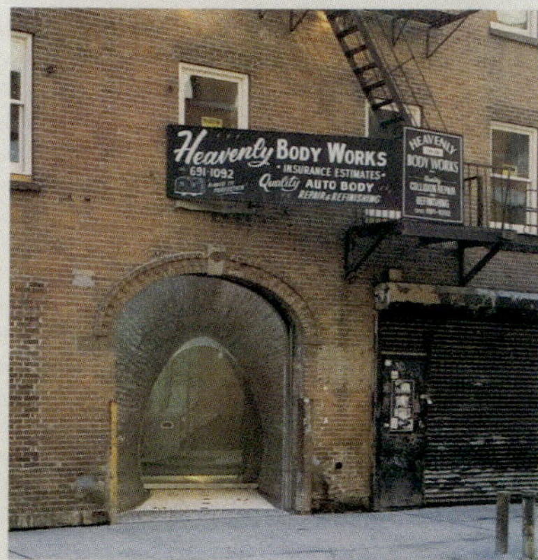

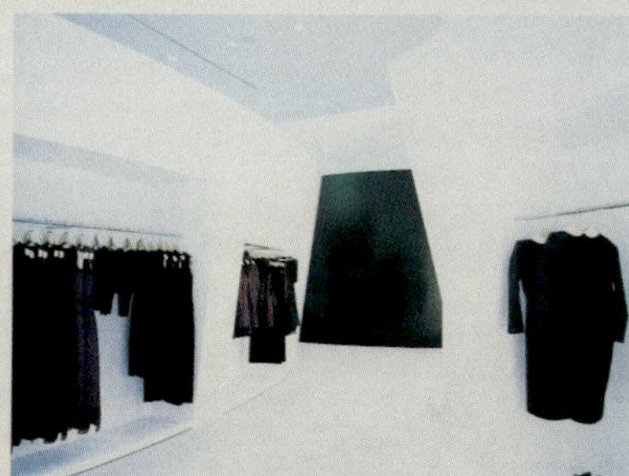

Outside Comme des Garçons' new shop in Chelsea (New York), an aluminium tunnel is all that suggests the special interior awaiting customers at the other end (top). Inside, a sculptural world of white immerses the senses in hidden luxury (left and above). Photography: Todd Eberle

SUBJECT
Comme des Garçons shops

LOCATIONS
Tokyo, Japan
New York, USA

DESIGNERS
Various, see page 097

PHOTOGRAPHERS
Various, see page 097

WRITERS
Asko Ahokas
Carolien van Tilburg

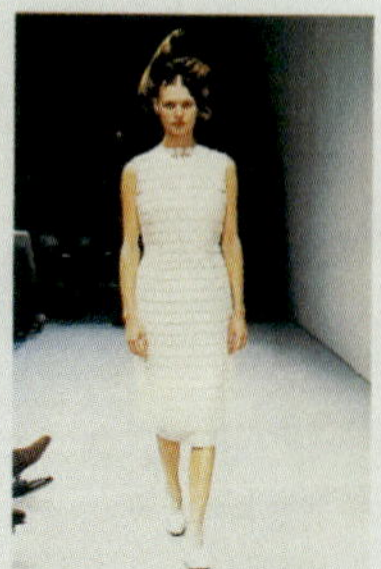

Comme des Garcons, Pret à porter collection, summer 1999. Photo: J. Francois Jose

You can't see the whole interior or all the clothes at once. The composition of sculptural elements is a seductive invitation to wander around and discover. After awhile you begin to realise how the space supports the fashions

of artists – a reinforcement of Kawakubo's ideology, which separates the designer and her work from the masses.

The century-old appearance and atmosphere of Chelsea has been maintained by its manufacturing-only zoning status. Since the mid-1990s, former industrial blocks on 22nd and 23rd Streets have enticed several art galleries away from the more retail-oriented Soho district. The same avant-garde arriving in Chelsea in the wake of these galleries are sure to be found browsing through the racks at Comme des Garçons.

Kawakubo's Chelsea store is the first of a series, all of which are to be based on a new, future-directed concept. According to the company, this concept expresses the pure will of the designer, uncompromising and unaccommodating. Its creation is the work of a closely co-operating international team, which includes Kawakubo, Japanese architect Takao Kawasaki, British architects Future Systems, Studio Morsa in New York, and contractors in Japan, England and the United States. Kawakubo and Kawasaki had also collaborated in developing the previous generation of the designer's stores.

Like an old-world geisha, the exterior is both modest and mysterious. The worn-out, unfinished look of the red brick facade has been left intact, and there are no plans to change it. Even the signage for 'Heavenly Bodyworks', a car-repair shop that formerly occupied the space, still hangs above the entrance. No window displays, billboards or trademarks identify the store from the outside – a strikingly different strategy from that of most designer labels, which boldly mark their territory and maximise display areas to attract the window-shopping crowd.

The original drive-in entrance to the former car-repair shop frames a lustrous aluminium tunnel, designed by Future Systems. The tunnel – manufactured in a shipyard in Cornwall, England – is visible from the street, but only to those passing directly in front of it. Mysteriously luminous, it subtly invites visitors to enter the hidden world of Kawakubo. The first bit of fun lies in discovering how to open the glass door to the tunnel. Having solved that puzzle, the customer moves through a passage flanked by shimmering curved walls and enters a heightened state, unlike the world outside.

At the other end, the space opens up into nearly 500 square metres of shiny white space. Gleaming white, meticulously enamelled steel structures are juxtaposed with black walls that curve around fitting rooms to create a space more like a sanctuary than a place to sell clothes.

The collections are enclosed in self-contained pods of steel, immaculately painted with white enamel and manufactured in Tokyo. Kawakubo added a softening element of black soundproof panels to create a space-age relief of sorts on the back wall. Monochromatic black, present throughout the store in shiny enamel and matte panelling, has been her signature in fashion for years. Kawakubo's black is distant and subtle rather than sensual or erotic. Viewing the shop's fixtures, several visitors have remarked on the similarity to sculptures done by Richard Serra, which were on display last season in a gallery across the street: the Dia Center for Contemporary Art.

You can't see the whole interior or all the clothes at once. The composition of sculptural elements is a seductive invitation to wander around and discover. After awhile you begin to realise how the space supports the fashions. Careful consideration and disciplined creativity permeate both clothes and interior. Just as ▶

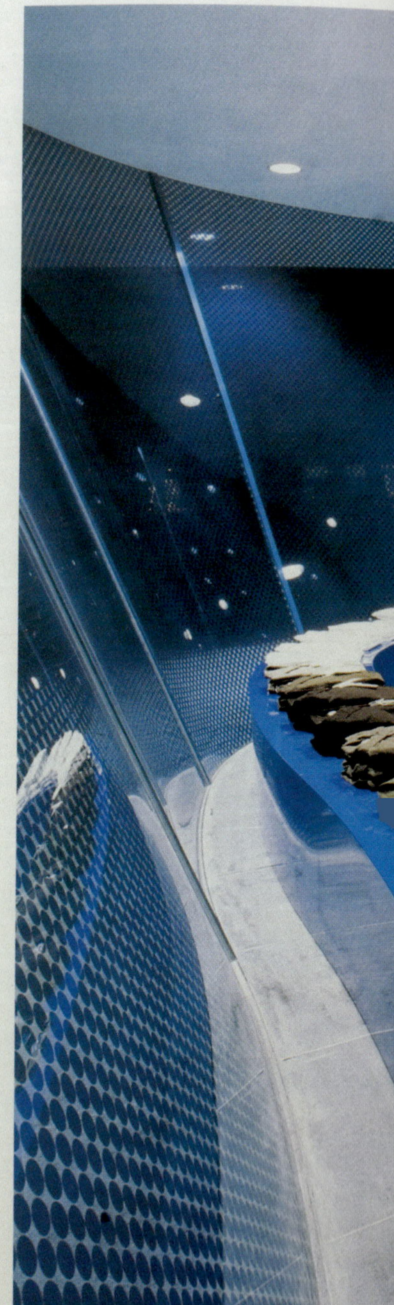

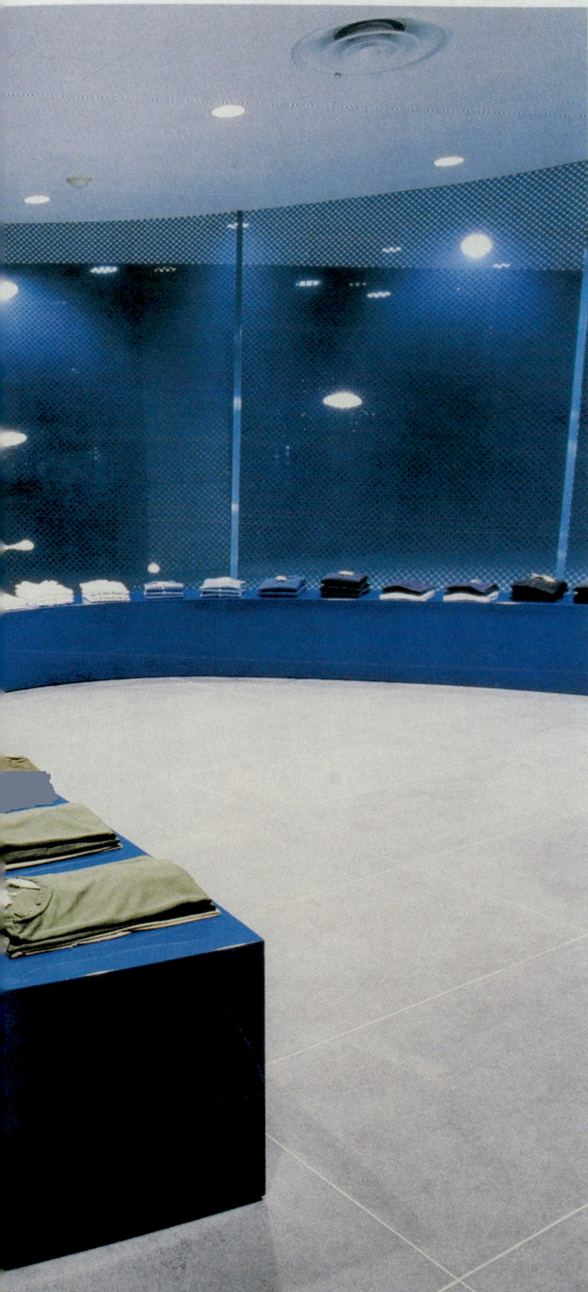

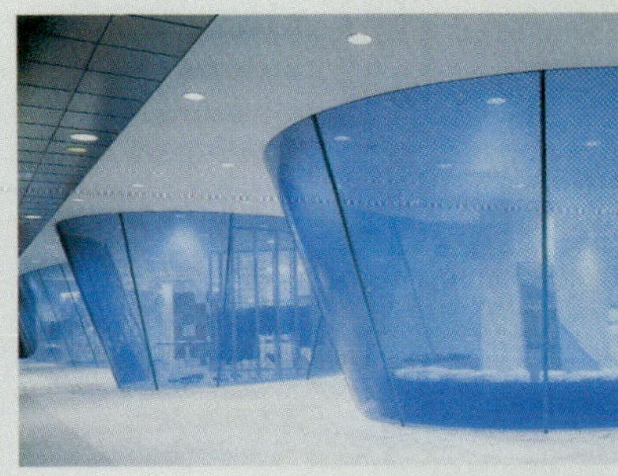

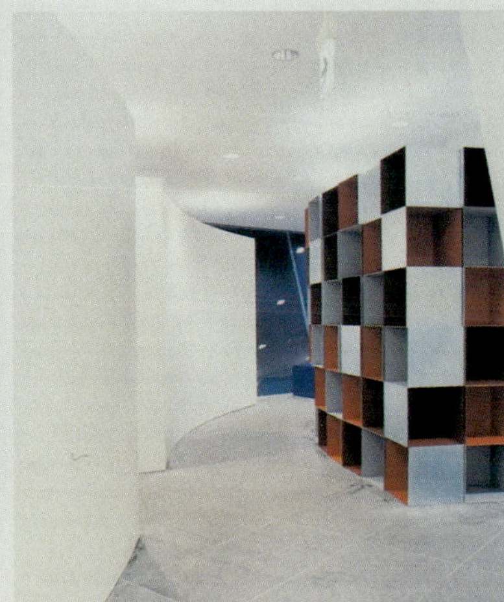

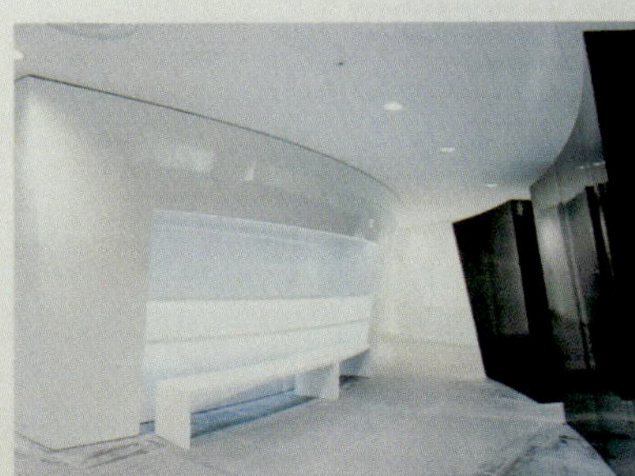

Comme des Garçons' main store in Tokyo lies concealed behind a partially transparent, curved facade of blue glass (top). On the opposite side of the glass, clothing is arrayed on a modest display element that follows the curve of the glass (left). Here, as in the New York shop, sculptural, enamelled components give the space its unique quality (above). Stacked metal boxes inject a note of colour into an interior dominated by white and, to a lesser extent, black (middle). Photography: Masayuki Hayashi

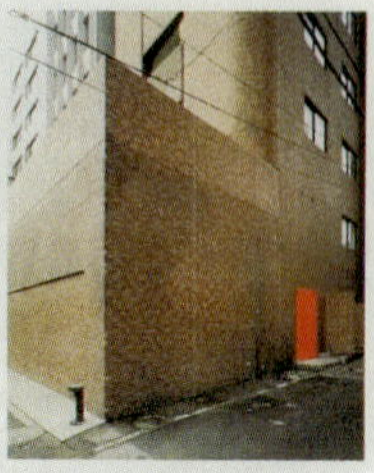

Fashion-world insiders have already nicknamed the establishment, affectionately referring to it as 'The Orange Door'

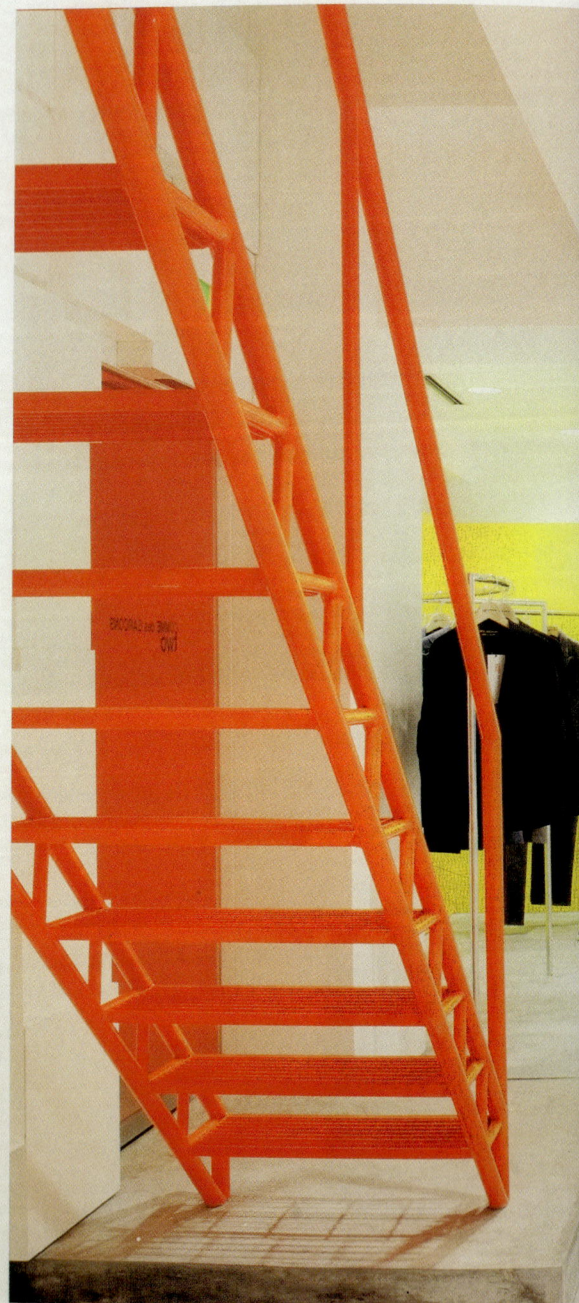

An orange door marking an otherwise blind facade is a whimsical introduction to Tokyo's second shop, laconically called Two (top). Many elements seen in the main shop are repeated here (below). New are the bright neon shades that form a contrast to a largely white environment (right). Photography: Masayuki Hayashi

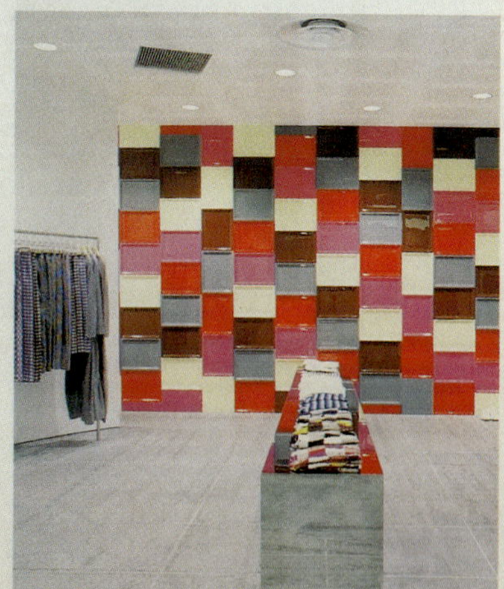

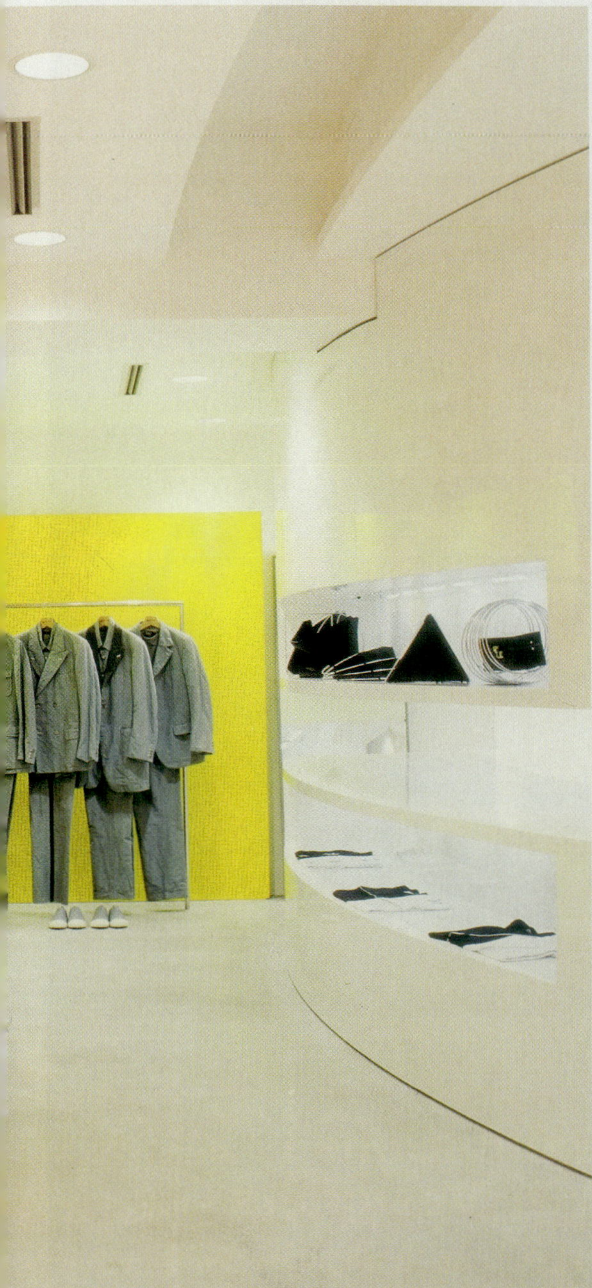

the entrance tunnel subtly draws the visitor into Kawakubo's world, the garments she designs seem to reveal, almost shyly, a hidden agenda. Take the shirred lining of a man's jacket, for example – although invisible from the outside, it reshapes the silhouette and protects the wearer from the bumps of urban life. New York's Comme des Garçons is not the only shop in the family that's been treated to a new outfit. In Tokyo's ultra-trendy Aoyama district, a drastic renovation has completely transformed Rei Kawakubo's main store. A second shop nearby – cleverly called Two – appears to be a base of operations for the expanding range of collections and labels marketed by Comme des Garçons. The main store – until recently, a transparent, sterile box – is now somewhat concealed behind a partially transparent, curved facade of blue glass. Designed by Future Systems, this slanting glass wall with bumpy blue protrusions extends a welcome, while subtly filtering the view of the interior. After entering the premises, you walk through a labyrinth featuring curved, white-enamelled walls and slanting, black-matte partitions. Just as in New York, visitors are not provided with an overall view. Each part of the collection has its own special section. Accessories are prominently displayed, while clothes are presented modestly, but clearly according to theme.

Dividers between these themes – shall we call them 'installations'? – include stacked metal boxes, soundproof panels and a cabinet with vases wrapped in fabric. The customer is manoeuvred through the space by means of views created to draw the eye to other areas of the store. Spacious fitting rooms are concealed within the walls.

Collections bearing the names of Comme des Garçons designers Junya Watanabe and Keiichi Tanaka have been moved to other premises. Two, as the second shop is called, is obviously an oriental cousin of the store in New York. Although the front of the Tokyo shop has not been left in its original state, it *has* been neatly walled up. The blind facade – without a single sign or logo indicating what's for sale inside – is a strange phenomenon in a shopping street lined with windows boasting the artfully arranged products of top fashion designers and exclusive antique dealers. Two's entrance, at the side of the building, is painted a fluorescent orange, a device that definitely arouses curiosity. Fashion-world insiders have already nicknamed the establishment, affectionately referring to it as 'The Orange Door'. Once inside the brightly illuminated white space, with its curving walls, you're immediately drawn by the unique charisma that typifies Comme des Garçons. Accents of colour are bolder here than those in the main store: one corner panel is fluorescent yellow, and steps leading upstairs radiate a brilliant orange. In comparison with the design of the main store and the fashions sold there, these more exuberant accents correspond well to the somewhat more baroque, experimental collections designed by Junya Watanabe, which are sold at ground-floor level. The stairway leads to 'casual' labels on the upper floor: Homme and Homme, homme. In the area of interior design, however, this floor offers nothing in the way of surprises.

Comme des Garçons' main store, which features the collections of Rei Kawakubo, is a trip through 700 square metres of intriguing destinations, while Two stirs the imagination with its minimalist exterior. Together with the shop in Chelsea, they've set the pace for the next generation of Comme des Garçons stores and, undoubtedly, for other brands eager to be *comme* Kawakubo. ◄

SUBJECT
Comme des Garçons shops

LOCATIONS
Tokyo, Japan
New York, USA

DESIGNERS
Various, see page 097

PHOTOGRAPHERS
Various, see page 097

WRITERS
Asko Ahokas
Carolien van Tilburg

Issue 11

November/December 1999

104-105

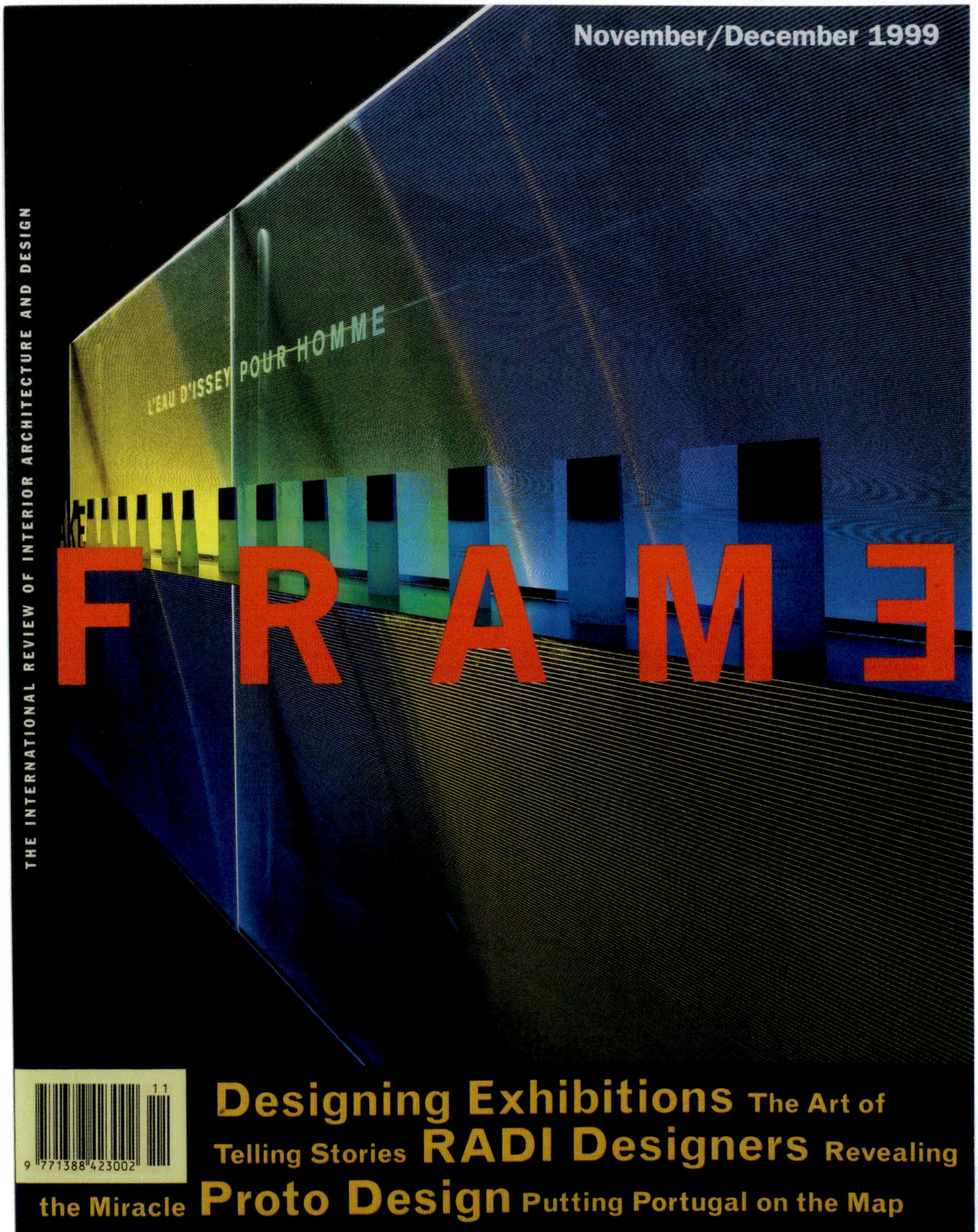

ART DIRECTOR
Ton Homburg,
Opera Designers

11

Interface for Cyberspace

Frame editor Robert Thiemann
about the project

As one of the few international interior-design magazines in existence, Frame has always paid attention to exhibition design. We look for shows and exhibitions that get their message across in particularly striking ways. Our idea of a well-designed exhibition is one that gives the visitor an architectonic experience that relates to the theme and one in which audiovisual media are kept to a minimum. A young Canadian outfit, (The) Fridge, amply satisfied both criteria with an exhibition design for the Canadian Centre for Architecture (CCA) in Montreal. (The) Fridge projected architectural photographs on the inner surfaces of Barrisol membranes covering tentlike structures. These 'aliens' gave visitors to the classical rooms of the CCA a curious and memorable experience, and their presentation did justice to the subject of the show: architectural photographs shown simultaneously in Montreal, in a Parisian gallery and on the internet. The contorted membranes provided a lucid interface that referred to the intangible nature of the internet. And a small exhibition, presented on only four pages of the magazine, has turned out to be the Frame 11 article that we most admire.

DESIGNERS
(The) Fridge

SUBJECT
Exhibition design for 'Interspace' at the Canadian Centre for Architecture

LOCATION
Montreal, Canada

PHOTOGRAPHER
Michel Legendre

WRITER
Robert Thiemann

Interface for Cyberspace

Architecture in cyberspace has an uneasy relationship with earthbound architecture. That which is fluent, elastic and constantly in motion on the Web is static, stiff and inflexible in the 'real' world, even though many architects are doing their best to unite the two. No one can deny that the phenomenon of virtual architecture is a growing source of inspiration for traditional architects. These thoughts formed the basis of an exhibition held last summer at the Canadian Centre for Architecture (CCA) in Montreal, an event that focused on the interface between the two types of architecture. Four photographers from the Canadian capital and four Parisian colleagues – a collaboration reflective of the age-old connection between Canada and France – were asked to illustrate the 'Interface' theme with digital photographs. The resulting images were shown simultaneously at the CCA, at an exhibition in Paris and on the Internet. A competition for the design of the CCA exhibition was won by fledgling firm (The) Fridge, consisting of architects René-Luc Desjardins, Régine Lafata and Francis Novak. Inspired by the fluidly elastic nature c cyberspace, they designed a structure of pliable forms covered b a stretchy transparent membrane with a contorted surface, upor which they projected photographs from within. These tent-like projection screens looked like extraterrestrials newly alighted in the classical rooms of the CCA – a brilliant thematic interpretation. *Robert Thiemann. Photography: Michel Legendre.* ◀

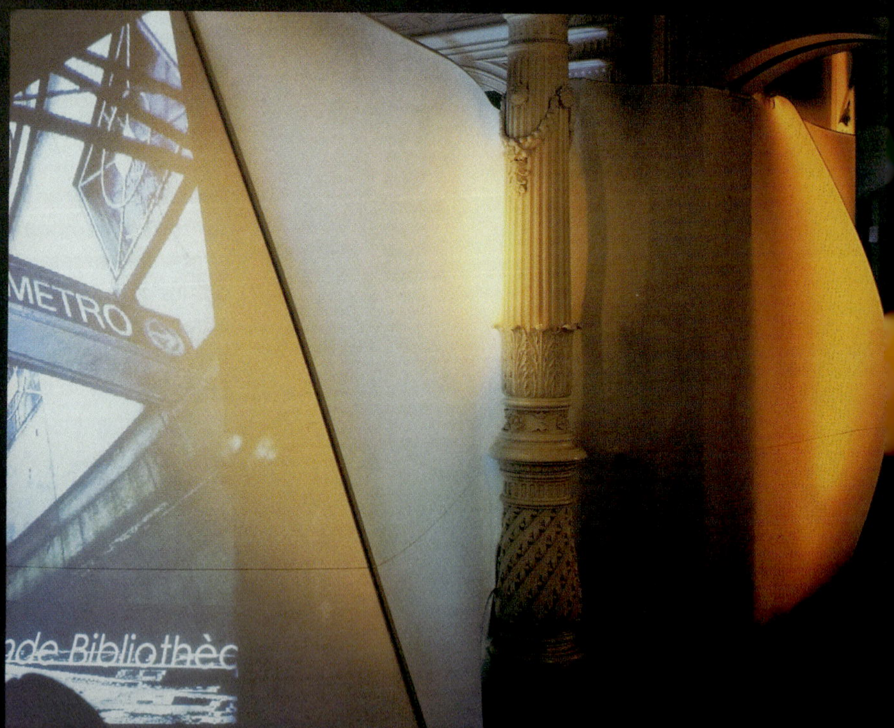

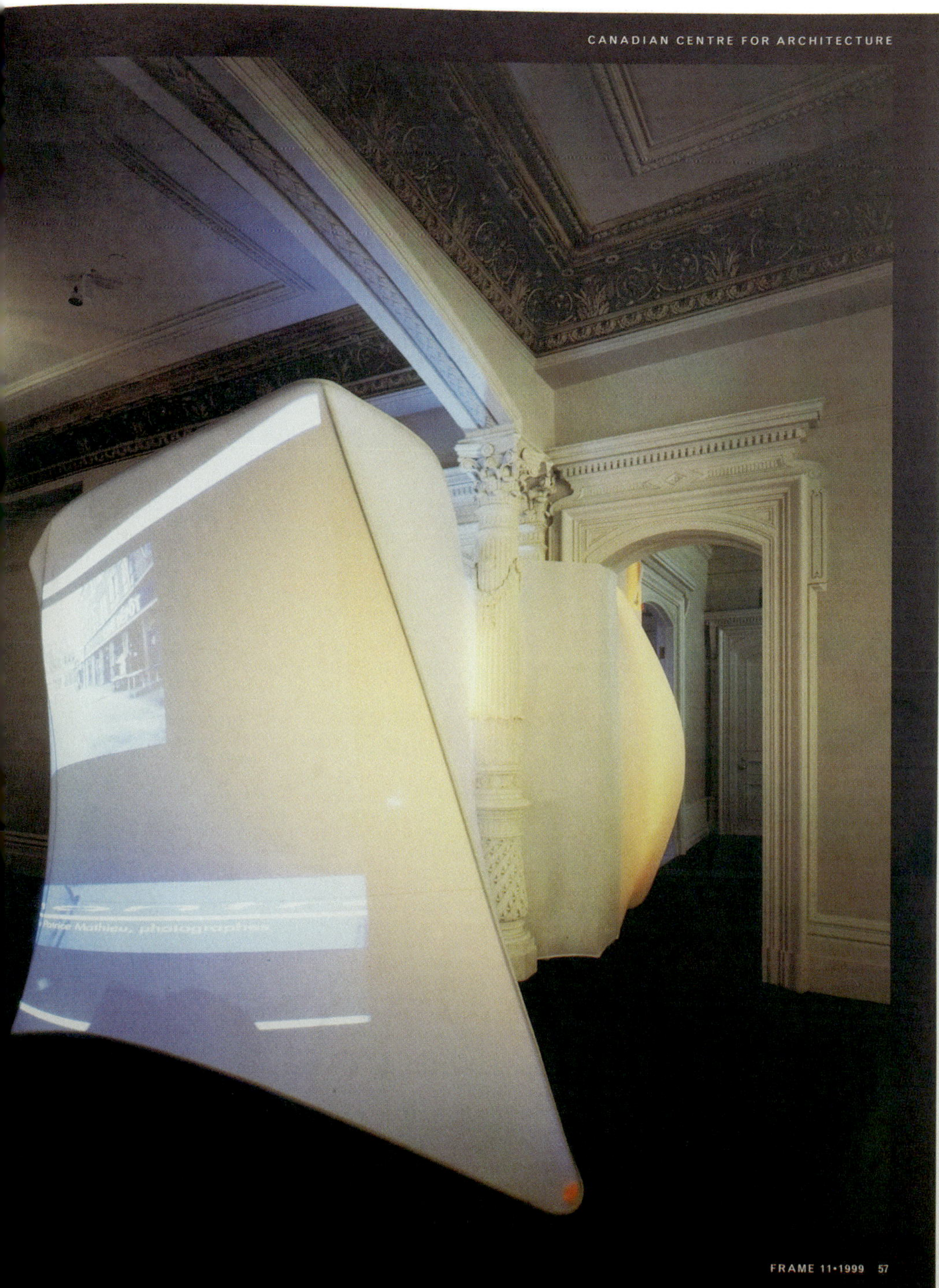

SUBJECT
Exhibition design for 'Interspace'

LOCATION
Montreal, Canada

DESIGNERS
(The) Fridge

PHOTOGRAPHER
Michel Legendre

WRITER
Robert Thiemann

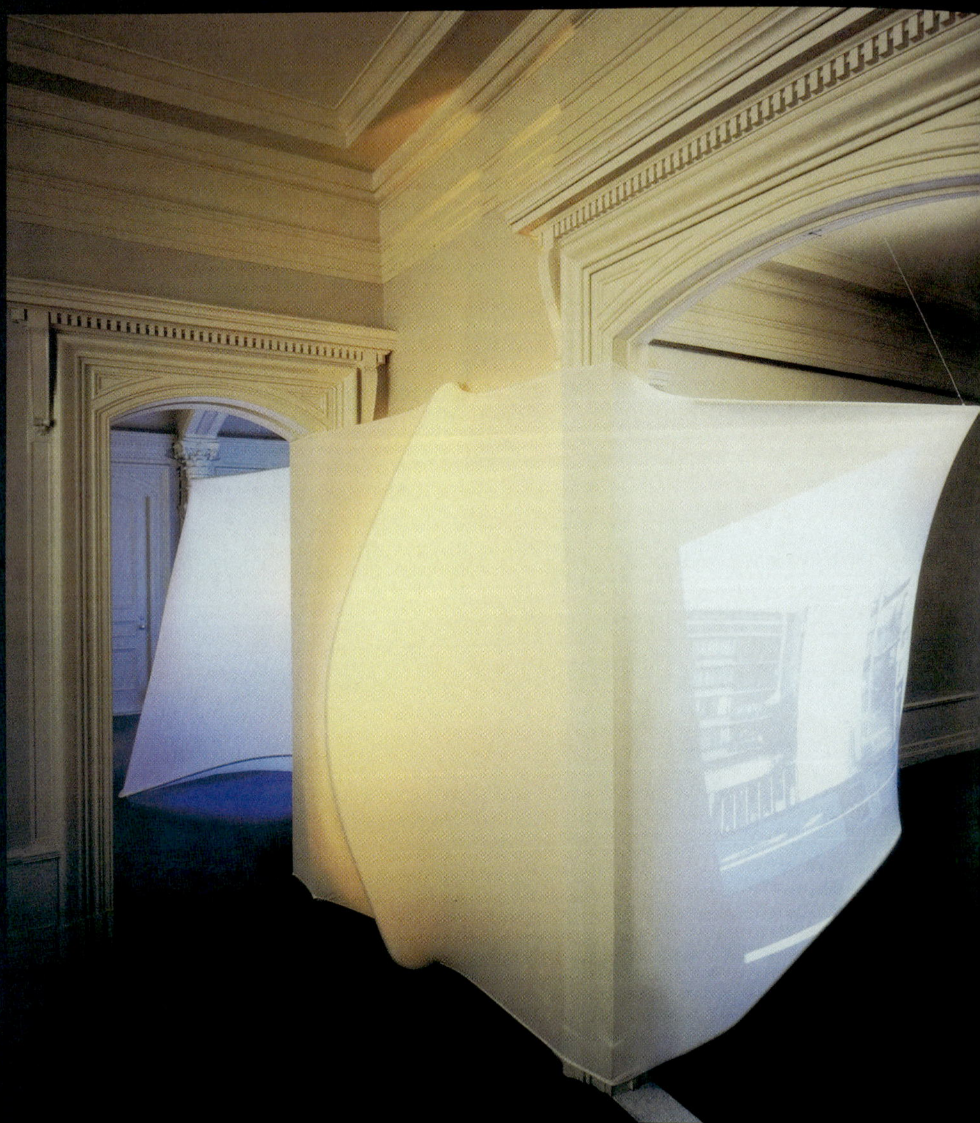

Tent-like projection screens look like extraterrestrials newly alighted in the classical rooms of the Canadian Centre for Architecture – a brilliant thematic interpretation

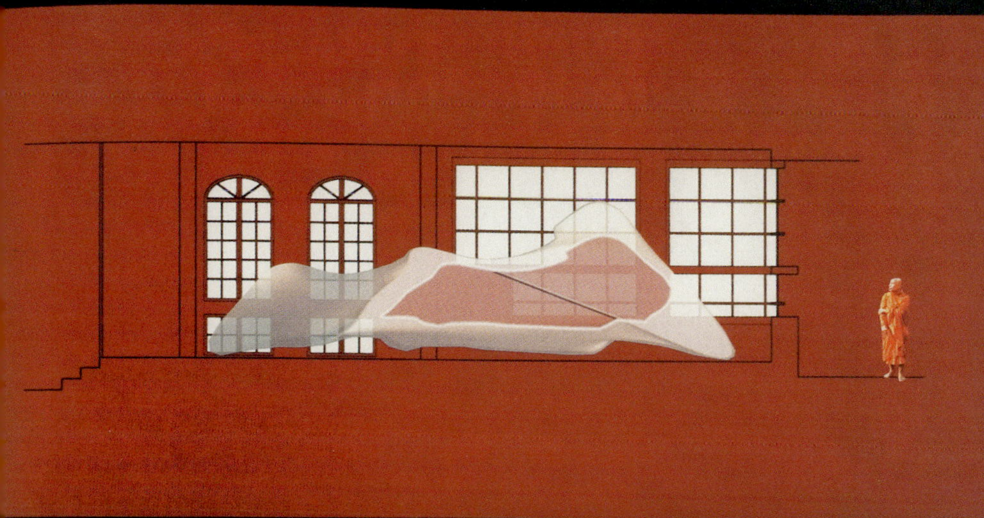

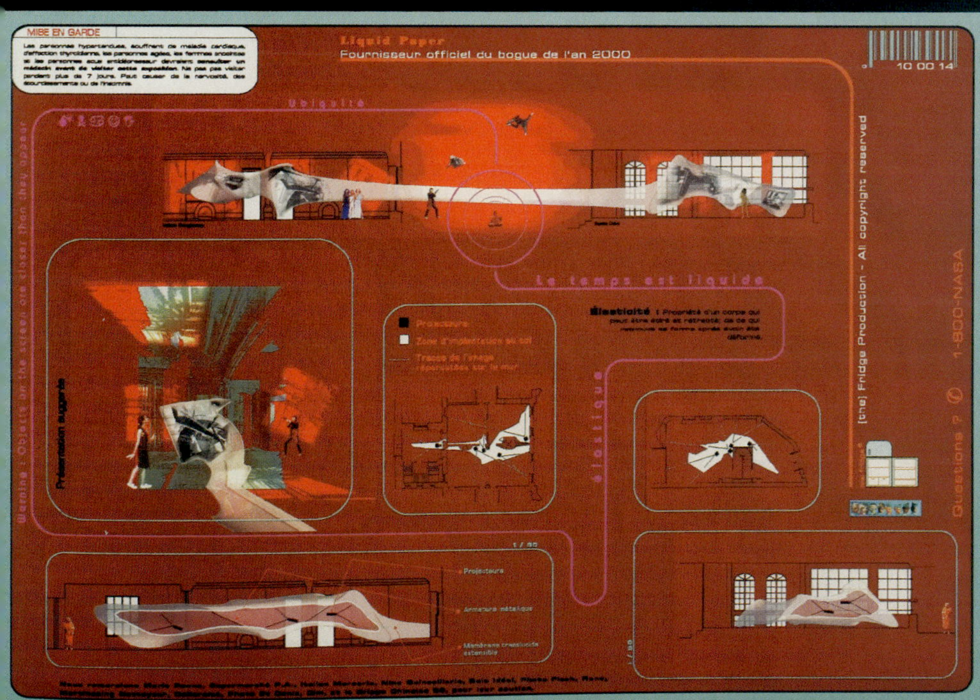

SUBJECT
Exhibition design for 'Interspace'

LOCATION
Montreal, Canada

DESIGNERS
(The) Fridge

PHOTOGRAPHER
Michel Legendre

WRITER
Robert Thiemann

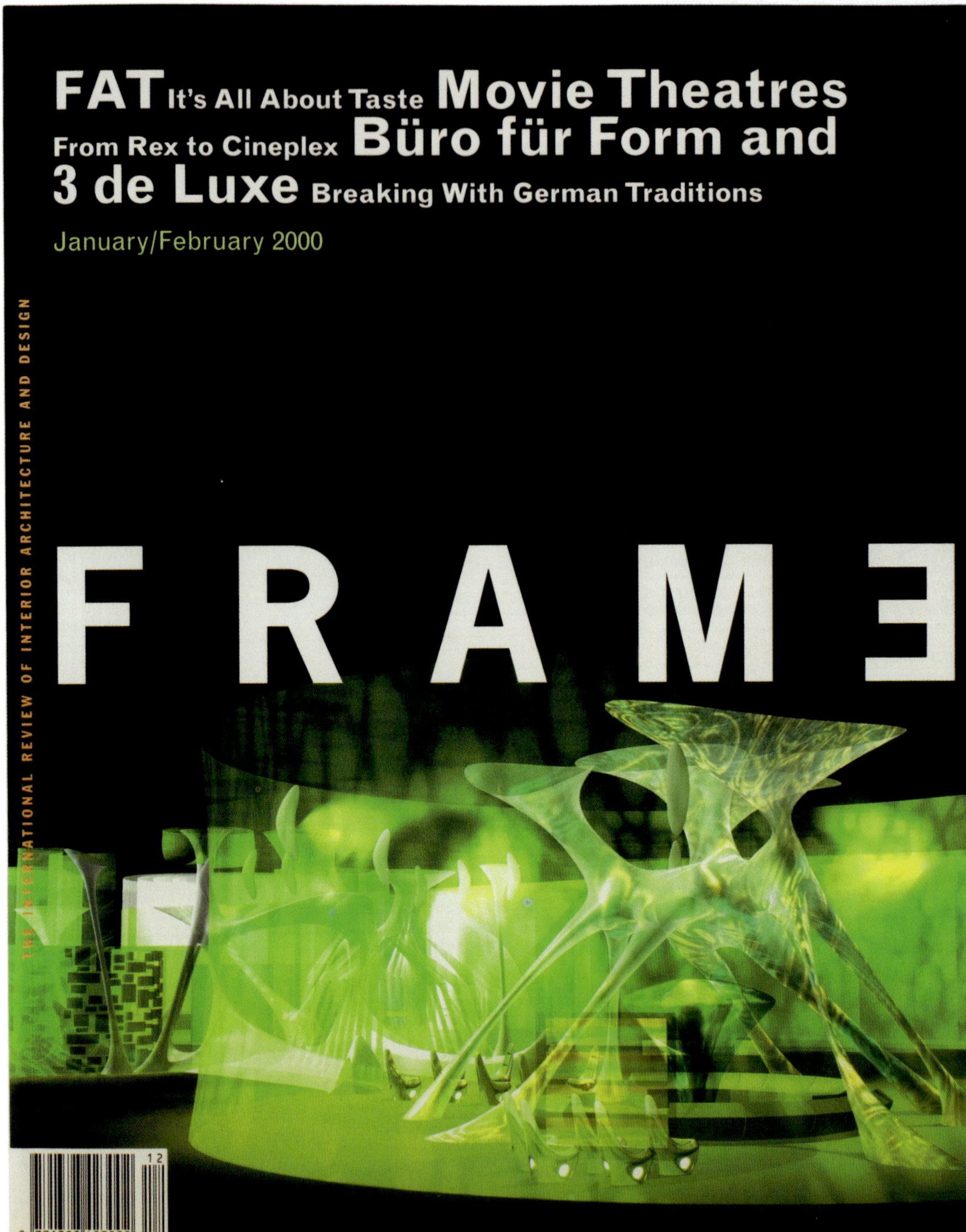

ART DIRECTOR
Ton Homburg,
Opera Designers

12

Virtual Reality de Luxe

3deluxe about the project

Scape, the Youth Media Pavilion for Expo 2000 in Hanover (on the cover of Frame 12) is – along with the Neue Räume installation for the 1996 Frankfurt Fair (p 86) – one of our key projects. Derived from the conceptual approach that we call 'genetic architecture', the organically shaped structures, now a trademark of 3deluxe, achieved their first realization within the interior 'scape' world. Our contact with Sven Väth, who played a DJ set at the closing event, led to our collaboration with him on CocoonClub, a project in Frankfurt that was completed in 2004.

Did the article have an effect?

A very positive one. Scape attracted international attention from those in the design scene. The cover story in Frame was an important medium for our increasing publicity.

Recent news or information about the project featured in *Frame*

Scape was a temporary project. The installation was dismantled after the last day of the World Exhibition: 31 October 2000.

Current projects

Closing ceremony (complete design and staging) of the 2006 FIFA World Cup // Salzzeitreise: Themed installation for a salt mine open to visitors in Berchtesgaden // Football Globe Germany, 2006 FIFA World Cup: Interactive information and light object that is touring cities such as Tokyo, Paris, Zurich and Milan as a cultural ambassador for the World Cup // Leonardo: Corporate architecture for a German glass manufacturer with a 2000-sq-m showroom // CocoonClub Summer Tour: Mobile event architecture // Aquasphere: Interior design of an aquarium at Hamburg's Hafencity.

DESIGNERS
3deluxe
www.3deluxe.de

SUBJECT
Portrait of 3deluxe

LOCATION
Wiesbaden, Germany

PHOTOGRAPHER
—

WRITER
Sven-Anwar Bibi

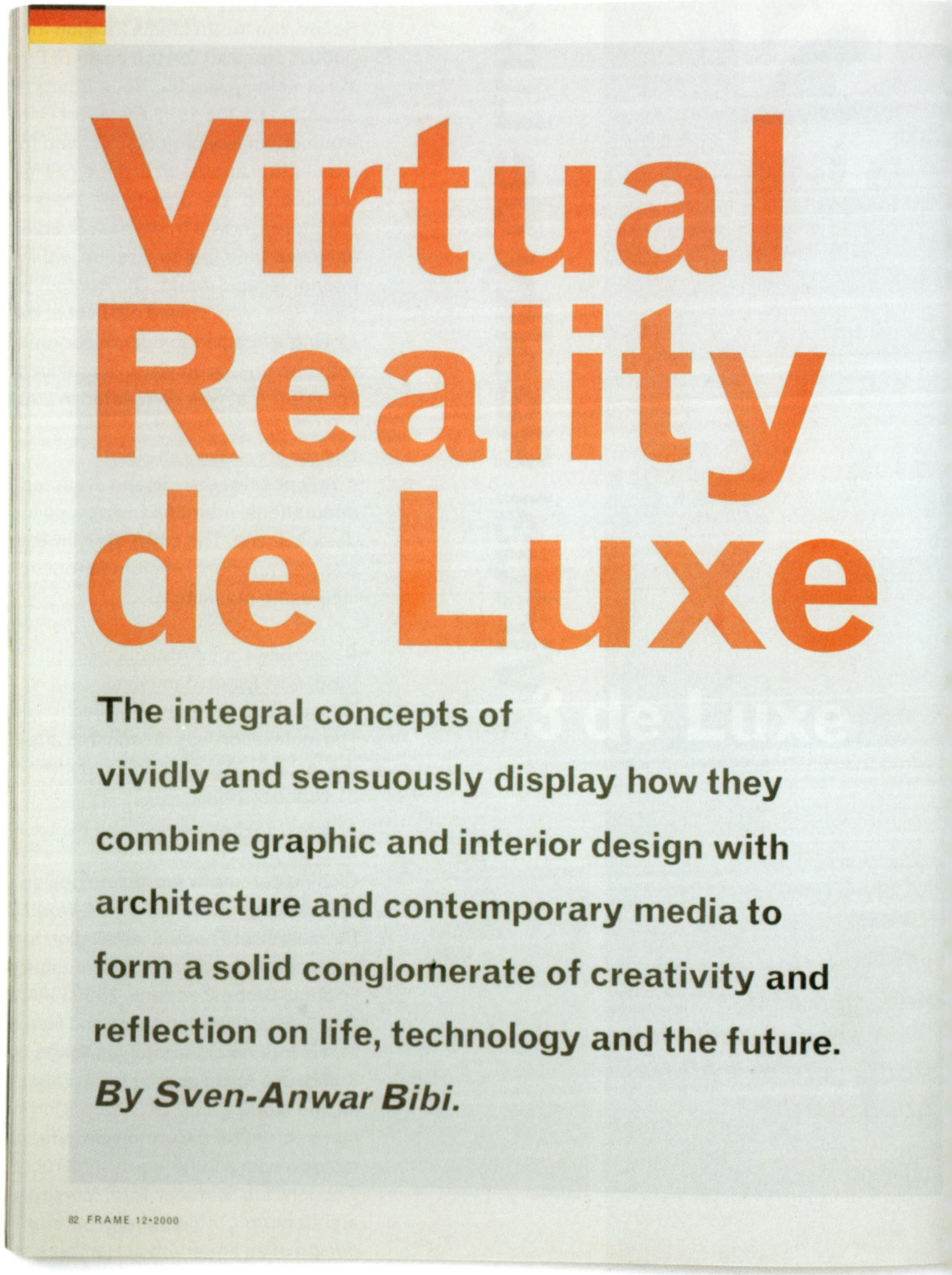

Virtual Reality de Luxe

The integral concepts of vividly and sensuously display how they combine graphic and interior design with architecture and contemporary media to form a solid conglomerate of creativity and reflection on life, technology and the future.

By Sven-Anwar Bibi.

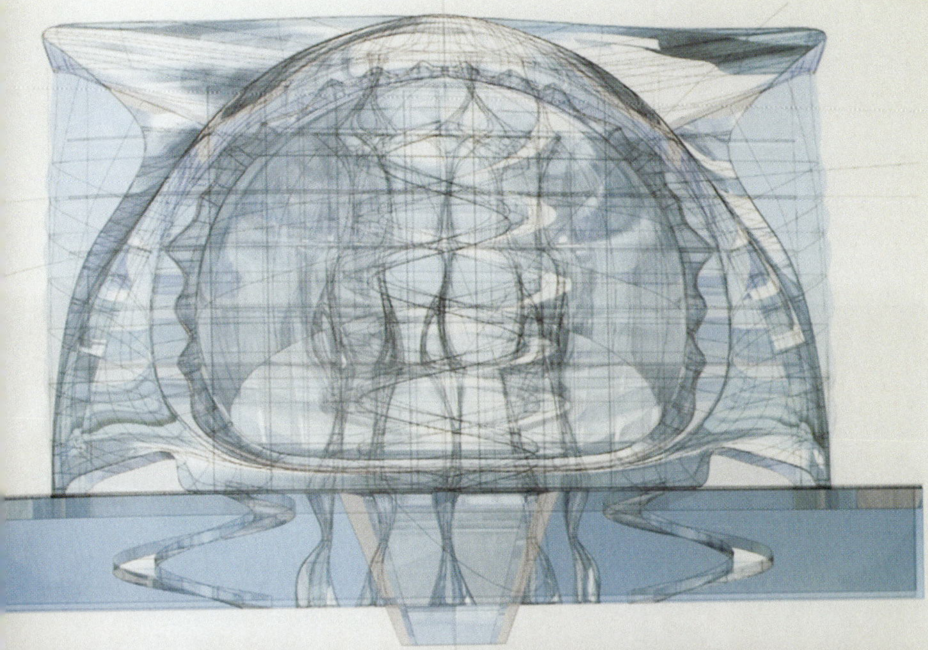

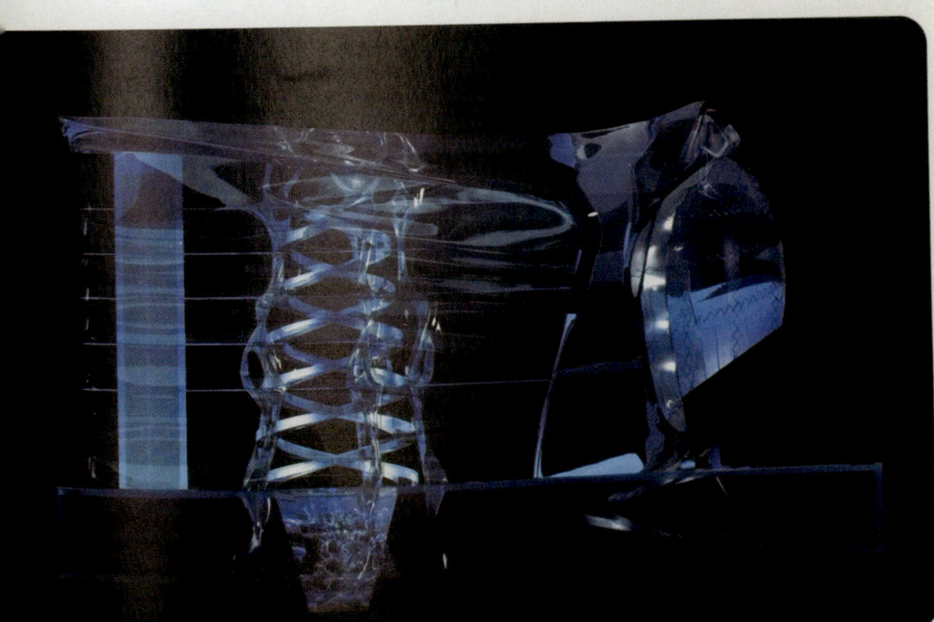

page: Bionic pavilion, a project for Expo 2000 in Hanover. The building is supposed to show what happens when natural mechanisms are applied to technology

SUBJECT
Portrait of 3deluxe

LOCATION
Wiesbaden, Germany

DESIGNERS
3deluxe

PHOTOGRAPHER
—

WRITER
Sven-Anwar Bibi

The 21st century is expected to unleash a flood of changes in the areas of design, architecture and music. The approach to the new millennium seems to be paved with futuristic visions, many of which also apply to housing and lifestyle. Although such concepts have been in the air for decades, however, relatively few revolutionary changes have shaken the surface of our lives. This can certainly be said of the music scene, and a good look at contemporary interior design reveals little more than a revised interpretation of futuristic objects popular in the sixties. Has innovation died of natural causes? Or do the cynical among us have to take a harder look at what's happening? On a recent visit to Wiesbaden, Germany, I found a design agency poised to challenge our wildest dreams.

Calling themselves 3 de Luxe, the Lauhoff brothers and their colleague Nik Schweiger established the agency about eight years ago. They were soon joined by Dieter Brell, and currently the firm's staff of 12 employees – representing the fields of design, architecture and media – focuses on projects in three primary areas: graphic design (they've worked for Fanatic Snowboards, Carhartt, and various record and fashion labels), interior design and architecture (the latter two areas are the province of Brell and Schweiger, both of whom I interviewed). The agency's diversity of

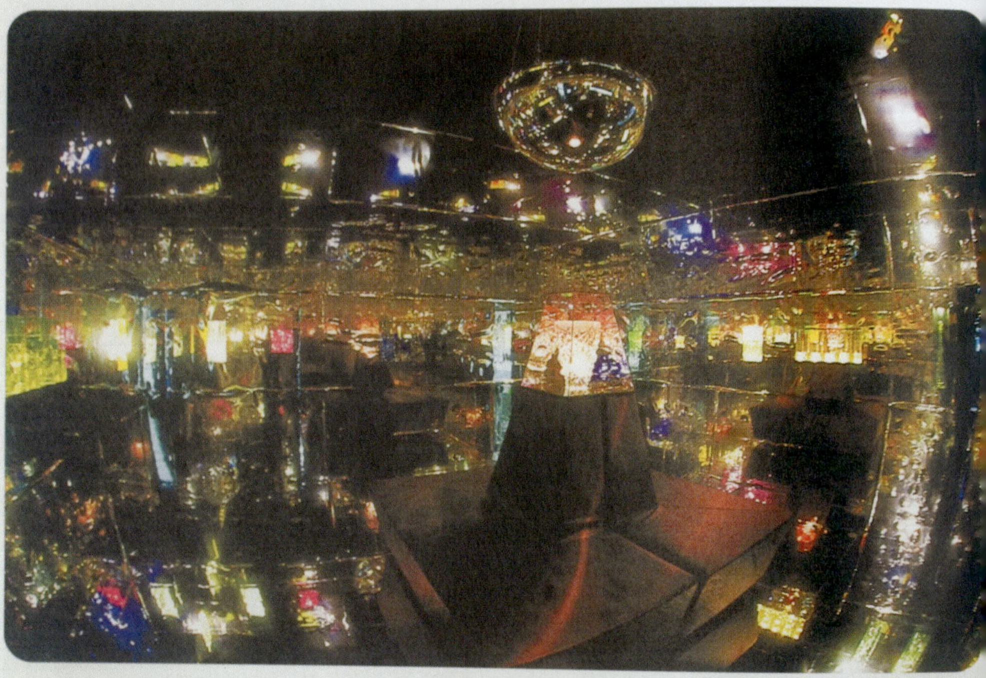

'German design and architecture often lack the ability to reflect on society. No longer can you melt down creative expression to a fixed point and expect it to match reality'

Extra-Heroes, a project for the Frankfurt Fair. In 16 virtual worlds, the effects of lighting lend vitality to glass, which 3 de Luxe calls 'the oldest and yet most futuristic material'

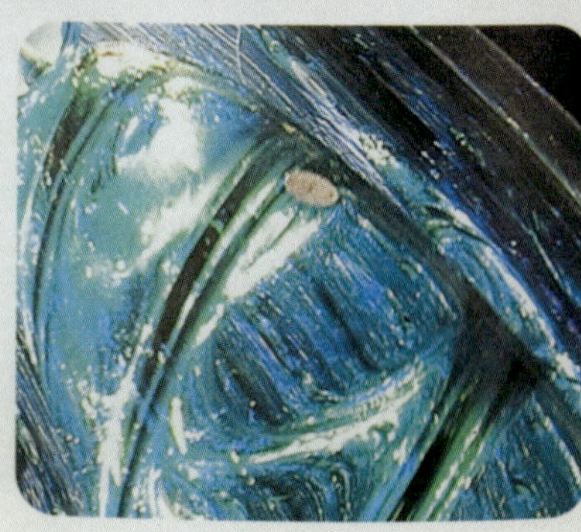

, which provides an opportunity to combine and share
essential to their modus operandi. What could be more
than this teeming pool of talent all located at one spot?
hat we've got here,' says Brell, 'is a healthy mix
ecture, interior design, spatial design and event planning
gh the term "event" often fails to tell the whole story.'
n has created installations for both the Frankfurt
e and Tendence, as well as a number of shop interiors.
designed TV studios for Pro 7, three MTV stands featured
mm in Cologne and a handful of powerful 'concept pro-
times they give free rein to ideas that they develop on

their own, without the input of a client. Schweiger speaks for the agency as a whole when he says: 'A dream that we've managed to realise is to work on projects that allow us to express spontaneous ideas in a language uniquely our own. We're not interested in being nothing but a member of a service industry. Planning projects without clients is a privilege, but ultimately, of course, we wouldn't mind seeing some of these designs manufactured.' Enthusiasm for the work of 3 de Luxe has resulted in more projects than the agency has the time and manpower to accept. Being forced to turn down jobs, however, has not led to an increase in employees. 'We don't want to get bigger,' says Brell. 'A larger staff ▶

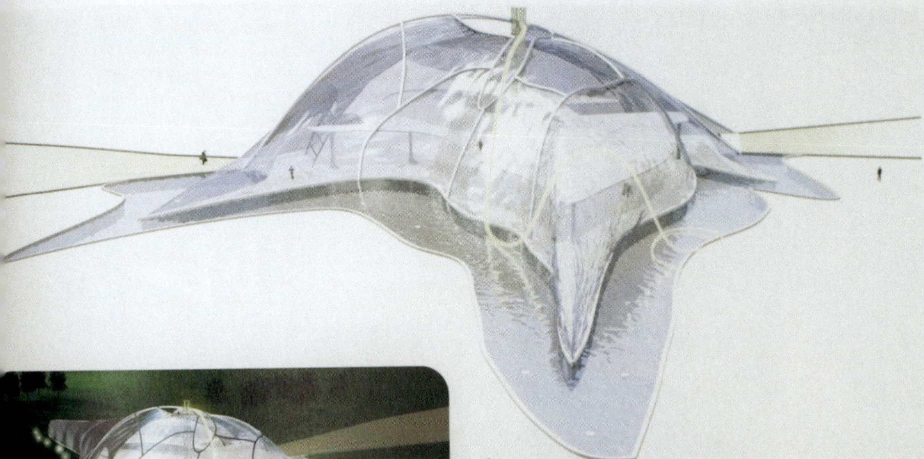

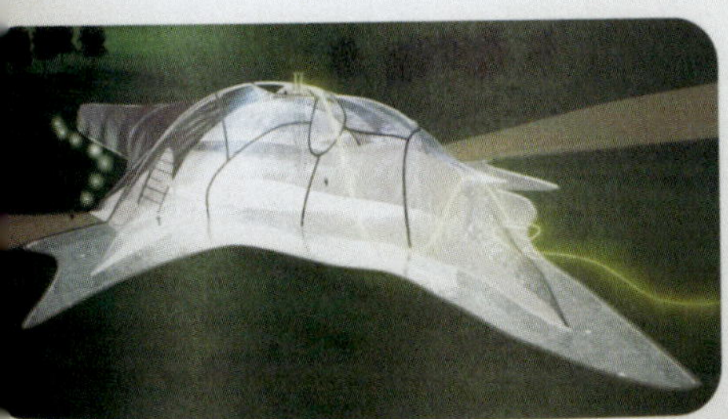

Genetic architecture: the study of a 'living' building. Having originated in the computer, genetic architecture is virtually enhanced and ultimately put through a metabolic cycle that produces an intelligent, self-reliant 'organism' able to communicate. At night, when it's not absorbing energy, the building becomes a fantastic light sculpture

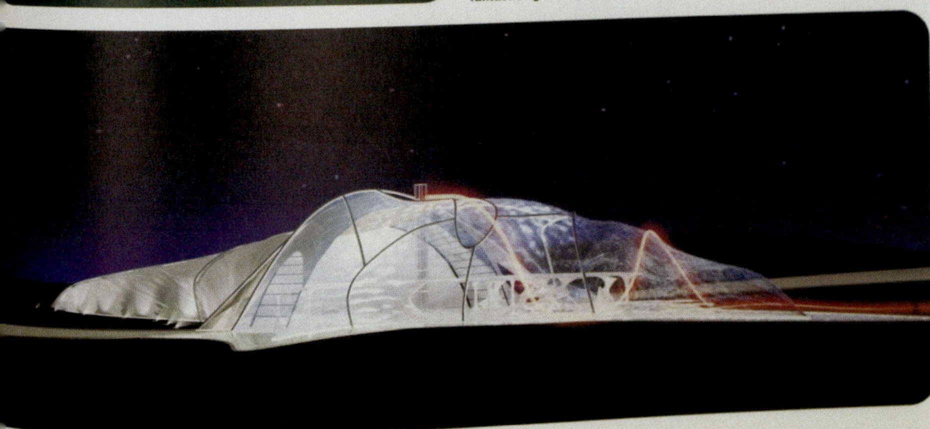

SUBJECT Portrait of 3deluxe
LOCATION Wiesbaden, Germany
DESIGNERS 3deluxe
PHOTOGRAPHER —
WRITER Sven-Anwar Bibi

Issue 12   Virtual Reality de Luxe   116-117

might lead to a change in atmosphere, and we're not willing to risk losing the way we work as a team and seeing the quality of our work diminish.'

The Frankfurt Fair, one of the agency's first clients, has given 3 de Luxe a free hand in designing their installations. Brell and Schweiger's 1996 creation, which revolved around the influence that technology and media have and will have on housing, gave visitors a glimpse into the future. Entering the installation, the viewer was invited to shrug off everyday reality and walk into a new temporal dimension where static objects were infused with mobility and light. Brell speaks of 'three-dimensional music in motion, aromatic space, visual perception reduced to a few glowing elements, and a blend of existing parameters'.

The installation in Frankfurt is one of several experimental designs. A more recent event, Cologne's Popkomm 1999, featured a design for MTV/VH1, which represented 3 de Luxe's third project for this client. Having created a futuristic club ambience in 1997 and a macrocosm of synthesised nature in 1998, the agency narrowed the gap between reality and virtuality in their 1999 design. The result was a multi-layered installation that evoked the type of stratification found in graphic programs. 'New information structures are producing a growing complexity with-

Set design for Max TV, 1998

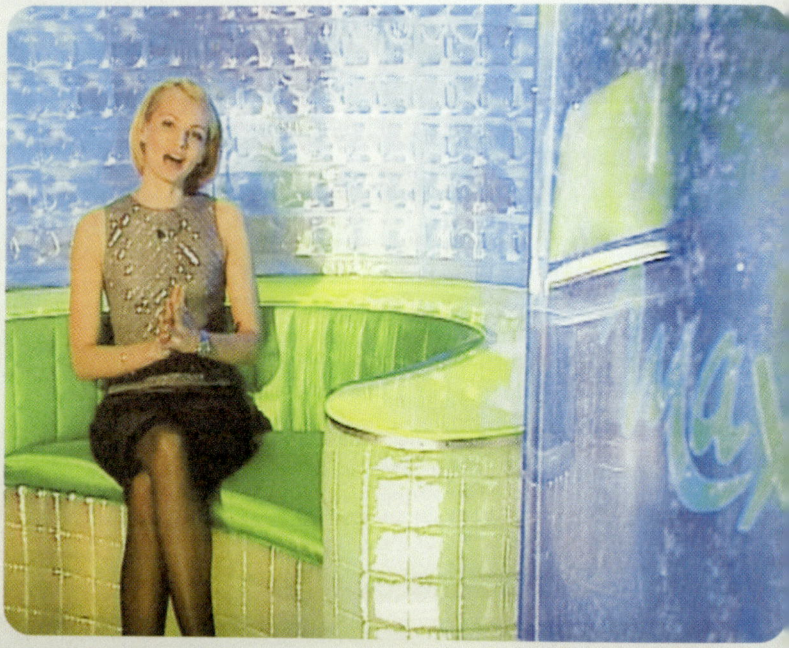

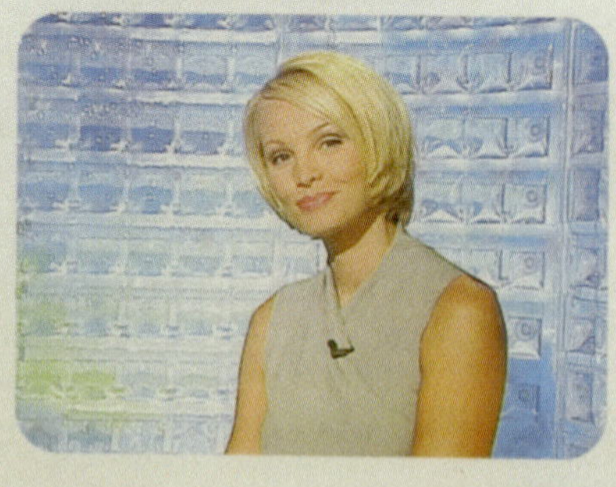

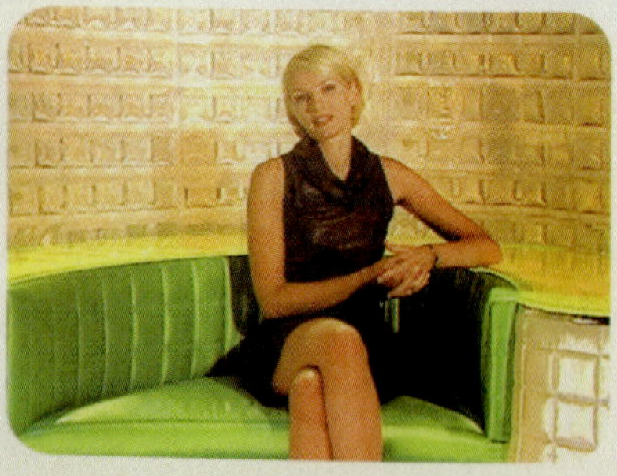

eal and virtual spaces,' says Schweiger. 'It's the new realities we have to concentrate on.' The layers – found in a real room taining glass panels – formed a backdrop for virtual images animated representations of organic forms and symbols. ni-transparent and thus providing a view of activities on both s of its surfaces, the installation was an interactive, three-ensional design that flirted with reality and virtuality simulta-usly. Music and sound selected by the designers underscored visual elements of the project.

sic is sometimes the primary source of the agency's designs, all their work relies heavily on music, which is an essential conduit to the perception of their message. Music indicates time and is a vital tool that they use to generate a creative atmosphere. They're quick to voice dissatisfaction with contemporary music, however, which they feel makes their work unpredictable, while also being incapable of producing euphoria.

Fitting 3 de Luxe into a category is no easy task. 'We don't feel at home in any single design niche,' Brell says. 'Sometimes that's a problem, but it's also an advantage.' The latter part of his statement is demonstrated by the team's integral concepts, which vividly and sensuously display how they combine graphic and interior design with architecture and contemporary ▶

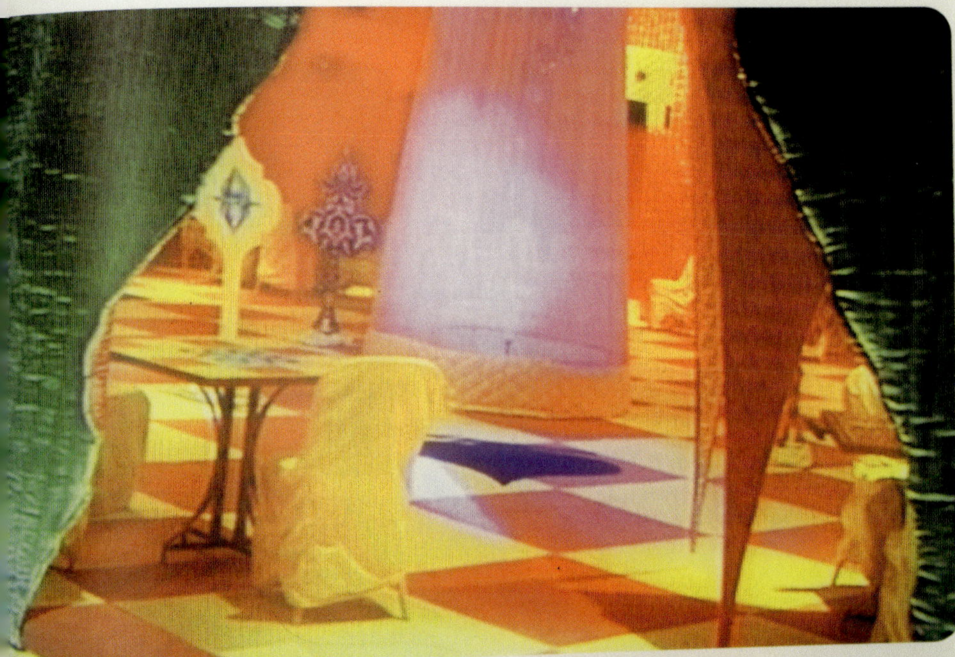

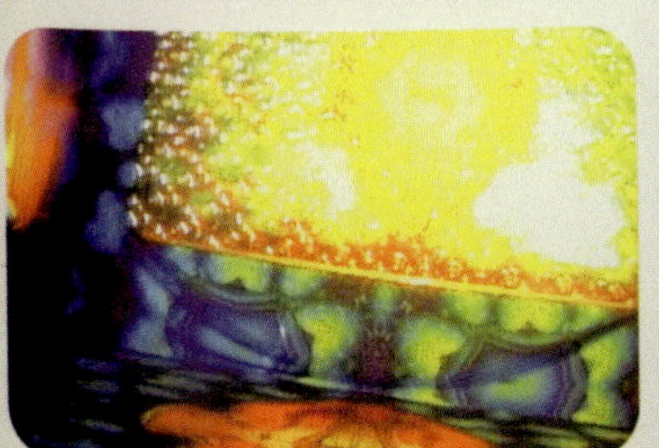

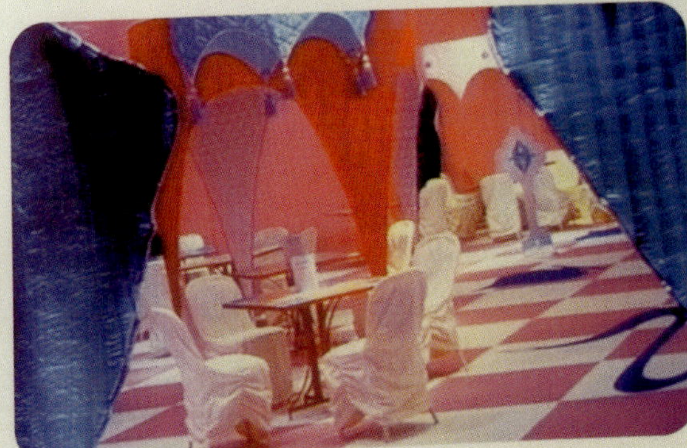

Oriental Café, a place with an 'Oriental-American' atmosphere ironically infused with the sounds of easy-listening melodies. Furniture and canopies revolve slowly around what 3 de Luxe calls 'virtual wells'

SUBJECT
Portrait of 3deluxe

LOCATION
Wiesbaden, Germany

DESIGNERS
3deluxe

PHOTOGRAPHER
—

WRITER
Sven-Anwar Bibi

media to form a solid conglomerate of creativity and reflection on life, technology and the future. It's this type of synergetic work that imbues their projects with such high quality. 'It's not all divided into fields and professions,' Brell adds. 'We're producing work that crosses boundaries, and we like the idea of being crossover designers.'

The agency's projects contain no formal aesthetic message. The breeding ground for their ideas is pop culture, with a special focus on music and fashion. Fashion, which presents an image of what's in at the moment, is often a harbinger of new technologies, trends, music styles and changes in society. 'German design and architecture often lack the ability to reflect on society,' Schweiger points out. 'No longer can you melt down creative expression to a fixed point and expect it to match reality. The reality presented and perceived has to be dynamic – not static.'

Looking for a message in today's architecture and design is often like looking for the proverbial needle in a haystack. And when the two men do find information that reflects on society, it's often part of a project that hasn't been realised. In their own work, Brell and Schweiger make every effort to recognise the existing situation and to stay abreast of the latest technological developments. In anticipation of more extensive collabora-

One phrase that Brell and Schweiger use to define their work is 'an attempt to visualise the interface between virtual and real spaces' an attempt that should allow the visitor to physically experience this newly created dimension

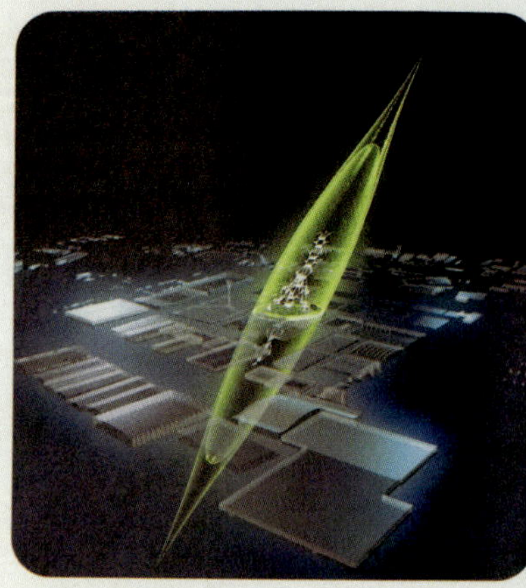

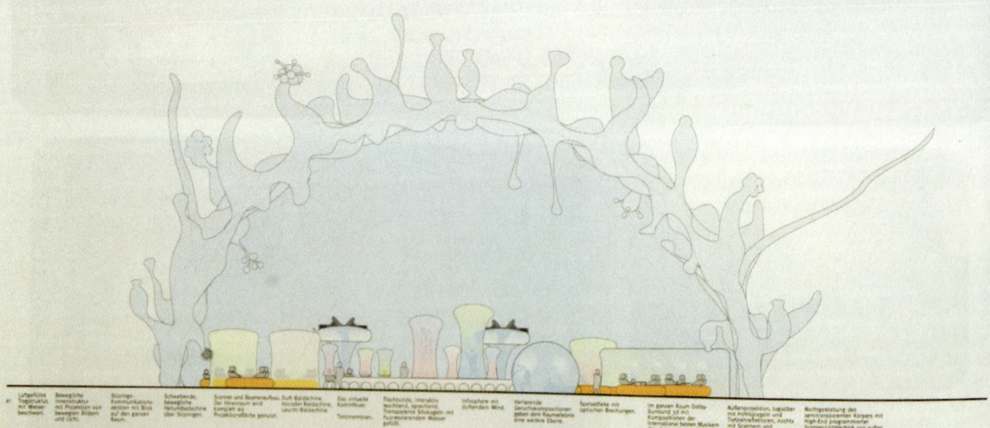

Below: Visionary pavilion, a project for Expo 2000. Computer-controlled pneumatic systems give 'life' to a transparent building that sways like an underwater plant and becomes an illuminated spectacle at night. The building's interactive installations feature biotechnology and bionics, as well as a lounge for relaxation, dining and enjoying music

n among artists, designers, scientists and technologists, 3 de
xe is also engaged in joint research projects with companies
t develop and manufacture new materials and technologies.
e phrase that Brell and Schweiger use to define their work is
attempt to visualise the interface between virtual and real
aces'. Their efforts are aimed at allowing the spectator to physi-
lly experience this newly created dimension. 'We try to be as vir-
ally realistic as possible,' Brell tells me. 'But we always run into
hnical details that keep us from realising the flawless virtual
lity that we have in mind. It's something we just have to live
th.' Their 'genetic architecture' combines the principles of organic entities and systems with technology and the media. It graphically yet concretely presents smart and innovative solutions for residential environments.

Although self-generating materials and 'living' architecture are still a thing of the future, 3 de Luxe's exhibition at Expo 2000 in Hanover allows them to introduce a revolutionary concept: Scape. This visionary experiential project offers visitors a multi-sensory journey through the world of futuristic media. Scape is the outcome of a long process that included three main stages of conceptual visualisation. The designers began with a concept for generative architecture, which evolved into a plan ▶

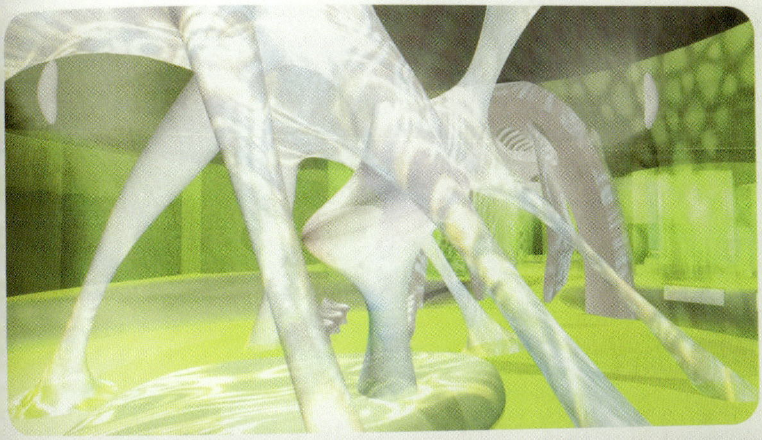

This page and opposite page top: Scape, another project for Expo 2000. Once more 'living' architecture invites visitors – teenagers in particular – to enter the magical world of 3 de Luxe

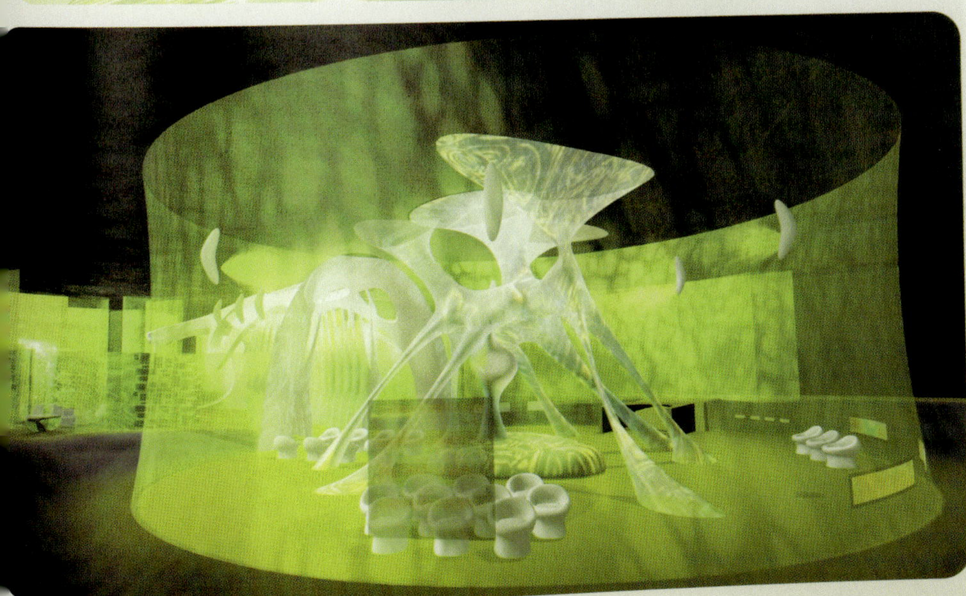

SUBJECT Portrait of 3deluxe
LOCATION Wiesbaden, Germany
DESIGNERS 3deluxe
PHOTOGRAPHER —
WRITER Sven-Anwar Bibi

for a bionic pavilion and, ultimately, progressed to Scape, which combines the two. The installation – a living, intelligent organism that undergoes a metamorphosis – pulls the visitor into a vision that provides an integral view of the diversity that technology has to offer. Scape's message is that designing a bright future requires not only technical and scientific knowledge but also – and above all else – the determination to experiment and to tackle each problem in an unbiased way.

Scape's target group is composed of kids from 12 to 15, a slice of the population often seen only as 'receivers'. When they enter Scape, they become 'transmitters' as well. Surrounded by architecture that uses translucent fabrics to reveal both the inside and outside of different systems, youngsters can opt to visit any number of rooms and interactive facilities. E-Cyas – a virtual pop star with an innate knowledge of new media, virtual reality and cyberspace – is an interactive part of the installation. A fashion studio invites visitors to use new fabrics to create (and wear!) computer-designed clothing.

The world that 3 de Luxe has shaped for visitors to Expo 2000 stands ready to catapult them into the next millennium, and the whole thing is packaged in generative – read 'living' – architecture, which carries the promise of a merger

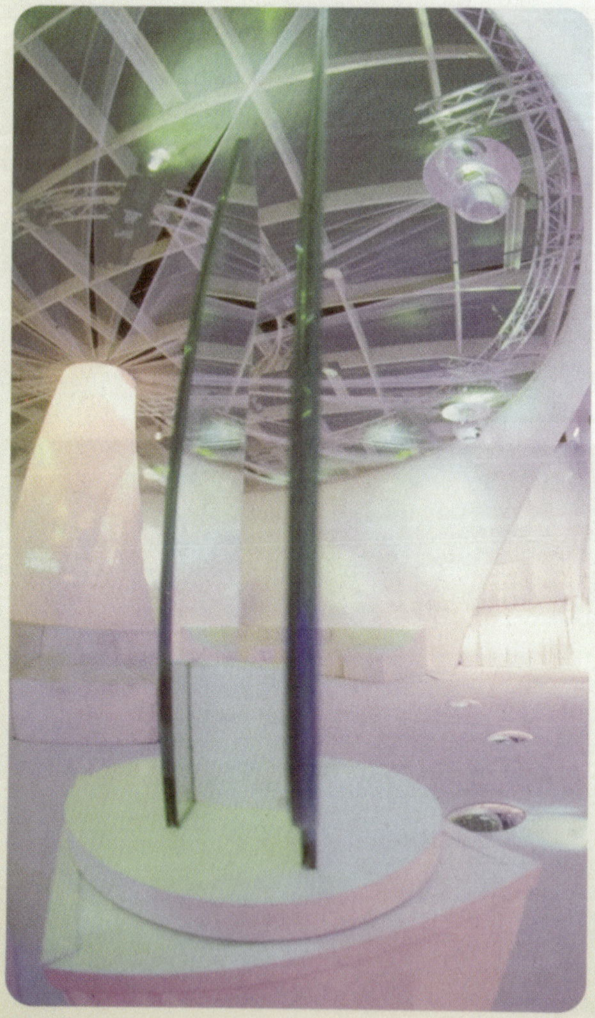

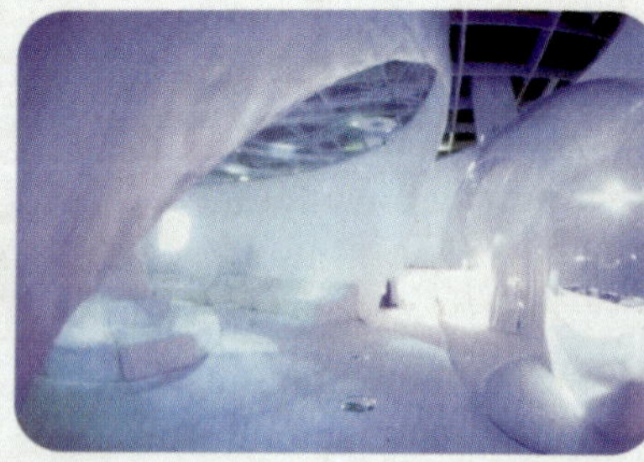

All their work relies heavily on music, which is an essential conduit to the perception of their message. Music indicates time and is a vital tool that they use to generate a creative atmosphere

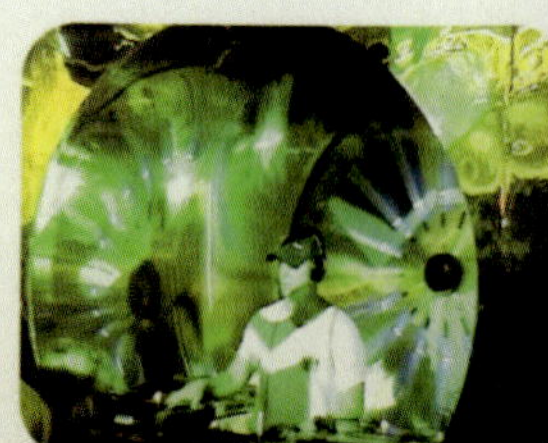

…ween technology and biology. Just as Haeckel's radiolarian …ches of organisms and microphytes inspired the design of …é Biret's pavilion for the World's Fair of 1900, held in Paris, … Luxe based Scape on an analysis of complex computer sys…s and other technological wonders of the late 20th century. …d like Biret, they used the structures, patterns and sometimes …otic forms they found to symbolise their project. A gigantic …me, visible on the skyline of Expo 2000 from some distance, …e symbol that draws visitors to the virtual reality of Scape. …lay's symbols cannot be static. The contemporary context asks … dynamics and change,' Brell explains. Mobile or not, however, Scape's symbol stands as a totem introducing both a new millennium and a generation of designers who recognise social change and are neither bogged down in conservative traditions nor blind to innovation. A sliver of the future that faces 3 de Luxe is already part of their present reality. ◄

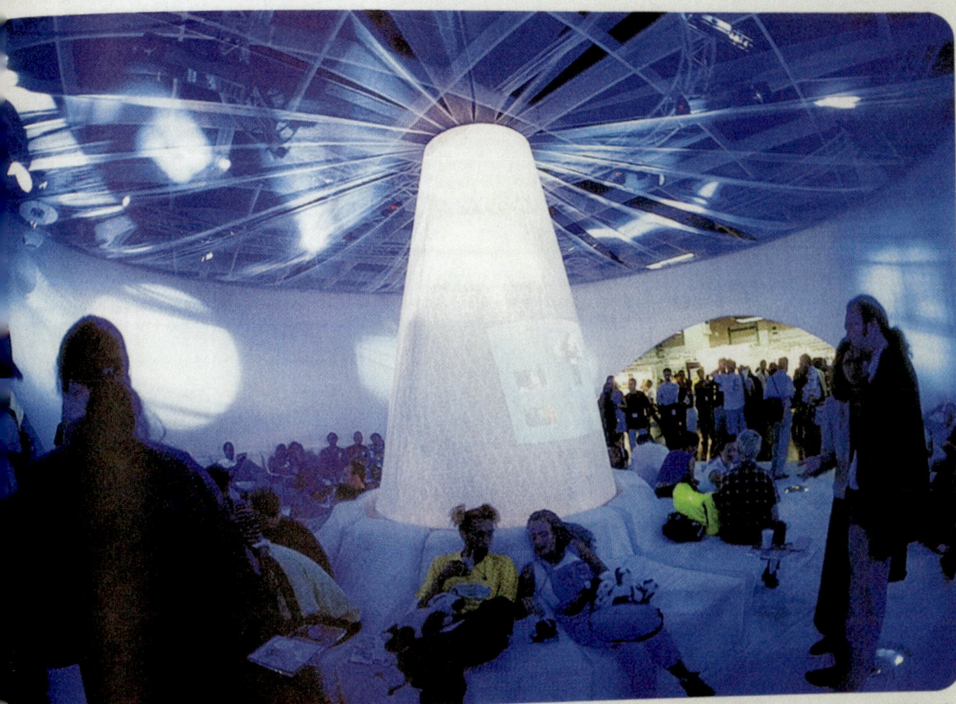

…his and opposite page: displays designed for MTV exhibits at Cologne's Popkomm. In 1997 (large photos) 3 de Luxe created a futuristic club ambience with DJs and …e acts from the drum 'n' bass scene. In 1998 (green images) the agency made an initial attempt to combine nature and technology: real plants vibrated in an artifi…ally produced breeze, while lighting effects created the colour of bright-green chlorophyll. The 1999 design was a blend of pop culture and 'artificial nature'

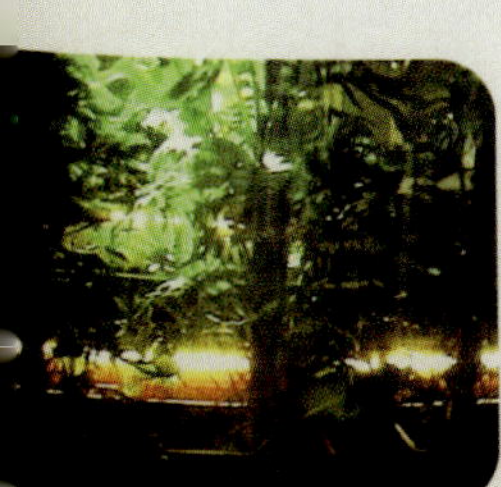

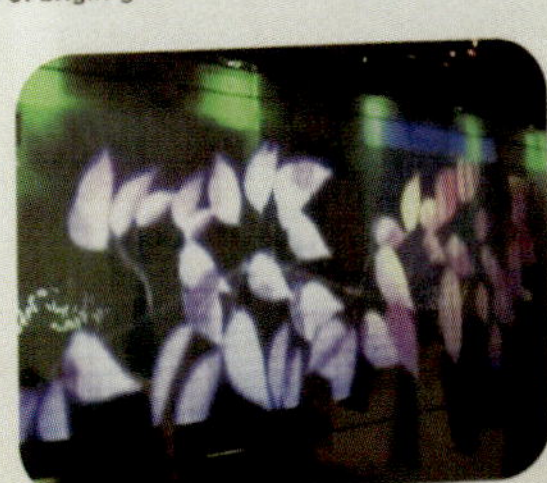

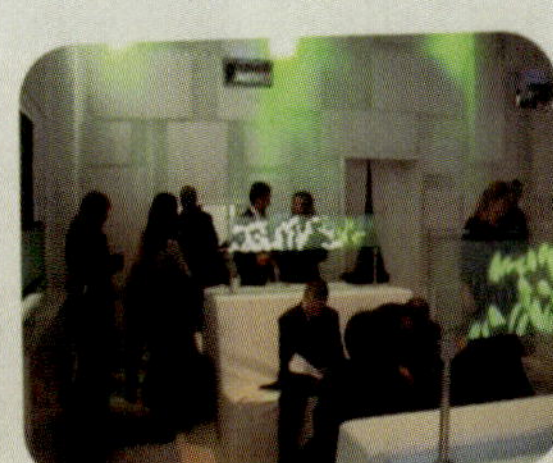

WRITER Sven-Anwar Bibi

PHOTOGRAPHER —

DESIGNERS 3deluxe

LOCATION Wiesbaden, Germany

SUBJECT Portrait of 3deluxe

Issue 13

March/April 2000

122-123

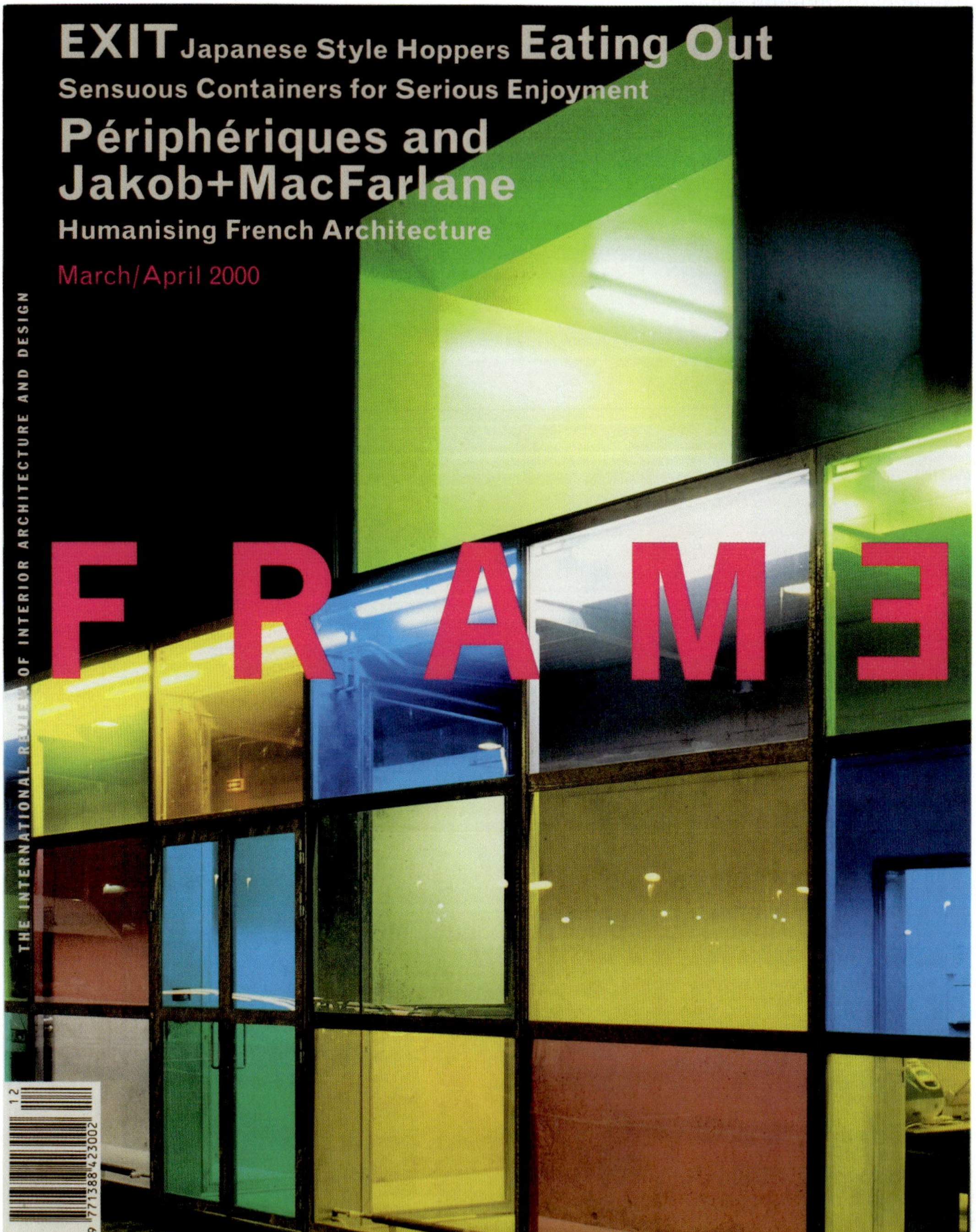

ART DIRECTOR
Ton Homburg,
Opera Designers

13

Absolut Nordic

Architects Kim Herforth Nielsen, Bo Boje Larsen and Kim Christiansen (3XN) about the project

The Royal Danish Embassy is one of our finest projects. It marked the beginning of a process of internationalization for 3XN and still stands as a poetic interpretation of a nation implemented through contemporary architecture. Friendly and inviting, the project acknowledges the constraints of space and the possible clash of formal exercises within the communal Nordic embassy complex.

Comments on the published article

Illustrations are sparse and could have been more informative. One photo that was supposedly shot in the Danish Embassy was actually taken in the communal building. The article tries to unite all five embassies, and with an eye to the many other articles on the complex that were published at that time, we feel the one in Frame is rather confusing. A thorough reading does reveal some interesting facts, however. We got a good laugh out of the author's (justifiable) opinion of planning director Stimmann.

Did the article have an effect?

It's hard to say, because the Nordic embassies were published in a wide range of international magazines that year. We did get some good feedback from the Netherlands, however, which probably can be credited to Frame.

Recent news or information about the project featured in *Frame*

Over the years, the Danish embassy has received a great deal of attention from a wide audience – so much so that the complex has introduced guided tours through the embassy area, which is normally off limits to the public.

DESIGNERS
Gert Wingårdh
www.wingardhs.se
3xN (formerly Nielsen, Nielsen & Nielsen)
www.3xn.dk
Snøhetta
www.snoarc.no
Pálmar Kristmundsson
Viiva Arkkitehtuuri
www.viiva.fi

SUBJECT
Five Nordic embassies

LOCATION
Berlin, Germany

PHOTOGRAPHERS
Jussi Tiainen
www.jussitiainen.com
Åke E:son Lindmann
Ibrahim El-Hayawan
Christian Richters

WRITER
Leo Gullbring
www.calimero.se

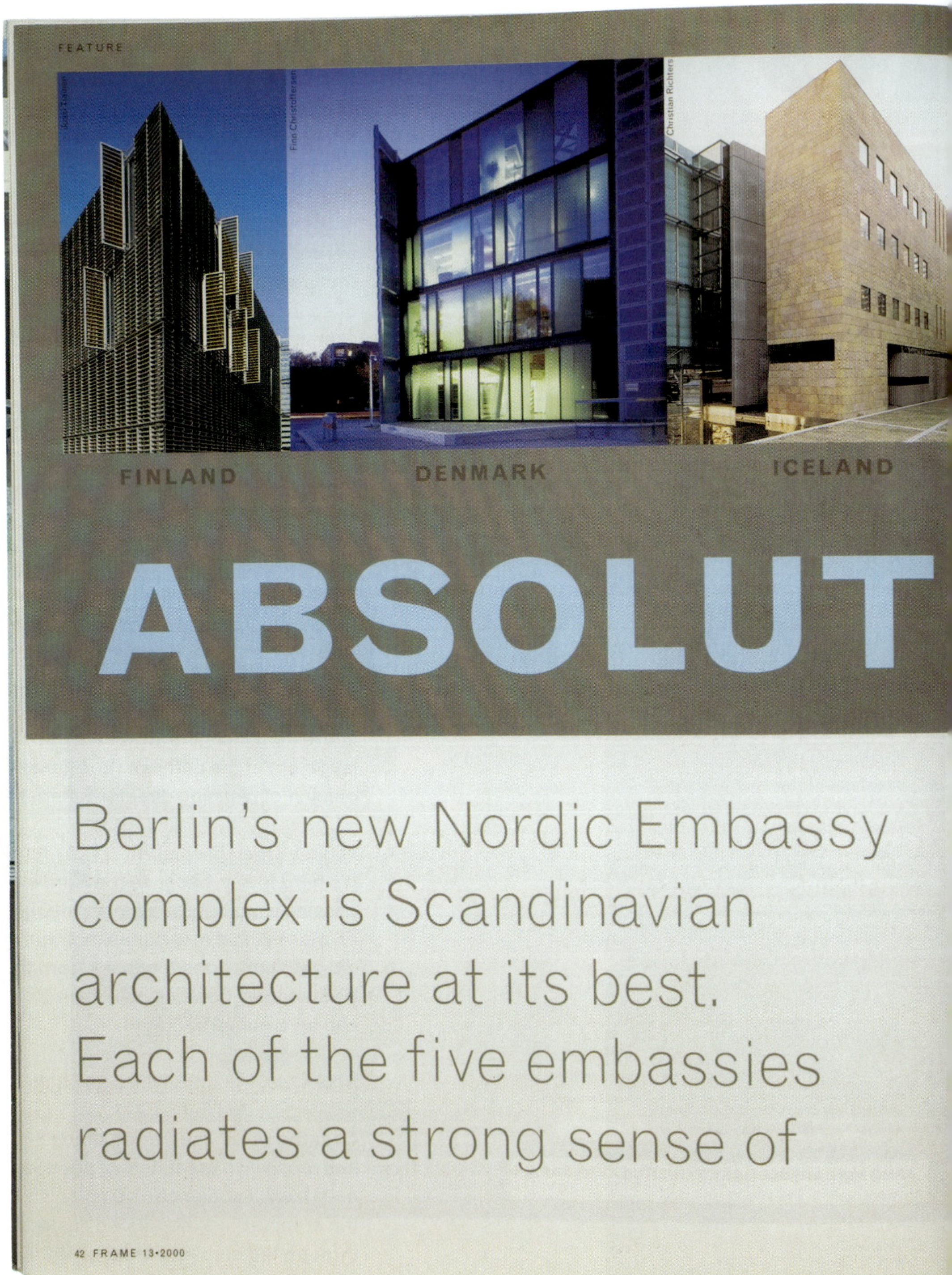
FEATURE

Finn Christoffersen

Christian Richters

FINLAND DENMARK ICELAND

ABSOLUT

Berlin's new Nordic Embassy complex is Scandinavian architecture at its best. Each of the five embassies radiates a strong sense of

42 FRAME 13•2000

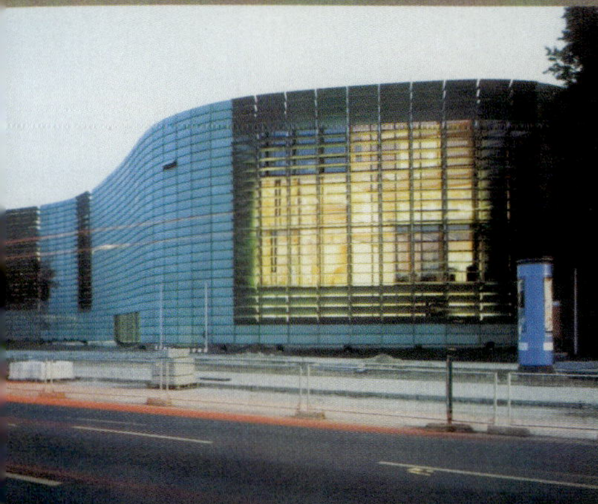

SWEDEN

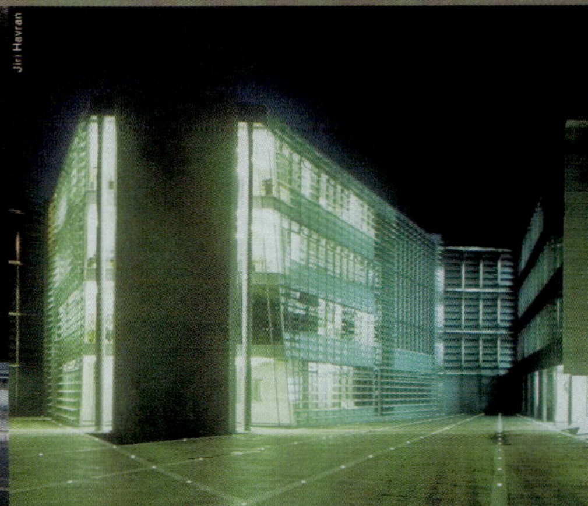

Jiri Havran

NORWAY

NORDIC

identity – a rare ingredient in the recent architectural history of Germany's new capital. *By Leo Gullbring.*

SUBJECT
Five Nordic embassies

LOCATION
Berlin, Germany

DESIGNERS
Various, see page 123

PHOTOGRAPHERS
Various, see page 123

WRITER
Leo Gullbring

What is it? A green frog? A turtle? Situated on Klingelhöferstrasse, a busy street in the southern part of Berlin's Tiergarten, the green copper-banded wall that billows along the Nordic Embassy complex has triggered various reactions. Although Berliners refer to it somewhat disparagingly as *Die Schildkröte* ('The Turtle'), the overall verdict is positive, especially after they've seen it at night with its large glass facades illuminated. The five embassies, each of which expresses a clear-cut identity, are attracting quite a crowd. To some extent their popularity is both a reaction to the 'critical reconstruction' of Berlin and a longing for the days of the International Building Exhibition (IBA), an urban construction program.

Well-defined identities – a rare ingredient in the reconstruction of Berlin as the new German capital – explain the success of the five embassies. This type of distinctive image is also expressed by Libeskind's Jewish Museum, another project that questioned the block-based character of the city and went on to expand it into something promisingly different. Can anyone explain why this city, with its visibly fractured history, pursues a cowardly, low-risk policy of 'critical reconstruction' that includes commissioning the most expensive and internationally well-known architects to do second-rate jobs?

Behind this reconstruction scheme stand architect Josef Kleihues – who deserves credit for his leadership of the IBA, although not for his work as an architect – and Berlin's former planning director Hans Stimmann. Libeskind's verdict is bitter: 'You can't restart history by pressing a button. It's a kitsch idea of history. Dullness is the bureaucratic result of what planners wanted to communicate: power and order, a new "old Berlin".' Instead of using the golden opportunity to create a unified Berlin bereft of its schizoid past and to attempt a new and healing architecture, the city has opted for blandness paired with an almost American commercialism. Taken to the extreme, the programme produces an inverted architecture, exemplified by Frank O. Gehry's DG Bank on Pariserplatz. This city square – with Hotel Adlon on one side and two banks designed by Kleihues on the other – was to be restored to its former pre-war splendour. Strict uniformity, limited heights and a restricted use of materials led Gehry to draw an austere but by no means bland exterior. The interior is the real work of art, however: his famous (though unrealised) 'Lewis House' is reincarnated here as a fish on a wave heading straight for the net.

Fortunately, the Nordic embassies are out of reach of planning director Stimmann. Here the spirit of the IBA lives on. During the public opening in October 1999, some 15,000 visitors viewed the complex, which is bound to spawn subsequent masterpieces in the central-Berlin area. It's something of a paradox, however, that the open plan designed by Austrian-Finnish architects Berger+Parkkinen is largely off limits to eager admirers. The joint entrance to the *Nordisches Botschaft* is located on a small side street where the exterior wall's patinated copper band ends. Inside Felleshuset – a shared building with exhibitions, auditorium and restaurant for embassy personnel – steel and glass dominate without creating an overcool environment. The incorporation of a large communal public area further accentuates the novelty of locating five embassies on one site. A hastily organised photography exhibition is on Felleshuset's upper level. The exhibition combines the idea of a reborn Nordic dream with the reality of five distinctly separate countries and, at the same time, offsets the hush-hush milieu normally associated with embassies. The semi-transparent staircase joining the two public storeys is too imposing. A different approach to the organisation of space could correct this problem, however. One such approach has been proposed: the vacant second-floor terrace can be transformed into a small cafeteria during the summer. The building's cosy auditorium, with seating for 150 visitors, has already been put to rewarding use.

Another problem is access. Security measures must be strict, of course, but bulletproof glass and a very low desk make it hard to communicate with the receptionist. Once through the glass doors, however, the visitor begins to fathom Berger+Parkkinen's plan. The undulating shape of the high exterior wall is partly explained by trees left on the site, which can also be interpreted as a concession to the importance of nature in the Nordic countries. All six buildings are equal in height and have the same flat roofs. And all share the surrounding copper-banded wall and the interior courtyard, which also separates the embassies. Intermediate space is open and clear – a foretaste of the clarity and purity of form that marks the individual embassies. Although the exterior as a whole radiates a rather cold image, a closer look at the facades produces a more intimate impression.

The Swedish Embassy faces its Norwegian partner with a black facade of overlapping diabase slate that fronts a glass wall. Here hand-hewn edges underline the metaphor for the frontier mountains between these two countries, while the wall that faces the Finnish Embassy looks out on a surface of water. Made of chalky white Norrvange limestone, this wall has an opal-tinted glass facade, which features a transparent 'viewing strip', suspended in front of it. 'Like Yin and Yang', explains the architect.

Being a Swede, I begin my tour with a visit to the **Swedish Embassy**. Inside I'm enveloped by both straight and voluptuously curved volumes clad in a warm-coloured birch rather than in the shades of white often associated with Swedish interiors. The staircase – with its curve and counter-curve – spirals upwards from an indoor pond. Architect Gert Wingårdh, who has opted for an almost baroque statement, cites Borromini as an inspiration. In a joking manner he explains that he started by putting the coffee room in the middle and building the rest around it: 'This is where we solve all the problems, you know.'

Horizontally adjustable antidazzle screens let daylight flood the interior for 'an attitude of transparency', according to the architect. The copper band on the exterior wall is repeated here on a smaller scale by means of horizontal stripes inset in wood, even inside the smoothly operating lift. A meeting room is set off by a low, white, Richard Serra-like arch, which ends outside, where it is punctuated by a column. Although I like the 2.5-tonne concrete door on its single hinge, the shape is one too many in my opinion. (Then, too, Serra would certainly have preferred steel.) The chairs by Mats Theselius are not to my taste either, although their spindly legs made of copper tubing do offset all the heavy volumes. The table – Wingårdh's own design – is a far better choice. That the architect was forced to let another firm do the working cubicles is a pity; their uninteresting interiors, which show no regard for furniture or detail, are disappointing when compared with the rest. The overall impression given by this unique architectural work is rewarding though, especially when viewed with ▶

Excessive ornament is *verboten* in the Finnish Embassy, and the resulting image is one of an almost vodka-clear architectural lucidity

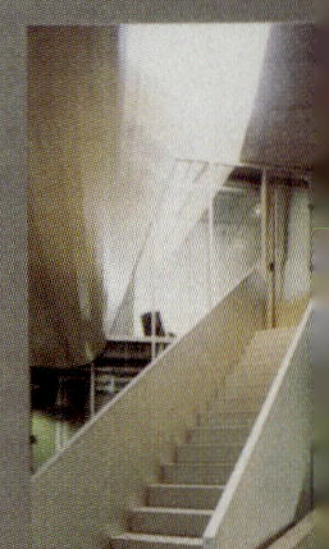

A blend of wood, glass, and concrete produces strongly defined charac

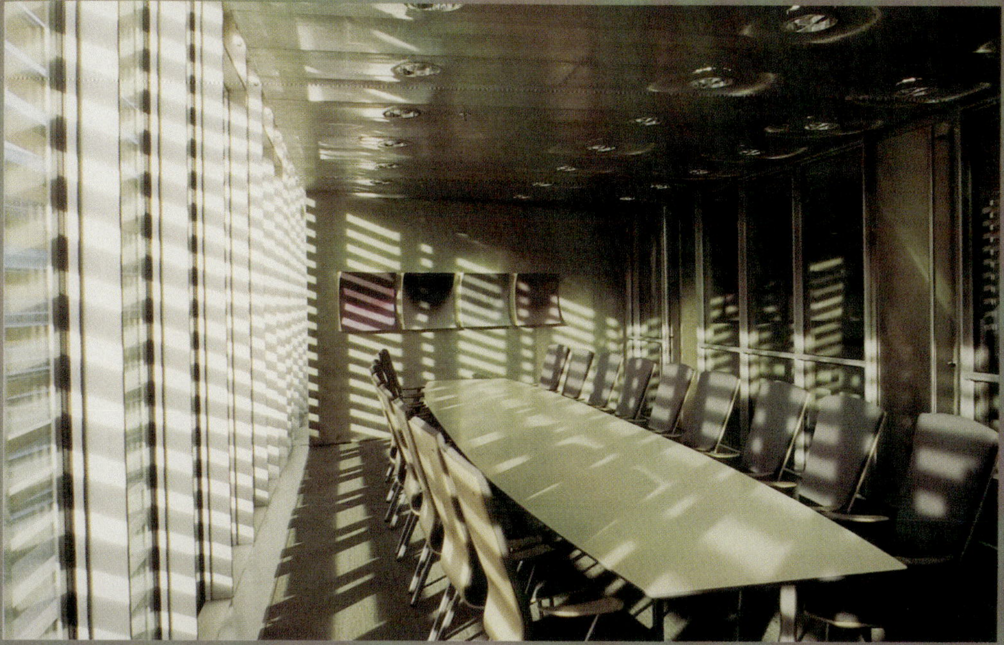

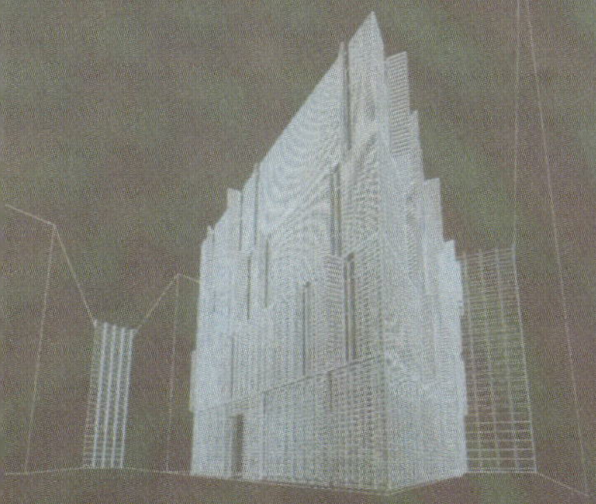

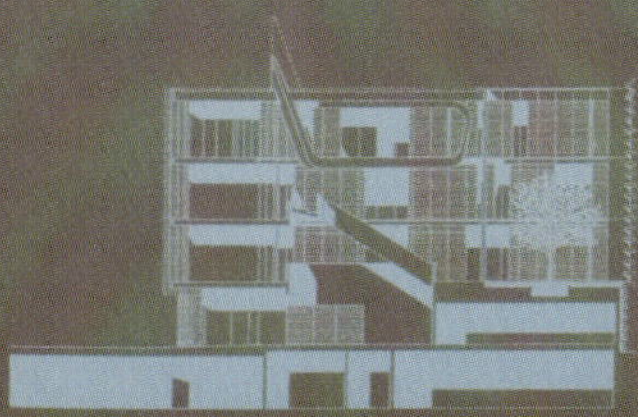

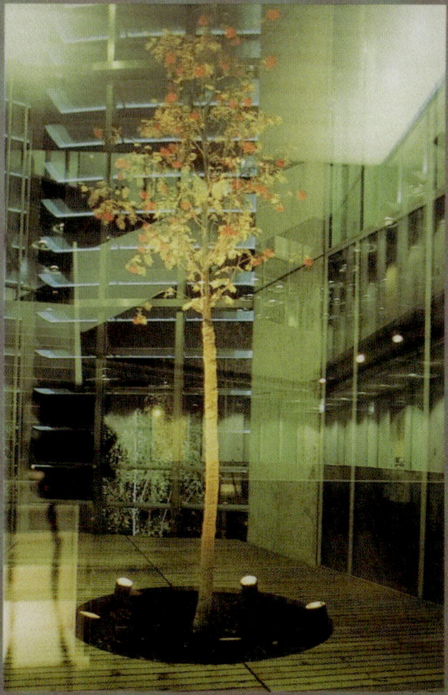

Above: Meeting room in the Finnish Embassy. The glass facade sports an adjustable screen of horizontal larch laminae

Left: The tree in one of the courtyards refers to Finland's national epic, Kalevala

Photography: Jussi Tiainen

SUBJECT
Five Nordic embassies

LOCATION
Berlin, Germany

DESIGNERS
Various, see page 123

PHOTOGRAPHERS
Various, see page 123

WRITER
Leo Gullbring

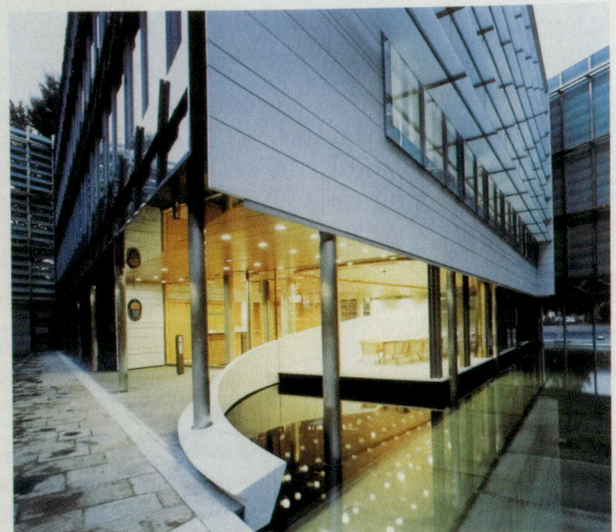

Inside the Swedish Embassy, you're enveloped by both straight and voluptuously curved volumes clad in a warm-coloured birch rather than in the shades of white often associated with Swedish interiors

This page: A meeting room on the ground floor of the Swedish Embassy is set off by a white arch which ends outside. The chairs are designed by Mats Theselius, the table by architect Gert Wingårdh

Photography: Åke E:son Lindmann

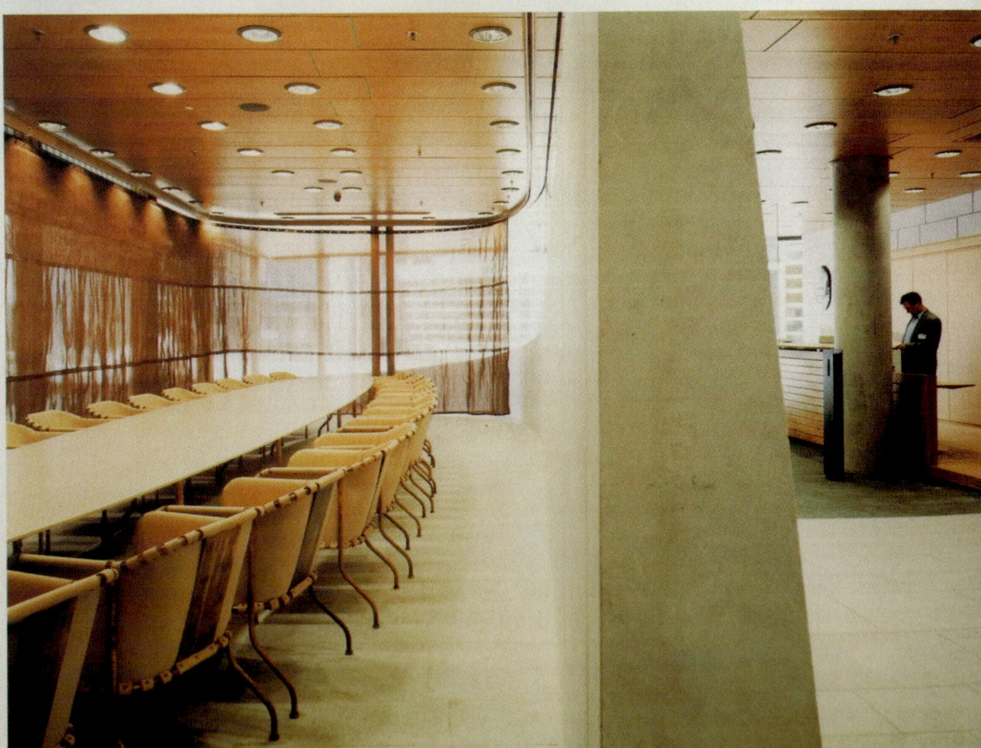

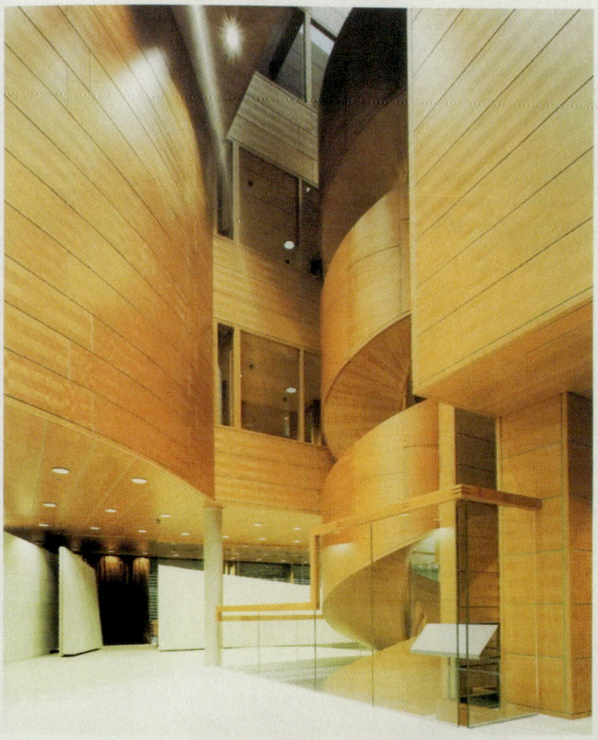

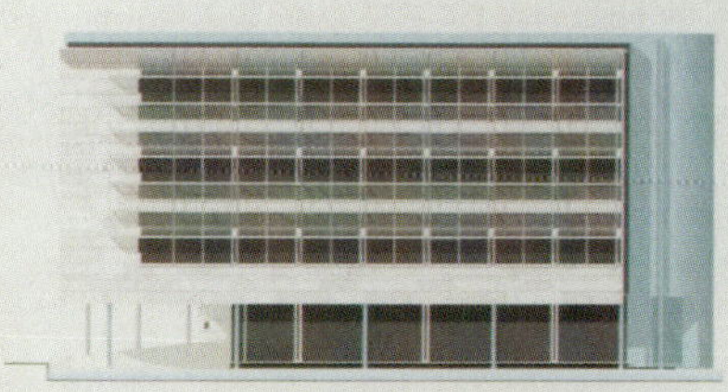

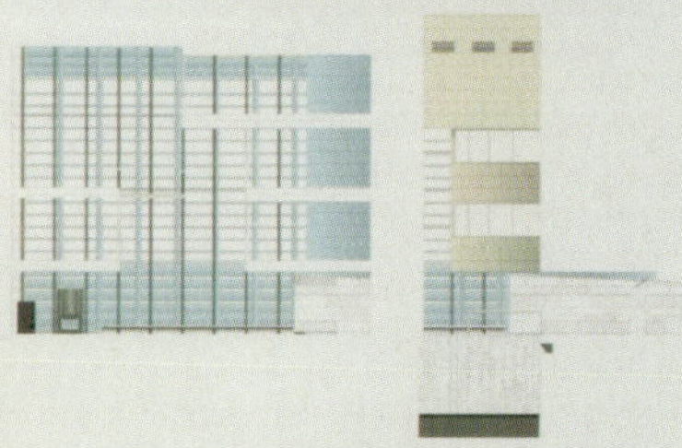

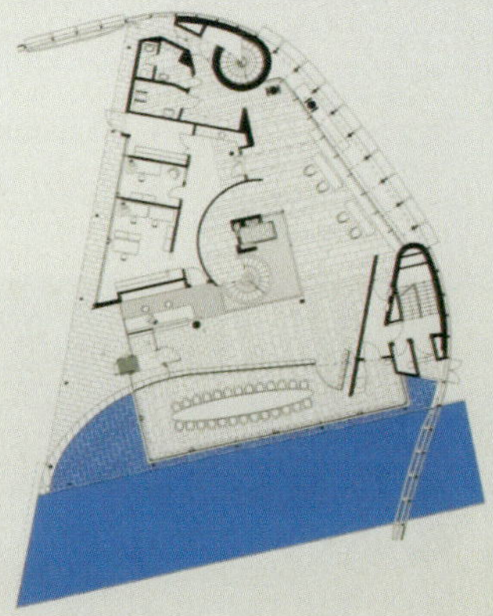

This page: The central staircase and its surroundings, clad in warm-coloured birch

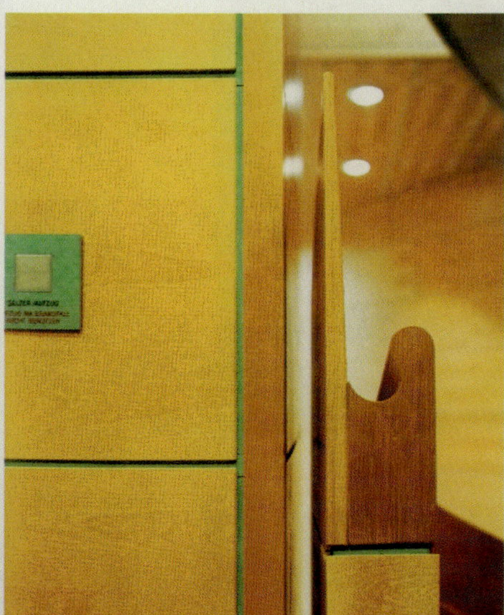

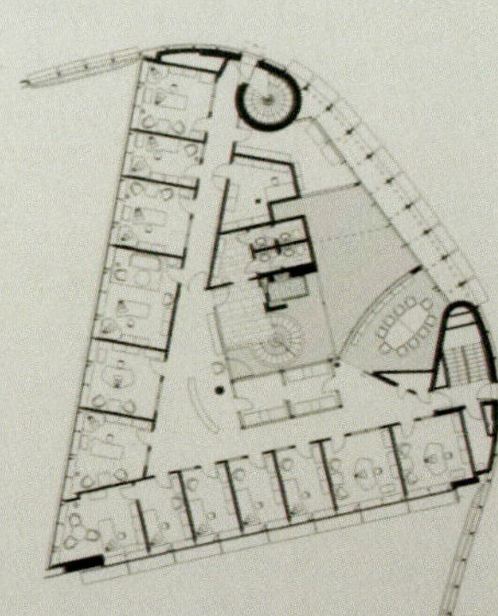

SUBJECT
Five Nordic embassies

LOCATION
Berlin, Germany

DESIGNERS
Various, see page 123

PHOTOGRAPHERS
Various, see page 123

WRITER
Leo Gullbring

the knowledge that officials in Stockholm are usually just as conservative as their colleagues in Berlin.

Nielsen, Nielsen & Nielsen took a different approach to the **Danish Embassy**. The panoptic lobby area with its glazed canopy welcomes visitors into a high-ceilinged, brightly illuminated space flanked by two opposing volumes. A screen of ash louvres runs along a slightly tilted organic volume, which takes its shape from the exterior wall and its wavy copper band. Set against the opposing wall with its straight, perforated-steel sheets, the volume creates a sense of positive tension, which increases as my gaze moves up between the narrowing building blocks. As I ascend the staircase, which lends access to footbridges at different levels, I experience an expanse of space and altitude, catch a glimpse of corridors behind the screen, and discover yet another space with ruby-red walls. Of the five, the Danish Embassy has the warmest ambience and is the most casual.

As one might expect from the Danes, materials are beautiful and details well executed. The embassy's floors are made of deep-brown, shimmering Portuguese stone. Arne Jacobsen's mat-polished stainless steel lamps, originally designed for the City Hall in Aarhus – the architects' hometown – fit perfectly into the interior, as do his chairs and Poul Kjærholm's tables. My one objection is to the heavy concrete core at the south end, which lacks the right touch of marbling. It leaves a nice space facing the outer wall, however, where the art and design objects exhibited can also be seen by passers-by.

The south wall of the new **Norwegian Embassy** is composed of a single, monumental slab of grey granite. Architect Craig Dykers, who works for Snøhetta in Oslo, explains that it was easy to extract the slab – 120 tonnes, over 14 metres high, 5 metres wide and up to 70 centimetres thick – from the surface of the quarry. Furthermore, this 900-million-year-old monolith has a dual personality. The natural condition of the untreated slab is exposed on the exterior side, where a series of glacial scars passes diagonally across its entire surface. On the opposite side, a diamond-surfaced cutting band has left a completely smooth surface with a convex curve. Unfortunately, the Germans have added two unnecessary pillars on either side of the stone block to prevent its collapse, and these seriously detract from the monumental sense of the structure.

While the monolith is an undeniable tour de force, the interior is less convincing. The lift shaft opposite the stone surface in the entrance area ties up the layout and leaves the corridors dark and uninspiring, even though the dimensions of this space match those of the Swedish Embassy. The ambassador's office, on the other hand, is filled with so much light that he needs a shade to block out the sun – a problem magnified by the room's slanted walls. Art is provided almost exclusively by Statoil, a Norwegian oil company. Plentiful outdoor space provides summertime compensation for the apparent lack of space inside the building.

The **Icelandic Embassy** may be the smallest of the five buildings, but here architect Pálmar Kristmundsson has produced a little ascetic gem surrounded by pieces of lava. Eye-catching whims – corrugated concrete details and a spiral staircase that climbs the entire height of the building – are part of an abstract language. No doubt Björk would like it.

The same ascetic elegance is found in the **Finnish Embassy**, which sports a glass facade behind an adjustable ▶

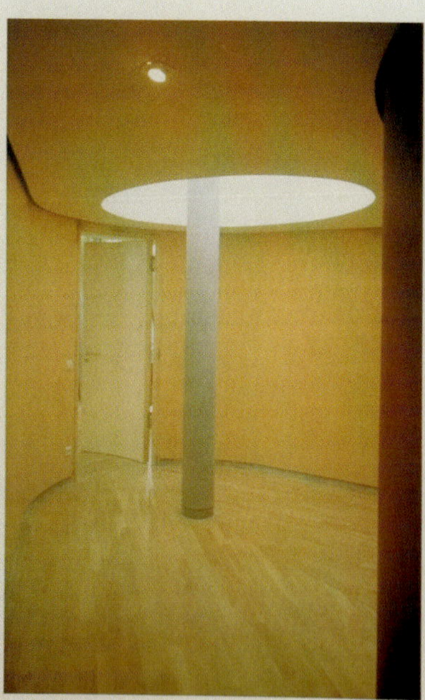

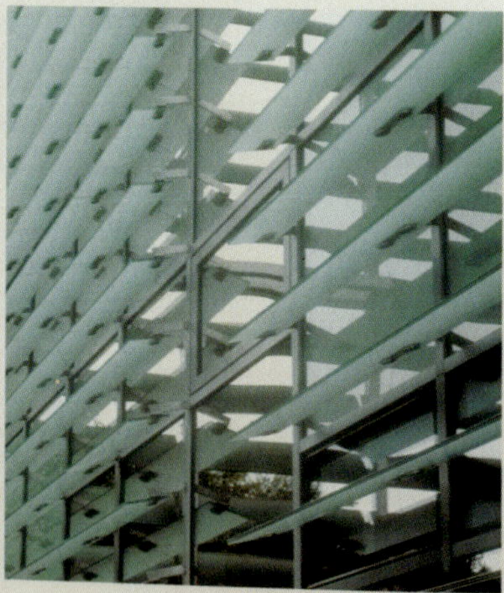

While the Norwegian Embassy's south wall is an undeniable tour de force, the interior is less convincing

Far left: One of the uninspiring corridors in the Norwegian Embassy. Photo: Jiri Havran
Left: Glass facade with adjustable screen
Photo: Ibrahim El-Hayawan

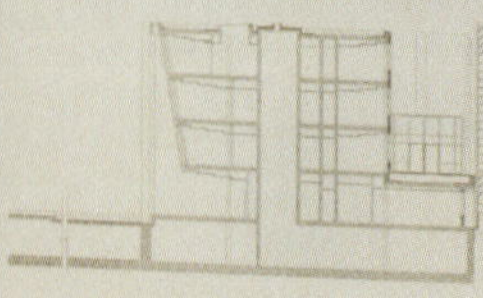

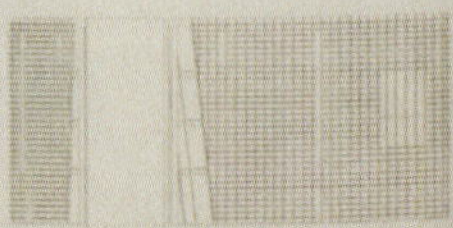

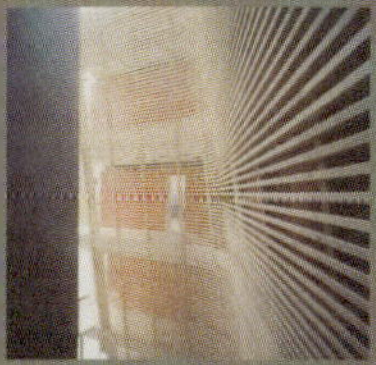

Above and right: The panoptic lobby area with its glazed canopy. A screen of ash louvres runs along a slightly tilted organic volume. The opposing wall is clad in perforated-steel sheets
Bottom: Exhibition space

Photography: Finn Christoffersen

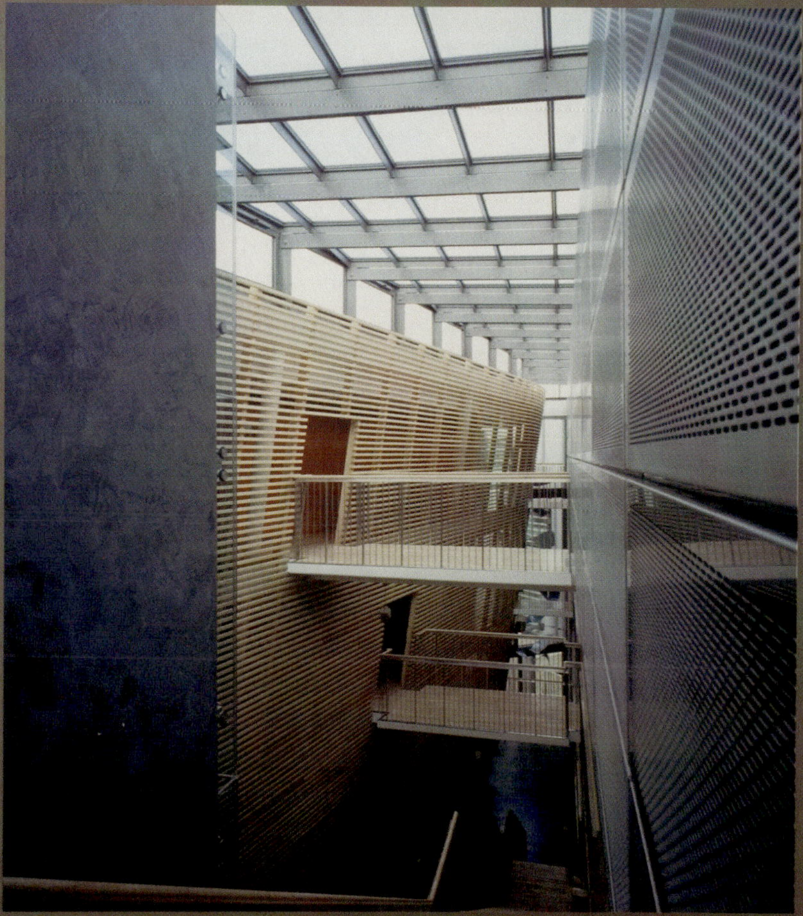

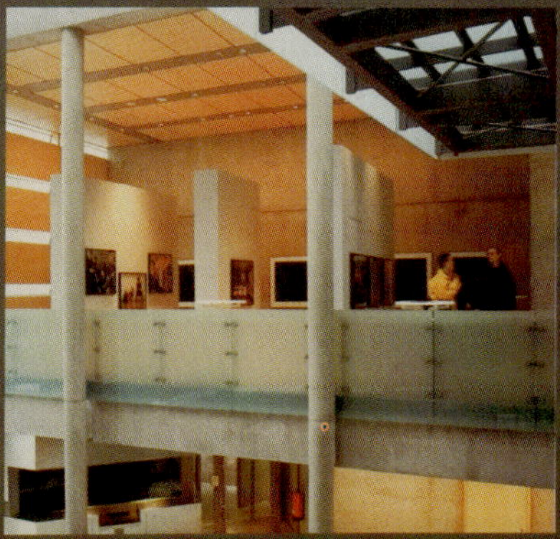

Of the five, the Danish Embassy has the warmest ambience and is the most casual

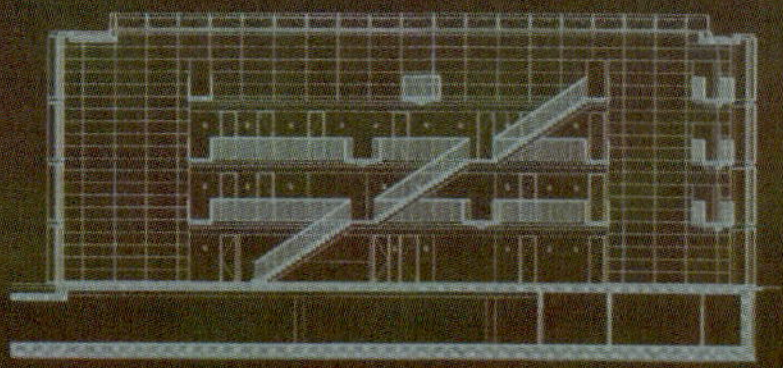

SUBJECT
Five Nordic embassies

LOCATION
Berlin, Germany

DESIGNERS
Various, see page 123

PHOTOGRAPHERS
Various, see page 123

WRITER
Leo Gullbring

screen of horizontal larch laminae. Having organised an open rather than an invitational design competition for its embassy, Finland awarded first prize to an unknown firm, Viiva Arkkitehtuuri, composed of young architects and students. Rauno Lehtinen, Pekka Mäke and Toni Peltola haven't used light as extensively as the other architects have, but the somewhat dim entrance-cum-lobby is illuminated by an exquisite work of colour plates by Silja Rantanen. A solemn atmosphere created by a blend of wood, glass, steel and polished concrete produces a space with a strongly defined character. Excessive ornament is *verboten*, and the resulting image is one of an almost vodka-clear architectural lucidity. Two saunas, which are more than token nods to Finnish culture, soften the sober impression created by the square layout. Another nice touch focuses on phrases from *Kalevala*, Finland's national epic, which are scattered here and there at the next level. Inspired by the same book, the embassy has planted a rowan tree in one of the small courtyards.

As far back as the early 1900s, embassies belonging to Finland and Sweden were located in this traditional ambassadorial area south of Tiergarten. As a result of the Allied bombardments of 1943, the site was abandoned. Only after the reunification did the area regain its former function, which is now augmented by the spirit of a united Germany and by the sense of togetherness inherent in a stronger EU. This collective feeling has not prevented the Finns from paying respect to their uniquely stylised modernism, however, nor has it made the Danes less loyal to their own contemporary architectural ideals. Both Sweden and Norway, on the other hand, have felt free to take a far more individualistic approach in designing their embassies. The complex is a showcase of Nordic architecture at its best, and each embassy expresses its national identity in a powerful way. Would it be overoptimistic to hope that these embassies and those under construction will prove that contemporary rather than anachronistic, *ersatz-nostalgia* architecture is the solution to a more beautiful Berlin? ◀

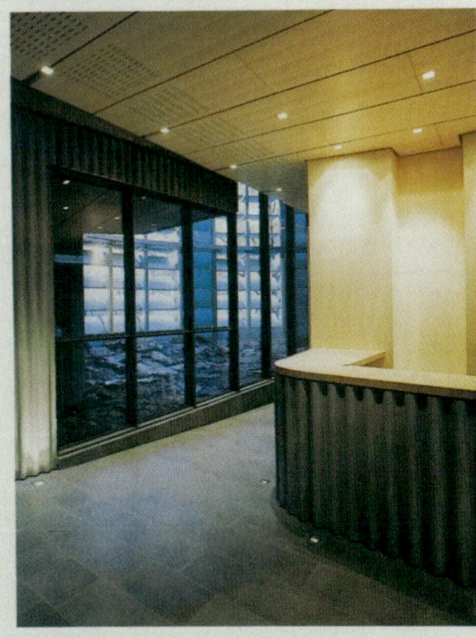

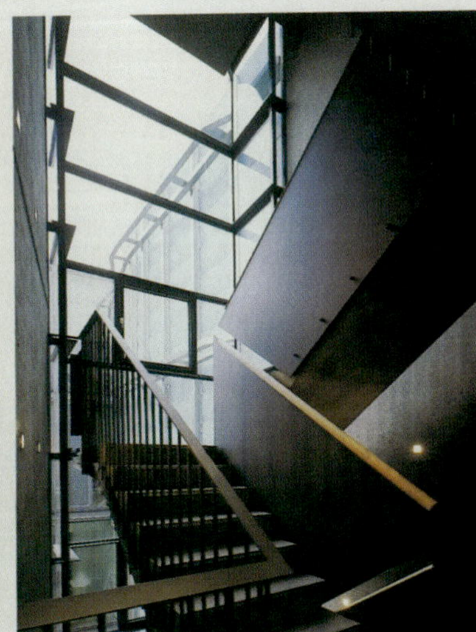

This page: The Icelandic Embassy features eye-catching whims, such as corrugated concrete details and a spiral staircase that climbs the entire height of the building. Opposite page: The embassy seems to be surrounded by lava

Photography: Christian Richters

The Icelandic Embassy is a small ascetic gem surrounded by pieces of lava. No doubt Björk would like it

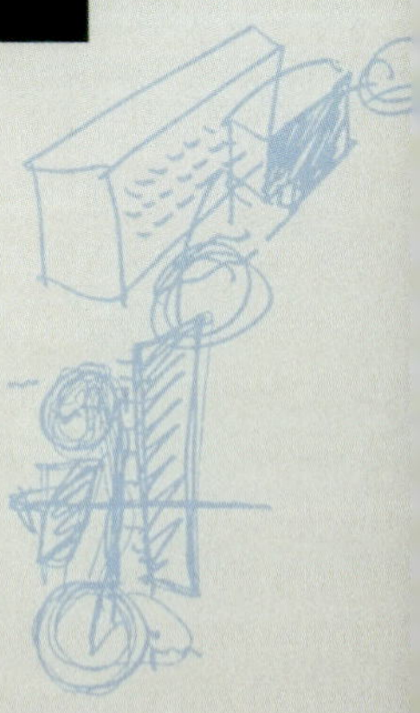

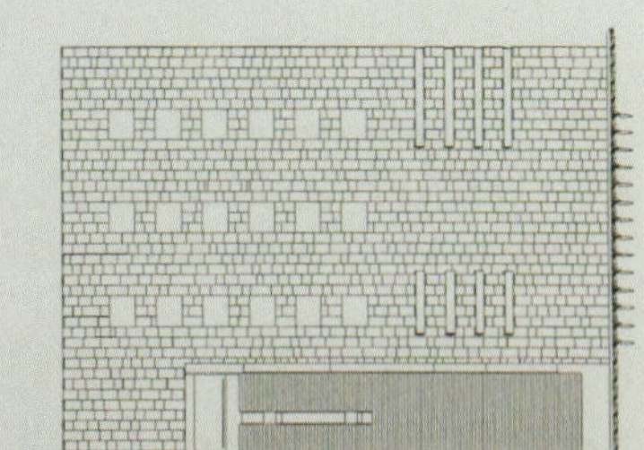

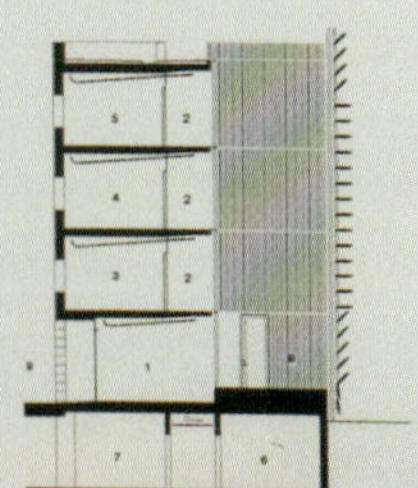

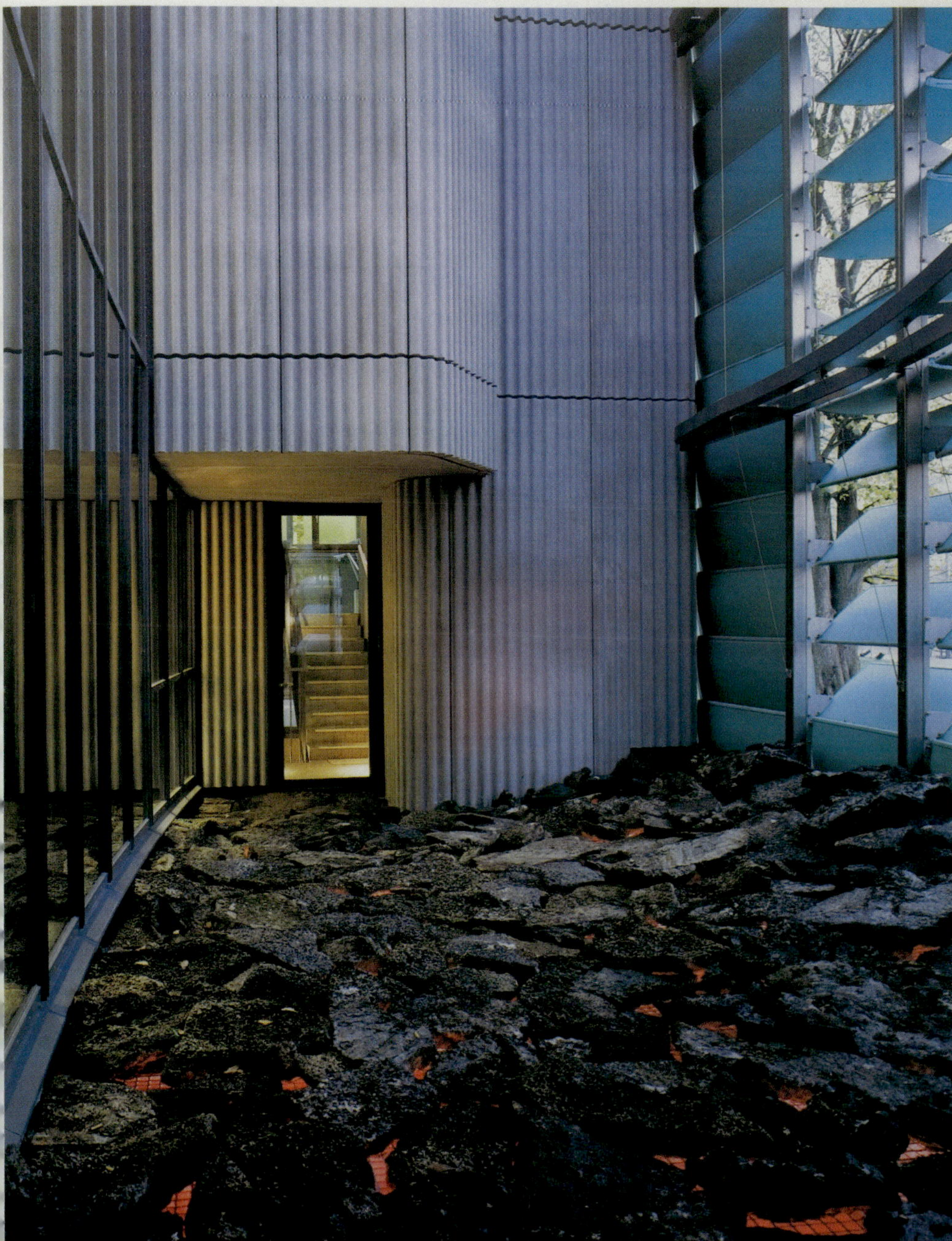

SUBJECT
Five Nordic embassies

LOCATION
Berlin, Germany

DESIGNERS
Various, see page 123

PHOTOGRAPHERS
Various, see page 123

WRITER
Leo Gullbring

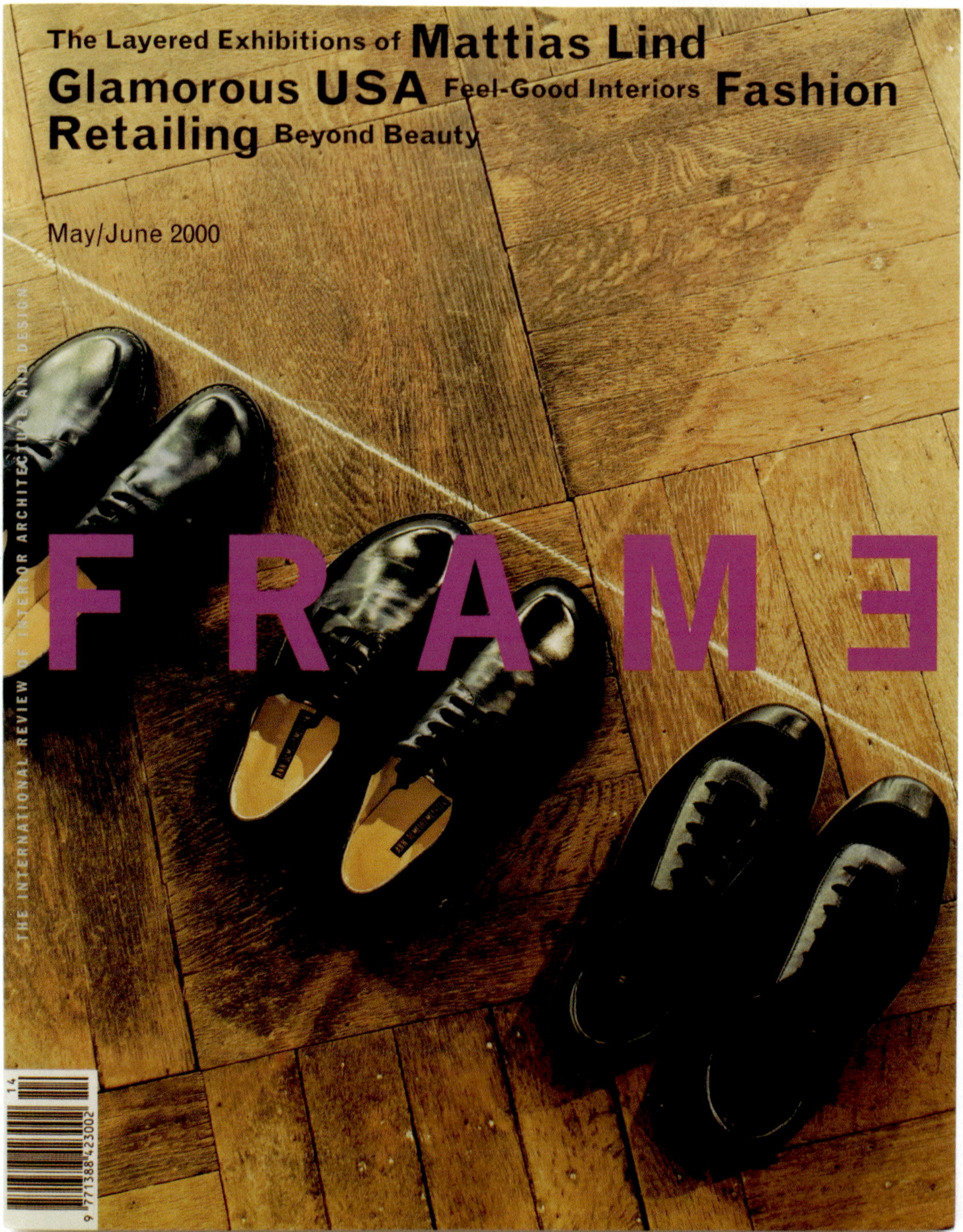

ART DIRECTOR
Ton Homburg,
Opera Designers

14

Worlds Apart

Architect Evert Crols of B-Architecten about the Walter Van Beirendonck shop

A pioneering project in search of a different approach to the shopping experience, in Antwerp and beyond.

Comments on the published article

It is a short, well-written article that captures the experience of the shop in a few words.

Did the article have an effect?

A lot of people got to know the project, as Frame was one of the first major magazines to publish this shop.

Current projects

We are working on shops for: Weekendesk, Antwerp, Belgium // Annemie Verbeke, Antwerp, Belgium // Missoni, Antwerp, Belgium // Ittala, Antwerp, Belgium // Kids by Women'secret, Madrid, Marbella and Almería, Spain

—

Architects Paul Robbrecht and Hilde Daem of Robbrecht and Daem Architecten about the project for Ann Demeulemeester

Timeless architecture.

Current projects

Concert hall, Bruges, Belgium // Opera, Antwerp, Belgium // St Felix warehouse, Antwerp, Belgium.

DESIGNERS
B-Architecten
www.b-architecten.be
Robbrecht and Daem Architecten
robbrechtendaem.com

SUBJECT
Fashion shops for Ann Demeulemeester and Walter Van Beirendonck

LOCATION
Antwerp, Belgium

PHOTOGRAPHER
Daniël Nicolas

WRITER
Chris Scott

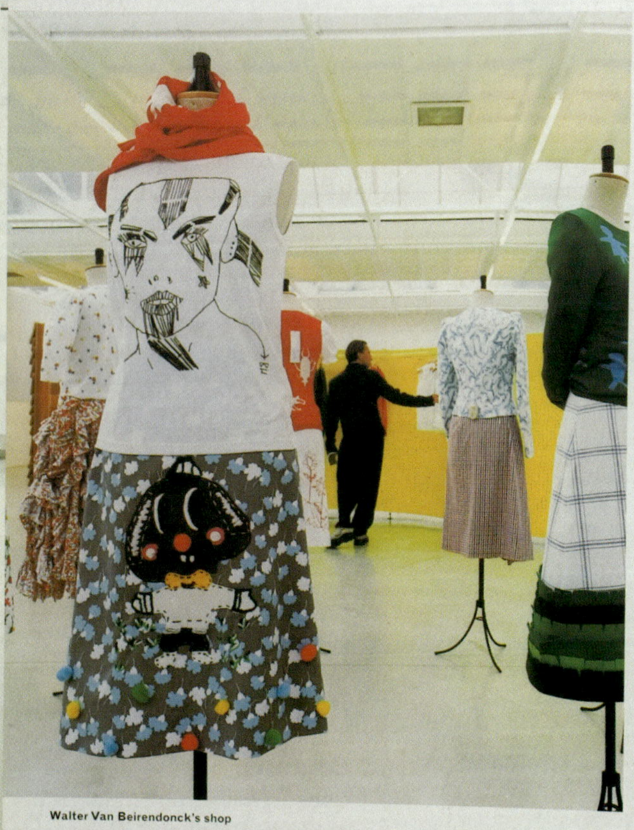

Walter Van Beirendonck's shop

Worlds

Looking around the shops of Belgian fashion designers **Ann Demeulemeester** and **Walter Van Beirendonck**, you get the feeling that clothes may have been an afterthought.

Apart

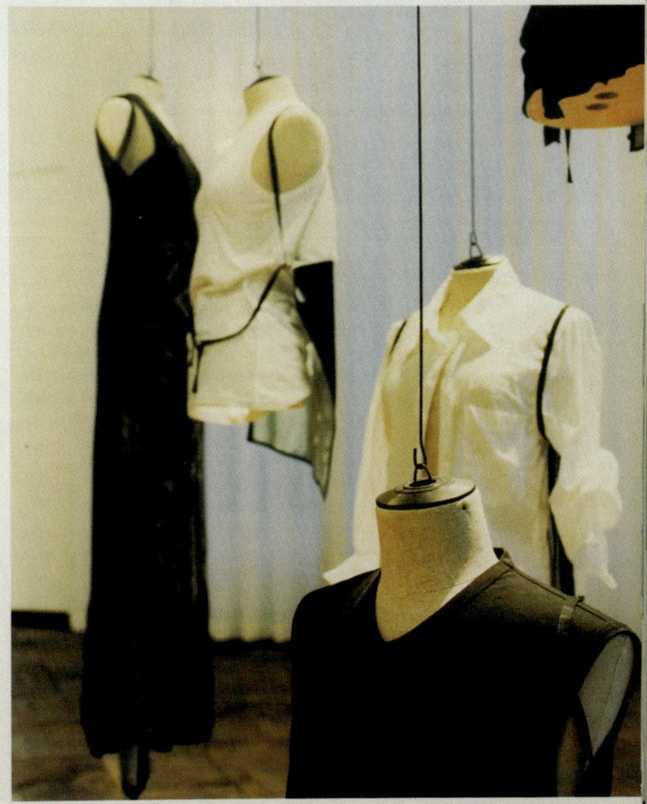

Ann Demeulemeester's shop

Both shops welcome you into the personal world of the designer, where you're persuaded to spend far more time than expected – and more money. *By Chris Scott. Photography by Daniël Nicolas.*

SUBJECT Shops for Ann Demeulemeester and Walter Van Beirendonck
LOCATION Antwerp, Belgium
DESIGNERS Various, see page 135
PHOTOGRAPHER Daniël Nicolas
WRITER Chris Scott

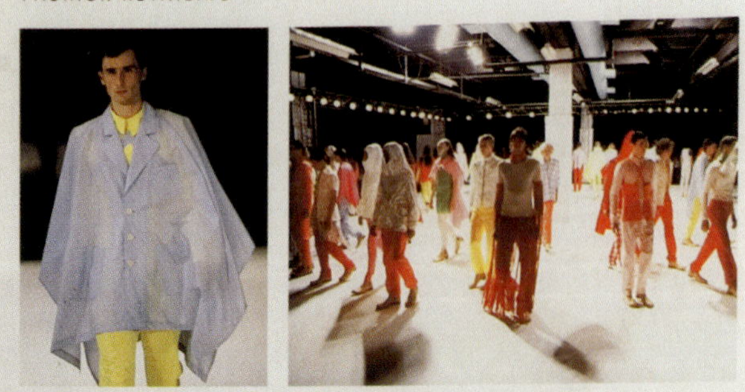

Fashion by Walter Van Beirendonck, summer 2000. Photography: Dan Lecca

Antwerp – apart from being famous for its beer and chocolate – has become a hot spot on the fashion map. It's home to a number of fashion designers, each with a highly distinctive identity. Two of them, Ann Demeulemeester and Walter Van Beirendonck, have recently opened their first shops in the city. Although very different designers with very different shops, the two exhibit certain similarities. While neither is keen on following the rules, both have found success in doing their own thing. By mixing fashion with architecture and focusing on aspects other than retailing clothes, Demeulemeester and Van Beirendonck have brought something extra to the fashion world. After seeing these shops, I'm convinced that minimalism is alive and breathing. When it's done well, the stripped-down look can be highly effective. The relatively stark ambience of both shops, in which each object has been placed with infinite care, creates an openness that is not in the least bit uncomfortable. I also had the intriguing impression that the clothes displayed here were almost an afterthought. Imbued with the mood of the designer, each space offers an insight into the world of an individual who – physically present or not – seems to be extending a warm welcome to all who enter. The personal touch puts customers at ease and gently persuades them to spend a lot more time – and money – than expected. The ultimate result is a desire to take home a 'souvenir' as a reminder of a pleasant experience. Now that's what I call successful retailing.

Walter

Fascinated, I approach Walter, a one-time car park with an unconventional drive-in entrance: a transparent plastic screen that lifts up to reveal Walter's wonderland, which can be likened to a funfair-cum-art exhibition. The 1000-square-metre area, all on one level, leaves me wishing for even more. I'm a little kid again, not knowing what to look at first.

The shop belongs to two designers – Walter Van Beirendonck and Dirk Van Saene – each of which has his own particular (often bizarre, frequently fun and always interesting) style. In designing the shop, they collaborated with Dirk Engelen of B-architecten, an Antwerp-based architecture firm. Lots of brainstorming, enlivened by constantly changing ideas and new proposals, m it an exciting, innovative project. The idea was to alter as little possible in order to maintain the original sense of space and emptiness that characterised the former car park.

Van Beirendonck had one initial requirement: 'I want a bear i the shop – a sleeping bear.' His request was honoured, and th big brown bear, prominently located, invites children to climb on and cuddle up. 'It should definitely not feel like a shop,' he further stipulated. 'Customers should have to find the fashion Walls smoothed down and whitewashed combine with a highl polished concrete floor to create a design completely in keepir with the original space. Filled with daylight pouring through industrial skylights, the uncluttered expanse is an excellent ba drop for brightly coloured objects of various shapes and textur that are distributed strategically throughout the shop. Wanderi around, I viewed these 'exhibits' as though I were in a gallery, realising only at second glance that many of them were displa for a number of fashion collections, including those of Van Beirendonck and Van Saene.

B-architecten were responsible for most of the furnishings an fixtures, including the rustic wooden chalet that houses the Dirk Van Saene collection. The 'floating counter' is the work o Marc Newson, who also collaborated with Van Beirendonck on the design of displays for his W&LT label.

Situated behind sliding, industrial-style panels, the fitting roor are yet another pleasant surprise. Floor, bench and walls are o continuous surface of cherry wood, which radiates a sense of l ury and comfort. Nothing is obvious here. My advice is to expe the unexpected. Leaving the premises in a pleasantly amused frame of mind, I resolve to come back soon. And I'm careful n to wake the bear. ▶

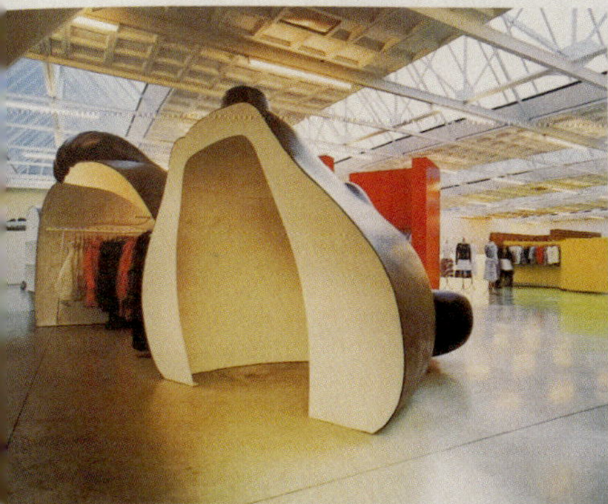

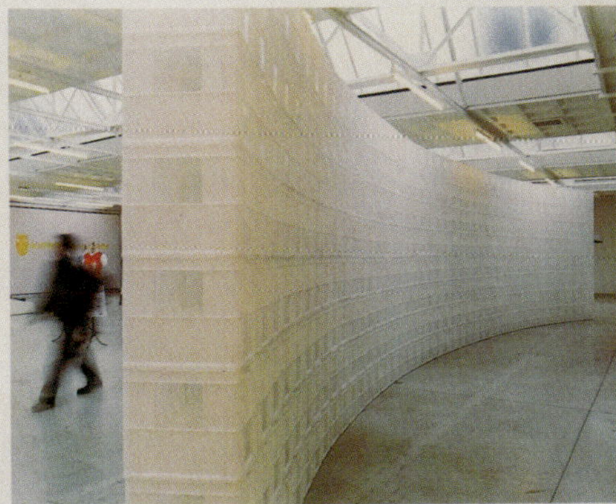

This page: Van Beirendonck's shop can be likened to a funfair-cum-art exhibition. Filled with daylight pouring through industrial skylights, the uncluttered expanse is a backdrop for brightly coloured objects that are distributed strategically throughout the shop. A big brown bear (top left) invites children to climb on and cuddle up

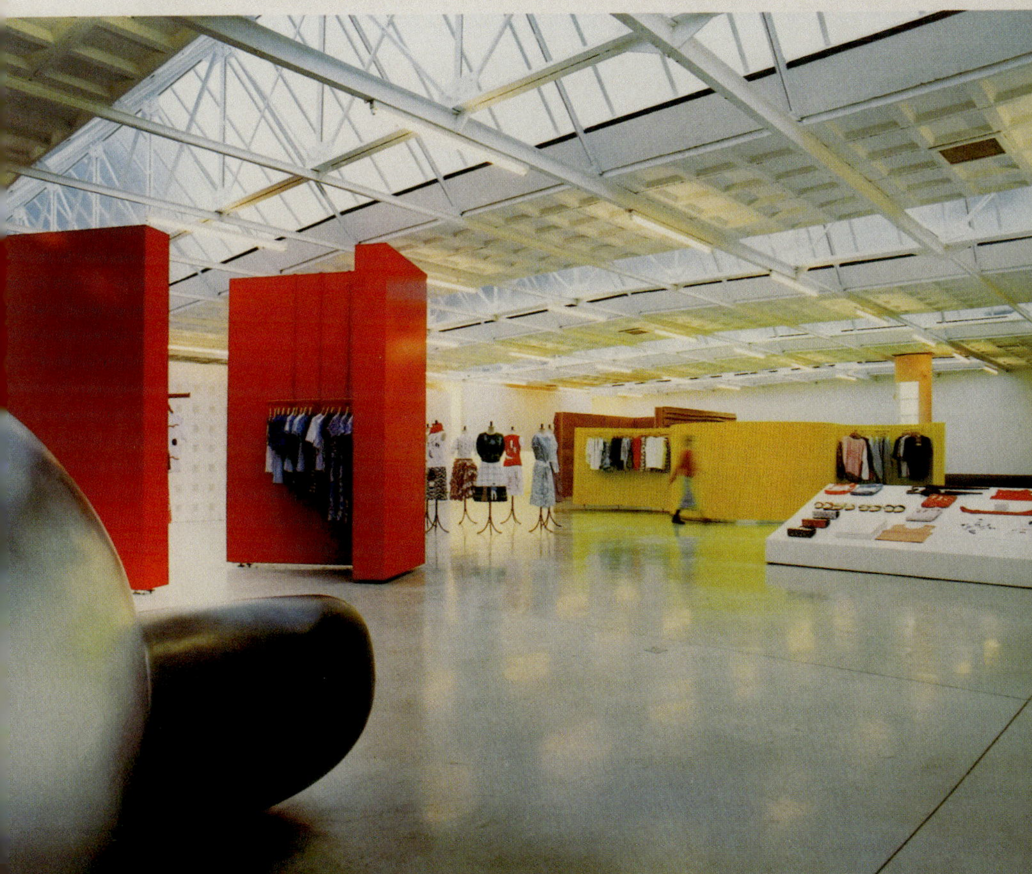

SUBJECT
Shops for Ann Demeulemeester and Walter Van Beirendonck

LOCATION
Antwerp, Belgium

DESIGNERS
Various, see page 135

PHOTOGRAPHER
Daniël Nicolas

WRITER
Chris Scott

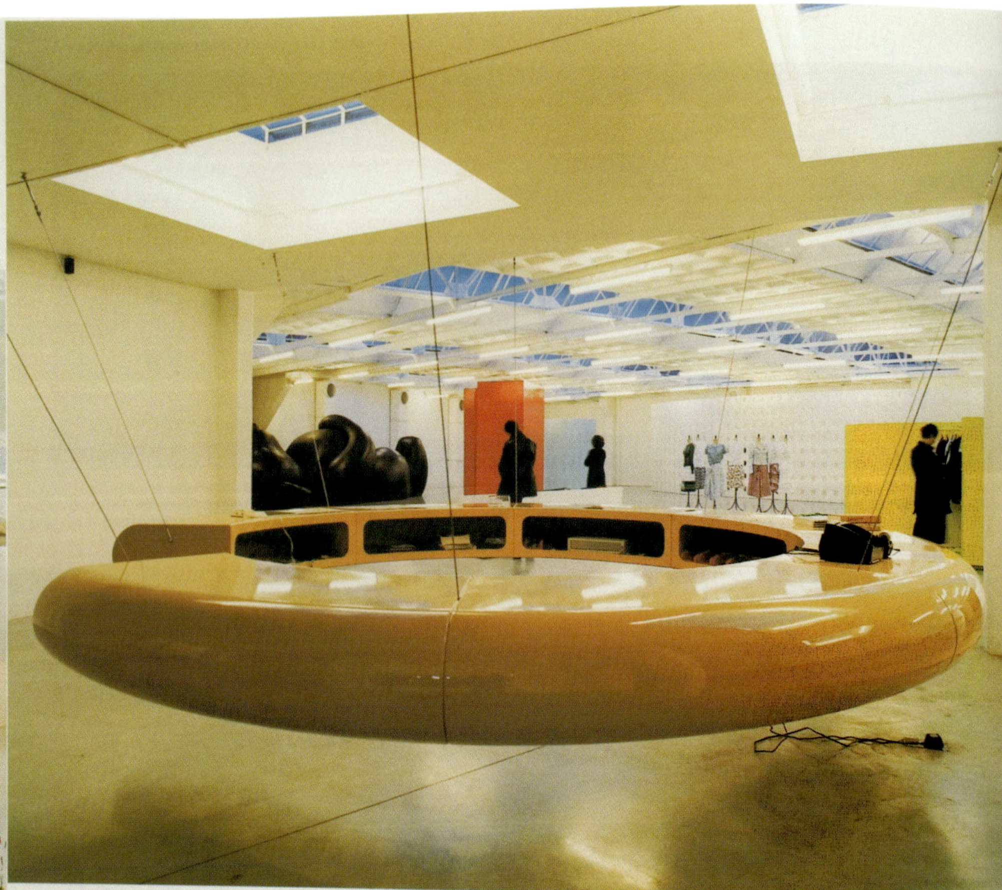

Van Beirendonck had one initial request: 'I want a bear in the shop – a sleeping bear'

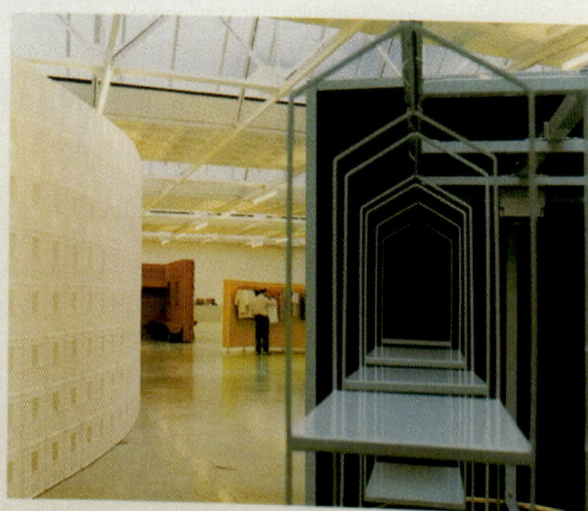

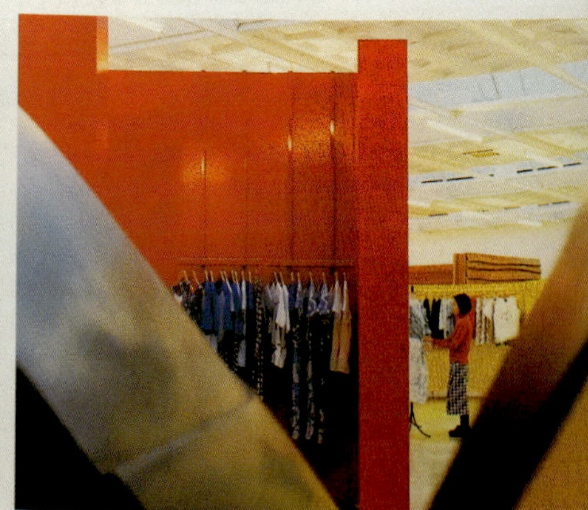

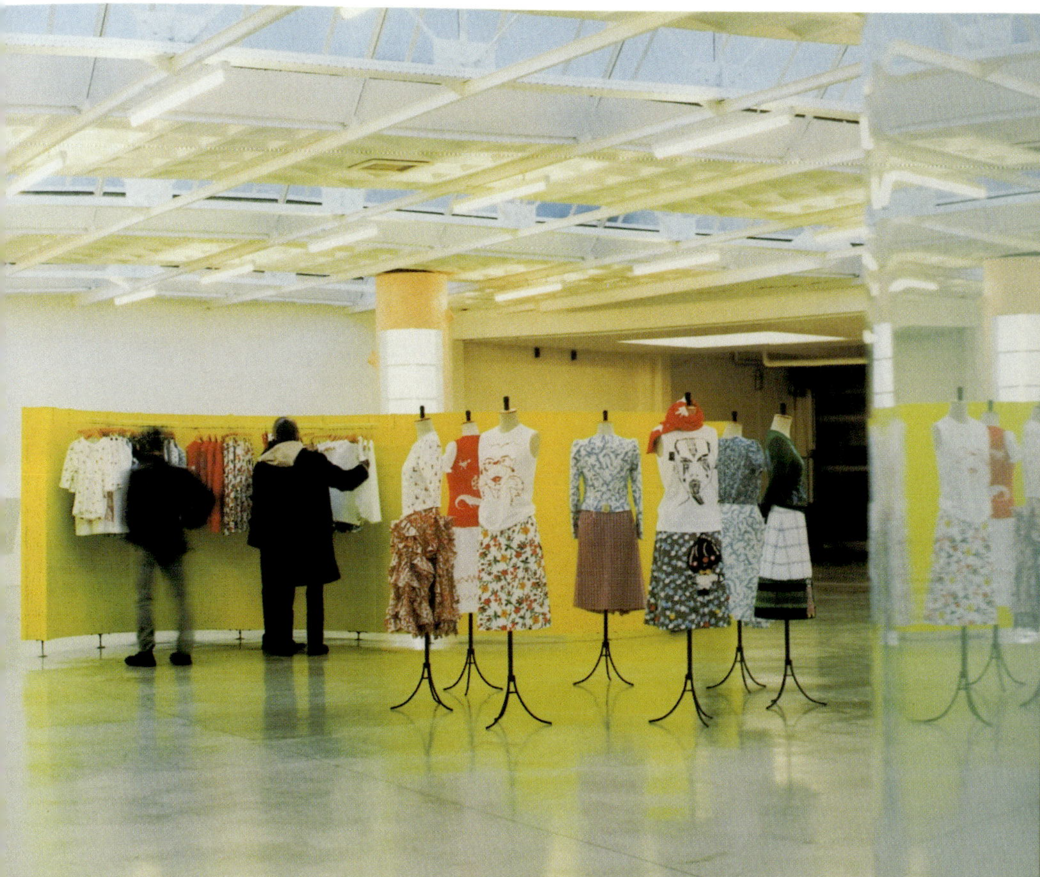

These pages: In Van Beirendonck's shop fashion displays often look like art exhibits. The 'floating counter' is designed by Marc Newson

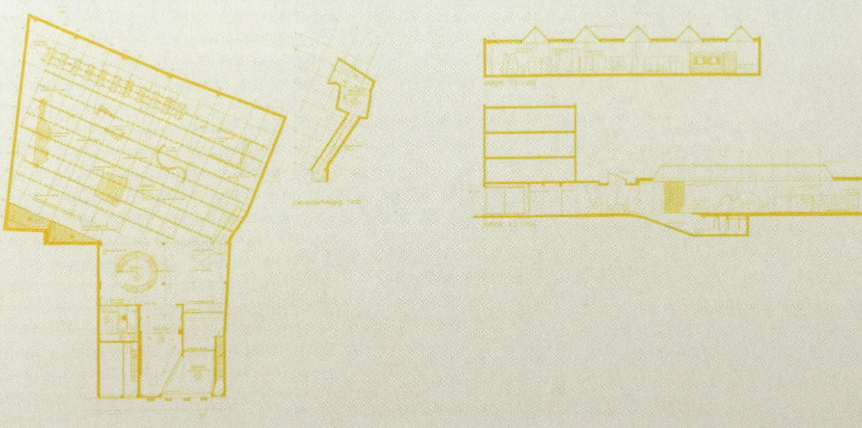

SUBJECT
Shops for Ann Demeulemeester and Walter Van Beirendonck

LOCATION
Antwerp, Belgium

DESIGNERS
Various, see page 135

PHOTOGRAPHER
Daniël Nicolas

WRITER
Chris Scott

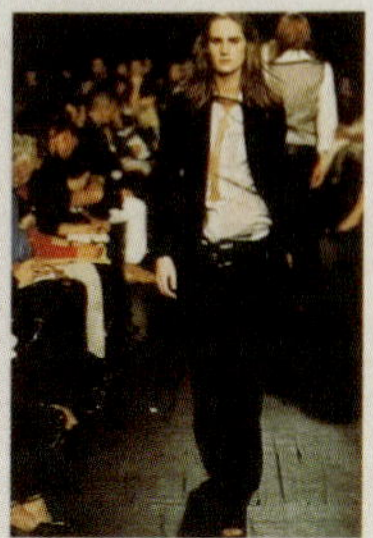

Fashion by Ann Demeulemeester, summer 2000. Photo: Dan Lecca

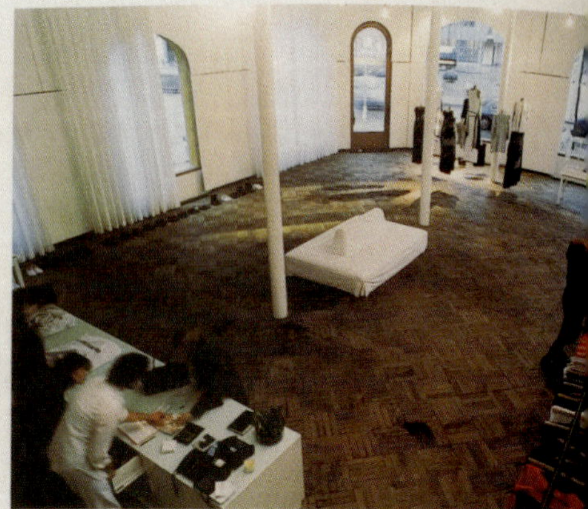

Above and opposite page: Demeulemeester's shop is locate[d] in a late-19th-century building on Antwerp's grand Museum Square. The interior was stripped and given a second skin: canvas-covered frames. Display windows that are virtually part of the streetscape form a courtyard of sorts, in which suspended items of clothing give the impression of figures mingling in the square

Ann Demeulemeester

Demeulemeester's shop is attractively located in a late-19th-century building on Antwerp's grand Museum Square, opposite the Museum of Fine Art. But Demeulemeester didn't select the building for its beautiful facade, which isn't really her style. Coming across the property unexpectedly, she was stimulated by the 'vibes' and pleased by the scarcity of shops nearby. She decided to create her own space, to divorce her establishment from the building's exterior and not to accentuate the existing architecture. Despite knowing these things, I cannot stifle a feeling of excitement as I near the charming surroundings. I'm certain that the shop will live up to – and perhaps exceed – my expectations. The custom-made Ann Demeulemeester sign outside is an indication of the individuality and style inside. Like her clothing, the shop has a studied simplicity achieved only after a great deal of thought and accomplished with such perfection and attention to detail that the result seems effortless and feels just right.

To Demeulemeester, fashion design is closely related to architecture. A unique atmosphere permeates each of her collections. Shapes evolve, collections are built and the outcome – fashions based on an architectural approach – can be seen in this shop. The term 'shop' is far too limiting to describe a place that is actually the designer's personal universe. The creation of this small world is rooted in Demeulemeester's concept of the ideal retail space, including everything from ambience and contact with sales personnel to the desired surroundings for trying on garments. She and her husband and partner Patrick Robyn collaborated with architect Paul Robbrecht, who recently redesigned Museum Square, to transform Demeulemeester's concept into reality. The interior of the building was stripped and given a second skin: canvas-covered frames. Display windows that are virtually part of the streetscape form a courtyard of sorts, in which suspended items of clothing give the impression of figures mingling in the square. Relaxing features include canvas-lined walls, gauze curtains, a pleasantly mild fragrance, cool music, white pigeons, gauze-wrapped hanging lamps, unpolished wooden floors and high black ceilings. The blend of stark simplicity softened by these calming notes produces a mood that both stimulates and sooth[es]. The look is basic and uncomplicated. Rows of hanging clothes complement rows of shoes – what could be simpler?

Womenswear, located on the ground floor, is linked to mensw[ear] upstairs by a plain white staircase. A few well-chosen and perf[ect]ly placed items of furniture (all furnishings are Demeulemeest[er] designs) complete the picture.

The aesthetically pleasing experience is further enhanced by fi[t]ting rooms that encourage customers to take their time. No ru[sh], no overattentive sales assistants, no cramped spaces with skim[py] curtains, but real rooms. Each of the five changing rooms exem[]plifies artless elegance. A peek inside reveals an occasional pie[ce] of furniture, a glass, a jug of water, a painting or two, several plants – and each space overlooks the courtyard garden, which provides an added element of nature. Like Demeulemeester's collections, the retail space changes and evolves. It reminds me [of] an artist's atelier, with its constant pulse of energy, its searchin[g] sense of creativity and its passionate enthusiasm.

Why do I like these shops so much? Why do they leave me wit[h] such a warm afterglow? Why am I planning to contact a few friends and go back? The two designers have very different – a[nd] very interesting – styles, but fashion is hardly the reason. Ther[e] lies a large part of the fascination. Unlike normal retail outlets, these shops offer access to the personal environments of two in[di]viduals. Refreshing and inspiring – words that define Walter's amusing surroundings as well as Ann Demeulemeester's relax[ed] setting – they are so much more than the latest in haute coutu[re]. They draw me into two uniquely different worlds suffused wit[h] authenticity – perhaps the most vital clue to their appeal. And I didn't manage to get away without a few 'souvenirs' of my ow[n]

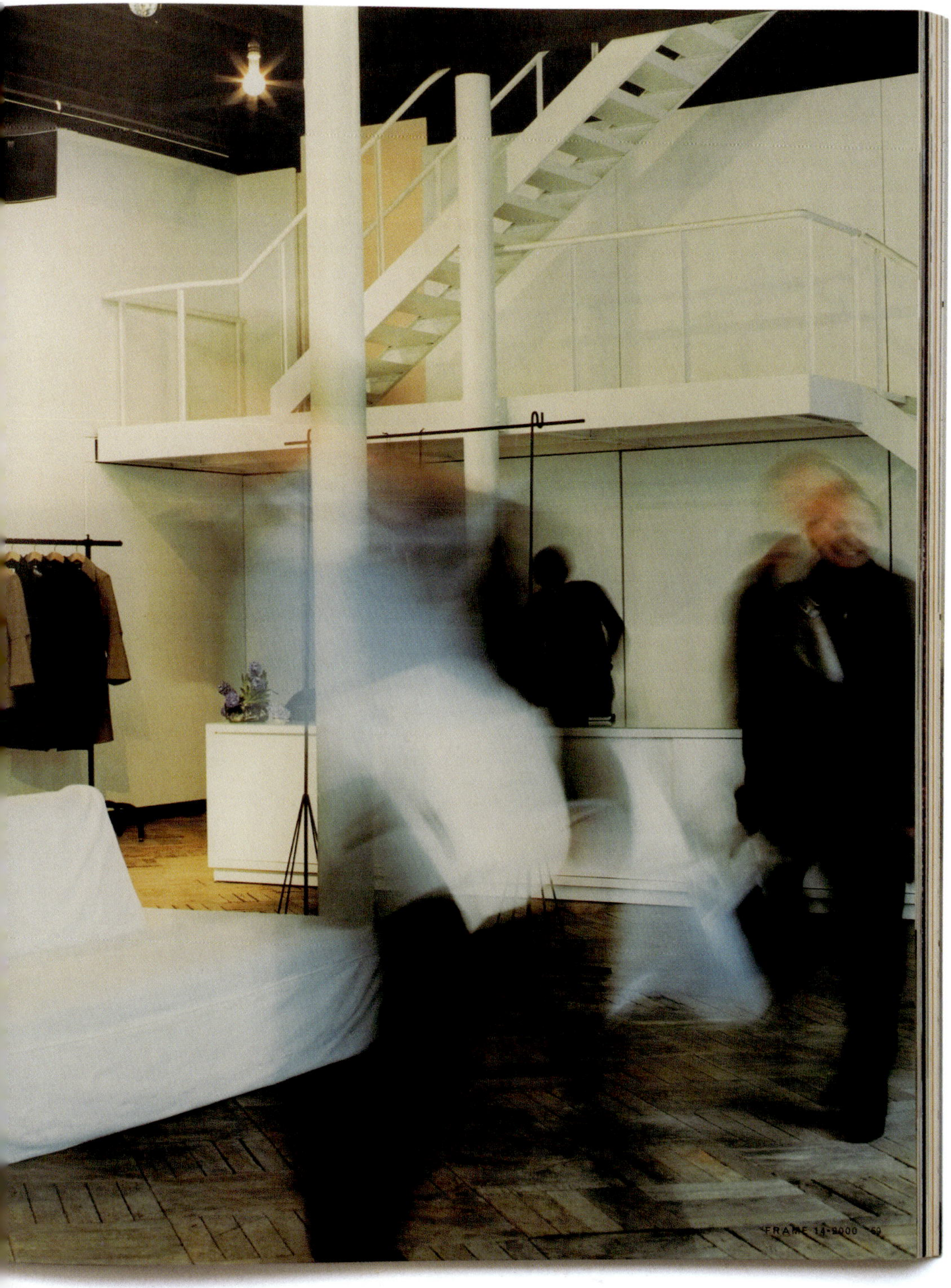

SUBJECT
Shops for Ann Demeulemeester and Walter Van Beirendonck

LOCATION
Antwerp, Belgium

DESIGNERS
Various, see page 135

PHOTOGRAPHER
Daniël Nicolas

WRITER
Chris Scott

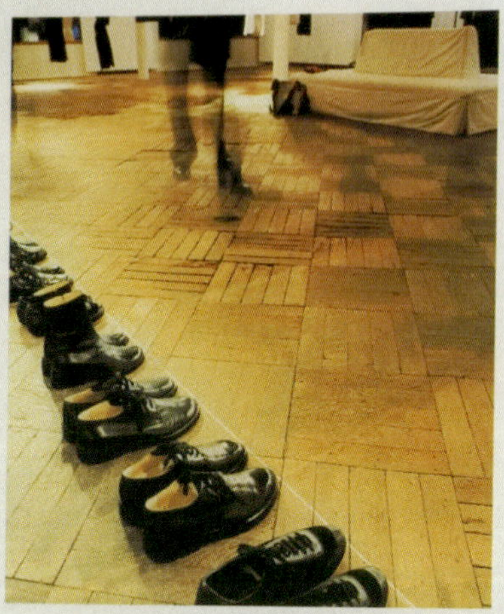

The term 'shop' is far too limiting to describe a place tha
is actually Ann Demeulemeester's personal universe

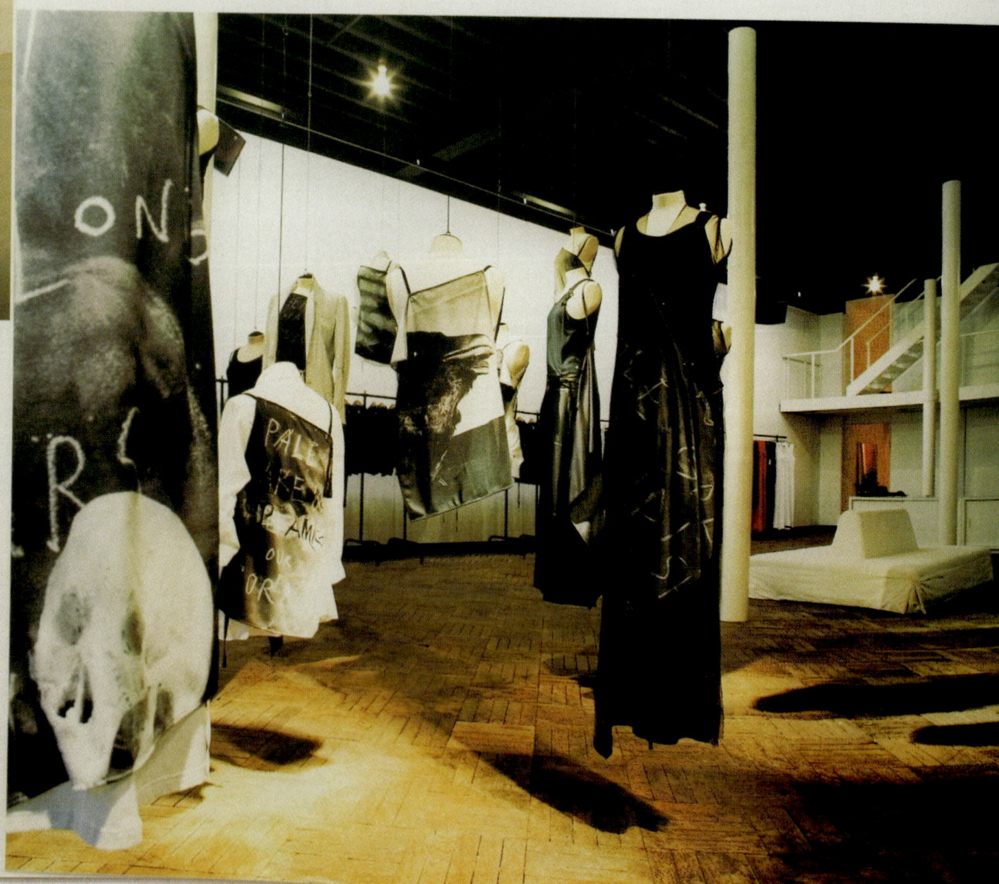

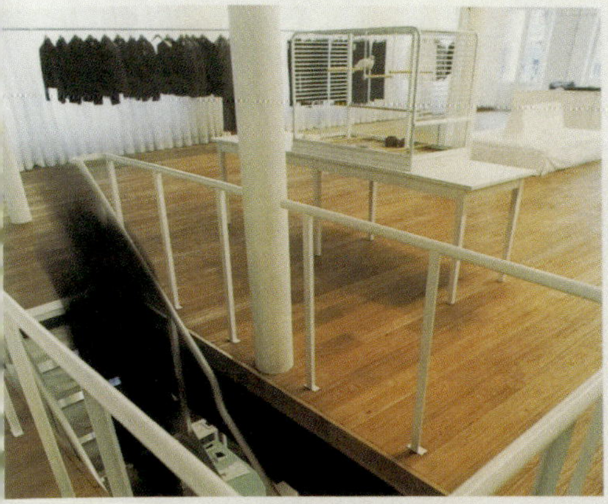

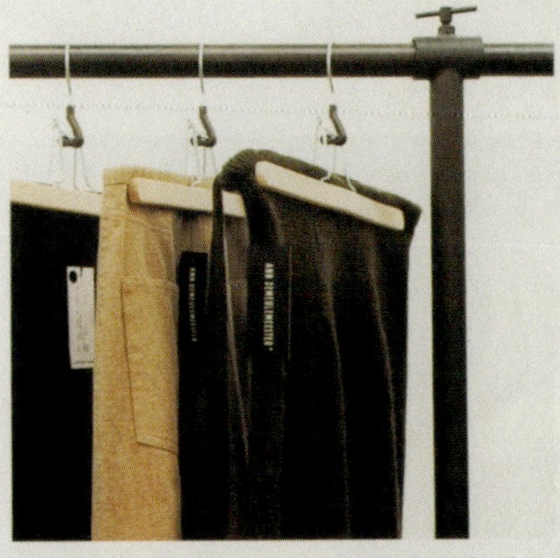

These pages: The look of Demeulemeester's universe is basic and uncomplicated. Rows of hanging clothes complement rows of shoes. Womenswear, located on the ground floor, is linked to menswear upstairs by a plain white staircase. A few items of furniture (all furnishings are Demeulemeester designs) complete the picture

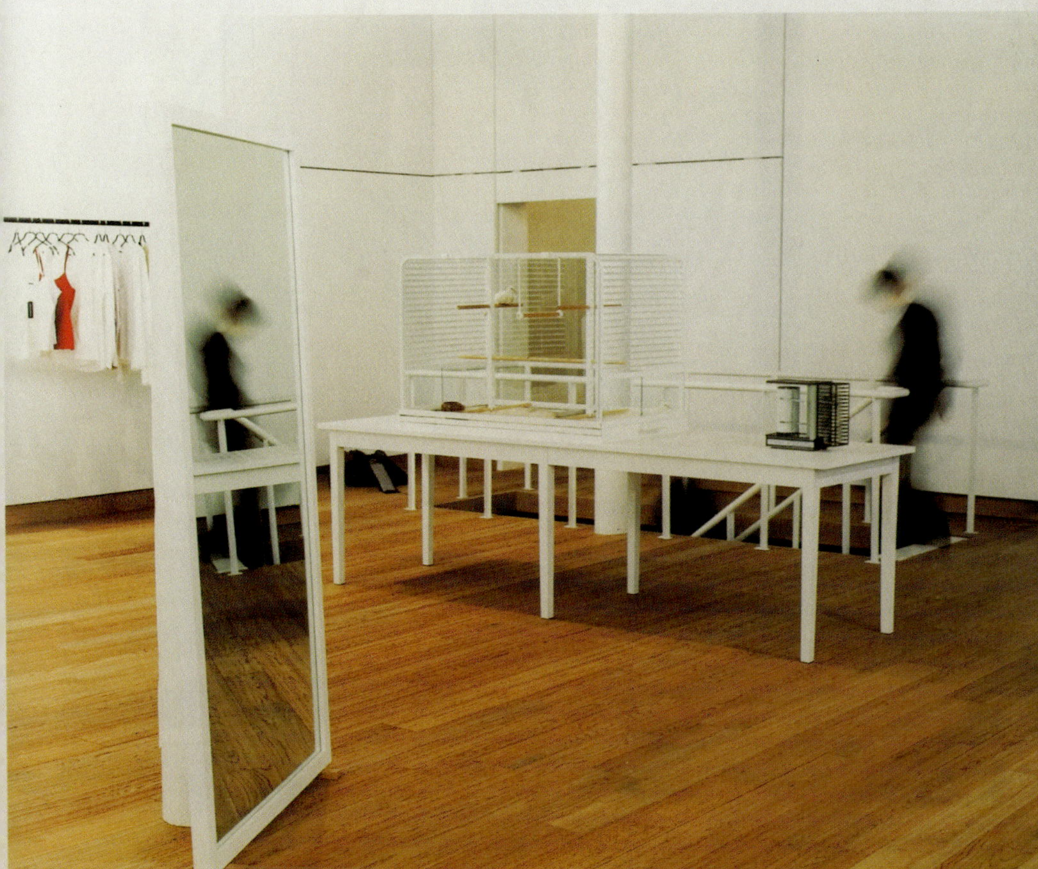

SUBJECT
Shops for Ann Demeulemeester and Walter Van Beirendonck

LOCATION
Antwerp, Belgium

DESIGNERS
Various, see page 135

PHOTOGRAPHER
Daniël Nicolas

WRITER
Chris Scott

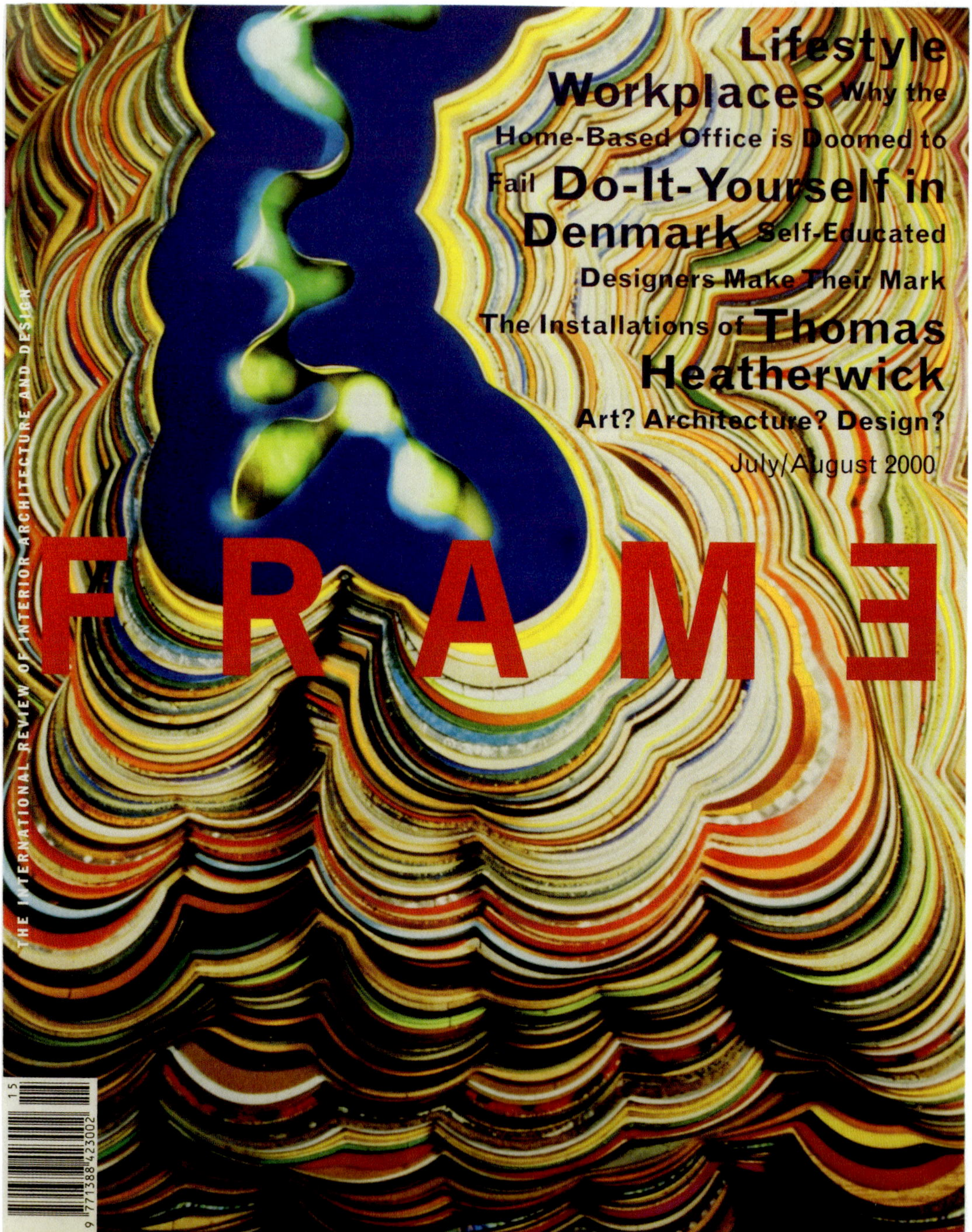

ART DIRECTOR
Ton Homburg,
Opera Designers

15

On the Cutting Edge

Architects Zoe Smith and Graeme Williamson (Block Architecture) about the project

We look at the project with a great deal of affection. It was one of our first, but we are still pleased with how it looks today. It was a very simple idea – to make a hybrid experience somewhere between hairdressing and cosmetic surgery.

Did the article have an effect?

It was great for us to have such a small project published at an early stage in our careers. It helped us to establish our interests in utilitarian design and developed awareness of our approach.

Current projects

Hussein Chalayan, Tokyo, Japan // Magasin 3 Gallery, Stockholm, Sweden // Office interior for Glue, London, England // Art space and accommodation for Blast Theory Art Collective.

DESIGNERS
Block Architecture
www.blockarchitecture.com
Tom Dixon
www.tomdixon.net
Jeremy Pitts
Jonathan Reed
Hideyuki Yamauchi
Jun Shirai

SUBJECT
Hair salons

LOCATIONS
London, England
Tokyo, Japan

PHOTOGRAPHERS
Leon Chew
Mitchell Sams
Shinbiyo Shuppan
Kazuya Shimokawa/
Nikkei Design

WRITERS
Chiharu Watabe
Chris Scott

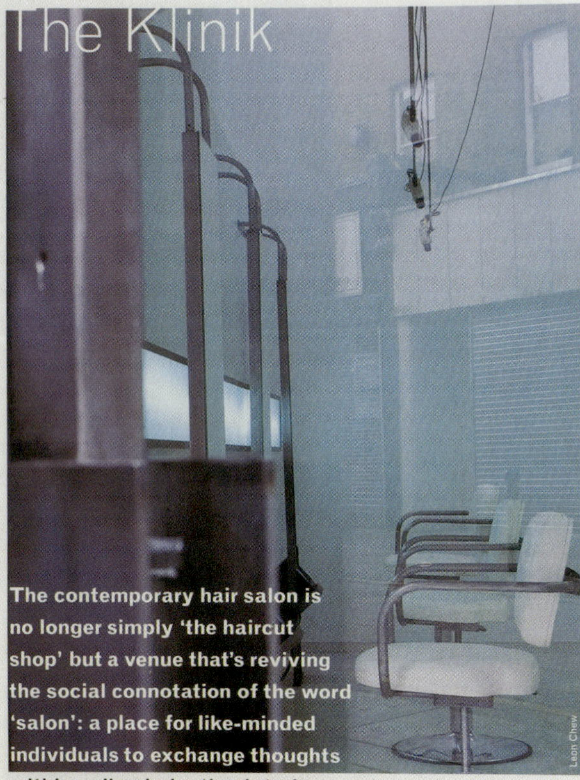

The Klinik

Leon Chew

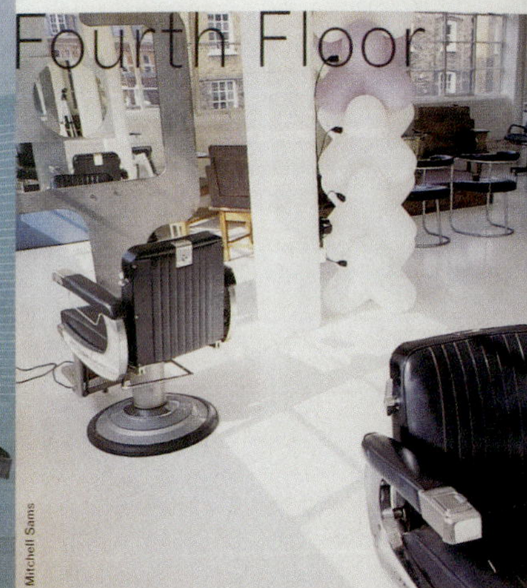

Fourth Floor

Mitchell Sams

London

The contemporary hair salon is no longer simply 'the haircut shop' but a venue that's reviving the social connotation of the word 'salon': a place for like-minded individuals to exchange thoughts within a discriminating interior. *Chiharu Watabe* reports from Tokyo and *Chris Scott* from London.

Japanese fashion magazines are publishing an increasing number of issues that feature hairstyles. What used to be an annual event gradually evolved to twice a year, and now we can expect a magazine presenting hot hair designs to turn up nearly every time the season changes, right in step with the latest fashion hits. Interestingly, hair salons seem to be attracting almost as much attention as writers pay to the hairstyles and their creators.

Many interior designers and architects in their 20s and 30s corroborate this observation when they report a steadily rising number of clients in the hairdressing sector, which appears to be equalling boutiques, restaurants and bars in terms of rapid growth. A look at the figures shows that the last 5 years have seen the emergence of 10,000 new hair salons throughout Japan. This incredible spurt, which is more obvious in big cities, is a major cause of overcrowded conditions in some parts of Tokyo. The most congested area spans the neighbouring districts of Harajuku and Aoyama, which accommodate no less than 600 hair salons. Not a large area, it's served by only two underground stations, which can be reached on foot in minutes. If we include the districts served by the following two stations – Shibuya and Daikanyama – the number of salons surely tops 1,000.

Though each has different characteristics, these four districts are home to the city's most trend-conscious shops and services.

The last 5 years have seen the emergence of 10,000 new hair salons throughout Japan

Heavens Link

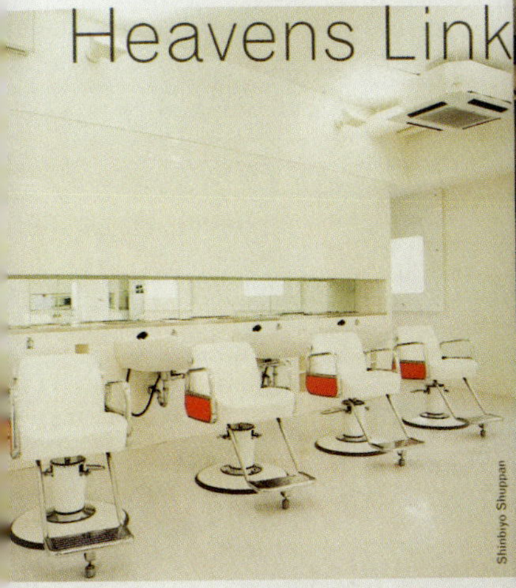

Shinbiyo Shuppan

Arcadia

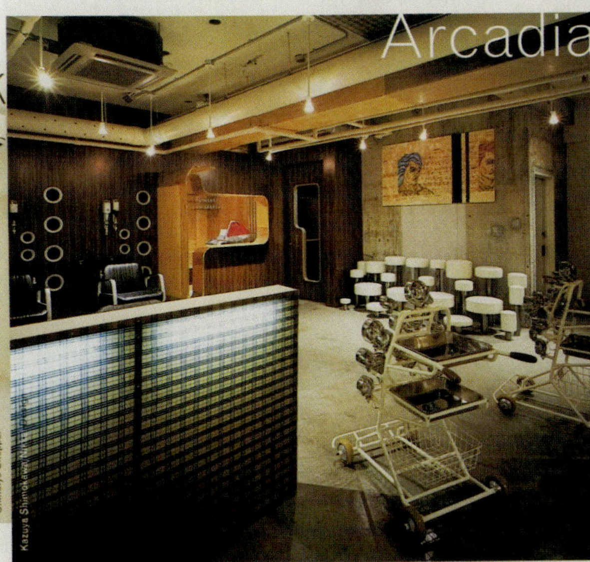

Kazuya Shimokawa

Tokyo

On the Cutting Edge

Densely packed with boutiques that offer everything from street-style apparel to haute couture, this part of Tokyo has its finger on the pulse of vogue. Success in this competitive market depends on grabbing the customer's attention and keeping it. Even the most talented hairstylist needs more than skill. The salon itself has to provide attractive surroundings, and the solution to this problem is good interior design.

Comfort is precious. With the exception of home or workplace, where do we regularly spend a couple of hours? For many of us, the answer is: at the hairdresser's. And what do we hope to find there? Relaxation, comfort and an atmosphere that radiates cleanliness. A recent trend in hair-salon design is an abundance of white, which conveys a sense of immaculate freshness. Today's haircut often involves a deeper level of communication. We're whisked back to the 'salons' of the 17th and 18th centuries, in which an exchange of ideas was an essential element of society. The hair-salon industry is investing in the revival of this concept, albeit a contemporary version that concentrates a bit less on culture and a lot more on fun. In the early '90s trendsetters (and their followers) were constantly asking, 'Which club is in?' Several years later the search was on for 'a posh restaurant', soon to be followed by 'the right cafe'. The latest question on Tokyo lips – and I'm not making this up – is: 'Have you found a really cool hair salon?' ▶

WRITERS
Chiharu Watabe
Chris Scott

PHOTOGRAPHERS
Various, see page 147

DESIGNERS
Various, see page 147

LOCATIONS
London, England
Tokyo, Japan

SUBJECT
Hair salons

The use of light boxes in the window has a strong visual impact, especially at night, when the place positively radiates. Photography: Leon Chew

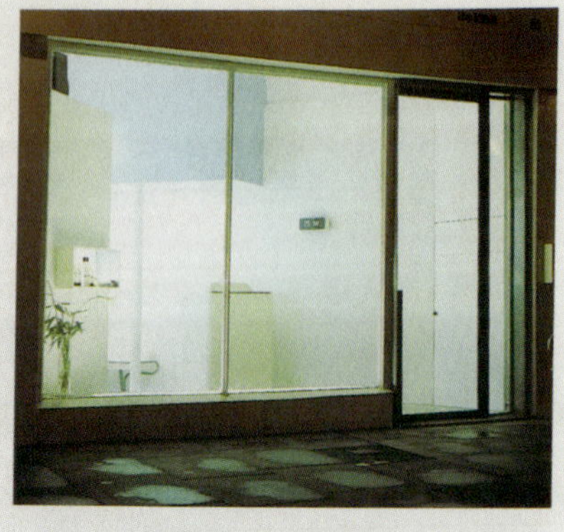

The Klinik, London

The Klinik's Swedish owner, Anna Forsling, turned to her native tongue for an eye-catching version of the word 'clinic'. Aware that a hair salon could and should be so much more than what she found listed in the Yellow Pages, she set out to create a visually interesting and entertaining venue, combined with efficient service. Persuaded by the architects 24/seven to take on a run-down space in the then up-and-coming Exmouth Market, Forsling worked closely with them throughout every stage of the project. (The firm later changed its name from 24/seven to Block Architecture, a label that embodies their down-to-earth approach to design.)

A major consideration in the plan for The Klinik, a salon that had to fit into a space of only 36 square metres, was the functional requirement. Block's point of departure was a rather medicinal and utilitarian aesthetic, which they developed into a well-functioning space. Thanks to its double height, clean graphic lines and a cleverly restrictive use of furnishings, the salon appears to be larger than it is. Daylight strip bulbs provide effective lighting, and the use of light boxes on opposite sides of the room and in the window has a strong visual impact, especially at night, when the place positively radiates.

Block designed the furniture and fittings. As they like to oversee all aspects of a project, the architects often create furnishings in their own workshop. For the most vital piece of furniture in the salon – the barber chair – they sought out some good second-hand frames and added their own white leatherette seating. Easy to clean, the chairs further emphasise the image of a sanitary environment. The same white upholstery covers the chaise longue-cum-hospital trolley, with its metal legs and casters. Everyday objects take on a completely different look when put together in this entertaining yet practical fashion – The Klinik's water sprays, for example, are actually upside-down soap dispensers.

Instead of staring in the beautifully designed mirror or thumbing through magazines, the client can watch her hairstylist at work on one of the salon's black-and-white, closed-circuit monitors – for a voyeuristic view of the back of her own head.

The clinical setting belies what is, in fact, a warm and welcoming place. At the high point of a party at The Klinik, the festive crowd spills out into the street. Friday evenings turn into social events, complete with wine and conversation, and images projected on the walls enhance the clubby atmosphere. ▶

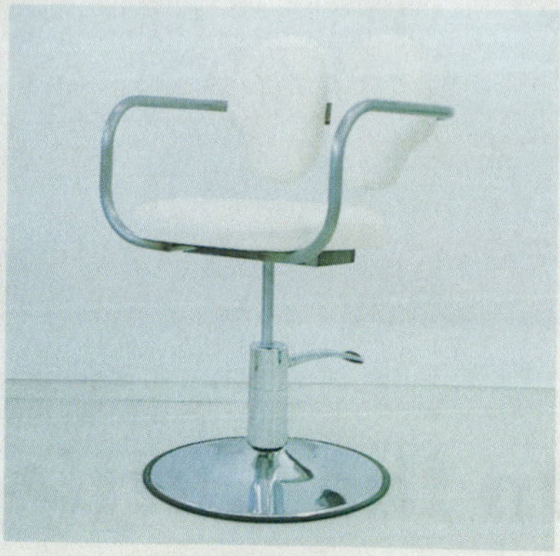

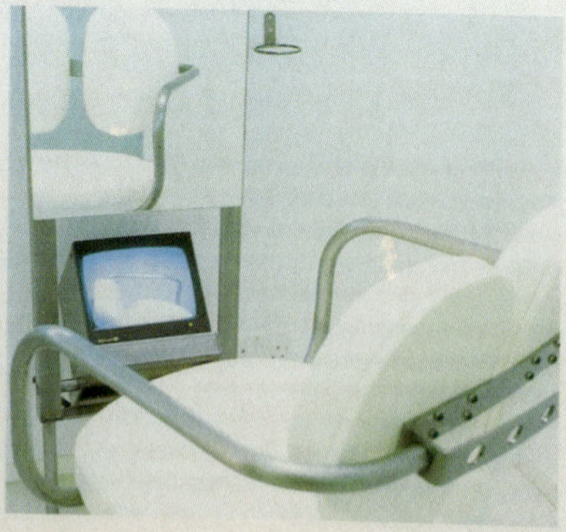

Block Architecture designed all furniture and fittings. For the barber chair, they sought out some good second-hand frames and added their own white leatherette seating. Easy to clean, the chairs further emphasise the image of a sanitary environment

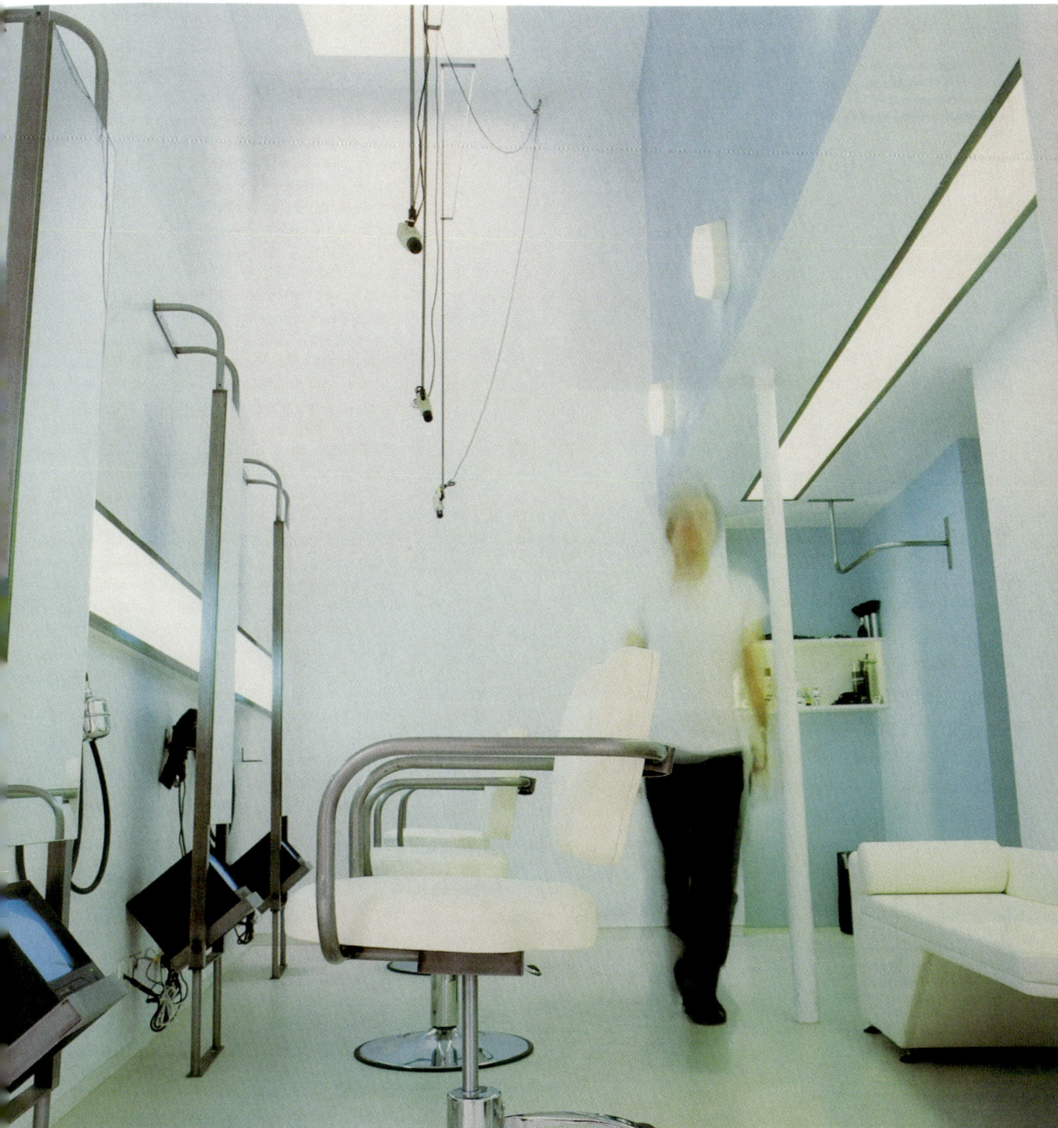

In designing The Klinik, Block's point of departure was a rather medicinal and utilitarian aesthetic, which the architects developed into a well-functioning space

Instead of staring in the beautifully designed mirror or thumbing through magazines, the client can watch her hairstylist at work on one of the salon's black-and-white, closed-circuit monitors – for a voyeuristic view of the back of her own head

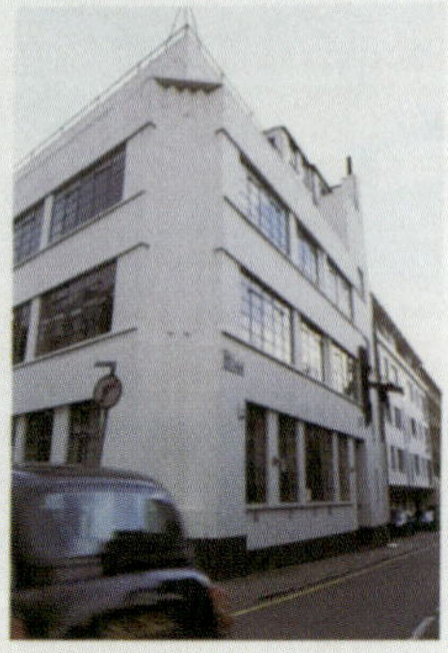

Fourth Floor is located in an industrial building on a deserted street in Clerkenwell. Photography: Mitchell Sams

Fourth Floor, London

Fourth Floor has to be acknowledged as a major forerunner of the 'new salon', which is rooted in a changing attitude within the industry. Owner and 'host' Richard Stepney went out on a limb in creating his business. Following his instincts he opted for Clerkenwell, an area of London not at all fashionable at the time, and subsequently moved into an industrial building on a deserted street. With no salon front, the result was a rather low-key, private space.

The idea behind Fourth Floor grew out of Stepney's talks with customers who were dissatisfied with the situation in conventional hair salons. The size of the establishment has increased from the original 75 square metres (fourth floor only) to 175 square metres (the combined area of the third and fourth floors). The original, low-budget premises were realised by Richard Stepney and Tom Dixon (now creative director of Habitat), whose mirrors, surrounds and lighting designs are found at both levels. Leaving many existing features unchanged, they washed rather than painted the walls and then allowed the space to develop 'organically'. The terrace looks out over the rooftops of central London. Sitting here, clients experience a sense of isolation while also feeling very much a part of the busy city.

The recently acquired third floor is a continuation of Stepney's understated concept. Keeping a hair salon pared down and uncluttered in the midst of clients and staff in seemingly constant motion and snippets of hair drifting to the floor – the activities that dictate the nature of the space – is no small achievement. The design's simplicity is based on an incredibly low number of hair stations: a mere six on each floor. In the world of hairdressing, 175 square metres and only 12 chairs is unheard of – the result is a veritable sea of space.

Stepney asked architect Jeremy Pitts (now in partnership with Jonathon Reed) to help him design the third floor. Pitts and Reed combine their architectural and interior-design talents to work primarily on residential properties. Many of their clients are in the arts/music business. A current project in New York involves David Bowie and his wife Iman.

Stepney wanted to make the most of the wonderful space and light that characterises the third floor. Walls came down and the basic, white space that emerged represents a blank canvas to which components can be added or changed and in which clients are part of the overall picture. Gearing the design to natural features found upstairs meant using wooden elements, such as a bench-table made of elm and chairs of solid oak.

Very few hair products are in view. Rather than containers of shampoo and conditioner, clients are encouraged to look at the art on display – work often done by the same people who come here to have their hair cut. News of the place seems to travel by word of mouth – rather like the grapevine that buzzes when an exclusive club opens it doors. Fourth Floor has evolved into a meeting point and a forum for discussion. New works of art replace the old 'exhibition', and a somewhat stale business card suddenly has a fresh graphic design. Sometimes a haircut can be the catalyst for a broad range of interesting experiences. ▶

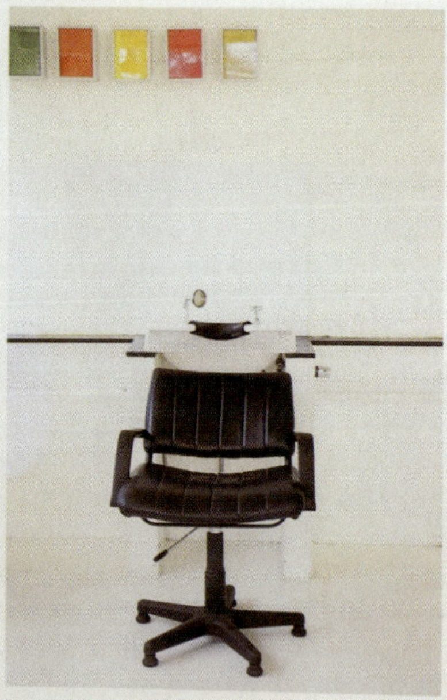

Right: Like the space upstairs, the recently added third floor has amazingly few hair stations: a mere six on each floor

Opposite page top: The terrace of the fourth floor looks out over the rooftops of central London

Opposite page bottom: The third floor is a basic white space – a canvas to which components can be added or changed and in which clients are part of the overall picture

Rather than containers of shampoo and conditioner, clients are encouraged to look at the art on display – work often done by the same people who come here to have their hair cut

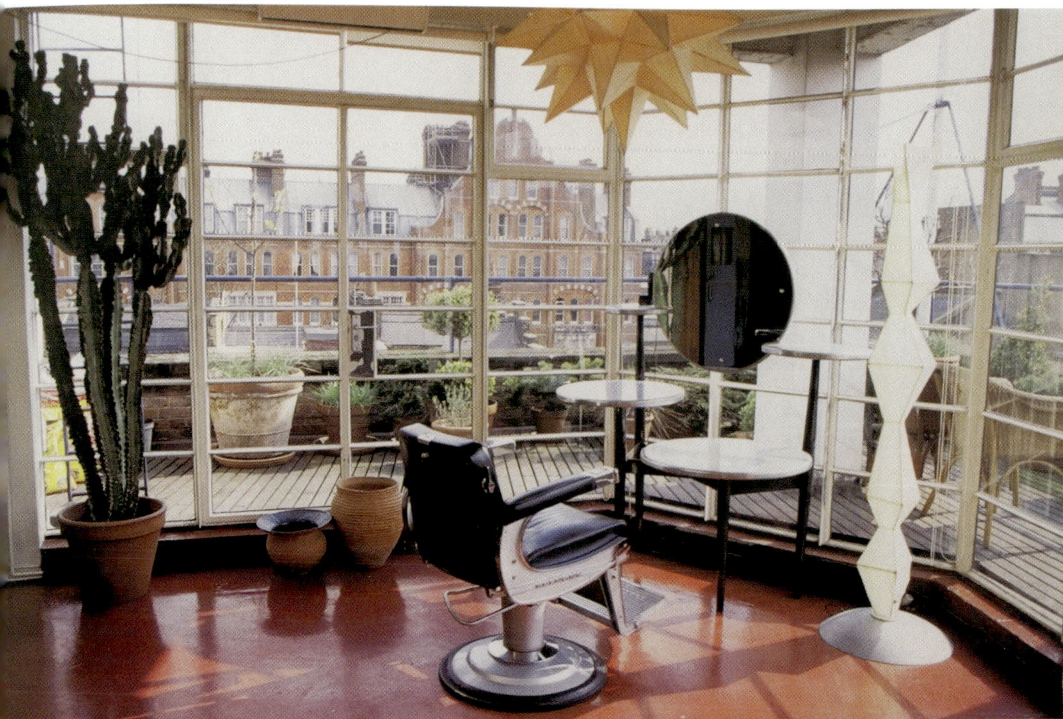

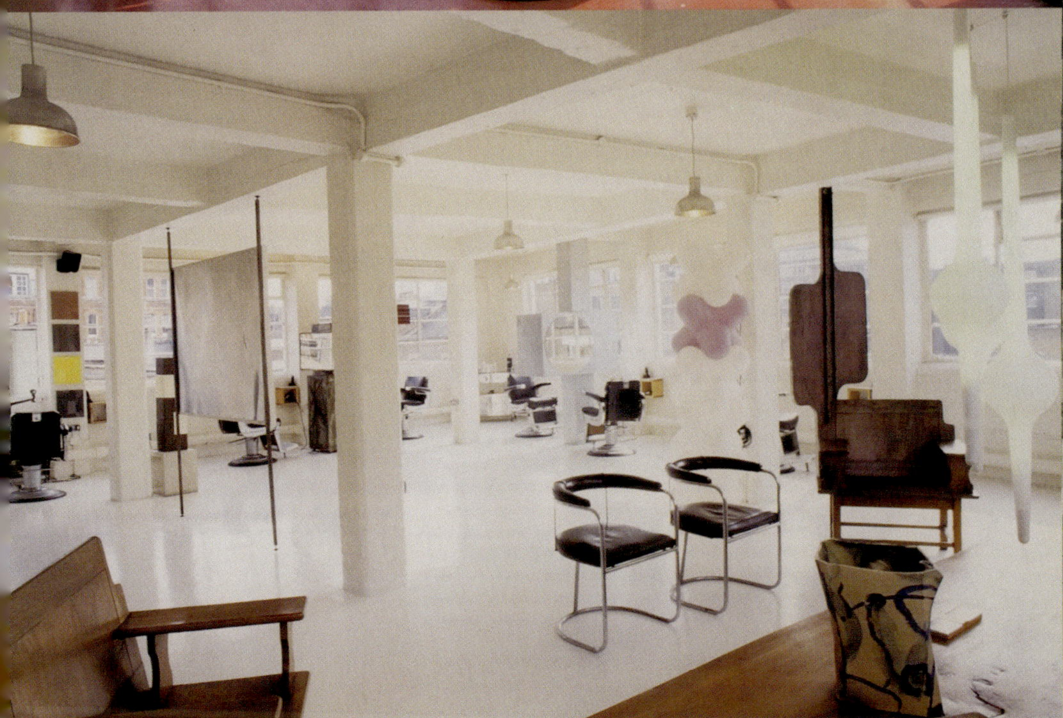

SUBJECT
Hair salons

LOCATIONS
London, England
Tokyo, Japan

DESIGNERS
Various, see page 147

PHOTOGRAPHERS
Various, see page 147

WRITERS
Chiharu Watabe
Chris Scott

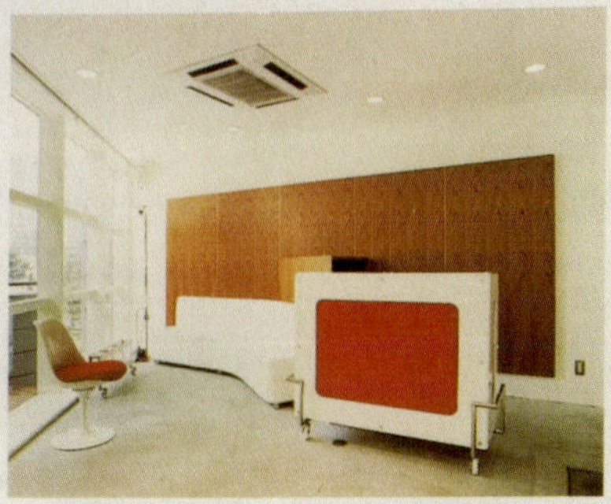

To avoid a cramped feeling, the designer used a movable cash desk and trolleys with legs especially designed for a 'light' look. Photography: Shinbiyo Shuppan

Heavens Link, Tokyo

Heavens Link in Shibuya opened in March 1998. Owner Atsushi Komatsu had already made a success of his first salon, Heavens, which is no more than a five-minute walk from the new establishment. Clients of both salons are primarily teenage girls and young women in their twenties, many of whom are in creative professions.

Interior designer Hideyuki Yamauchi, who's known Komatsu for at least ten years, designed the salon. The concept emerged from discussions between the two, often at bars. Although Yamauchi used white to create the 'clean' effect desired, to avoid producing a hard, overly inorganic impression he added red accents for a 'cutie' image. The combination of red and white is repeated in the logo on the shop sign. The concept's retro touch – shades of the '50s and '60s – materialises in the salon's outdated curved-arm chairs, which were ordered from a catalogue. The mixture of red and white and the blend of old and new were part of a deliberate strategy to transcend the trendy look and create a space with a respectable life span.

The floor area is only 76 square metres. To allow sufficient space for professional equipment without producing a cramped feeling, the designer used a movable cash desk and trolleys with legs especially designed for a 'light' look. Space for towels, located behind the shampooing sinks, can be stocked from the rear to avoid disturbing customers. A shiny white floor is the basis of the clean image. Not often seen in hair salons, white flooring is easily stained by the chemicals that beauticians use. At the end of each working day, potential stains are thoroughly removed at Heavens Link. Thanks to the staff's efforts, the two-year-old floor is still pristine. ▶

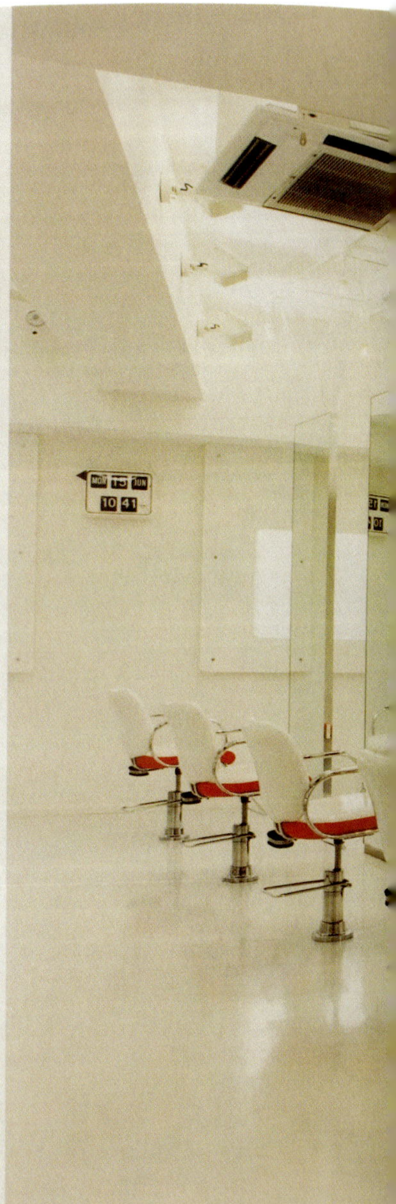

Interior designer Hideyuki Yamauchi used white for a 'clean' effect and red accents for a 'cutie' image

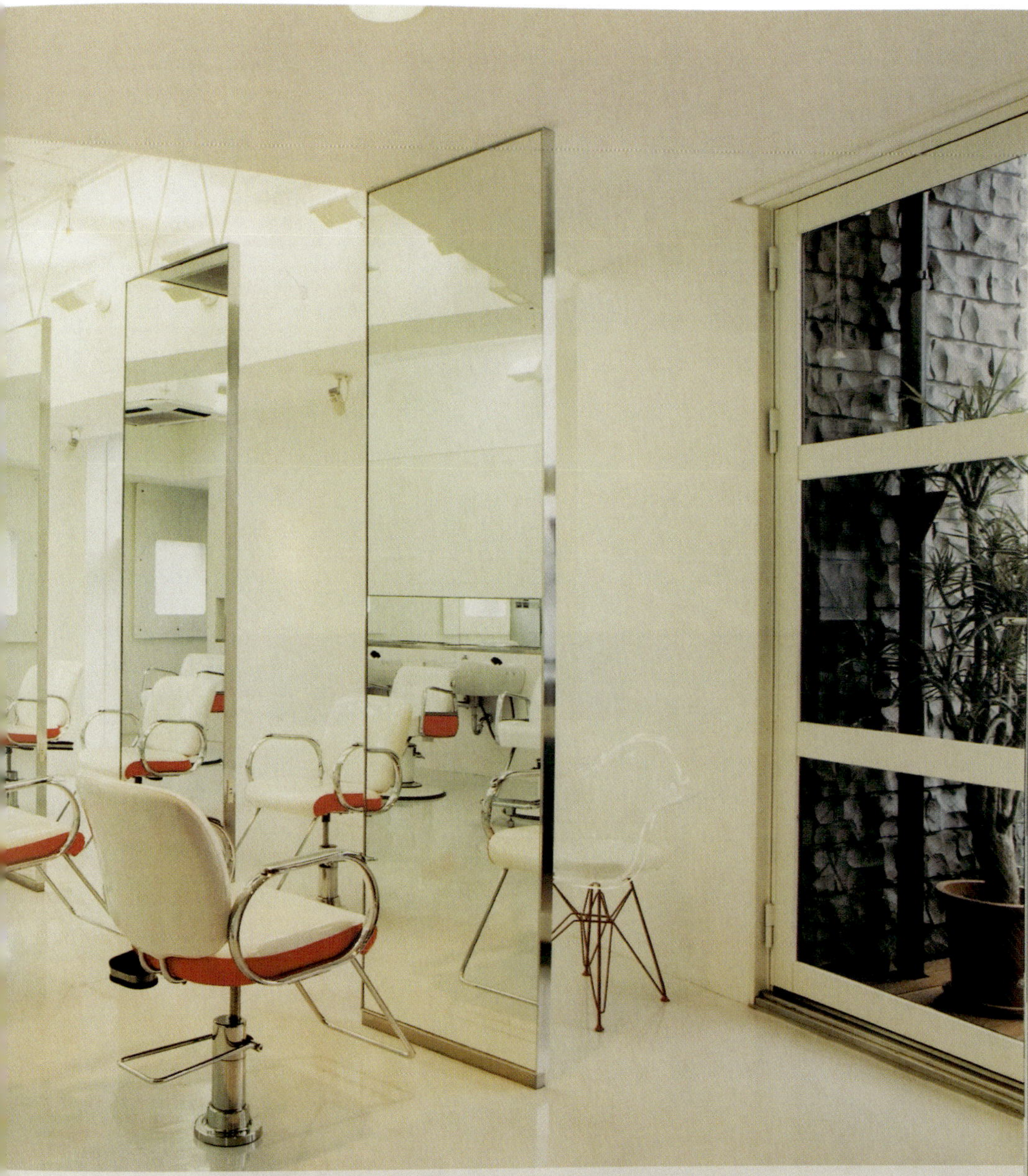

The mixture of red and white and the blend of old and new were part of a deliberate strategy to transcend the trendy look and create a space with a respectable life span

SUBJECT
Hair salons

LOCATIONS
London, England
Tokyo, Japan

DESIGNERS
Various, see page 147

PHOTOGRAPHERS
Various, see page 147

WRITERS
Chiharu Watabe
Chris Scott

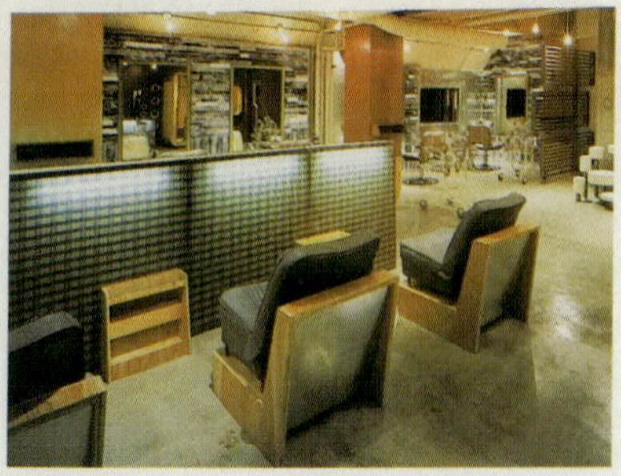

Sculptor-artist Jun Shirai created the salon's interior design. Bypassing white, he opted for a brown-based space. Photography: Kazuya Shimokawa/Nikkei Design

Arcadia, which recently opened its doors in Aoyama, is not easy to find. Clients must snake into a back street to find their destination on the first floor of a building nearly hidden in a secluded block. Owner and hairstylist Masafumi Matuura commissioned sculptor-artist Jun Shirai to create the salon's interior design. Bypassing white, Shirai opted for a brown-based space. Shampoo holders are the same bottle holders used in bars. Chairs that resemble the seats of a Mini Cooper and stools upholstered in artificial snakeskin give the salon a masculine ambience, despite its largely female clientele of chic young women.
'I didn't think about the salon's clients,' declares Shirai. 'A client's opinion of the space is irrelevant to me. I wanted to break out of the mould.' He says that the rough plywood panel on one wall causes people to ask if the room is still under construction: 'That's my intention. Matuura and I didn't talk much about details. Most of my clients appreciate the way I work and don't try to impose their ideas on me. I'm free to do almost anything I like. I love heavy stuff – making a white space would never even occur to me.' ◄

Arcadia, Tokyo

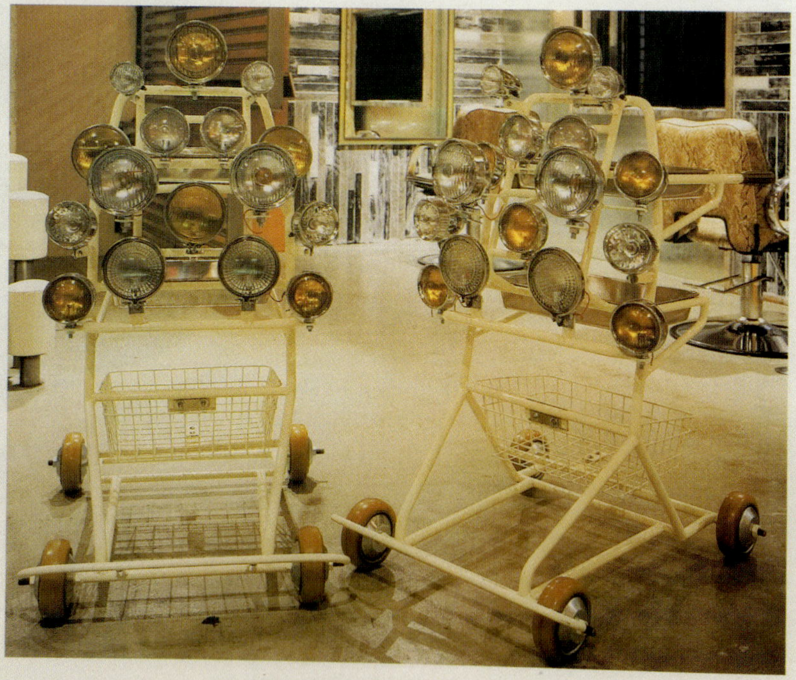

Left: Trolleys look like works of art

Opposite page: Chairs that resemble the seats of a Mini Cooper and stools upholstered in artificial snakeskin give the salon a masculine ambience. Shampoo holders are the same bottle holders used in bars

'A client's opinion of the space is irrelevant to me. I wanted to break out of the mould'

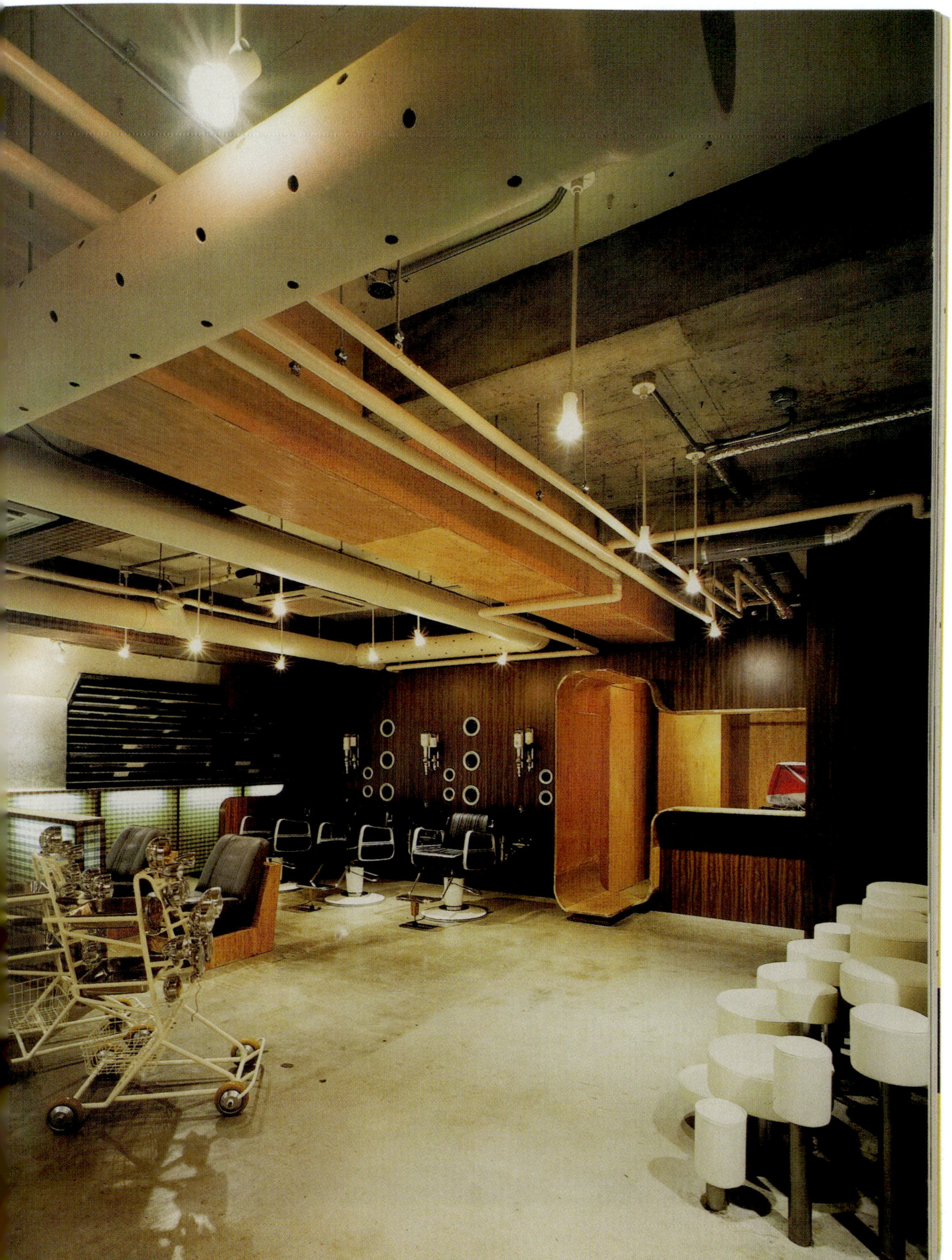

SUBJECT
Hair salons

LOCATIONS
London, England
Tokyo, Japan

DESIGNERS
Various, see page 147

PHOTOGRAPHERS
Various, see page 147

WRITERS
Chiharu Watabe
Chris Scott

September/October 2000

Issue 16

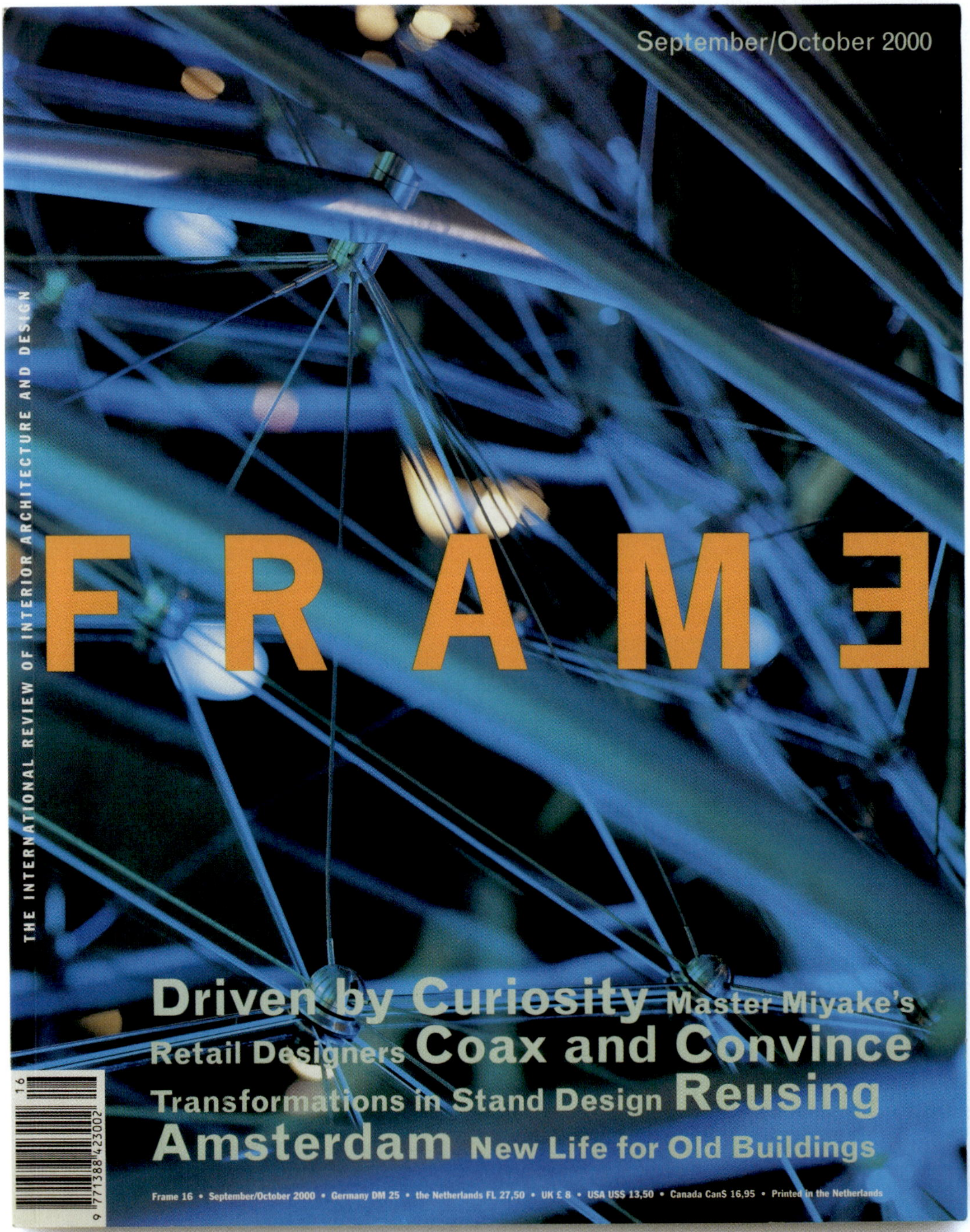

ART DIRECTOR
Ton Homburg,
Opera Designers

Commercial Tate

DESIGNERS
Herzog & de Meuron
www.herzogdemeuron.ch
Jasper Morrison
www.jaspermorrison.com
Lumsden Design
Partnership
www.ldp.co.uk

SUBJECT
National gallery
Tate Modern

LOCATION
London, England

PHOTOGRAPHER
Rolant Dafis

WRITER
Daniela Mecozzi

Designer Callum Lumsden about the project

This project has become our 'flagship'. No matter which country in the world I go to, most people seem to have heard of Tate Modern and those who have actually visited it have bought something in one of our shops. We still remain very proud of the design for this project.

Comments on the published article

The article was extremely well written, very accurate and covered every aspect of the project.

Did the article have an effect?

The Tate Modern project as well as the article really did put Lumsden Design Partnership on the map. It was the first piece of international recognition which my design agency received and we are very grateful to Frame for their attention!

Recent news or information about the project featured in *Frame*

The design still works extremely well, although the whole space is now in need of some refreshment. The best comment we have received about the Tate Modern bookshop was from Graham Crowley, the Professor of Painting at the Royal College of Art. He described it as 'the best modern art bookshop in the world' in an interview on the BBC. This was music to my ears as that was our design brief from Sir Nicholas Serota for the project!

Current projects

A new concept for a chain of mobile phone stores in Saudi Arabia // The Science Museum Store, London, England // The London Transport Museum store and restaurant // A chain of fruit and juice bars to be rolled out throughout Europe for Chiquita // A showroom for fashion designer Peter Werth.

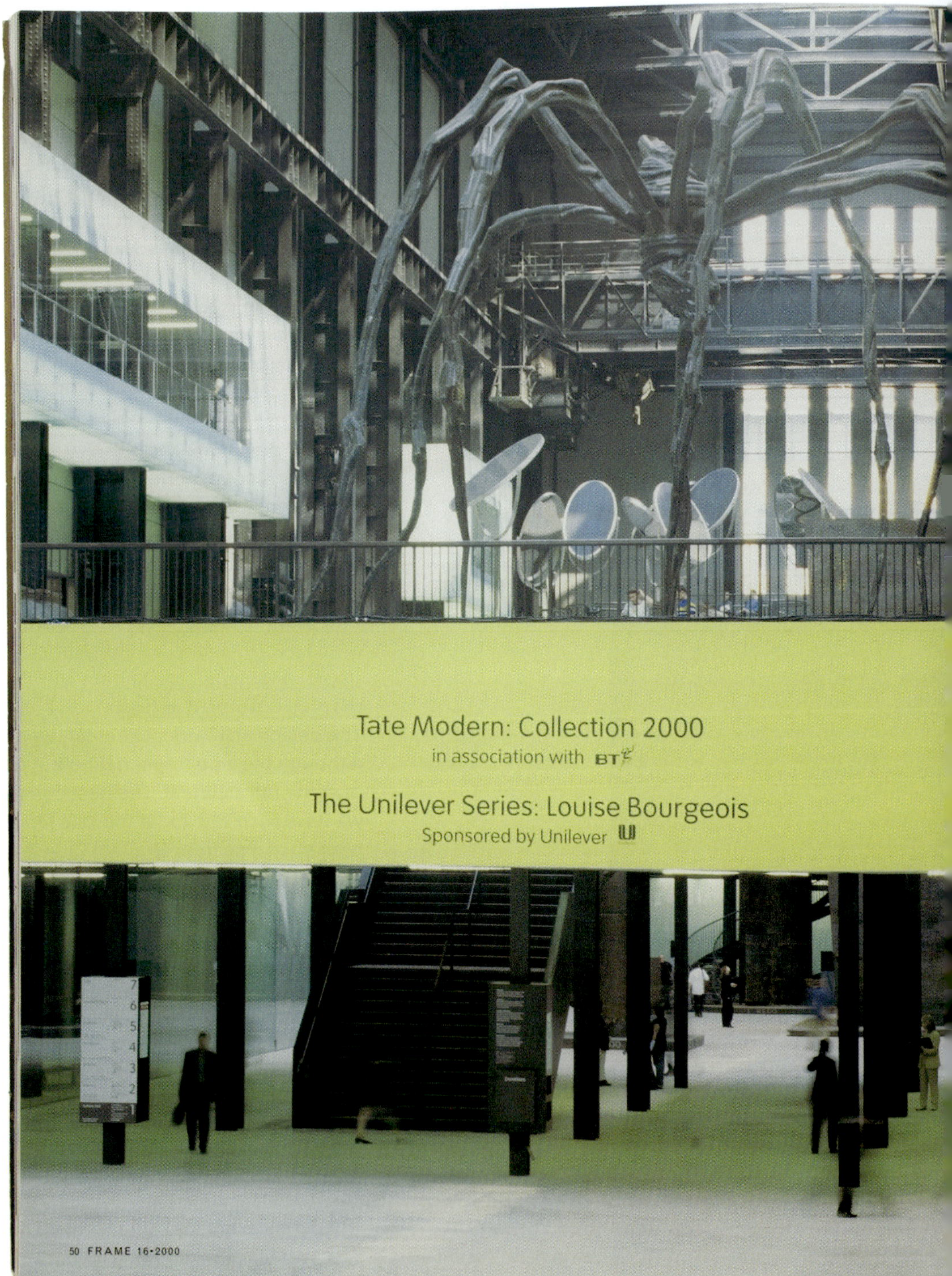
Tate Modern: Collection 2000
in association with BT
The Unilever Series: Louise Bourgeois
Sponsored by Unilever
50 FRAME 16•2000

Opposite page: The entrance is a feature of the enormous turbine hall, at one end of which two towering sculptures by Louise Bourgois determine the scene

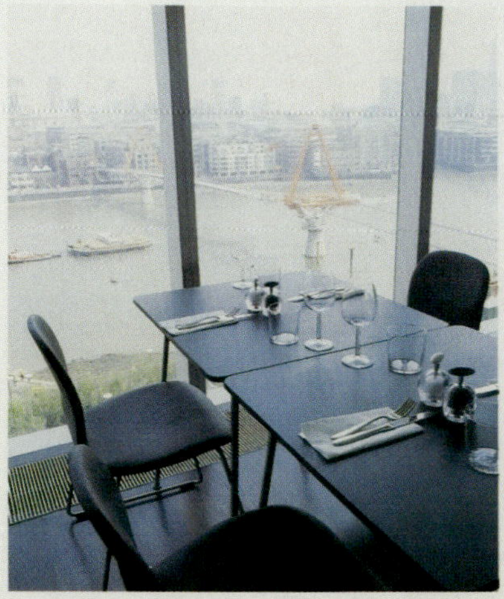

Commercial Tate

First and foremost, of course, the new Tate Modern is a museum. But it's one that requires, at least in part, a degree of self-sufficiency. Under Herzog & de Meuron's strict supervision, Jasper Morrison and the Dennys Lumsden Partnership designed a number of remarkably unremarkable commercial spaces. ***By Daniela Mecozzi. Photography by Rolant Dafis.***

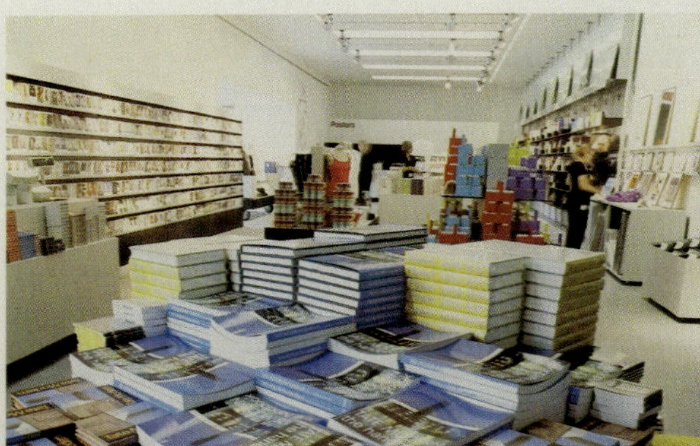

Top: An imposing view of London dominates the design of the restaurant on the seventh floor

Right: The second-floor bookshop, designed by the Dennys Lumsden Partnership, gives centre stage to the Tate Modern catalogue

SUBJECT National gallery Tate Modern

LOCATION London, Enlgand

DESIGNERS Various, see page 159

PHOTOGRAPHER Rolant Dafis

WRITER Daniela Mecozzi

'Architecture that serves art': with these words Nicholas Serota, director of the Tate galleries, introduced the initial stage of Herzog & de Meuron's refurbishment of the Tate Modern, located on London's South Bank. The design of the first large-scale museum of the new millennium benefits much more than art, however. Perhaps a more insightful introduction would have honoured 'architecture that serves the city and the public'. Housed in the smaller of two massive power stations designed by Sir Giles Gilbert Scott, the Tate Modern boasts a collection of international, 20th-century art previously in the possession of the old Tate. The latter, renamed Tate Britain, is finally free to resume the role for which it was established in 1897: the museum of British art from 1500 to the present day.

The Bankside power station was designed in 1947, built in 1963 and decommissioned in 1981. For decades the austere edifice facing St Paul's Cathedral and the corporate buildings of the City of London has formed the architectural, economic and ideological counterbalance to development on the opposite bank of the Thames. Today, as the Tate Modern, it represents the last step in the regeneration process targeting this area. Together with the recently completed Shakespearean Globe Theatre and the London Eye, the new gallery is part of a cultural walk that starts at the South Bank complex and continues to the Designmuseum. The project also has repercussions, in the form of fresh opportunities, for the working-class people who have traditionally lived on this side of the river. The old polarity between the two shores has evolved into a dialectical situation, symbolised by the construction of a pedestrian bridge designed by Sir Norman Foster and Anthony Caro.

Even surrounded by gardens, the Tate Modern retains its severe character, a quality left untouched by Herzog & de Meuron. The architects' understanding of the meaning and role of the building, an industrial landmark in the centre of the city, was the ground on which they won the competition to design this museum in 1995. On the outside, the architects avoided any attempt to disguise the building's former function or eradicate the signs of age. They preserved the soaring chimney and left the original brickwork intact. Their only addition is a dual-level glass structure that runs the length of the roof and provides the upper galleries with an extra source of light. This sculptural element is a welcome contrast to the massive structure below.

Internally, the museum is divided into two distinct but interconnecting areas that run parallel to each other: the turbine hall, which serves as the access to the museum, and the seven-storey volume that accommodates galleries, transport systems and services. The impressive, dark-grey entrance hall, 155 metres long and 35 metres high, is illuminated by skylights overhead and cathedral-like windows at both ends. In the democratic tradition of the South Bank, the hall is a public space that receives both visitor and passer-by with equal cordiality. Currently the long ramp of the turbine hall is dominated by the first art installations specially commissioned for this site. Louise Bourgeois' *Maman*, an enormous steel spider, stands over the bridge connecting the hall to the galleries. Beyond it, three towers by Bourgeois – *I Do, I Undo and I Re-Do* – feature gigantic swivel mirrors that reflect encounters between observers and the architecture around them. These large installations establish the vanishing point of the hall, and cantilevered light-box windows indicate the museum's different levels. Both elements offer points of reference with which to read this otherwise overwhelming space.

Illuminated bay windows link the turbine hall to the galleries. Along with circulation areas at each level and a massive stainless-steel escalator, these windows serve as spaces that encourage social, visual and aesthetic interaction. In contrast to all this activity, the atmosphere of galleries at the third, fourth and fifth levels is one of calm and contemplation. To avoid the monotony of the rows of rooms found in traditional museums, Herzog & de Meuron designed an itinerary characterised by rooms of various sizes and with differing wall heights. Mood and ambience rely mainly on light: both daylight and artificial sources in the form of slim, flush-fitting light boxes designed by the architects in collaboration with Ove Arup. Stripped of any detail deemed inappropriate, such as skirtings, and with dark oak doors flush to the walls and unsanded oak flooring, these rooms provide an ideal setting for the display and appreciation of art.

Diversity, which makes each room a unique experience, is combined with a curatorial approach that gives a thematic presentation of art precedence over a traditional chronological display. The result is a juxtaposition of works from different periods. The interpretation of art as a provider of experience rather than as a set of acquired notions has dictated the layout of the art, which is divided into four categories: Landscape/ Matter/ Environment, Still Life/ Object/ Real Life, History/ Memory/ Society, and Nude/ Action/ Body. Although such an approach has previously been adopted for temporary exhibitions, this is the first time that a permanent collection has been presented following this criterion. The aim is to remove people's preconceptions or expectations and to allow them to experience art in new contexts. Then, too, by breaking with tradition and providing a dynamic display, the museum hopes to broaden its audience.

People and their needs seem to be of paramount concern at the Tate Modern. The presence – at different levels of the museum – of three bookshops, a cafeteria, a restaurant, a café and a members' room reflects this concern. In the overgenerous provision of such environments, however, the Tate Modern's emphasis on accessibility reveals, more than a social or cultural agenda, a highly commercial approach to the management of this space. Good examples of the sales-based strategy are a new Tate logo by Wolffs Olins, the creation of a Tate brand and, should one want to avoid the museum experience completely, the opening of a Tate bookshop in Selfridges department store.

The design of the commercial and restaurant areas is the result of Herzog & de Meuron's collaboration with the Dennys Lumsden Partnership, a firm specialising in retail design, and Jasper Morrison, brought in to help with the design and selection of free-standing furniture. Dennys Lumsden's brief for the largest of the three bookshops, a 700-square-metre space on the first level of the gallery, was to design 'the best art bookshop in the world'. He points out that 'the main architectural challenge of this space was its position off the main entrance . . . the whole of this space opens up gradually as you walk down.' Although Lumsden was eager to take advantage of the situation, Herzog & de Meuron imposed a number of restrictions on both the layout and the design of the furniture, which prohibited the insertion of any visual element that might affect the adjacent ramp, a major feature of the turbine hall.

A large glass door off the main entrance to the gallery floors leads visitors into this vast space, where 10,000 books and magazines ▶

White calico on the ceiling of the seventh-floor restaurant acts as a light diffuser. Morrison's Tate tables and chairs contribute to the welcoming ambience

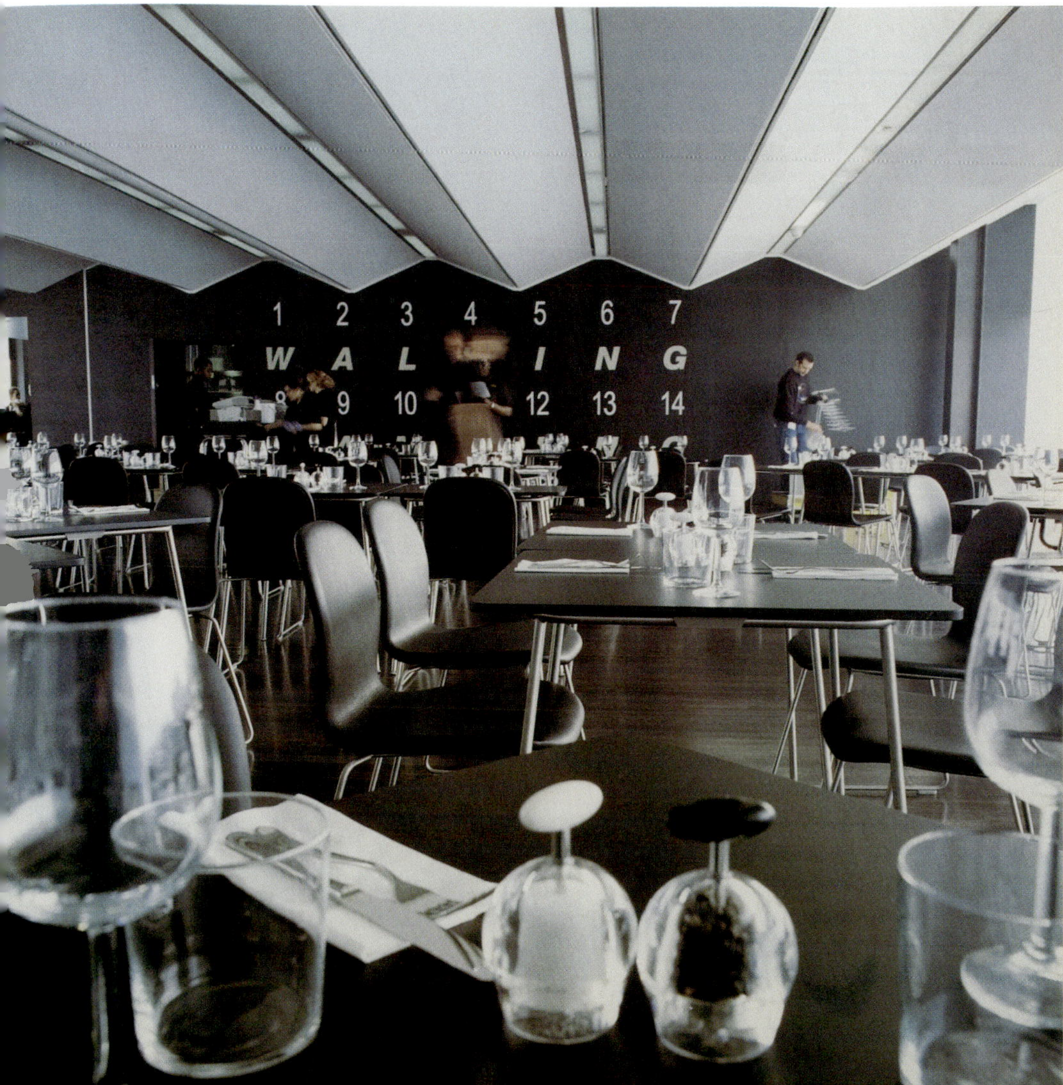

The ability to create different, site-specific moods by introducing minor changes into the same elements testifies to the success of the collaboration between the architects and Jasper Morrison

SUBJECT
National gallery
Tate Modern

LOCATION
London, Enlgand

DESIGNERS
Various, see page 159

PHOTOGRAPHER
Rolant Dafis

WRITER
Daniela Mecozzi

View from the east room on the seventh floor

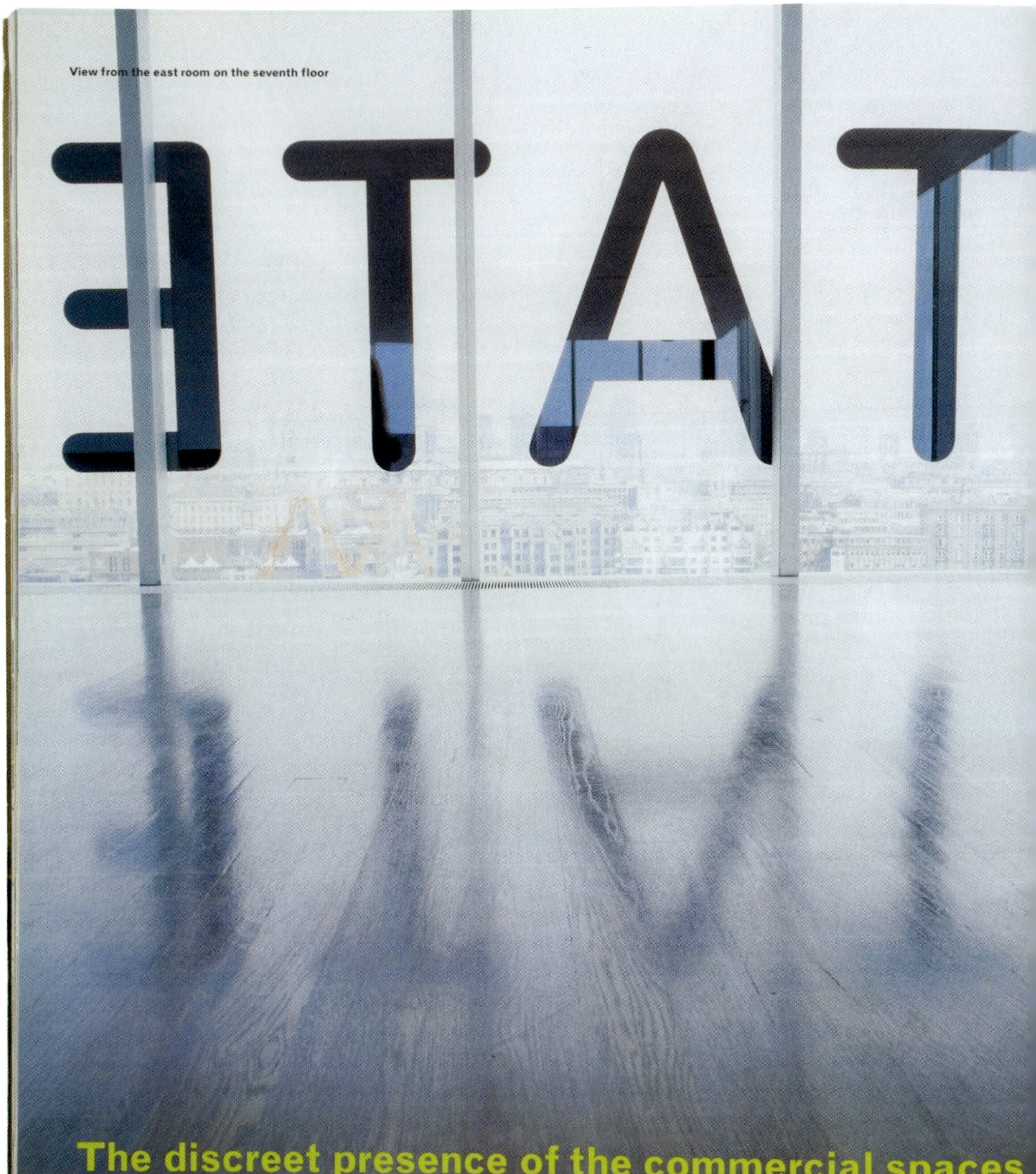

The discreet presence of the commercial spaces creates the freedom needed to experience and enjoy the art and architecture

From top to bottom: the cafeteria on the second level opens directly onto the gardens; the clean lines of Morrison's designs join with rows of hanging Eclipse lamps by DK1; heavy benches are complemented by black-topped tables and lightweight Tate chairs

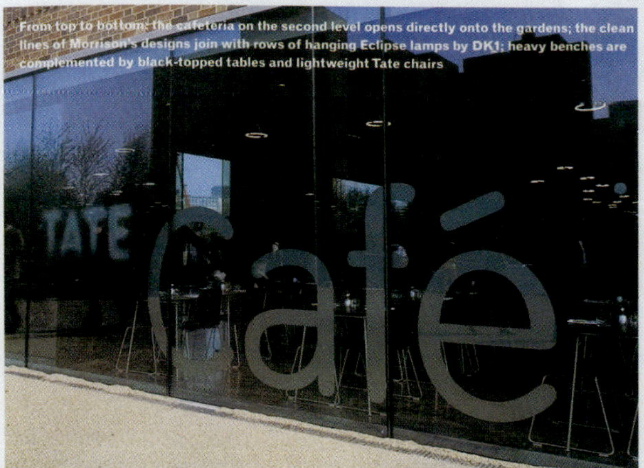

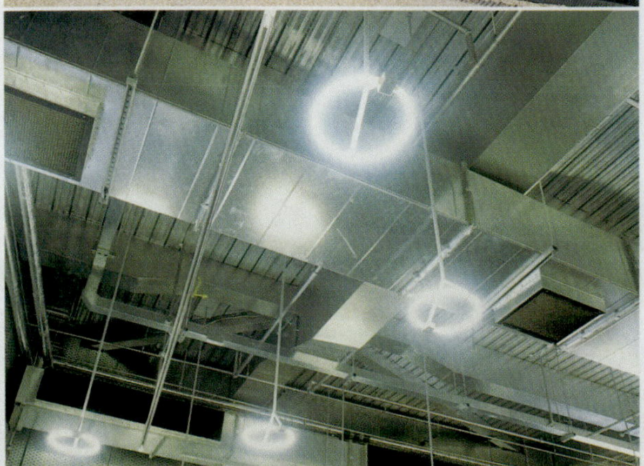

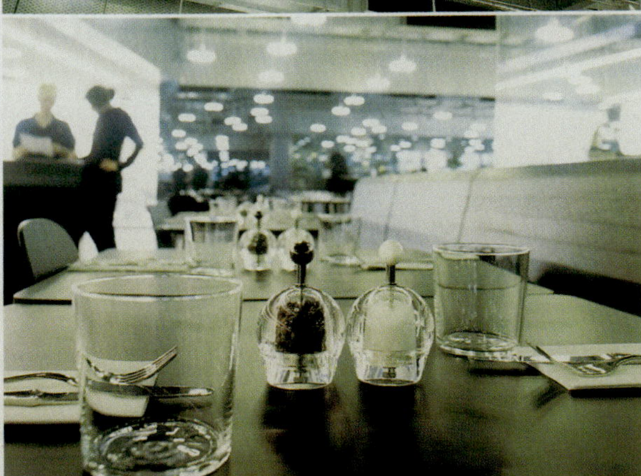

fill a Vitsoe bookshelf that spans the entire length of the main wall. A light box above this shelf accommodates temporary works of art. A track of oak trolleys running parallel to the long bookshelf holds additional books and separates this space from the gift area. Trolleys in the latter support display cases containing items such as jewellery, ornaments, T-shirts, framed prints and furniture. Positioned both at right angles and parallel to the ramp, the mobile display cases never rise above the ramp wall, thus ensuring the visual clarity of the line formed by this element. To further maintain architectonic continuity in the bookshop area, oak drawers for storing prints are part of the ramp wall, while tall revolving stands for art videos replicate the lines of the broad steel columns next to them.

The use of materials – unsanded oak flooring, custom-made oak furniture and metal surfaces of shot-blasted aluminium – creates the relationship between this space and other parts of the gallery. Despite the array of items for sale, the result is an open and ordered space in which the crisp and contemporary shapes of the furniture counterbalance the raw quality of iron columns and a ceiling with ventilation and electric ducts in full view.

The same restraint and sobriety applies to the other two shops, each of which is 200 metres square. The shop at the second level has the advantage of a large window that opens onto the gardens and Foster's pedestrian bridge. Here, too, a glass entrance has been positioned off the main route in line with project architect Harry Gugger's view that 'you don't want to be forced into a bookshop; you choose to enter it.' Again, materials used in this retail space refer to those in adjacent areas: dark-stained oak furniture and a polished concrete floor reflect the look of the large staircase, as well as that of bar counters and benches. The third shop, dedicated to temporary exhibitions, is completely panelled in dark-stained oak. Glass openings allow browsers to see the narrow espresso bar next door, designed to complement the shop. Although Lumsden is pleased with the end result, he acknowledges that the architects' restrained strategy, coupled with their lack of interest in the problems that accompany the design of an effective commercial outlet, led to a 'difficult and challenging' relationship.

As restraint and sobriety are Jasper Morrison's trademarks, it's unsurprising to hear that his association with the architects – the first time he ever worked on this type of interdisciplinary project – was a 'very nice collaboration'. He adds that it 'worked very well, because there was no competition really'. In addition to selecting furniture for the gallery, Morrison designed the large digital clocks and the donation box. He also acted as a consultant for the design and furnishing of the cafeteria, restaurant and members' room.

Located on the second level, the cafeteria, which accommodates 220 people, opens directly onto the gardens and thus provides an independent entrance to the gallery. The architects replaced much of the wall area with windows, leaving only the structural pilasters. Painted black and wrapped in glass, these provide highly reflective surfaces. As a consequence, the perimeter of the café is an alternating surface of mirrored images and open views of the gardens and the river. The principles of reflection, which multiplies visual stimuli, and serialisation, which reiterates identical elements, combine to create a rich breeding ground for accidental encounters as well as an environment conducive to privacy. Inside, the sturdy appearance of a dark-stained counter of slatted ▶

SUBJECT National gallery Tate Modern

LOCATION London, Enlgand

DESIGNERS Various, see page 159

PHOTOGRAPHER Rolant Dafis

WRITER Daniela Mecozzi

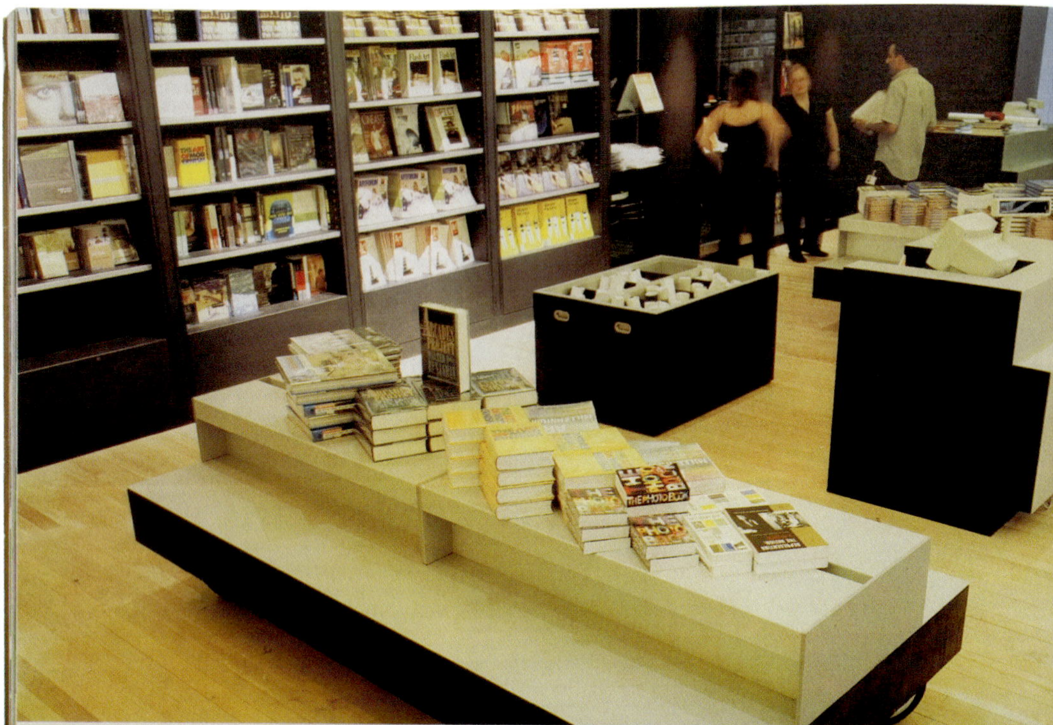

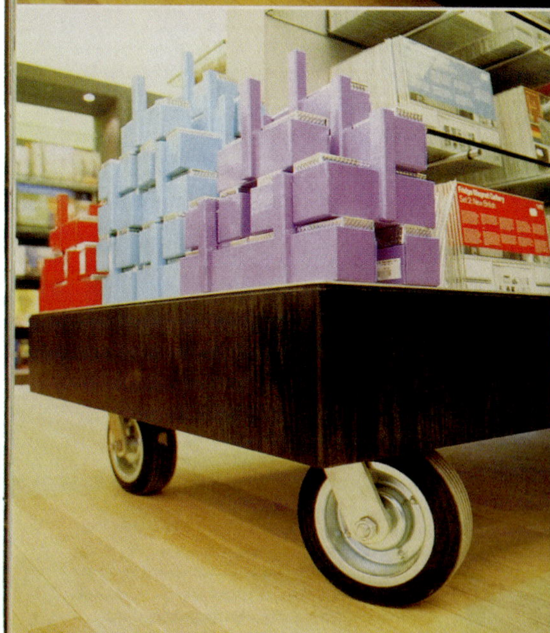

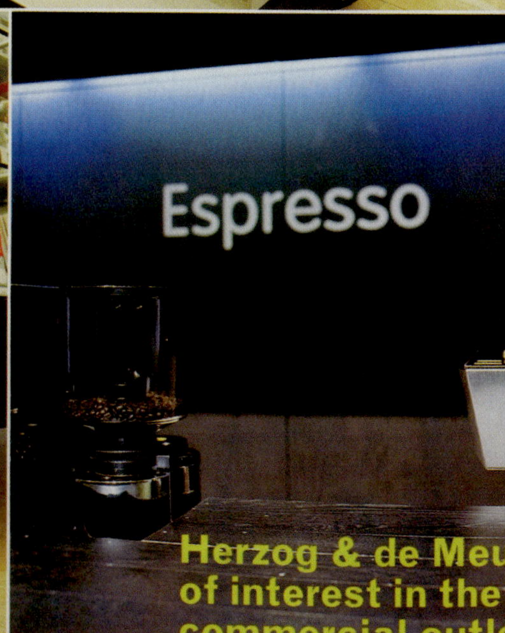

Herzog & de Meu
of interest in the
commercial outle

Opposite page: The fourth-floor bookshop, designed by the Dennys Lumsden Partnership, is completely panelled in dark-stained oak. Glass openings allow browsers to see the narrow espresso bar next door, designed to complement the shop

oak is balanced by glass partitions that provide a view of the kitchen area. Similarly, heavy benches are complemented by black-topped tables and lightweight Tate chairs of black plywood and stainless steel, all designed by Morrison – as individual units that can be rearranged to suit groups of different sizes – and manufactured by Cappellini. The clean lines of Morrison's designs join with rows of round, hanging Eclipse lamps by DK1 and kitchen walls tiled in a repetitive cookery-related pattern to convey a communal atmosphere of informality.

An imposing view of London dominates the design of the restaurant on the seventh level, where a long counter backed by the glass façade invites customers to sit and enjoy the panorama. The restaurant area, which seats 200 people, has a double view – the West End in one direction and Southwark in the other. The white calico on the ceiling, which is also used for the blinds, acts as a light diffuser to create a bright environment. Morrison's Tate tables and chairs, now upholstered in black leather, contribute to the welcoming ambience.

Another space that functions as an outlook across the city is the members' room. Lining the walls, Morrison's large brown leather settees, manufactured by Elam, give visitors the opportunity to relax in comfort while contemplating the world outside. Despite the size of the room and the white walls, the seating elements combine with Morrison's Alessi tray tables and Rondo floor lamps to give the space a domestic, intimate feeling. The ability to create different, site-specific moods by introducing minor changes into the same elements testifies to the success of this collaboration. The substance and harmony of the areas described indicate the amount of thinking and care that went into the planning of these spaces. At every stage and even under pressure, Herzog & de Meuron take time to concentrate on the smallest detail. Their ethical approach to design is shared by Morrison, whose final comment on the project expresses his respect and admiration: 'I think they are exceptionally brave when it comes to experimenting. They really push to get things done in a new and different way.'

The Tate Modern is proof of the success of this daring and uncompromising architectural strategy. A constant source of surprise, the building engages the eye and the mind with its interior and exterior views, its promenades and its tasteful presentation of 20th- and 21st-century art. In a conservative country like Britain, the realisation of a leading museum of contemporary art is a great achievement. Moreover, this genuinely public institution is determined to maintain a free access policy. Although it can count on partial government funding, the Tate Modern is relying on a commercial approach to fill the coffers, as well as to respond to the public demand for services. Based on principles of simplicity and restraint, Herzog & de Meuron's exacting design of the gallery extends to all the museum's commercial outlets, each of which has its own unique atmosphere. Their discreet presence, which incorporates none of the commercial pressure found in urban shopping areas, creates the freedom needed to experience and enjoy the art and architecture of this place. ◄

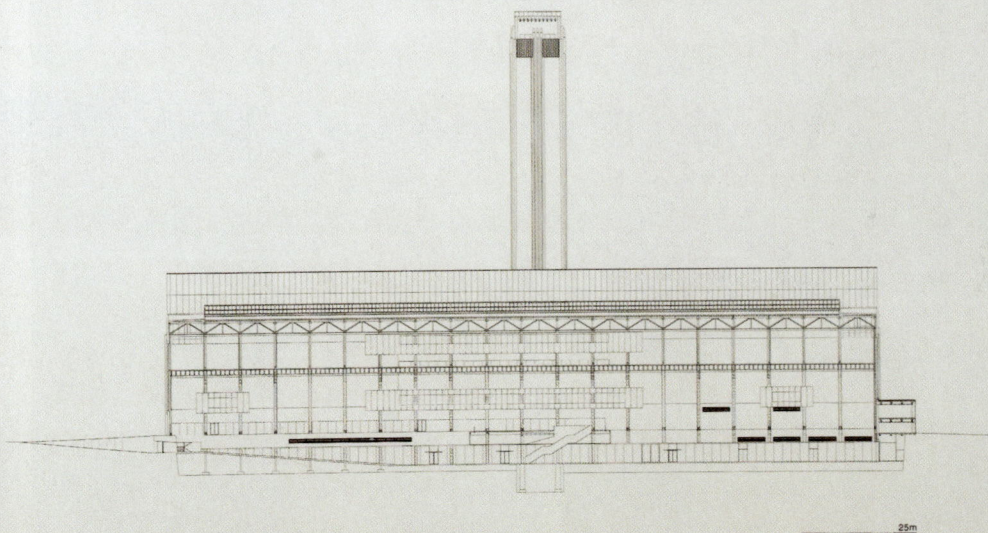

Section of the Tate Modern

restrained approach to architecture and their lack
ems that accompany the design of an effective
to a difficult relationship with the retail designers

SUBJECT National gallery Tate Modern

LOCATION London, Enlgand

DESIGNERS Various, see page 159

PHOTOGRAPHER Rolant Dafis

WRITER Daniela Mecozzi

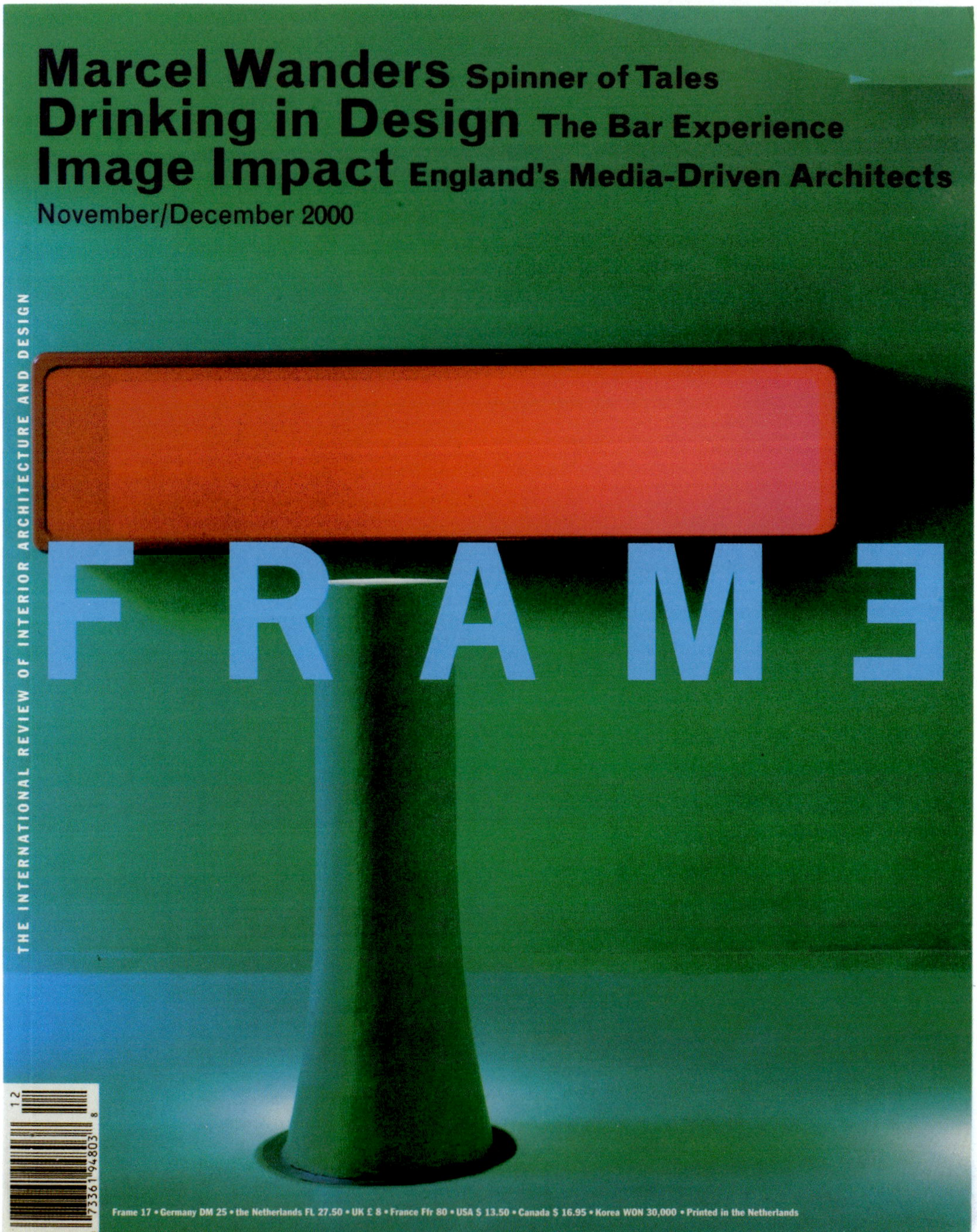

ART DIRECTOR
Ton Homburg,
Opera Designers

17

Chill Out

Designer Michael Young about the project
Surreal, but to really understand it you had have to been there, it was like being inside a dream.

Comment on the published article
One of the best.

Did the article have an effect?
Yes, but by accident it was on the front covers of Domus and Monitor, too, so it is hard to pinpoint. Anyway, I got asked to do 100 interiors after that and I declined them all.

Recent news
My new office is open in Hong Kong.

Current projects
An interior in Florence, Italy // A bike for Giant // An MP3 player for Kuro.

DESIGNER
Michael Young
www.michael-young.com

SUBJECT
Astro Bar nightclub

LOCATION
Reykjavik, Iceland

PHOTOGRAPHER
Ari Magg
www.arimagg.com

WRITER
Robert Thiemann

Chill Out

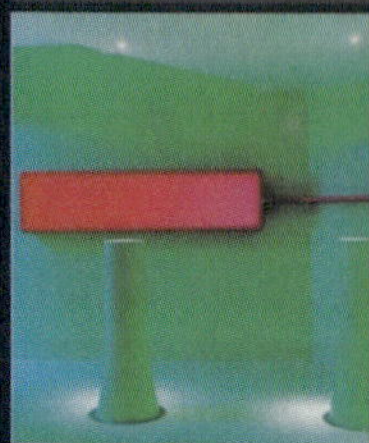

The surreal club that Michael Young created in Reykjavik looks like a cross between a swimming pool and a picnic ground. *By Robert Thiemann. Photography by Ari Magg.*

'I'd never felt more prepared for a project,' says Michael Young. 'And I couldn't have had a better client – youthful and enthusiastic, with an understanding of the younger generation.' He's talking about a client in Reykjavik, Iceland, who recently asked him to convert one of the oldest buildings in the city into a club with four bars and two dance floors.

Young (1966) moved to the island republic in 1998, more or less as an act of self-defence. Previously, the Englishman had lived and worked in London, where a 'very sociable job', as he describes it, got him 'embroiled in the many other activities that one encounters in a big city'. Iceland has proved to have fewer distractions. But living on the periphery hasn't stopped him from working for clients like Cappellini, Magis and Sawaya & Moroni (three Italian-based firms), Idée (Japan) and Telia Info Media (Sweden). Ten days of globe-hopping every five or six weeks allow him to maintain his contacts. Thanks to the Internet, colleagues in other parts of the world are able to develop the sketches he makes during his travels. 'It works well,' he says, 'though at times I do miss the human interaction.'

Young wanted the Astro to be a 'fun house'. He considers it a joke that parts of the 90-year-old building are protected by Iceland's listed-building legislation. 'It's not that special,' he claims. 'There wasn't a straight line anywhere.' His first task, therefore, was to get the interior in working condition by 'sorting out the geometry'.

The designer's point of departure was to introduce a slight touch of Iceland's chill, windy, but breathtakingly pure landscape into the club. To implement his plan, he hired a local firm that specialises in building geothermal environments. Young is a big fan of the building skills he's found in this country, as well as of the attitude of the construction workers: 'They're extremely good, especially at concrete and steel work, which are used a lot in swimming pools and geothermal areas in Iceland. Unlike your average British builder, who seems to work only in 30-minute shifts, they have a genuinely enthusiastic nature.' The interior does indeed evoke an image that's part swimming pool, part glacier and part picnic ground. A smooth skin gives every surface a seemingly slippery sheen, and the ground floor is immersed in ice-cool colours. Bars and shelves have been finished in Corian laminate, which reinforces the club's surreal atmosphere. In Young's words, 'It's almost like stepping into the virtual.'

Small clusters of furniture and lamps interrupt this austere ambience. These mini-picnic sites offer a brief historical summary of Young's oeuvre. The club features all his furniture designs for Cappellini and Sawaya & Moroni, plus a number of new models. The Stick lamp that he created for Eurolounge is here as well, not to mention earlier work such as the Woven Steel lamps and Smarty seat cushions. The latter add a comfortable layer to the club's waterless Smarty Pool. In explaining his strategy, Young says, 'I've treated the building more as a reflection of my approach to design than as a single piece of architecture.'

The mood upstairs is warmer and more laid-back. Here an ambient floor by Jeremy Lord shares the space with a private area for members only: the Red Room. Thermo-formed walls contain a lighting system that reacts to movement; a minimum of physical activity elicits a pale pink light, which deepens to a vivid crimson as the dancing comes to a boil.

Young's clever division of the club into zones with different atmospheres makes Astro by far the most contemporary place of entertainment on the island (this venue would even turn heads in London). It's this very exclusivity, however, that poses a problem. As Young points out, Iceland's population (270,000 in 1996) is probably too small to attract a weekly crowd of party people properly appreciative of what the club has to offer. If bankruptcy does eventually bring Astro to its knees, however, let's hope that the club's ultra-cool interior is added to Iceland's inventory of listed buildings. ◀

Opposite page: Detail of the window covering and Young's Stick light

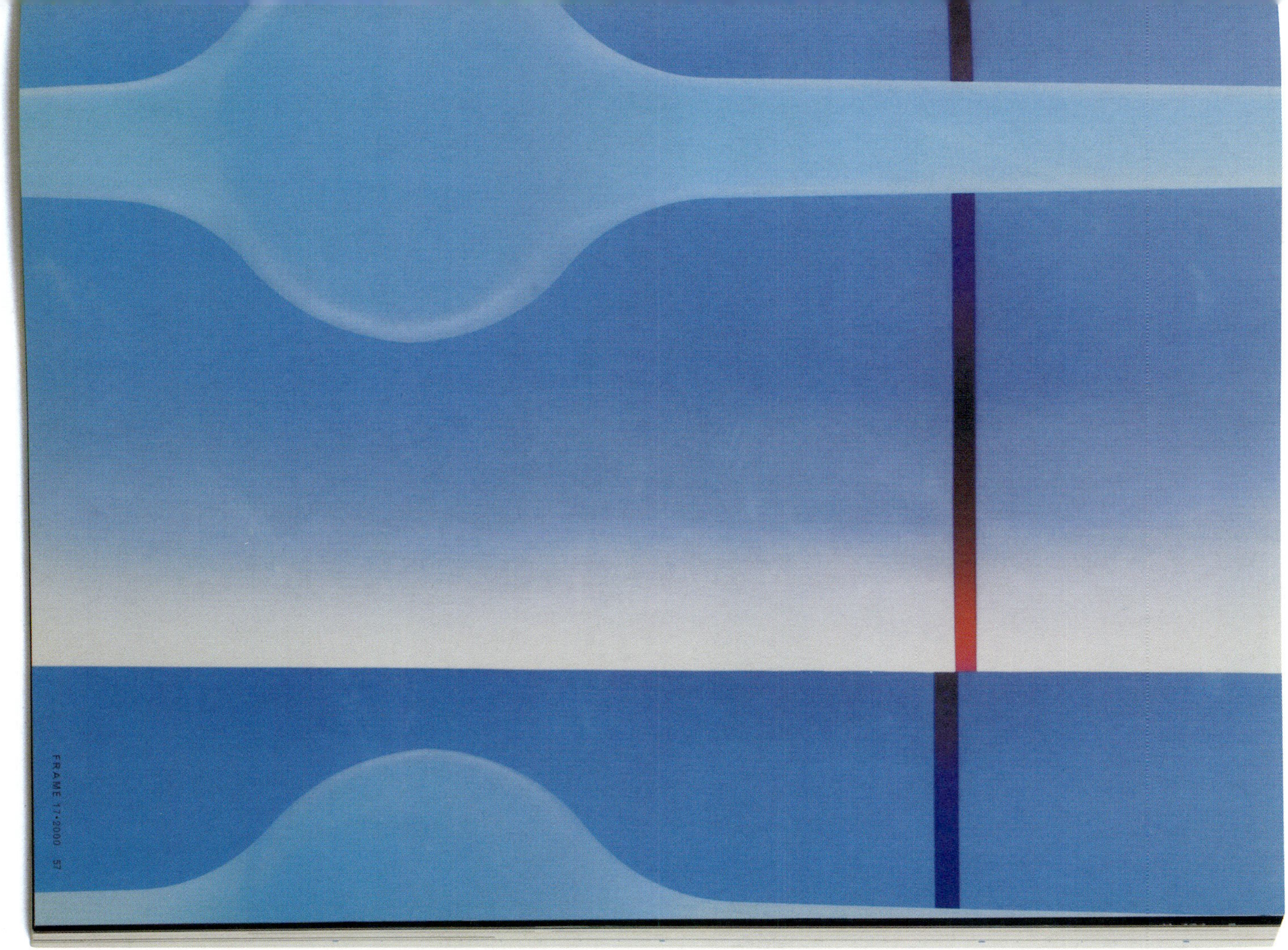

SUBJECT
Astro Bar nightclub

LOCATION
Reykjavik, Iceland

DESIGNER
Michael Young

PHOTOGRAPHER
Ari Magg

WRITER
Robert Thiemann

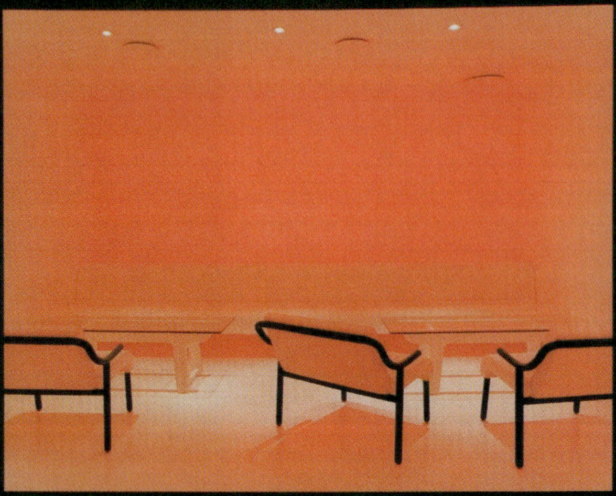

Below, bottom and opposite page: The ground floor is immers in ice-cool colours, and a smooth skin gives every surface a seemingly slippery sheen. Small clusters of furniture and lamps interrupt this austere ambience. Smarty seat cushions add a comfortable layer to the club's waterless, jacuzzi-like Smarty Pool

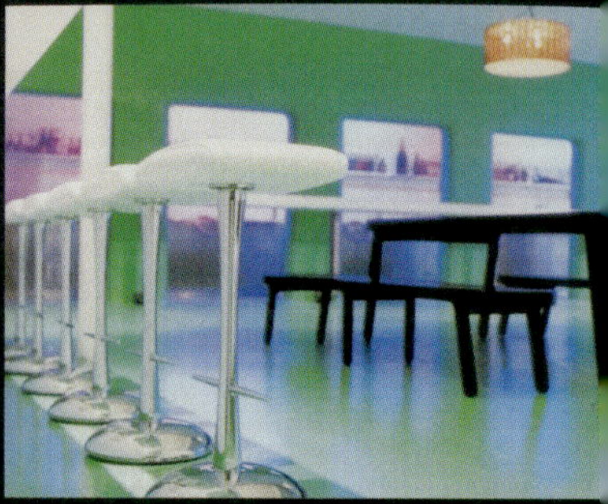

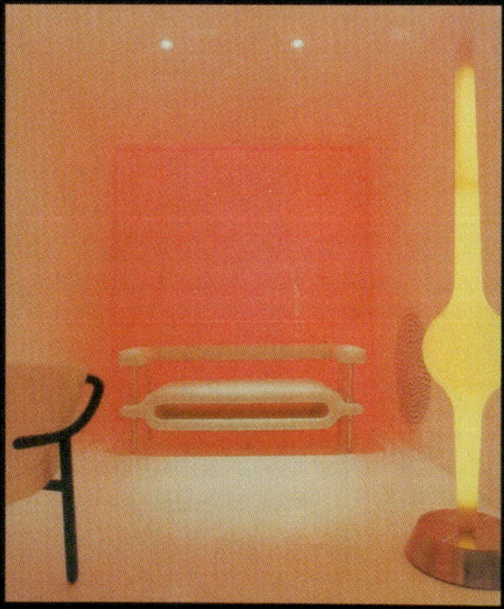

Mini-picnic sites offer a brief historical summary of Young's oeuvre

Top and above: The Red Room is a private area for members only. Thermo-formed walls contain a lighting system that reacts to movement; a minimum of physical activity elicits a pale pink light, which deepens to a vivid crimson as activities come to a boil

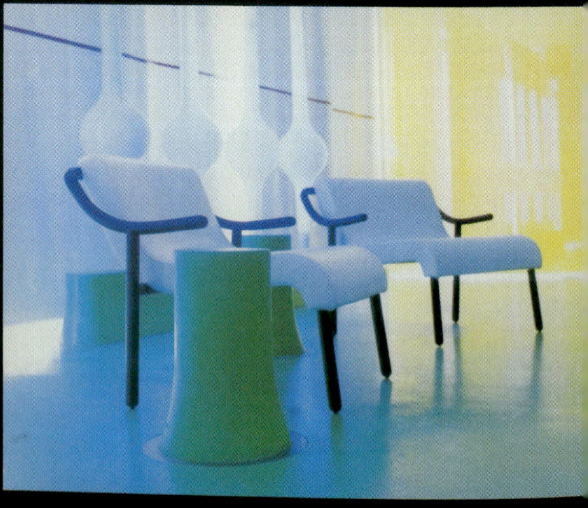

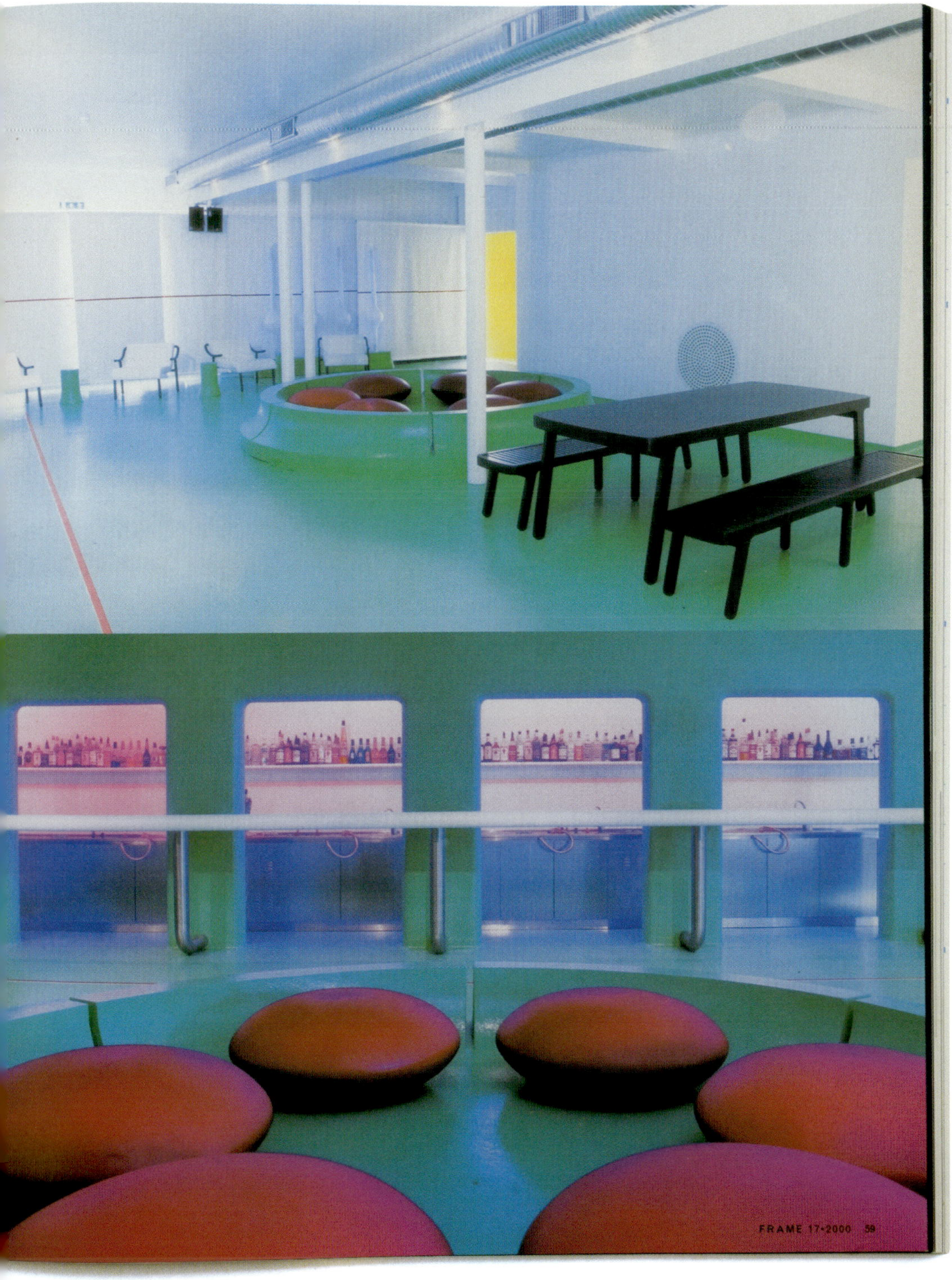

SUBJECT
Astro Bar nightclub

LOCATION
Reykjavik, Iceland

DESIGNER
Michael Young

PHOTOGRAPHER
Ari Magg

WRITER
Robert Thiemann

Issue 17 Chill Out 174-175

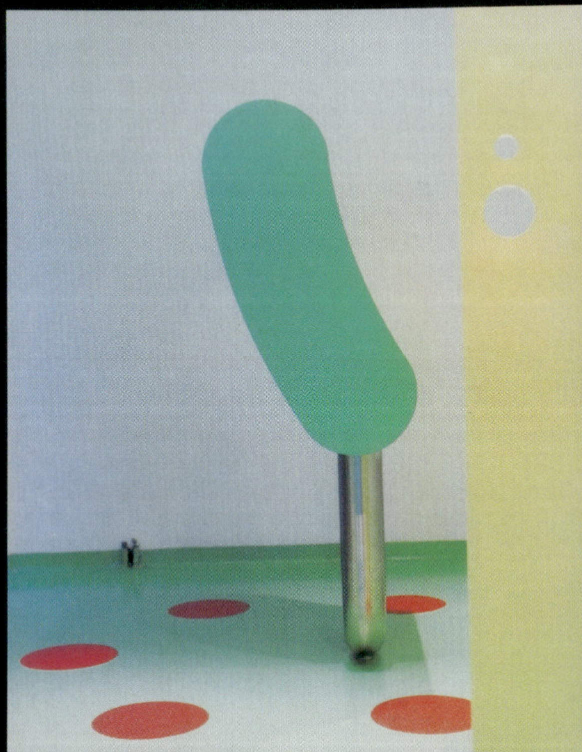

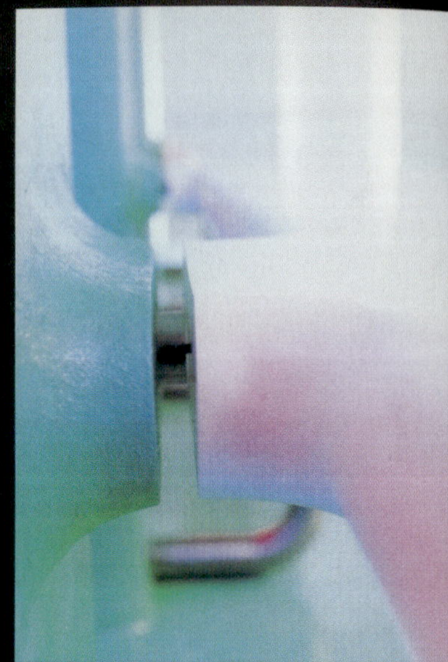

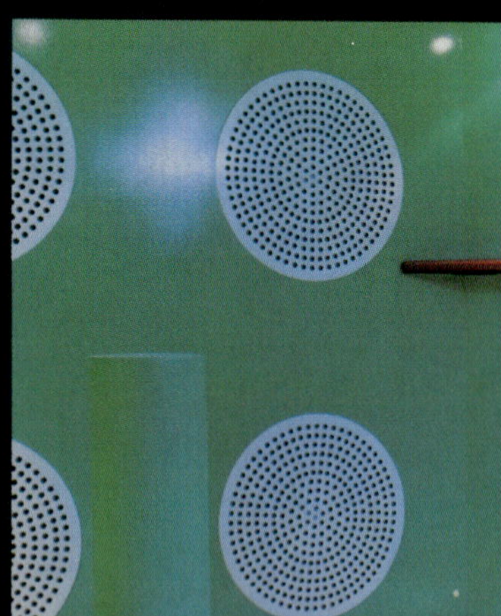

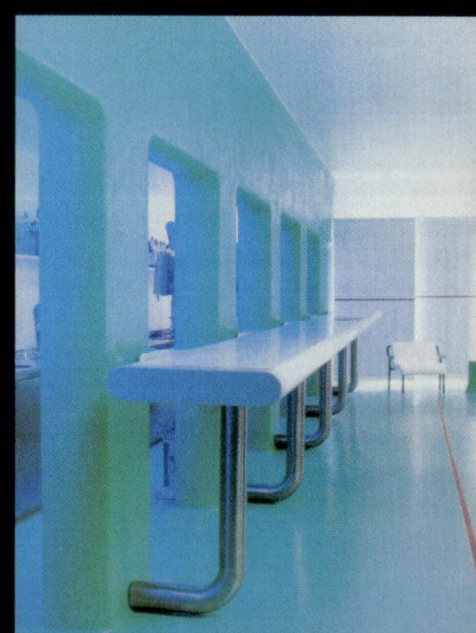

Clockwise from top left: The kiosk at the entrance; bars and shelves are finished in Corian laminate, which reinforces the surreal atmosphere; loudspeakers are integrated into the walls

Michael Young's point of departure was to introduce a slight touch of Iceland's chill, windy, but breathtakingly pure landscape into the club

back

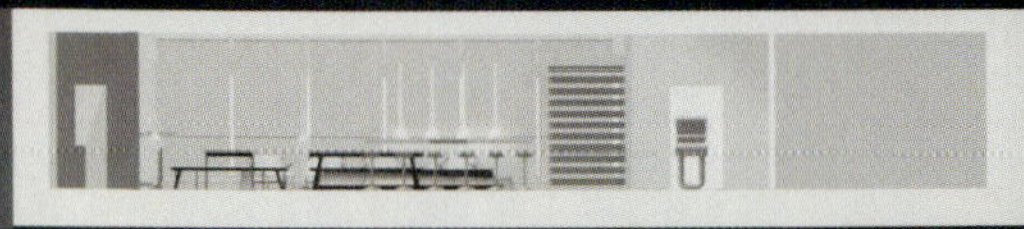

front

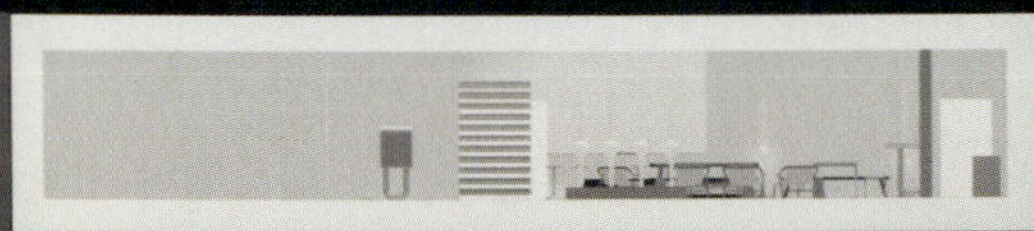

left

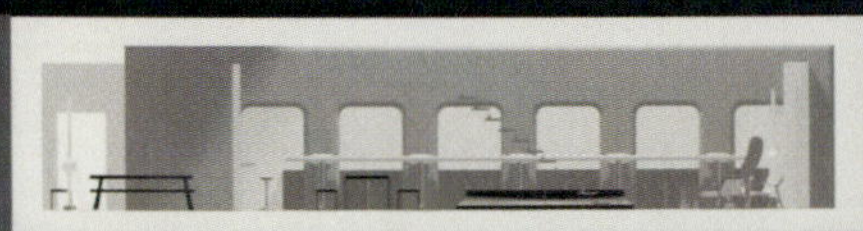

right

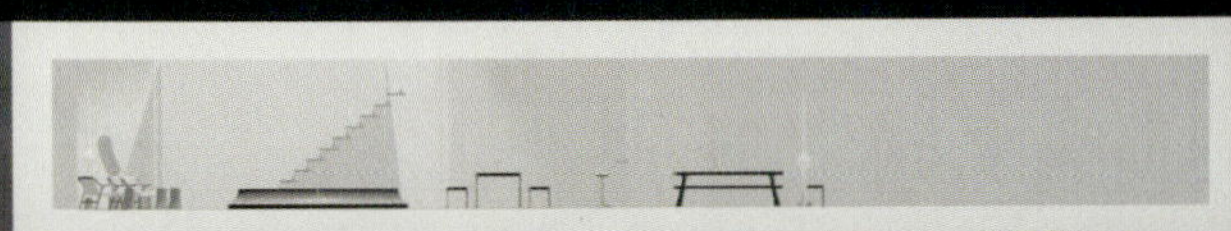

top

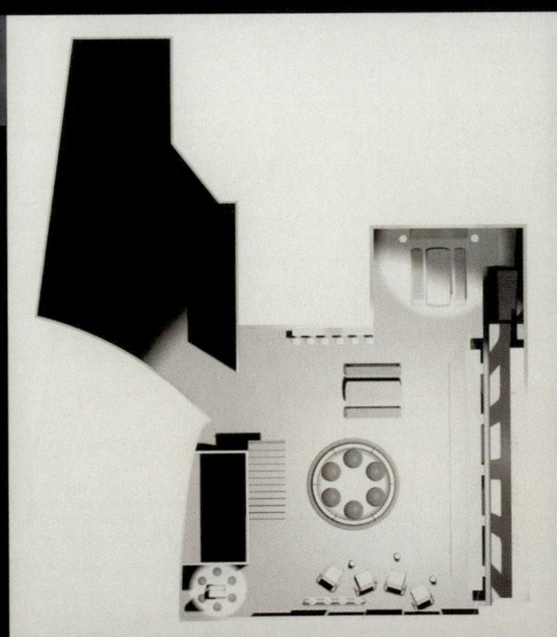

WRITER
Robert Thiemann

PHOTOGRAPHER
Ari Magg

DESIGNER
Michael Young

LOCATION
Reykjavik, Iceland

SUBJECT
Astro Bar nightclub

Issue 18 January/February 2001 176-177

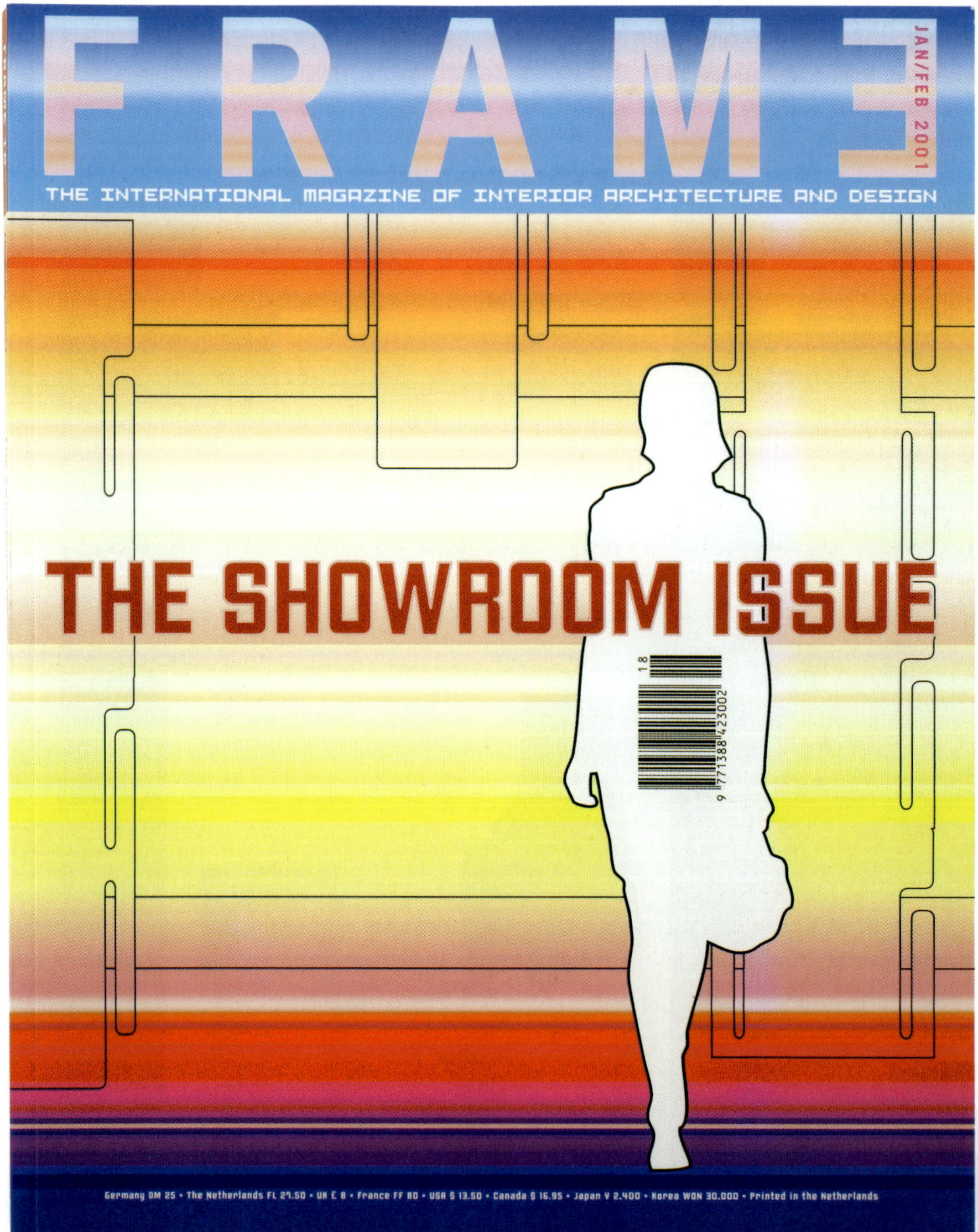

ART DIRECTOR
Roelof Mulder

18

Soft Machine

DESIGNERS
Erwan and Ronan Bouroullec
www.bouroullec.com

SUBJECT
Issey Miyake's A-POC shop

LOCATION
Paris, France

PHOTOGRAPHER
Wouter van den Brink

WRITER
Chris Scott

Designers Ronan and Erwan Bouroullec about the project

The A-POC project bears all the marks of its location. It was constructed not from the space to the detail, but rather from the detail to the space. First we designed the hangers, then a support for the hangers, and so on. That process was aimed at the creation of an array of tools that would allow Issey Miyake to play with the image of his space over time. Unlike a demonstrative proposition in terms of architecture, our design is an evolutionary system that tends to disappear behind the clothes.

Comments on the published article

The article published in Frame is by far one of the most complete articles featuring the A-POC store. For us, it was a great opportunity to have Frame show the main steps of the design: from the first conception to the drawings to the modelling to the very building of the space. We also appreciated very much the serious demand for accuracy, the relevance of the questions and the accessibility of the journalist.

Did the article have an effect?

We have the feeling that those who read the article gained a deeper understanding of the project.

Current projects

We are working on several projects for Vitra, Magis and Roset, among others. We recently designed a new showroom for Kvadrat in Stockholm (The opening was in February 2006).

THE SEAMLESS SPACE THAT ERWAN AND RONAN BOUROULLEC DESIGNED FOR ISSEY MIYAKE'S A-POC SHOP IN PARIS PROVIDES A SOFT, WHITE BACKDROP FOR COLOURS FOUND ALMOST EXCLUSIVELY IN THE FASHIONS DISPLAYED THERE. BY CHRIS SCOTT. PHOTOGRAPHY BY WOUTER. STYLING BY ERJAN BORREN.

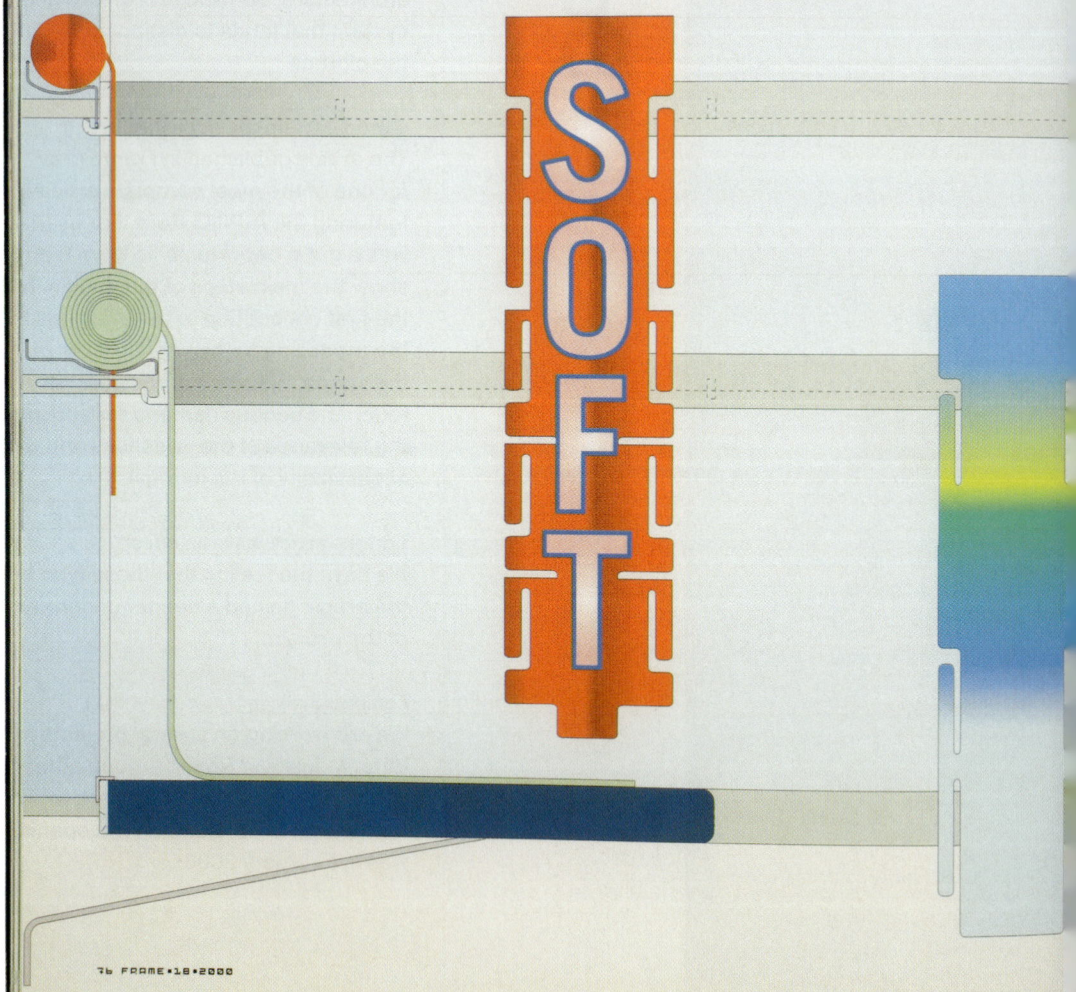

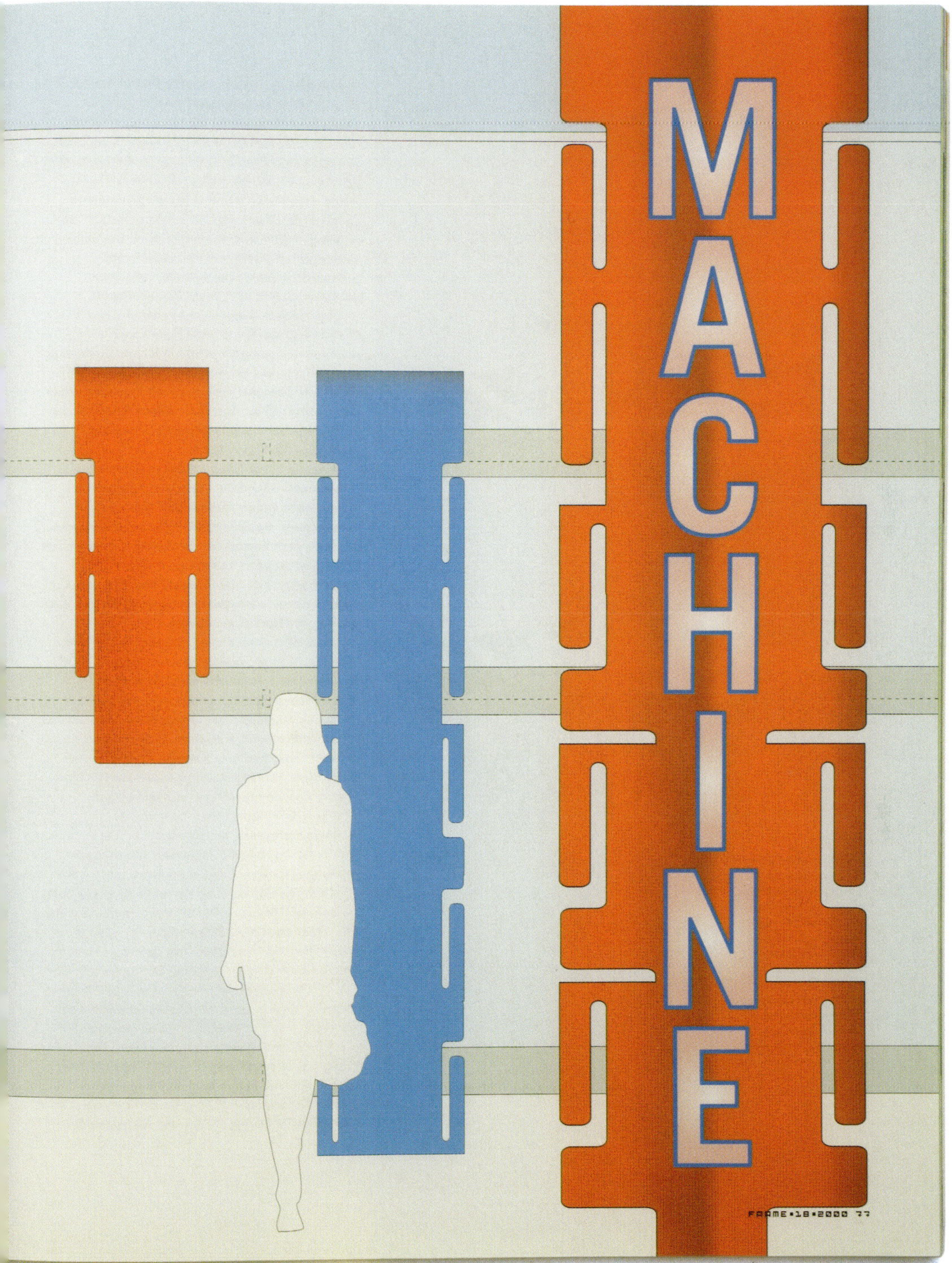

SUBJECT
Issey Miyake's A-POC shop

LOCATION
Paris, France

DESIGNERS
Erwan and Ronan Bouroullec

PHOTOGRAPHER
Wouter van den Brink

WRITER
Chris Scott

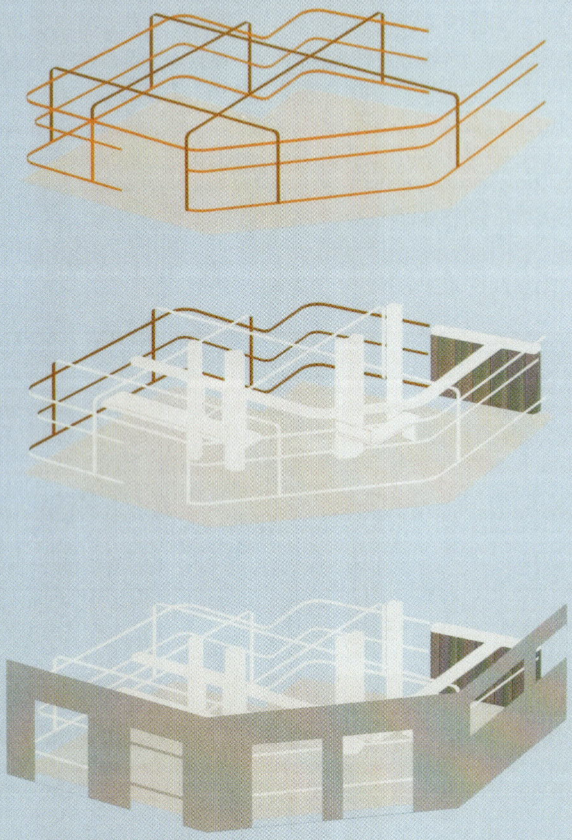

According to the designers, the space resembles a body, a living box in which the layers of skin have been peeled back to reveal the 'organs' and 'bones'. A module that circulates through the space is attached invisibly to existing columns to allow for the dispensation of 'energy', the space's life source. The conventional separation between window display and selling place has been deleted.

Following the success of Issey Miyake's A-POC shop in Tokyo, the designer recently opened a second outlet in Paris. The two are remarkably different interpretations of 'A Piece of Cloth', the concept that Miyake created in collaboration with Dai Fujiwara. Differences aside, however, both shops rely on a constantly changing, interactive space that makes customers major players in the creation of their own garments. This revolutionary idea results in clothes with a certain 'mass produced' quality, yet each article is custom designed to meet the buyer's requirements.

The Paris boutique was a dream project for French brothers Ronan and Erwan Bouroullec, who were responsible for the total look. The young duo has won several awards, including the ICFF Best New Designer Award, which they received in New York in 1999. The Bouroullecs' impressive list of clients boasts such names as Cappellini, Authentics, Boffi and Iittala.

A big fan of their work, Miyake identifies with the brothers' thoughtful, sensitive approach, not to mention their freshness of mind and an interrogative, disciplined attitude that often carries an element of humour. A keen analytical insight into the project at hand forms a good basis for reinterpreting and reworking a concept to meet the client's needs, while also producing a result clearly bearing the Bouroullec stamp.

Confident in his choice, Miyake held no major briefings. A short, general discussion on A-POC and the clothes was sufficient. He asked the Bouroullecs to submit a proposal, and even though the first draft failed to win his approval, Miyake was determined to work with them. A second proposal, for which they had only a week to prepare, was much more to his liking. They proceeded with no restrictions, no stipulations, no set budget – Miyake's only request was: 'Surprise us.' An intimidating task, as this was their first architectural project. Enthusiastic about the new challenge, however, the Bouroullecs set to work.

Three months on – having had only occasional contact with Miyake – the brothers completed the job. While other new shops rushed to meet the crucial deadline that marks the opening of autumn salons and shows in Paris, the Bouroullecs finished the A-POC project right on time, ready for the first invasion. What's more, Miyake was delighted with the result. His confidence in the pair had paid off.

A-POC has a prime corner location. The façade along one street is lined with milk-white windows, which allow not even a glimpse of the interior. Passers-by are intrigued by this air of mystery, while customers inside experience the wall as a gift of

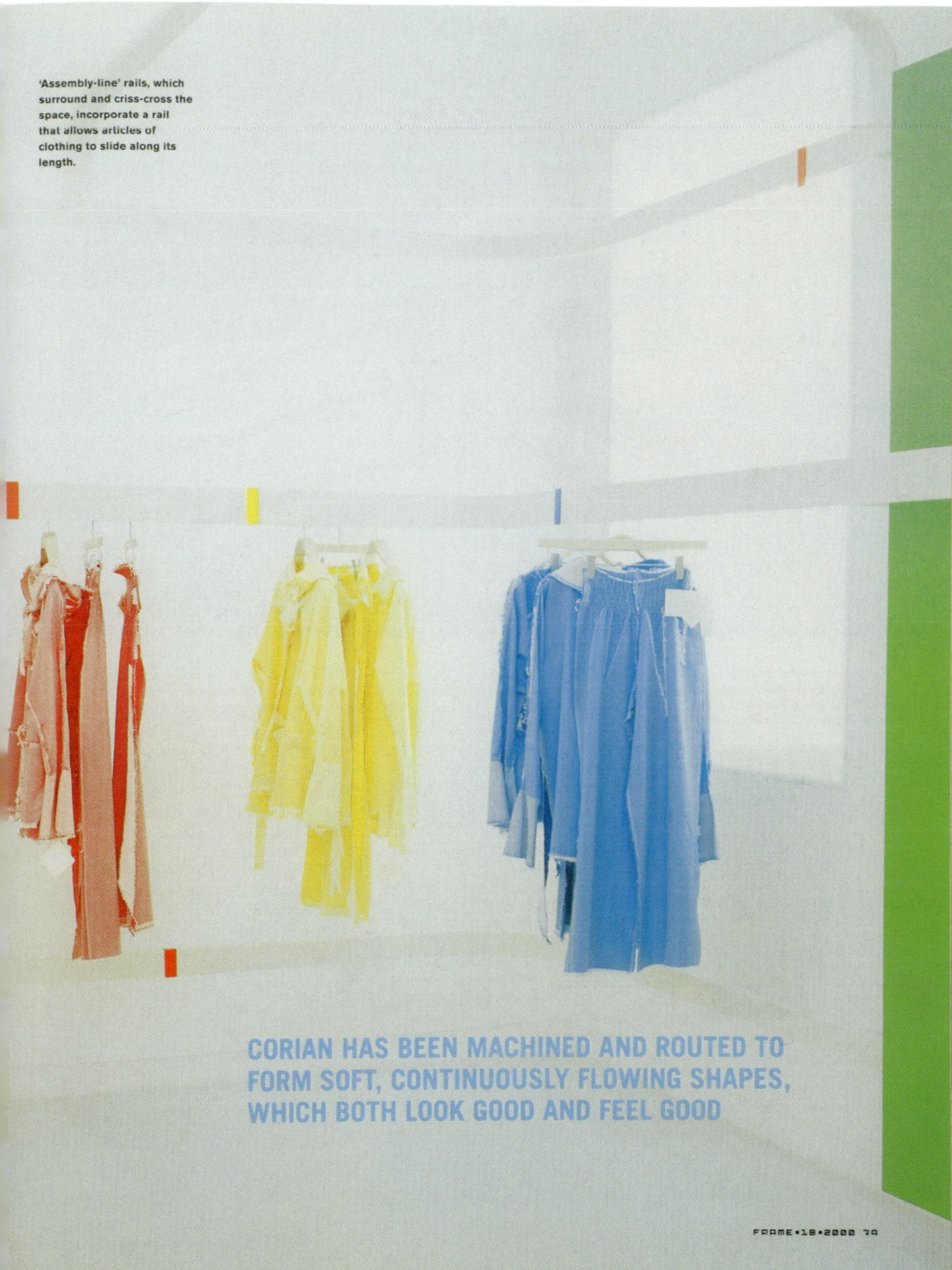

'Assembly-line' rails, which surround and criss-cross the space, incorporate a rail that allows articles of clothing to slide along its length.

CORIAN HAS BEEN MACHINED AND ROUTED TO FORM SOFT, CONTINUOUSLY FLOWING SHAPES, WHICH BOTH LOOK GOOD AND FEEL GOOD

SUBJECT
Issey Miyake's A-POC shop

LOCATION
Paris, France

DESIGNERS
Erwan and Ronan Bouroullec

PHOTOGRAPHER
Wouter van den Brink

WRITER
Chris Scott

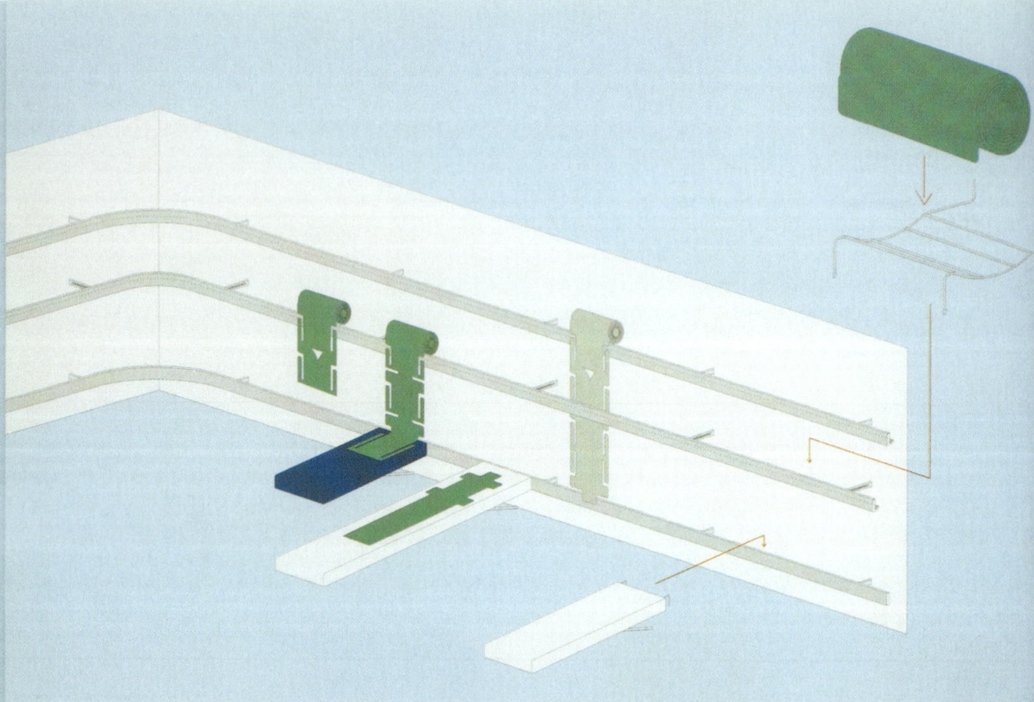

Adjustable boards attached to the Corian strips display bolts of fabric, which can also be moved as required.

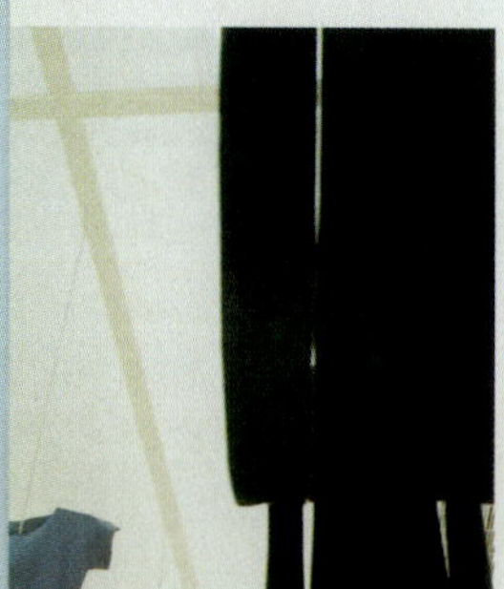

privacy. The only exterior sign of life is the cryptic A-POC inscription on the wall. Turn the corner, however, and a starkly contrasting, clear-glass shop front draws the eye inside, where beautiful graphic shapes, lines and colours play gleeful games with the imagination. Lacking even the simplest of window displays, this is obviously no ordinary retail outlet. A concept without compartmentalisation, the Paris A-POC is a single organic, integrated, interactive space – a showroom, gallery, atelier, factory and boutique in one.

Inside, 'assembly-line' rails display a variety of fashions and fabrics. These strips, which surround and crisscross the space, are made of Corian; each strip incorporates a rail that allows articles of clothing to slide along its length. Adjustable boards attached to the strips display bolts of fabric, which can also be moved as required. Apparel and fabrics that hang like paintings and pose like sculptures can easily make the observer forget that this is a boutique devoted to fashion.

The 100-square-metre shop is a largely white environment, whose main source of colour is the collection of clothes displayed within its confines. Most of the furniture and fittings (tabletops, display surfaces, rails and coat hangers) are made of Corian, a material put to optimum use in this setting. It has been machined and routed to form soft, continuously flowing shapes, which both look good and feel good. Counters, cutting tables and ironing boards that slip into tabletops all merge together without a single visible seam or join. Cast in polyurethane resin, the floor also forms an uninterrupted surface. The space is a fluid, organic entity.

The rails support long metal rods that are adjustable in length. They resemble the picture-hanging rails at an exhibition. Attached to the rods are Corian hangers, which are designed to let the clothes hang freely without slipping, sliding or falling to the ground. Garments not displayed on hangers lie flat against coloured metal panels. Here again, no hooks, pegs or strings are to be seen. The secret is magnets, which fasten the items to the panels without producing an unwanted fold or wrinkle.

After a customer has selected a style or colour from the displays, she watches as one of the staff moves a huge bolt of fabric into place. It fits perfectly into a predetermined position on the moulded tabletop, from which a length of fabric is rolled out on the illuminated surface in preparation for the next

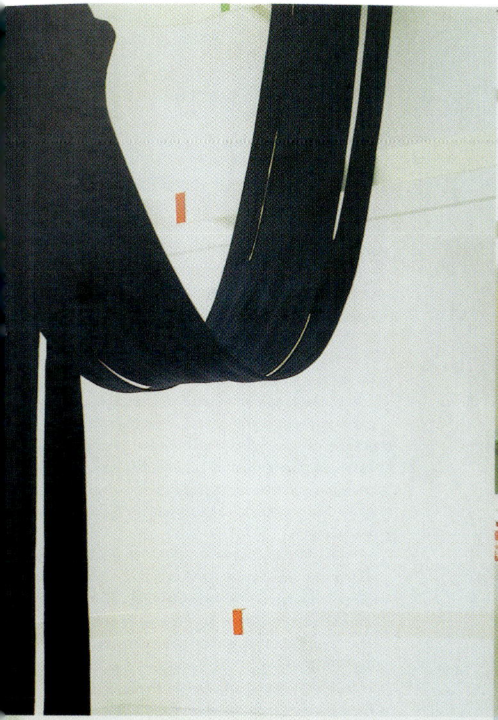

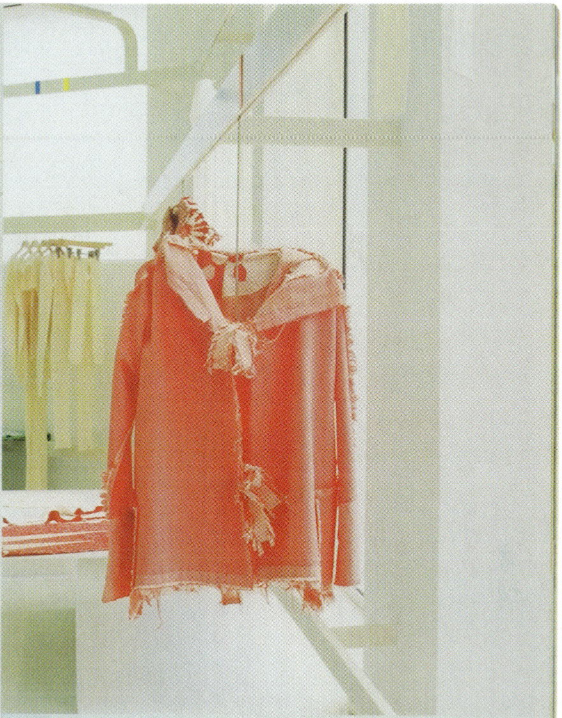

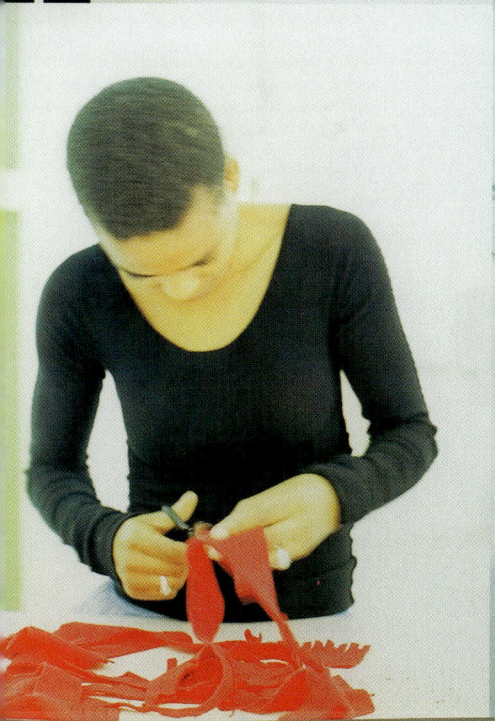

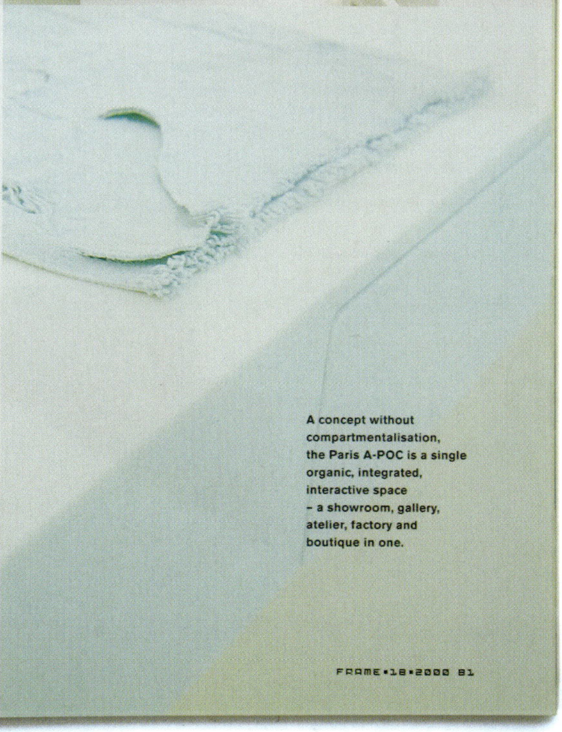

A concept without compartmentalisation, the Paris A-POC is a single organic, integrated, interactive space – a showroom, gallery, atelier, factory and boutique in one.

SUBJECT
Issey Miyake's A-POC shop

LOCATION
Paris, France

DESIGNERS
Erwan and Ronan Bouroullec

PHOTOGRAPHER
Wouter van den Brink

WRITER
Chris Scott

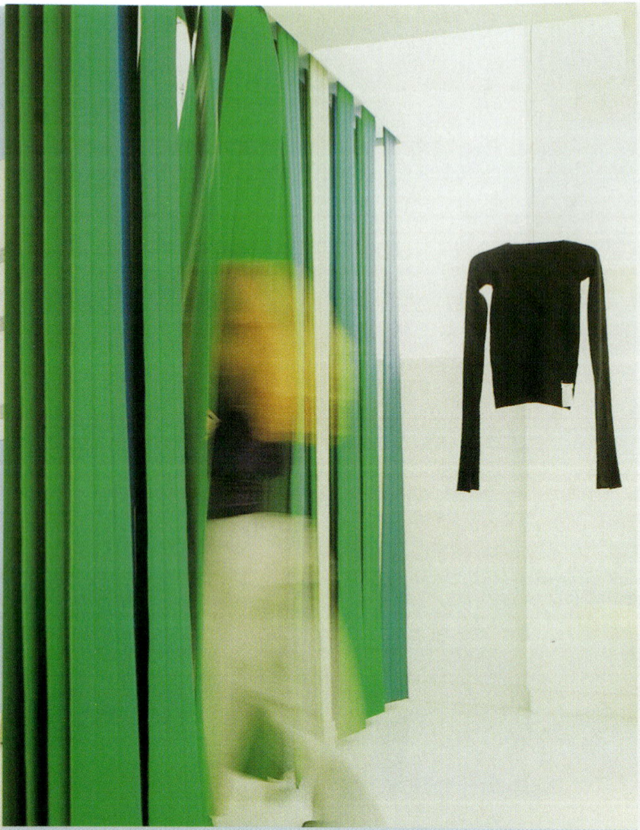

Above: In the fitting rooms green panels of quilted mousseline de laine block out noise and offer privacy.

Right: Bolts of fabric fit into a predetermined position on the moulded tabletop, from which a length of fabric is rolled out and crafted into a tailor-made creation.

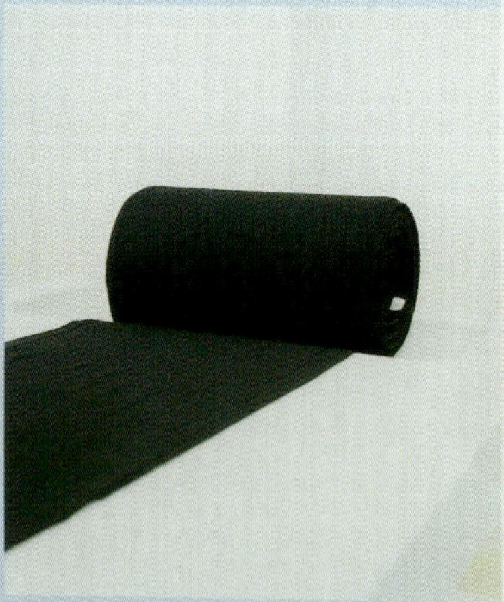

step. A pre-cut outline already marks the tubular jersey material, which awaits the customer's instructions. Sleeves, neckline, length – alterations to this non-fraying fabric are simple. As the scissors snip away, the garment undergoes a metamorphosis. The operation, which blends modern technology with a hands-on approach, glides along with apparent ease and efficiency.

A-POC's fitting rooms are another good example of a well-thought-out design. Green panels of quilted *mousseline de laine* (literally 'muslin of wool') block out noise and offer the privacy needed to decide if a purchase should show a bit more knee, take a deeper dip at the neckline or be a sleeveless style after all. The customer truly participates in crafting a tailor-made, and thus unique, creation. The fashion-designing fun can even continue at home. The only drawback is that once the material is cut, there's no turning back. A-POC is an innovative approach to shopping for a new outfit.

The lighting scheme evolved from the collaborative efforts of the Bouroullecs and a team of theatre and exhibition specialists. A combination of halogen and tungsten lamps, along with fluorescent tubes, produces illumination that closely resembles daylight. The aim of the designers was a smooth transition between artificial and natural light. Their choices enhance the daylight that enters the shop, and the result does not alter the colours of the merchandise.

The brothers gave the interior a final touch by adding a few objects of their own design: good examples are the Cappellini chairs and Valauris ceramics. The shop combines the talents of Miyake and Bouroullec in a complementary fusion of East and West.

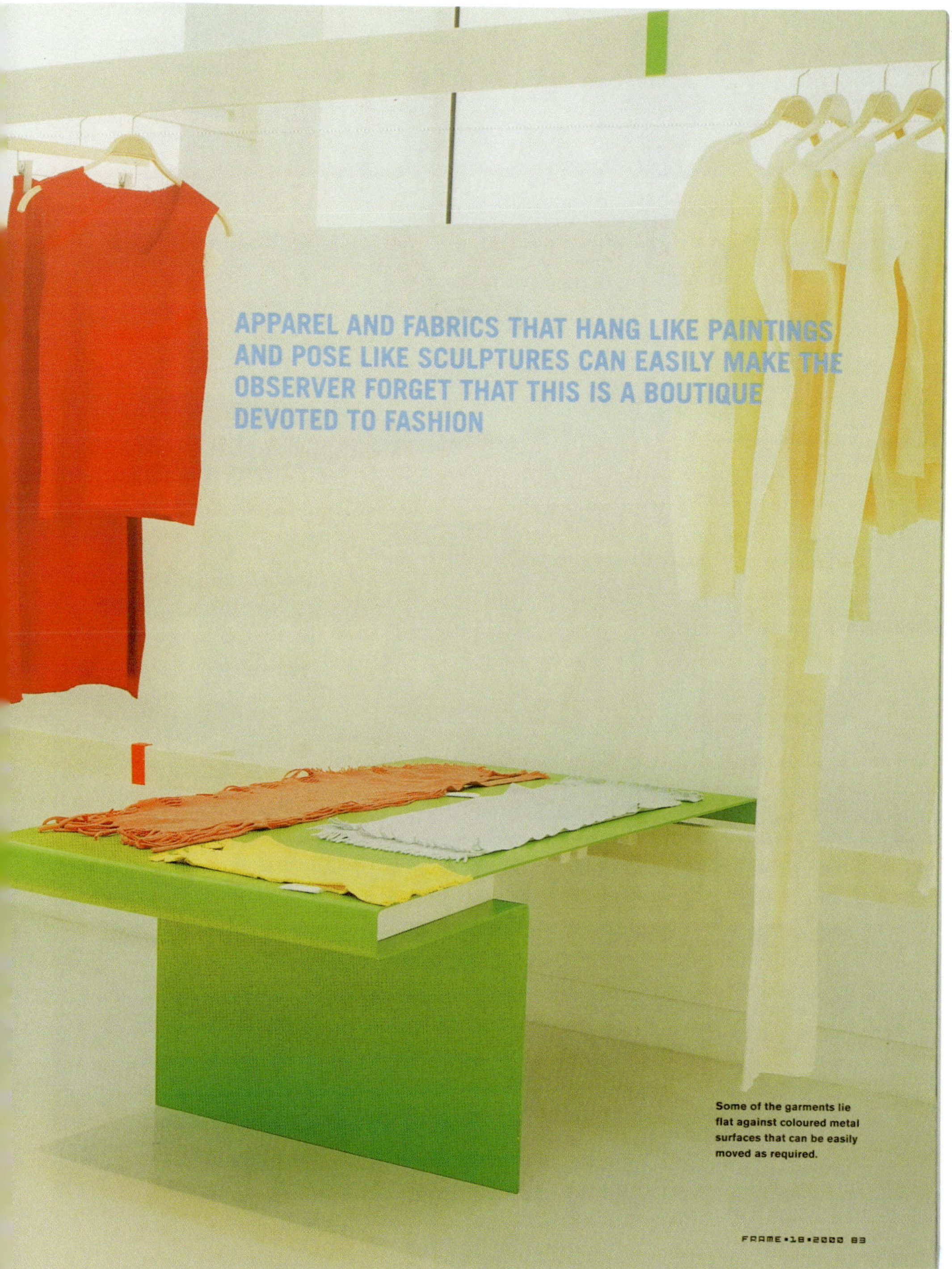

APPAREL AND FABRICS THAT HANG LIKE PAINTINGS AND POSE LIKE SCULPTURES CAN EASILY MAKE THE OBSERVER FORGET THAT THIS IS A BOUTIQUE DEVOTED TO FASHION

Some of the garments lie flat against coloured metal surfaces that can be easily moved as required.

Issue 19 March/April 2001 186-187

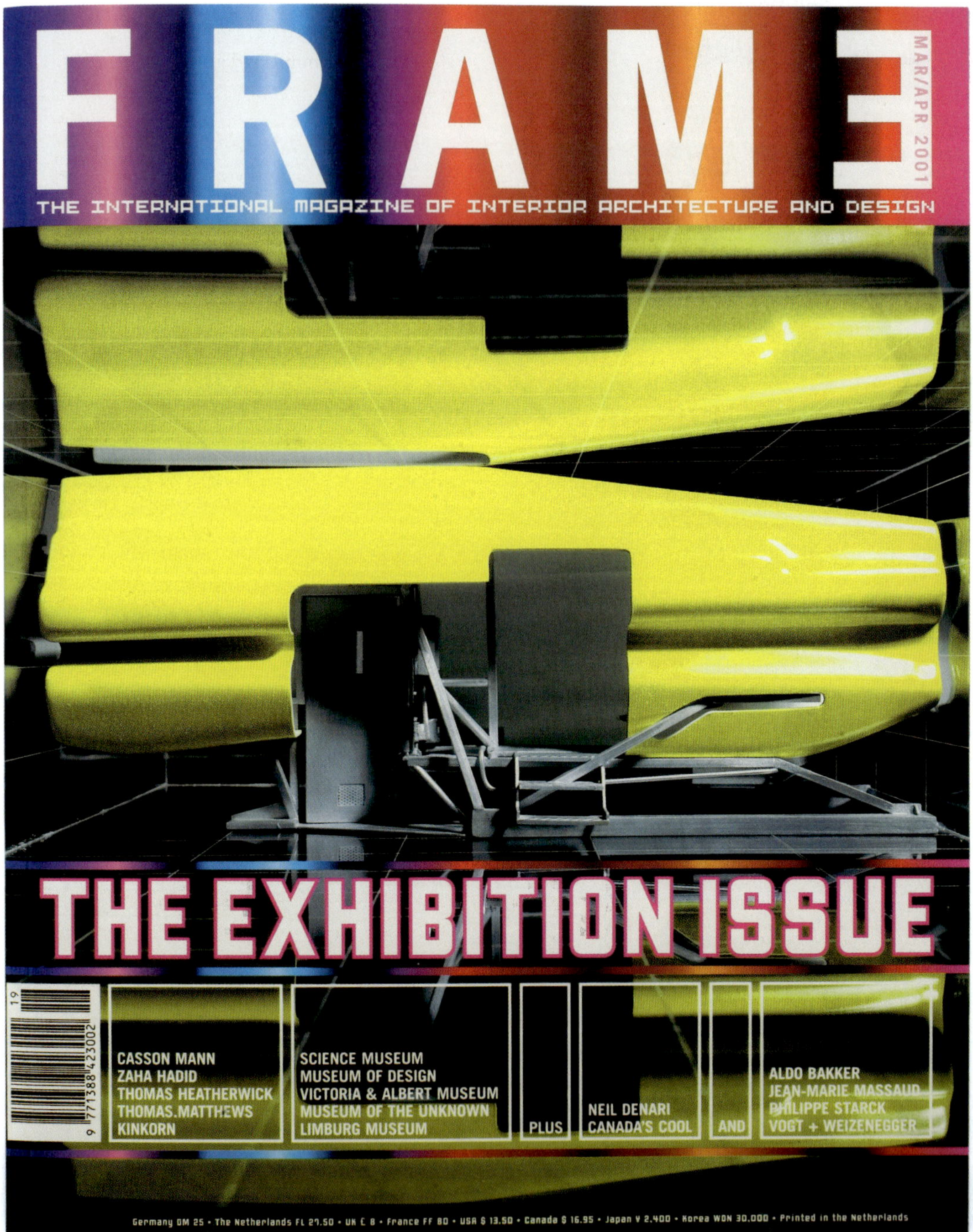

ART DIRECTOR
Roelof Mulder

The Folding Game

DESIGNER
Neil M. Denari
www.nmda-inc.com

SUBJECT
Portrait of Neil Denari

LOCATION
Los Angeles, USA

PHOTOGRAPHERS
Benny Chan
Fujitsuk Mitsumasa

WRITER
Leo Gullbring
www.calimero.se

Architect Neil M. Denari about the published article
Very informative. Great layout.

Did the article have an effect?
We are not able to quantify the effect.

Current projects
High Line 23: Mid-rise residential design (13-storey loft building), completion date August 2007, New York, USA // NEU Development: Large-scale, eco-sustainable, urban- and landscape-design project featuring loft housing (new-build and conversion). Final master plan to be completed winter 2006; first phase of site development scheduled for 2007, Nashville, USA // Mitsubishi / UFG Bank, completion date April 2006, Nagoya, Japan // Alan / Voo House: Residence for family of five, construction in progress, Los Angeles, USA // International Center for Living Watersheds (ICLW): Cultural/ educational complex including exhibition spaces, educational facilities and offices, Nashville, USA // Ideal House: 1:1-scale mock-up of 45 sq m of an 'ideal house' to be exhibited at the Megaron Museum in the autumn of 2006, Athens, Greece.

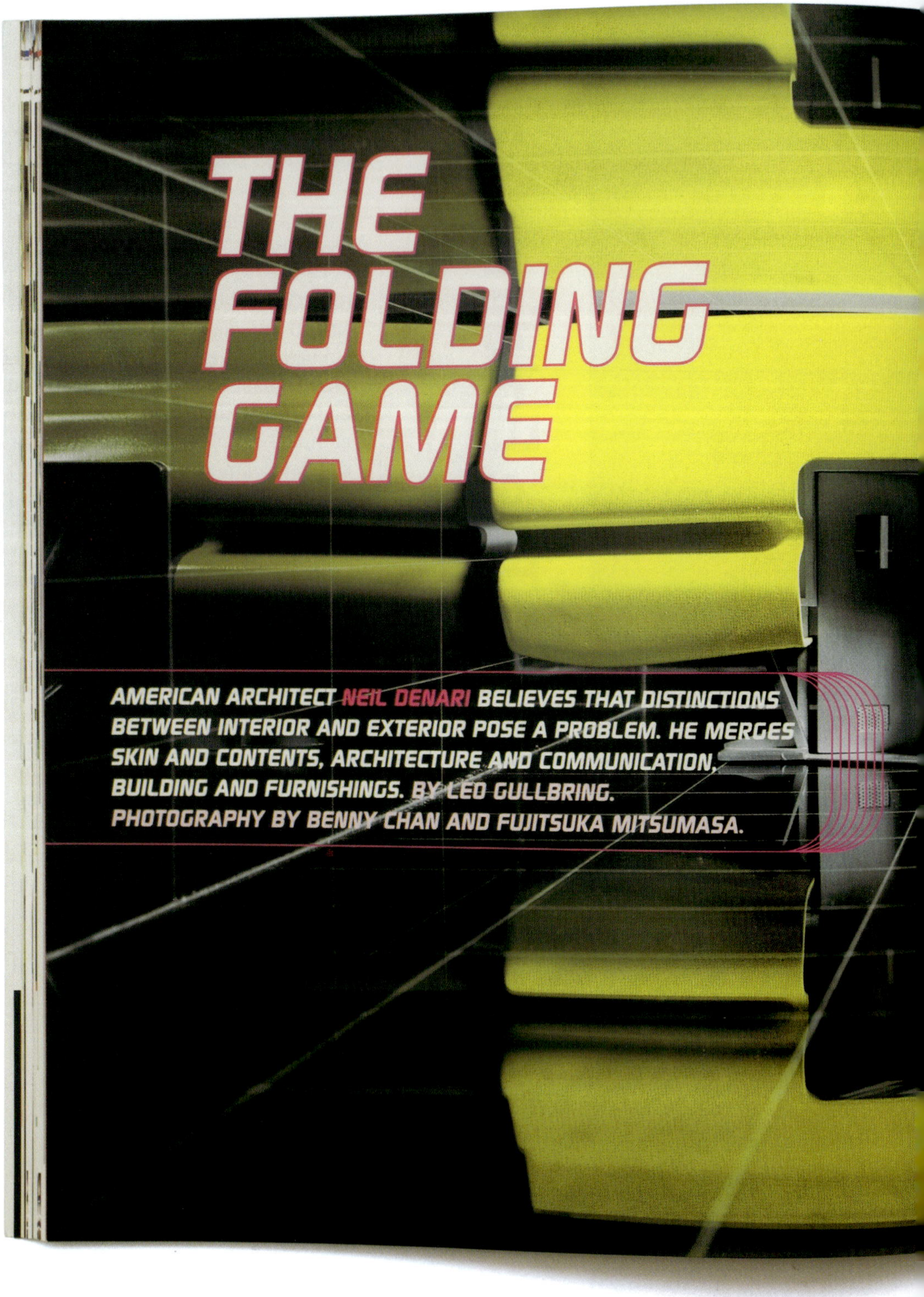

THE FOLDING GAME

AMERICAN ARCHITECT NEIL DENARI BELIEVES THAT DISTINCTIONS BETWEEN INTERIOR AND EXTERIOR POSE A PROBLEM. HE MERGES SKIN AND CONTENTS, ARCHITECTURE AND COMMUNICATION, BUILDING AND FURNISHINGS. BY LEO GULLBRING. PHOTOGRAPHY BY BENNY CHAN AND FUJITSUKA MITSUMASA.

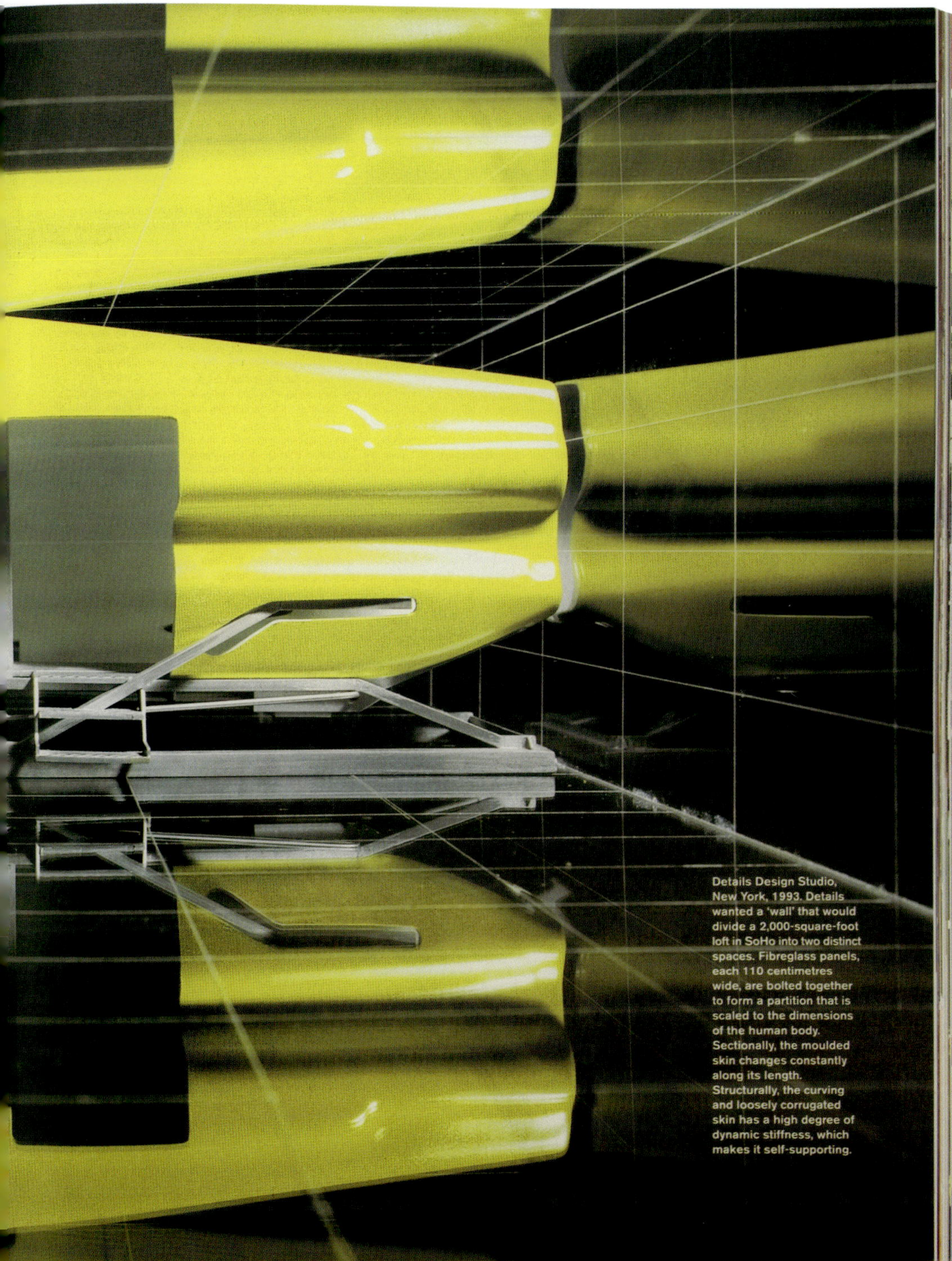

Details Design Studio, New York, 1993. Details wanted a 'wall' that would divide a 2,000-square-foot loft in SoHo into two distinct spaces. Fibreglass panels, each 110 centimetres wide, are bolted together to form a partition that is scaled to the dimensions of the human body. Sectionally, the moulded skin changes constantly along its length. Structurally, the curving and loosely corrugated skin has a high degree of dynamic stiffness, which makes it self-supporting.

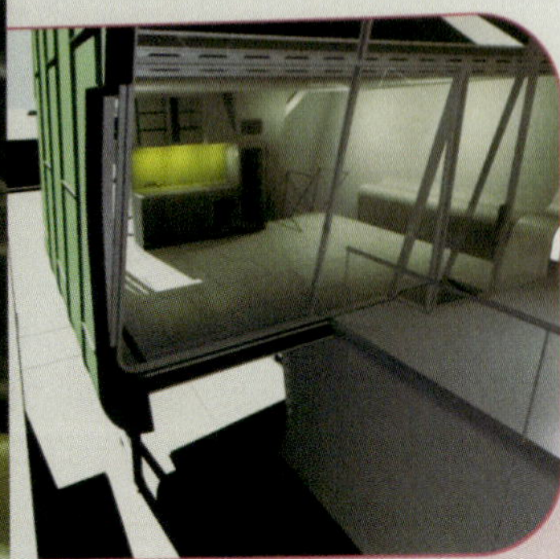

'How do I get to the school?' Having finally reached Denari by phone, I still have to drive from downtown Los Angeles to the Southern California Institute of Architecture (SCI-Arc) for the interview. Listening to his directions, I'm already cruising the streets of LA in my mind. Blue sky. Windows wide open. Dr Dre on the radio. A cityscape of houses in melting ice-cream pastels. Billboards promoting fine liquors next to signs announcing the Second Coming and ads for psychic readings. Bland bungalows interspersed with more recent designs by celebrities like Frank Gehry, Eric Owen Moss, Thom Mayne and Franklin D. Israel. Not a trace, however, of anything by Neil Denari.

A dilapidated industrial zone is the setting for SCI-Arc's temporary home: a gigantic white tent rising hazily from its waterfront site. I spot Denari, who heads the institute, as he walks towards a makeshift office with several students. When I identify myself, he suggests that we talk in 'the café'. We climb onto the loading platform of a former warehouse that's little more than floors, columns and a roof. A couple of students are surveying the narrow skeleton of the building, which must be at least a half-mile long. As Denari explains that we're standing on the future premises of SCI-Arc, I have no trouble recognising the extraordinary potential of this structure.

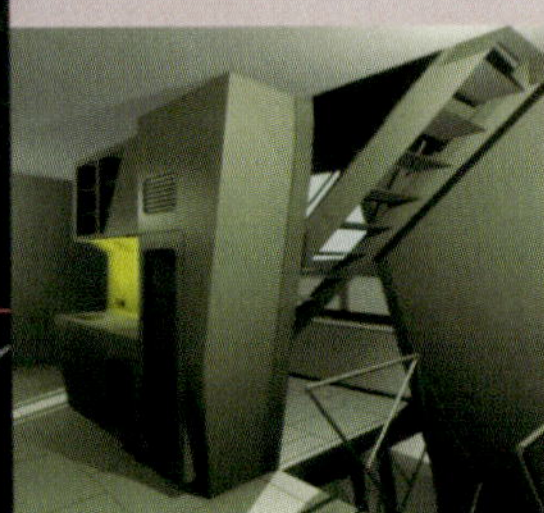

The café forms a fashionably understated backdrop for the activities going on outside. Its lofty and somewhat minimalist interior offers a cool contrast to the sizzling asphalt surroundings. Waiting for espresso and cheesecake, I contemplate the man who – together with Neil Spiller, Marcos Novak and Gregg Lynn – is credited for establishing the 'transarchitectural movement'. My initial interest in Denari was triggered by his beautiful book, Gyroscopic Horizons (1999), a volume filled with references to a number of French and Italian philosophers and cinematographers whom I find fascinating as well. When I express my admiration for Gyroscopic Horizons, he says, 'That book represents the summation of one period and the beginning of another. One project that I included, a house in Palm Springs, dealt with climatic issues and changing aesthetics. I was focusing on the transformation of a roof envelope, which corresponds to my interest in things commercial and to the phenomenon of the image. The book addresses the inability of architecture to demand attention and to control the articulation of space. It talks about architecture that aspires to something other than a Frank Gehry-like expression of the sublime. Architecture that holds the attention and becomes useful – or even more than useful.'

He goes on to explain that he no longer puts iconography above pure function, as he did when writing the book: 'I'm back to a more normative way of thinking, which implies a better understanding of how performance and ideas are rooted in culture. It's not all about creating obscure devices that allow architects to express ideas. Our work should relate to more contemporary pressures.'

The conversation shifts to social issues and, in particular, the vast number of people who cannot afford housing of any kind. This long-standing situation is one of Denari's major concerns. 'Our culture is based strictly on profit,' he says. 'Only a small percentage of local and federal funds goes into the building of low-cost housing. How can architects introduce new spatial ideas, different planning methods or alternative economic models within such a climate?'

He gazes out the windows at the cloud-like tent. 'I'd like to explore new types of housing and new forms of density. It's not that I want to make Los Angeles into some sort of Manhattan. With the exception of certain downtown areas overcrowded with immigrants, the situation here represents a more dispersed reality. A big problem is functional segregation: housing here, commercial properties there, offices somewhere else. With everything spread out like it is, LA reminds me of Garnier's linear city project. I want to explore more oblique or diagonal relationships among the various functions. My thoughts go beyond housing and employment to the merging of commercial spaces and circulation systems, not to mention the effect of the Internet on travel and shopping, for example.'

How many architects do you need to remove the perpetually burning light bulb of decentralisation in LA? A bad joke with no clever punch line deserves to remain unspoken. Instead of responding to his concerns with humour, I direct my attention to a sheaf of computer-generated drawings that he's placed on the table. Included are images of Denari's amazing Multisection Office Block, a proposal reminiscent of Archigram's mega-structures yet with folds, splits and splices that evoke the work of Peter Eisenman. Three similar but not identical eight-storey buildings, designed for Pico Boulevard, reflect the surrounding commercial and residential areas. The immense scale of the buildings is geared to that of the nearby San Diego Freeway. The shearing effect produced by the exterior walls and interior circulation spaces of this design can be compared to analogous effects in structuralist film-making. The basic structure resembles certain undulating designs by Rem Koolhaas, but its graphically charged surfaces tell quite another story. Denari remarks that most contemporary architecture seems to be stuck in another century. Few architects have what it takes to cope with current events. 'We see no definitive

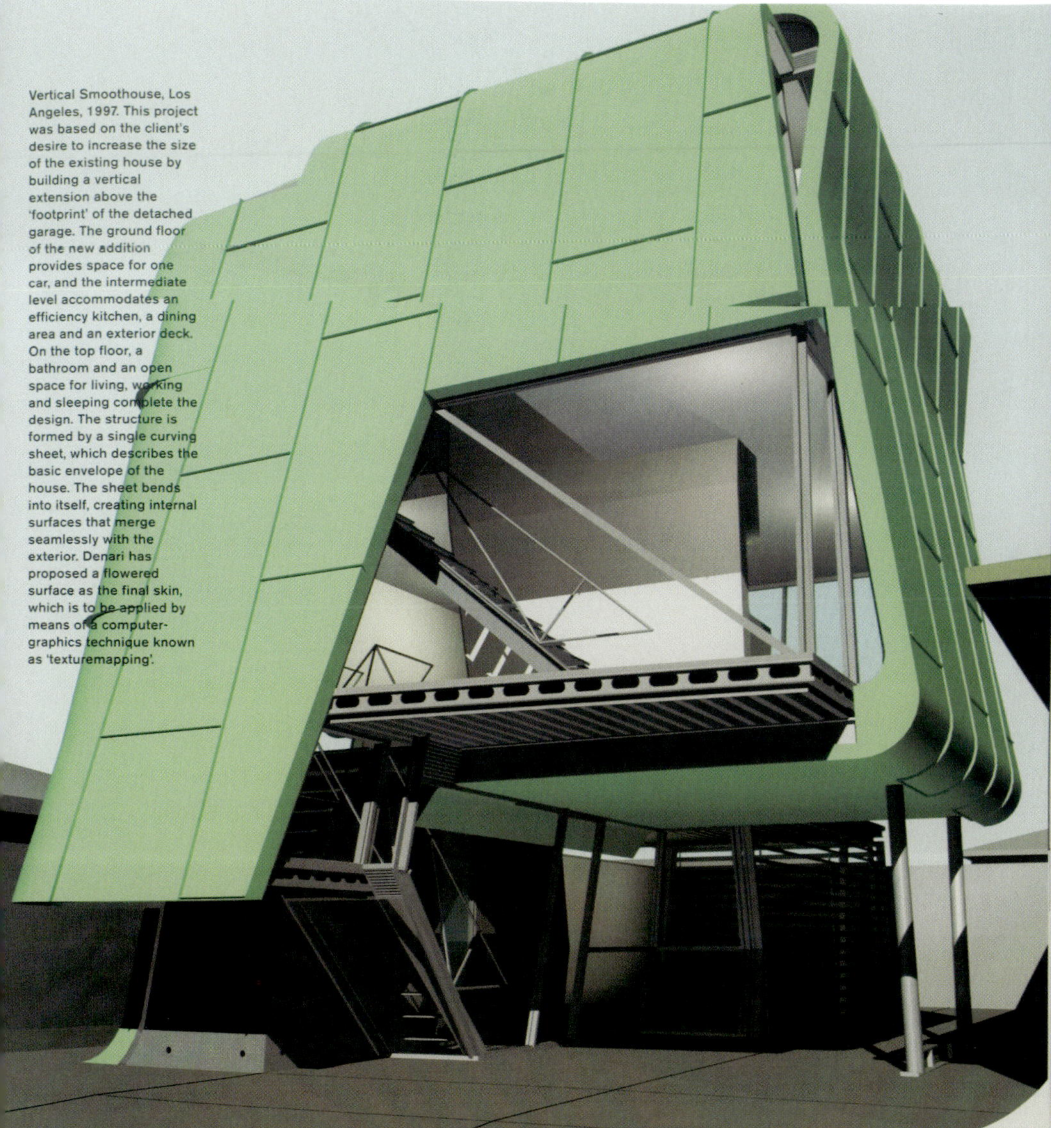

Vertical Smoothouse, Los Angeles, 1997. This project was based on the client's desire to increase the size of the existing house by building a vertical extension above the 'footprint' of the detached garage. The ground floor of the new addition provides space for one car, and the intermediate level accommodates an efficiency kitchen, a dining area and an exterior deck. On the top floor, a bathroom and an open space for living, working and sleeping complete the design. The structure is formed by a single curving sheet, which describes the basic envelope of the house. The sheet bends into itself, creating internal surfaces that merge seamlessly with the exterior. Denari has proposed a flowered surface as the final skin, which is to be applied by means of a computer-graphics technique known as 'texturemapping'.

'THINK ABOUT A PORTABLE ARCHITECTURE – AN ARCHITECTURE THAT CAN BE COMPARED TO CAMERAS, WALKMANS AND SHOES'

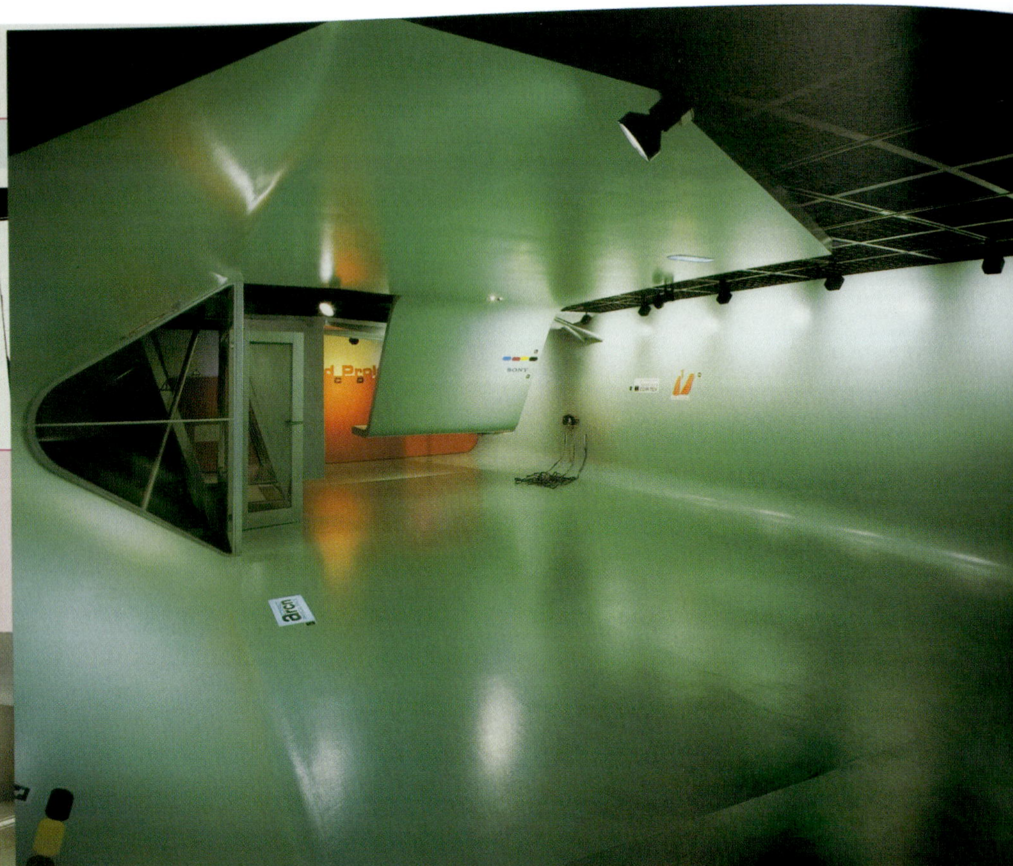

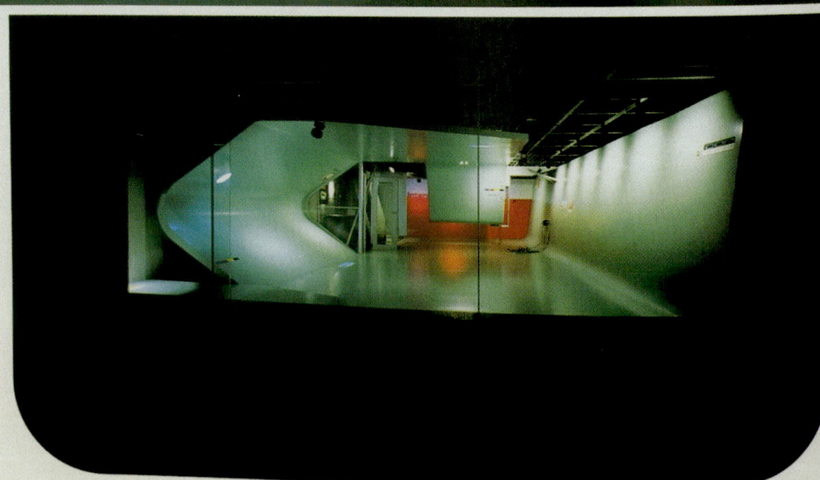

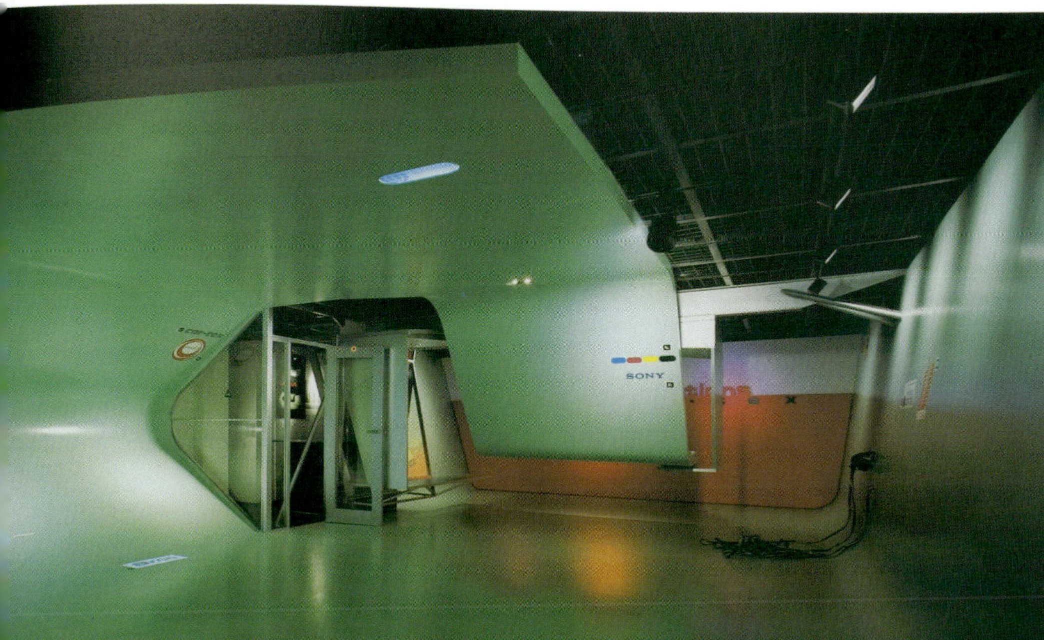

'THE NOTION OF AN ENVELOPE THAT HIDES THE COMPLEXITY OF THE INTERIOR IS REFLECTED IN NATURE AS WELL AS IN INDUSTRIAL DESIGN. MECHANISMS INSIDE, SKIN OUTSIDE. BUT SEPARATING THE SHELL FROM THE CONTENTS POSES A PROBLEM'

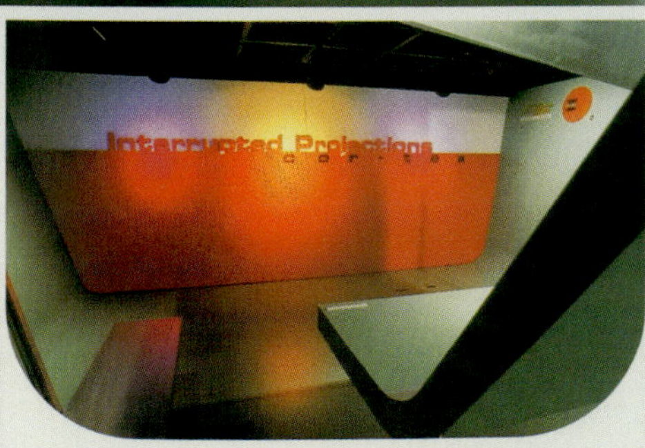

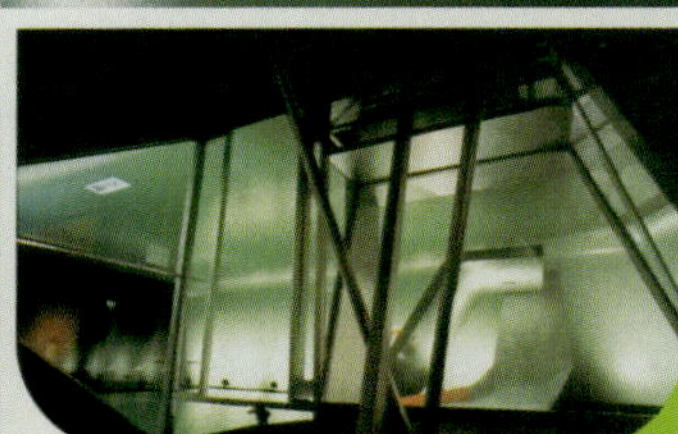

Interrupted Projections: Another Global Surface, Tokyo, 1996. Exhibited at Gallery MA, this project is based on the Homolosine Interrupted Projection Mapping System. Here the word 'map' is interpreted as 'sheet': a surface on which to record territories. Because the Homolosine Projection depicts the world as a series of sheared ellipses, the inner green surface of the gallery is an interrupted projection, which uses an empty, flattened surface to form space. This surface bends and loops to create a three-dimensional geometry, both smooth and complex, which is capable of merging with the graphically 'logoised' world of visual codes and conventional signs.

SUBJECT Portrait of Neil Denari

LOCATION Los Angeles, USA

DESIGNER Neil M. Denari

PHOTOGRAPHERS Benny Chan, Fujitsuk Mitsumasa

WRITER Leo Gullbring

'THINK ABOUT NEW LIFESTYLES AND ABOUT HOW ECONOMY INFLUENCES THEM MORE THAN BUILDINGS DO'

statements about the form of the city anymore,' he says, 'because the city itself is inarticulate. Fixed models have got to make way for procedural models. The question is how to go about building the phenomenon of the city into architecture.' Denari has moved beyond a techno-scientific approach to architecture. He's looking for a new aesthetic and for a way to blend real sustainability with his commitment to ecology. 'Let's try to break down the sort of binary thinking that puts high-rises into a category separated from the ground and small dwellings into a category that's part of the ground. Up in the sky, far from the benefits of nature, you're immersed in a harsh urban lifestyle that seems to be fabricated. On the other hand, a beautiful inside-outside environment is possible in small, single-family homes like those designed by Neutra. Can't an urban matrix accommodate something more continuous than what we see in most cities today? Can't public and private spaces be interwoven by finding new treatments for envelopes, different solutions for the use of glass or innovative acoustical applications? What about collaborating with industry for the purpose of creating new types of plant materials?'

Adding vertical structures to LA's horizontal sprawl is another of Denari's ideas, but he approaches the subject by claiming that space is just as important to global ecology as biodegradable materials and solar energy are. As a solution to a host of environmental problems, space holds implications for themes like water, greenery, building envelopes, architectural voids, circulation and physical comfort.

'If we concentrate all our efforts on these issues,' he says, 'we can get some healthy results. Such an attempt involves a whole new aesthetic, however, and an even greater degree of performance.' Warming to the subject, he finds an analogy in the profession that occupied his father when Neil Denari was growing up in the Texas metropolis of Dallas-Fort Worth: aeronautics. He names the enigmatic Stealth Fighter as an example of the powerful and ideal world he envisions. 'I'm imagining, of course, that the Stealth had been the success it was meant to be,' he concedes. 'Nevertheless, the aircraft represents a complete transformation. It's made of a different material. It becomes invisible. But it still flies. It still works. A whole paradigmatic shift is involved here. If we could apply the same rules to the levels of performance in architecture – if we could demand buildings that use less energy, that are cheaper to build, that can be mass produced – just think of the consequences. Yes, these ideas have been around for a long time. But what have we actually achieved?'

Denari says that most people see architecture as a means of shelter, a superstructure that provides a place in which to live and work. 'The notion of an envelope that hides the complexity of the interior is reflected in nature as well as in industrial design,' he explains. 'Mechanisms inside, skin outside. But separating the shell from the contents poses a problem, in my opinion. Rather than making the inside the outside, as the modernists did with glass and very open designs, I'm trying to extend the potential of a building. So far, however, I haven't gone beyond concept diagrams.'

I look at the photographs of an experimental space that he exhibited in Tokyo a few years ago. 'My point of departure was to create something more than a shelter. In an attempt to make the envelope of the building come alive, I tried to have the vessel and its interior merge. This particular interpretation of merging is rooted in the New Economy. Spatial and material ideas are laminated, as it were. My goal is to merge architecture with communication, as well as with interior furnishings.' Examples of merging, he says, are the first Schindler House in Los Angeles and, in Prague, the Bata Shoe Store, with its logo-covered glass façade. Although he admits that this idea is no more innovative than his position on global ecology, he emphasises that the time has never been so ripe for the union of architectural content and surface or for the penetration of design and media into deeper programmatic layers. Denari's obsession with an envelope that represents both outside and inside carries certain Miesian qualities and underlines architecture as a spatial medium. But the concept of merging cannot be construed as a secluded room for cultivating modernist ideas. A better definition focuses on the possibilities of cultural conditions as yet missing from the language of architecture.

The pervasive California sun welcomes us as we exit the café, but Los Angeles – a city whose social and economic climate should embrace his ideas – is still without a Denari. If not LA, I wonder aloud, then why not Europe, the utopia of cultural exploration? 'I'm interested in pragmatic experimentation in undefinable utilitarian zones,' he replies cryptically. 'My objective is to portray architecture as a highly impressive encounter, not to strip architecture of the possibility of experience, thus making it undefinable and irrelevant. But this particular goal implies a new discourse, because building in a city starts with politics, a power directed by social and economic forces. Architecture is nearly always at the bottom of the list. Rather than waiting patiently for commissions, architects should be shaping these very forces. Think about new lifestyles and about how economy influences them more than buildings do. Think about a portable architecture – an architecture that can be compared to cameras, Walkmans and shoes. It's an avenue that's being explored. But where are the answers?'

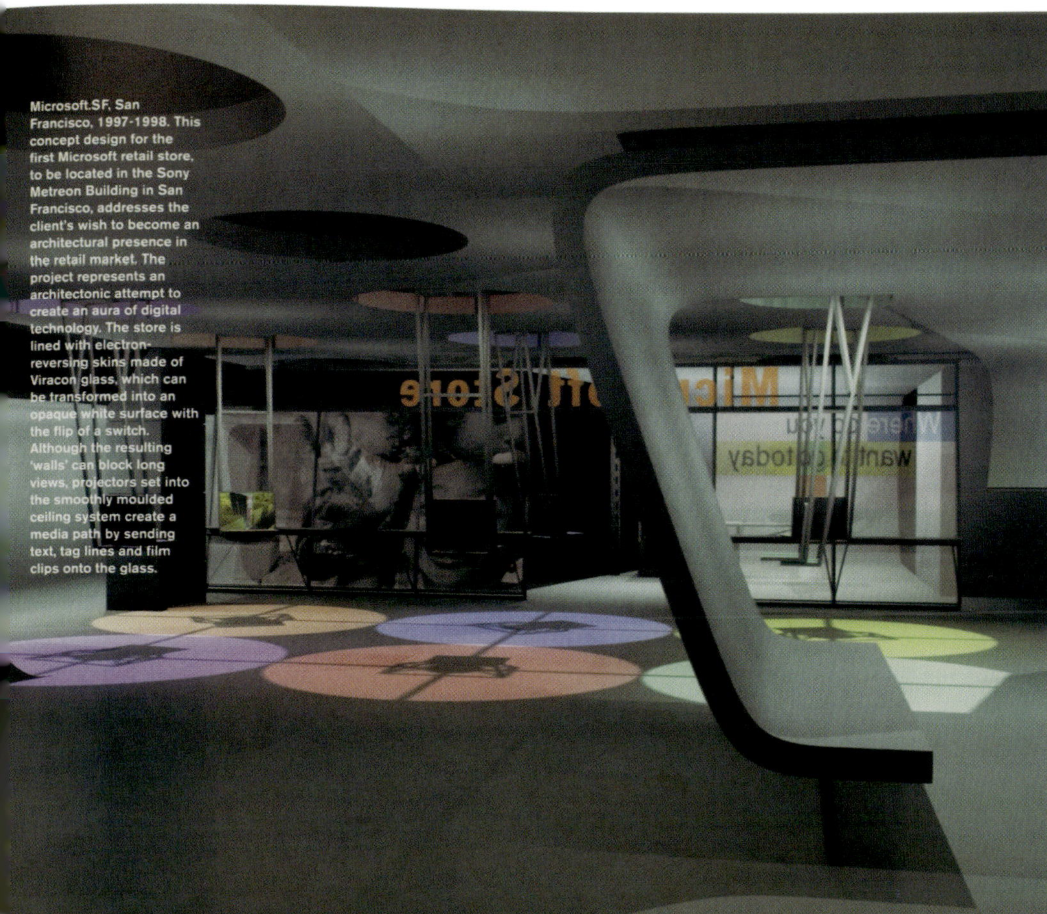

Microsoft.SF, San Francisco, 1997-1998. This concept design for the first Microsoft retail store, to be located in the Sony Metreon Building in San Francisco, addresses the client's wish to become an architectural presence in the retail market. The project represents an architectonic attempt to create an aura of digital technology. The store is lined with electron-reversing skins made of Viracon glass, which can be transformed into an opaque white surface with the flip of a switch. Although the resulting 'walls' can block long views, projectors set into the smoothly moulded ceiling system create a media path by sending text, tag lines and film clips onto the glass.

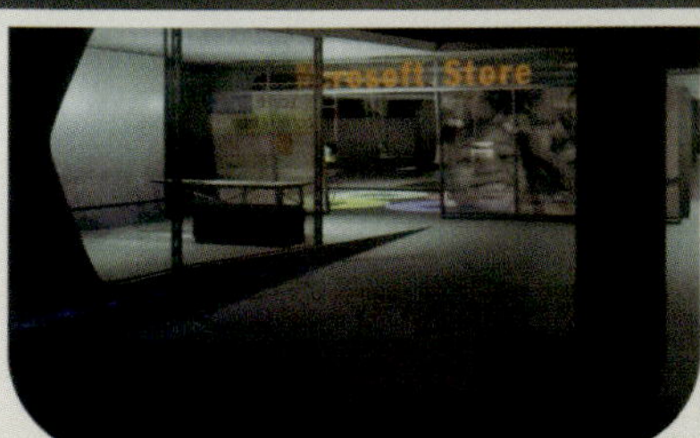

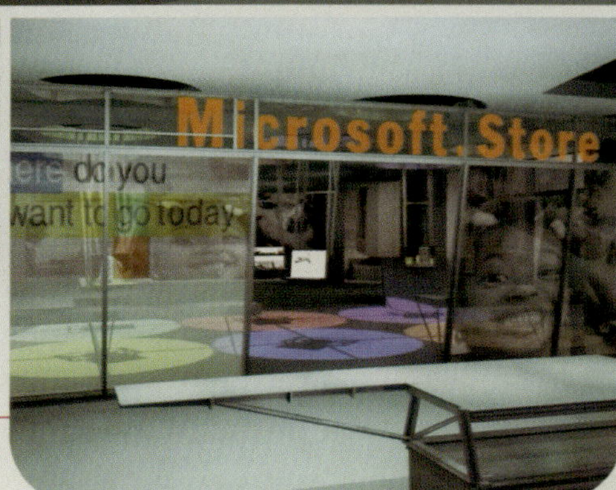

SUBJECT Portrait of Neil Denari

LOCATION Los Angeles, USA

DESIGNER Neil M. Denari

PHOTOGRAPHERS Benny Chan, Fujitsuk Mitsumasa

WRITER Leo Gullbring

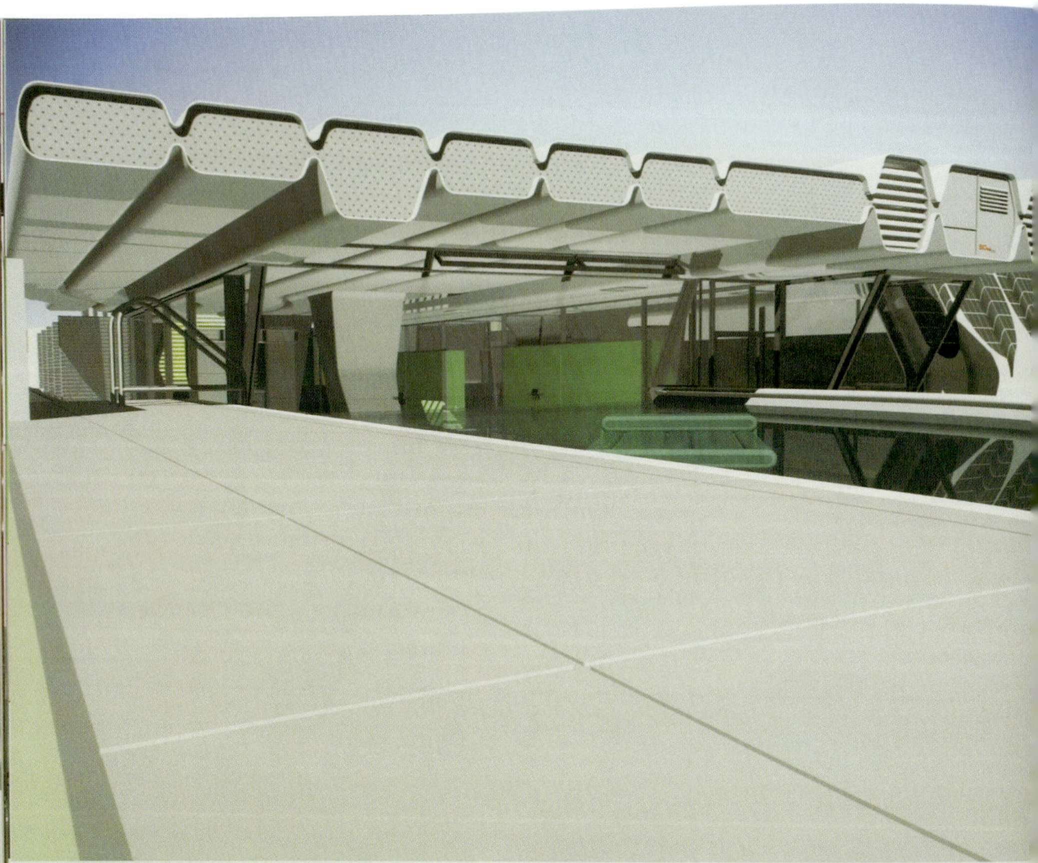

Corrugated Ducts House, Palm Springs, 1998. Desert mountain ranges in the Coachella Valley and the dramatic heat and sunshine of Palm Springs, California, form the context for this experimental house: 2,000 square feet of conditional space with a flexible interior. The scheme revolves around a large, white cavity roof whose mirrored, double-skin construction reflects heat and allows cool air and water to penetrate the interior of the house. Air-handling units are attached to the edges of the roof.

The corrugations are vital to the design of the roof, but they also twist and turn to form random columns/shafts, as well as exterior wall fragments, internal partitions and built-in furniture. Other features are a terrazzo floor and curtain walls made of aluminium and glass.

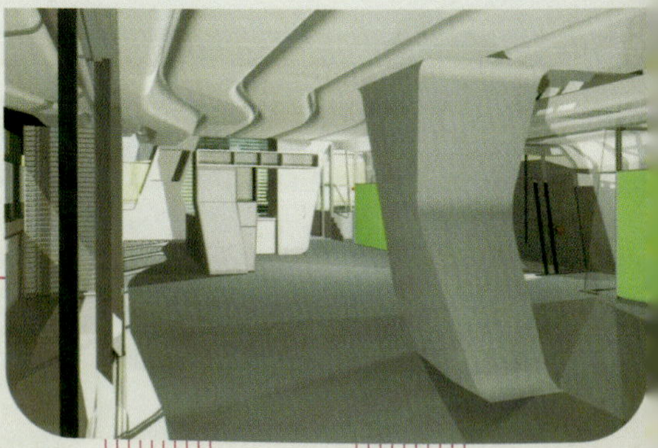

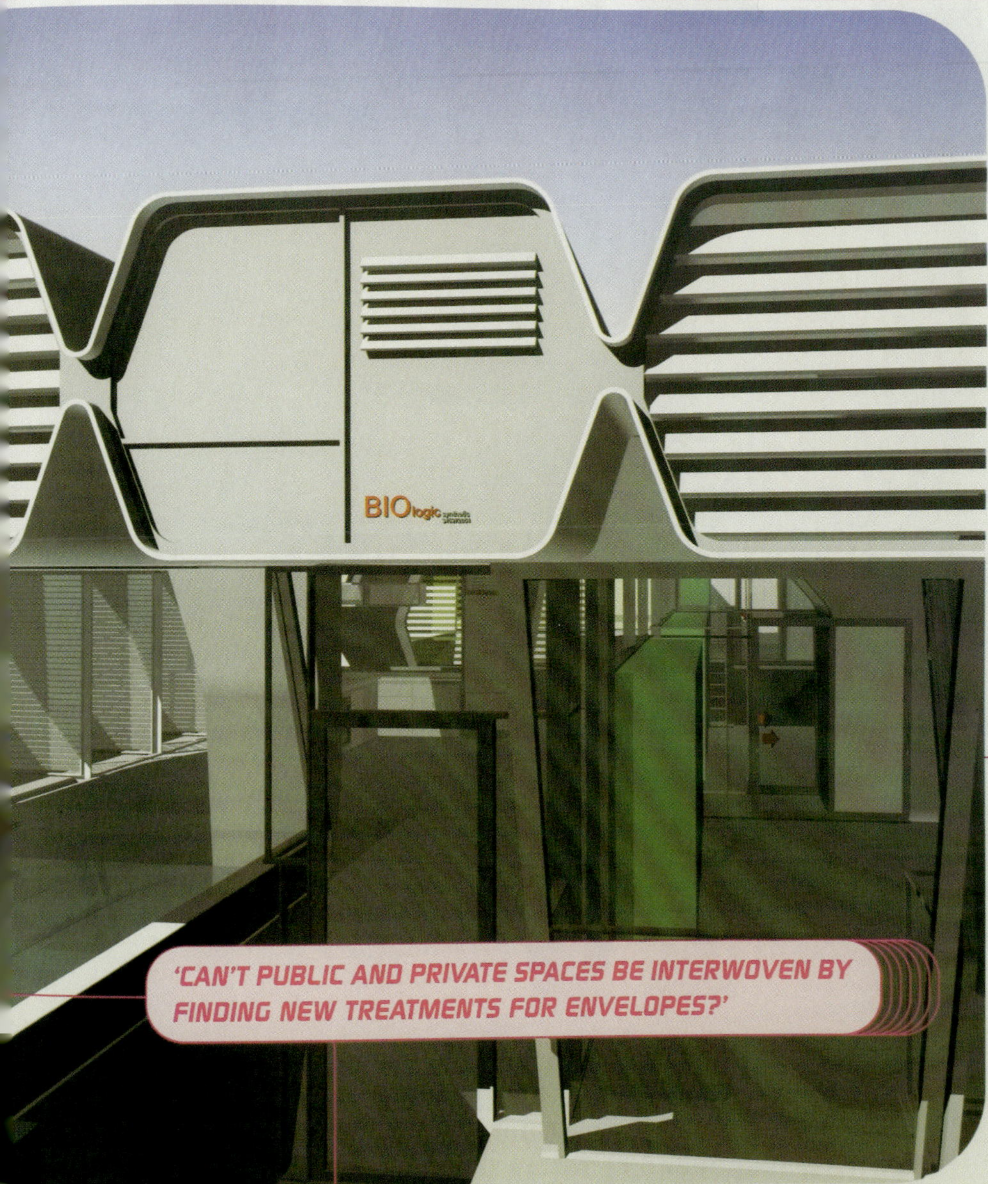

'CAN'T PUBLIC AND PRIVATE SPACES BE INTERWOVEN BY FINDING NEW TREATMENTS FOR ENVELOPES?'

SUBJECT
Portrait of Neil Denari

LOCATION
Los Angeles, USA

DESIGNER
Neil M. Denari

PHOTOGRAPHERS
Benny Chan
Fujitsuk Mitsumasa

WRITER
Leo Gullbring

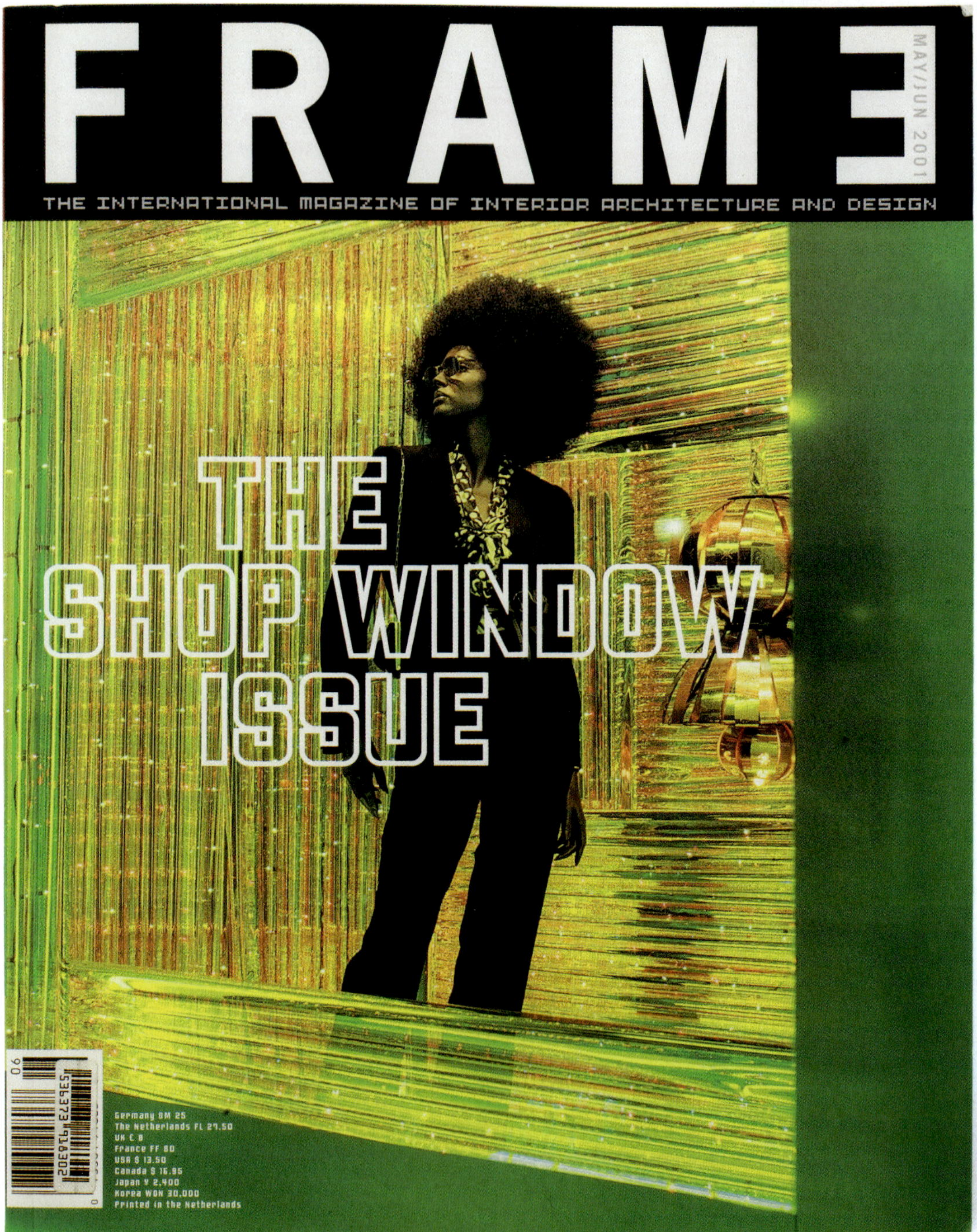

ART DIRECTOR
Roelof Mulder

20

Shop Windows

Frame editor Tessa Blokland about the project

It was the first time we did a feature on shop-window design. The shop window is as important for a retail environment as any other part of the store; it all starts with trying to attract your customer to enter the shop, in order to have them eventually leave the premises with bags of purchased products. The brief we gave to the photographers was that all shop windows should have 'stopping power'.

Did the article have an effect?

A new world of interiors opened for us, and the result was several features in Frame magazine about shop windows (Frame 27 and Frame 38), as well as a book on shop-window design: Forefront: The Culture of Shop Window Design, published in 2005. I have no idea what the effect has been on our readers, but the book got good response from those involved in this field. Finally, some attention!

DESIGNERS
Various

SUBJECT
Twenty shop-window designs in three European cities

LOCATIONS
Amsterdam, Netherlands
London, England
Milan, Italy

PHOTOGRAPHERS
Wouter van den Brink
Ramak Fazel

WRITERS
Jeroen Junte
Daniela Mecozzi
Roberto Monelli

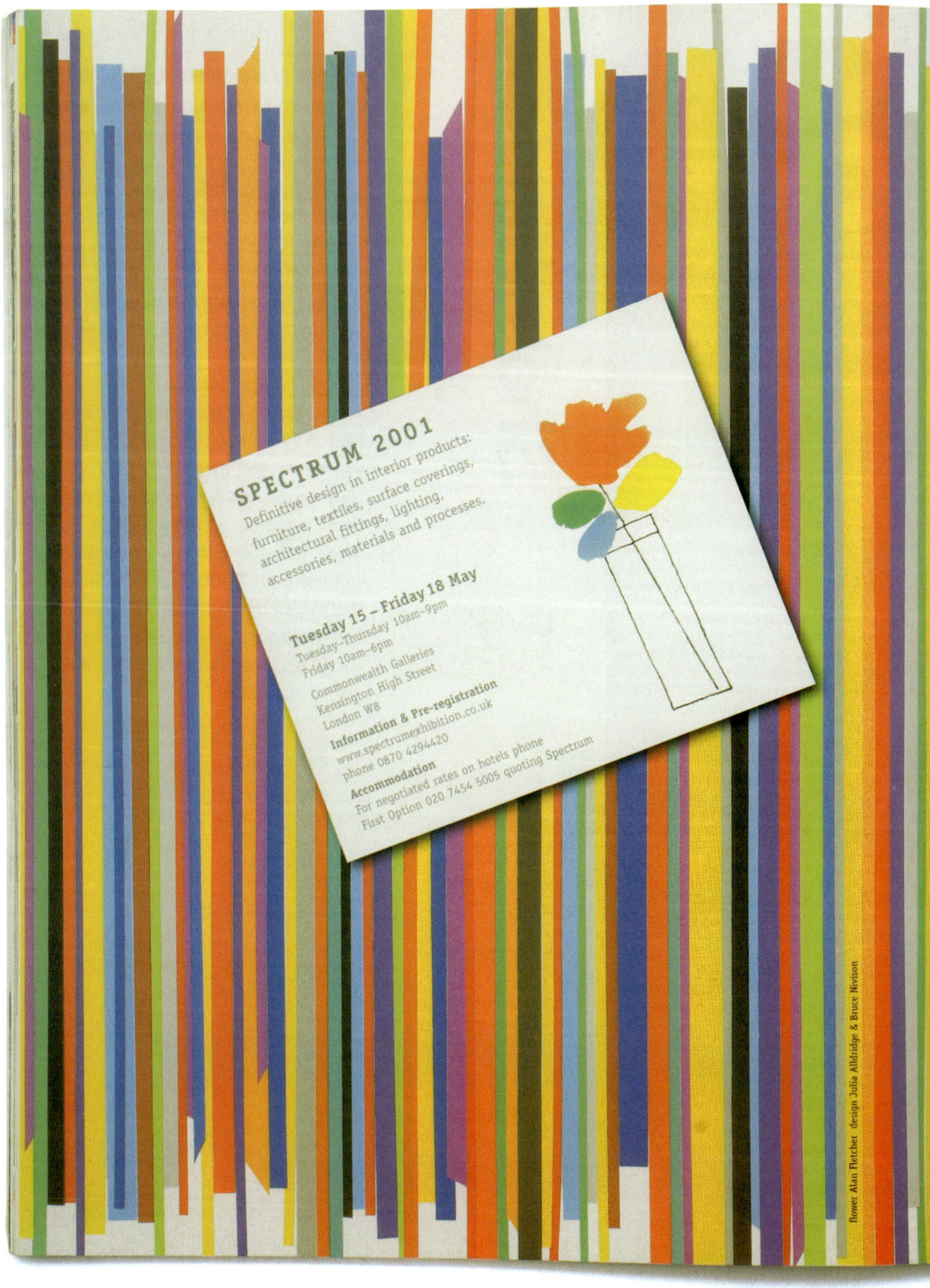
SPECTRUM 2001
Definitive design in interior products:
furniture, textiles, surface coverings,
architectural fittings, lighting,
accessories, materials and processes.
Tuesday 15 – Friday 18 May
Tuesday–Thursday 10am–9pm
Friday 10am–6pm
Commonwealth Galleries
Kensington High Street
London W8
Information & Pre-registration
www.spectrumexhibition.co.uk
phone 0870 4294420
Accommodation
For negotiated rates on hotels phone
First Option 020 7454 5005 quoting Spectrum
flower Alan Fletcher design Julia Alldridge & Bruce Nivison

shop windows

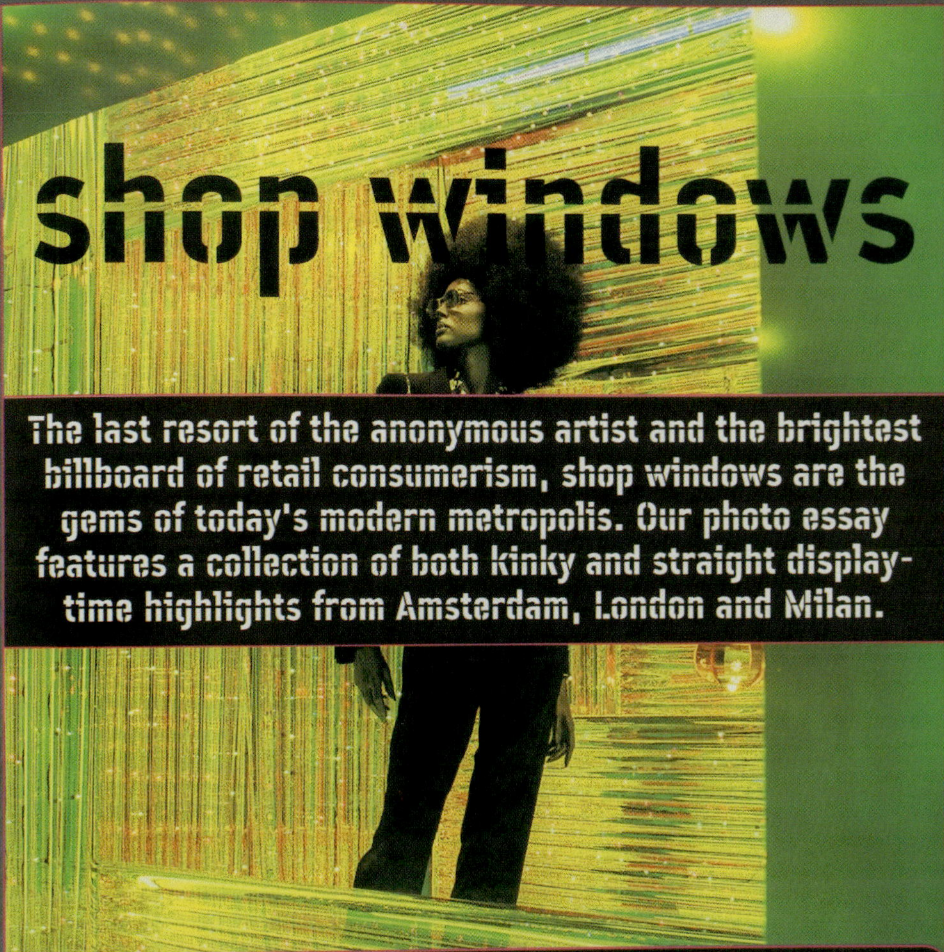

The last resort of the anonymous artist and the brightest billboard of retail consumerism, shop windows are the gems of today's modern metropolis. Our photo essay features a collection of both kinky and straight display-time highlights from Amsterdam, London and Milan.

Amsterdam
Words by Jeroen Junte
Photography by Wouter

London
Words by Daniela Mecozzi
Photography by Wouter

Milan
Words by Roberto Monelli
Photography by Ramak Fazel

SUBJECT
Twenty shop-window designs in three European cities

LOCATIONS
Various, see page 199

DESIGNERS
Various

PHOTOGRAPHERS
Wouter van den Brink
Ramak Fazel

WRITERS
Various, see page 199

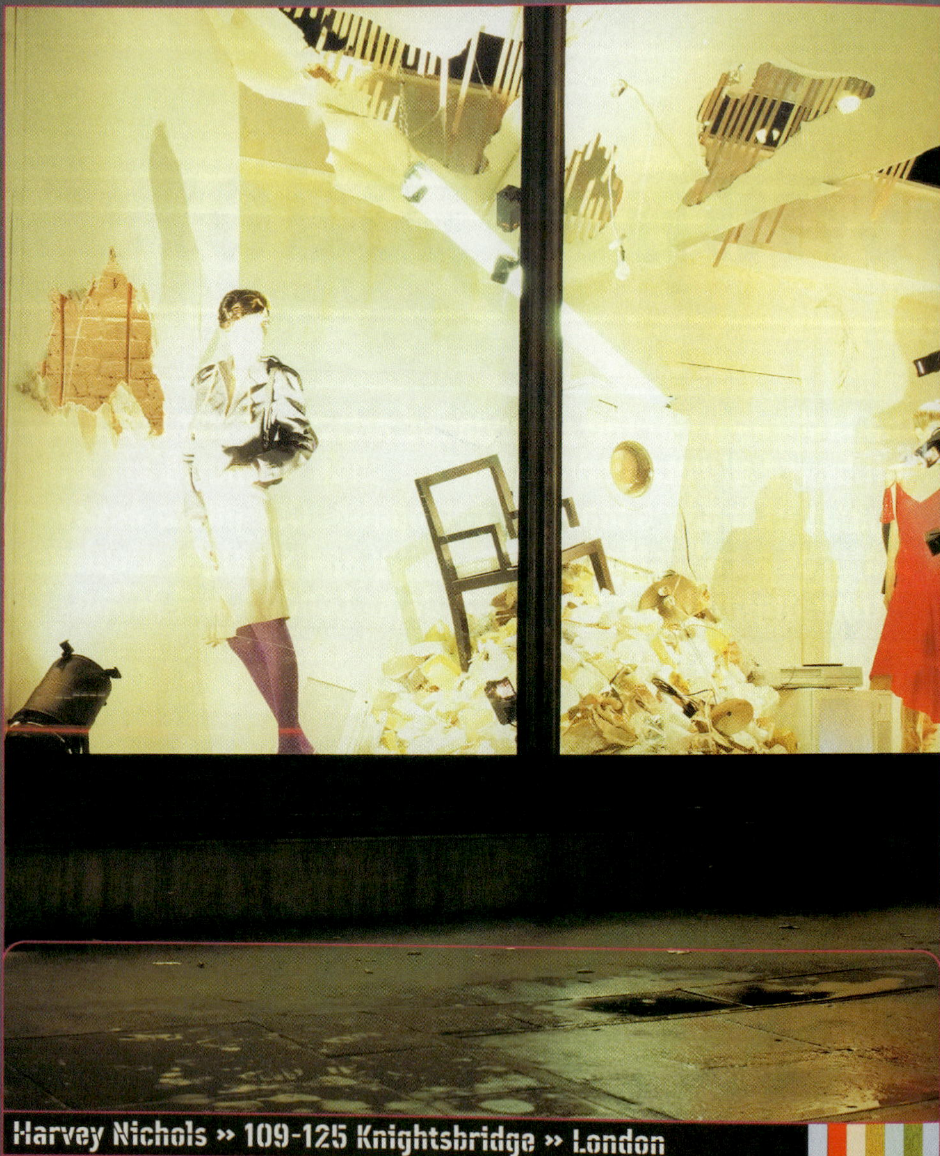

Harvey Nichols » 109-125 Knightsbridge » London

It takes events on a grand scale to surprise or provoke Londoners, so why do the conceptual window displays at Harvey Nichols often manage to jolt their sensibilities? Perhaps it's the design policy, which often plays down any association with glamour, preferring to draw inspiration from the broader social context. Litter and debris provide the setting for the display shown here. Empty fast-food containers, a collapsing ceiling and obsolete machinery reflect the neurotic way in which we live and consume. The message is reinforced by the sound of leaking pipes, the frantic spinning of a washing-machine drum and throbbing flashes of light from a TV screen. By displaying Harvey Nichols fashions and furniture among the chaos, the retailer points an accusing finger at the contrast between quality and quantity.

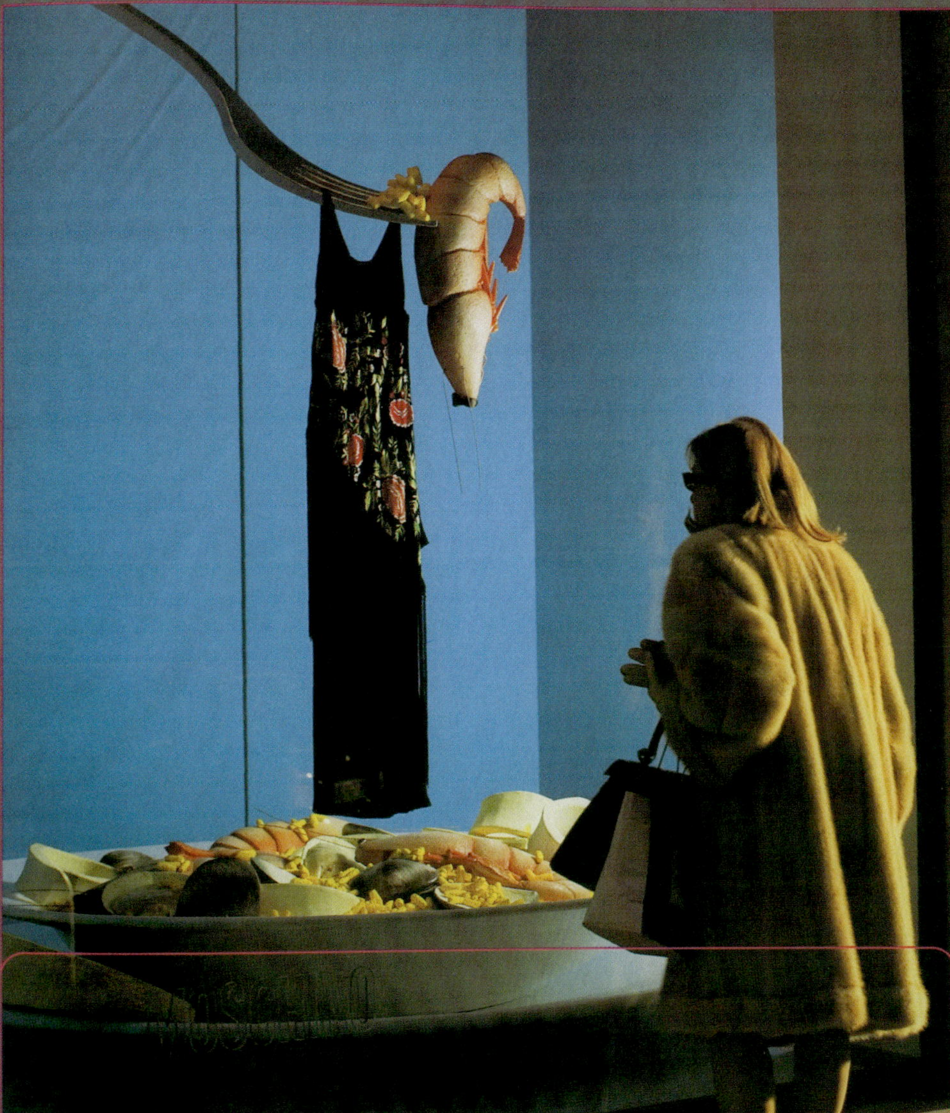

Moschino » Via S. Andrea 12 » Milan

Featured on the Moschino menu: Italian dressing. Take a classic concept like the theatrical window display built around a three-dimensional still life and go from there. High-tech? No. Polystyrene and silicone? Hardly. Nouvelle cuisine? Negative. Mediterranean fare? Take another look. It's a dress. If Moschino wasn't able to 'Stop the Fashion System', Studio Moschino – the label's in-house design team – certainly stops pedestrians dead in their tracks with the most irreverent window display on Via S. Andrea – just steps away from Armani's austere nod to shoppers.

SUBJECT
Twenty shop-window designs in three European cities

LOCATIONS
Various, see page 199

DESIGNERS
Various

PHOTOGRAPHERS
Wouter van den Brink
Ramak Fazel

WRITERS
Various, see page 199

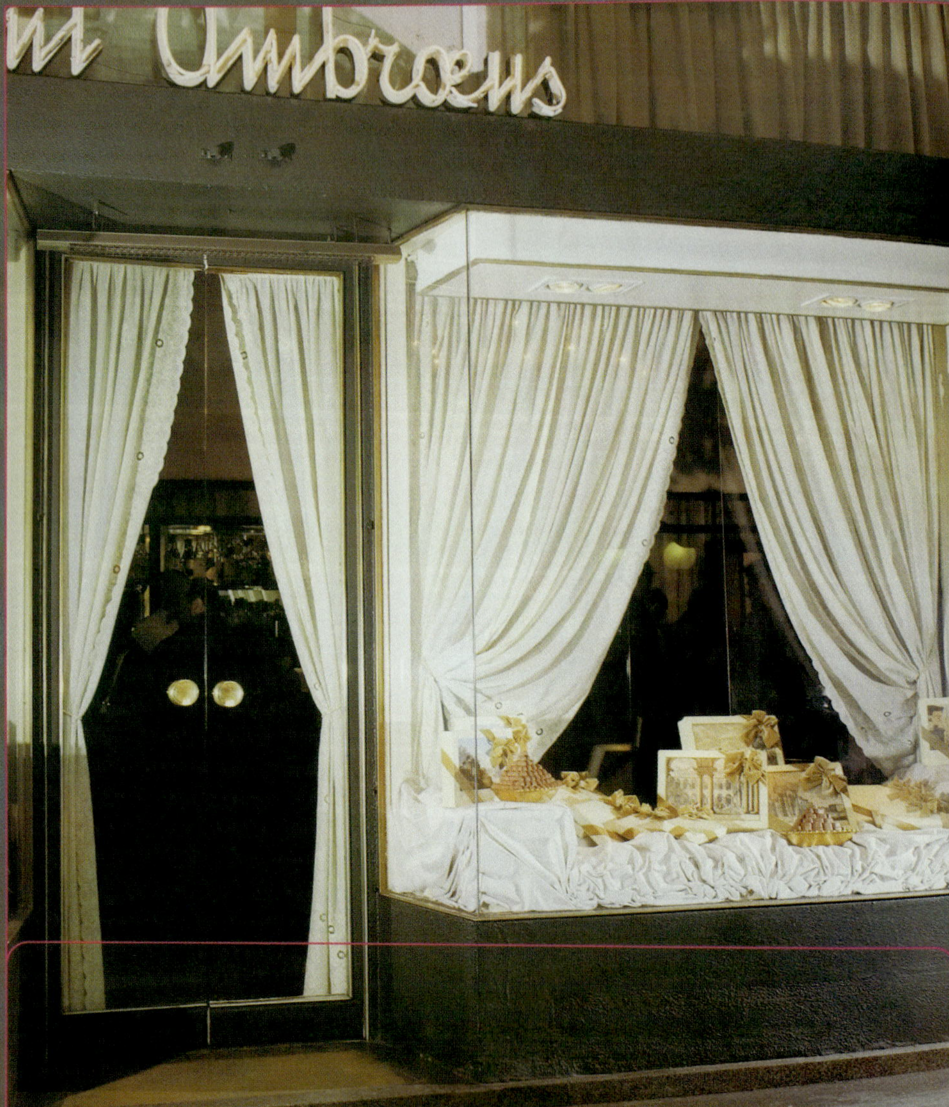

Sant Ambroeus » Corso Matteotti 7 » Milan

This pastry-shop window presentation designed by Luciano Vismara (chocolate sculptures) and Simona Merlin (boxes and displays) is a blend of tradition and theatre. Stage curtains, satin drapery and etched mirrors are featured in a series of six windows, each with a different theme. Chocolate sculptures, sweets and gift boxes snuggle together like precious eggs in a nest. The baroque impression made by the window display continues inside, where briarwood boiserie (woodwork) and salmon-pink walls form the setting for two grand Murano chandeliers that would not be out of place at La Scala.

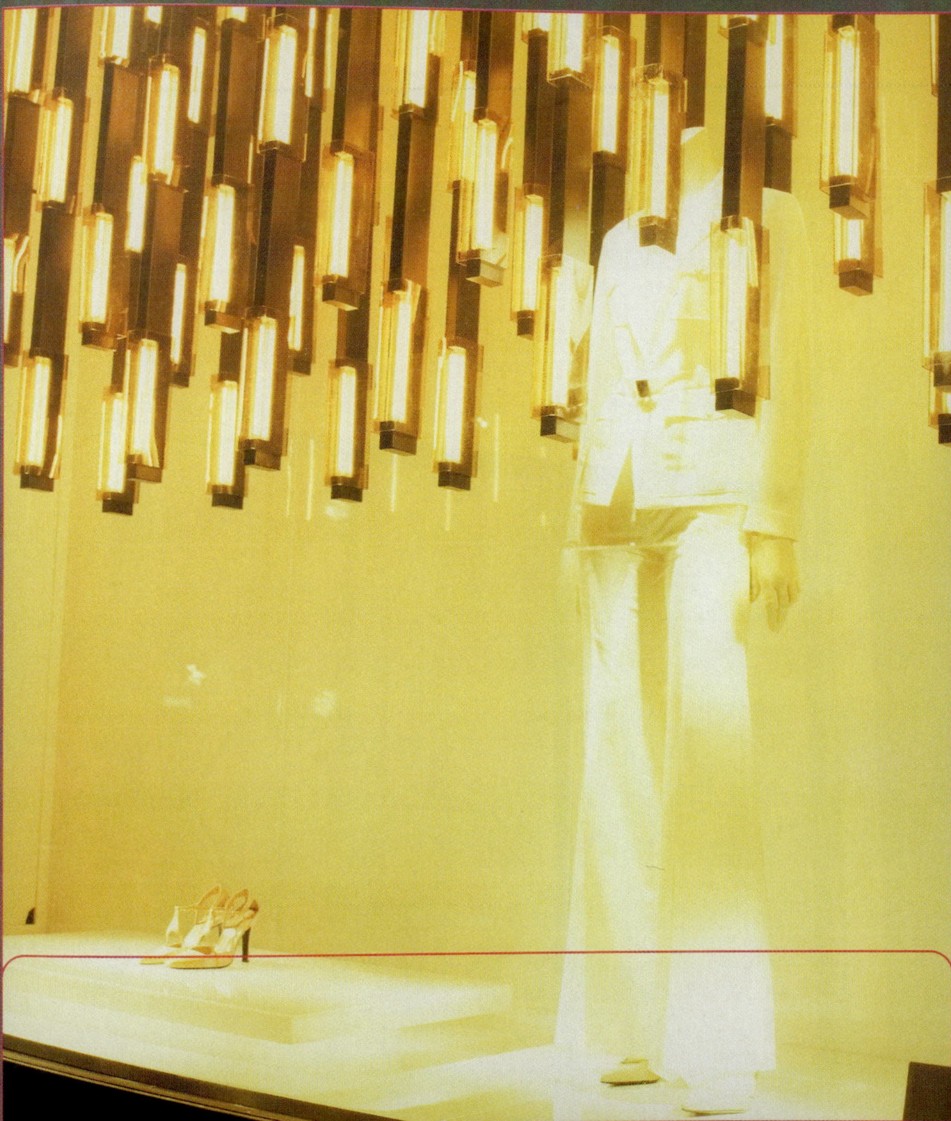

Gucci » 32 Old Bond Street » London

A sea of whiteness surrounds a handful of fashions and accessories on pedestals – the embodiment of minimalism? Not when the Gucci team adds bold, '70s-style, vertical brass lighting, which transforms the deceivingly simple backdrop into a glamorous stage set. Flat chandeliers provide a reverberating frame for the products on display, thus converting the snow-white setting into a highly reflective environment. Though each fashion and accessory has a glow of its own, the result is a coherently unified design.

SUBJECT
Twenty shop-window designs in three European cities

LOCATIONS
Various, see page 199

DESIGNERS
Various

PHOTOGRAPHERS
Wouter van den Brink
Ramak Fazel

WRITERS
Various, see page 199

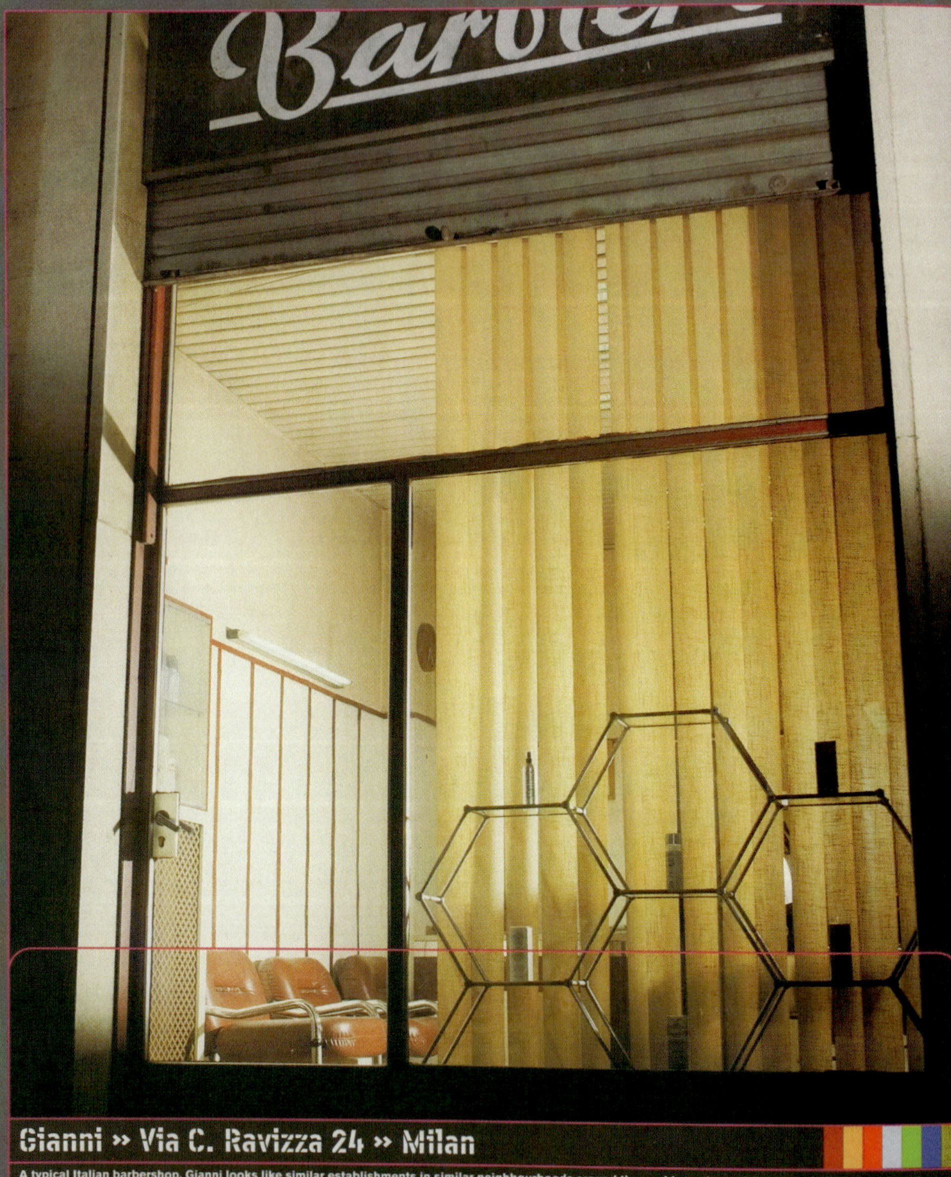

Gianni » Via C. Ravizza 24 » Milan

A typical Italian barbershop, Gianni looks like similar establishments in similar neighbourhoods around the world – a place where guys of all ages go to get a haircut, to chew the fat, and to thumb through magazines that laud the beauty of cars and the female body. The furniture is timeless and the window display basic: a bottle of shampoo, a can of hairspray and a dubious container of anti-balding lotion. Add to this minimalist version of iconographic imagery a photo of James Dean or a traditional print of the early-20th-century tonsorial technician, and you've reached the pinnacle of barbershop design. Why are spaces dedicated to 'the look' almost always devoid of 'look appeal'?

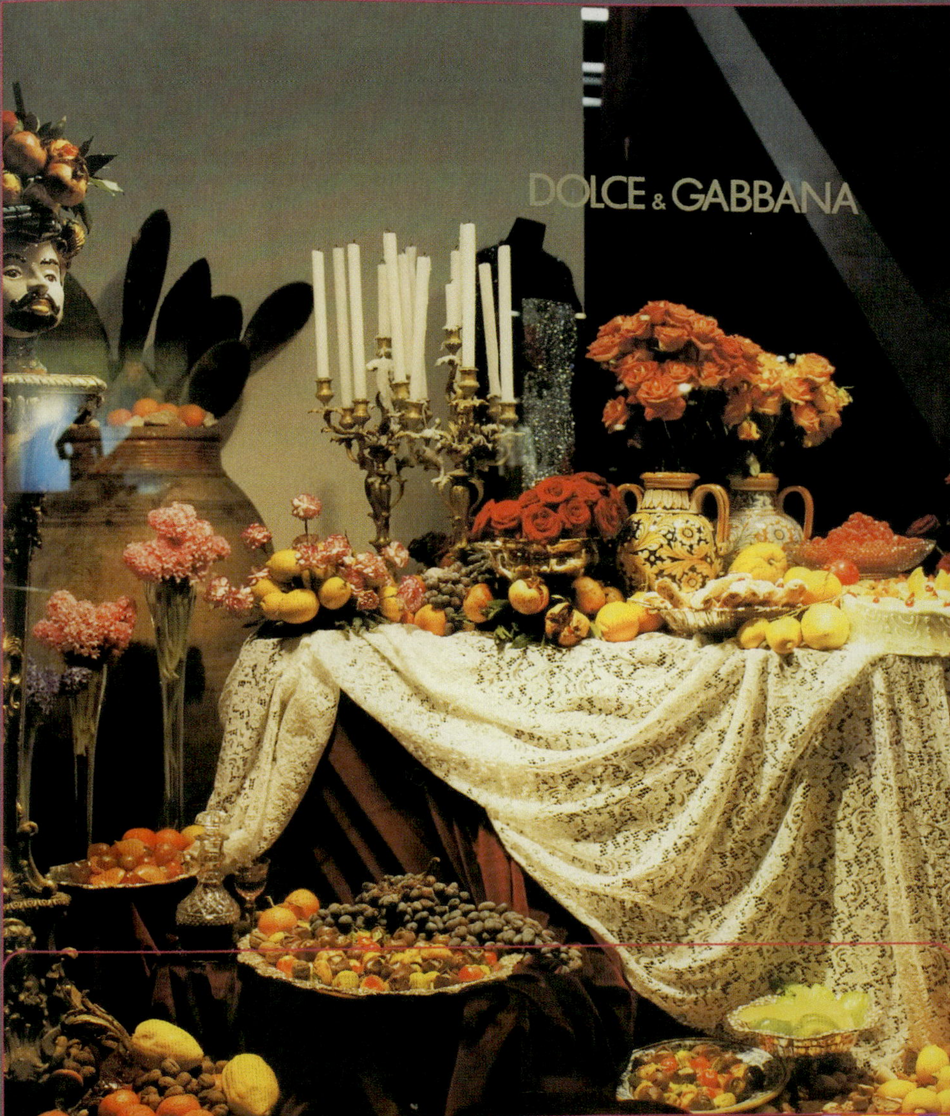

Dolce & Gabbana » 175 Sloane Street » London

The Christmas display at Dolce & Gabbana celebrates the typical Mediterranean atmosphere surrounding this holiday, which revolves around the gathering of family and friends. The shop window focuses not on clothes but on a luxurious table decked out in a white lace tablecloth. Glistening chandeliers, floral arrangements and opulent baskets of fruit complete the colourful picture. Like a Renaissance tableau vivant, the ornate scene conveys a message of tradition and festivity: powerful references to the distinctive style of these two designers.

SUBJECT
Twenty shop-window designs in three European cities

LOCATIONS
Various, see page 199

DESIGNERS
Various

PHOTOGRAPHERS
Wouter van den Brink
Ramak Fazel

WRITERS
Various, see page 199

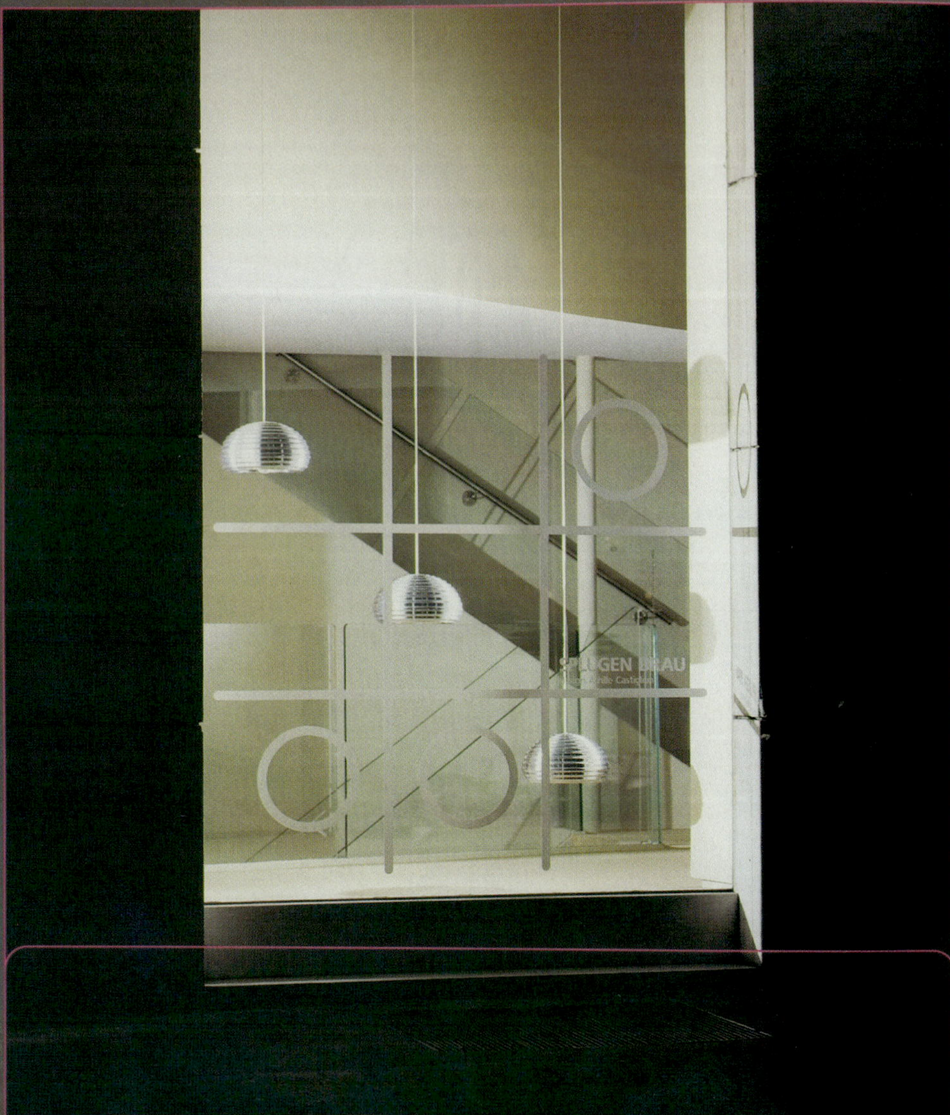

Flos » Corso Monforte 9 » Milan

'Few and simple' frequently describes elements that get results. In his design for shop windows at the Flos showroom, Ferruccio Laviani envisioned the lamps as pieces in an elementary game of noughts and crosses. In the display shown here, Laviani suspended lamps created by Achille Castiglioni at different heights, while making no reference to the setting and no attempt to demonstrate lighting effects. Visitors entering the showroom become part of the display, which has a stairway leading up to the office area as its backdrop.

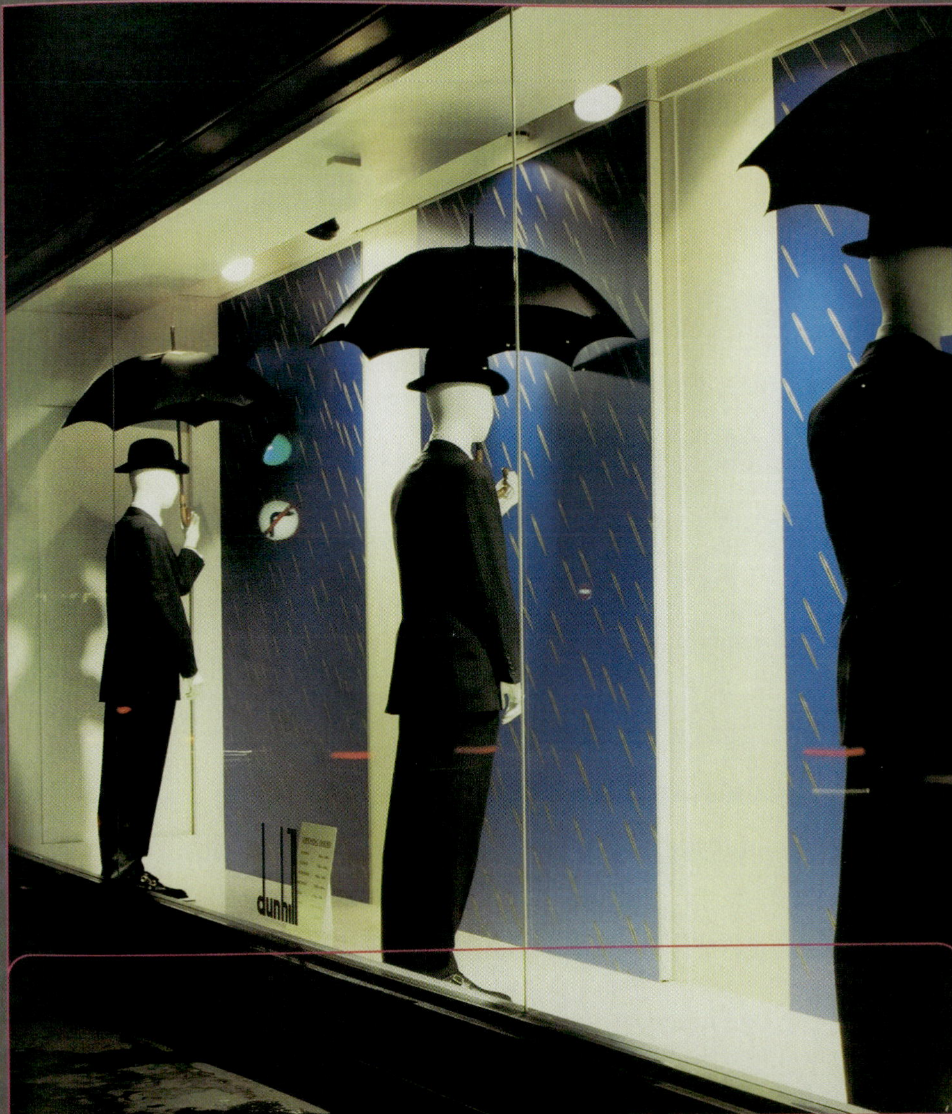

Alfred Dunhill » 30 Duke Street » London

Building a winter window display around the theme of rain takes more than an ounce of courage, particularly in London. Clearly inspired by the work of Magritte, the team at Alfred Dunhill created a surreal setting for the display of traditionally tailored men's suits. Standing in a row with their backs to the street, as if to underline the introverted mood of the display, mannequins in bowlers hold umbrellas as rain falls from a graphically rendered sky. The intriguing background animates the whole display and subtly emphasises the fine quality of the various suit fabrics.

SUBJECT
Twenty shop-window designs in three European cities

LOCATIONS
Various, see page 199

DESIGNERS
Various

PHOTOGRAPHERS
Wouter van den Brink
Ramak Fazel

WRITERS
Various, see page 199

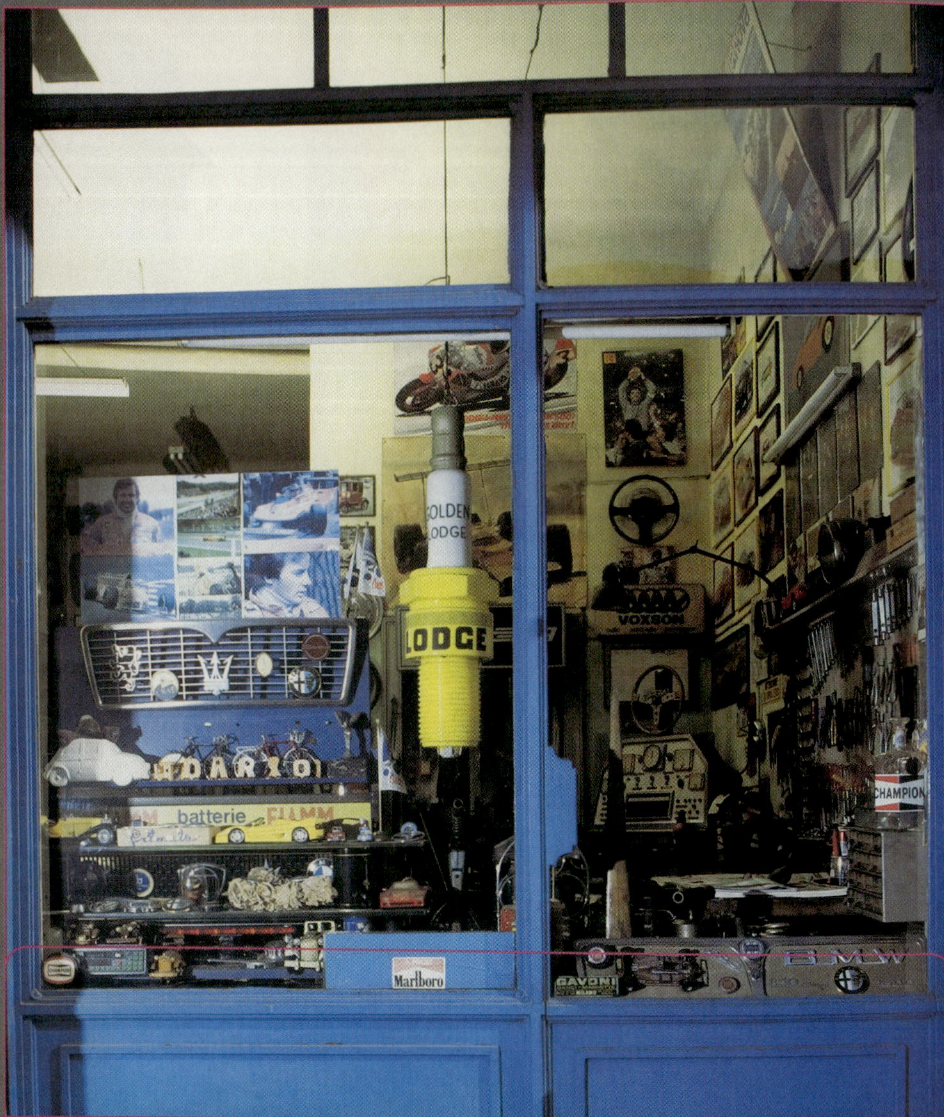

Dario Cerquoni » Via A. Verga 22 » Milan

You're not on Route 66 in the Arizona desert, but outside a small Italian service station carved out of a residential building. Over a span of 35 years these premises have become a museum of automotive memorabilia. A veritable altar to the world of cars, the shop window offers a harvest of calendars, photos of Formula One champions, model cars and a crankshaft. The luminosity of a giant spark plug is the only indicator that a product is for sale inside the shop, where the visitor is greeted by tools, tires, Pirelli calendars and simple paintings featuring automobiles. The proprietor is more than just a collector; he uses his booty to build mechanical illuminated sculptures for fellow artists and collectors, not to mention the occasional fascinated newcomer to this address.

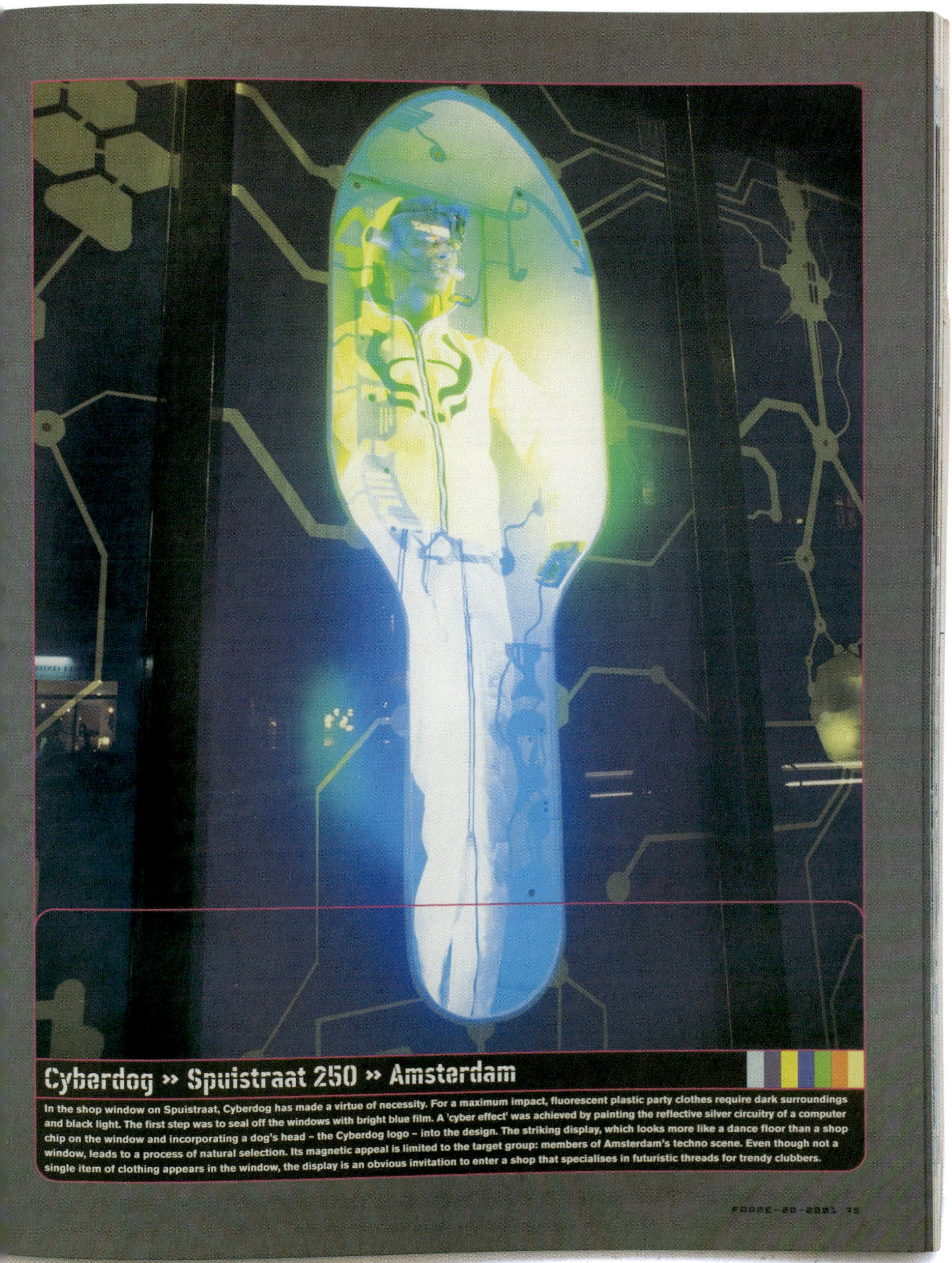

Cyberdog » Spuistraat 250 » Amsterdam

In the shop window on Spuistraat, Cyberdog has made a virtue of necessity. For a maximum impact, fluorescent plastic party clothes require dark surroundings and black light. The first step was to seal off the windows with bright blue film. A 'cyber effect' was achieved by painting the reflective silver circuitry of a computer chip on the window and incorporating a dog's head – the Cyberdog logo – into the design. The striking display, which looks more like a dance floor than a shop window, leads to a process of natural selection. Its magnetic appeal is limited to the target group: members of Amsterdam's techno scene. Even though not a single item of clothing appears in the window, the display is an obvious invitation to enter a shop that specialises in futuristic threads for trendy clubbers.

FRAME-20-2001 75

SUBJECT
Twenty shop-window designs in three European cities

LOCATIONS
Various, see page 199

DESIGNERS
Various

PHOTOGRAPHERS
Wouter van den Brink
Ramak Fazel

WRITERS
Various, see page 199

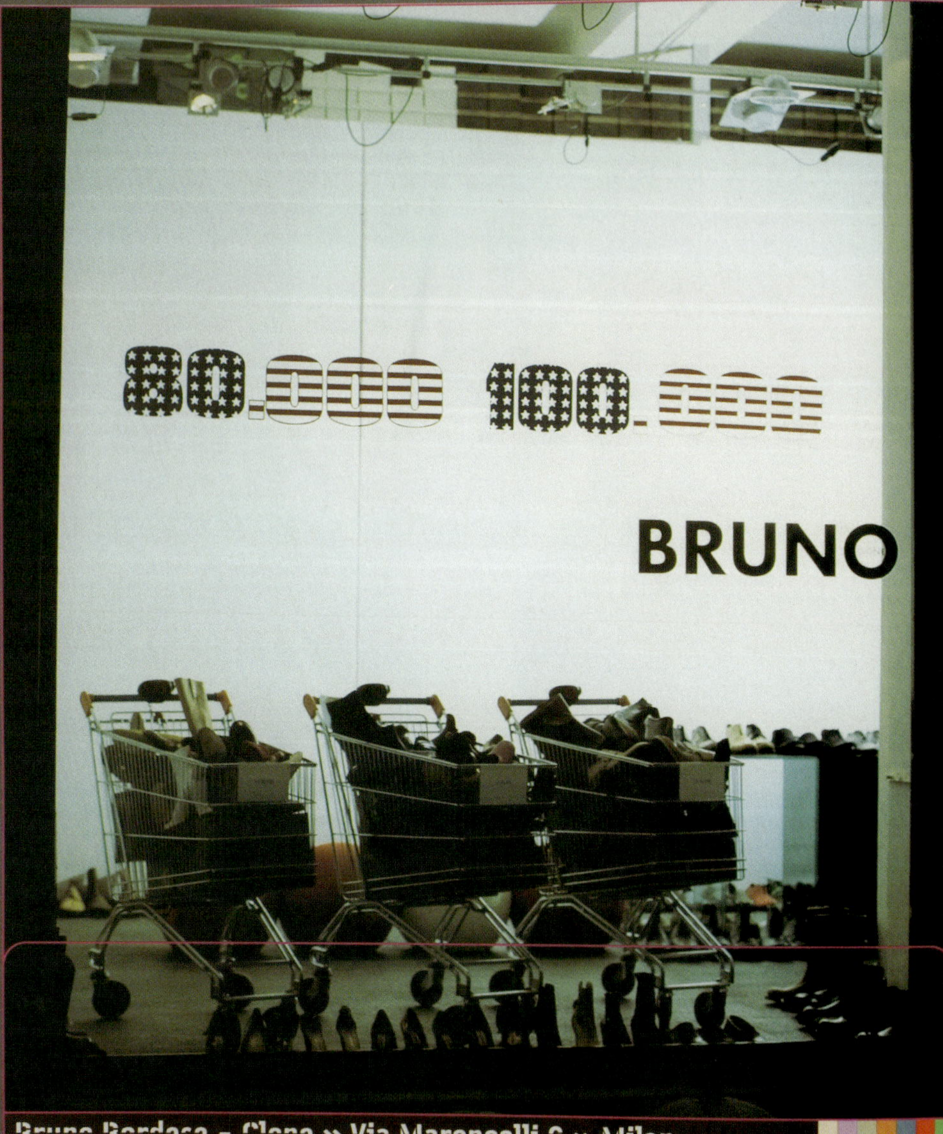

Bruno Bordese - Clone » Via Maroncelli 2 » Milan

This fashion-conscious neighbourhood is the perfect setting for two shop windows with a rather 'tame trash' display of shoes and trendy accessories. Designed by architect Massimiliano Raggi, the outlet carries the Clone collection: footwear created by Bruno Bordese of Milan. The two men pooled their talents to come up with the window display, which revolves around an experimental combination of materials and colours. It somewhat ironically promotes mass consumption of products available only in limited quantities: models from last season's collection, one-of-a-kind pieces (presented in shopping carts) and special creations by 'guest designer' Vivienne Westwood. Potential customers also find an orderly arrangement of shoes on the grey concrete pavement.

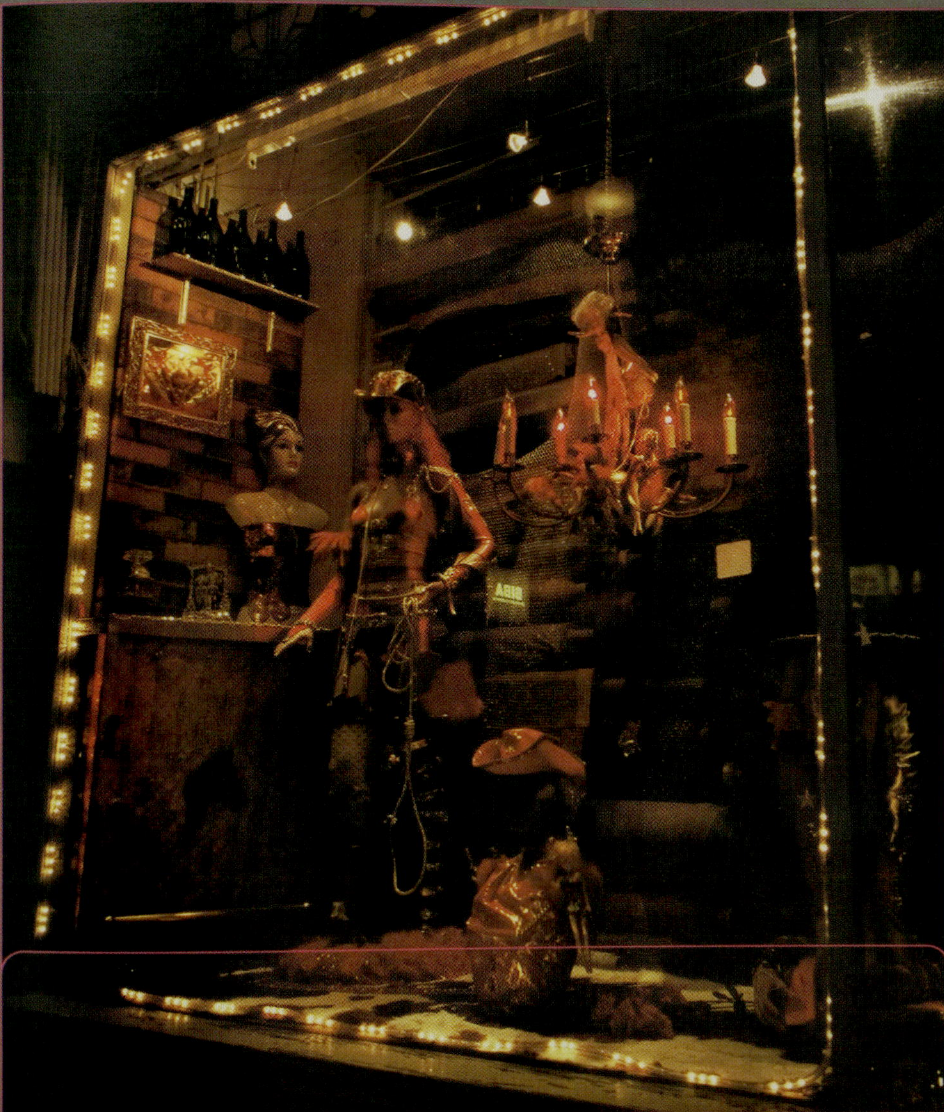

A. Boeken » Nieuwe Hoogstraat 31 hs » Amsterdam

Stylist Thea Lichtenberg's display for A. Boeken Fabrics & Haberdashery (only a block away from Amsterdam's notorious Red Light District) is a titillating still life. 'It refers to Madonna's cowboy look, but when we mention Madonna, we're talking about the biggest sex symbol around today. That's the idea behind cowgirls in a red brothel bar.' Although there's less excitement inside, where bolts of fabric share shelf space with buttons, ribbons and embroidery silk, Lichtenberg is quick to point out the connection. 'The display implies more than just the merchandise. It relates to what can be made with it. Everything in the window came from the shop. I want to tempt customers to pick up a needle and thread and create something themselves.'

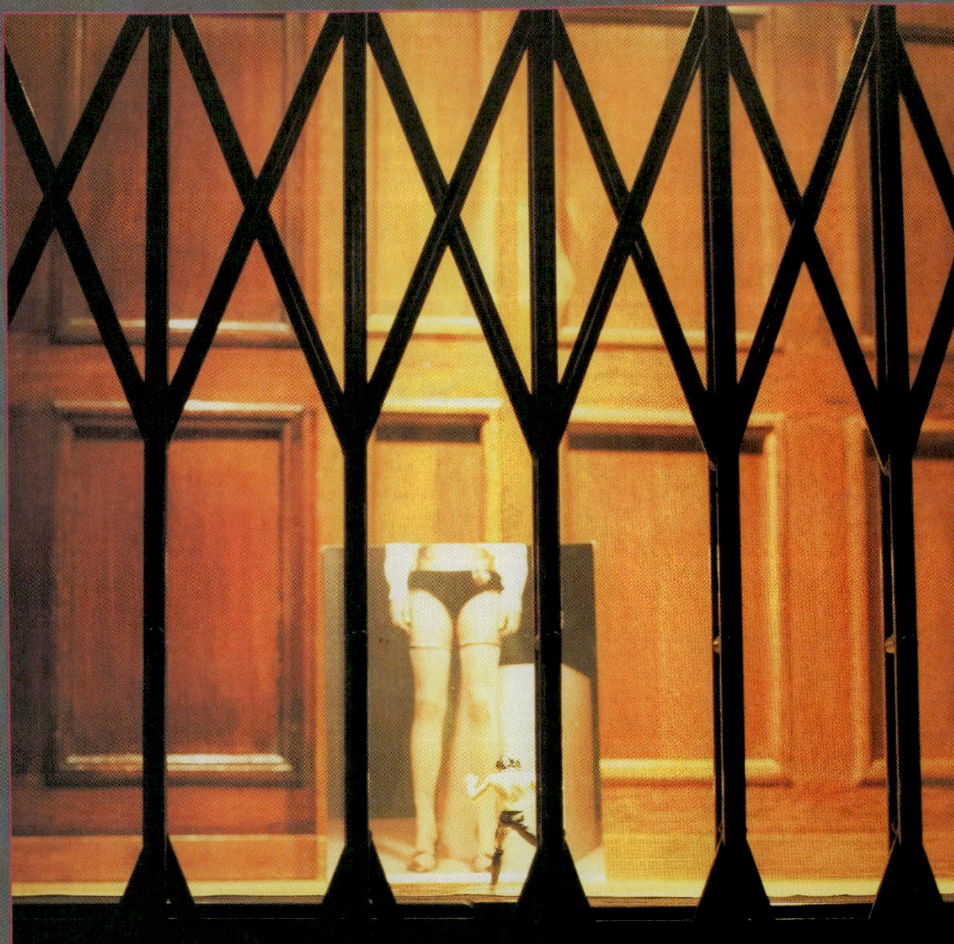

Paul Smith » 40-44 Floral Street » London

The display window at Paul Smith links the entrances of two shops: Men and Women. This link often becomes an ironic comment on the relationship between one half of our society and the other. In this case, the cover of Tank magazine provides both scale and theme. A woman photographed from the waist down, her tights at half-mast, menacingly confronts a small, partly naked man whose masculine features – from muscles to moustache – are neutralised by his diminutive size. Viewers of both sexes are left to draw their own conclusions.

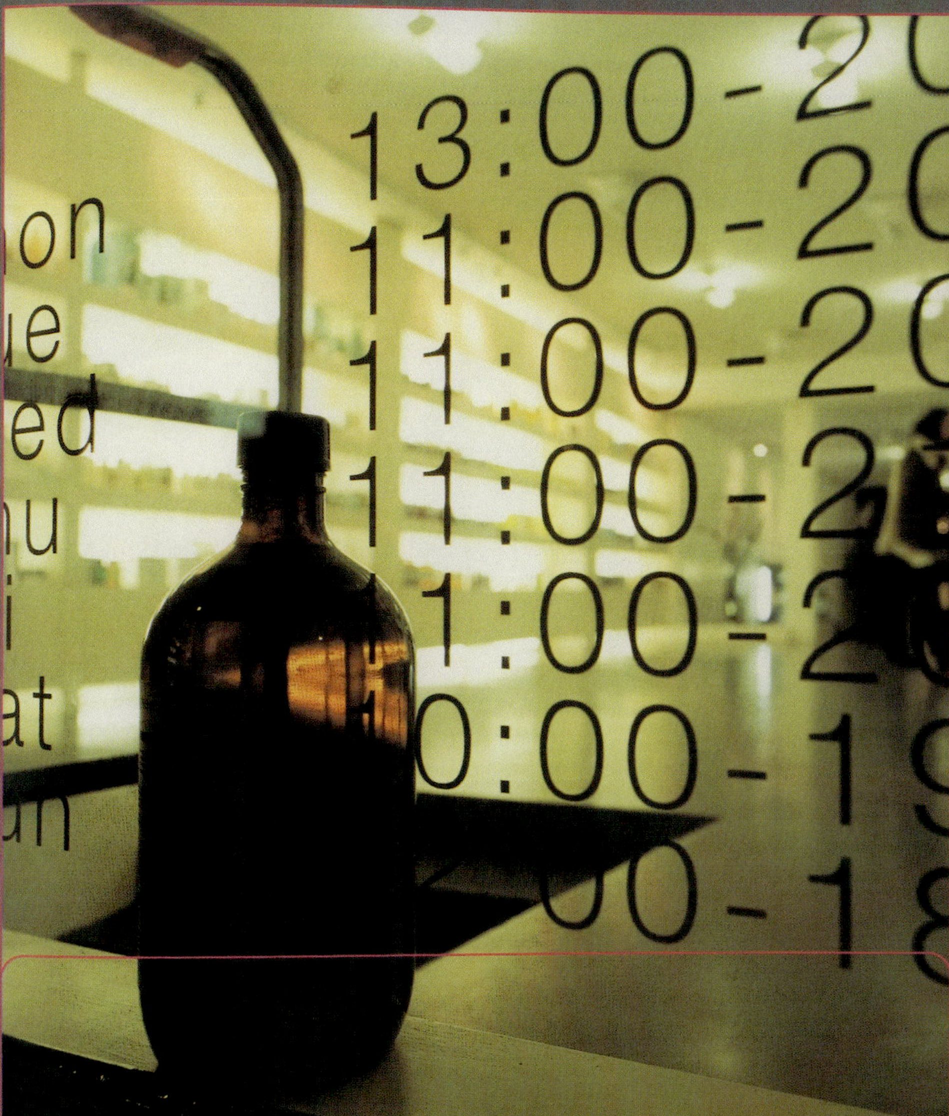

Skins » Runstraat 9 » Amsterdam

The display window at Skins is the shop itself. The concept that a 'cosmetics lounge' like this one doesn't need extra frills is expressed by the shop's clean, contemporary look. Both front and rear façades are fully glazed, giving the interior the benefit of warm, natural lighting. Lettering on the windows is minimal but clear, and the sober, functional interior is equally up-to-date. The message: Our products satisfy the demands of the 2000s. At the same time, the shop radiates the ambience of an old-fashioned apothecary. Orderly rows of jars, tubes and boxes line the long shelves. The only items displayed in the window are the classic brown bottles once used to dispense medicines. The message: Products sold here are still made using time-honoured methods.

SUBJECT
Twenty shop-window designs in three European cities

LOCATIONS
Various, see page 199

DESIGNERS
Various

PHOTOGRAPHERS
Wouter van den Brink
Ramak Fazel

WRITERS
Various, see page 199

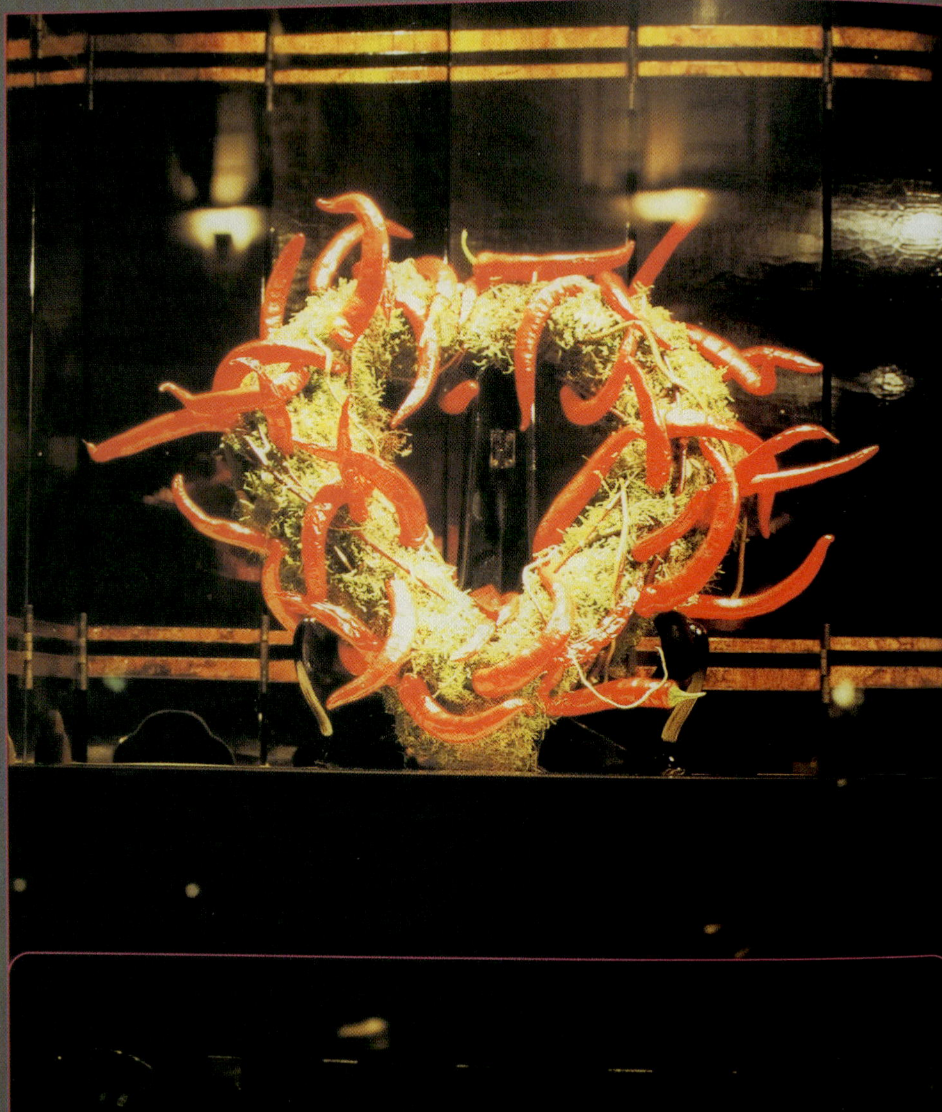

Chanel Fine Jewellery » 173 New Bond Street » London

Among the array of interpretations on the romantic theme of St Valentine's Day is a little gem at Chanel Fine Jewellery, whose shop window presents a piquant image with a decidedly French touch. Dark wood panelling frames a heart-shaped wreath featuring hot chilli peppers. With its subtle sense of humour, the display forms an interesting contrast between the contained elegance of the setting and the wild exuberance of the colourful arrangement. The message is that the secret of enduring love and the thrill of passion is plenty of spice.

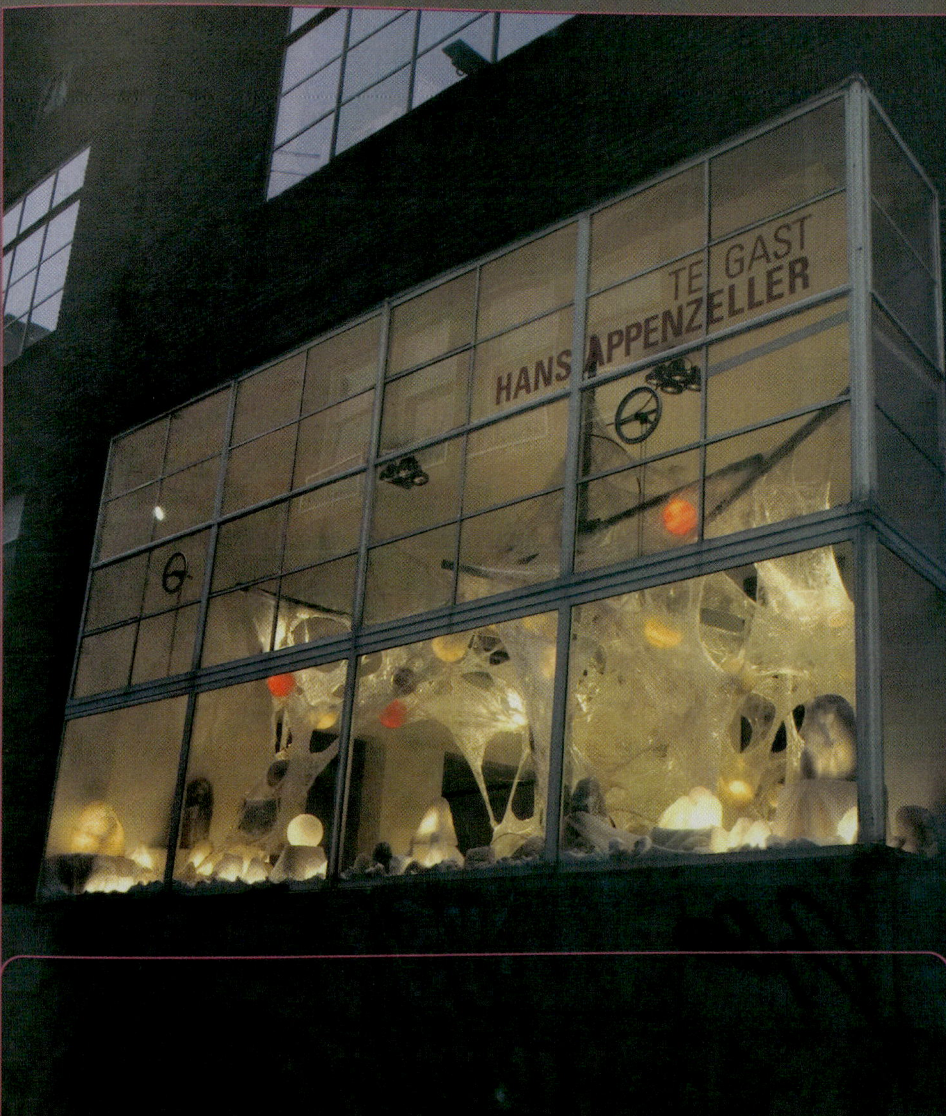

Hans Appenzeller » Grimburgwal 1 » Amsterdam

Jewellery maker Hans Appenzeller calls on a different designer every time he wants to update his shop window-cum-office space. This plastic-tape cobweb is the work of Marsel Pott, a graduate of the Design Academy in Eindhoven. Pott previously used the same material for lamps and stage sets. Here his intricate structure is the décor for a product sold in Appenzeller's jewellery shop: rock-salt lamps. Pott has more or less woven the lamps into the plastic filaments of the web and, at the same time, integrated the office furniture into the design. Passers-by may not realise exactly what's being promoted, but the enchanting light of the lamps in this sci-fi entourage will certainly whet their curiosity. And that's what creating window displays is all about.

SUBJECT
Twenty shop-window designs in three European cities

LOCATIONS
Various, see page 199

DESIGNERS
Various

PHOTOGRAPHERS
Wouter van den Brink
Ramak Fazel

WRITERS
Various, see page 199

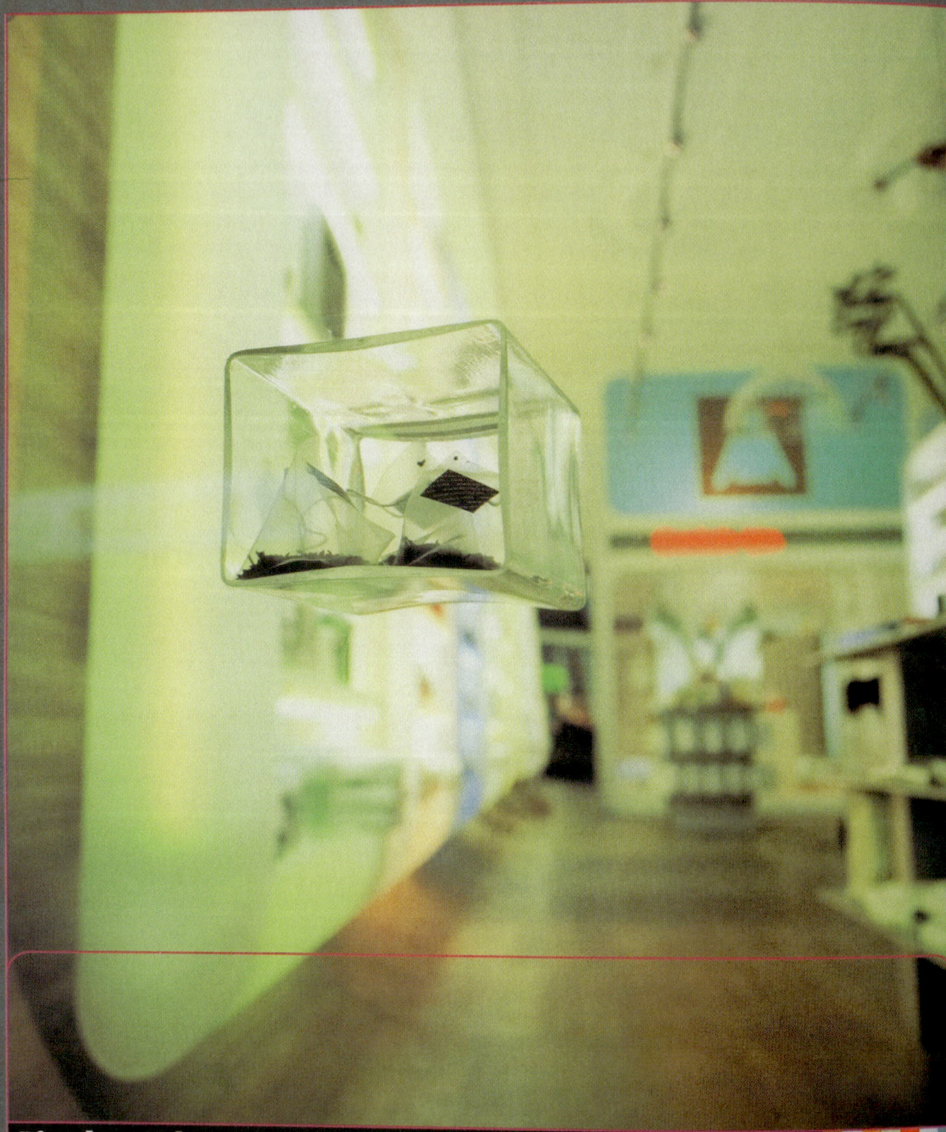

Rituals » Kalverstraat 73 » Amsterdam

Rituals on Kalverstraat supports its virtual counterpart – the e-shop featured on the Rituals website – as well as the company catalogue. Customers with questions and complaints like the idea of a physical contact point, with real people in a tangible setting. 'We want the shop to convey a sense of openness,' says Rob Wagemans of Concrete, the outfit responsible for Rituals. 'A wide-open shop window allows people to see the entire interior from the street. And what you see is what you get.' Rituals sells what it calls 'care products', a term that includes everything from tea and shaving soap to towels. Small glass containers displaying the firm's products sparkle in the window like drops of rain. 'All the products are linked to water,' Wagemans explains, 'which is transparent, just like glass.'

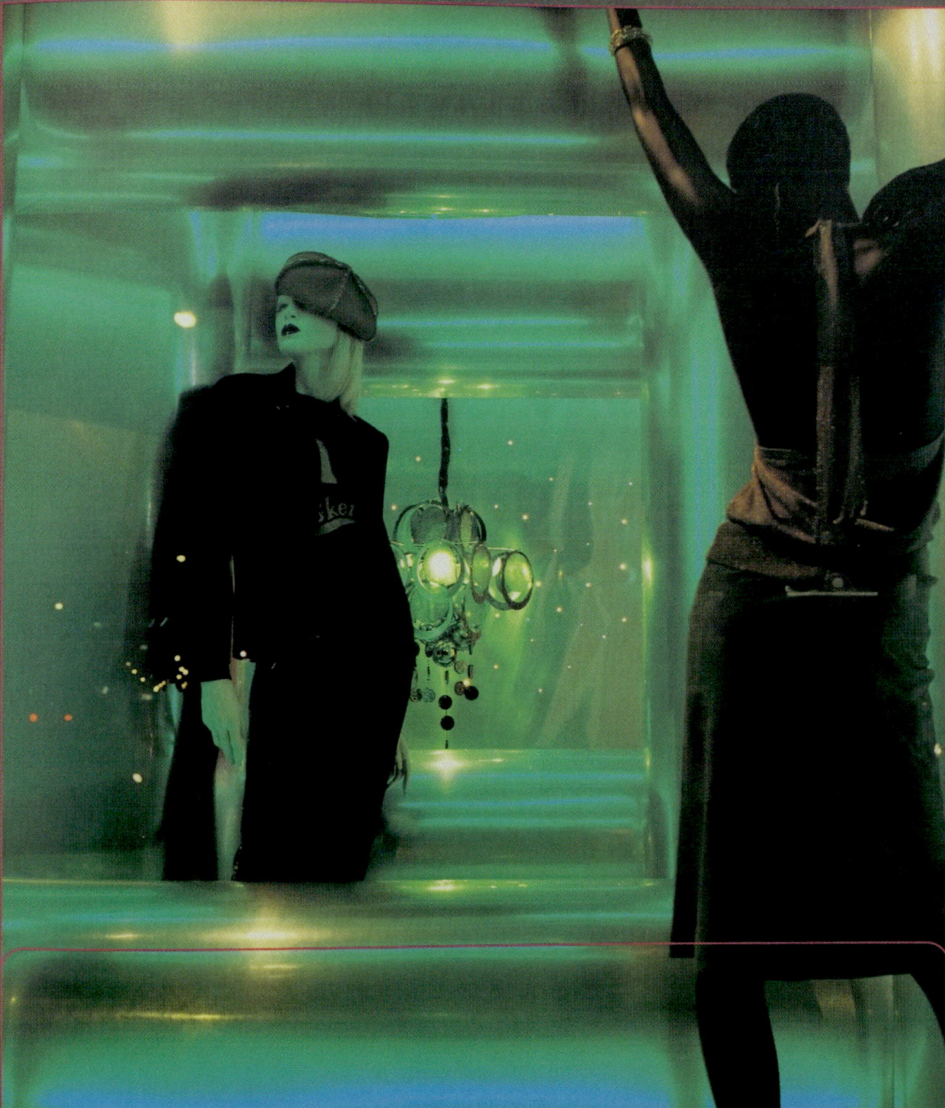

Dickins & Jones » 224 Regent Street » London

The theme explored by the Dickins & Jones design team is the month of December, conceived as one long stretch of parties that culminates in the Christmas season. Framed by curved sheets of metal in matt silver, each window is a stage for mannequins dressed to dance the night away at a '70s disco. The layered effect of the design creates an illusion of depth, which is reinforced by the reflective surfaces of stretched plastic wrap, metal sheets and glittered boards. Holiday fashions can step right out of this festive scene and go clubbing.

SUBJECT
Twenty shop-window designs in three European cities

LOCATIONS
Various, see page 199

DESIGNERS
Various

PHOTOGRAPHERS
Wouter van den Brink
Ramak Fazel

WRITERS
Various, see page 199

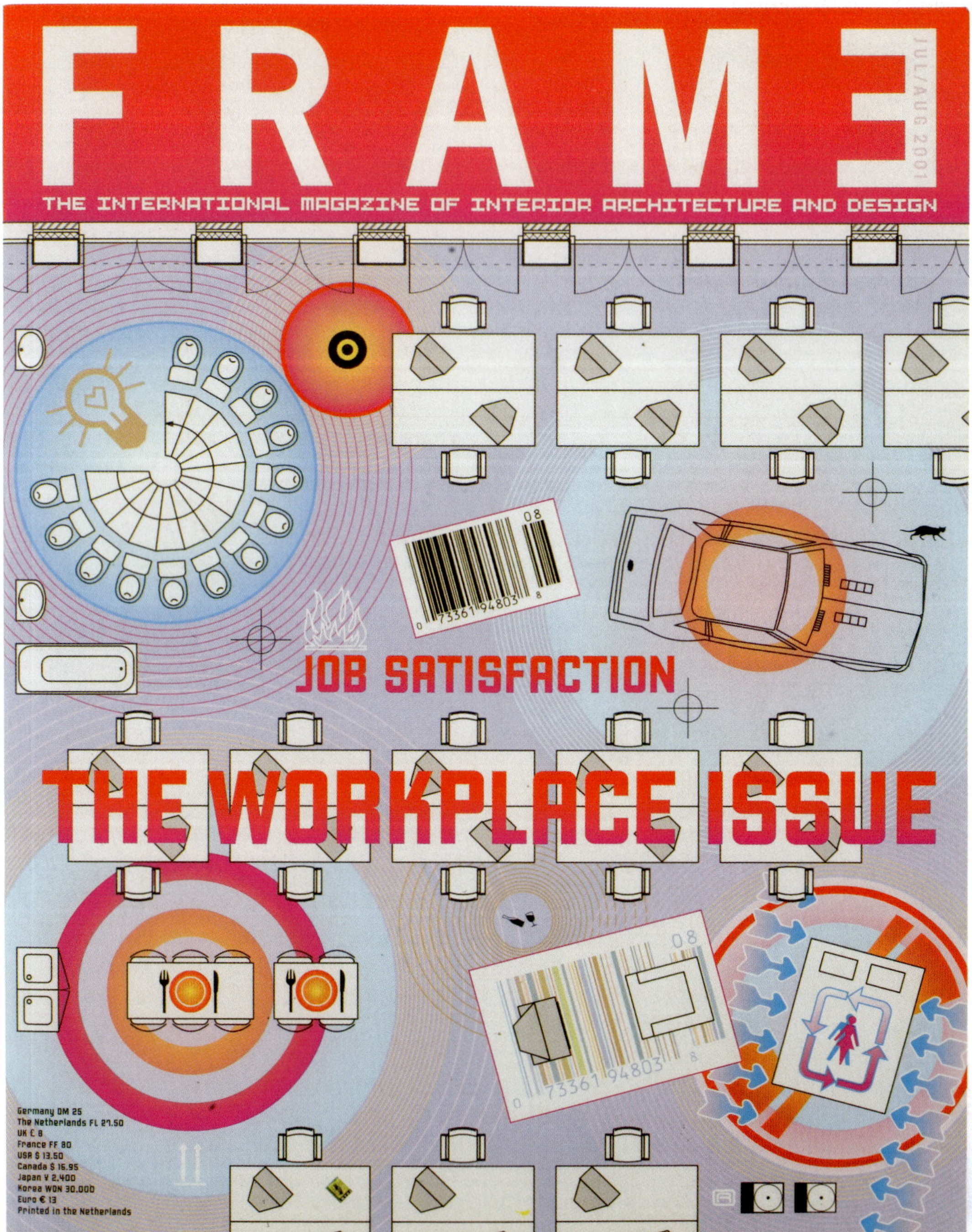

ART DIRECTOR
Roelof Mulder

21

Pixel Pop

Designer Fabian Hofmann (formgeber-berlin) about the project

We were given a free hand in many of the aspects of this project – a direct result of the Internet-boom at the time. The idea was to incorporate the 'Zeitgeist' of that period – the web-designers' life-style and the brave new world of Internet – within a corporate situation. I feel that we succeeded in creating a dramaturgy for the spaces which works, without being intrusive or disturbing.

Comments on the published article

The article was beautifully done.

Did the article have an effect?

We received a lot of positive feedback on the article.

Recent news or information about the project featured in *Frame*

Pixelpark has since downsized sub-stantially and no longer uses many of the rooms that we designed.

Current project

**formgeber-berlin disbanded in 2003.
I work freelance.
Two other partners now run an interior-furnishings shop in Berlin, Germany.**

DESIGNERS
formgeber-berlin
www.formgeber-berlin.de

SUBJECT
Pixelpark headquarters offices

LOCATION
Berlin, Germany

PHOTOGRAPHER
Frank Hülsbömer

WRITER
Owen Dunne

PIXEL POP

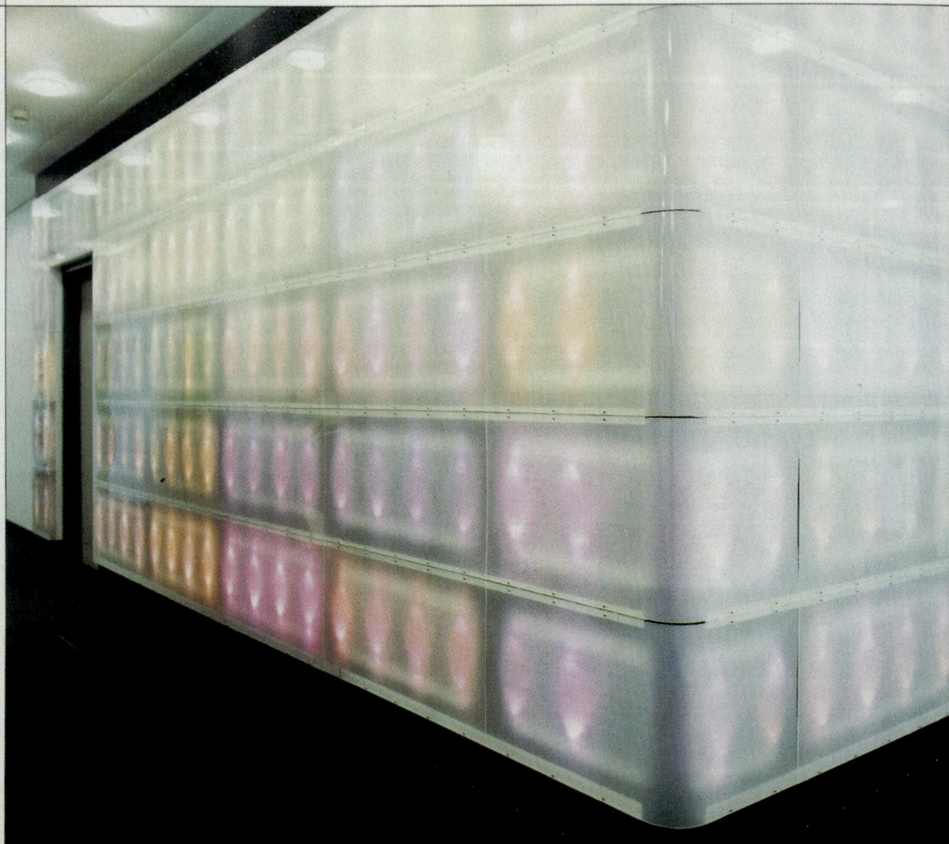

FORMGEBER-BERLIN REWORKED THE STERILE OFFICE SPACE OF AN IT FIRM UNTIL IT BECAME AN INTERACTIVE GALLERY OF HANDS-ON INSTALLATIONS AND KINETIC EXHIBITS.

BY OWEN DUNNE. PHOTOGRAPHY BY FRANK HÜLSBÖMER.

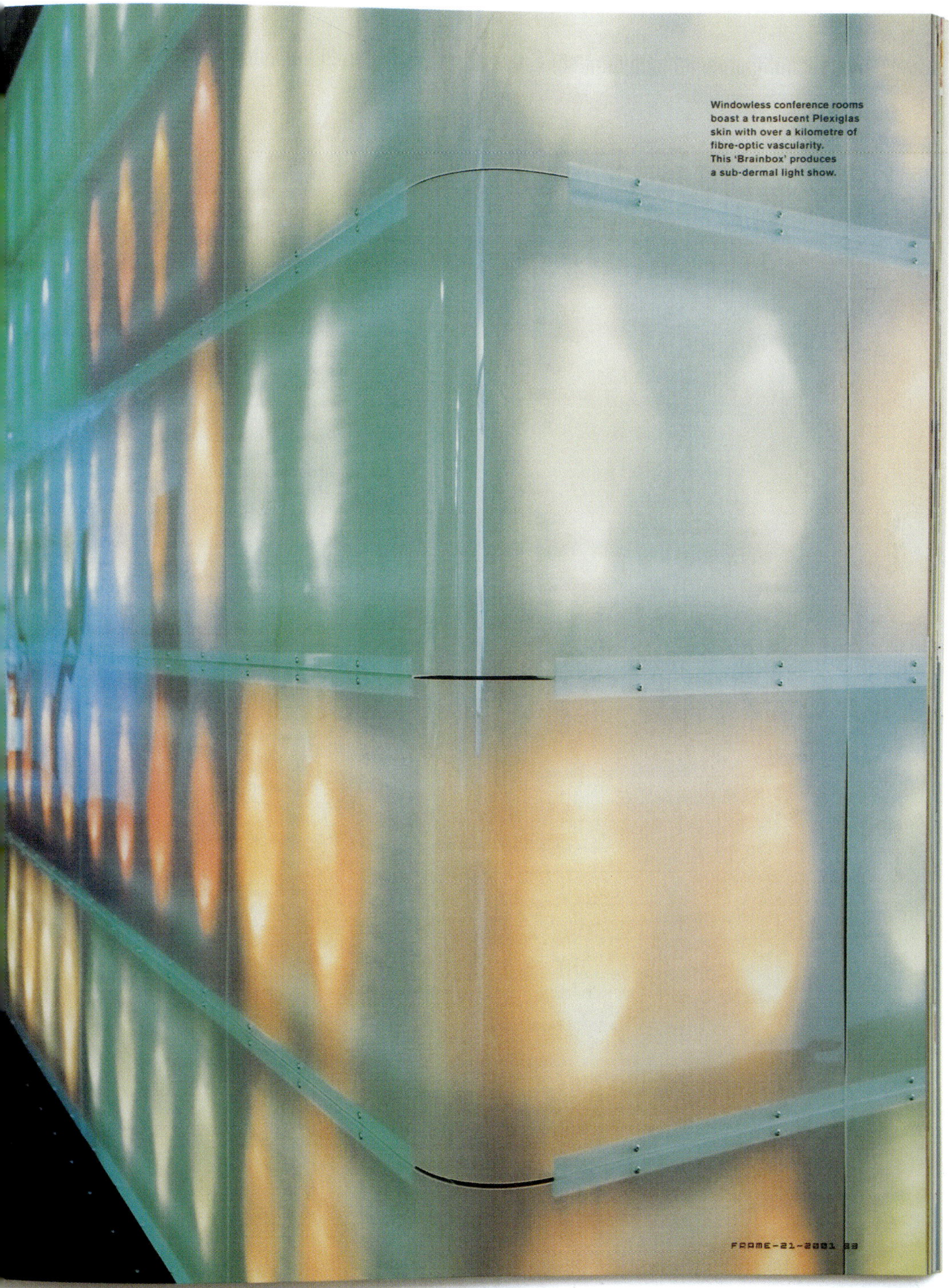

Windowless conference rooms boast a translucent Plexiglas skin with over a kilometre of fibre-optic vascularity. This 'Brainbox' produces a sub-dermal light show.

SUBJECT Pixelpark headquarters offices

LOCATION Berlin, Germany

DESIGNERS formgeber-berlin

PHOTOGRAPHER Frank Hülsbömer

WRITER Owen Dunne

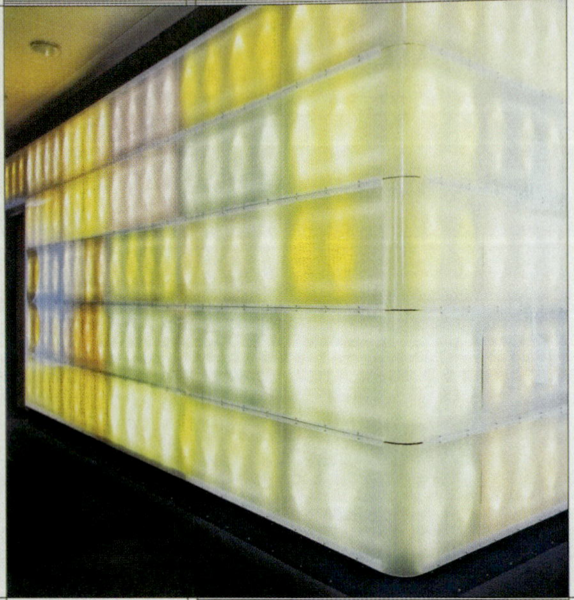

The 'Brainbox' displays computer-generated patterns that alter in colour and brightness at 20-second intervals, creating a hypnotic show for passers-by.

A CENTRAL THEME OF ALL THE INSTALLATIONS IS THE VISUALISED PIXEL

Corporate identity is reflected not only in the congruity of a company's work, but also in the coherent design of its workspace. When Pixelpark moved into a refurbished light-bulb factory in Oberbaum City last year, the goal was a thematically oriented space based on an in-house motto: Move first. A fast-growing multimedia giant with 13 affiliates worldwide, Pixelpark wanted its Berlin headquarters to be a visual interpretation of the company philosophy.

The brief was crystal clear: Translate IT into a tangible design and make it fun. Convinced that a favourable environment stimulates creativity, Pixelpark was eager to pamper its employees, some of whom spend as many as 12 hours a day at their workstations. Rob Houst, André Eid and Fabian Hofmann of formgeber-berlin opted to rework the sterile office space into an interactive gallery of hands-on installations and kinetic exhibits. Established in 1999, the three-man office specialises in product design and concept development. With a collection of 18 products under its belt, in addition to conceptual work for several small shops and a design for the Echo Awards, formgeber-berlin was looking forward to the firm's initial confrontation with a large-scale interior.

'Working with finished spaces,' says Houst, 'meant our work was more or less cosmetic.' Fundamental to the Pixelpark concept was the use of reflected or projected colour juxtaposed with neutral surfaces and the introduction of interactive elements. Visitors entering the foyer encounter 26 king-size Plexiglas panels, which fill the hall with a dozen hues. Hung on dual tracks parallel to a 30-metre-long expanse of fenestration, these 'colourpanes' virtually beg to be manoeuvred; the subsequent colour combinations, clusters and voids underline the volatility of the internet and create a light that merges Gothic cathedral with Star Trek.

Eighteen projects on seven floors invite Pixelpark's young designers and programmers to interact with the working environment. A central theme is the visualised pixel in a variable chromatic setting. Windowless third- and fourth-floor conference rooms boast a translucent Plexiglas skin with over a kilometre of fibre-optic vascularity, which produces a sub-dermal light show. Activated when lights in the meeting room are turned on, the 'Brainbox' outside the room mirrors the neural impulses of the conferees. Light panels displaying random, computer-generated patterns alter in colour and brightness at 20-second intervals, creating a hypnotic show for passers-by.

Asked to design the interior of another branch of Pixelpark located in the same office complex – the Zentrum für Logistik und Unternehmungsplanung (ZLU) – formgeber-berlin focused on movement. They propelled the core of the office, a simple horseshoe-shaped hall, into a new dimension. 'The idea,' says Hofmann, 'revolves around the dissolution of spatial relations and the game of perspective.' The hall, a metaphor for the widespread activities of a logistics centre, has no hard corners. Linoleum flooring running up the wall to a height of 1.20 metres meets a battery of computer-controlled fluorescent bulbs, which pulsate an illusion of movement in the hallway.

'ZLU is less a design than a walk-through work of art based on colourful pop art produced in the late '60s,' Hofmann explains. 'We're not talking about inspiration here; the idea is to recapture the lifestyle that marks the world of Verner Panton, but in a way that fits our fast-forward civilisation. A hall – the walk from one meeting to another – can offer a moment of relaxation.'

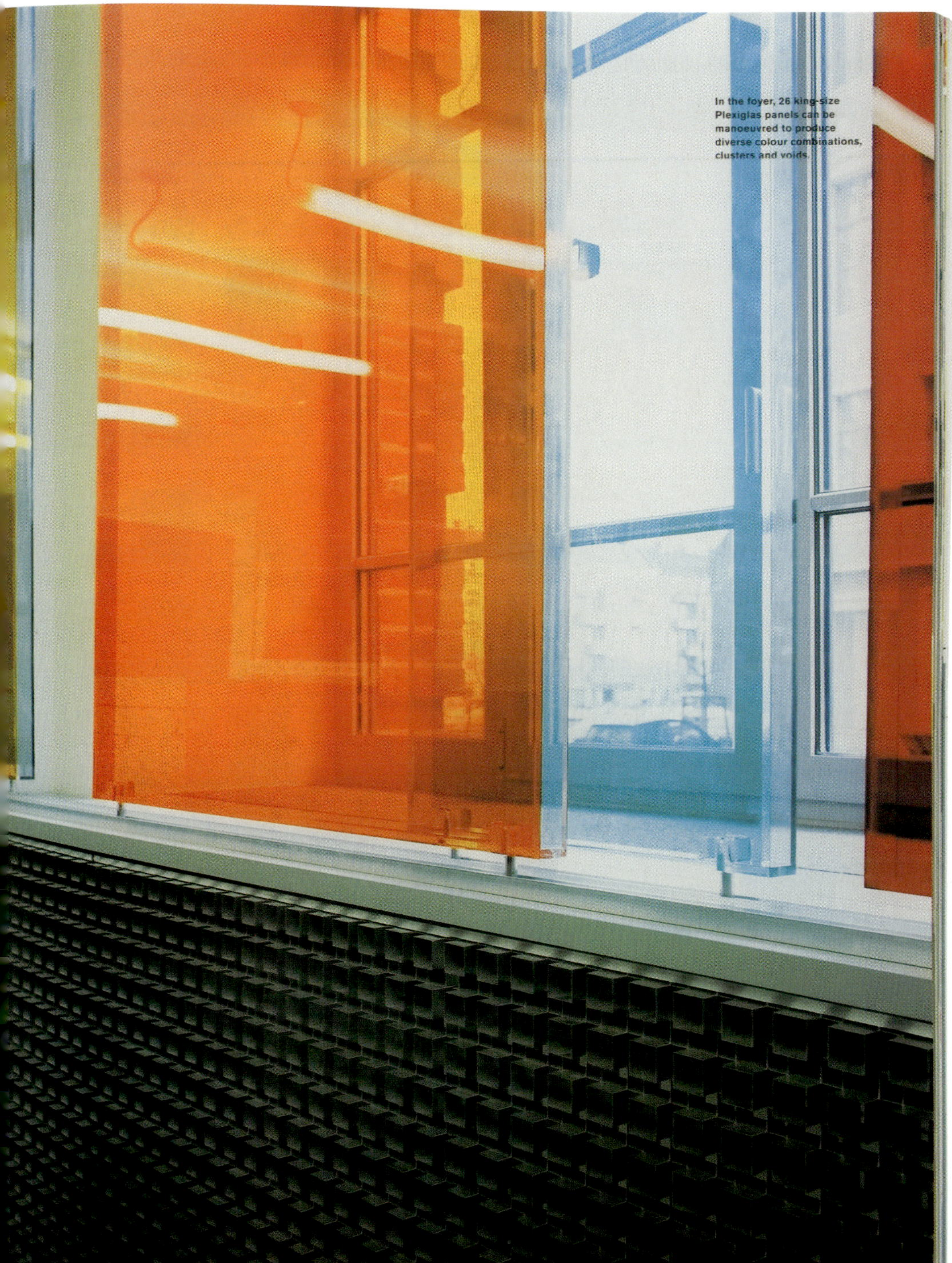

In the foyer, 26 king-size Plexiglas panels can be manoeuvred to produce diverse colour combinations, clusters and voids.

SUBJECT
Pixelpark headquarters offices

LOCATION
Berlin, Germany

DESIGNERS
formgeber-berlin

PHOTOGRAPHER
Frank Hülsbömer

WRITER
Owen Dunne

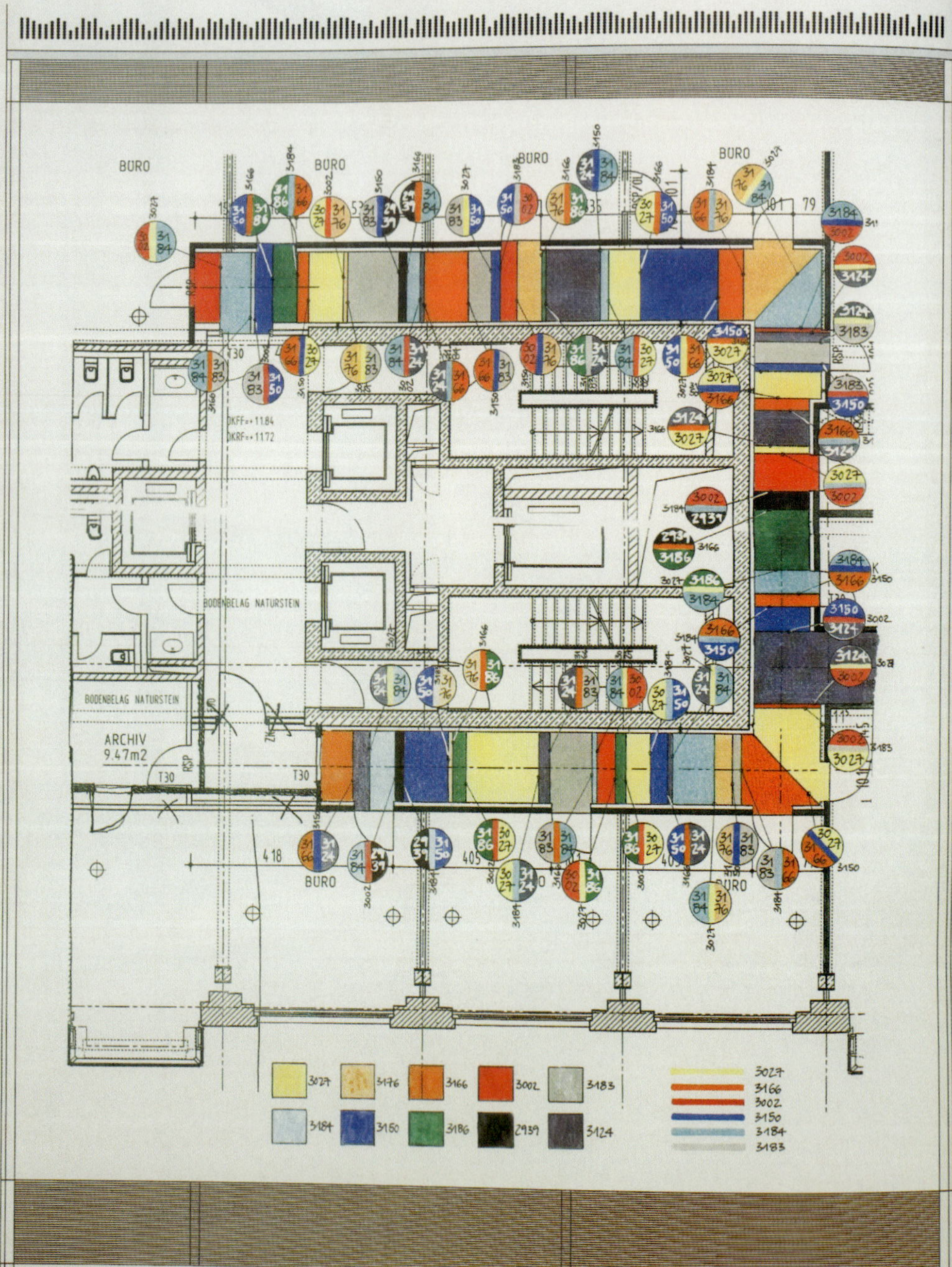
BÜRO
BÜRO
BÜRO
BÜRO
BÜRO
BÜRO
BODENBELAG NATURSTEIN
BODENBELAG NATURSTEIN
ARCHIV
9.47m2
T30
T30
T30
3027
3176
3166
3002
3183
3184
3150
3186
2939
3124
3027
3166
3002
3150
3184
3183

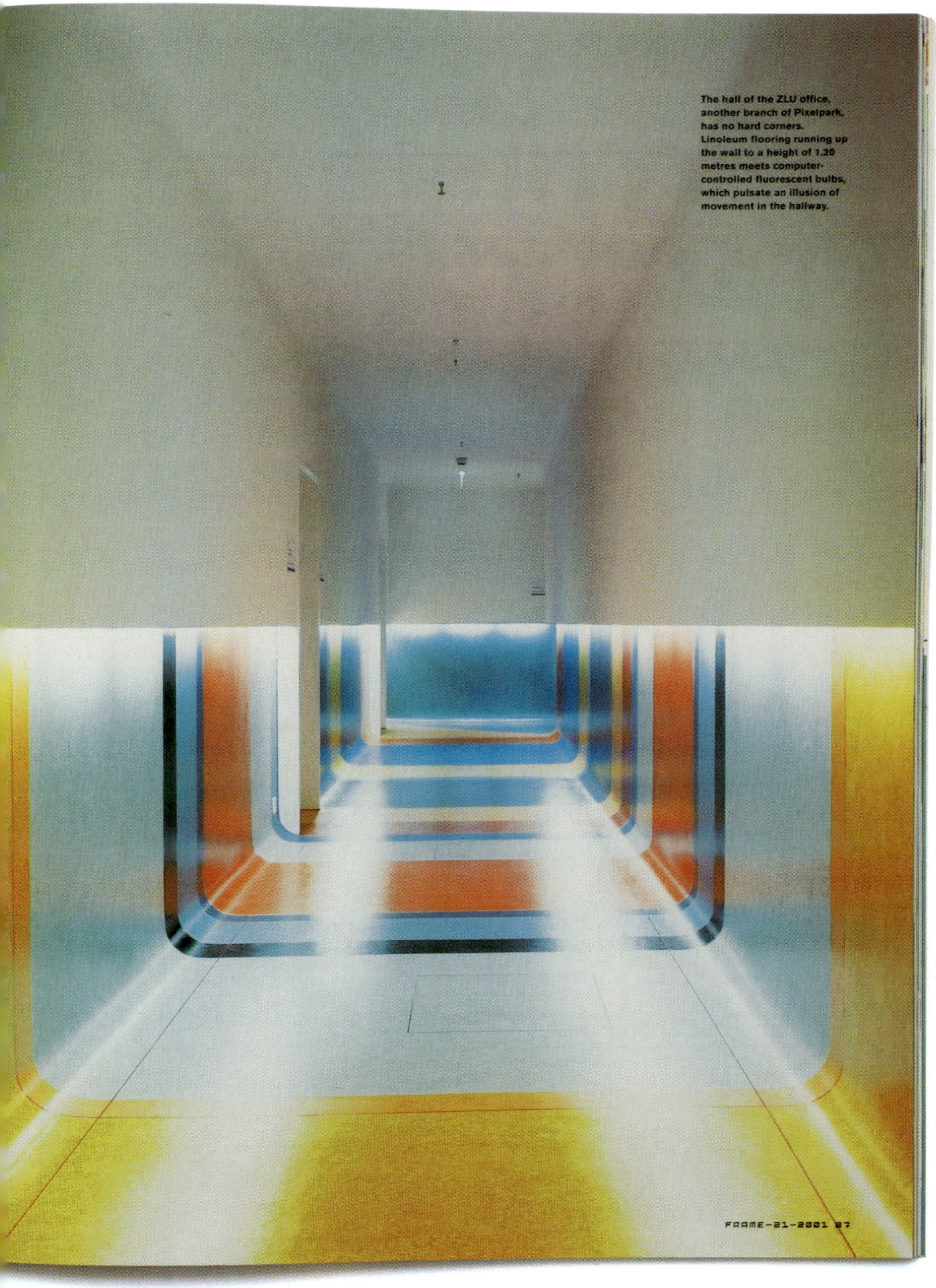

The hall of the ZLU office, another branch of Pixelpark, has no hard corners. Linoleum flooring running up the wall to a height of 1.20 metres meets computer-controlled fluorescent bulbs, which pulsate an illusion of movement in the hallway.

SUBJECT
Pixelpark headquarters offices

LOCATION
Berlin, Germany

DESIGNERS
formgeber-berlin

PHOTOGRAPHER
Frank Hülsbömer

WRITER
Owen Dunne

228-229

September/October 2001

Issue 22

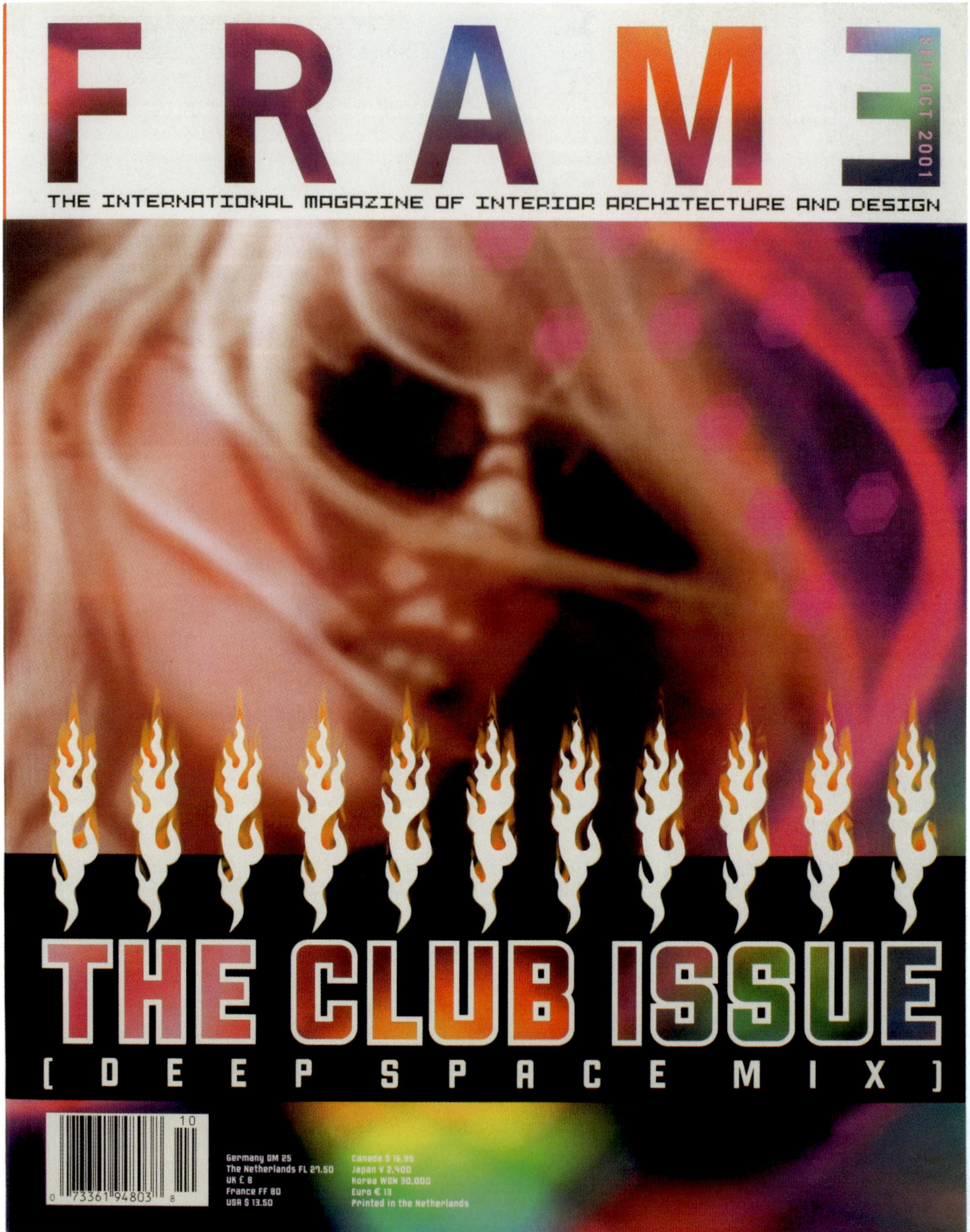

ART DIRECTOR

Roelof Mulder

Gay in Gotham

Designer Sam O'Donahue (Desgrippes Gobé Group) about the project

It was a unique opportunity to work with a client who was as committed to the integrity of the design as we were. This gave us the opportunity to indulge ourselves creatively and to design every detail from the furniture right down to the ashtrays. It was also a fascinating challenge to consider in detail how the architecture could serve to manipulate the behaviour of the patrons and control their experience in specific ways in each area and also at different times of the night. For example, the use of dynamic lighting, video cameras and even bar stools designed to prevent slouching, created an atmosphere of almost uncomfortable energy, which makes you feel alive and flirtatious.

Comments on the published article

It was great that the article included several pages and many photos which presented a real picture of the space. So many other publications limit coverage to one photo and a sound bite which often results in a misleading presentation of the work.

Did the article have an effect?

Clients and colleagues considered the selection of the project by Frame to be a real testament to the high quality of our work.

Recent news or information about the project featured in *Frame*

Still going strong.

DESIGNERS
d/g* Worldwide
(now called Desgrippes Gobé Group)
www.dga.com
John Salibello

SUBJECT
XL club

LOCATION
New York, USA

PHOTOGRAPHER
John Horner

WRITER
Asko Ahokas

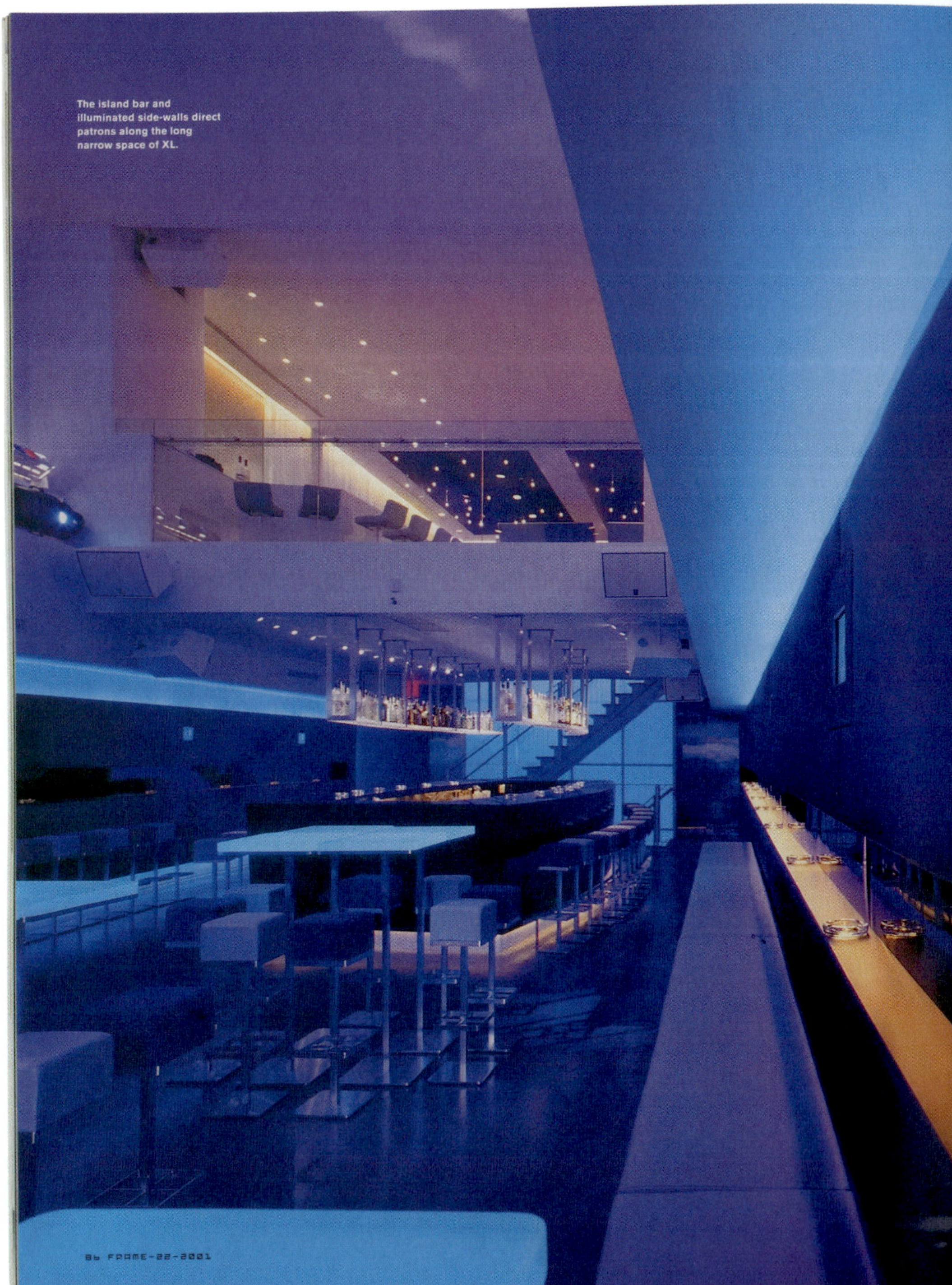

The island bar and illuminated side-walls direct patrons along the long narrow space of XL.

GAY IN GOTHAM

BEHIND THE MINIMALIST FRONTAGE OF AN EXISTING BUILDING IN DOWNTOWN NEW YORK LIES THE TRULY BLUE ENVIRONMENT OF XL.

BY ASKO AHOKAS. PHOTOGRAPHY BY JOHN HORNER.

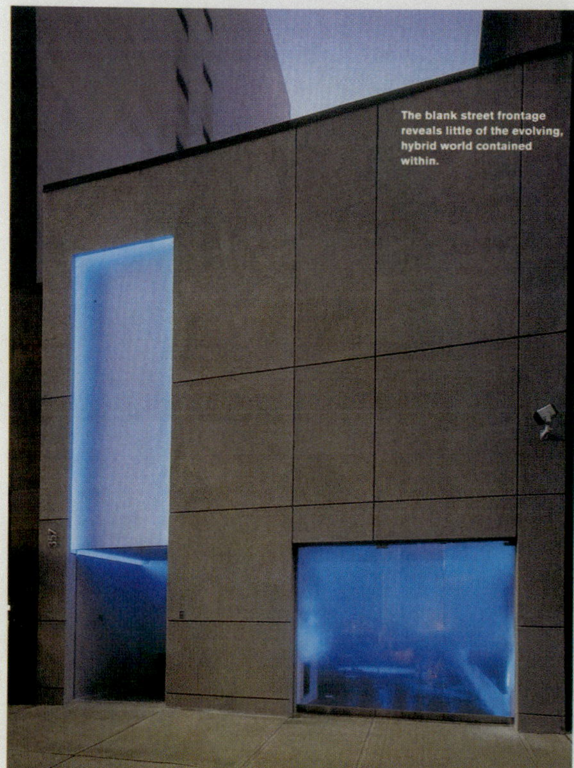

The blank street frontage reveals little of the evolving, hybrid world contained within.

Successful nightclubs are kinetic environments that can accommodate many different activities simultaneously. To avoid the traditionally static setting associated with bars and lounges, the designers of the XL club – architect David Ashen of New York firm d/g*Worldwide and interior designer John Salibello – created an evolving, hybrid space that fuses elements of both nightclub and bar.

After a fourteen-month refurbishment costing 2.5 million dollars, the existing venue tucked away behind a minimalist frontage in New York's industrial neighbourhood of Chelsea has been transformed into one of downtown's premium gay entertainment spots. The entrance corridor, which initiates clubbers into the ambience of the world within, leads to the main level of XL, where an island bar and a glowing wall direct the circulation of patrons. The primary design consideration was to define the long, narrow space so that it could cater to a varied clientele – from those dropping in for after-work drinks to serious aficionados of late-night clubbing. State-of-the-art lighting effects adjust to the time of day, ensuring a constantly evolving environment.

A floating, theatrical staircase rises to the mezzanine level. Compared with the darker and more intense energy downstairs, the atmosphere above is almost ethereal, with lighting fixtures imitating a starry sky and endlessly reflecting mirrors shrouded in white veils that seem to make the tight space bigger.

On the third level a VIP lounge with a one-way mirrored wall creates a sense of voyeurism, which is taken further with monitors that display the scene at other levels. The warm tones of dark wood and marble, mixed with animal skins and tiles, accentuate the feeling of privacy.

But the truly memorable feature of XL has got to be the men's room to the rear of the main level. A cool aquatic blue, an intense red, and a palpable air of voyeurism make for a highly sensual, even confrontational climate. The central accent of the space is the massive fish tank that separates the wash basins from the urinals. A final touch is the digitally mastered sound of water emanating from the fish tank.

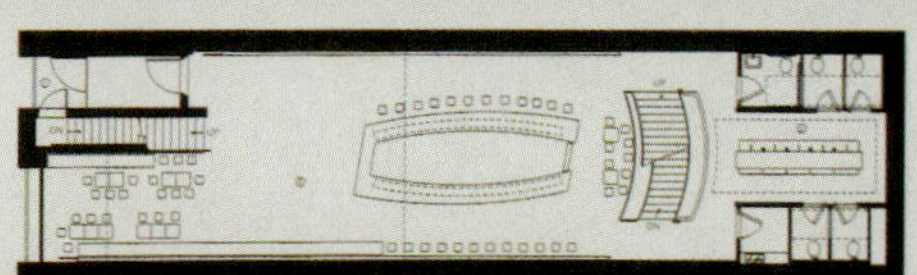

Mezzanine.

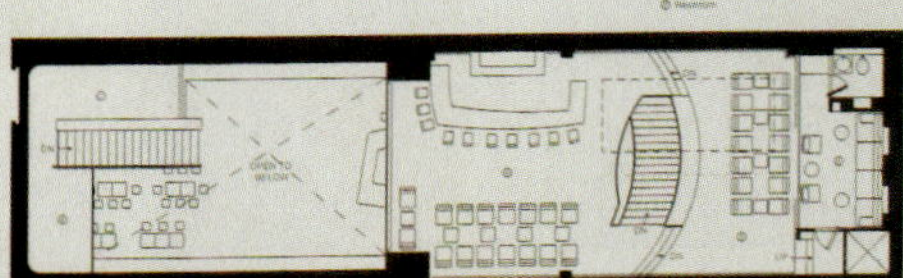

Ground floor.

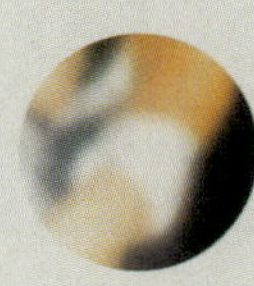

SUBJECT XL club
LOCATION New York, USA
DESIGNERS d/g* Worldwide John Salibello
PHOTOGRAPHER John Horner
WRITER Asko Ahokas

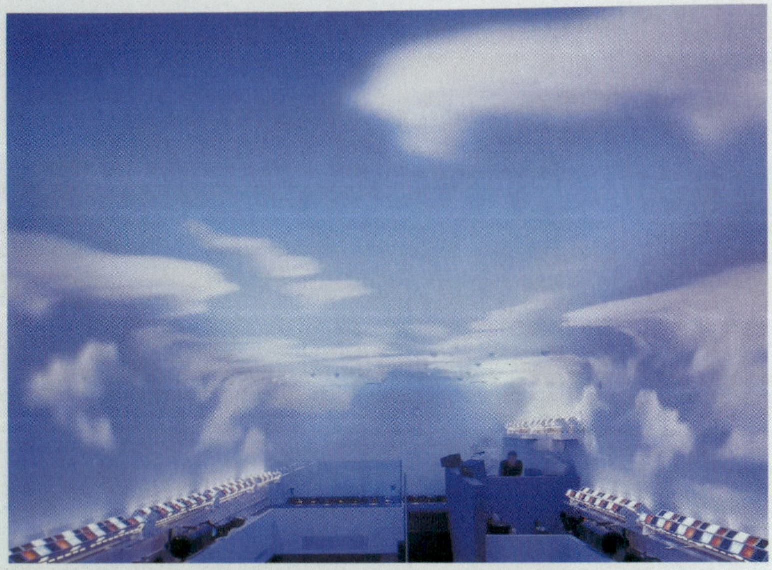

A COOL AQUATIC BLUE, AN INTENSE RED, AND A PALPABLE AIR OF VOYEURISM MAKE FOR A HIGHLY SENSUAL, EVEN CONFRONTATIONAL CLIMATE IN THE MEN'S ROOM

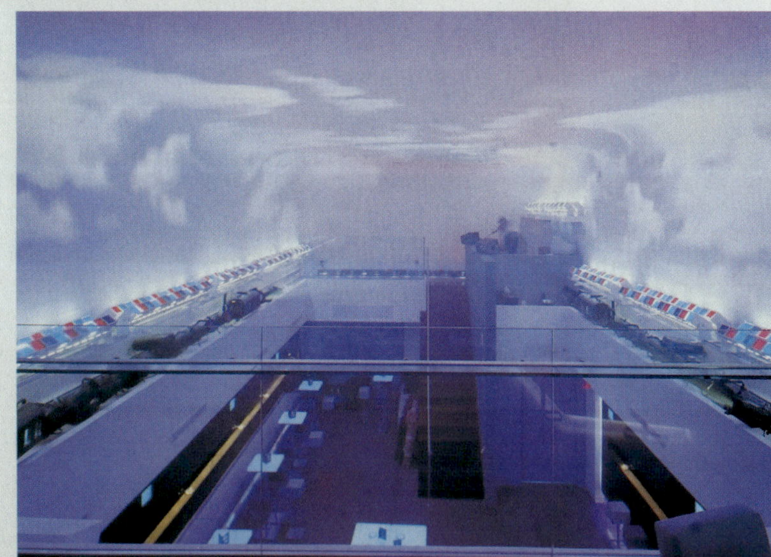

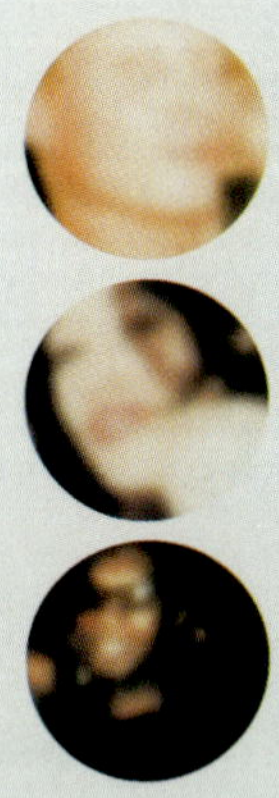

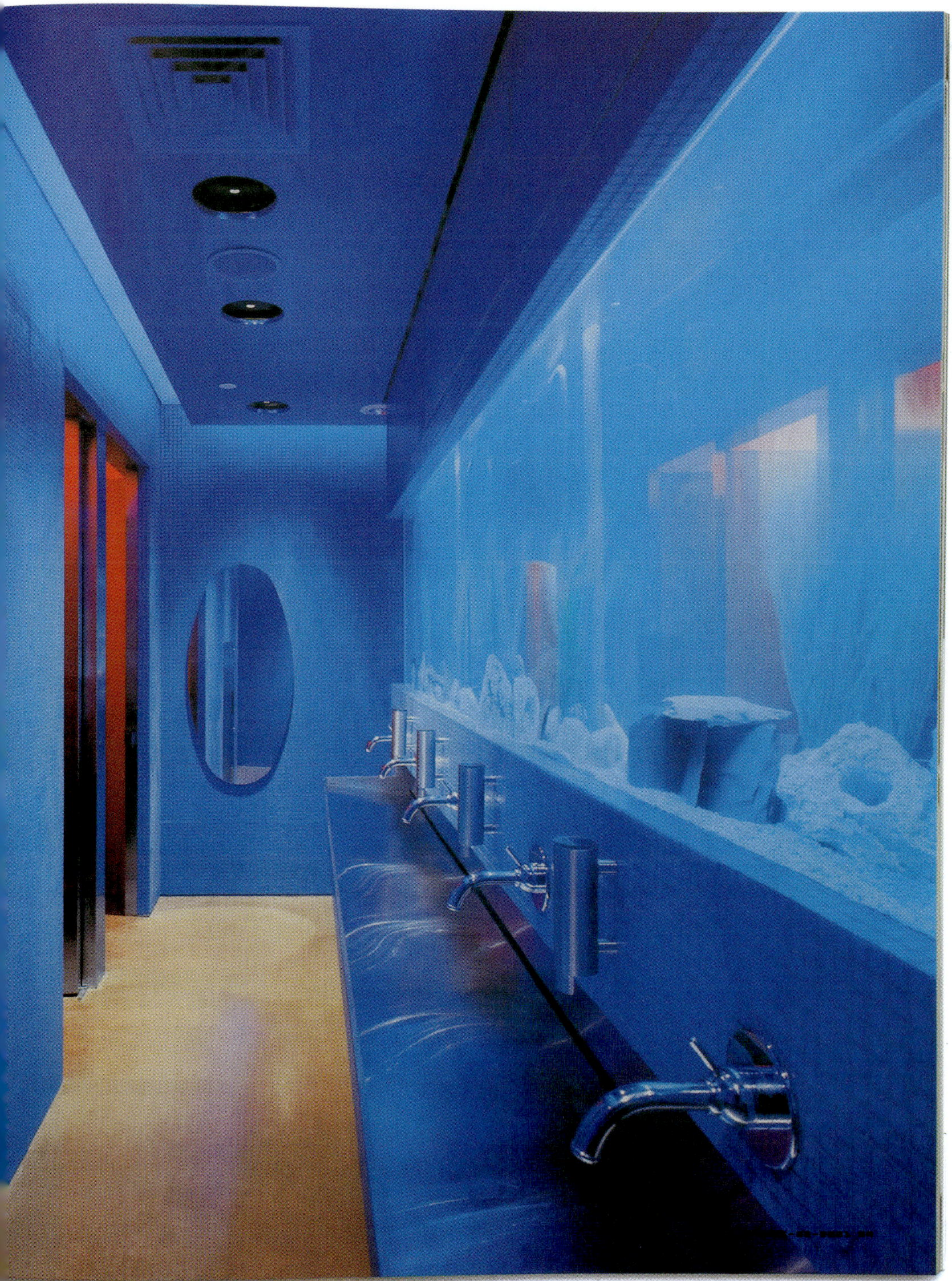

SUBJECT
XL club

LOCATION
New York, USA

DESIGNERS
d/g* Worldwide
John Salibello

PHOTOGRAPHER
John Horner

WRITER
Asko Ahokas

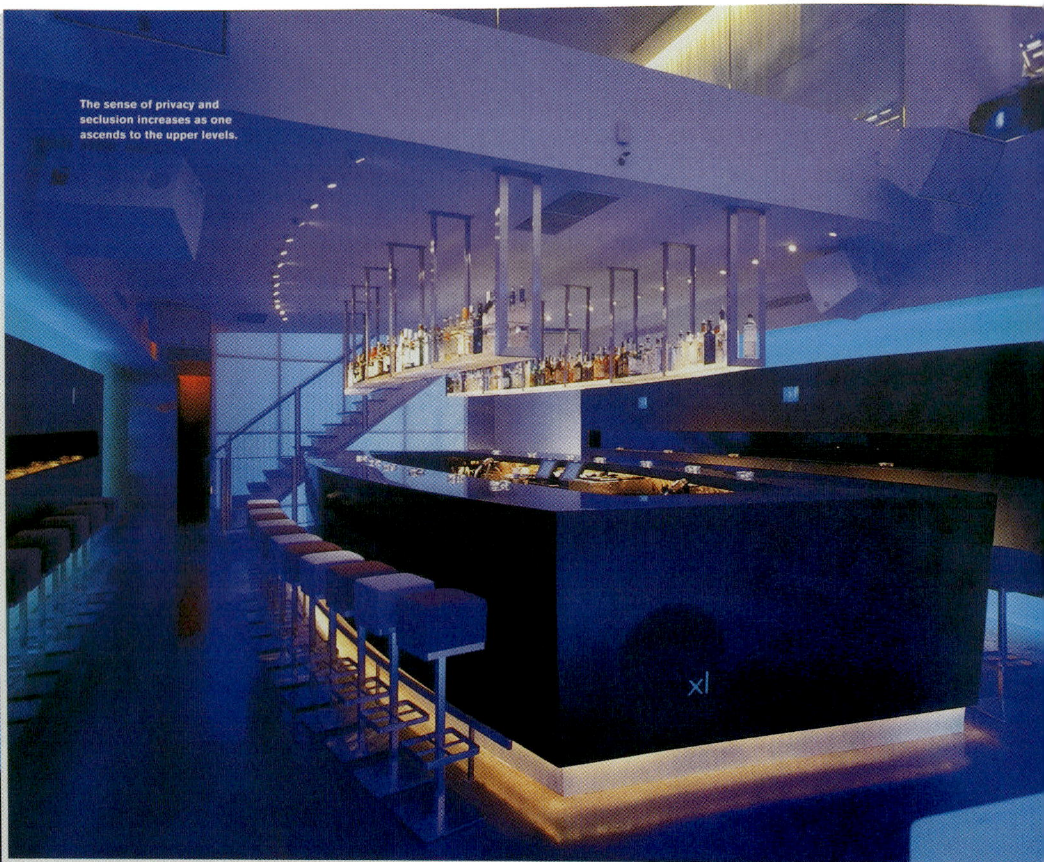
The sense of privacy and seclusion increases as one ascends to the upper levels.

STATE-OF-THE-ART LIGHTING EFFECTS ADJUST TO THE TIME OF DAY, ENSURING A CONSTANTLY EVOLVING ENVIRONMENT

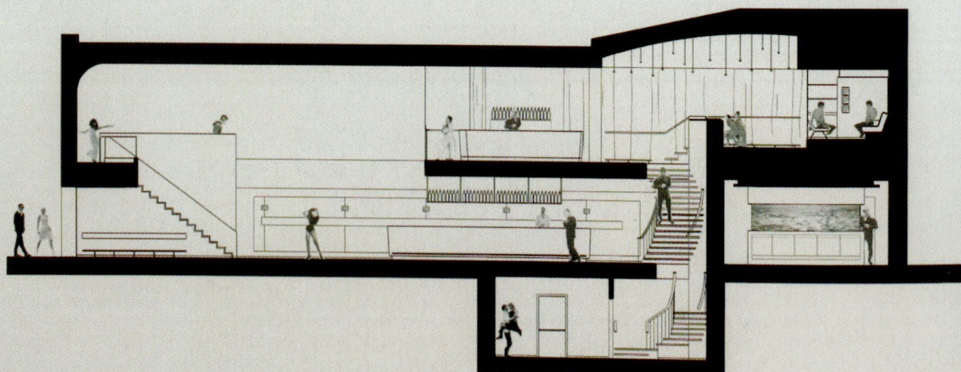

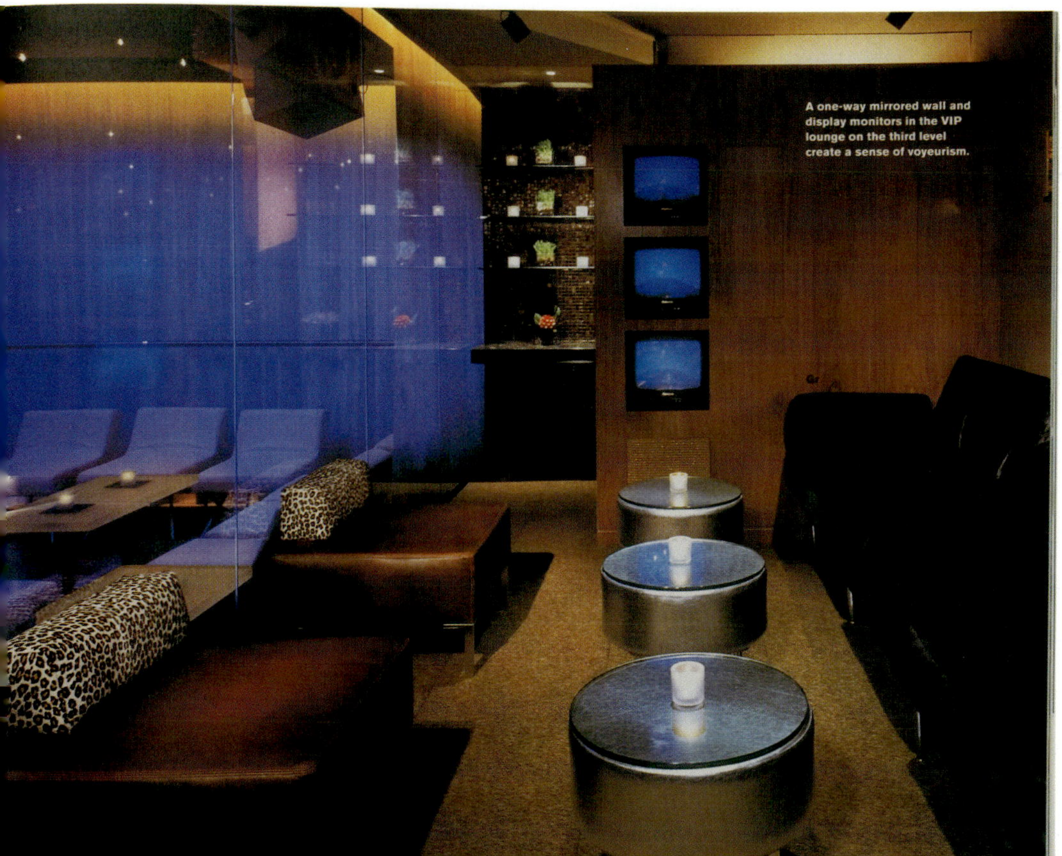

A one-way mirrored wall and display monitors in the VIP lounge on the third level create a sense of voyeurism.

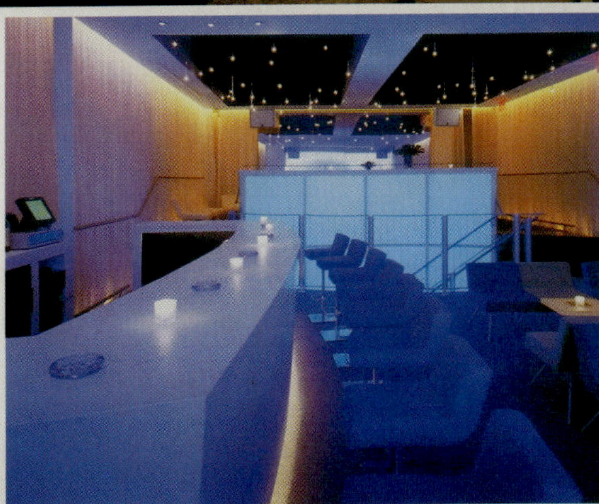

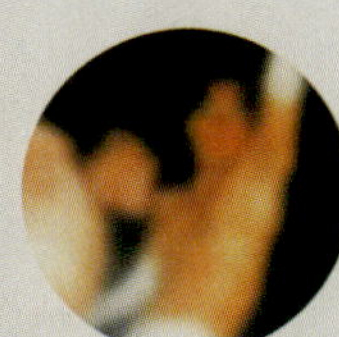

SUBJECT
XL club

LOCATION
New York, USA

DESIGNERS
d/g* Worldwide
John Salibello

PHOTOGRAPHER
John Horner

WRITER
Asko Ahokas

Issue 23 November/December 2001 236-237

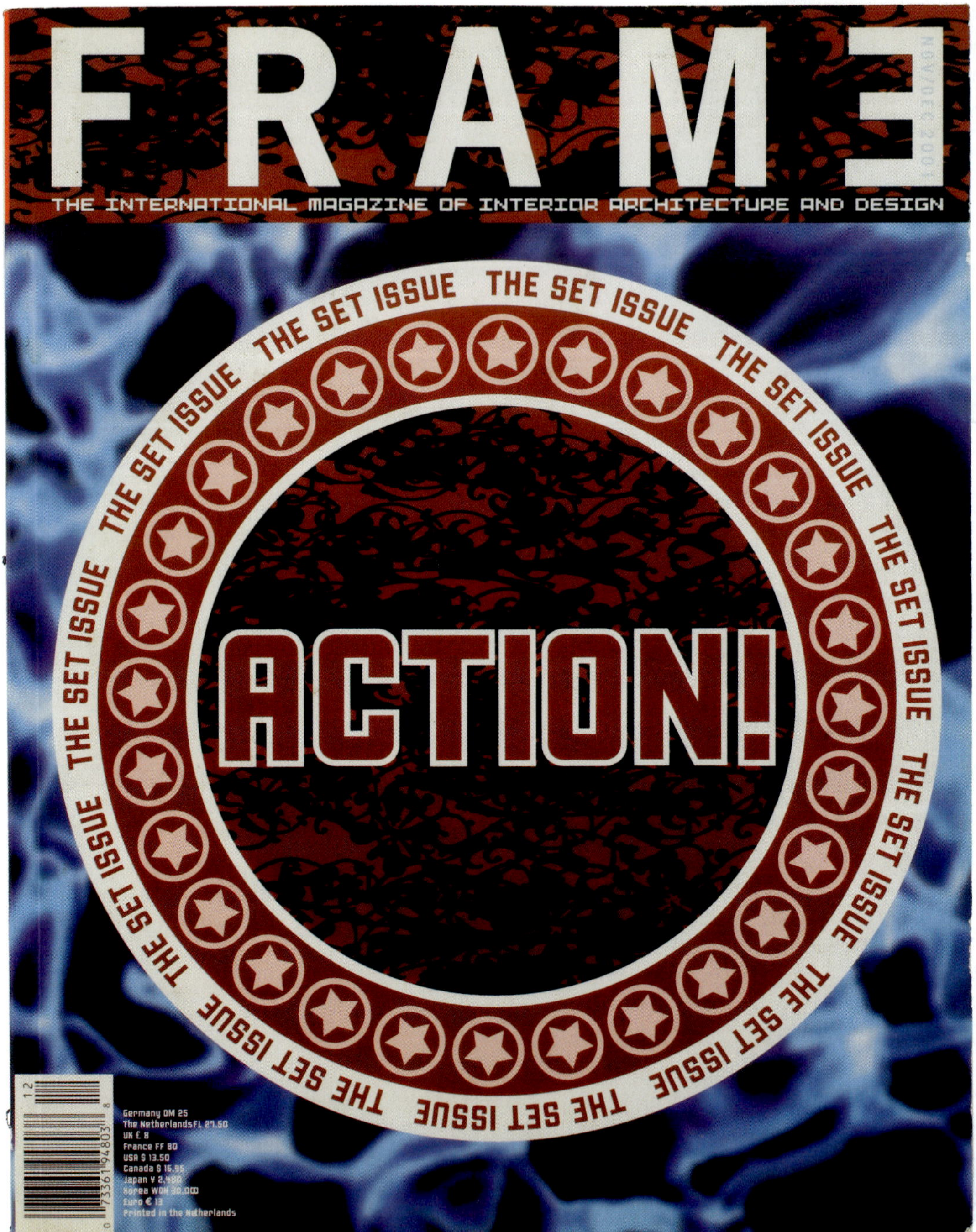

ART DIRECTOR
Roelof Mulder

23

New Is Now

Architects Astrid Klein and Mark Dytham about their current projects

Heidi House – a unique and light-hearted work environment for a young web design company // Brillare – a shining party space in the Risonare resort where our Leaf Chapel is situated // Uniqlo – Ginza flagship store about to feature in your Frame publication Dress Code.

DESIGNERS
Klein Dytham architecture
www.klein-dytham.com

SUBJECT
Portrait of Klein Dytham architecture

LOCATION
Tokyo, Japan

PHOTOGRAPHERS
Katsuhisa Kida
www.katsuhisakida.com
Kozo Takayama

WRITER
Carolien van Tilburg

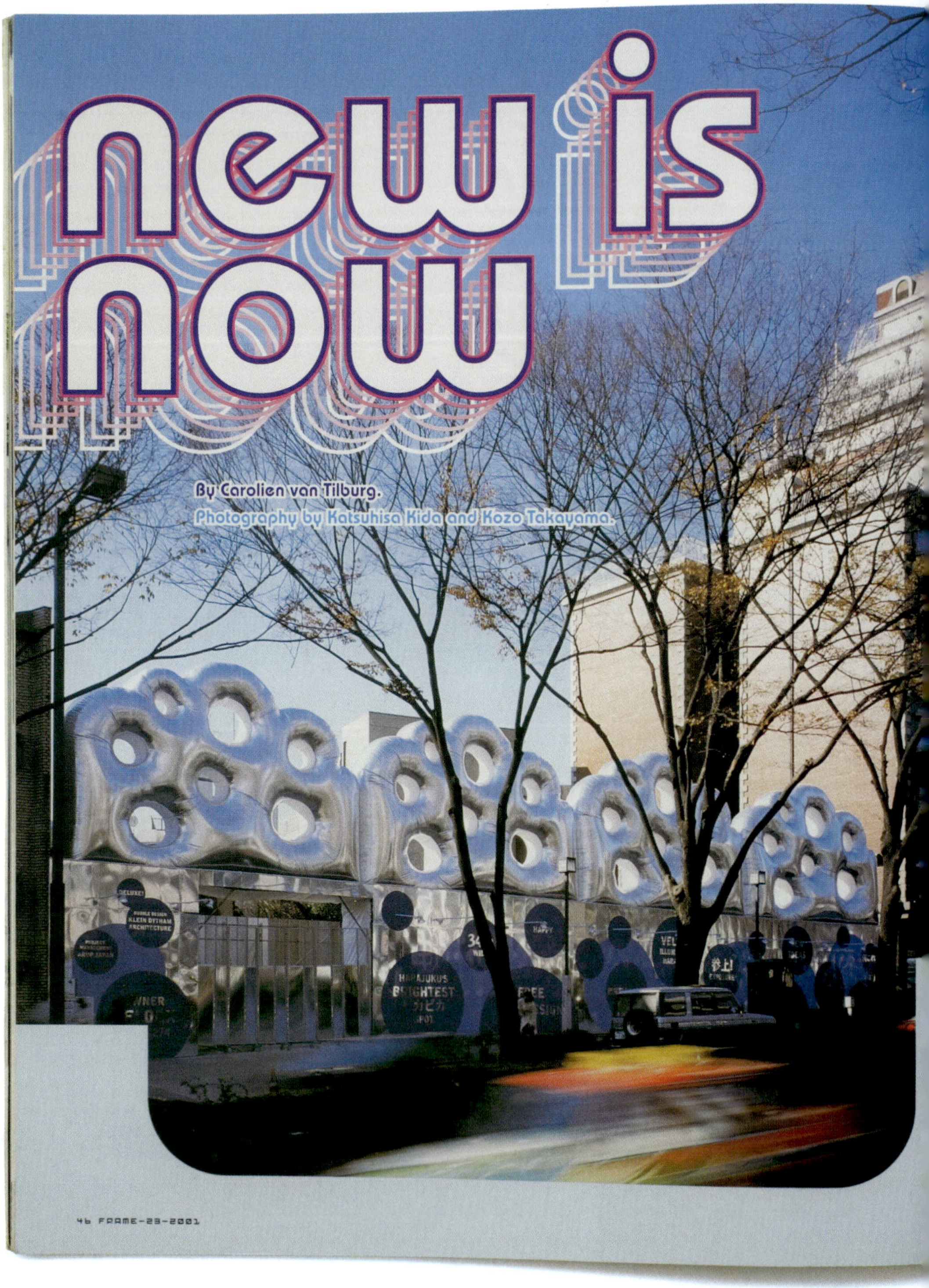

new is now

By Carolien van Tilburg.
Photography by Katsuhisa Kida and Kozo Takayama.

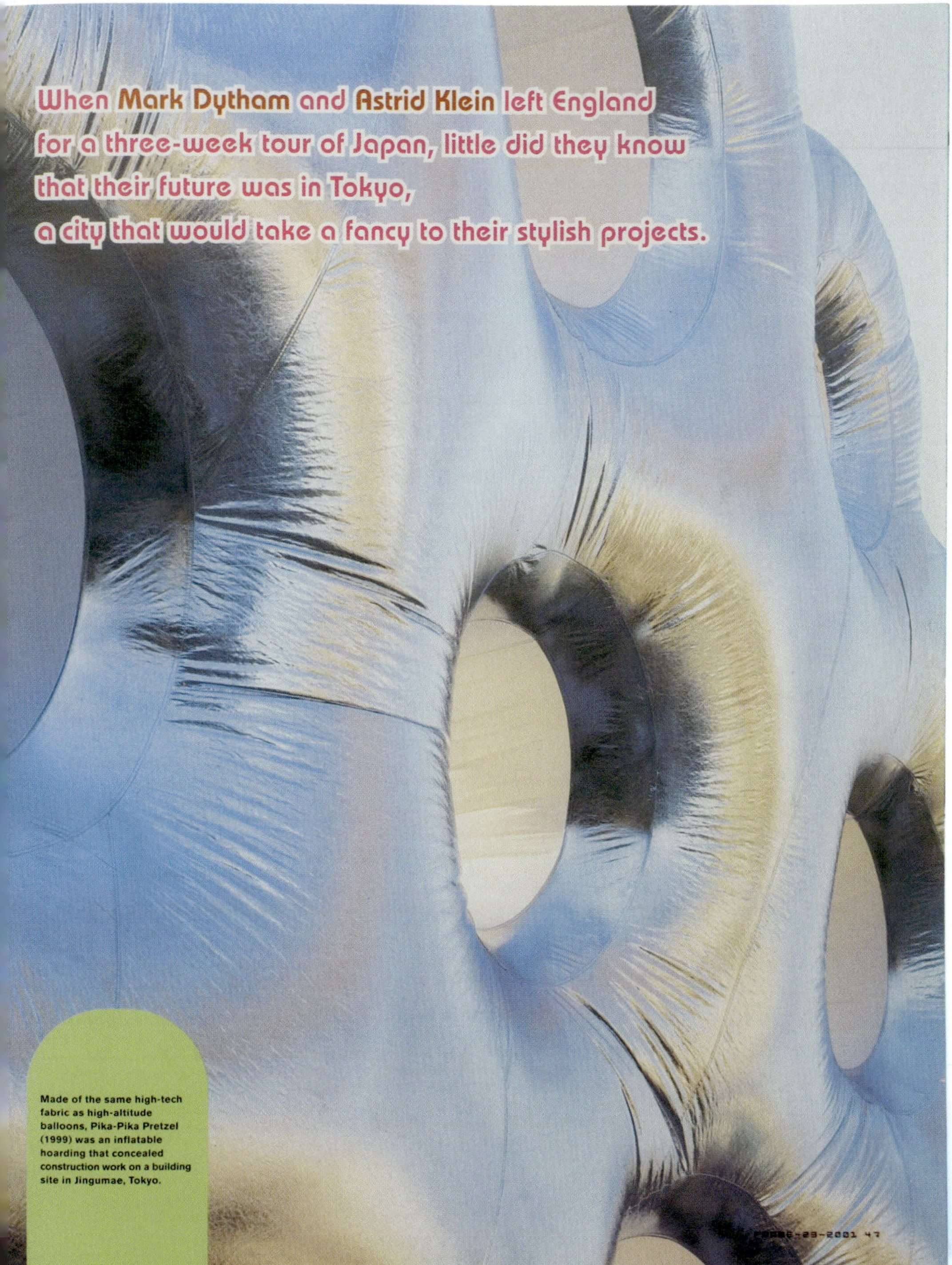

When Mark Dytham and Astrid Klein left England for a three-week tour of Japan, little did they know that their future was in Tokyo, a city that would take a fancy to their stylish projects.

Made of the same high-tech fabric as high-altitude balloons, Pika-Pika Pretzel (1999) was an inflatable hoarding that concealed construction work on a building site in Jingumae, Tokyo.

SUBJECT
Portrait of Klein Dytham architecture

LOCATION
Tokyo, Japan

DESIGNERS
Klein Dytham architecture

PHOTOGRAPHERS
Katsuhisa Kida
Kozo Takayama

WRITER
Carolien van Tilburg

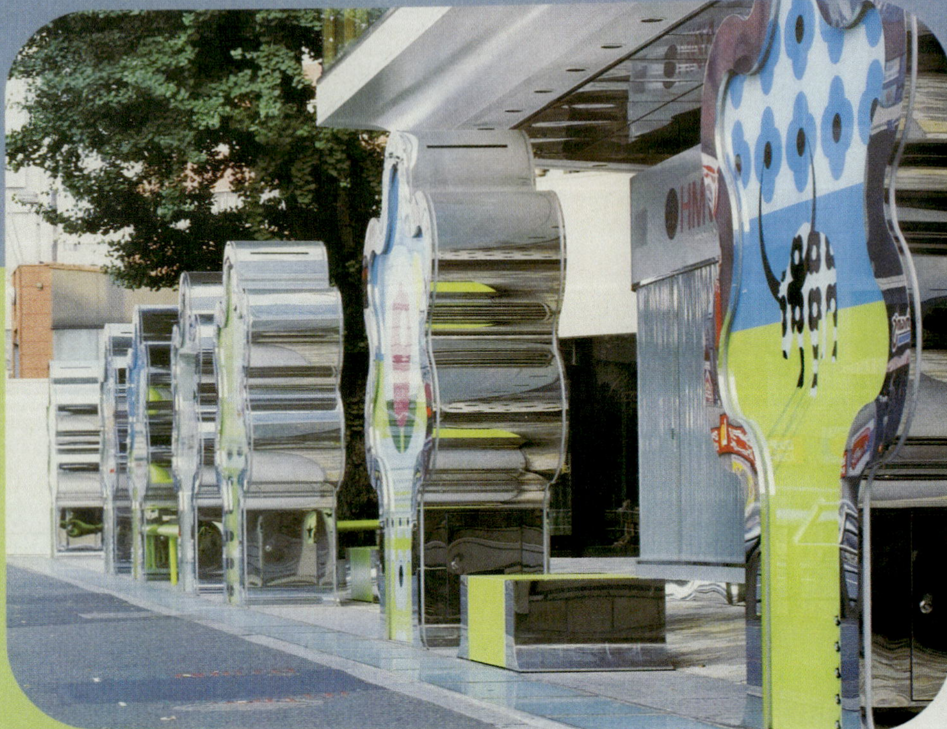

When asked to attract fashionable youth to Tokyo's trendy fashion palace Laforet, KDa 'planted' a row of tree-shaped objects along the building's 50-metre-long street frontage. The line of polished-stainless-steel trees, called Rin-Rin (2001), is emphasised by a glass strip set in the ground and illuminated from below.

After studying architecture at The Royal College of Art in London, Astrid Klein (1962) and Mark Dytham (1964) made plans for what they thought would be a three-week tour of Japan in 1989. Their trip was preceded by a flurry of letters and portfolios, all addressed to the various Japanese architects they hoped to meet. No one replied. Refusing to take this lack of response lying down, the persistent youngsters turned to the telephone for a more direct form of contact. Imagine their surprise when enthusiastic voices at the other end of the line assured them they were expected. 'People like Arata Isozaki, Fuhimiko Maki, Toyo Ito, Kenzo Tange, Tadao Ando and both the Suzukis were looking forward to our visit,' says Mark Dytham. 'It was amazing! They hadn't replied to our letters simply because they couldn't write English very well.'

They recall landing in Tokyo and immediately being caught up in the energetic pace and exotic atmosphere of the city. Within weeks of their arrival, a series of coincidences led to their first project, the renovation of a hair salon, which more or less fell into their laps thanks to the kindness of Toyo Ito. It was the beginning of a collaborative relationship, during which the young designers helped Ito to contact foreign clients and he guided and supported their work in Japan. 'I was struck by their complete and utter sense of style,' Ito says of Klein and Dytham.

Having extended their three-week tour into two full years, the duo set up their own office but continued to work for Ito on a part-time basis. To make ends meet, KDa (Klein Dytham architecture) took on all sorts of jobs, particularly interiors. Born of necessity, this broad-based strategy continues to characterise work produced by a firm that currently boasts a staff of eight.

The two gaijins did and still do well in Japan. In 1993 KDa walked off with the Kajima Space Design Award for best young practice in Japan. Three years later, their Idee workstation won both the Asahi Glass Design Award and the National Panasonic Design Award. In 1998 KDa's interior for a new information centre at the Tokyo British Council was honoured as the nation's best library. And the popularity of their website

'When you're a young architect, dying to do wild projects, Tokyo is the perfect city. Just follow the building regulations, and your design can look like anything you want it to'

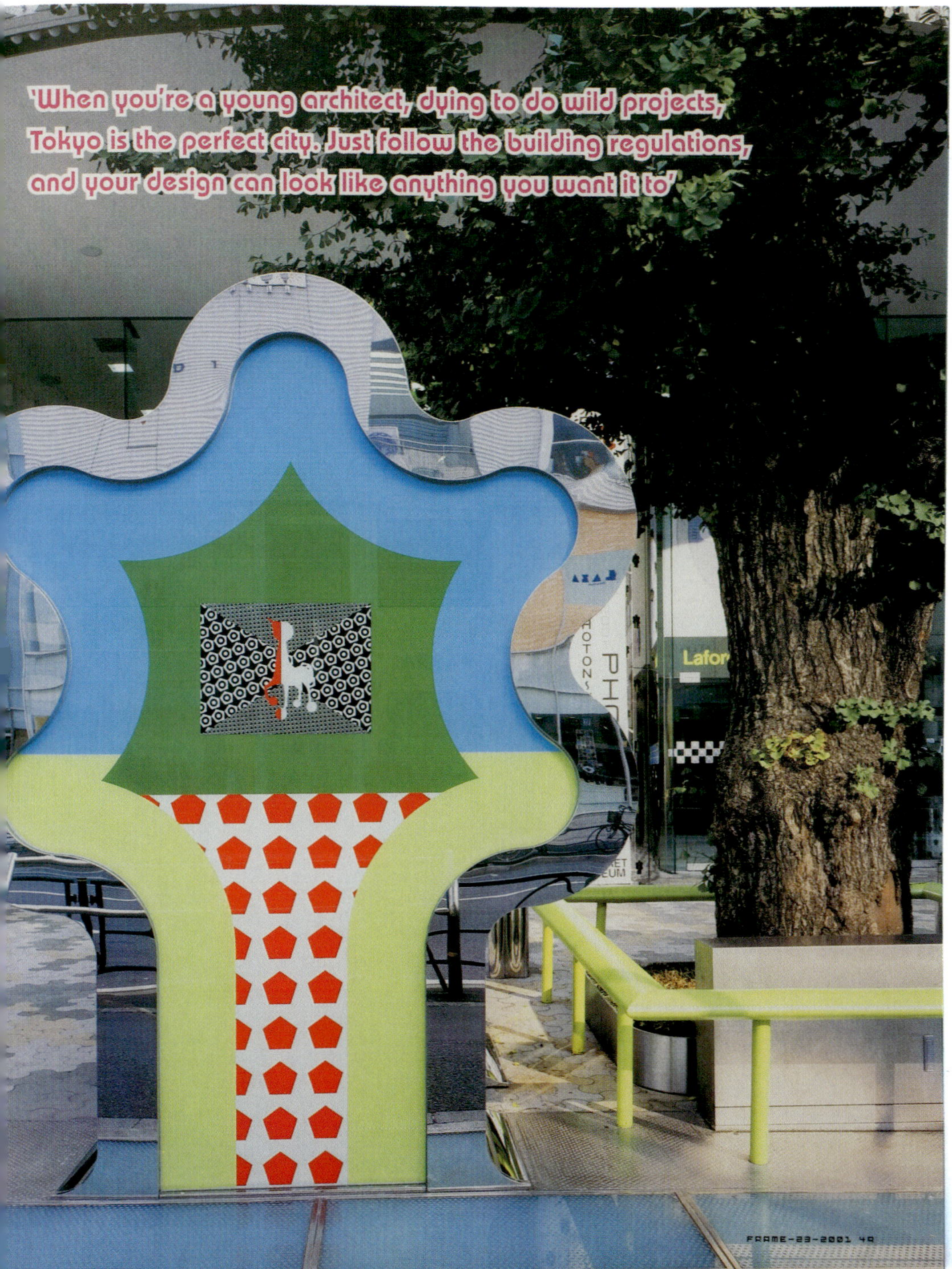

SUBJECT
Portrait of Klein Dytham architecture

LOCATION
Tokyo, Japan

DESIGNERS
Klein Dytham architecture

PHOTOGRAPHERS
Katsuhisa Kida
Kozo Takayama

WRITER
Carolien van Tilburg

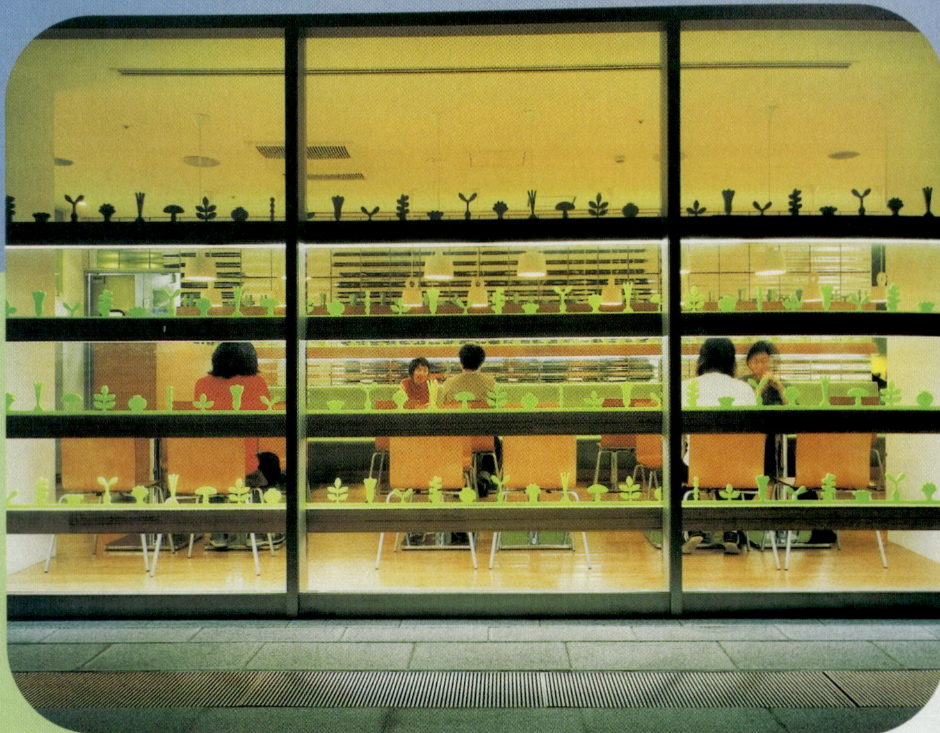

Spongy, bright-green pictograms of vegetables are eye-catching features at Vegie-To-Go (2000), a café-deli serving healthy meals for the young and trendy crowd of Daikanyama. Clad entirely in wavy mirror panels, the rear wall reflects daylight and distorts reflections.

– www.klein-dytham.com – which generates over 10,000 page hits a week, is a prize in itself. A fun feature is a collage of the Tokyo landscape composed of ads, billboards and a potpourri of products.

Astrid Klein – born and raised in Italy by her German parents – has a nose for trends, colours and materials. Mark Dytham, a witty Briton, is the engineering half of the pair, a guy with a fascination for new gadgets and technologies. 'I'm an early adopter,' he says. 'I don't mind if products have growing pains. I like to tackle the flaws and find a solution. The structure of a project and the logic behind it are my main contributions to KDa. I analyse the concept from every possible angle. Astrid is into colour and the kind of quirky stuff that I wouldn't dare put into a project. If I don't like something and can't come up with a better idea, however, I trust Astrid's judgement. We don't compromise.'

Mark and Astrid like to create projects that make people smile. They want to stir up the kind of fun that promotes a positive image of their client. A prime example is the inflated silver walls they created for a construction site on Omotesando – a dynamic, out-of-the-box design that's easy to understand. 'We try to make things that everyone can enjoy,' Astrid explains. 'We don't want to be elitist, and our approach is far from academic. Architecture is all about the visual; you never find a note inside that describes the concept. Just look at it. See it.'

A project that 'speaks for itself' – that's the catch phrase at KDa. 'Years ago, when we couldn't explain our ideas in proper Japanese, we used drawings and sketches. We gesticulated.' Mark grins at the memory. 'It's that sort of visual communication which is still so essential to our work.'

Toyo Ito appreciates this down-to-earth approach. 'I do believe that design in Japan, both now and in the future, needs the kind of lively feeling, rooted in direct experience, that these two represent. The type of society they propose is not one in which a sharply transparent or super-flat design freezes the user, but one in which we find an expression of maturity and friendliness. Klein and Dytham show us, time and again, what truly stylish design is.'

A flair for trends and fashions sometimes appears in their work as a retro look – a flashback to the '60s and '70s.

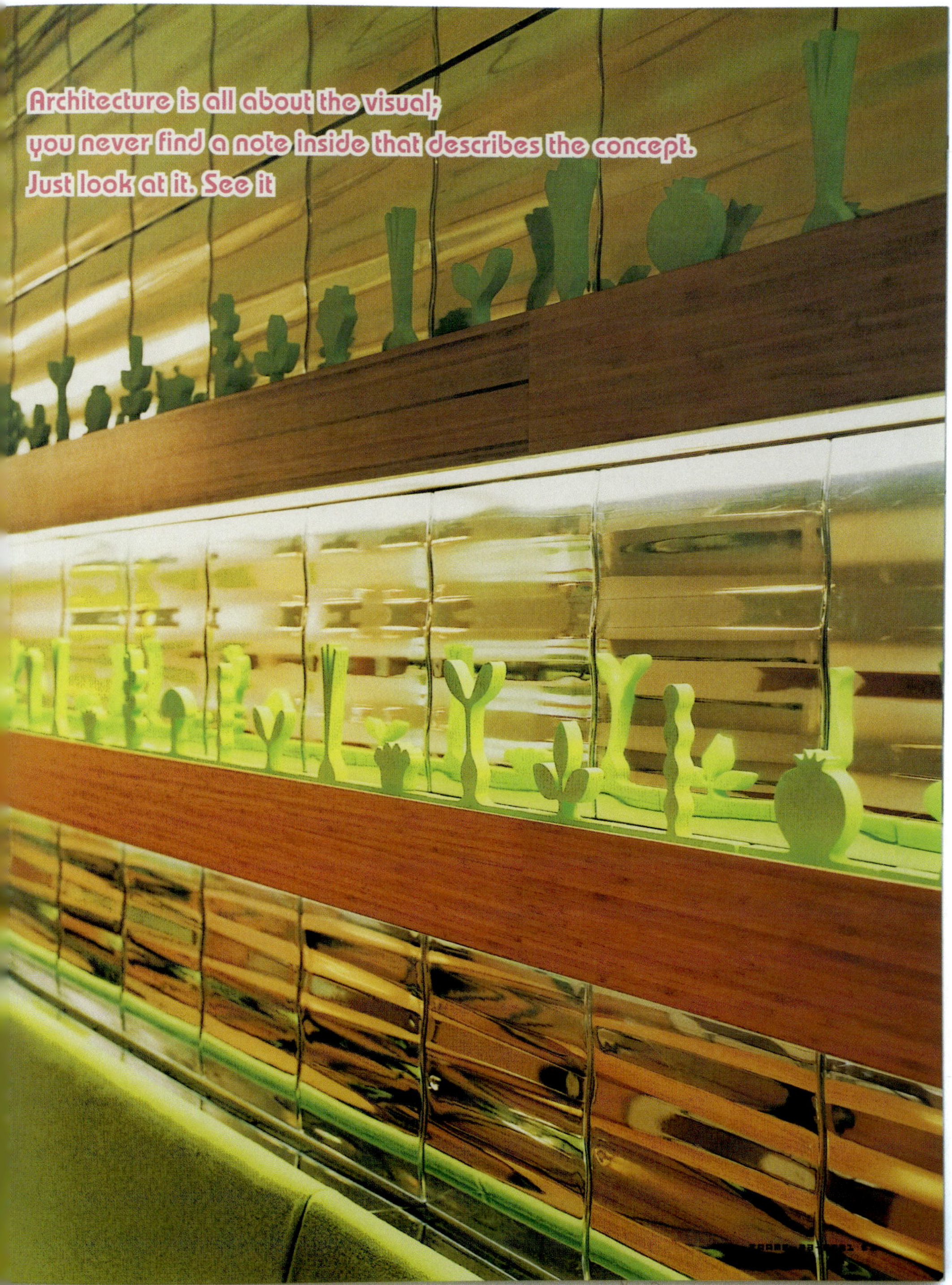

Architecture is all about the visual;
you never find a note inside that describes the concept.
Just look at it. See it

SUBJECT
Portrait of Klein Dytham architecture

LOCATION
Tokyo, Japan

DESIGNERS
Klein Dytham architecture

PHOTOGRAPHERS
Katsuhisa Kida
Kozo Takayama

WRITER
Carolien van Tilburg

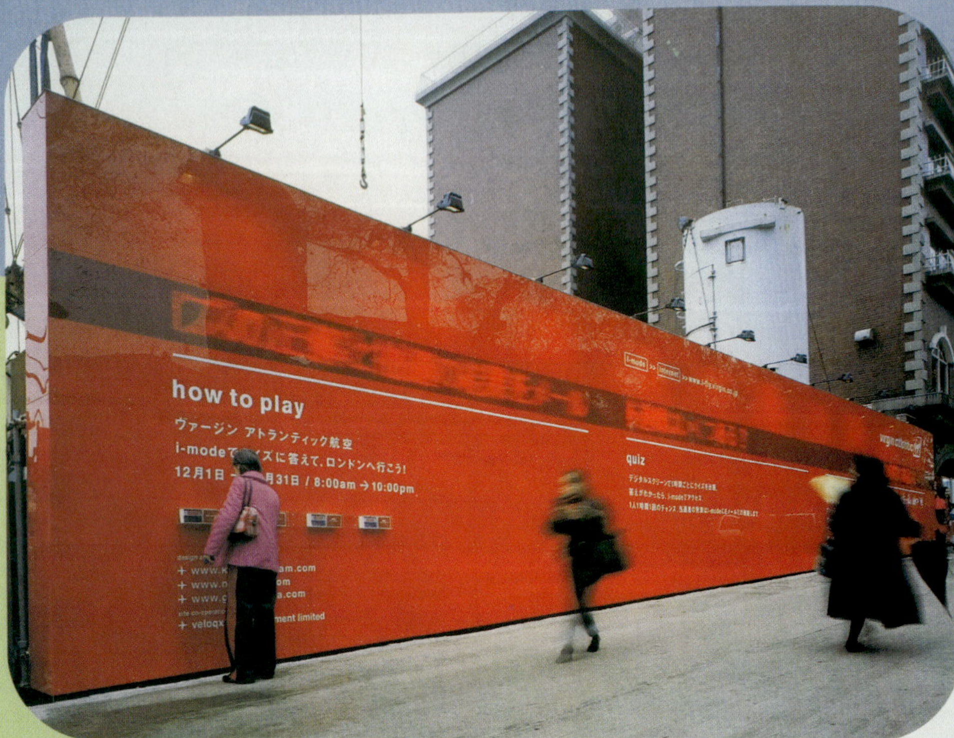

In 2000 KDa developed an interactive, red-acrylic billboard for Virgin Atlantic Airways. A 20-metre-long LED ticker-tape display set into this Wonderwall flashed questions that passers-by with i-mode phones could answer by logging onto the campaign website.

They often opt for the most obvious solution, like the artificial trees and other forest-related objects that characterise KDa's design for the La Foret Building. Some critics may consider their unambiguous approach superficial, but others, like myself, label it light-hearted and unpretentious.

Given the Japanese thirst for novelty, the team knows that anything new is sure to catch on immediately. New materials are no exception. 'Supporting Klein Dytham's designs is a sensitivity to the materials in which they work,' says Ito. 'Their projects and their use of building materials are always infused with discovery.' The La Foret project, for example, boasted a world premiere: a wall of optical fibres. Despite the sky-high cost of installation – countered to some extent by the low cost of maintenance – the client loved the idea of being identified with something brand new.

In designing the Vrooom! garage, Klein Dytham used a new insulating paint developed by NASA. The client encouraged them to try it. But novelty alone isn't the main criterion. 'We want each design to suit the particular project involved,' Astrid explains. 'Our goal is a design that responds well to the brief, doesn't exceed the budget and respects the allotted time frame. We feel the same about materials. We try to use them in a relevant way, to let them express their special qualities. Take this new artificial wood by Mizawa. It looks incredibly organic. Our idea would be to use it in a way that would never work if it were natural wood, such as in a counter made of one piece, for example, with lots of curves. People would do a double take before finally realising that it's fake wood.' Humour, technology, colour and materials are all key elements of KDa's work.

Another strong point is a knack for promoting the name of the firm, as well as that of Deluxe, the six-faceted collective they helped to establish in Tokyo. A day-to-day association with five other creative outfits – all six share a communal workplace-cum-exhibition space – has brought a number of new interests into the lives of Mark and Astrid, from beer-brewing to sound design. Although the pair certainly can't be pigeonholed as 'architects to the core', they're quick to point out that their peripheral

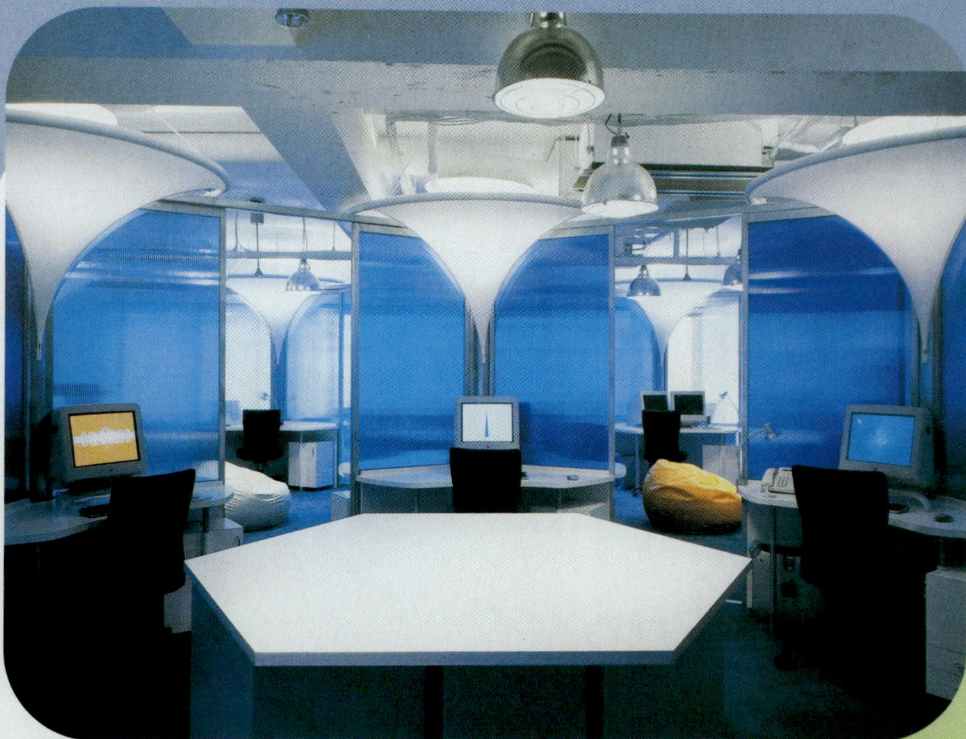

The hexagonal centre of the office design for Spike Cyberworks (2000) features a cluster of three desks in each corner. Separating the clusters are twin-wall polycarbonate screens finished in different hues of transparent blue film.

activities are 'mostly for fun'. Nevertheless, brewing beer and organising art-related events are part of building a widespread public image as a creative team par excellence. Clients have come to think of them not simply as the umpteenth architecture firm to set up shop in Tokyo, but as a smart, exciting group of people that get things done.

With no official policy on the appearance of buildings, Tokyo presents a bewildering array of modern architecture, a cityscape rich in diversity and visual discord. 'When you're a young architect, dying to do wild projects, Tokyo is the perfect city,' Astrid laughs. 'Just follow the building regulations, and your design can look like anything you want it to.'

Mark finds the city inspiring. 'There's a lack of context. Whatever you see today may be knocked down by next year. Everything is more or less disposable, and change is inevitable. It's a refreshing way to look at architecture.' This train of thought leads him to convenience stores, which 'can't leave anything as it is. It's all about improvement, about creating a new look. Change the chocolate bar, renew the beer can, buy the latest mobile phone.'

The English, he claims, want to keep things for life. 'Think of Burberry and of all those classic items made of leather. Guys use their leather briefcases forever – they love the idea. From the perspective of architecture, bricks get better as they age. Maybe because so many ancient European structures still exist, Europeans want products that will last just as long.' Not so in Japan, where the average building is only 25 years old at the time of its demolition and subsequent replacement.

'The demand for new things does keep us on our toes,' Astrid remarks. 'We can't be seen turning out the same old recipe. Our clients want innovative designs, which means we have to reinvent ourselves as well. A fresh face and a different approach to analysing the latest project can result in that unique – and uniquely relevant – solution that we're always waiting to discover.'

This is a revised version of Carolien van Tilburg's introduction to the book Klein Dytham architecture: Tokyo Calling, published by Frame in collaboration with Birkhäuser.

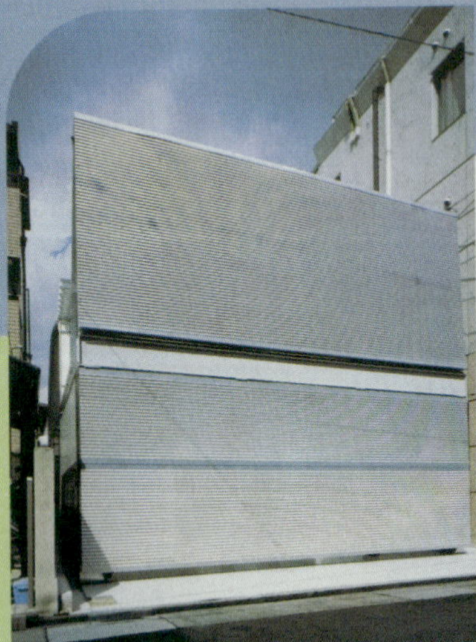

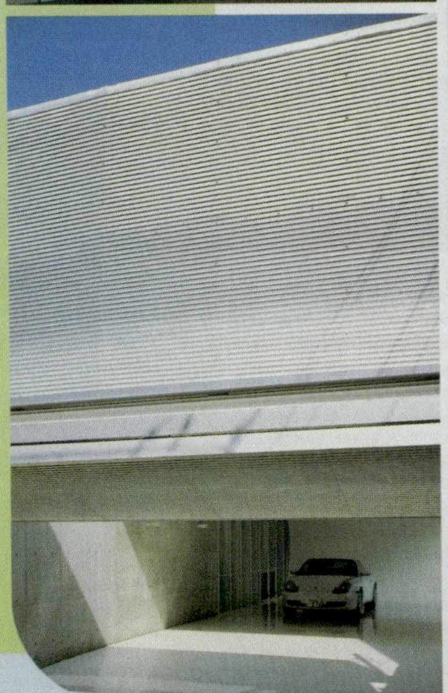

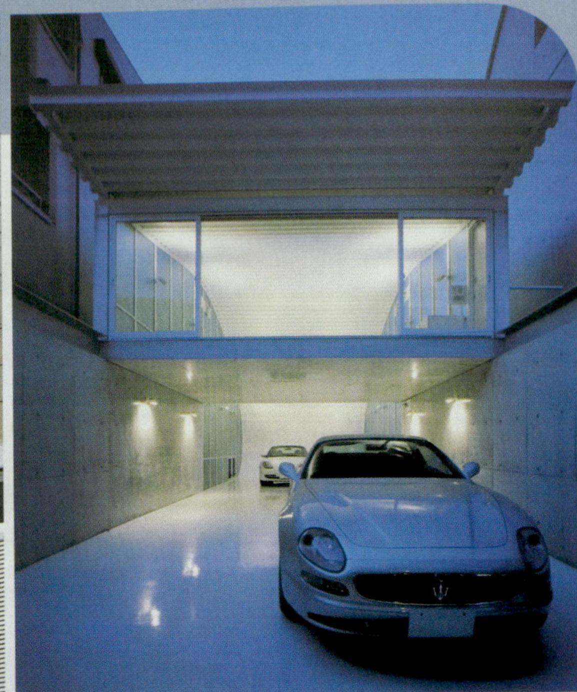

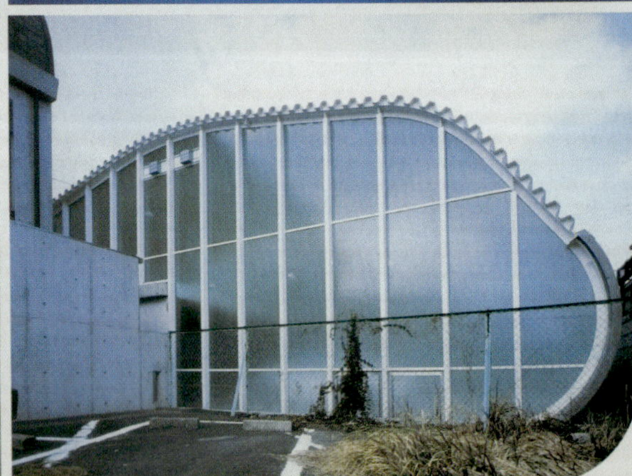

'There's a lack of context in Tokyo.
Whatever you see today may be demolished by next year.
It's a refreshing way to look at architecture'

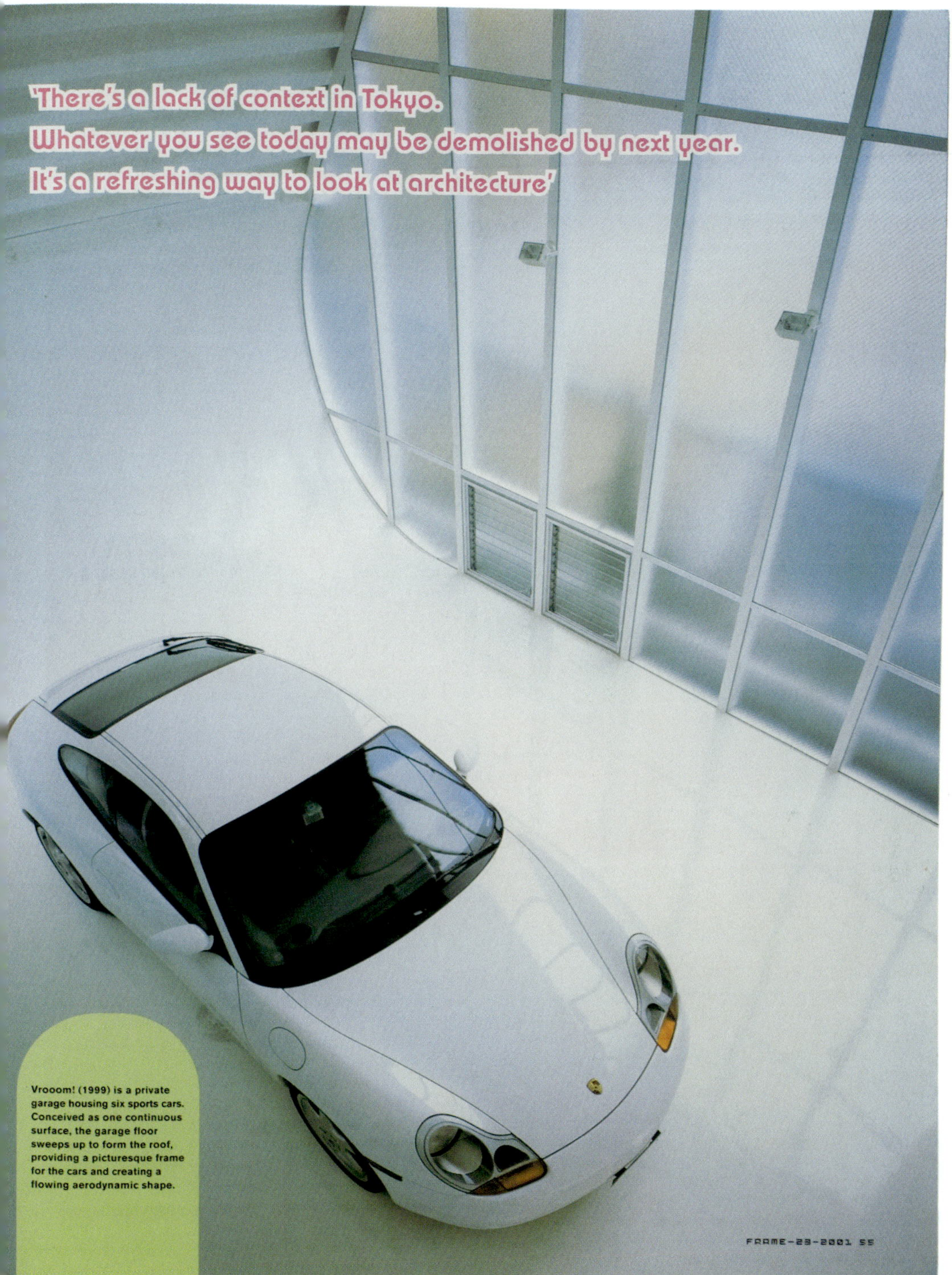

Vrooom! (1999) is a private garage housing six sports cars. Conceived as one continuous surface, the garage floor sweeps up to form the roof, providing a picturesque frame for the cars and creating a flowing aerodynamic shape.

SUBJECT
Portrait of Klein Dytham architecture

LOCATION
Tokyo, Japan

DESIGNERS
Klein Dytham architecture

PHOTOGRAPHERS
Katsuhisa Kida
Kozo Takayama

WRITER
Carolien van Tilburg

Issue 24 January/February 2002 248-249

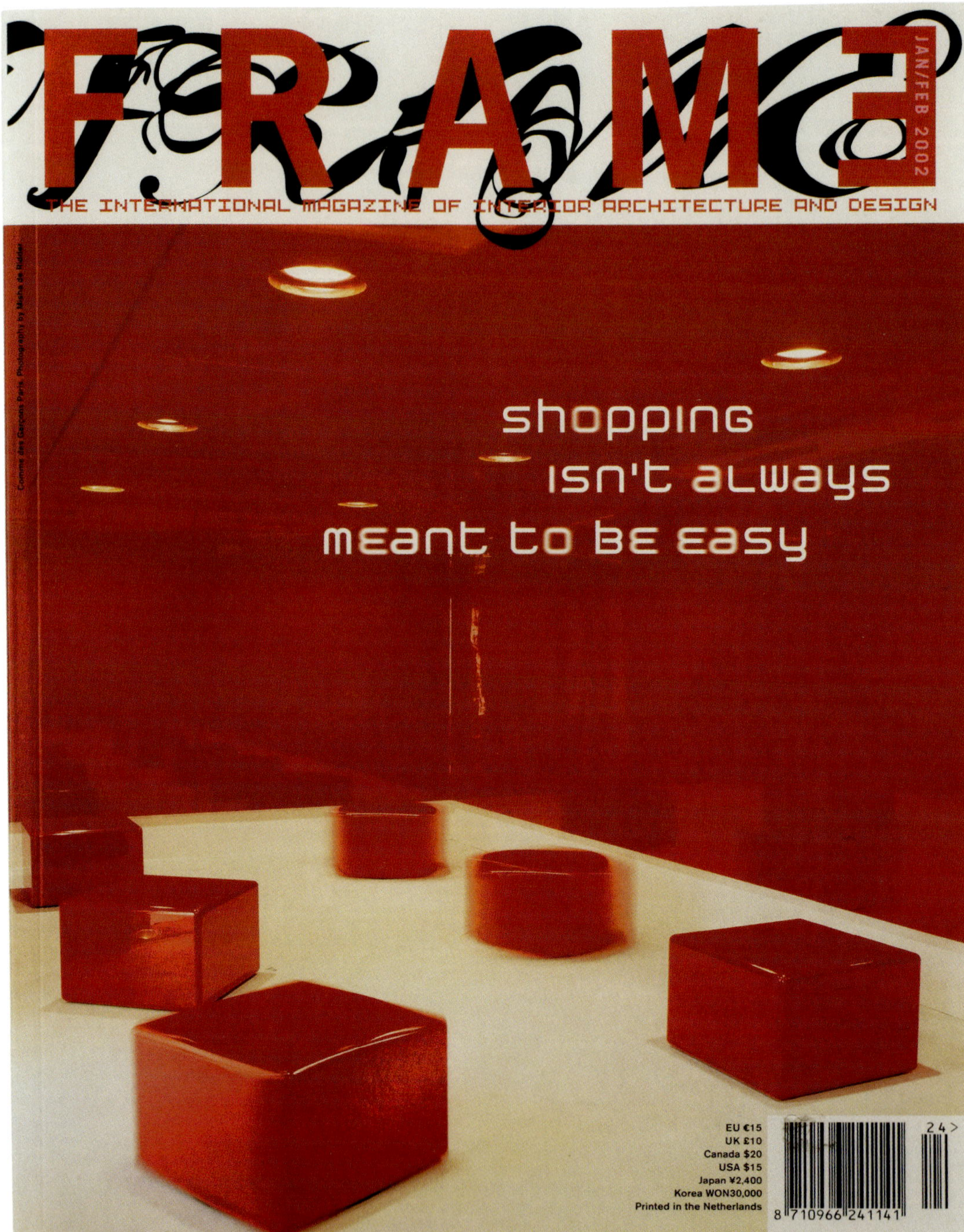

ART DIRECTOR
Roelof Mulder

Simply Red

Frame editor Tessa Blokland about the project

Comme des Garçons is a fashion brand we watch closely. Rei Kawakubo continually pushes the boundaries of all design disciplines. In her eyes, fashion is never 'just fashion', and a shop can double as a gallery. Why not? Because we knew that other magazines would also be featuring this fantastic shop, we asked Misha de Ridder to go to Paris and make an exclusive photo report for Frame. We were really excited when we saw the results.

Did the article have an effect?

It added to an increasing realization that, in terms of design, the world of fashion is often one step ahead of the rest. It was a lesson that we at Frame took to heart: being one step ahead is vital, not only with respect to content, but also to graphics and to our exploration of a wide range of design-related disciplines.

DESIGNERS
Rei Kawakubo
(Comme des Garçons)
Takao Kawasaki
KRD
www.krd-uk.com
Architectures Associés
Red Wave

SUBJECT
Comme des Garçons
shop

LOCATION
Paris, France

PHOTOGRAPHER
Misha de Ridder
www.mishaderidder.nl

WRITER
Chris Scott

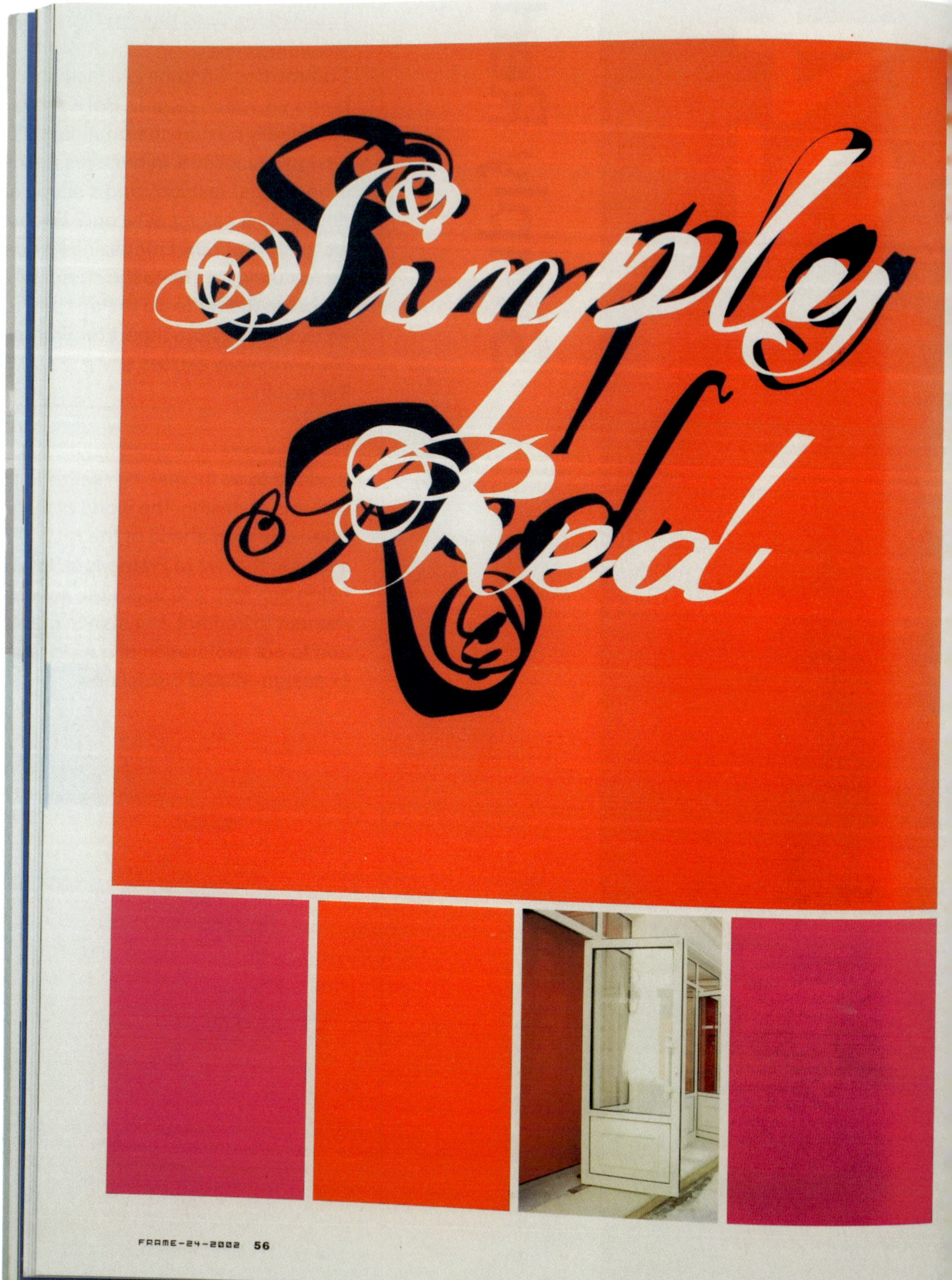
Simply Red
FRAME-24-2002 56

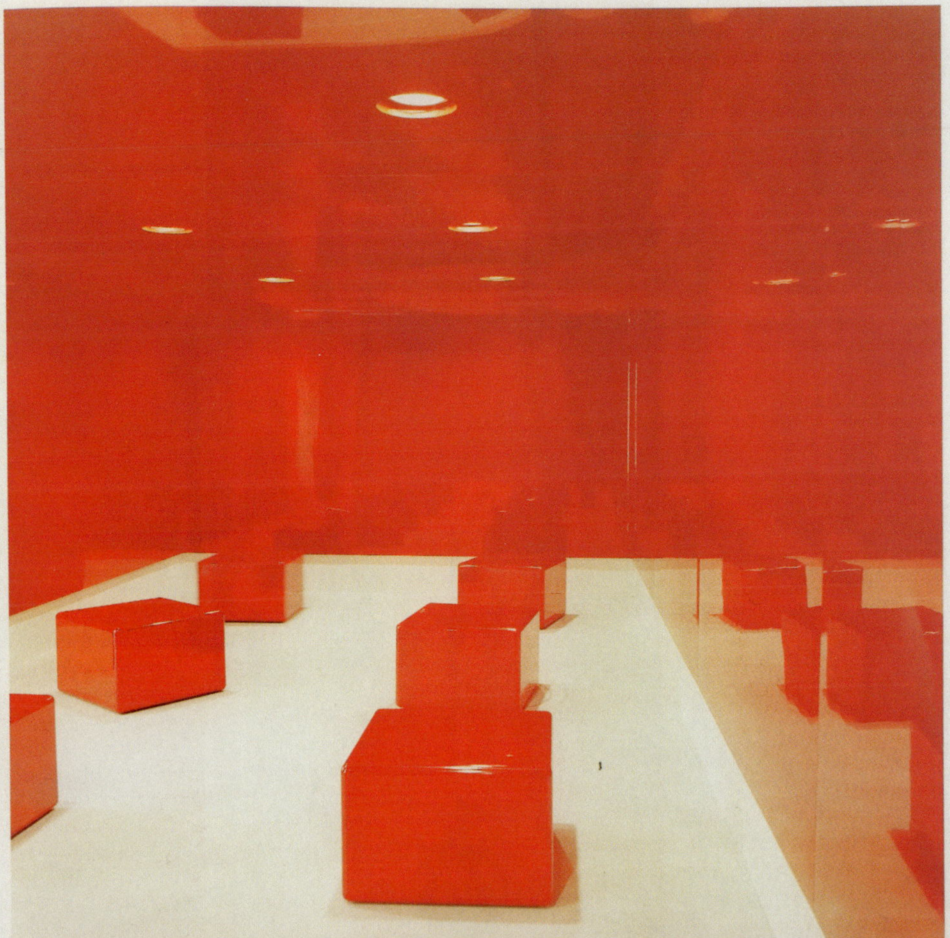

Not only has Rei Kawakubo been at the forefront of fashion for nearly two decades; the founder of Comme des Garçons also sets the pace in retail interiors, as her latest Paris shop proves once again.

By Chris Scott. Photography by Misha de Ridder.

SUBJECT
Comme des Garçons shop

LOCATION
Paris, France

DESIGNERS
Various, see page 249

PHOTOGRAPHER
Misha de Ridder

WRITER
Chris Scott

I vividly recall my very first Comme des Garçons purchase. Buying it was like being initiated into a select club and given a stamp of approval that signified credibility, like having some higher force declare me artistically sensitive but no slave to fashion – a feeling I'd never had before in a boutique.

An article about Comme des Garçons is an article about Rei Kawakubo, the woman who is Comme des Garçons. Renowned for shunning the limelight, for failing to show up when expected, for seeming aloof, this doyenne of style has always insisted that gathering insight into her personality is as simple as looking at her work. Arrogance has nothing to do with it. She genuinely believes that her work is more interesting than any other aspect of her life. Those who have had the privilege of meeting her agree that she comes across a good deal more sincere than most fashion bigwigs do.

The unmistakable Comme des Garçons style has penetrated many areas. Kawakubo currently sets the pace in perfumes and toiletries, made irresistibly attractive by their packaging and colour schemes. Unusual fragrances like tea, lily and mint come with candles and chocolate to match – products that not only smell good but look good and taste good too.

Another prime example is the Comme des Garçons fashion show, a recent example of which was seen at Landed, a fashion and cultural event held in Antwerp. By showing her Autumn/Winter collection at five totally different venues (museum, school, church, zoo and stock exchange), Kawakubo offered five equally different interpretations of one and the same collection. Here, as always, the designer's creative process – forever moving forward, never resorting to the past – started from scratch. Although she declines to name any influencing factors, objects, places or people, she does express her admiration for anyone who works with heart and soul. And she's perfectly willing to mention a few of her favourite

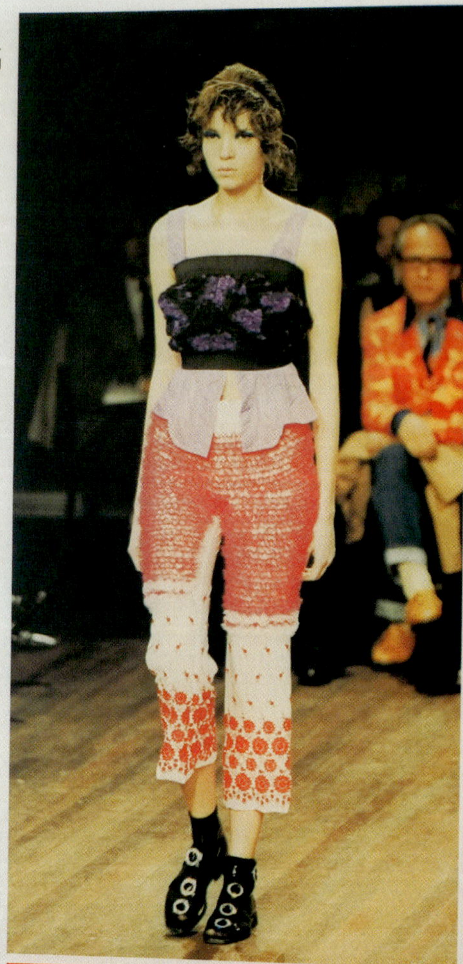

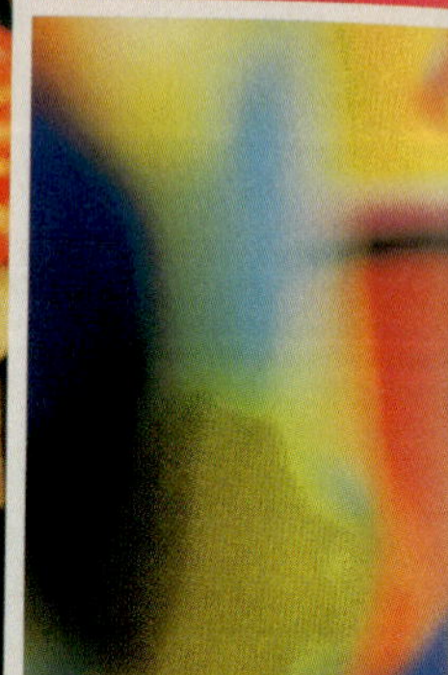

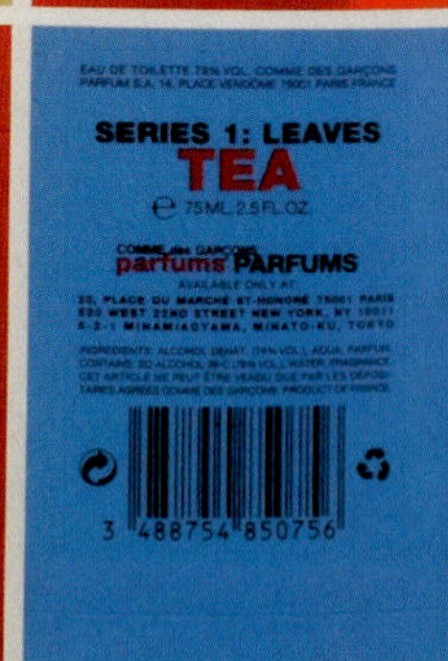

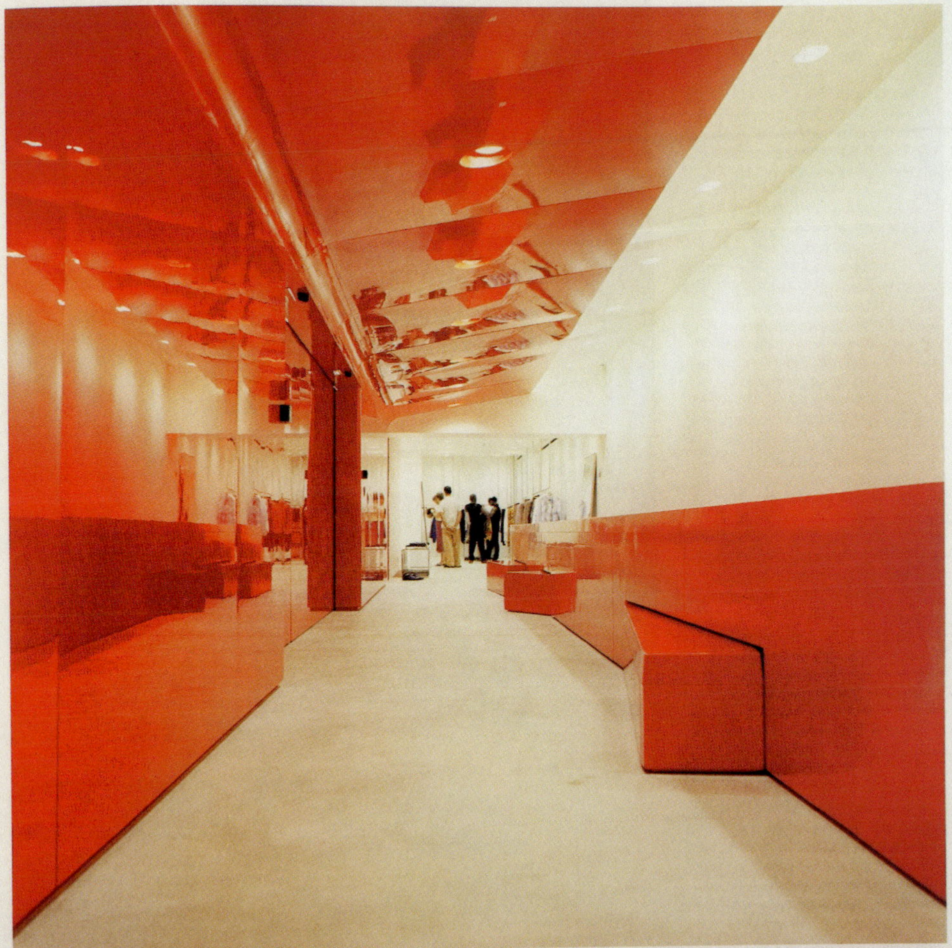

Kawakubo claims that 'red is as strong as black' – quite a statement from the woman who helped instigate a seemingly endless 'black period' in fashion

SUBJECT
Comme des Garçons shop

LOCATION
Paris, France

DESIGNERS
Various, see page 249

PHOTOGRAPHER
Misha de Ridder

WRITER
Chris Scott

things, among which are trees, animals and diamonds. In the absence of Comme des Garçons (heaven forbid!), she sees herself enjoying the company of frolicking cats and dogs.

All facets of her work are accorded equal consideration. 'Comme des Garçons cannot express all it wants to say through clothes alone,' she acknowledges. 'Environment, space and printed matter are subjected to the same treatment.' The position of a logo on a page, the width of a shop door, the stitching of a hemline: each is infused with the firm's sense of fresh, aesthetically provocative values, which challenge prevailing notions. That said, it is still difficult to explain the enduring success of this fashion label, which has acquired a visionary and virtually cult status that has not come without a price. 'As time goes by, the pressure to maintain a certain quality only mounts,' says Kawakubo. 'The weight of expectation grows with each success.' The demands on the designer and her employees are severe. Known as a stickler for perfection, she is self-critical and seldom satisfied with results. Believing that complacency is the enemy of creative achievement, she's always on the lookout for the next barrier to break through.

Set up in 1973 with just 6 employees, today Comme des Garçons has a staff of around 450. Despite such a large and skilled workforce, however, Kawakubo personally oversees all areas of design. On occasion she grants a certain degree of freedom to talents she admires, among them Keiichi Tanaka, who's involved with the company's 'Homme' designs, and Junya Watanabe, designer of the 'Tricot' line as well as a personal collection sold by Comme des Garçons.

Kawakubo's involvement in architecture and interiors is something she particularly enjoys, perhaps because it's less demanding than designing a new collection every six months. She participates in all aspects of an interior project and firmly rejects anything that

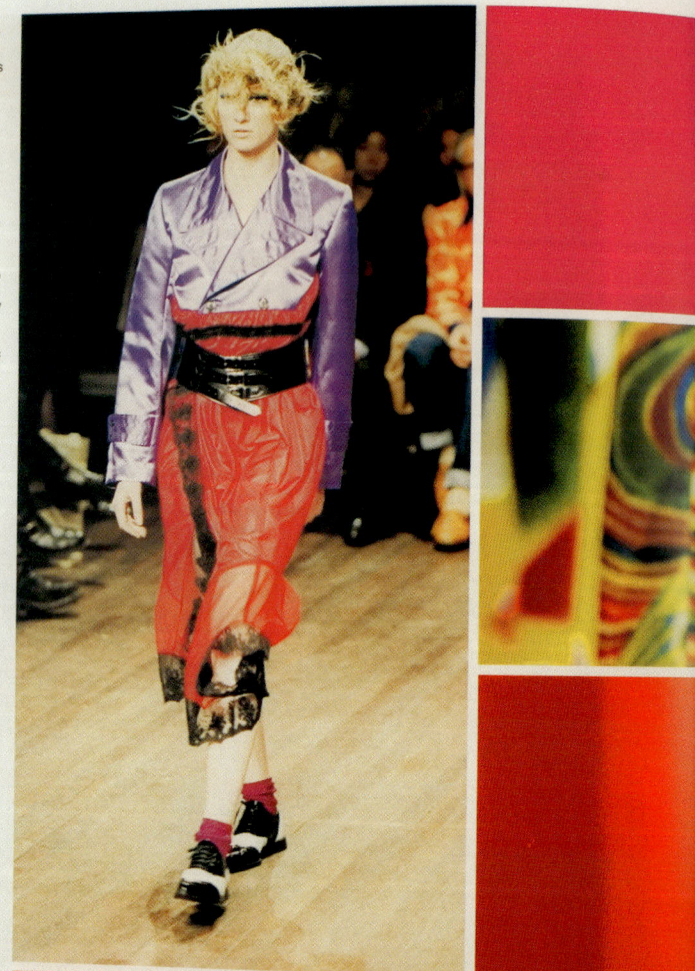

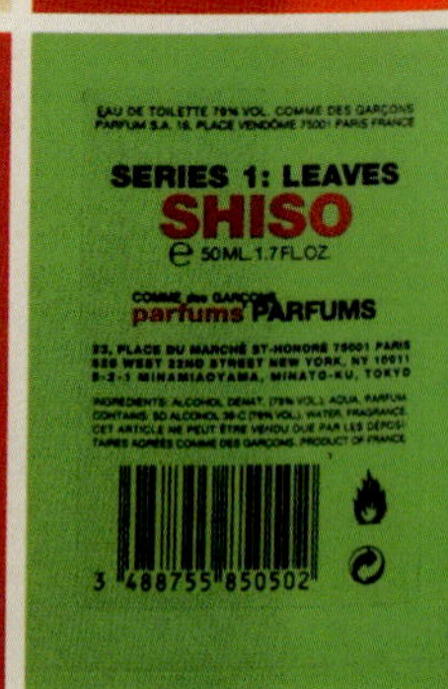

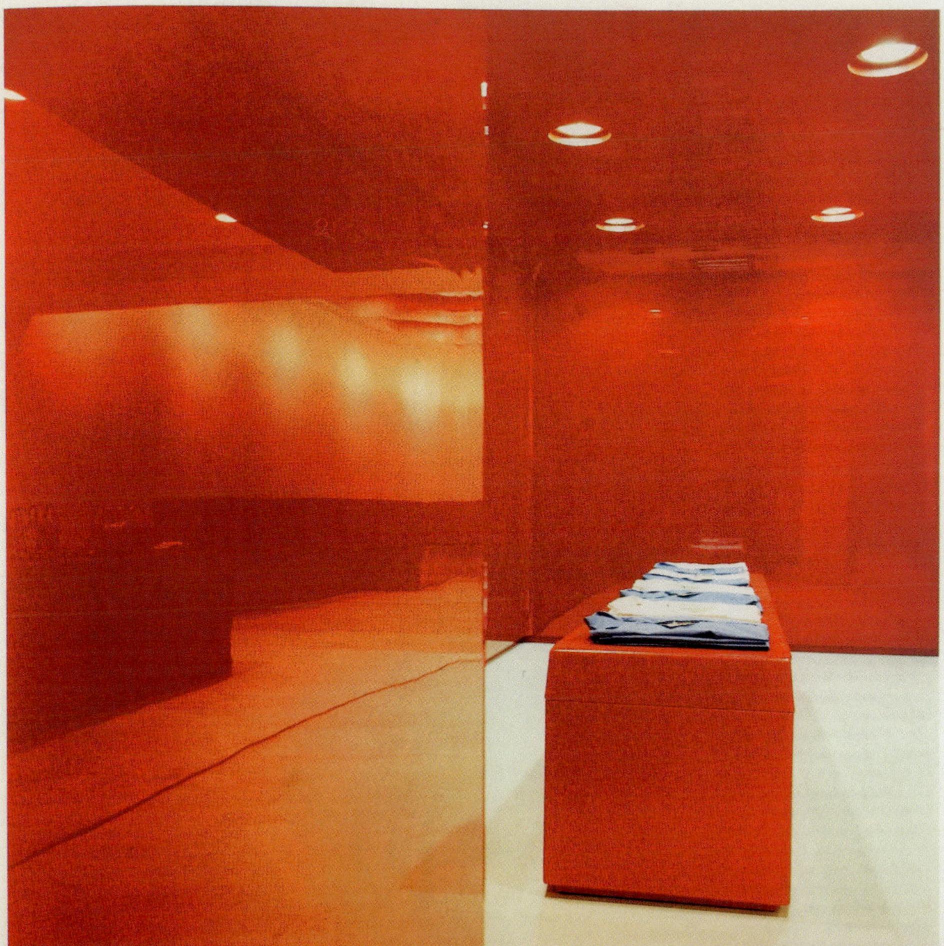

Shopping at Comme des Garçons isn't meant to be easy. Getting to grips with the sparsely furnished interior requires determination

SUBJECT Comme des Garçons shop

LOCATION Paris, France

DESIGNERS Various, see page 249

PHOTOGRAPHER Misha de Ridder

WRITER Chris Scott

is not redolent of Comme des Garçons. Though one of the first designers to work with a variety of big-name architects – a practice now common in retail design – she has collaborated repeatedly with some of her favourites, such as Japanese architect Takao Kawasaki.

Opting for locations off the beaten track is a Kawakubo strategy first applied in New York and Tokyo and, more recently, in Paris, where her latest shop is found on Faubourg St Honoré. A surprising choice, but Comme des Garçons is nothing if not surprising. 'The shops are aimed at customers who already know us,' she explains, 'Locations are not exactly secret, but private, and easily accessible to those in the know. I wanted to create a sense of discovery and exploration, to transport customers to another world.'

Subtly inserted into its architectural context, the shop in Paris proves her point. A discreet red plaque on the façade is the only exterior hint that a shop exists within. On reaching the courtyard at the rear, the first-time visitor wonders if she's on the right track. Is there a connection between these opaque red windows and the red sign out front? The mystery intensifies as she enters a second, completely red, courtyard. Compelled to continue the exploration, she's drawn to the red panels ahead. Suddenly, as if by magic, they swing open with a soft swish to reveal an almighty, blood-red interior. The effect is startling, to put it mildly.

Shopping at Comme des Garçons isn't meant to be easy. Getting to grips with the sparsely furnished interior, which appears to recede into infinity, requires determination. In Paris, as the shock of the red subsides, the first objects to demand attention are ceiling-mounted, chromium-steel bars hung with garments against a backdrop of white-lacquered walls. Although dramatic, the scene invites the browser to look around and make a selection. Summoning up the courage to take a few items to the fitting room is the next challenge,

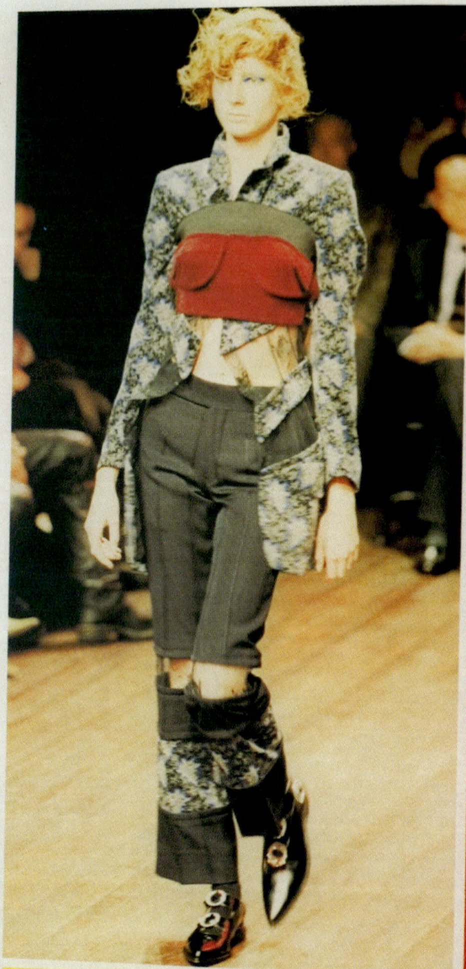

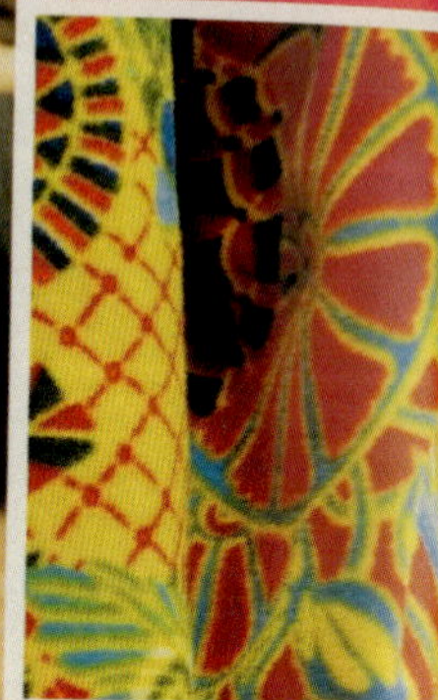

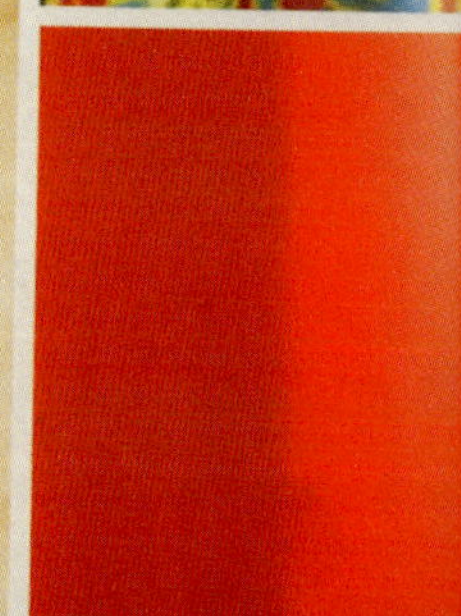

SERIES 1: LEAVES
CHOCOLATE
TEA

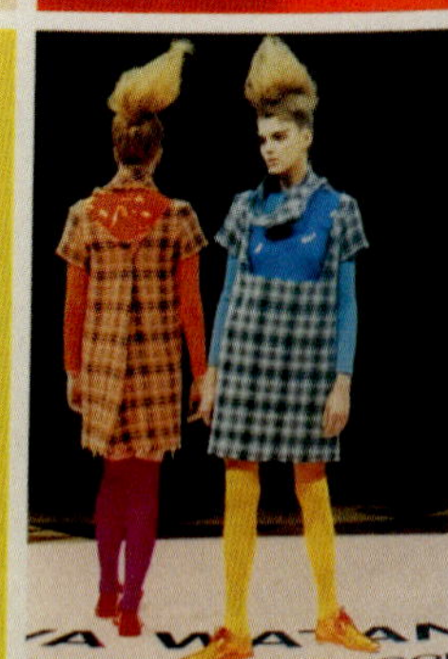

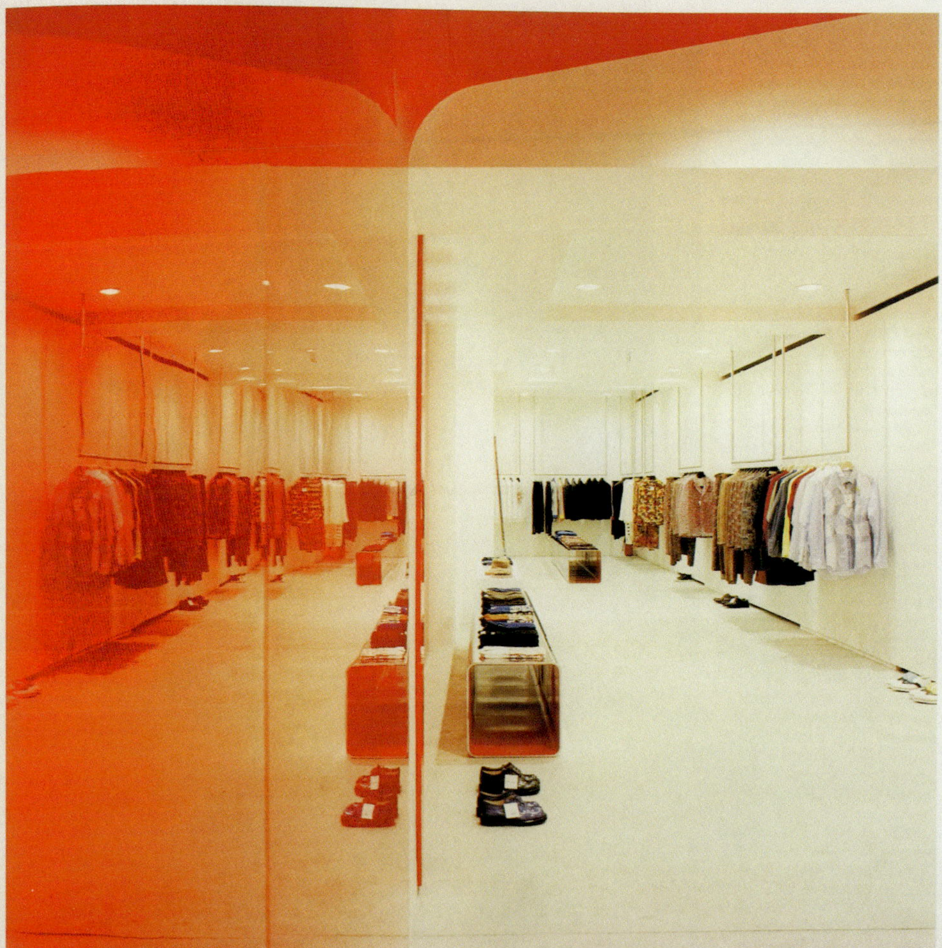

'I wanted to create a sense of discovery and exploration,' says Kawakubo of her latest shops, 'to transport customers to another world'

SUBJECT
Comme des Garçons shop

LOCATION
Paris, France

DESIGNERS
Various, see page 249

PHOTOGRAPHER
Misha de Ridder

WRITER
Chris Scott

followed in turn by a glance at the price tag. Whether overwhelmed or unruffled, the customer can take a relaxing break in the resting room across the courtyard, an area furnished with red cube seats that glide around the floor when unoccupied. Considered 'the essence of the shop', the space is the result of Kawakubo's collaboration with KRD duo Shona Kitchen and Ab (son of Richard) Rogers. Also involved in the project were Kawasaki, Architectures Associés and Red Wave, an English outfit. Other than the fabric-covered table by Christian Astuguevieille, all furniture was designed by Kawakubo.

Why red? Why so red? Apart from the fact that it looks fantastic on the fibreglass surface, Kawakubo claims that 'red is as strong as black' – quite a statement from the woman who helped instigate a seemingly endless 'black period' in fashion. Bold accents of colour now enliven much of her work. Moving forward and evolving are vital to her sustained success. 'I'd like to undertake a complete overhaul of the shops more often, but that's simply not practical,' she says. 'Instead, I alter the feel of the shops every ten years or so, though new departures must have a familiar feel to them.'

Her next venture into retail design will be a collaborative effort with another very talented female force in design: Carla Sozzani, founder of Milan's super-stylish 10 Corso Como. Scheduled to open in February 2002, the establishment will be located in Minami-Aoyama, Tokyo. Here Kawakubo wants the shop design to question where the fashion industry is going. She admires Sozzani's strong visual sense and marvels at her ability to select merchandise intended, in this case, to moderate the minimalist elements that are the hallmark of Comme des Garçons. Responsible for the overall design of the 840-square-metre space, Kawakubo is working with Kris Ruhs, Sozzani's partner, on interior details. Will this be the umpteenth lifestyle

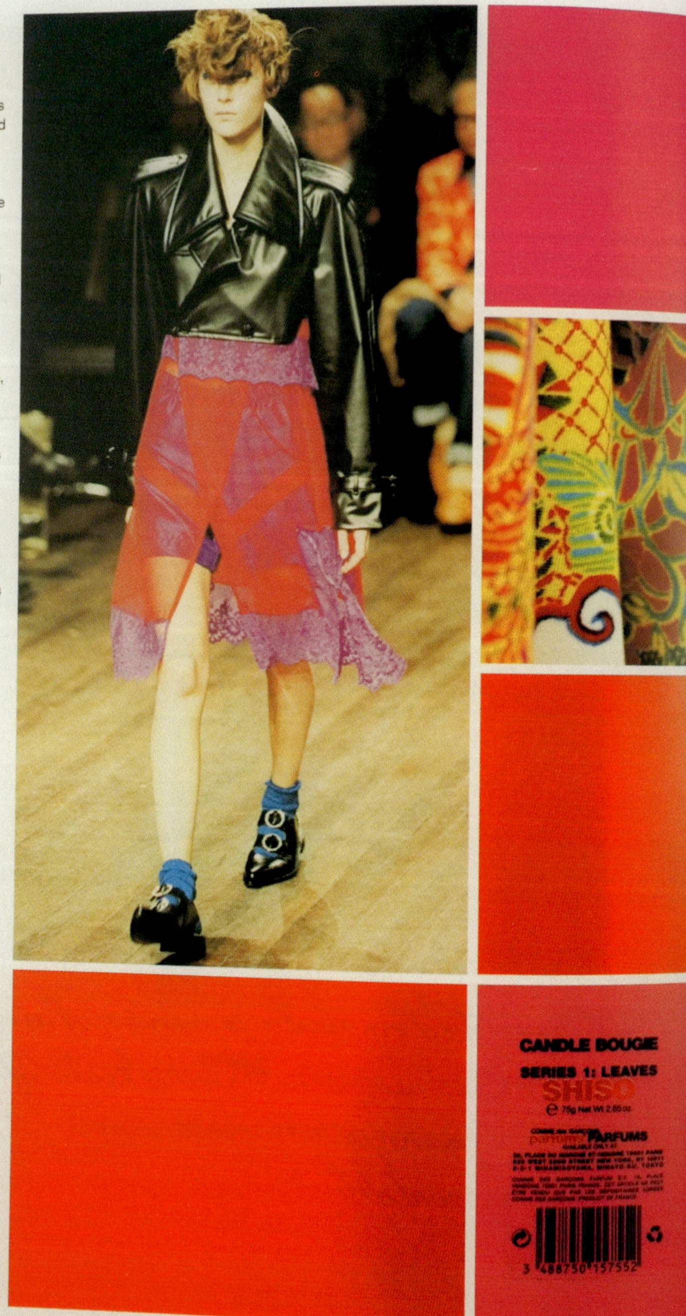

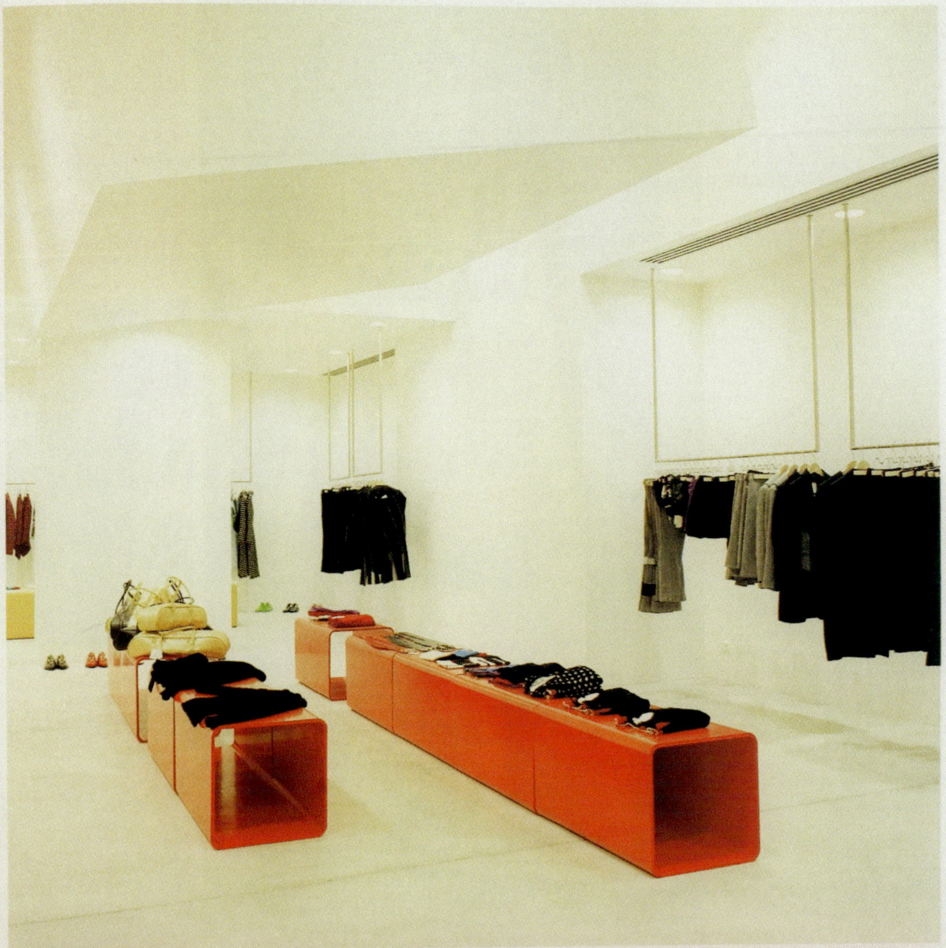

'The shops are aimed at customers who already know us. Locations are not exactly secret, but private, and easily accessible to those in the know'

SUBJECT
Comme des Garçons shop

LOCATION
Paris, France

DESIGNERS
Various, see page 249

PHOTOGRAPHER
Misha de Ridder

WRITER
Chris Scott

shop, albeit a highly stylised one? 'I do the shops I like to do,' she replies. 'I follow my instincts. I don't pay attention to or follow trends.' Not only fashion aficionados but also designers in general eagerly await her latest projects.

Never intent on provoking shock for shock's sake, Kawakubo does like to make things that surprise, excite and energise people. 'I constantly search for new ways to express unexpected beauty. Each piece is an attempt to raise spirits and to stimulate. I want to make clothes that liberate and pose questions, that allow people to express themselves.' A colleague sums her up succinctly: 'She makes interesting products, has a sound business acumen and is independent – a unique combination.' This is only too true in a world in which personal feelings are often smothered beneath an avalanche of commercialism and in which major corporations crush all traces of individualism and free spirit. Rei Kawakubo cherishes the independence that allows her to do what she sets out to do. Her achievements are those of a woman many of us would love to emulate.

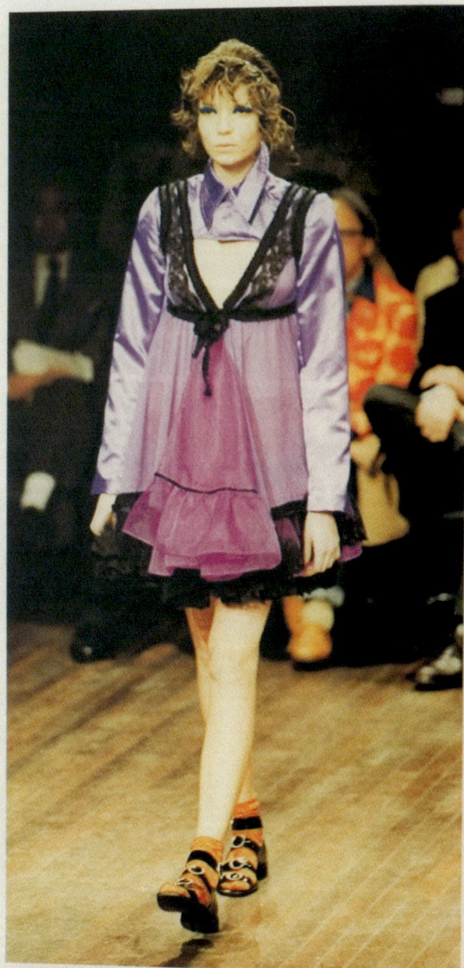

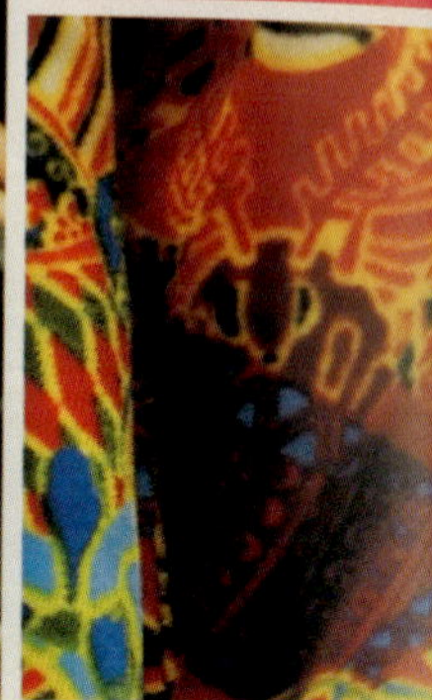

EAU DE TOILETTE 75% VOL. COMME DES GARÇONS PARFUM S.A. 16 PLACE VENDÔME 75001 PARIS FRANCE

SERIES 1: LEAVES

MINT

℮ 25ML 0.80FL.OZ

COMME des GARÇONS parfums PARFUMS

23, PLACE DU MARCHÉ ST-HONORÉ 75001 PARIS
520 WEST 22ND STREET NEW YORK, NY 10011
5-2-1 MINAMIAOYAMA, MINATO-KU, TOKYO

INGREDIENTS: ALCOHOL DENAT. (75% VOL.), AQUA, PARFUM CONTAINS: SD ALCOHOL 39-C (75% VOL.), WATER, FRAGRANCE. CET ARTICLE NE PEUT ÊTRE VENDU QUE PAR LES DÉPOSITAIRES AGRÉÉS COMME DES GARÇONS. PRODUCT OF FRANCE

3 488752 850253

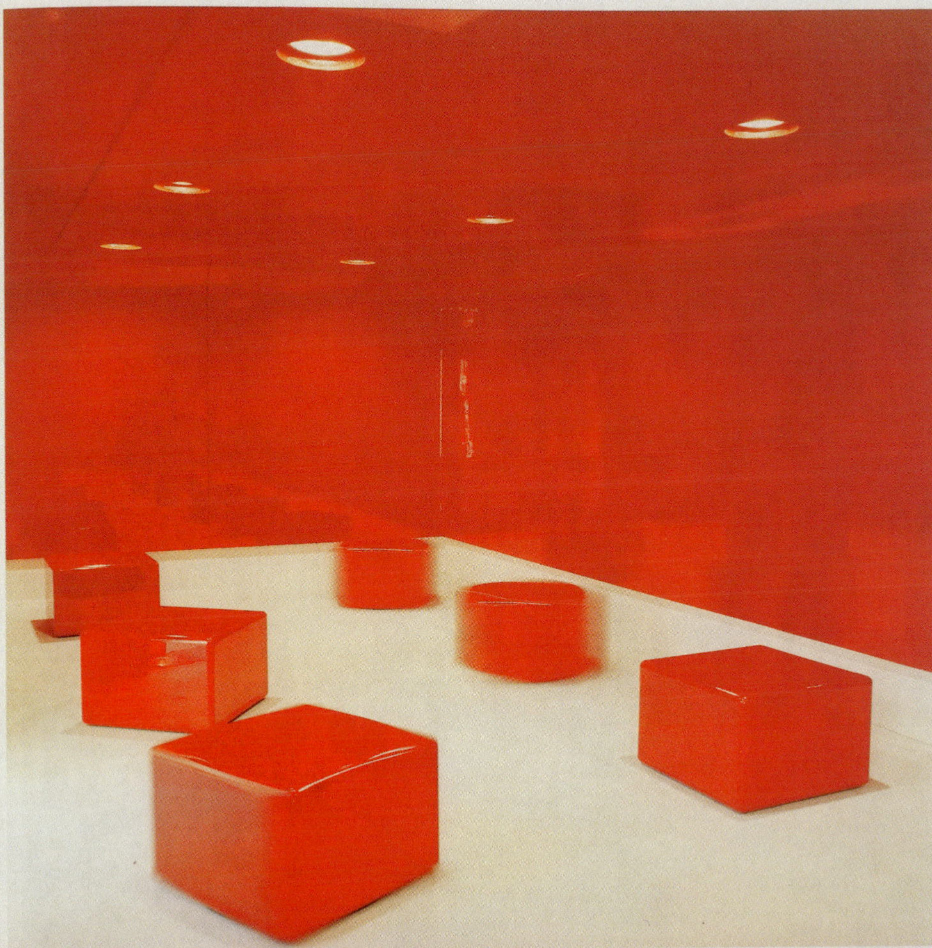

'I want to make clothes that liberate and pose questions, that allow people to express themselves'

SUBJECT
Comme des Garçons shop

LOCATION
Paris, France

DESIGNERS
Various, see page 249

PHOTOGRAPHER
Misha de Ridder

WRITER
Chris Scott

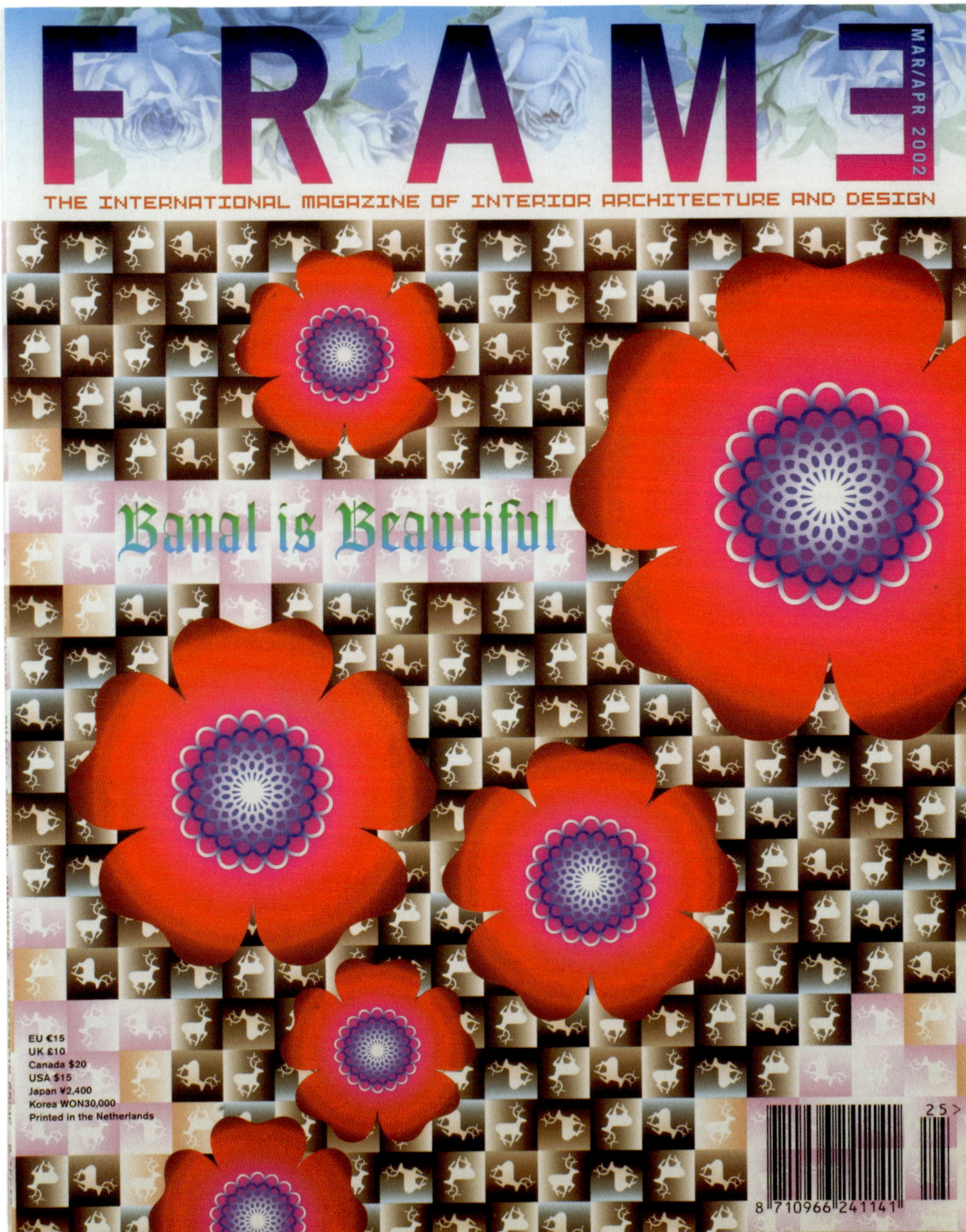

ART DIRECTOR
Roelof Mulder

25

Car Culture

Designer Wiebe Boonstra about the project

Great thinking on great masculine themes – cars and cosmetics.

Did the article have an effect?

A surprising effect.

Current projects

'Op bezoek bij de overheid' (A Visit to the Government): a study on the quality of semi-public space, commissioned by the Public Fund for the Promotion of Architectural Quality // Noncommercial projects // Retail projects // Development of a new collection of furniture.

DESIGNERS
Wiebe Boonstra
Martijn Hoogendijk
Marc van Nederpelt
(formerly Dumoffice)
www.dumoffice.com

SUBJECT
Visual essay on cars and their relationship to architecture, interiors and furniture

LOCATION
Amsterdam, Netherlands

PHOTOGRAPHER
—

WRITERS
Dumoffice

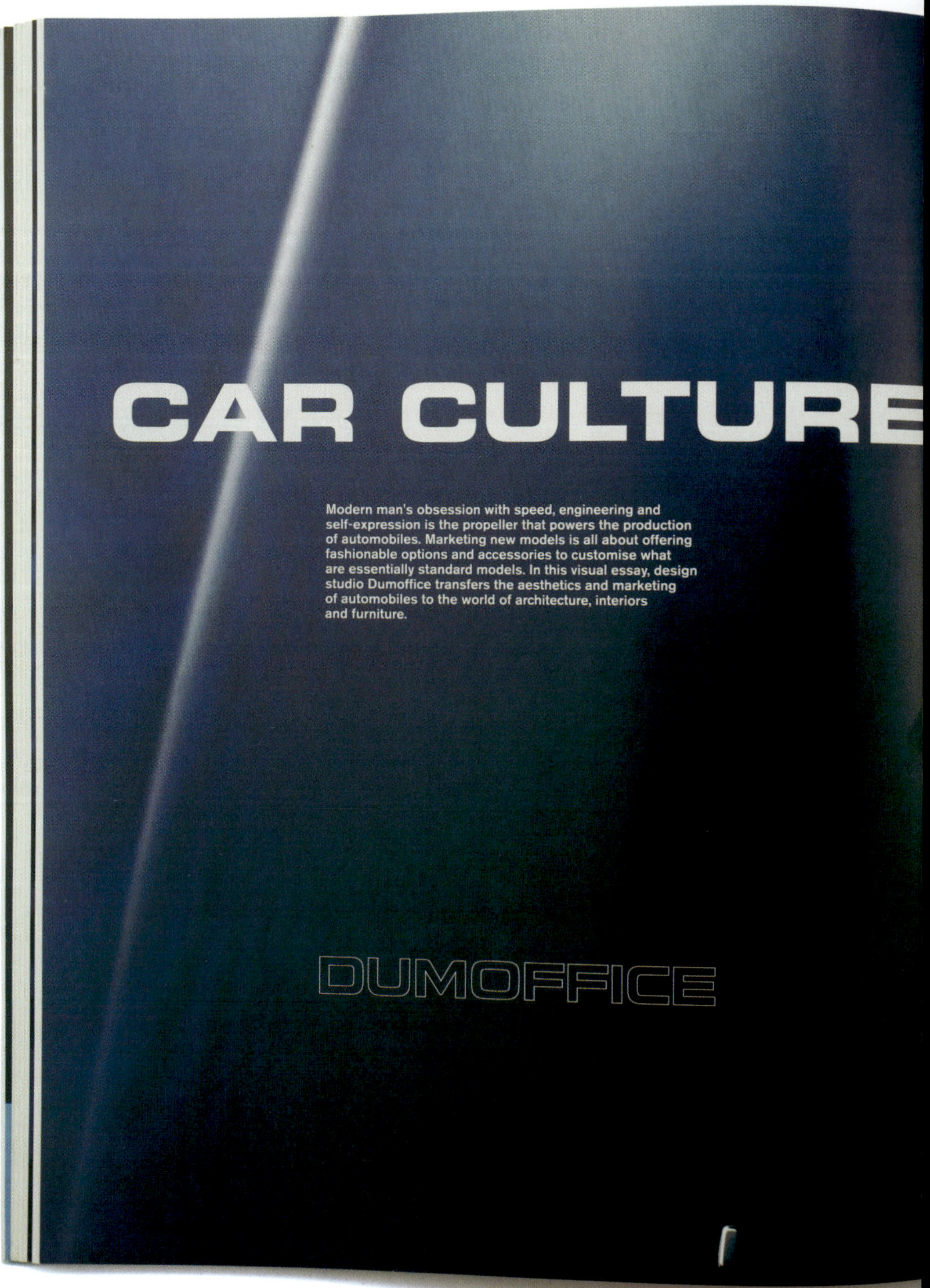

CAR CULTURE

Modern man's obsession with speed, engineering and self-expression is the propeller that powers the production of automobiles. Marketing new models is all about offering fashionable options and accessories to customise what are essentially standard models. In this visual essay, design studio Dumoffice transfers the aesthetics and marketing of automobiles to the world of architecture, interiors and furniture.

DUMOFFICE

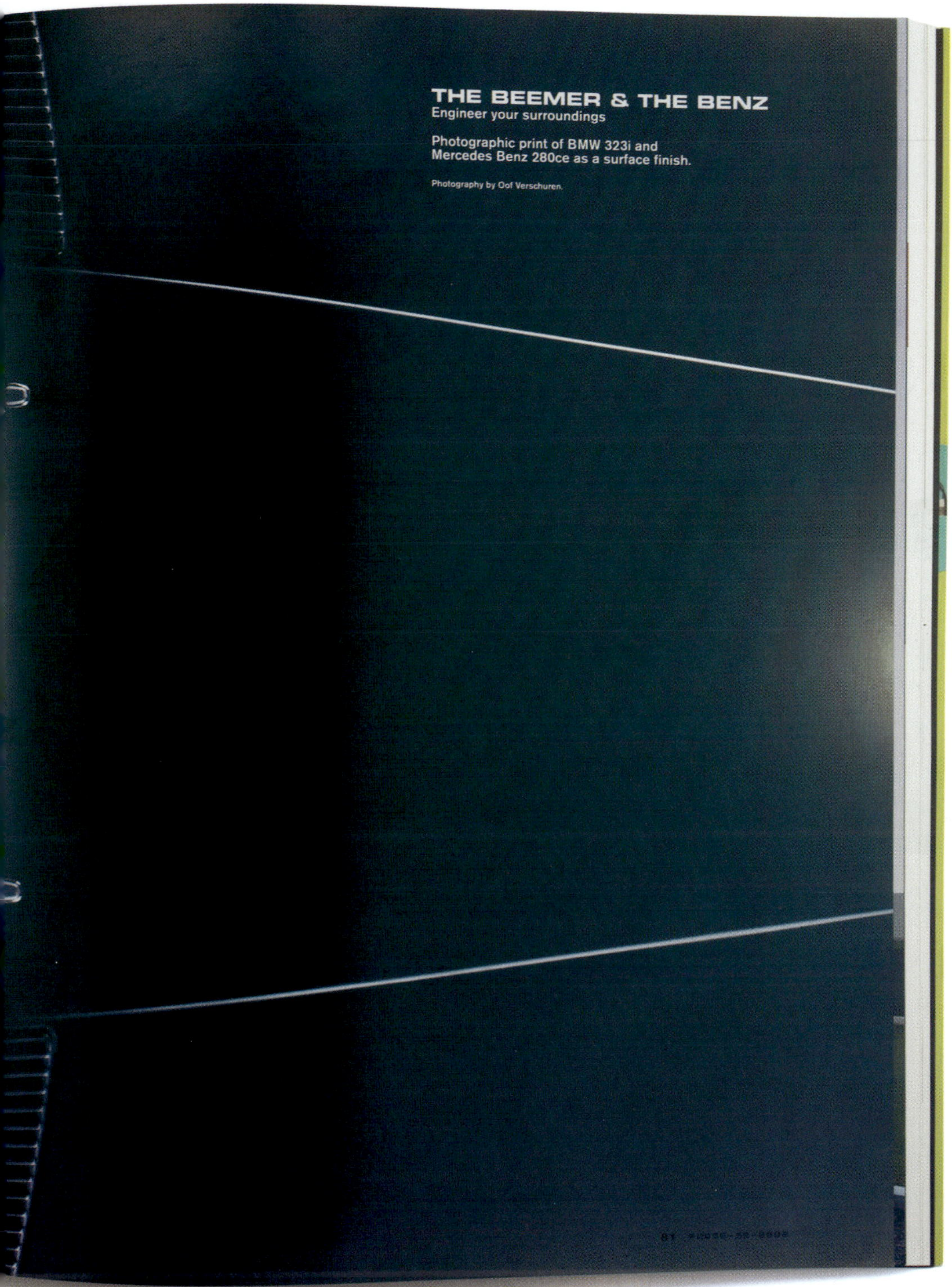

SUBJECT
Visual essay on cars

LOCATION
Amsterdam, Netherlands

DESIGNERS
Dumoffice

PHOTOGRAPHER
—

WRITERS
Dumoffice

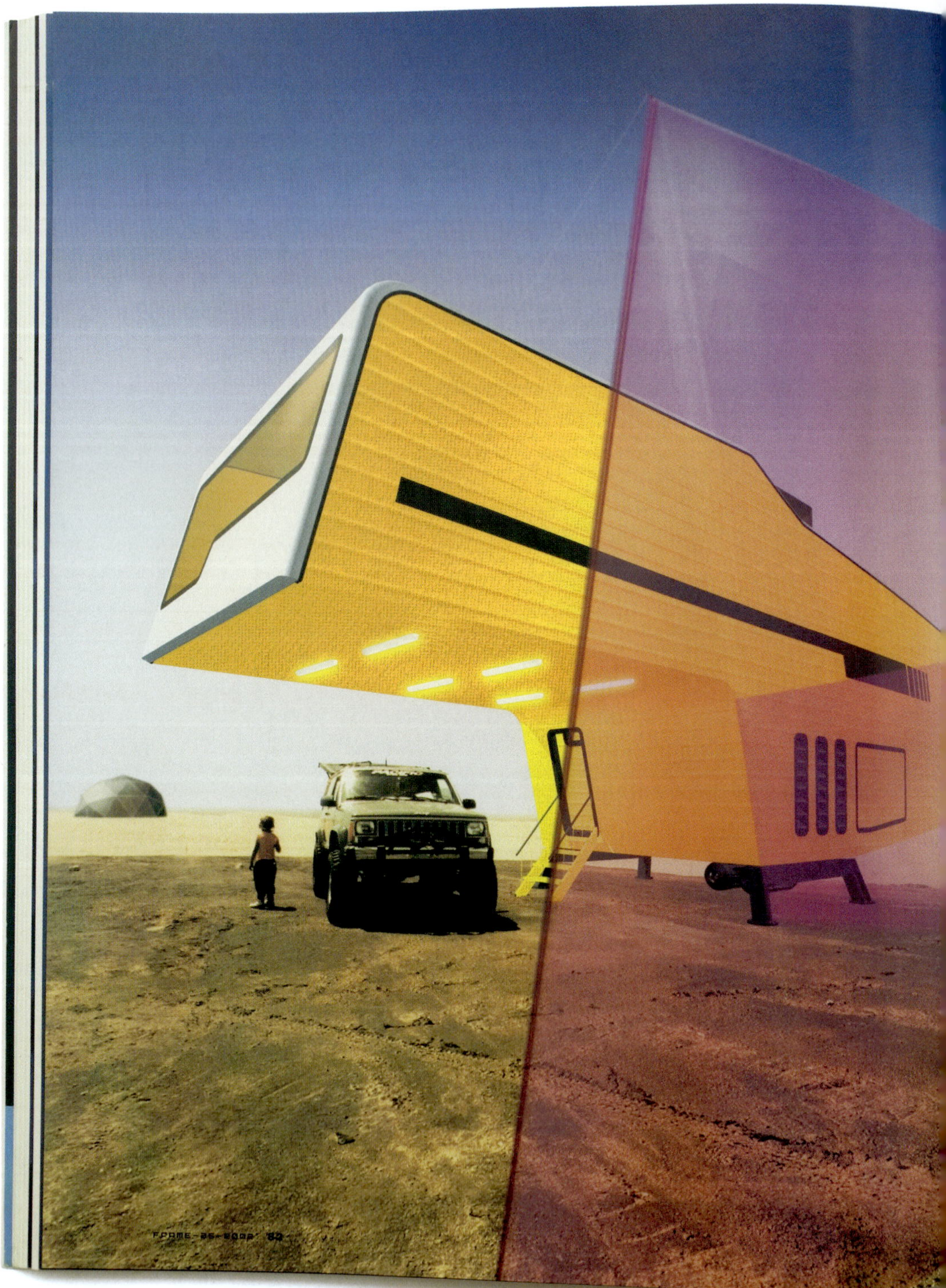

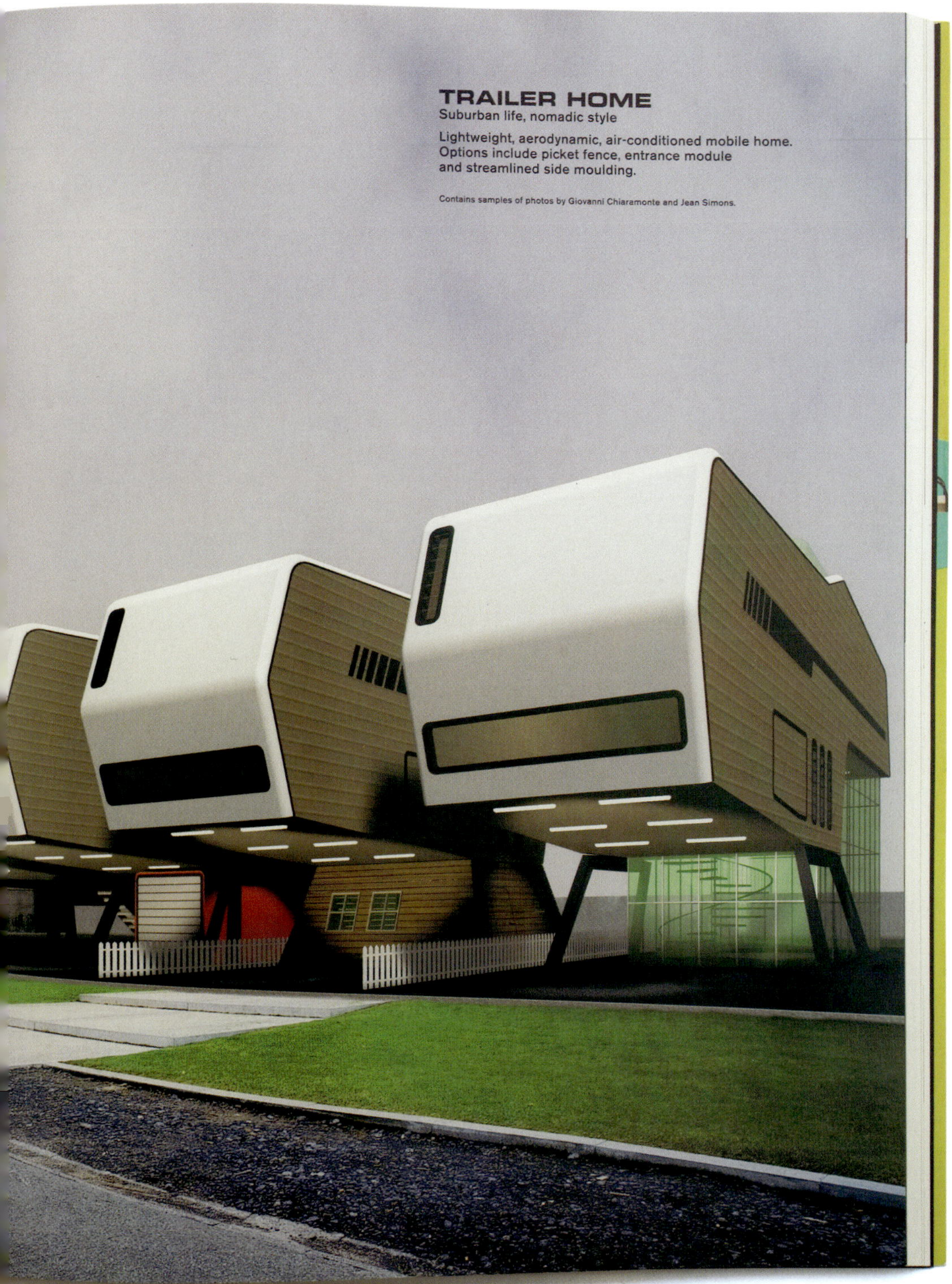

TRAILER HOME

Suburban life, nomadic style

Lightweight, aerodynamic, air-conditioned mobile home.
Options include picket fence, entrance module
and streamlined side moulding.

Contains samples of photos by Giovanni Chiaramonte and Jean Simons.

SUBJECT
Visual essay on cars

LOCATION
Amsterdam, Netherlands

DESIGNERS
Dumoffice

PHOTOGRAPHER
—

WRITERS
Dumoffice

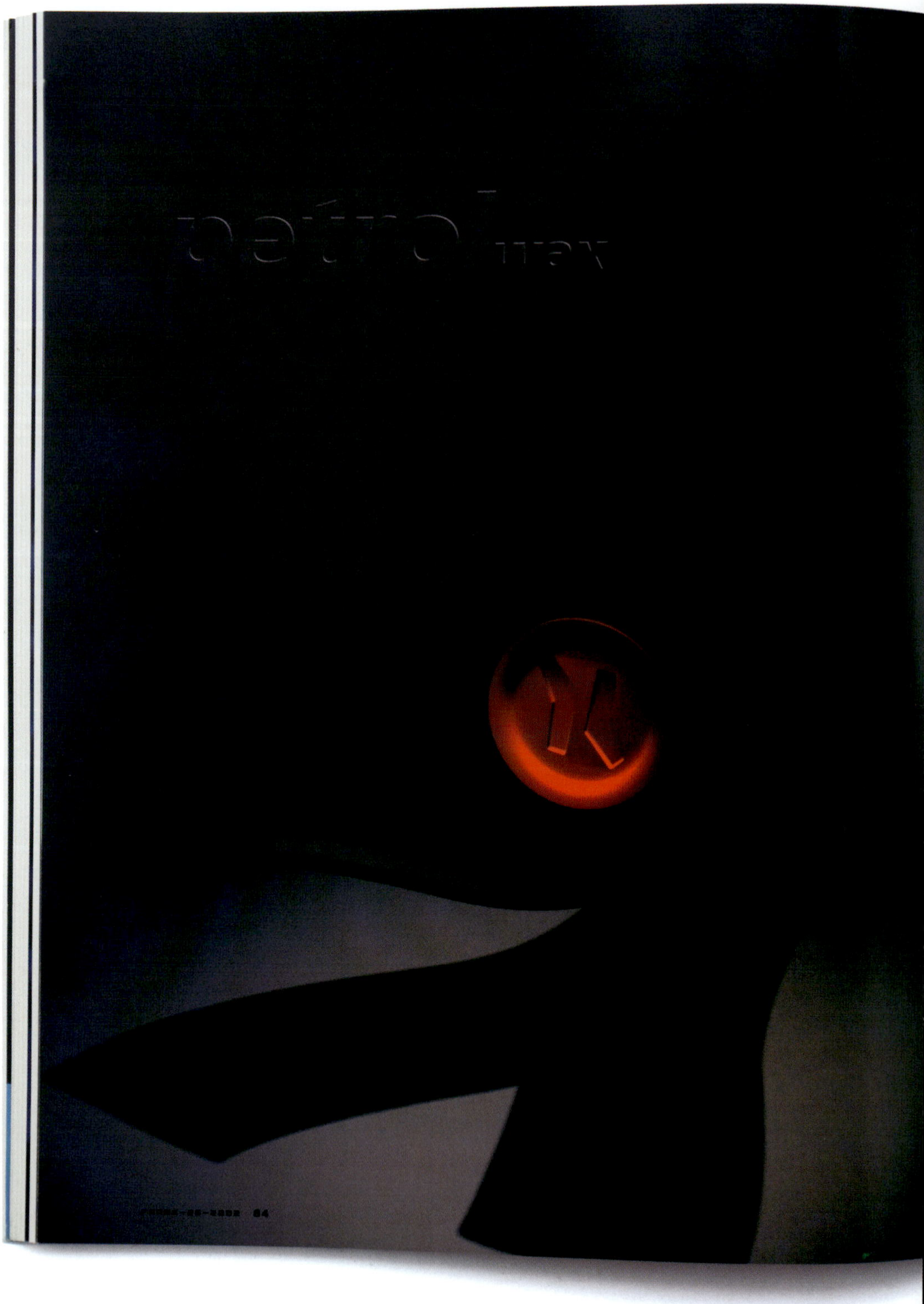

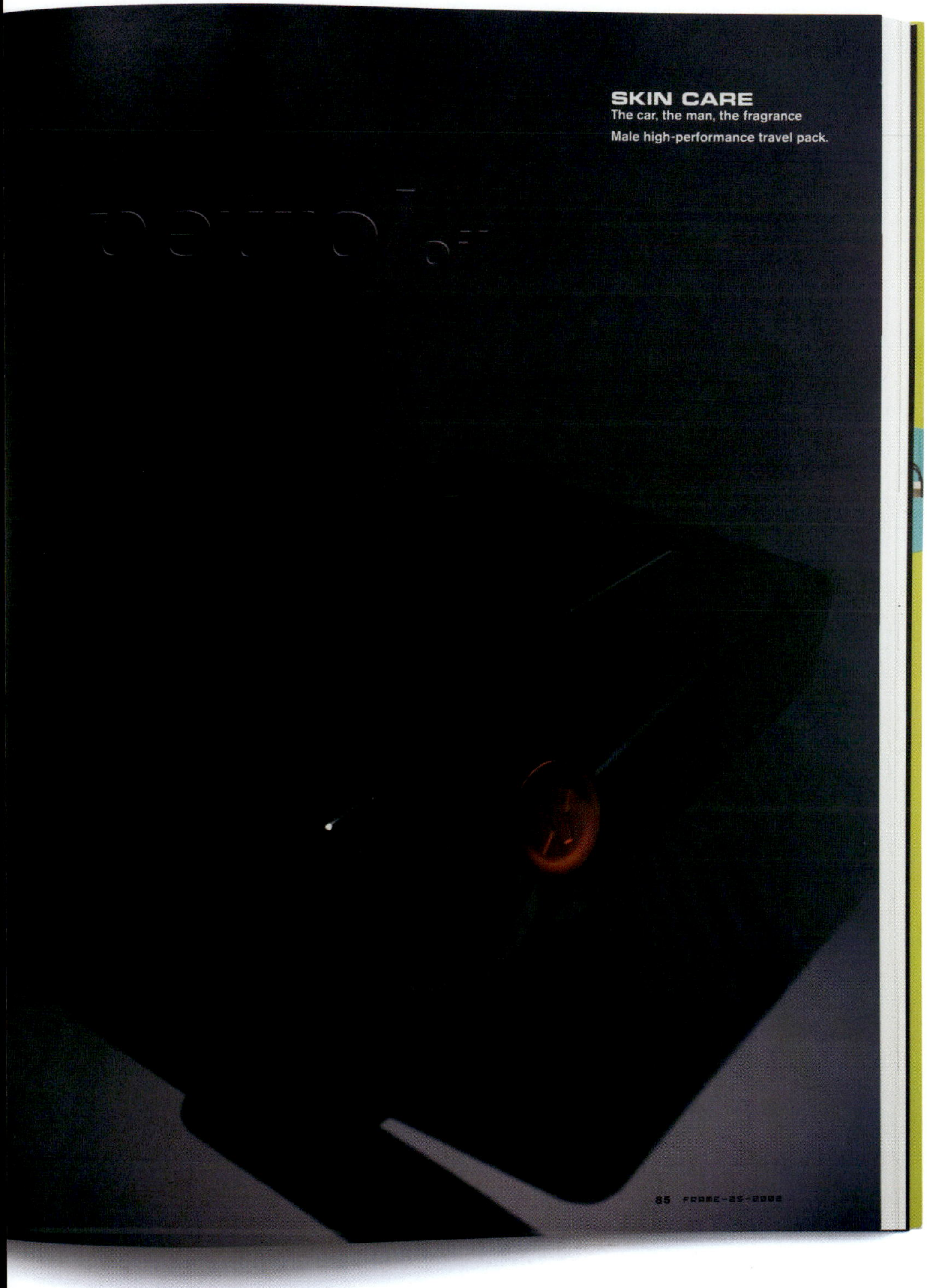

SUBJECT
Visual essay on cars

LOCATION
Amsterdam, Netherlands

DESIGNERS
Dumoffice

PHOTOGRAPHER
—

WRITERS
Dumoffice

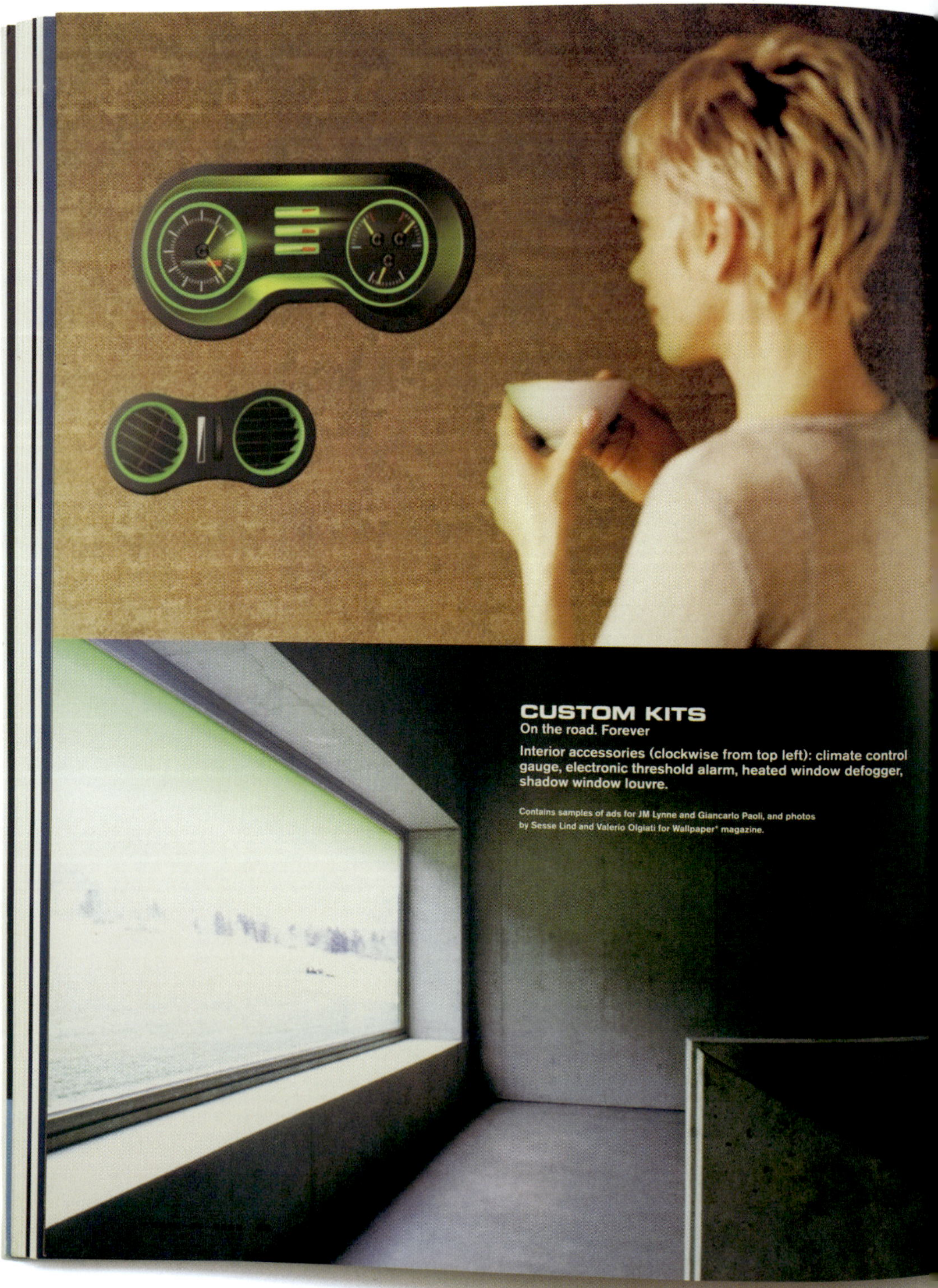

CUSTOM KITS

On the road. Forever

Interior accessories (clockwise from top left): climate control gauge, electronic threshold alarm, heated window defogger, shadow window louvre.

Contains samples of ads for JM Lynne and Giancarlo Paoli, and photos by Sesse Lind and Valerio Olgiati for Wallpaper* magazine.

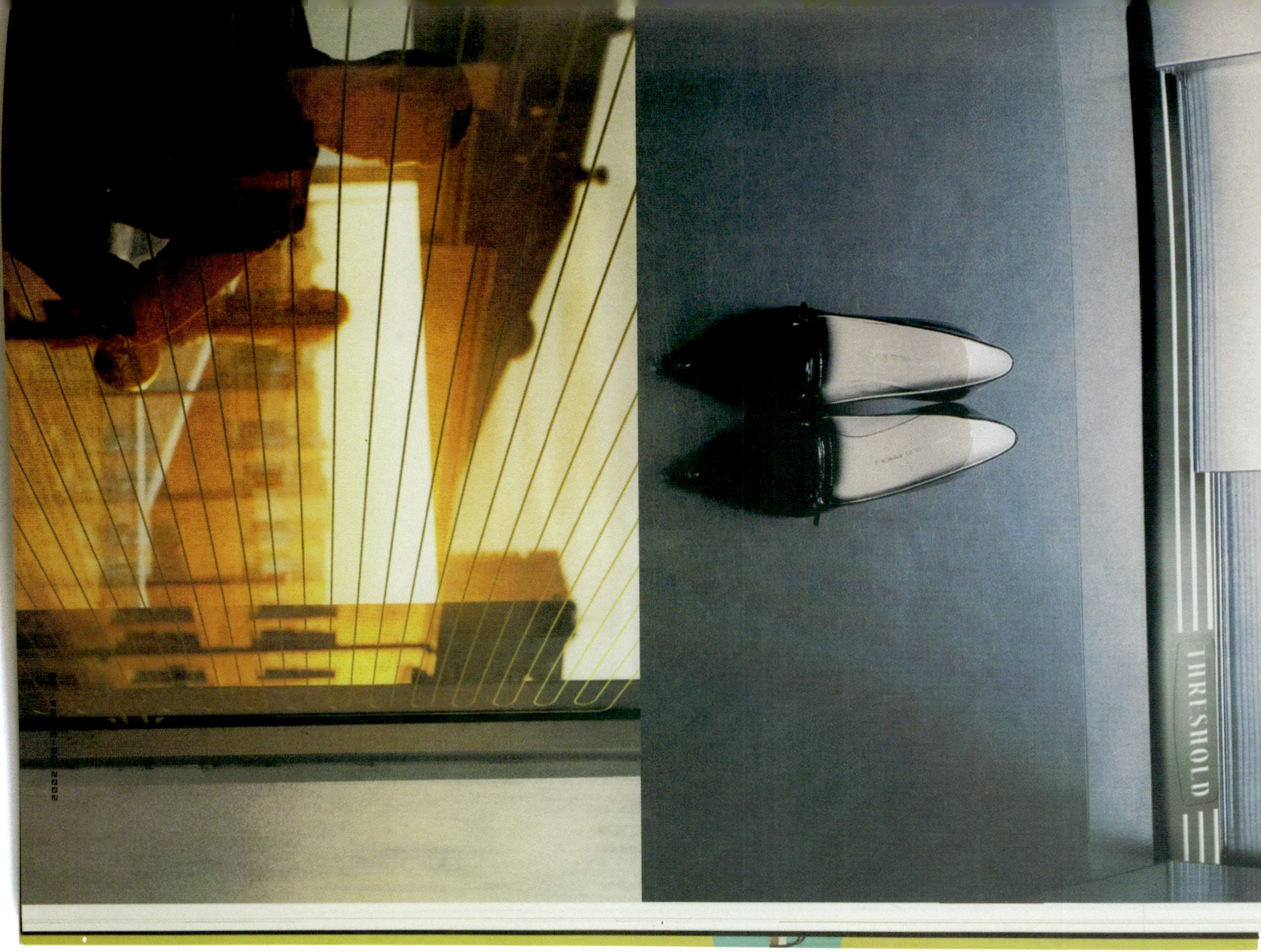

SUBJECT
Visual essay on cars

LOCATION
Amsterdam, Netherlands

DESIGNERS
Dumoffice

PHOTOGRAPHER
—

WRITERS
Dumoffice

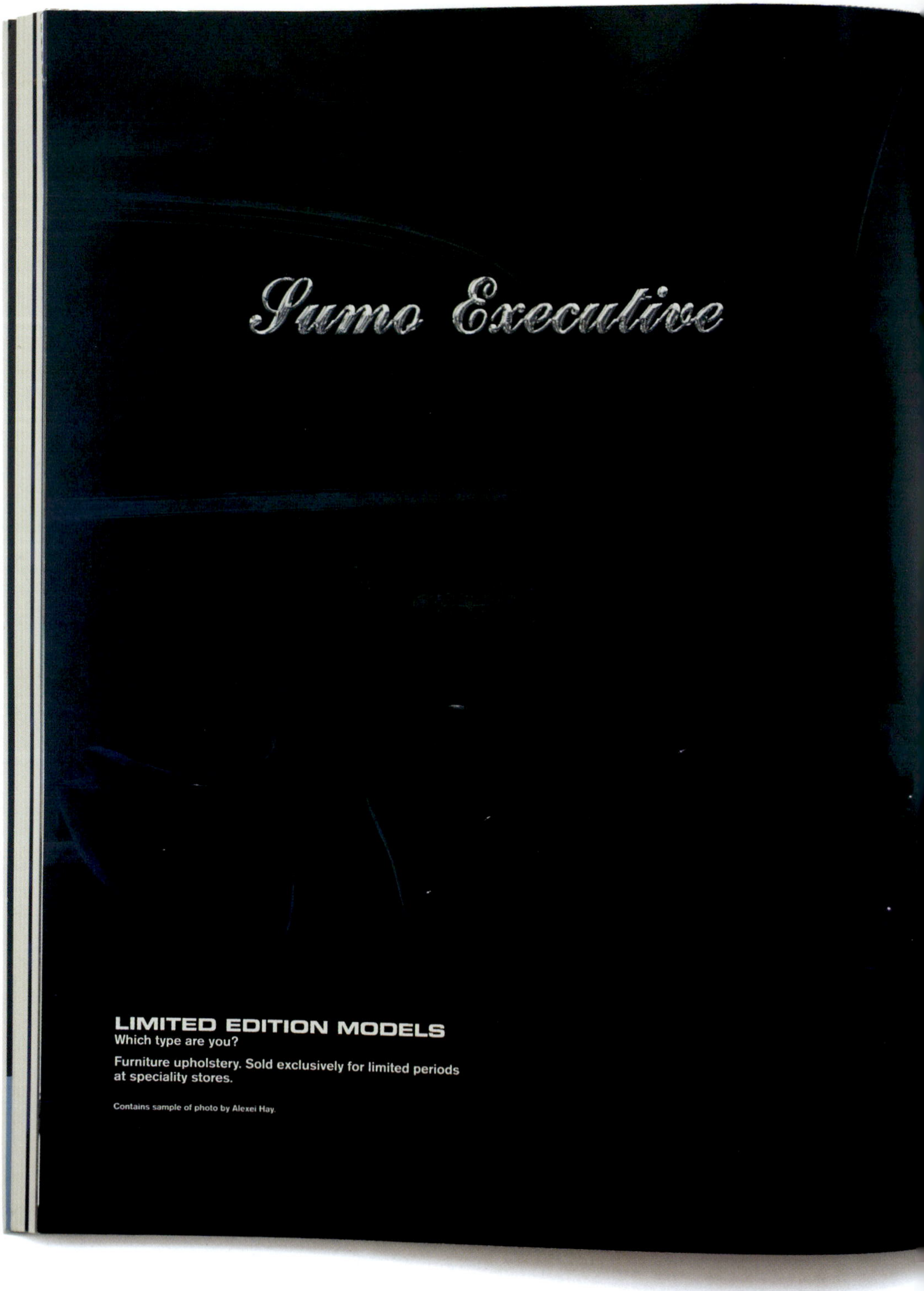
Sumo Executive
LIMITED EDITION MODELS
Which type are you?
Furniture upholstery. Sold exclusively for limited periods at speciality stores.
Contains sample of photo by Alexei Hay.

sumo sport

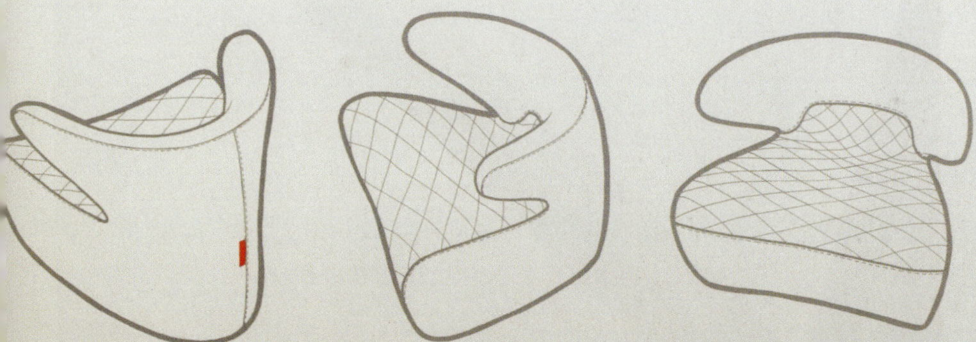

SUBJECT
Visual essay on cars

LOCATION
Amsterdam, Netherlands

DESIGNERS
Dumoffice

PHOTOGRAPHER
—

WRITERS
Dumoffice

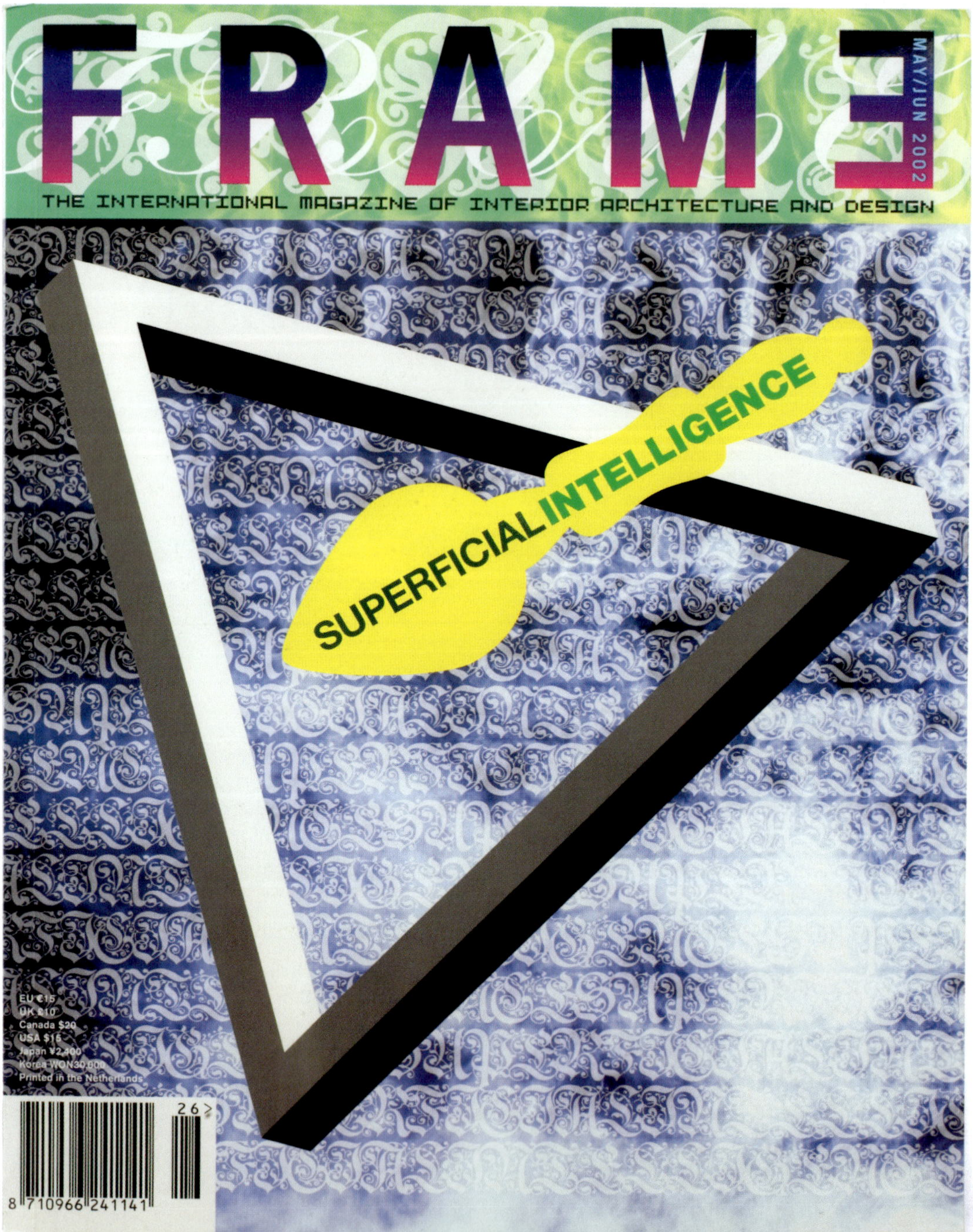

ART DIRECTOR
Roelof Mulder

Alcoholics Anonymous

DESIGNERS
Visomat
www.visomat.com

SUBJECT
Private club Automaten

LOCATION
Berlin, Germany

PHOTOGRAPHER
Rob Houst

WRITER
Rob Houst

Designers F.R.E.d Rubin (Rotations-recycling) and Michael Weinholzner (visomat inc.) about the project
An extraordinary experience.

Did the article have an effect?
It had a positive effect on our self-conception. Automaten was an unenslaved project – based on our concept, realized within our network. An article in a Frame issue feels good.

Recent news or information about the project featured in *Frame*
We closed Automaten in September 2004. It was a temporary idea with no commercial aspect from the start in 2001. In 2002 we build a mobile version of Automaten with similar design elements and an audio-video jukebox. The mobile Automaten (www.automatenbar.de) fits in a small truck and is travelling to festivals and exhibitions.

Current projects
Our favourite current project is a public room called M12 (http://M12.visomat.com), located at Berlin-Alexanderplatz inside a shopping mall. The interior design is based on chrome alloyed MERO M12. The concept is to offer a fair system to an interested public with around 900 nodes and tubes, that is totally flexible. With the continuous variation of the space, the space itself will become content. Audio- and Video Performances, exhibitions, lectures, screenings, etcetera in a loose structure will fill the programme-part of M12. The 120-sq-m space is equipped with 60 small loudspeakers, various screens and projections.

ALCOHOLICS

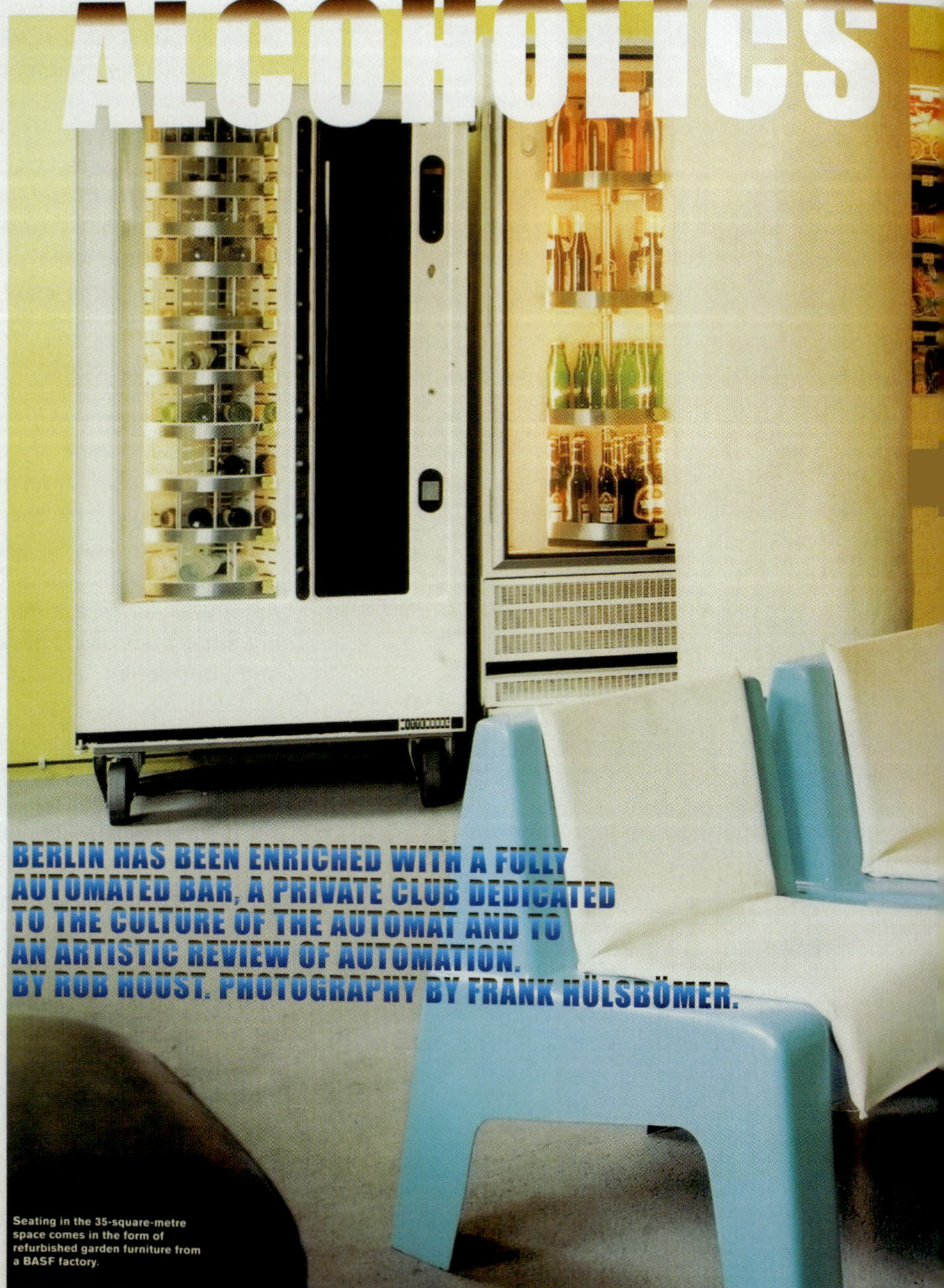

BERLIN HAS BEEN ENRICHED WITH A FULLY AUTOMATED BAR, A PRIVATE CLUB DEDICATED TO THE CULTURE OF THE AUTOMAT AND TO AN ARTISTIC REVIEW OF AUTOMATION. BY ROB HOUST. PHOTOGRAPHY BY FRANK HÜLSBÖMER.

Seating in the 35-square-metre space comes in the form of refurbished garden furniture from a BASF factory.

ANONYMOUS

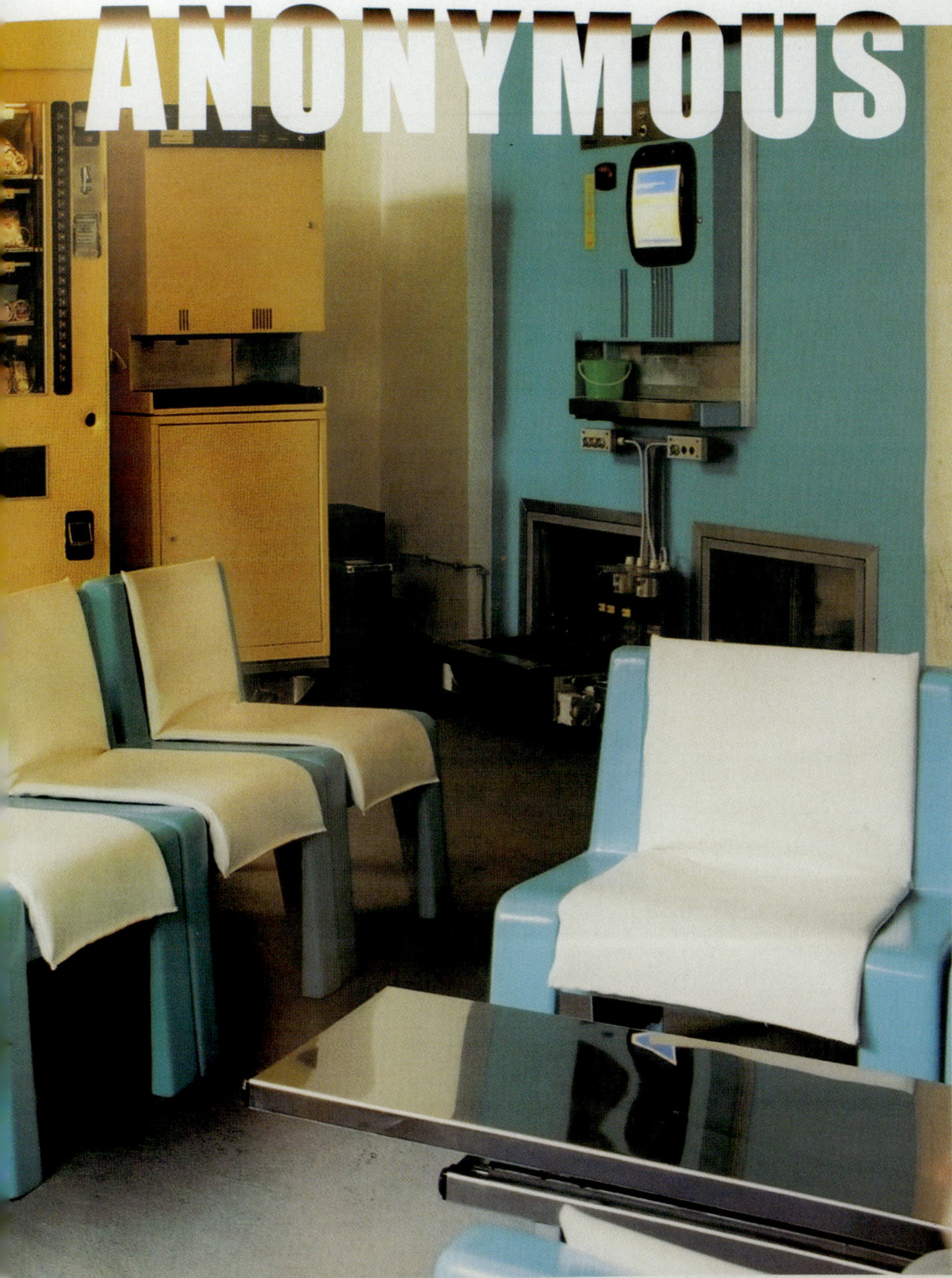

SUBJECT
Private club Automaten

LOCATION
Berlin, Germany

DESIGNERS
Visomat

PHOTOGRAPHER
Frank Hülsbömer

WRITER
Rob Houst

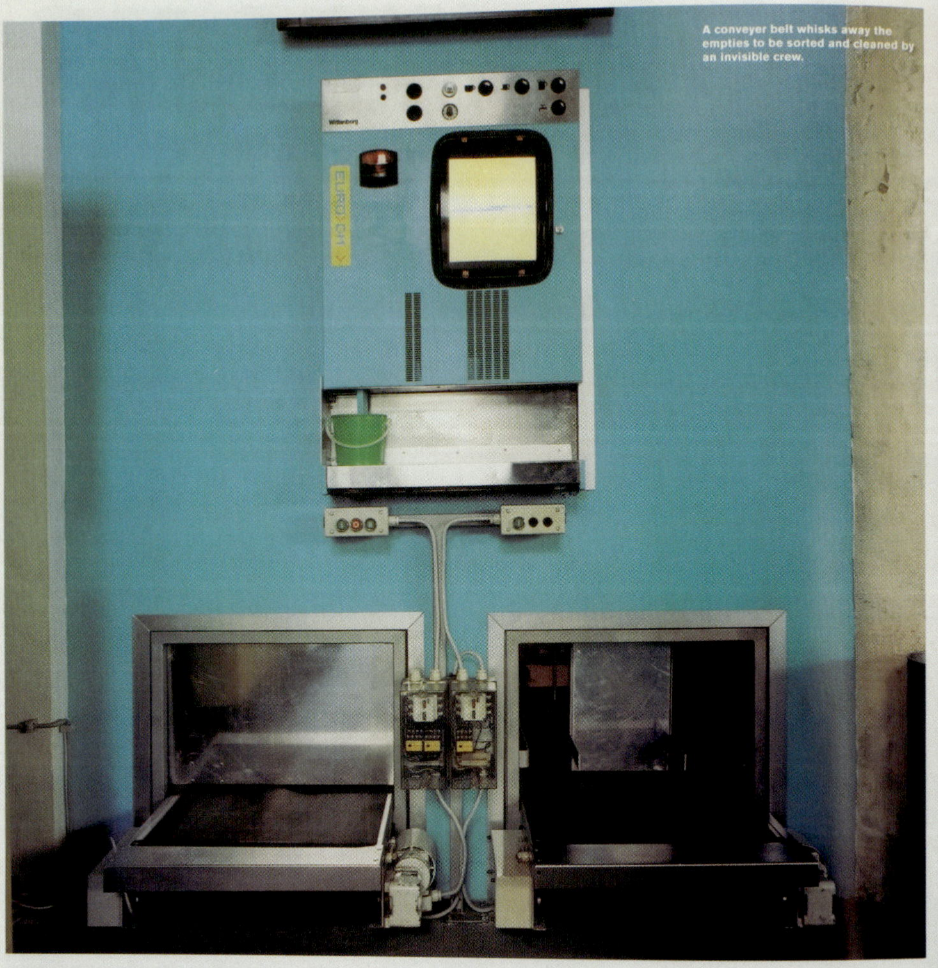

A conveyer belt whisks away the empties to be sorted and cleaned by an invisible crew.

Exactly one hundred years ago, German technology and American ingenuity gave birth to the fast-food revolution. In 1902 the first Horn & Hardart Automat premiered in Philadelphia. Ten years later the Automat had reached Manhattan, and by the 1930s Horn & Hardart was the world's largest restaurant chain, with 180 branches serving 800,000 people a day. Boasting chrome and glass coin-operated machines, these cavernous, waiterless establishments brought inexpensive, high-tech dining to a low-tech era.

Founded by American Joseph Horn and German-born Frank Hardart, the restaurants were a new concept in food service, borrowed from a successful German eatery. Automats immediately captured America's interest and imagination. As the restaurant industry's first attempt at emulating the assembly line, they reflected society's belief in an automated future.

Eventually, however, Automats fell victim to consumers' changing tastes. No longer able to compete in a jungle of contemporary fast-food outlets, New York City's last Automat closed its doors in 1991. Nonetheless, the concept survives as a vision of 20th-century futurism – a relic of a modern world of robots and space-age idealism that failed to materialise.

For the founders of Automaten, relics and nostalgia are not enough. Conceived by media-space designers Visomat Inc., Automaten is a private club, chartered with the aim of 'preserving automated culture and encouraging the artistic analysis of automation'.

'It's a childhood dream,' explains Visomat's Gereon Schmitz. 'Modernism is automatic. Robots should take care of all the aggravations of daily life. This is the key to unlimited leisure time – the world of the Jetsons and their utopia.' In the clubroom in Berlin-Mitte, Automaten has set up a fully automated bar. For an annual fee of 18 euros,

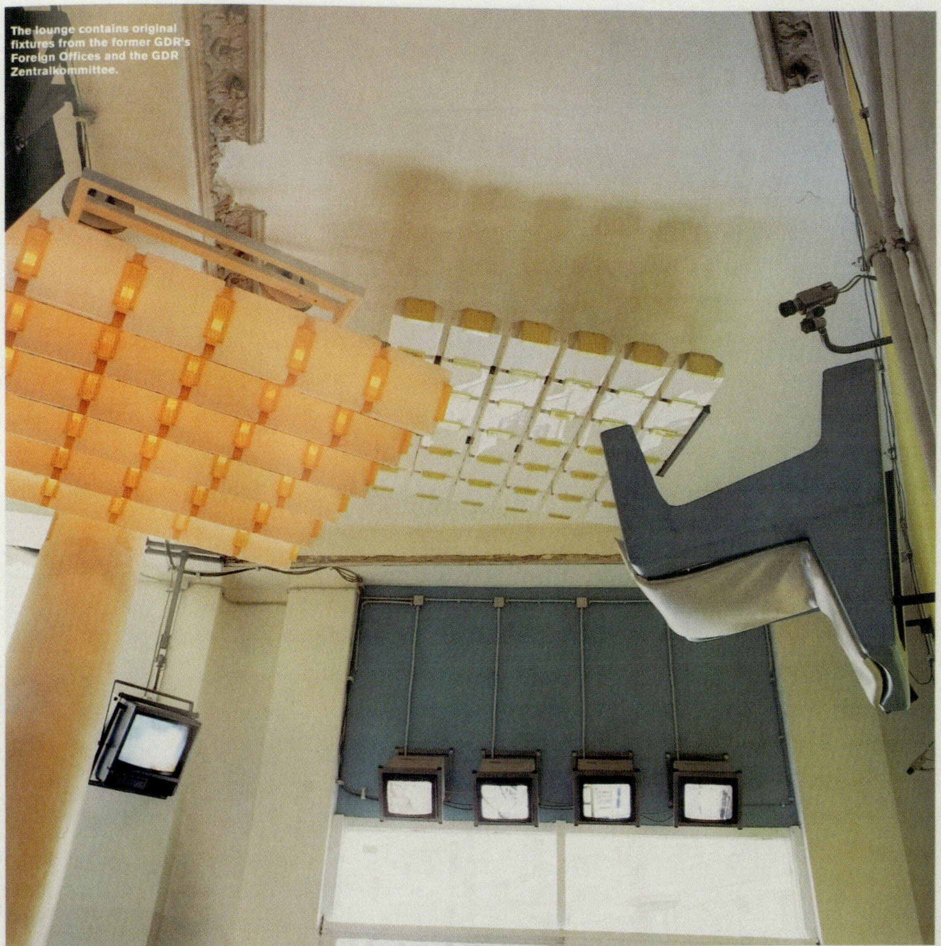

The lounge contains original fixtures from the former GDR's Foreign Offices and the GDR Zentralkommittee.

FOR AN ANNUAL FEE OF 18 EUROS, 'MEMBERS CAN SURRENDER TO THEIR AUTOMATED FANTASIES, WITHOUT THE ANNOYING SMALL TALK OF A BARTENDER'

SUBJECT Private club Automaten

LOCATION Berlin, Germany

DESIGNERS Visomat

PHOTOGRAPHER Frank Hülsbömer

WRITER Rob Houst

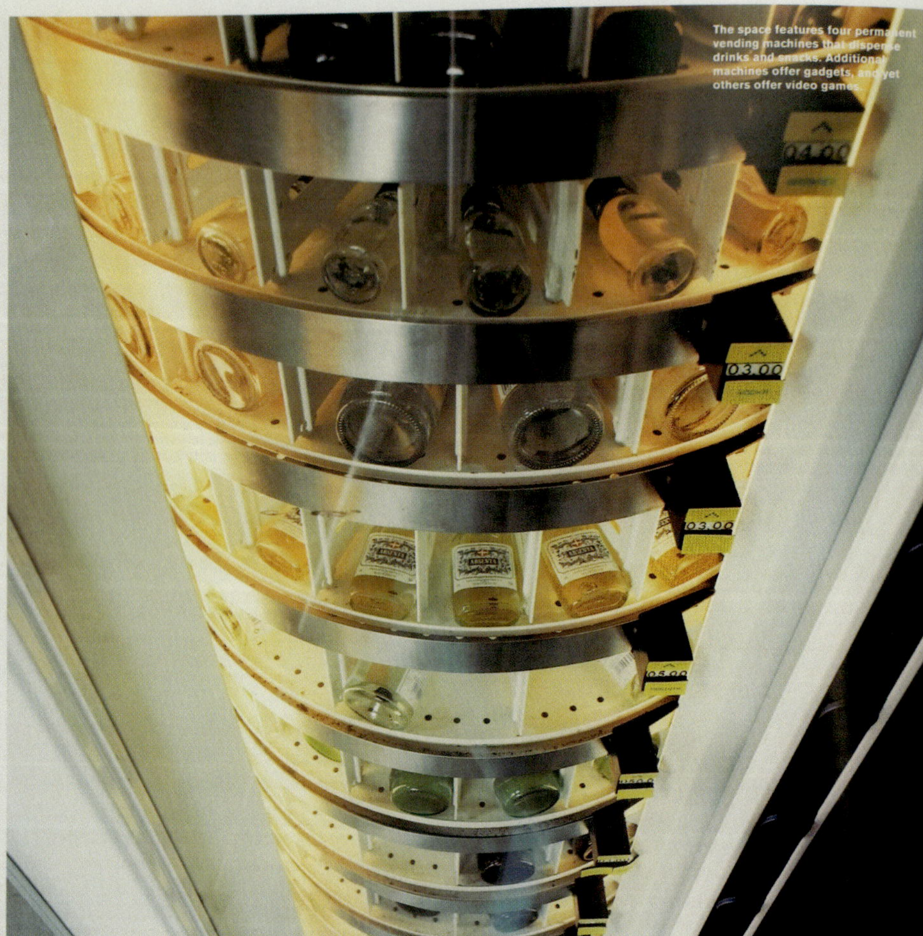

The space features four permanent vending machines that disperse drinks and snacks. Additional machines offer gadgets, and yet others offer video games.

'Members can surrender to their automated fantasies, without the annoying small talk of a bartender.' But it's not only the bartender that's missing. Don't look for a doorman, waiters, busboys or bouncers at Automaten. Even the entertainment is automated. Designed by Alexander Baumgardt, the bar is managed by a central computer that controls the door, monitors the room and functions as a multimedia jukebox. Members gain access to the bar with a simple magnetic card. Inside the 35-square-metre lounge, they can exchange euros for old-fashioned Deutschmark coins (in an automatic machine, of course) and browse the bank of windows and coin slots for drinks and snacks. Although ready-made cocktails are not available, guests can buy the ingredients for most long drinks and mix their own favourites. A conveyor belt whisks away the empties to be sorted and cleaned by an invisible crew. Additional machines offer gadgets, and yet others feature video games.

A blend of styles, the lounge contains original fixtures from the former GDR's Foreign Offices and the GDR Zentralkommittee, as well as refurbished garden furniture from a BASF factory. The central computer is a remodelled and reprogrammed coffee automat.

In addition to its permanent collection of four machines, Automaten has a programme of changing exhibitions, lectures and 'guest automats', the first of which – a vintage 1980s Atari Centipede video console – made its debut in March.

'*Deus ex machina* is the ultimate prophecy of post-socialist society,' says Schmitz. But is contemporary Berlin ready for a Fritz Lang-style fantasy? In a world of internet, mobile phones and Palm Pilots, where personal contacts are becoming less and less common, that annoying bartender can be a nice change of pace.

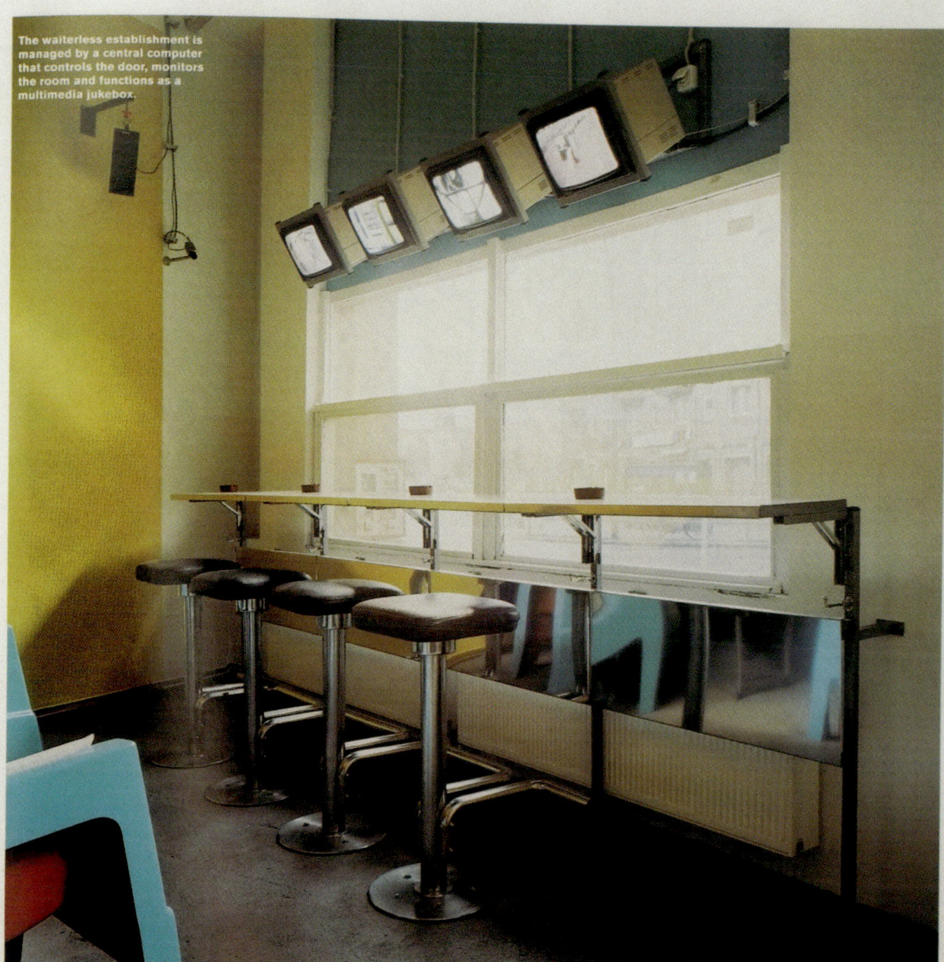

The waiterless establishment is managed by a central computer that controls the door, monitors the room and functions as a multimedia jukebox.

AUTOMATEN'S PROGRAMME OF 'GUEST AUTOMATS' FIRST FEATURED A VINTAGE 1980S ATARI CENTIPEDE VIDEO CONSOLE

SUBJECT
Private club Automaten

LOCATION
Berlin, Germany

DESIGNERS
Visomat

PHOTOGRAPHER
Frank Hülsbömer

WRITER
Rob Houst

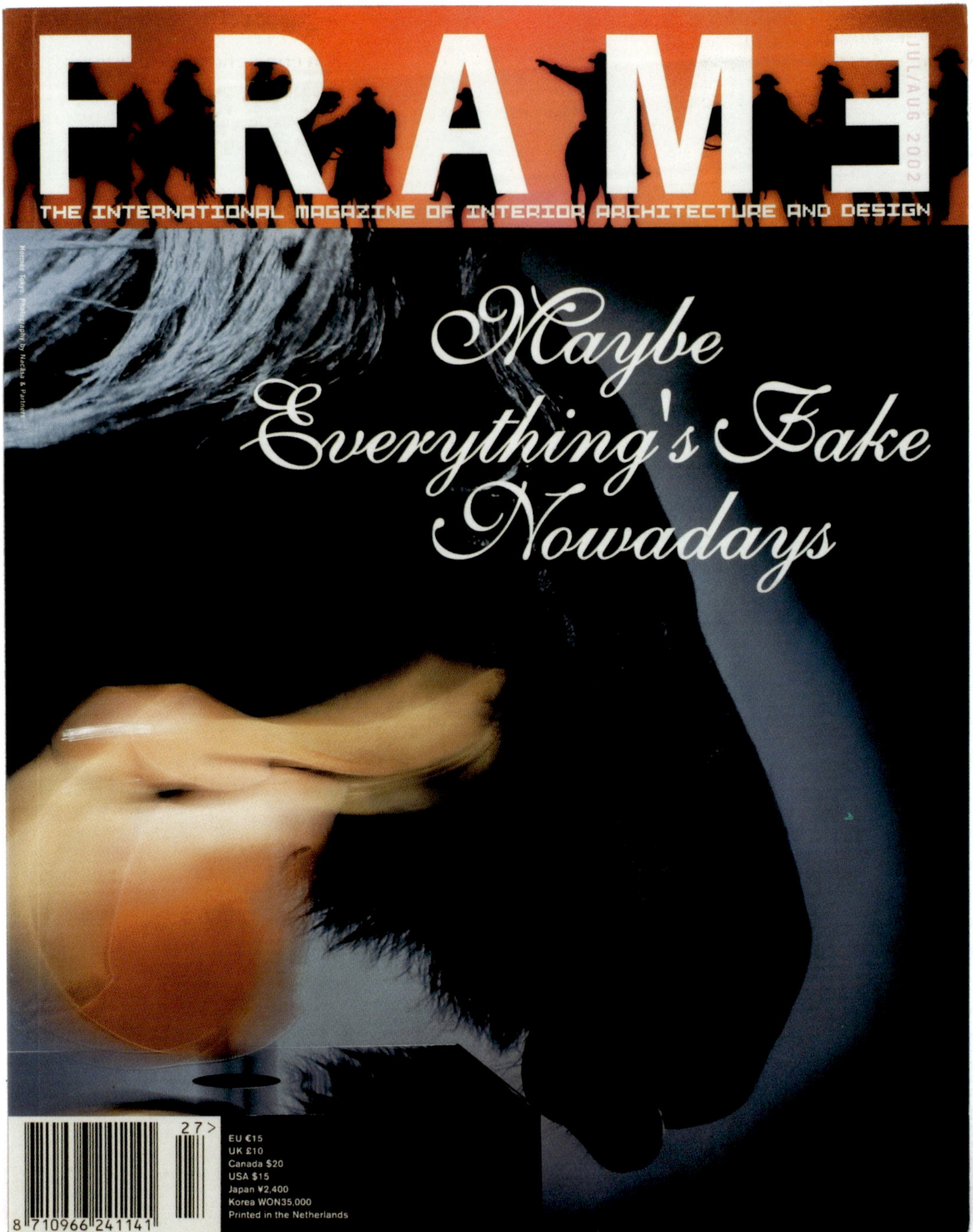

ART DIRECTOR
Roelof Mulder

27

The Last Chance Saloon

DESIGNERS
Union North
www.unionnorth.co.uk

SUBJECT
Bar Multi-Purpose Venue (MPV)

LOCATION
Leeds, England

PHOTOGRAPHER
Chris Cascoigne/View
www.viewpictures.co.uk

WRITER
Will Jones

Union North about the project

MPV was a confection of integrated cross-disciplinary design, occupying several grey areas around architecture, interior, product and identity.

Comments on the published article

We have a high regard for Frame and enjoyed the irreverent way in which it was featured.

Did the article have an effect?

It is always a boost to morale for the team to see projects in print, we are not aware of any direct effects but the accumulative effect of being published brings recognition and jobs.

Recent news or information about the project featured in *Frame*

Recently re-painted two-tone blue by the client in a serious lapse of judgement!

Current projects

Parade & Drive – conversion of '60s social housing, Leeds, England // Irk Valley – conversion of '60s tower blocks, Manchester, England // Manchester Regiment – new build sustainable housing, Manchester, England // Midland Hotel – refurbishment and extension of listed Art Deco hotel, Morecambe, England // Rochelle – conversion and extension of Victorian school for gallery and artists residential, London, England // Montreal – five storey hotel, bar, restaurant and club, Liverpool, England // New Quay – new build shoreline restaurant and beach-houses, County Clare, Eire // Point of Ayr – private house, Oxton, England // New York – apartment refurbishment, New York, USA.

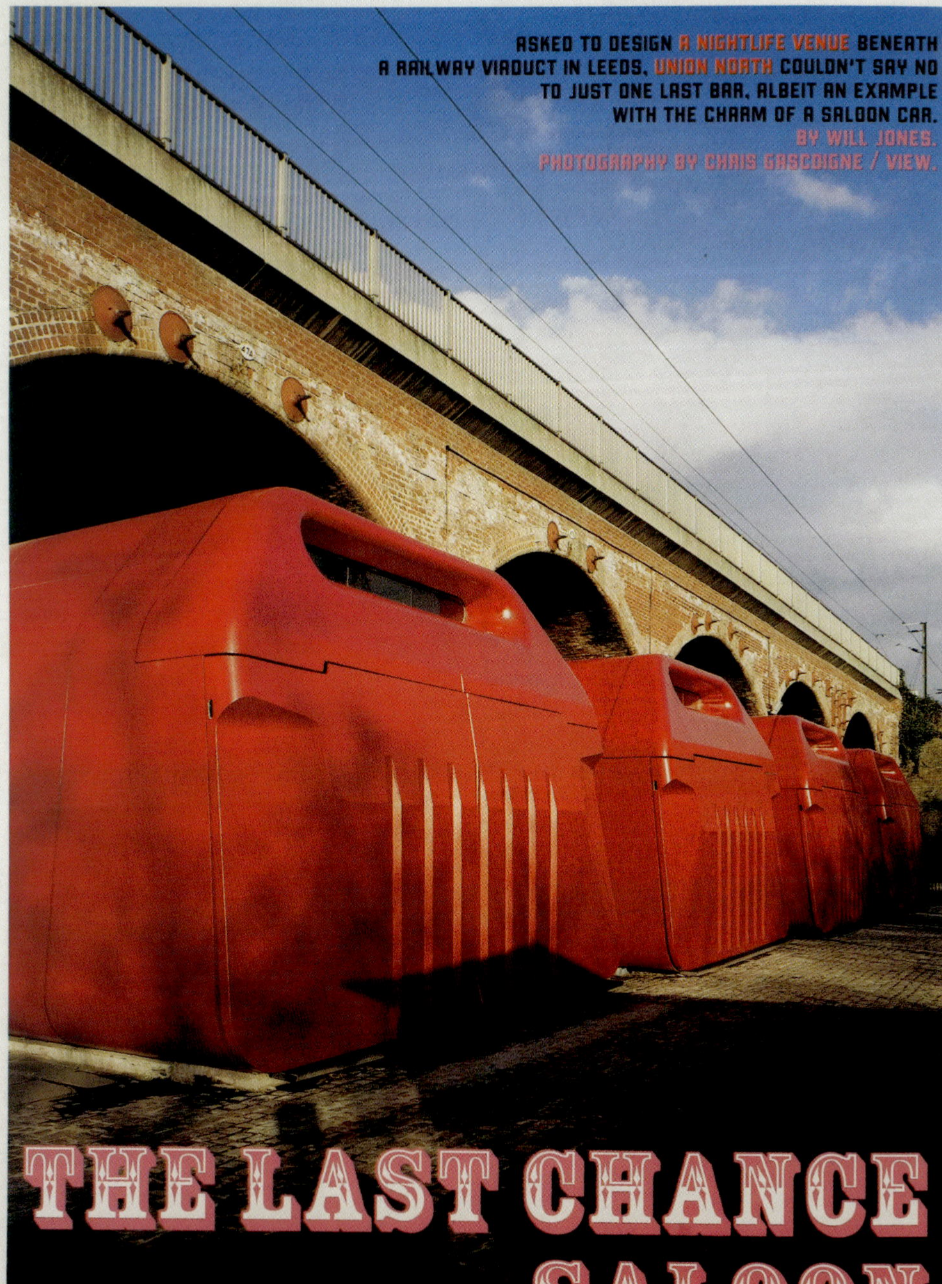

ASKED TO DESIGN A NIGHTLIFE VENUE BENEATH A RAILWAY VIADUCT IN LEEDS, UNION NORTH COULDN'T SAY NO TO JUST ONE LAST BAR, ALBEIT AN EXAMPLE WITH THE CHARM OF A SALOON CAR.

BY WILL JONES.

PHOTOGRAPHY BY CHRIS GASCOIGNE / VIEW.

THE LAST CHANCE SALOON

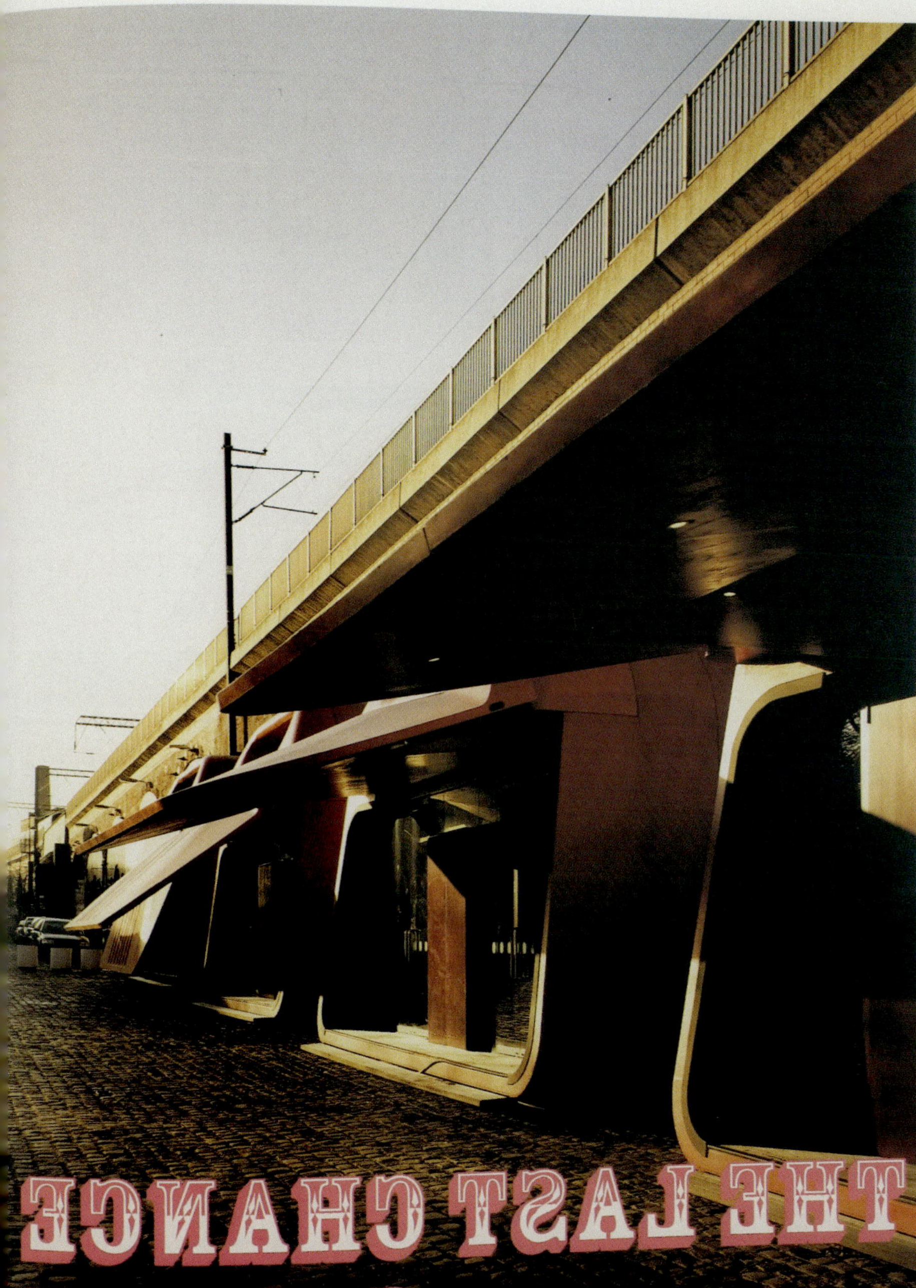

THE LAST CHANCE SALOON

SUBJECT
Bar Multi-Purpose Venue (MPV)

LOCATION
Leeds, England

DESIGNERS
Union North

PHOTOGRAPHER
Chris Cascoigne/View

WRITER
Will Jones

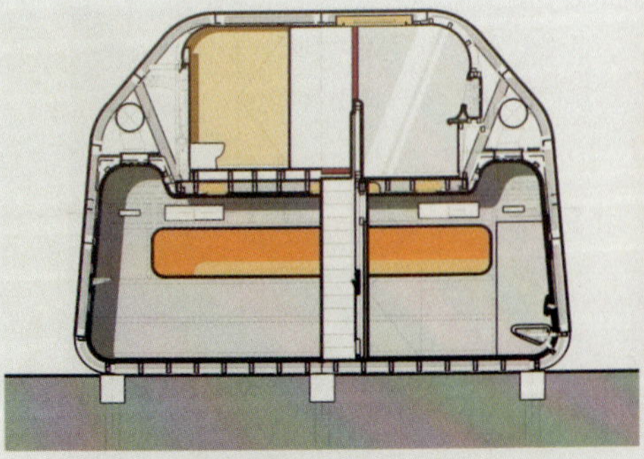

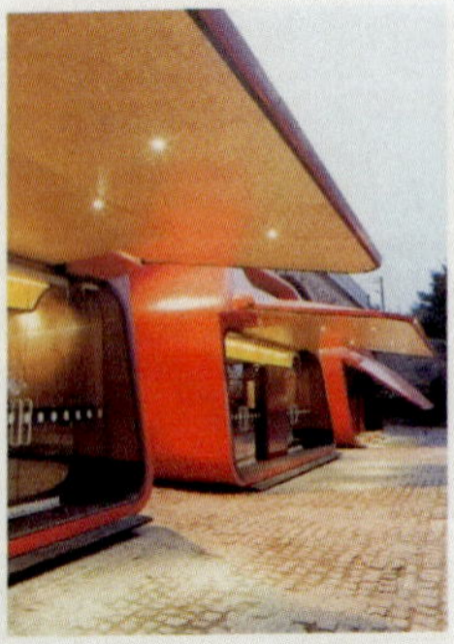

Previous pages: Located beneath the 19th-century arches of a railway viaduct, the four steel-framed pods are clad in a continuously welded 3-mm mild-steel skin. By day the pods remain firmly shut.

The front of each pod is a steel-frame shutter door clad in GRP. The opening mechanism is a motorised winch that pulls the door upwards to form a canopy. Spotlights in the birch soffit of the canopy illuminate the cobbled overflow area in front.

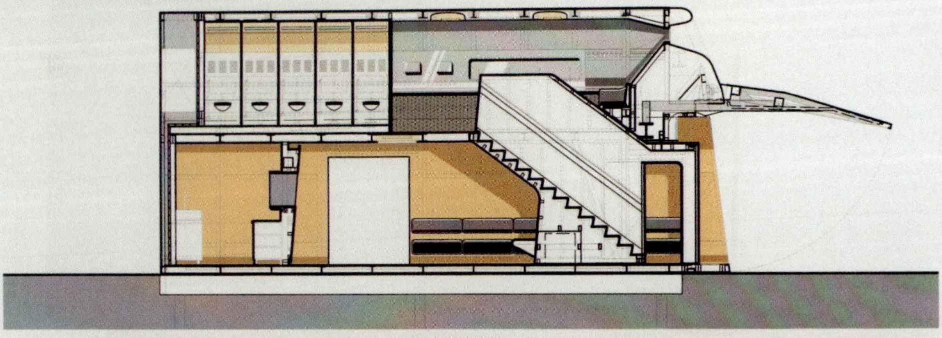

THE RED BOXES COULD BE GIANT HANDBAGS, RECYCLING CONTAINERS, STAR WARS TROOP CARRIERS, OR PERHAPS SOME FUTURISTIC DESIGN FOR A TRAIN CARRIAGE

A row of brick arches supports the Trans-Pennine railway line as it pushes its way through the city of Leeds en route to other centres in the north of England. Similar to thousands of others, the arches are familiar landmarks in what was once the industrial centre of the world. They show all the signs of wear and tear from years of freight rumbling across their backs. Rainwater still leeches the brickwork before gathering in puddles on the cobbled surface below.

The ageing process is even more striking now that four bright red steel boxes have been inserted into the spaces beneath four arches. All smooth skin and rounded edges, the boxes look incongruous and a bit unearthly next to the weathered brick. They could be giant handbags, recycling containers, Star Wars troop carriers, or perhaps some futuristic design for a train carriage shelved for fear of upsetting conservative souls.

In fact, these strange capsules have an altogether different function. By day they stay shut, but come nightfall all is revealed. With one smooth action the fronts of all four open upwards, leaving gaping mouths that seem ready to disgorge a clamouring army of aliens, with Darth Maul leading the way. Concealed within are four slick bar and club rooms that go by the name MPV, short for Multi-Purpose Venue. The four pods house a lounge, two bars and a club, linked to one another laterally by passageways through jack arches in the massive piers of the railway arches. MPV can accommodate around 550 people inside and a similar number outside, underneath the canopies formed when the fronts are open.

MPV is the work of Union North, a team of architects from Liverpool. The practice is made up of six individuals with backgrounds in architecture and in fields like metalwork, upholstery and fine art. For founding member Miles Falkingham, the combination of disciplines is exactly what enables the group to come up with imaginative, off-the-wall design schemes. 'We're not hung up about what makes a legitimate approach to a project, and the same goes for processes and materials. To come up with the best solution, we'll use a range of techniques and materials borrowed from anywhere, from 2-D design to engineering.'

The outfit was formed after four of the current team of six split from Shed KM Architects. Since then, Union North has gained a reputation in the north of England for innovative bar and club designs. Its creations are scattered around Liverpool's nightlife district, and MPV was not the first venture into the Leeds entertainment scene. Though the plethora of bar and club designs was welcomed at first, the practice now senses a danger of typecasting if it doesn't move on to other things. But when Townhouse Life – a client for whom Union North had already converted a disused mill into a bar and restaurant – came along with the

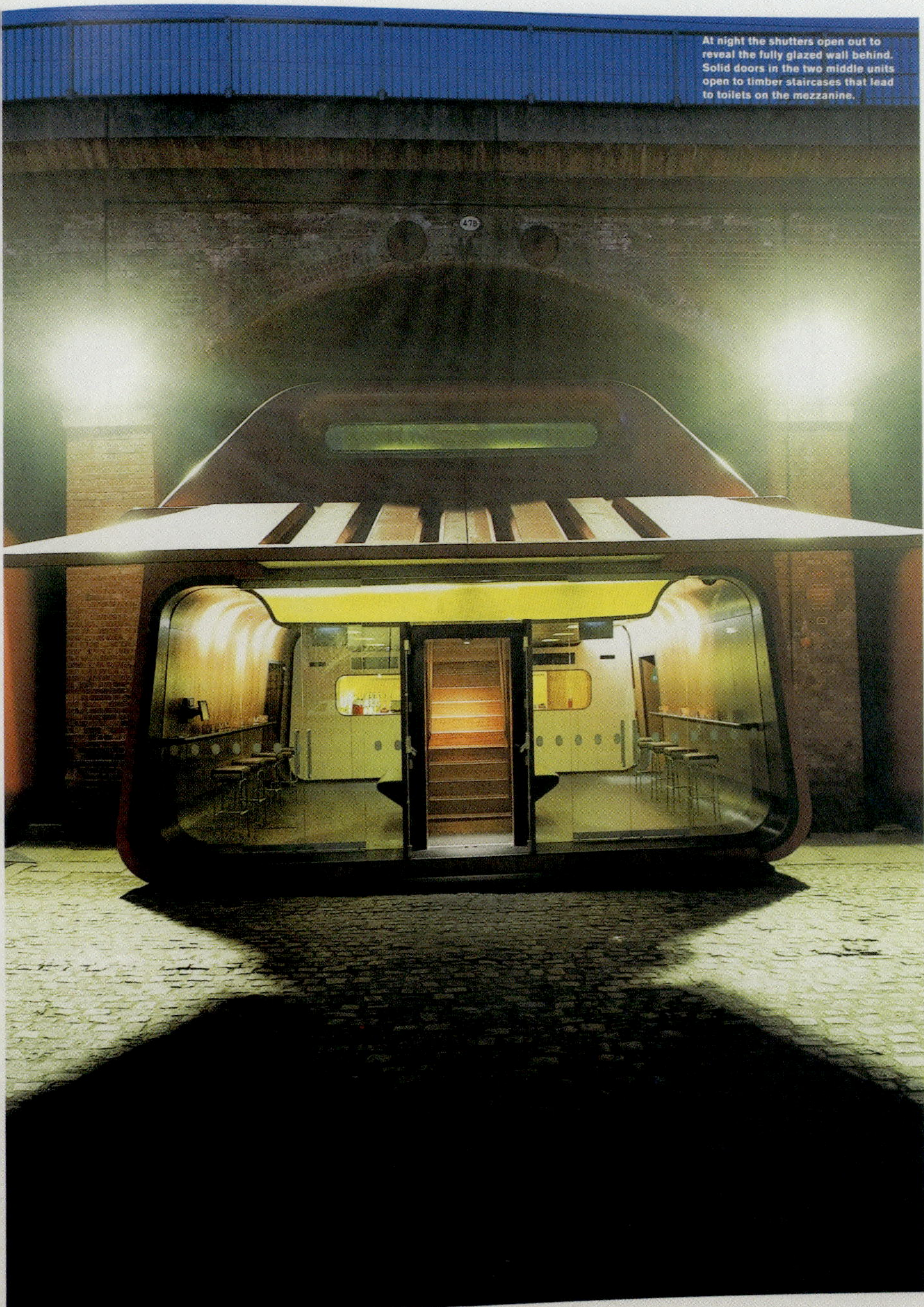

At night the shutters open out to reveal the fully glazed wall behind. Solid doors in the two middle units open to timber staircases that lead to toilets on the mezzanine.

SUBJECT
Bar Multi-Purpose Venue (MPV)

LOCATION
Leeds, England

DESIGNERS
Union North

PHOTOGRAPHER
Chris Cascoigne/View

WRITER
Will Jones

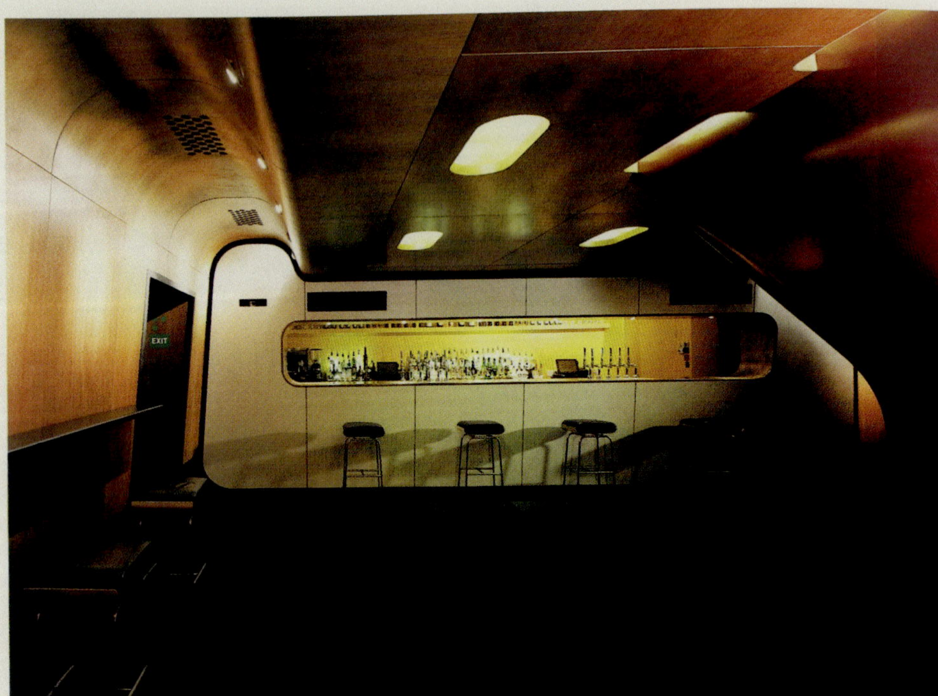

'MPV IS AS MUCH ABOUT PRODUCT DESIGN AS ARCHITECTURE,' SAYS FALKINGHAM. AND THE STYLISH VEHICULAR AESTHETIC OF THE SPACES BRINGS HOME THE POINT

railway-arch site, the designers couldn't turn down just one more bar job. MPV's distinctive look is not the result of the site alone. The success of the finished project has as much to do with Union North's ability to borrow skills from other design disciplines and industries, as well as with a highly democratic system that gives each player, regardless of the area of expertise, a say in the design. All six form a think-tank that tosses ideas around with the aim of establishing a 'proactive relationship' with the site. In this particular case, it soon became clear that the aesthetic and rhythm of the viaduct structure lent itself to a modular solution: the insertion of separate but related entities beneath the arches.

From the outside, the four capsules are identical. Shells are made of rolled-steel sections with a continuously welded mild-steel skin, a structure commonly used in shipbuilding. Manufactured by the appropriately named Merseyside Ship Repairers, this tough, heavyweight exterior provides an impenetrable hull that protects the richly detailed interiors. The two central pods house identical bars. Visitors enter under the canopy and through glass doors into a space that at first sight looks like a Scandinavian bar morphed with a burger vendor's van. Panels of self-coloured Meranti ply flow seamlessly from wall to ceiling. The curved ceiling profile conceals from view a row of spots that splash illumination off the ply wall and into the space. To the rear, behind a slot cut into the ply wall, is the drinks counter. In the centre of the space, the ceiling drops to create headroom for a washroom above, accessed from a stairway that rises from a solid door in the otherwise glass façade. At just 2 metres high, the tight space leaves no room for decoration. Sparse stainless-steel latrines and cubicles line the walls.

The cramped confines are the designers' own doing, however. 'We created those conditions ourselves,' says Falkingham. 'Dimensions were always going to be tight, because we had to leave access room to the railway arch above. But the scheme took on more of a vehicular aesthetic that an architectural one, so we just exploited that cosy, close-fit feeling.' The lowered ceiling upstairs gives a pent-up feel similar to a toilet block on a train or ferry. 'Wanting to be consistent in our approach, we lowered the ceiling as much as we dared.'

Two outer pods, accommodating lounge and club areas, are accessed through glass façades or along a circulation route through existing jack arches. Though different from the bars, the lounge and club spaces are a pair. Again, a ply lining adorns the walls. Here, however, these surfaces have been sprayed with a deep-plum lacquer. The furniture is simple: plum sofas with steel legs and curved ply backs, complemented by low tables with ply tops that curve up at their short

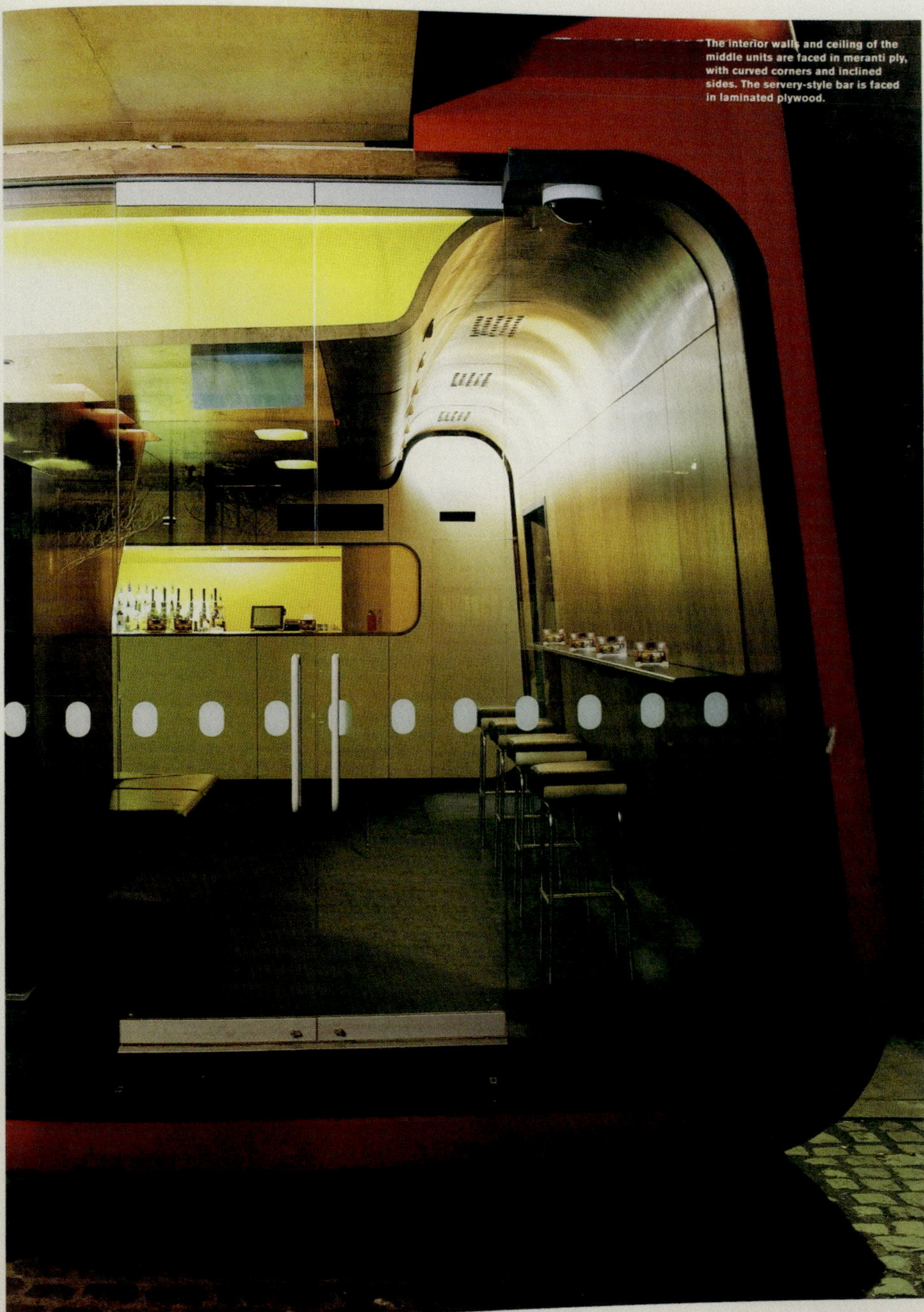

The interior walls and ceiling of the middle units are faced in meranti ply, with curved corners and inclined sides. The servery-style bar is faced in laminated plywood.

SUBJECT
Bar Multi-Purpose Venue (MPV)

LOCATION
Leeds, England

DESIGNERS
Union North

PHOTOGRAPHER
Chris Cascoigne/View

WRITER
Will Jones

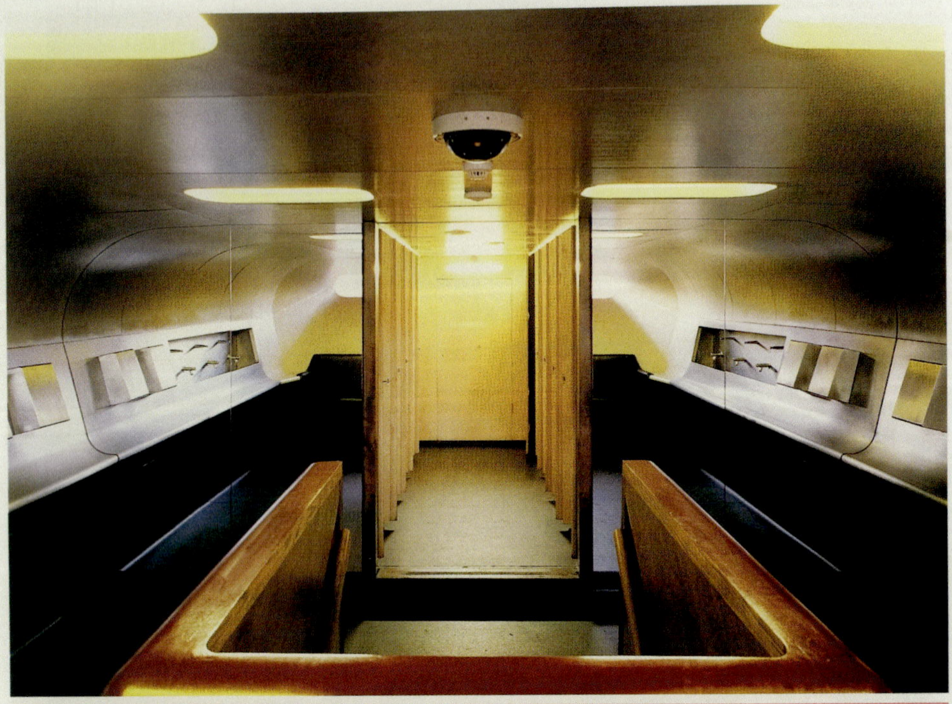

BY DAY THE CAPSULES STAY SHUT. BUT COME NIGHTFALL ALL IS REVEALED WHEN THE FRONTS OF ALL FOUR TILT UPWARDS, THE INTERIORS IN FULL VIEW THROUGH THE GLASS

ends. Bar stools, too, are closely related. A conventional ply-faced bar – no burger-van slot here – stands in one corner of a space that, compared with the two low-slung bars, feels positively cavernous. The lounge and club spaces utilise the full height of the pods, giving a completely different ambience. A glittering disco ball hanging from the ceiling – a virtual impossibility in the bar area – scatters flecks of colour and light across the dark walls. Rods that hang from the ceiling support large speakers, and recessed light fittings illuminate the counter. To come up with a vehicular analogy here is more difficult; these would have to be the funkiest armoured personnel carriers ever seen.

It's important to realise, however, that the recurring vehicular aesthetic was not predetermined. Though the scheme has taken on a decidedly train-boat-plane form (well, alien spacecraft) in both construction and atmosphere, Falkingham denies any thematic approach. 'Theming has more to do with style than content. MPV has a vehicular aesthetic because it was built using methods normally confined to that field. An architectural commission is about solving problems. That sets up the form of response, which in turn produces the style of architecture. We didn't come to the project with any preconceived notion of what it should look like. 'MPV is as much about product design as architecture,' he adds. And the stylish vehicular aesthetic of the spaces brings home the point. 'Crossovers between fields are what interest us. We're not obsessed with whether or not our peers think we do highbrow architecture. We simply apply our collective experience to the design process to create the best possible solution.'

It's this desire to come up with creative answers to interesting architectural conundrums that is pushing Union North on to fresh pastures away from bar design. The practice is currently involved in refurbishing the Blue Coat Art Centre in Liverpool, but its real aspirations lie in social housing. Various schemes in Liverpool and Manchester are on the drawing board, as is an imaginative project of 'creative demolition' for an inner-city developer that involves stripping a block and temporarily re-skinning it.

From designing bars using shipbuilding techniques to knocking down tower blocks ever so stylishly, Union North will supply the most left-field scheme imaginable. Who says the Beatles were Liverpool's greatest export?

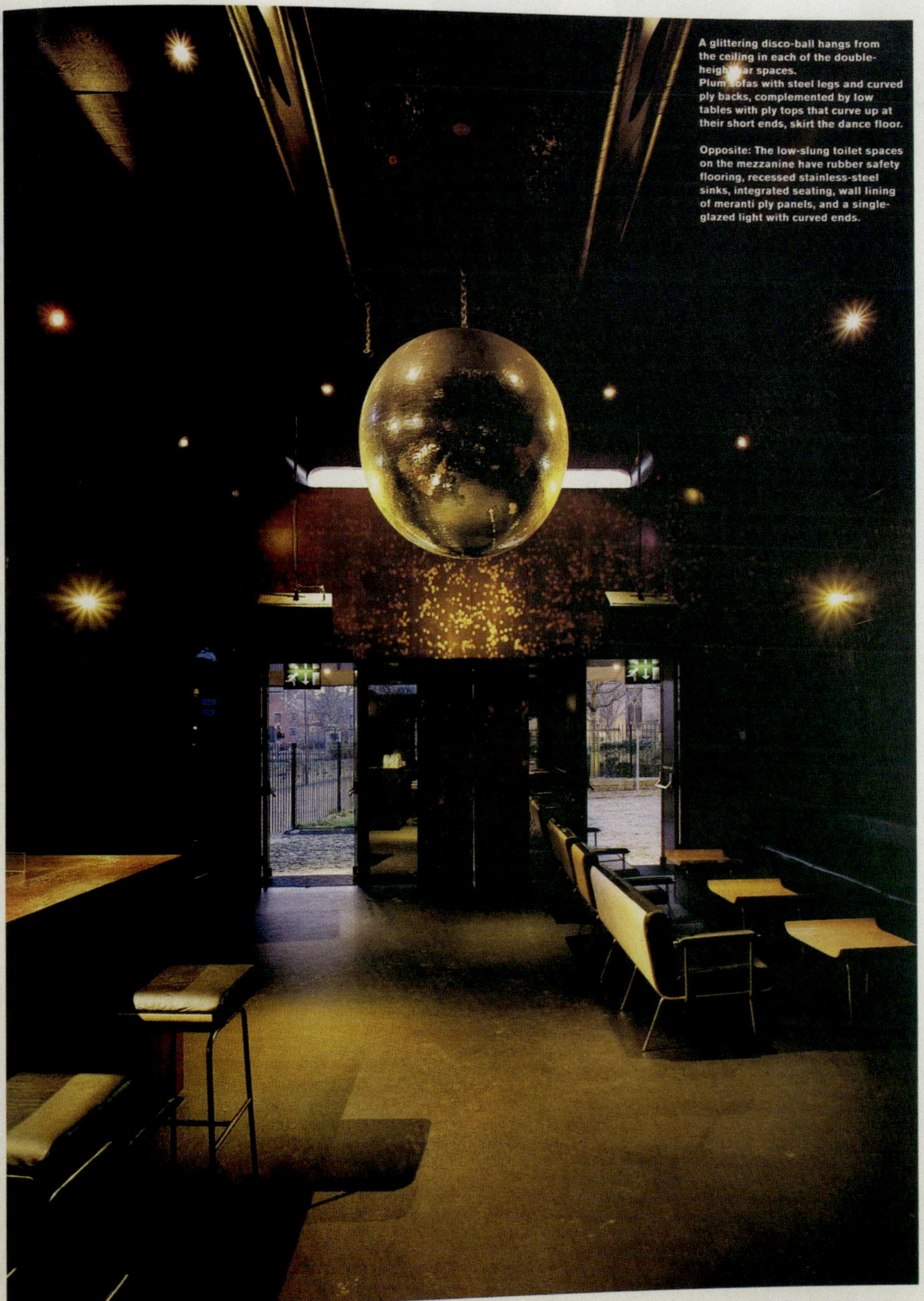

A glittering disco-ball hangs from the ceiling in each of the double-height bar spaces.
Plum sofas with steel legs and curved ply backs, complemented by low tables with ply tops that curve up at their short ends, skirt the dance floor.

Opposite: The low-slung toilet spaces on the mezzanine have rubber safety flooring, recessed stainless-steel sinks, integrated seating, wall lining of meranti ply panels, and a single-glazed light with curved ends.

SUBJECT
Bar Multi-Purpose Venue (MPV)

LOCATION
Leeds, England

DESIGNERS
Union North

PHOTOGRAPHER
Chris Cascoigne/View

WRITER
Will Jones

292-293

September/October 2002

Issue 28

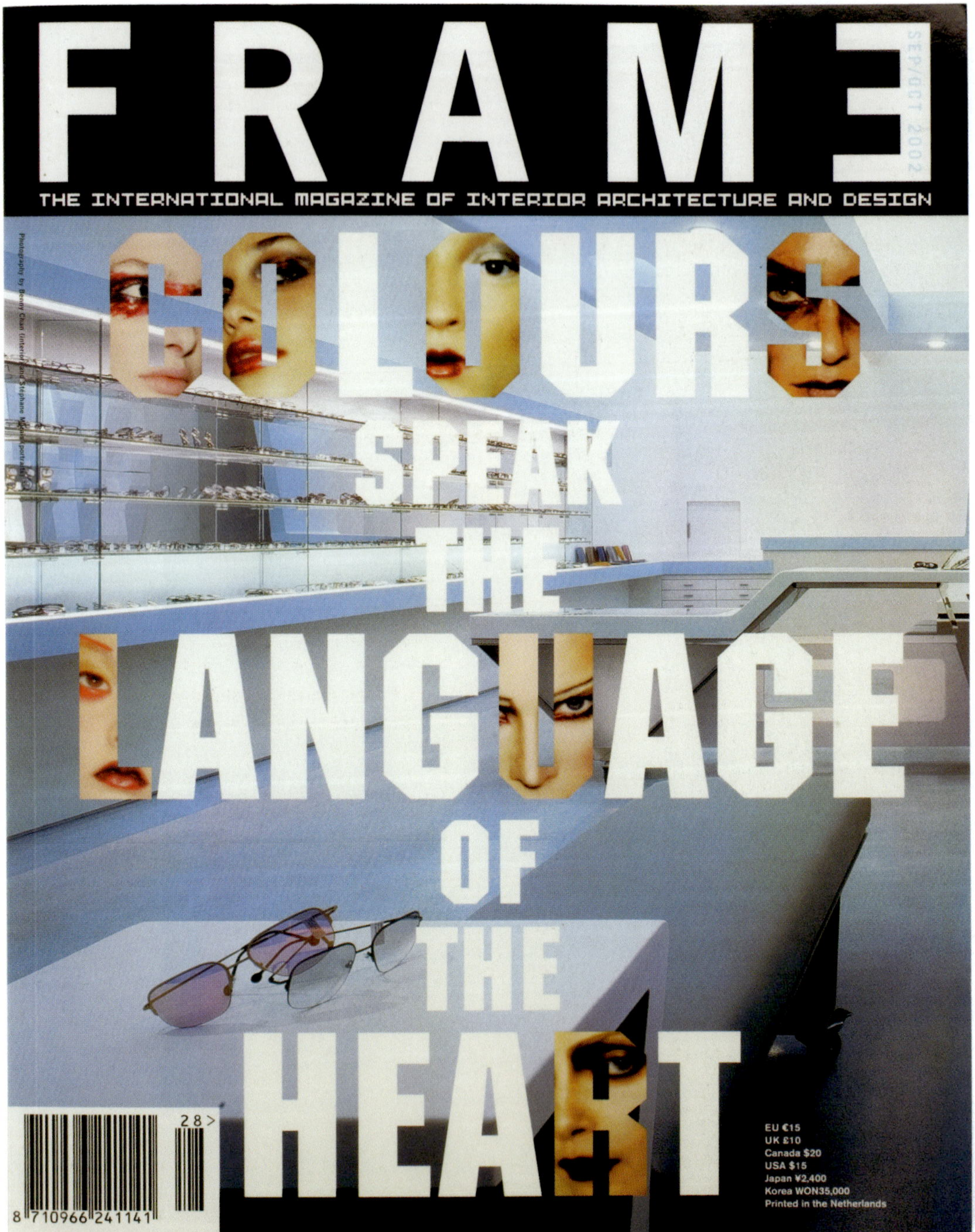

ART DIRECTOR
Roelof Mulder

Lost And Found

Designer Hella Jongerius (JongeriusLab) about the project

The occasion was the design of Repeat for Maharam (and the launch at Moss), which was one of our greatest projects of the past years and one with a very positive impact.

Comments on the published article

A clear, bright article which gives a good overview of Hella Jongerius's work in words and pictures.

Did the article have an effect?

Good publicity.

Recent news or information about the project featured in *Frame*

The Repeat fabric turned out to be quite a good seller.

Current projects

At the moment we are mainly working on a new design for Maharam and on the Home collection for Vitra.

DESIGNERS
Hella Jongerius
www.jongeriuslab.nl

SUBJECT
Portrait of Hella Jongerius

LOCATION
Rotterdam, Netherlands

PHOTOGRAPHER
Frank Oudeman

WRITER
Louise Schouwenberg

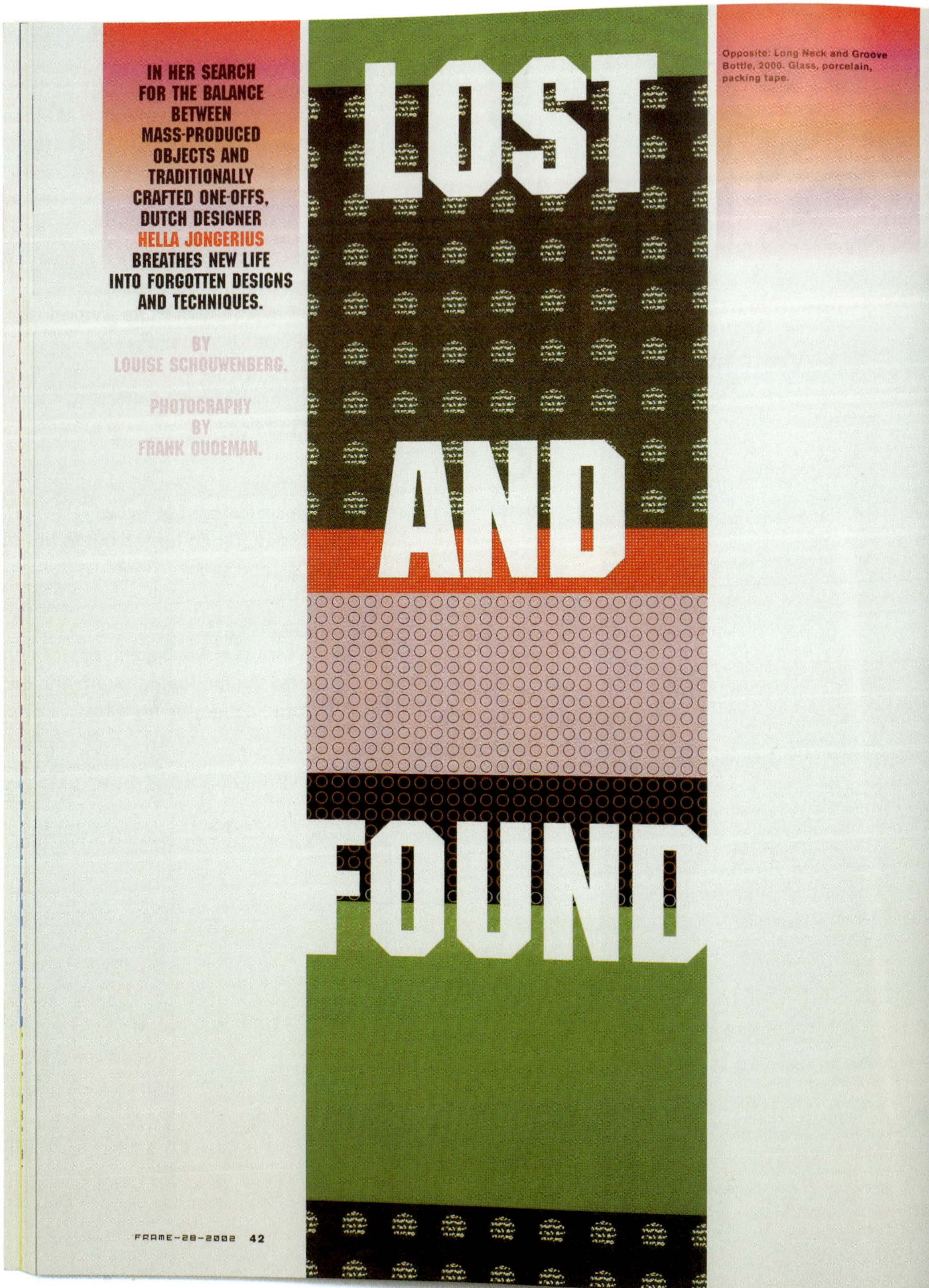

LOST AND FOUND

IN HER SEARCH FOR THE BALANCE BETWEEN MASS-PRODUCED OBJECTS AND TRADITIONALLY CRAFTED ONE-OFFS, DUTCH DESIGNER HELLA JONGERIUS BREATHES NEW LIFE INTO FORGOTTEN DESIGNS AND TECHNIQUES.

BY LOUISE SCHOUWENBERG.

PHOTOGRAPHY BY FRANK OUDEMAN.

Opposite: Long Neck and Groove Bottle, 2000. Glass, porcelain, packing tape.

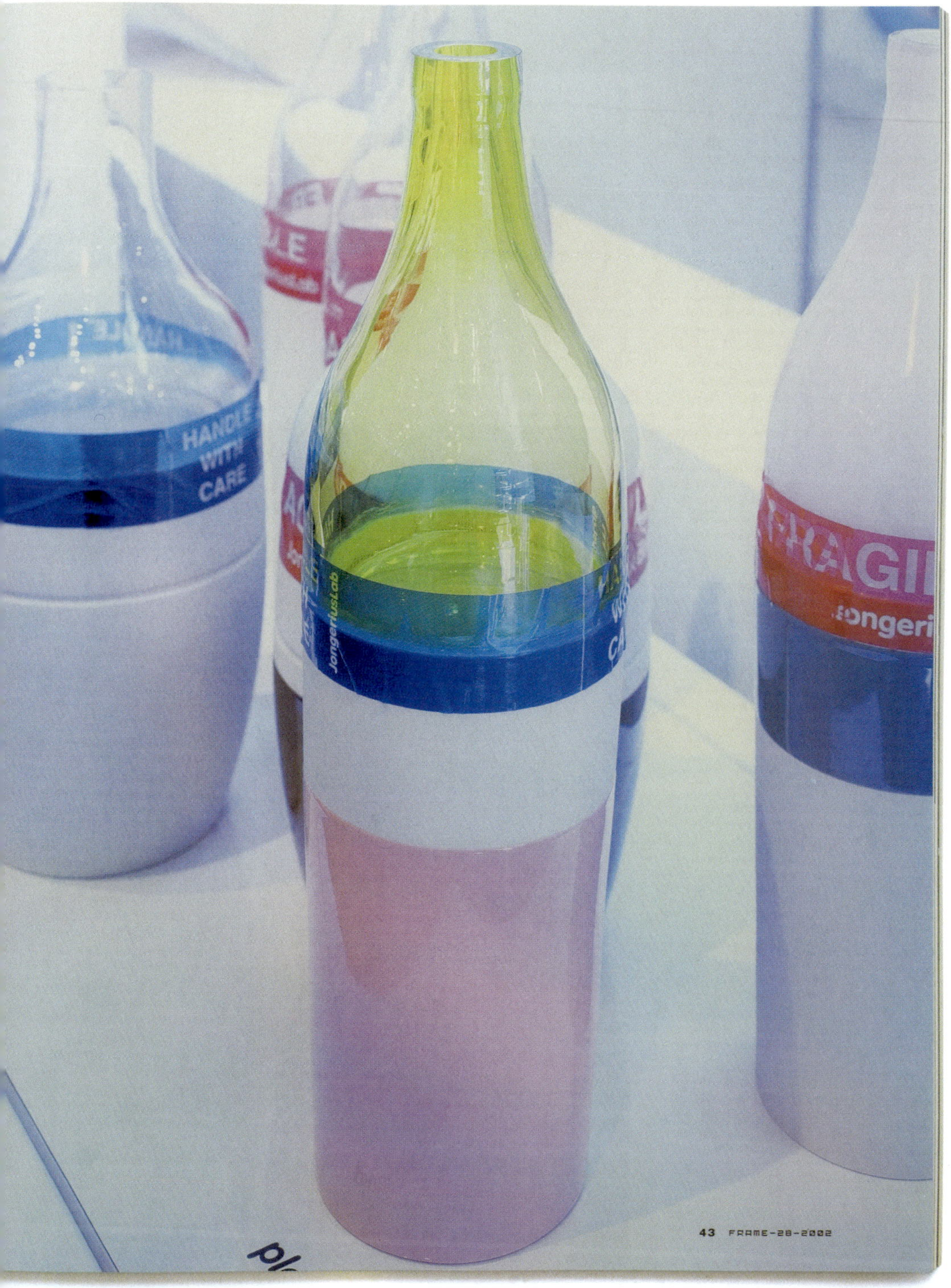

SUBJECT
Portrait of Hella Jongerius

LOCATION
Rotterdam, Netherlands

DESIGNER
Hella Jongerius

PHOTOGRAPHER
Frank Oudeman

WRITER
Louise Schouwenberg

'I REALLY DON'T TRY TO IMAGINE HOW EVERYTHING COULD BE BETTER OR NICER LOOKING. DESIGN IN ITSELF DOESN'T INTEREST ME'

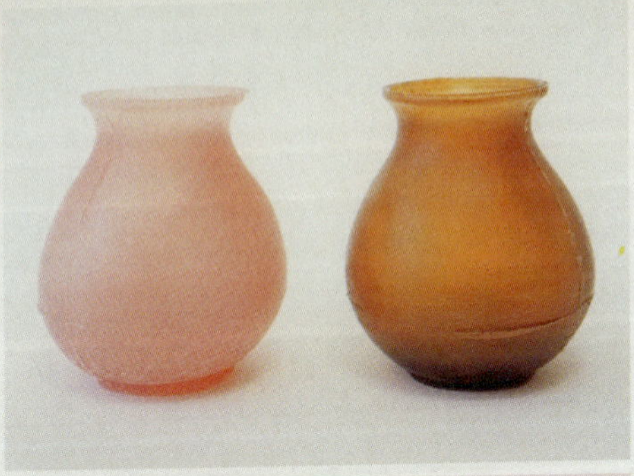

Soft Vase, 1994-1999. Polyurethane. Photography by Bob Goedewaagen.

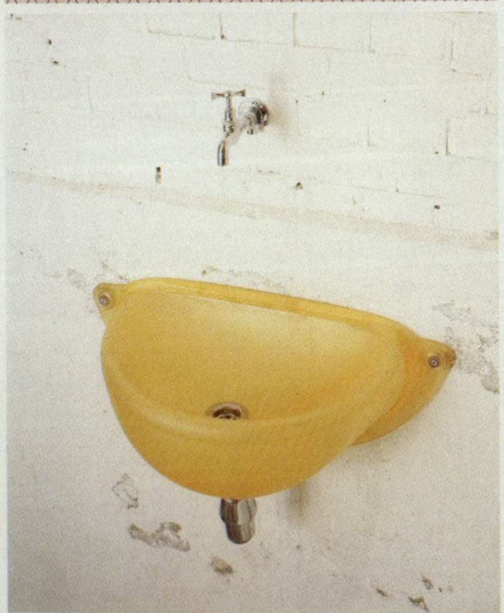

Pushed Soft Washtub, 1996. Polyurethane. Photography by Bob Goedewaagen.

Opposite: Crystal Frock, 2002. Designed for Swarovski, this crystal chandelier looks like a ball gown. The skirt is made of pink rubber letters that ask searching questions about the world of product design.

Despite the current plethora of cross-disciplinary work, differences between artists and designers still exist. Most artists, for example, are buried in a sea of flowers at the openings of their exhibitions. Every available pail, wine bottle and tin can is called into service as a temporary vase. As a visitor to many contemporary design shows, however, I recall only one or two instances in which a guest was brave enough to bring the exhibitor an aesthetically correct spray of blossoms. Designers know one another. They might *design* vases, but they don't plague their colleagues with annoying bouquets that would destroy the character of the ultimate urn. Although they play with practical objects, they often place the resulting designs on a pedestal like works of art, stripped of their functional nature.

Patently aware of this phenomenon, designer Hella Jongerius can chuckle at the humour of the situation. How does it relate to her own work? Her products, though exceptionally beautiful, show little evidence of 'design'. On the contrary, her vases are archetypes that exist by the thousands; the form of her Kasese chair replicates that of an African prayer chair, and her dinnerware has no intention of surpassing the splendour of a Wedgwood service. And let's not forget the fabrics that Jongerius recently designed for textile manufacturer Maharam, which feature existing motifs from the factory archives.

Jongerius has no problem with occupational disability. In the minds of the general public, the stockbroker lies awake at night thinking of dollars, the plastic surgeon scans every face in search of warts and wrinkles, and the designer is offended by the sight of a badly designed object. Hella Jongerius doesn't fit this description, however. 'I really don't try to imagine how everything could be better or nicer looking,' she insists. 'Design in itself doesn't interest me. What fascinates me is the story that a product tells.'

Trendsetter. A word that would surely horrify her, but a term nonetheless attributable to Jongerius. Her work appears regularly in nearly every interior-design magazine in the world. The marketplace teems with rip-offs of her rubber vases, embroidered ceramics and knitted lamps. What special quality does she see in her work that makes it an obvious prey for imitators? 'I guess I have a good nose for what's going on. Evidently my personal preferences transcend who I am. Whatever I intuit seems to be, sooner or later, in line with the times. Sounds arrogant perhaps, but up to now that's been my experience.'

She hastens to point out that, like all design-related terms, words such as 'intuition' and 'trendsetter' suffer from inflation. They no longer mean anything. The reason, according to Jongerius – who admits, when pressed, that she is offended by what often passes for design – is that we're drowning in a flood of bad design: rubbish presented as 'hip design' on televised DIY programmes, insipid gift-shop gags and cheap mass-produced articles. 'Every corner of the world has been "designed", but the quality of that design is appalling. We're fouling our own designer nest.

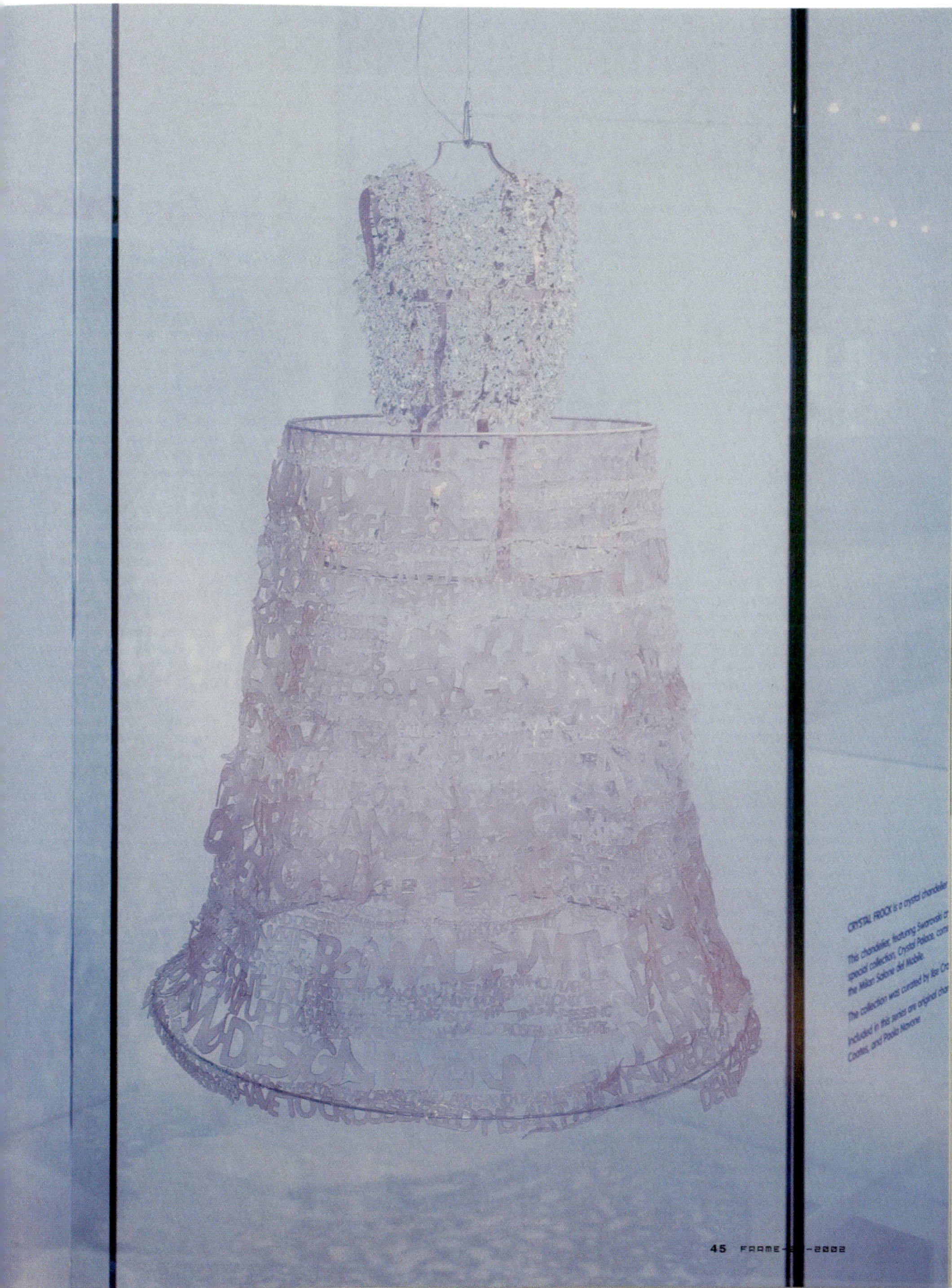

SUBJECT
Portrait of Hella Jongerius

LOCATION
Rotterdam, Netherlands

DESIGNER
Hella Jongerius

PHOTOGRAPHER
Frank Oudeman

WRITER
Louise Schouwenberg

'I GUESS I HAVE A GOOD NOSE FOR WHAT'S GOING ON. EVIDENTLY MY PERSONAL PREFERENCES TRANSCEND WHO I AM'

Roughly ten years ago, designers still played a major role in the world, but now they seem to be mired in their own success. The story they tell becomes shorter by the day.'

Jongerius's ambition is nothing less than a reassessment of the design profession. 'I'm looking for new meanings. In creating a single product, you can change the way people think about products in general. That's something I genuinely believe. It explains why I'm not satisfied with my work unless it makes a statement.'

Her Slightly Damaged Dinner Service refers to the importance of imperfection as a precondition for becoming attached to an object. Observing the people around her, she noted that they cherished the chipped cups and cracked saucers in grandma's cupboard far more than the pristine service for 12 that every designer of china burns his fingers making. Consequently, she crafted a second-rate service with minor flaws that make each piece unique. Lengthy experimentation with types of porcelain and various kiln temperatures produced the desired result. The next step was to persuade the ceramics industry (Royal Tichelaar Makkum) to fire her pieces at the 'wrong' temperatures.

'An incredible amount of intelligence goes into the creative process. Getting your hands dirty is essential to being a good designer. Concocting something on a purely rational basis provides no new insights, and concepts alone are much too bare.' Jongerius would rather delve into materials and manufacturing methods, milking them until they become her own. 'Only when the material pushes the envelope do new possibilities and unexpected concepts emerge. I find the creative process more important than the end result.'

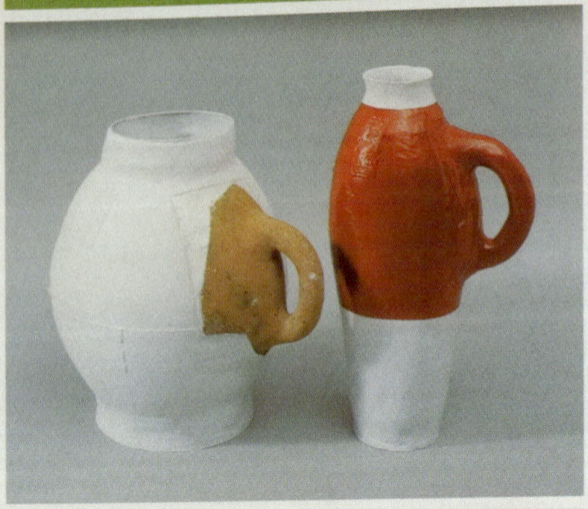

7 pots / 2 centuries / 3 materials, 1998. Porcelain, epoxy, shards of ceramic.

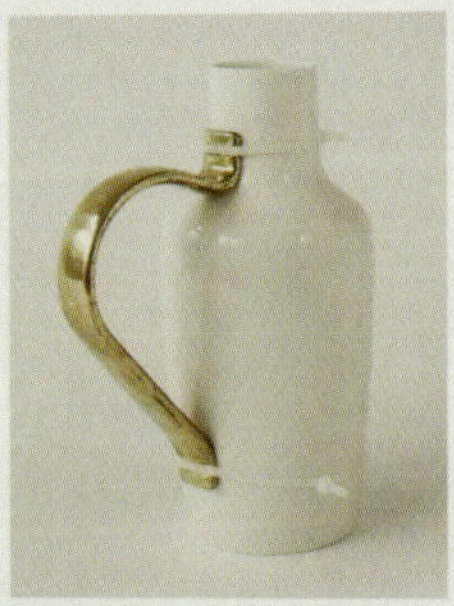

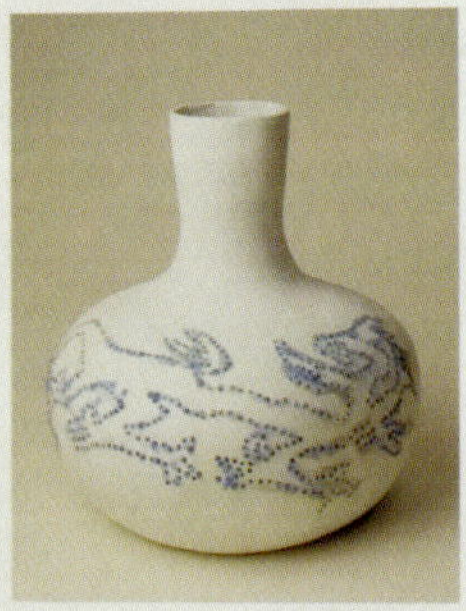

Left: Delft Blue Jug, 2001. Porcelain, bronze, tie wraps.
Right: Prince, 2000. Porcelain, silicone.

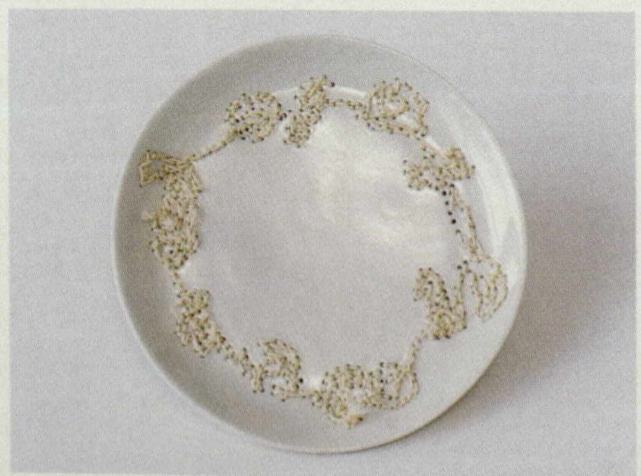

Delft Blue Plate, 2001. Porcelain, gold wire.

Opposite: Repeat Porcelain, 2002. Porcelain, thread.

To become fully immersed in this process, she has to carry out her experiments completely undisturbed. Material and intuition dictate the direction. Outside issues are irritating, as are rational explanations that arise too early in the process. 'Words follow much later, and even more time passes before they're strung into sentences and explanations – often only during conversations with others.' Despite her emphasis on process rather than final result, Jongerius feels she's a designer in heart and soul. Apart from the odd exception, such as Embroidered Tablecloth (2000), a project in which she embroidered a tablecloth to the dinnerware, her work is functional.

A brief history. Jongerius graduated from the Design Academy in Eindhoven in 1993. She 'wasn't a brilliant student'. But it's not all about talent. 'An intuitive sense of what's happening in the world of design, an eye for refined culture as well as street life, social intelligence and a huge passion for the profession are preconditions for success in this business.'

From the word go, Droog Design has promoted Jongerius and a number of other young designers, among whom Jurgen Bey, Jan Konings and Marcel Wanders. Jongerius's work was notable for its obvious attention to the study of materials. She came up with concepts, even during her years at the academy, by means of an almost alchemistic approach to testing materials: the 'dirty hands' method. In making her first

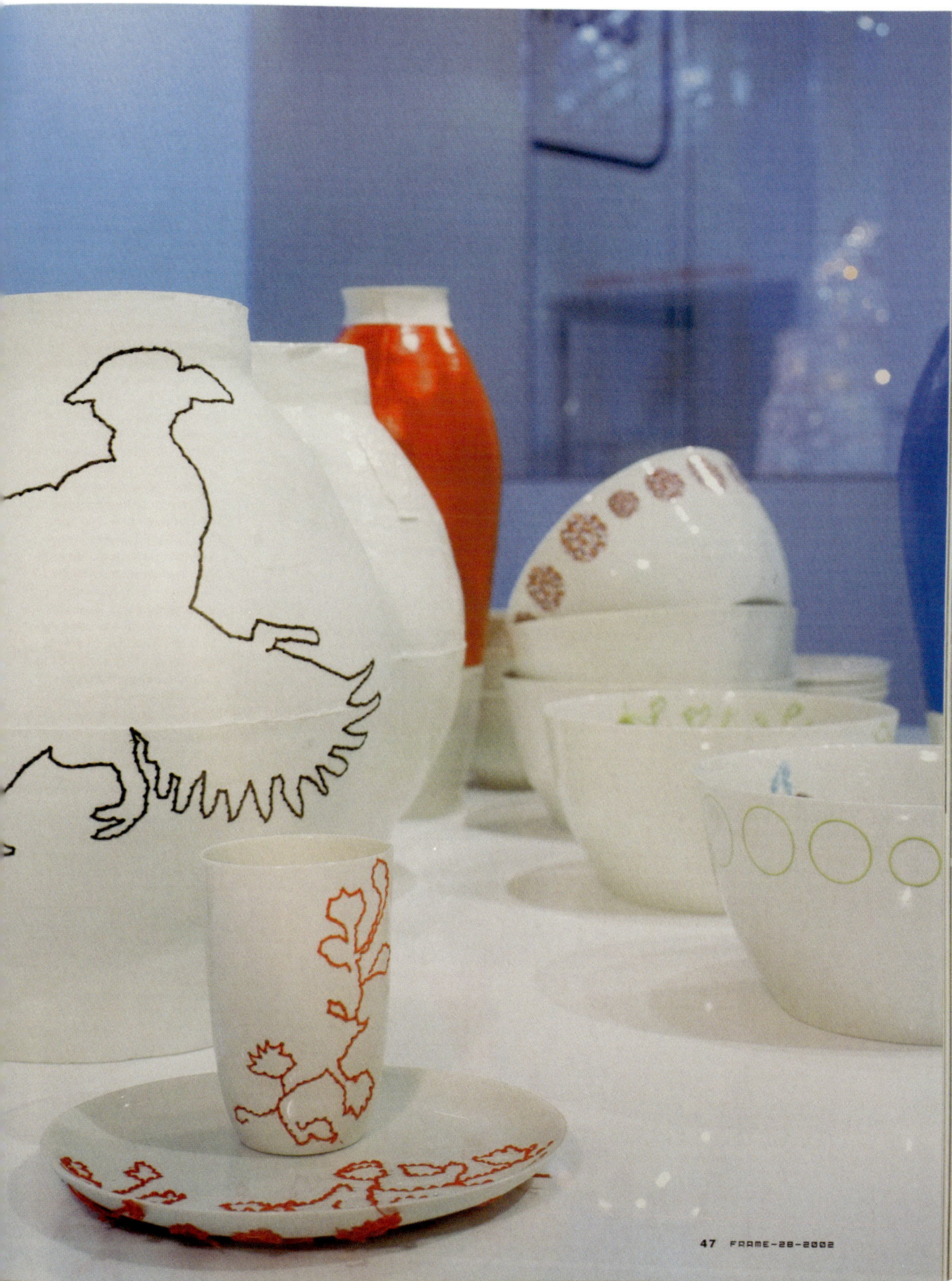

SUBJECT
Portrait of Hella Jongerius

LOCATION
Rotterdam, Netherlands

DESIGNER
Hella Jongerius

PHOTOGRAPHER
Frank Oudeman

WRITER
Louise Schouwenberg

'GETTING YOUR HANDS DIRTY IS ESSENTIAL TO BEING A GOOD DESIGNER'

products – rubber bathmats (1994), Soft Vases (1994) and Pushed Washtub (1996) – she forced the properties of polyurethane beyond the usual boundaries. The shapes she chose were, to use her own term, 'archetypal'. 'The world has enough vases. I don't feel the need to come up with a new form. What *is* important is how materials can add a new dimension to existing forms while combining beauty and concepts.' Later work required research into high-tech synthetic fibres and traditional crafts involving ceramics, glass, textiles and the like.

From the outset, Jongerius's work has revolved around the search for a balance between the mass-produced object and the traditionally crafted one-off. A search that lifts both spheres to a higher level: the industrial product is given a soul, and the one-off sheds its frumpish image. In addition to undisturbed working conditions in her own lab, Jongerius needs strong powers of persuasion to convince manufacturers that her ideas are worthy of production.

Sometimes the puzzle pieces fit together perfectly. A prime example is the 'dream project' that Jongerius realised for textile manufacturer Maharam, who knew exactly what they wanted. Maharam asked her to create an upholstery fabric that would turn a suite of chairs into 'cousins – each unique, but all related'. Her solution was Repeat: 16 fabric designs available by the roll, each of which has an unusually long repetitive pattern. In the textile industry, a motif repeated every 30 centimetres is the norm. Such fabrics are easier for manufacturers to use and thus

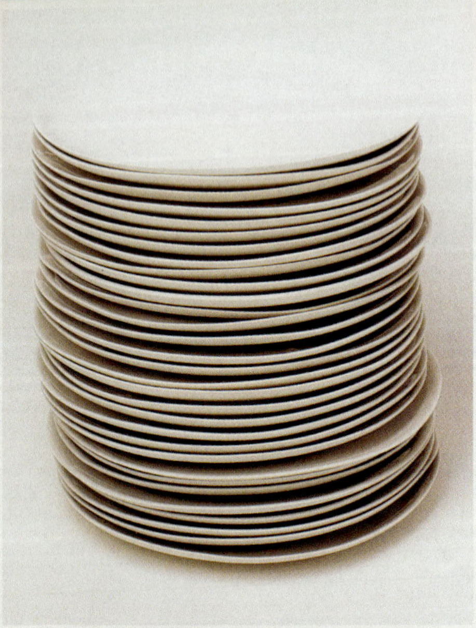

Slightly Damaged Dinner Service, 1998. Manufactured by Royal Tichelaar Makkum, the plates are all deliberately flawed. Photography by Bob Goedewaagen.

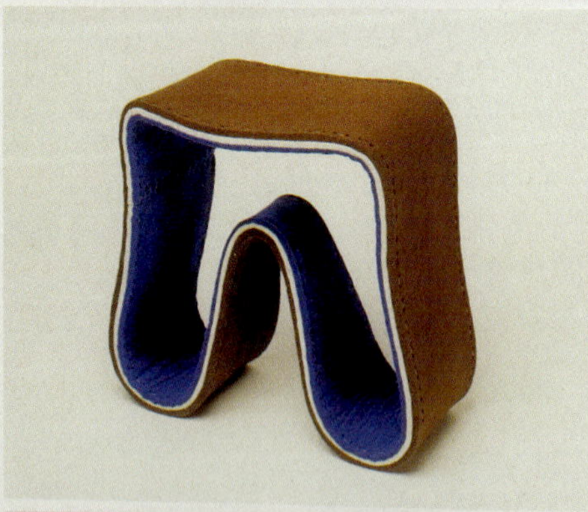

Felt Stool, 2000. Felt, metal. Distributed by Cappellini.

Opposite: Repeat, 2002. Upholstery textile designed for Maharam. Large-scale pattern sequences yield exaggerated repeats of up to 3 metres. When applied to furniture, each piece takes on an individual character, while remaining in the vocabulary of the grouping.

easier to sell. The Repeat pattern – a full 3 metres long – allows manufacturers to cover several pieces of furniture, none of which are identical to any of the others. Individual pieces within a single family.

Jongerius didn't turn to the drawing board, as might be expected, to design a trendy pattern for Repeat. She turned to the archives, walked through the factory and gathered archetypal patterns like pied-de-poule, stripes, and birds and flowers ('bird and vine' in technical jargon). Other discoveries incorporated into her patterns were perforations in punched cards for the looms, and numbers and codes referring to technical data. By changing sizes and colourways, and by distributing prints of enlarged dots and initials – hers and Maharam's – across the patterns, Jongerius created a range of overtones and motifs. The resulting fabrics display a wonderful marriage of old and new, of tradition and street culture.

Jongerius: 'It's a celebration of a product in all its richness. This is about more than the concept and the treatment of materials. It's about the industrial process and making something on a large scale, not to mention marketing. This project, like the Slightly Damaged Dinner Service, requires a user who is expected to contribute his own brand of personal creativity to the work. After all, the fabric invites him to determine which parts of the pattern will appear at which places on the furniture. More remnants than usual make this an expensive material. The salespeople will have to come up with a darn good pitch.' She praises the vision of

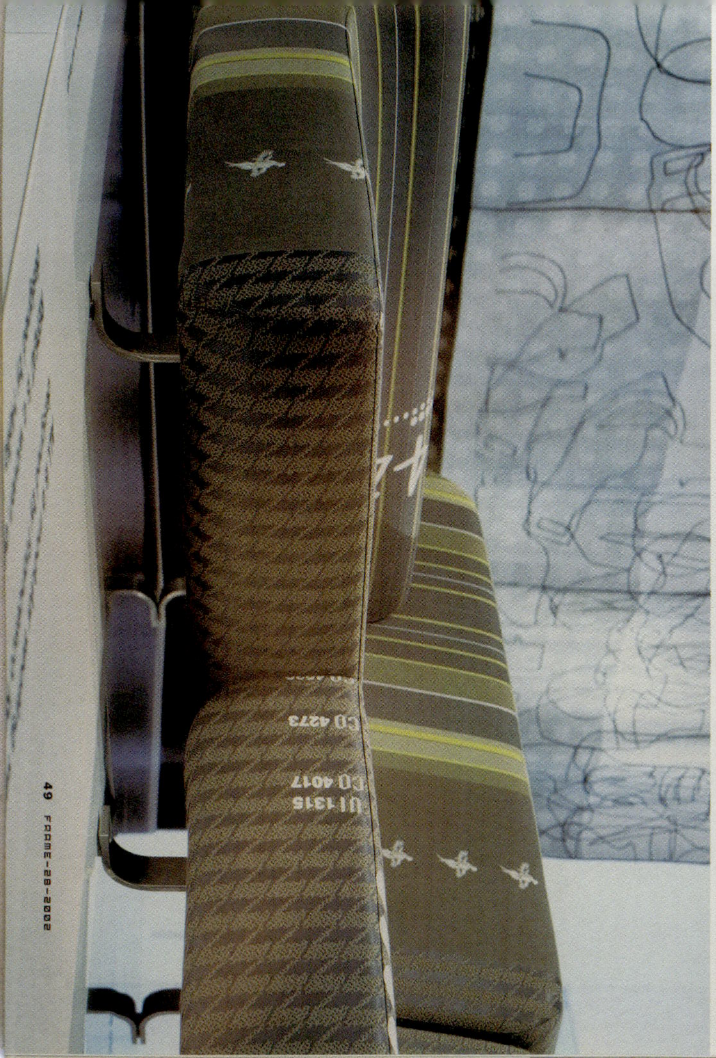

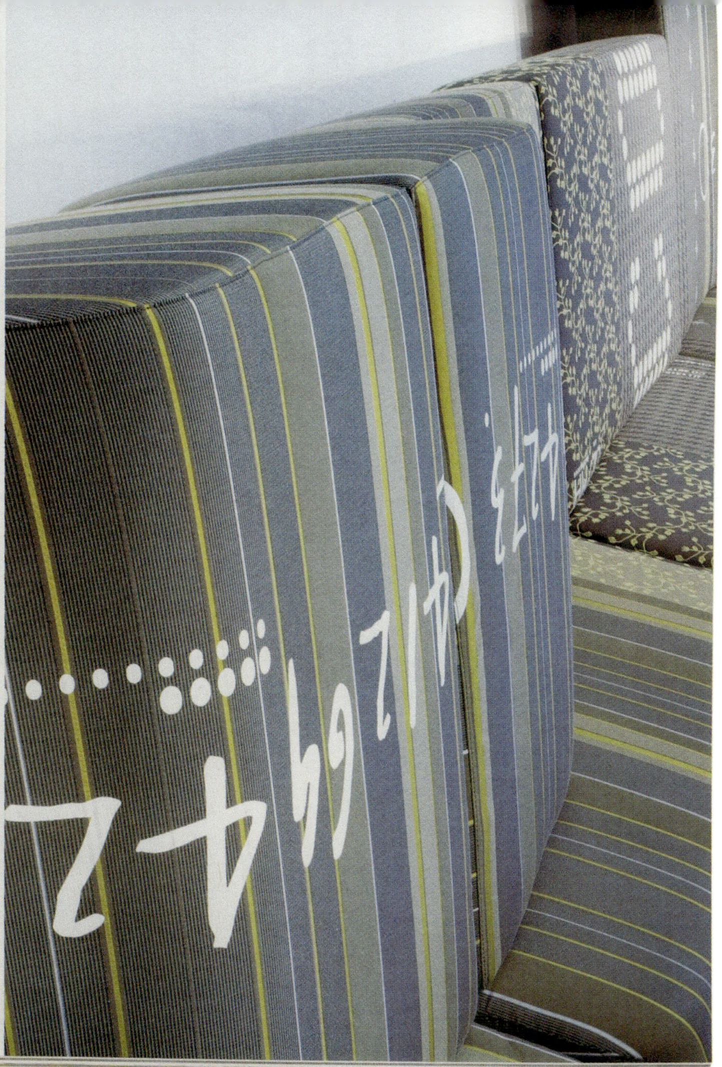

SUBJECT
Portrait of Hella Jongerius

LOCATION
Rotterdam, Netherlands

DESIGNER
Hella Jongerius

PHOTOGRAPHER
Frank Oudeman

WRITER
Louise Schouwenberg

'ONLY WHEN THE MATERIAL PUSHES THE ENVELOPE DO NEW POSSIBILITIES AND UNEXPECTED CONCEPTS EMERGE'

Maharam and the Moss Design Gallery in New York, where she first exhibited the fabric at a popular one-woman show in May 2002. They took a considerable risk in backing this project.

I remind her of the freedom she claims is essential for working in an undisturbed atmosphere. 'You're right. Working with industry always carries a risk. Throughout the process, you're constantly forced to make concessions. Fortunately, having done the preliminary work – mainly by experimenting with porcelain, strangely enough – I had a lot of self-confidence going in. For years I've been working with the ingredients that come together so amazingly in this project: old versus new, classic tradition versus banal symbols, and the power of decoration, which can transcend the visual to take on a different meaning.' She's referring to 'mix and match' designs in which she put rubber patterns on porcelain jugs and used ordinary plastic tape to make vases with ceramic bases and glass necks. Like the upholstery fabrics, these too were strange marriages that united old family members without stripping them of their individuality.

What Jongerius found so exciting about the Maharam project was the additional aspect of mass production. Indeed, for Jongerius, whose work so often teeters on the brink of function and practicality, the project formed a field of tension. To what extent is industry prepared to permit experimentation? Ultimately, even the most enthusiastic manufacturer thinks in terms of supply and demand, makes decisions that involve potential target groups, subjects the concept to a feasibility test, considers ease of use and contemplates possible marketing strategies. Although a designer is not an artist in the traditional sense of the word, Jongerius largely avoids such questions. She does describe her own target group, however, as consumers equipped with vision, creativity and the courage to opt for the audacious. She leaves it up to other designers to make functional vases and tablecloths that are not embroidered to the dinnerware.

Another industrial commission, this time for crystal king Swarovski, provided an ideal vehicle for Jongerius's powerful statements. Swarovski asked her and several other designers for innovative chandelier designs. Aware that chandeliers are nearly impossible to modernise, Jongerius seized the opportunity to question the design profession itself. From a distance, her opulently glittering ball gown fits right into the exclusive world of crystal chandeliers. A closer look, however, reveals a pink rubber skirt covered with letters that spell out discerning queries: Can industrial be affectionate? Can crafts be contemporary? Will design have to crossbreed? Can I only translate what's in the air? Can quality be made without affection? In this design, theory becomes decoration, and decoration asks critical questions.

Jongerius presumably knows the answers all too well.

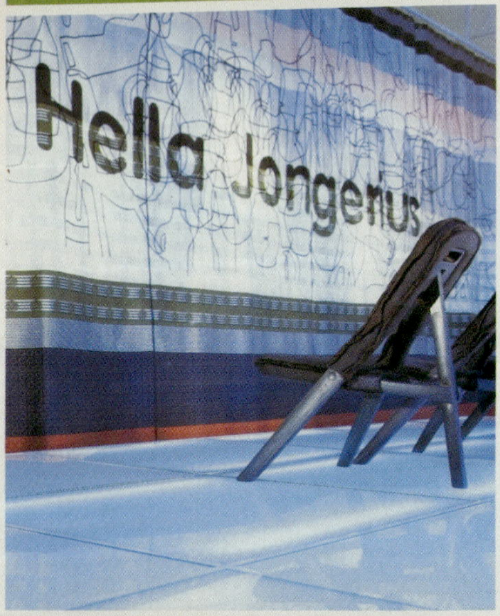

Opposite: For her exhibition earlier this year at the Moss Gallery in New York, Jongerius reworked a large piece of fabric from the Repeat collection with a stitched pattern that traces the outlines of her earlier work.

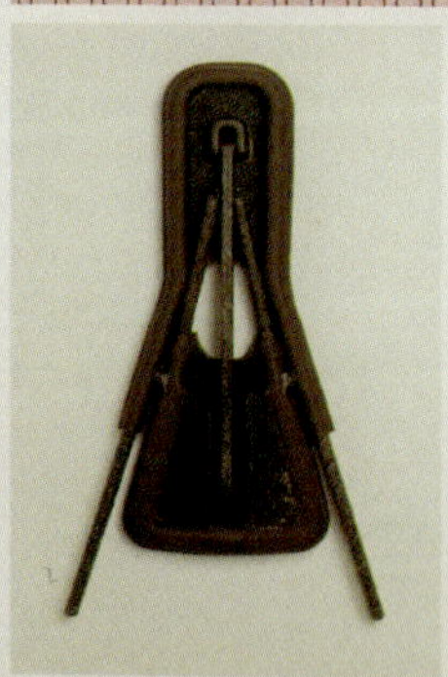

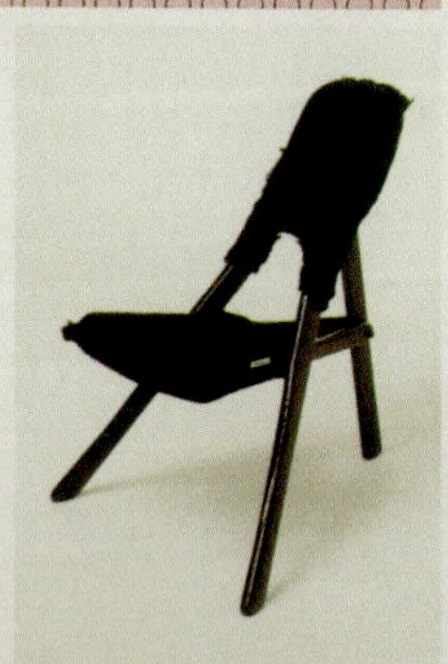

Kasese chair, 1999. Carbon-fibre frame with felt or foam cover. Jongerius based her design on a wooden chair that she saw in Africa. Distributed by Cappellini. Photography by Bob Goedewaagen.

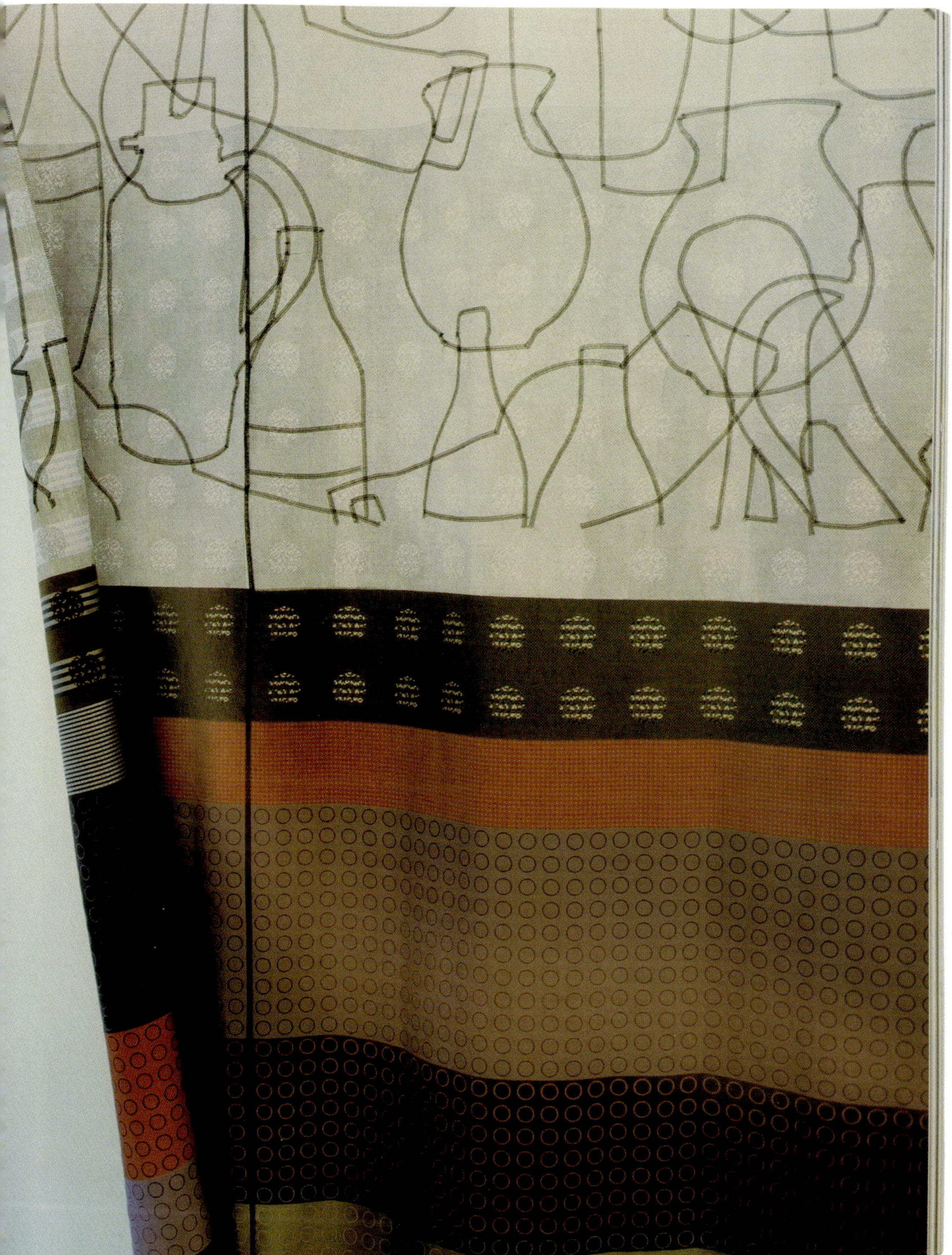

SUBJECT
Portrait of Hella Jongerius

LOCATION
Rotterdam, Netherlands

DESIGNER
Hella Jongerius

PHOTOGRAPHER
Frank Oudeman

WRITER
Louise Schouwenberg

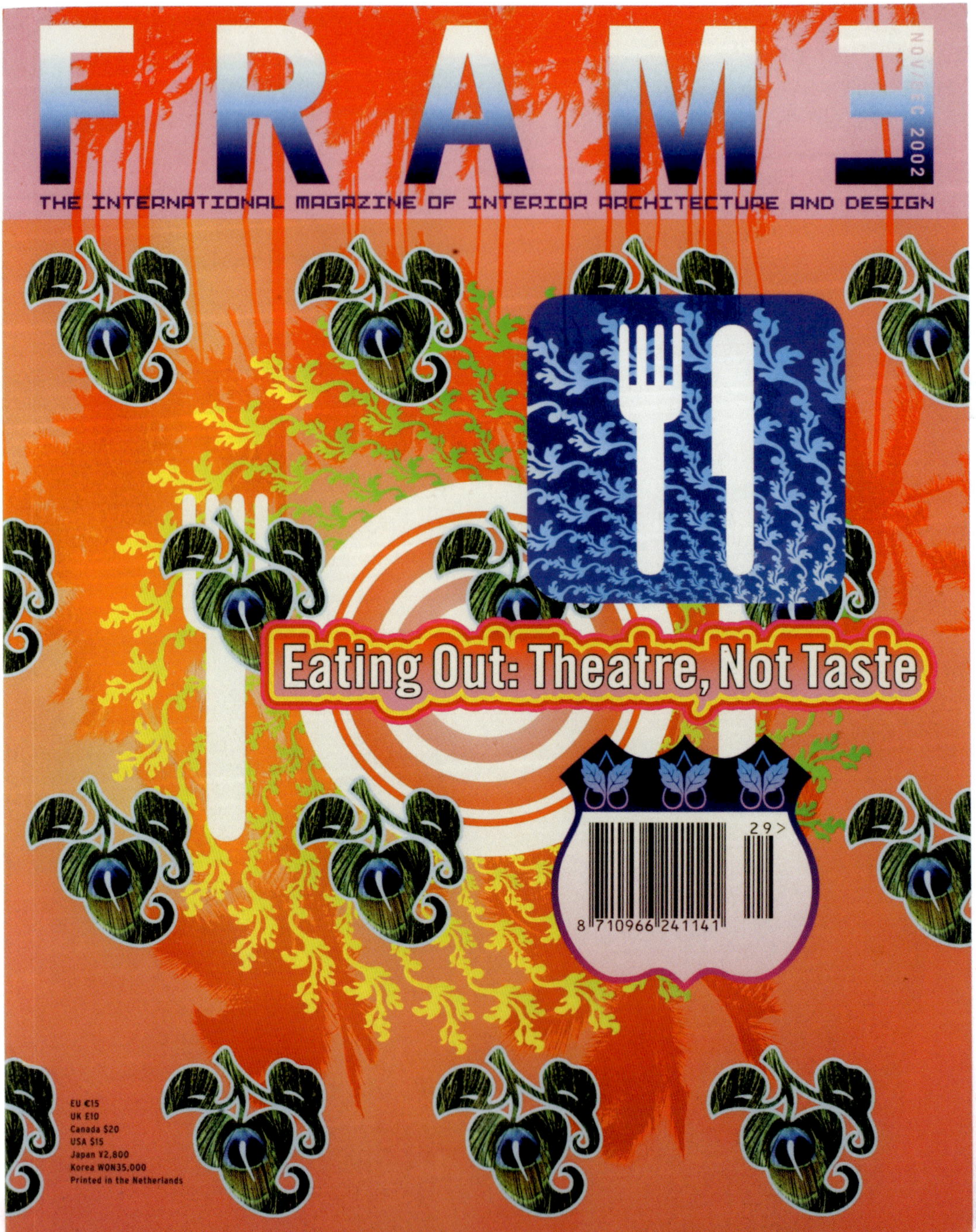

ART DIRECTOR
Roelof Mulder

29

Make the Earth Move

DESIGNERS
Foreign Office Architects
www.f-o-a.net

SUBJECT
Yokohama Port Terminal

LOCATION
Yokohama, Japan

PHOTOGRAPHER
Daichi Ano/Nacása
& Partners
www.nacasa.co.jp

WRITER
Shigekazu Ohno

Architects Farshid Moussavi and Alejandro Zaera-Polo (Foreign Office Architects) about the project

The Yokohama International Port Terminal was a determining project in our career as architects. It launched our office, and its construction was an intense and complex process that resulted in what we consider a very successful and interesting architectural experience.

Comments on the published article

The article illustrates well the richness and complexity behind the building and its story.

Did the article have an effect?

The Yokohama International Port Terminal was widely published in numerous magazines and books since its opening in June 2002. Obviously, every article counts towards increasing the exposure of the project.

Recent news or information about the project featured in *Frame*

The Yokohama International Port Terminal has been open and running since its opening in 2002.

Current projects

Oleiros Shopping Complex, La Coruña, Spain // Uemraniye Retail Complex and Multiplex, Istanbul, Turkey // Trinity Complex, London, England // Ravensbourne College of Design & Communication, London, England // Shires West Department Store, Cineplex and Pedestrian Bridges, Leicester, England // Villa in Pedralbes, Barcelona, Spain // Durango Residential Tower, Durango, Spain // Torre Forum, Barcelona, Spain // BBC Music Box, London, England.

AT THE YOKOHAMA PORT TERMINAL, A PROJECT BY FOREIGN OFFICE ARCHITECTS, ARCHITECTURE, INFRASTRUCTURE AND LANDSCAPE DESIGN COALESCE TO FORM A UNIQUE URBAN LANDMARK.

'What on earth is that?' gasps my taxi driver in astonishment as he pulls up to the drop-off point. And inside the terminal, a space that evokes the sensation of having been swallowed by Moby Dick, a woman exclaims, 'I never expected anything like this!' Devoid of columns and capped by a roof resembling the folded forms of origami, the hall is a spatial wonder not easily forgotten.

Inaugurated this year on June 2 (the same day the port was officially opened in 1859), the Yokohama Port Terminal is one of the most eagerly awaited structures to be completed in the past decade. It was designed by Alejandro Zaera Polo and Farshid Moussavi, who hail from Madrid and Tehran respectively. With offices in London and Tokyo, their aptly named firm – Foreign Office Architects (FOA) – has realised work in Asia, America and Europe, and will represent Britain at the Venice Architecture Biennale in 2002. In 1995 their design proposal was selected from 753 anonymously submitted schemes assessed by a heavyweight jury that included Toyo Ito, Arata Isozaki and Rem Koolhaas (a former employer of both architects). Virtually overnight the duo, still in their twenties, was catapulted into the architectural stratosphere. But it hasn't all been plain sailing. The fairytale start to their careers soon gave way to harsh reality and the colossal responsibility of guiding

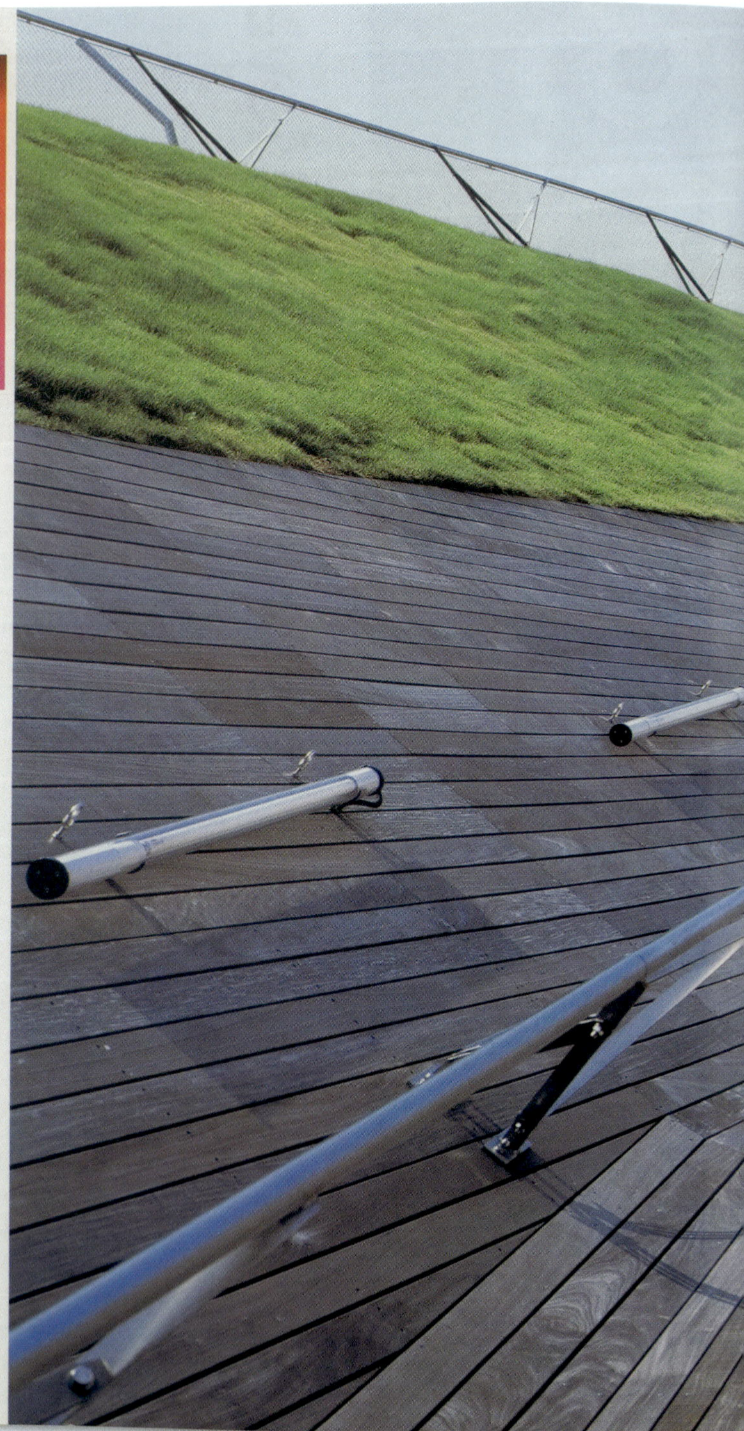

making the earth MOVE

BY SHIGEKAZU OHNO.
PHOTOGRAPHY BY DAICHI ANO/NACÁSA & PARTNERS.

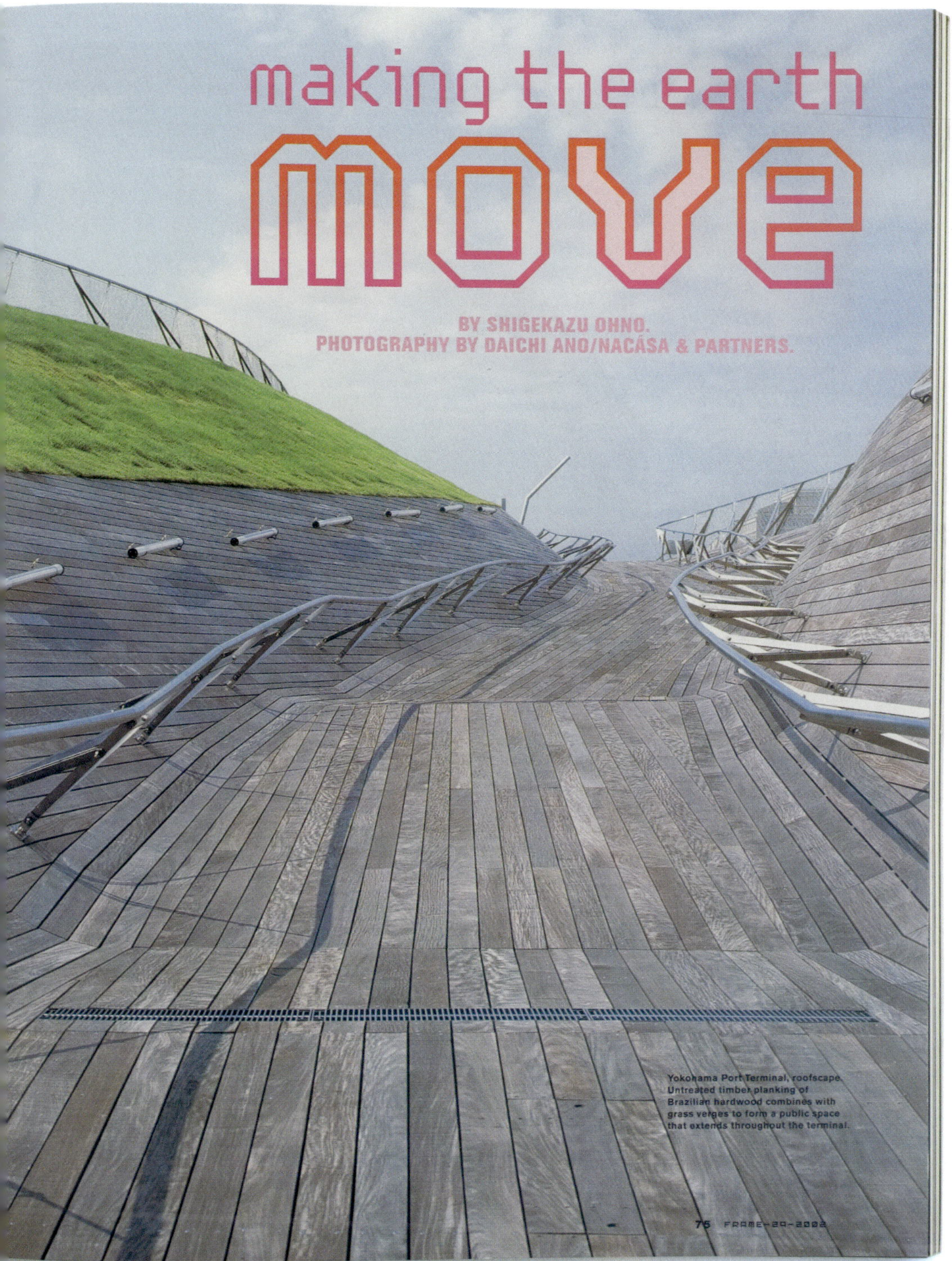

Yokohama Port Terminal, roofscape. Untreated timber planking of Brazilian hardwood combines with grass verges to form a public space that extends throughout the terminal.

SUBJECT Yokohama Port Terminal

LOCATION Yokohama, Japan

DESIGNERS Foreign Office Architects

PHOTOGRAPHER Daichi Ano/Nacása & Partners

WRITER Shigekazu Ohno

THE TERMINAL REPRESENTS A SEAMLESS MELDING OF ARCHITECTURE, URBAN DESIGN AND LANDSCAPE ARCHITECTURE

such a complex work of infrastructure to a successful conclusion. To make matters worse, the Japanese economic slump of the late '90s threatened to ruin the party. Lingering doubts about the feasibility of the project were dispelled only when Yokohama was chosen to stage the 2002 World Cup Final.

The original terminal had been constructed to mark an earlier sporting event, the 1964 Olympic Games in Tokyo, but by the early '90s it had ceased functioning. Given the opportunity to redevelop, city authorities were quick to stress the civic nature of a new structure. They weren't looking for just a replacement terminal, but for 'a square where city and port come together, a place of welcome and farewell for passengers from all over the world'. FOA's winning scheme does more than satisfy these requirements. It represents a radical rethink in terms of the ferry terminal programme and a seamless melding of architecture, urban design and landscape architecture. The complex network of interconnected and folded planes forms a public hard-surface park that connects to the neighbouring Yamashita Park. Housed beneath this topsy-turvy landscape are 17,000 square metres of terminal facilities (check-in, luggage handling, customs and immigration, administration), 13,000 square metres of amenities (conference spaces, shops, restaurants, multipurpose hall) and 18,000 square metres

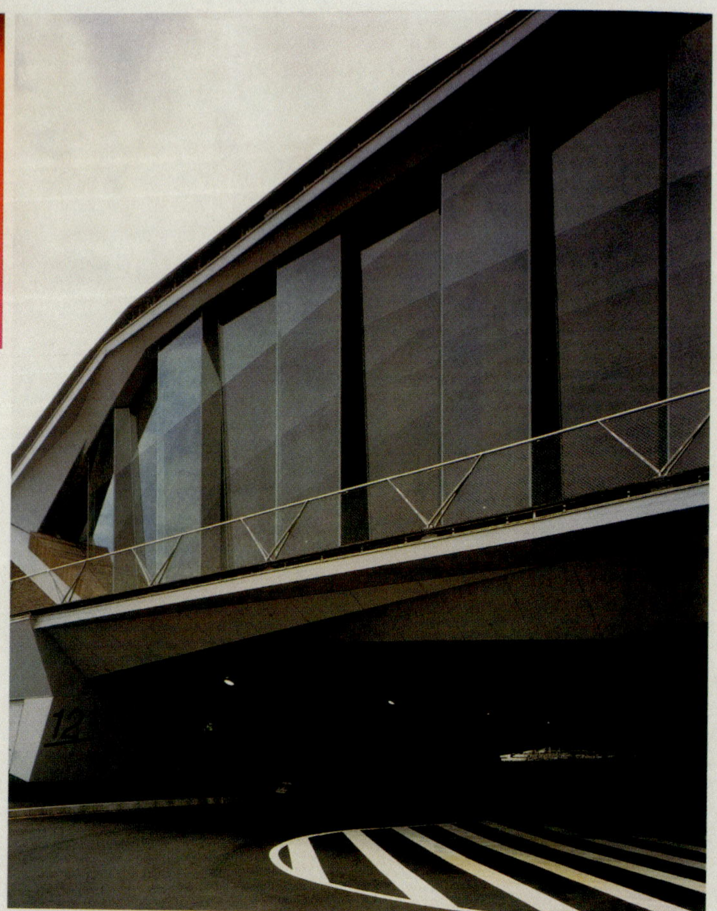

Above: Entrance to parking level.

Opposite top: The cruise terminal, 70 metres wide and 430 metres long, has a maximum height of 15 metres.

Opposite bottom: Foreign Office Architects describes the building as a ground plane that swells to become an active surface from which architectural elements emerge.

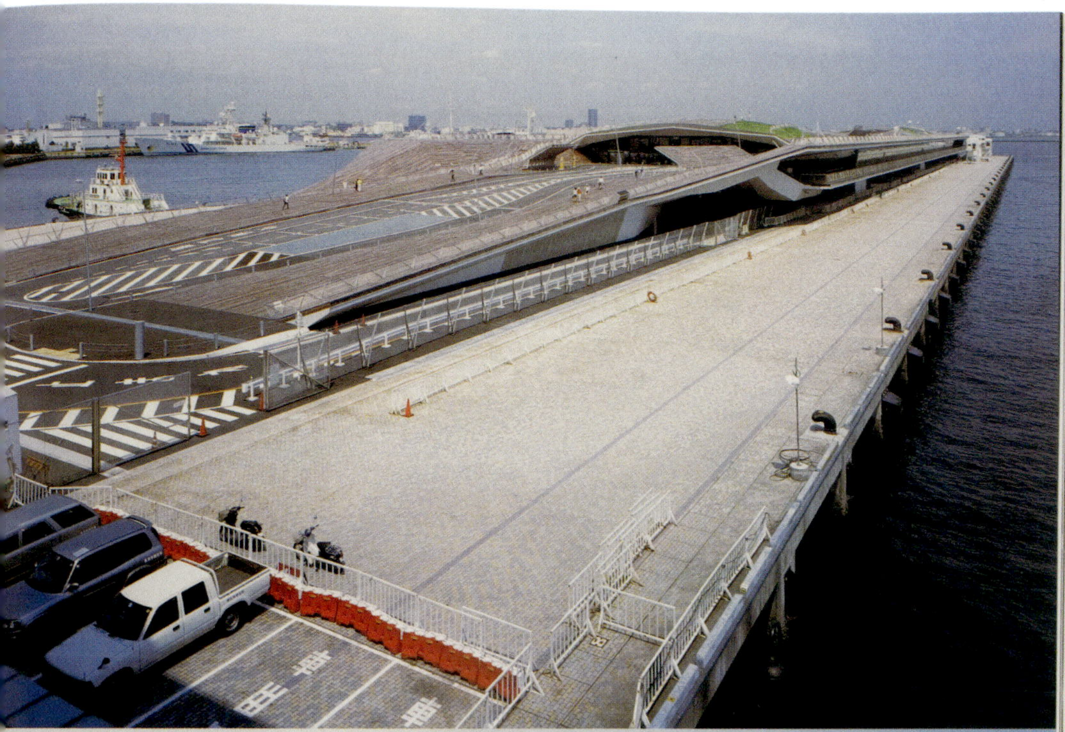

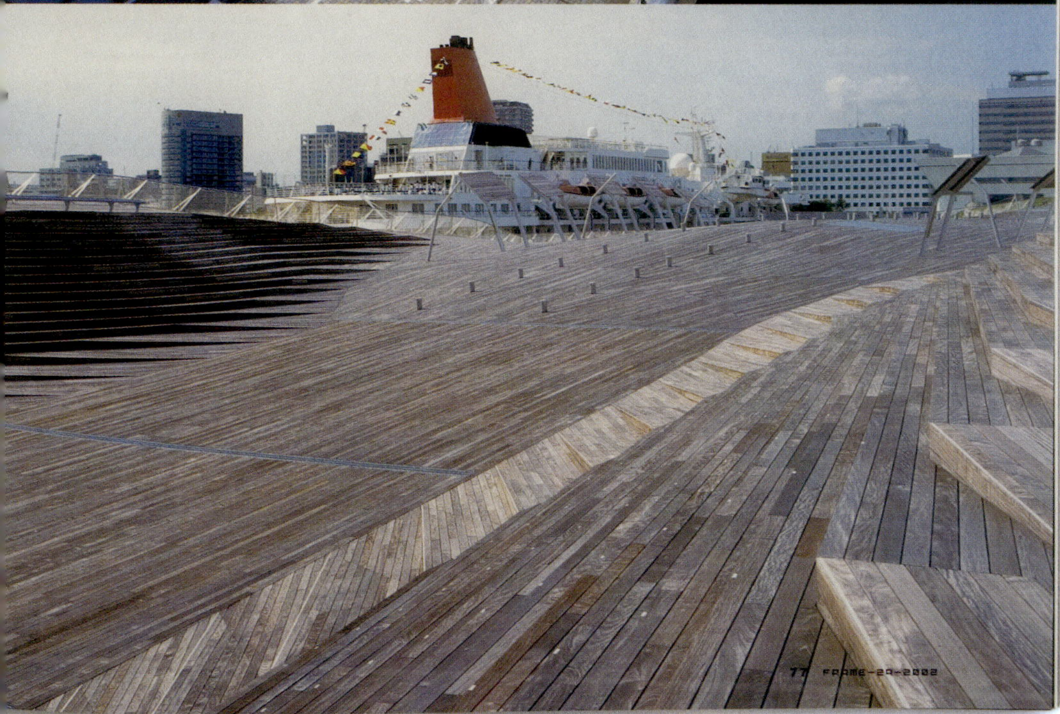

SUBJECT
Yokohama Port Terminal

LOCATION
Yokohama, Japan

DESIGNERS
Foreign Office Architects

PHOTOGRAPHER
Daichi Ano/Nacása & Partners

WRITER
Shigekazu Ohno

PEDESTRIANS ARE LEFT TO WONDER WHAT COULD LIE HIDDEN BENEATH THIS STRANGE LANDSCAPE OF SURFACES, SLOPES AND RAMPS MANIPULATED INTO AN ARTIFICIAL TOPOGRAPHY

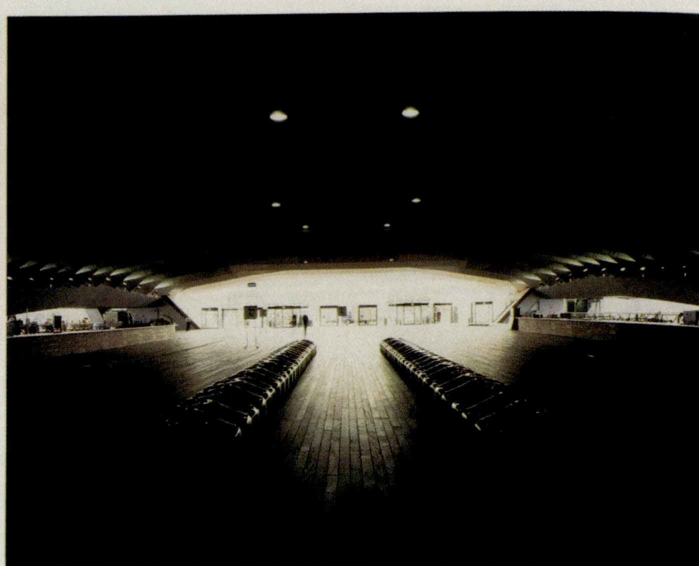

of traffic facilities (car and coach park, pick-up and drop-off plaza). Smoothly connecting the three levels – ground floor (traffic), first floor (facilities and amenities) and roof (park and observation platform) – are ten ramps and five glass lifts. The whole structure is a veritable groundscraper, 70 metres wide and 430 metres long, jutting out into Tokyo Bay.

Up on the roof, winding metal handrails border the looped circuit of gangways that meander across timber-planked decks. The route twists and turns as it journeys into steep-walled cuttings, runs through dark tunnels and emerges onto manicured lawns. Pedestrians are left to wonder what could lie hidden beneath this strange landscape of surfaces, slopes and ramps manipulated into an artificial topography. Inside and outside merge into one monolithic piece. At the same time, untreated timber planks, curving metal handrails and sea views bring to mind the deck of an ocean cruiser being tossed on a stormy sea. Rising from below, deep inside the building, is the echo of your footsteps.

Such sound effects are surpassed by the visual experience inside. The plate glass flanking both sides of the terminal has no frames. Each pane is fixed at the bottom only to withstand the impact of an earthquake, and the frameless fenestration opens up unobstructed views of the expanse of water outside.

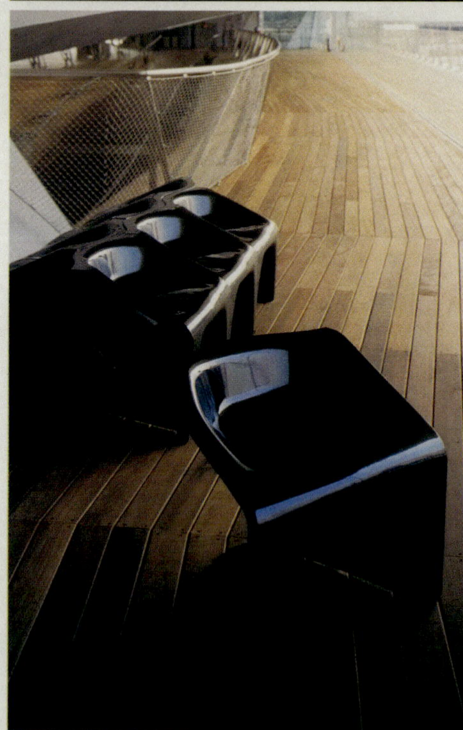

Above: The main level houses arrival and departure facilities, as well as meeting and waiting areas, restaurants and shops. The column-free space is spanned by steel trusses that rest on box girders.

Left: The fibre-reinforced plastic chairs on the deck were designed by Foreign Office Architects.

Opposite: Like the main level, the parking deck is spanned by an origami-like structure of trusses. The structure was fabricated in Korean shipyards.

SUBJECT
Yokohama Port Terminal

LOCATION
Yokohama, Japan

DESIGNERS
Foreign Office Architects

PHOTOGRAPHER
Daichi Ano/Nacása & Partners

WRITER
Shigekazu Ohno

FOA'S DESIGN FOR YOKOHAMA IS A PRIME EXAMPLE OF THE MUSCLE-FLEXING ARCHITECTURE THAT DEFINED THE '90S

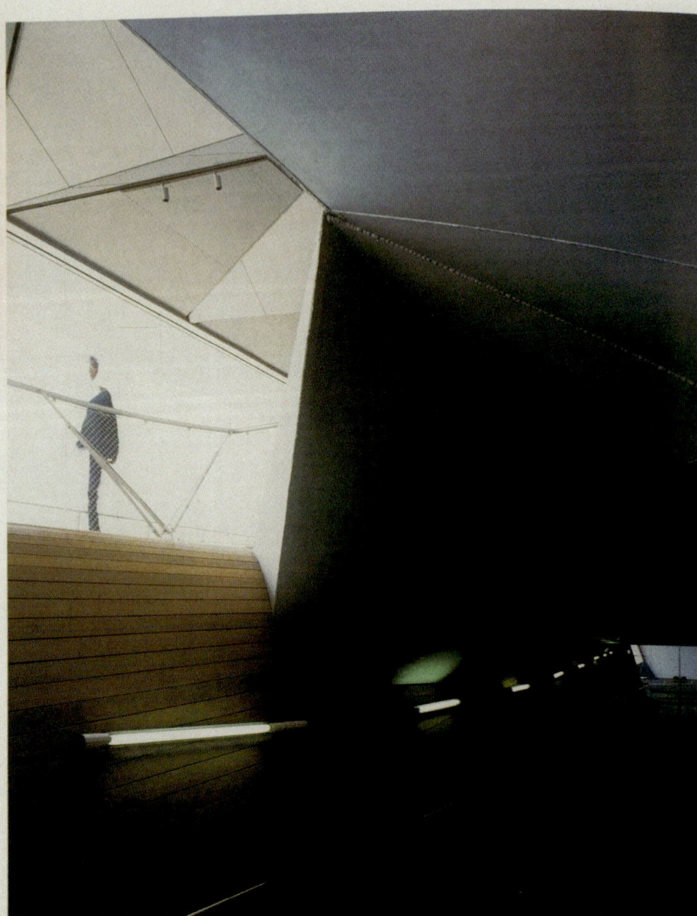

Spanning the column-free space are steel trusses that rest on box girders, a daring structure evocative of maritime construction and elaborated in collaboration with Structural Design Group. Focal points in the passenger halls include neat rows of Seven Chairs by Arne Jacobsen. Not to be outdone, FOA came up with a chair of their own, without legs. Inspired by Verner Panton, this cantilevered seat made of fibre-reinforced plastic is both organic and spatial, an object in perfect sync with the architectonic décor. The wavy pattern created by long rows of these chairs arranged back to back reminds me of an army of Darth Vaders frozen in space.

Even more remarkable, however, is the sheer otherworldly appearance of the entire complex. Figures like Gunther Domenig and Santiago Calatrava have previously shown that highly expressive architectural landmarks can capture the attention of architecture critics and general public alike. And whether the team at FOA would admit it or not, they are following the example of their architectural elders. The terminal is very much one of today's unprecedentedly popular 'look-at-me' buildings. And if towers, city halls and churches are entitled to stand out, then why not an seaport building? Here, however, the architects have carefully avoided a monumental, civic character. Zaera Polo and Moussavi have restricted their

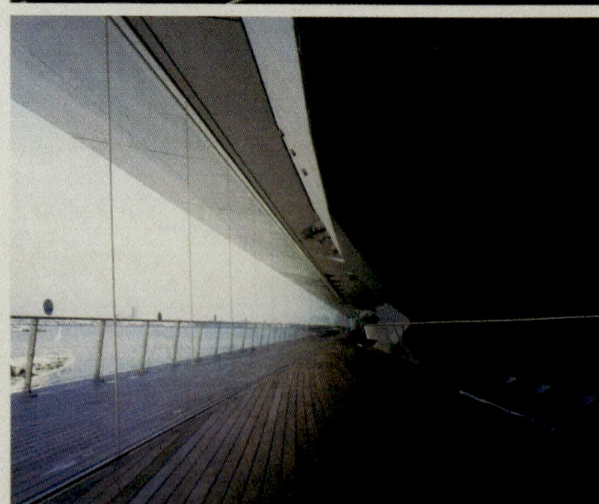

Above: Inclined gangway connecting two levels.

Left: Timber-planked gangways, elegant handrails and continuous views of the sea emphasise the nautical imagery. The palette of materials is limited to steel, glass and wood.

Opposite: Entrance to the terminal lobby. Lifts and shallow slopes facilitate movement between floors in a space without stairways.

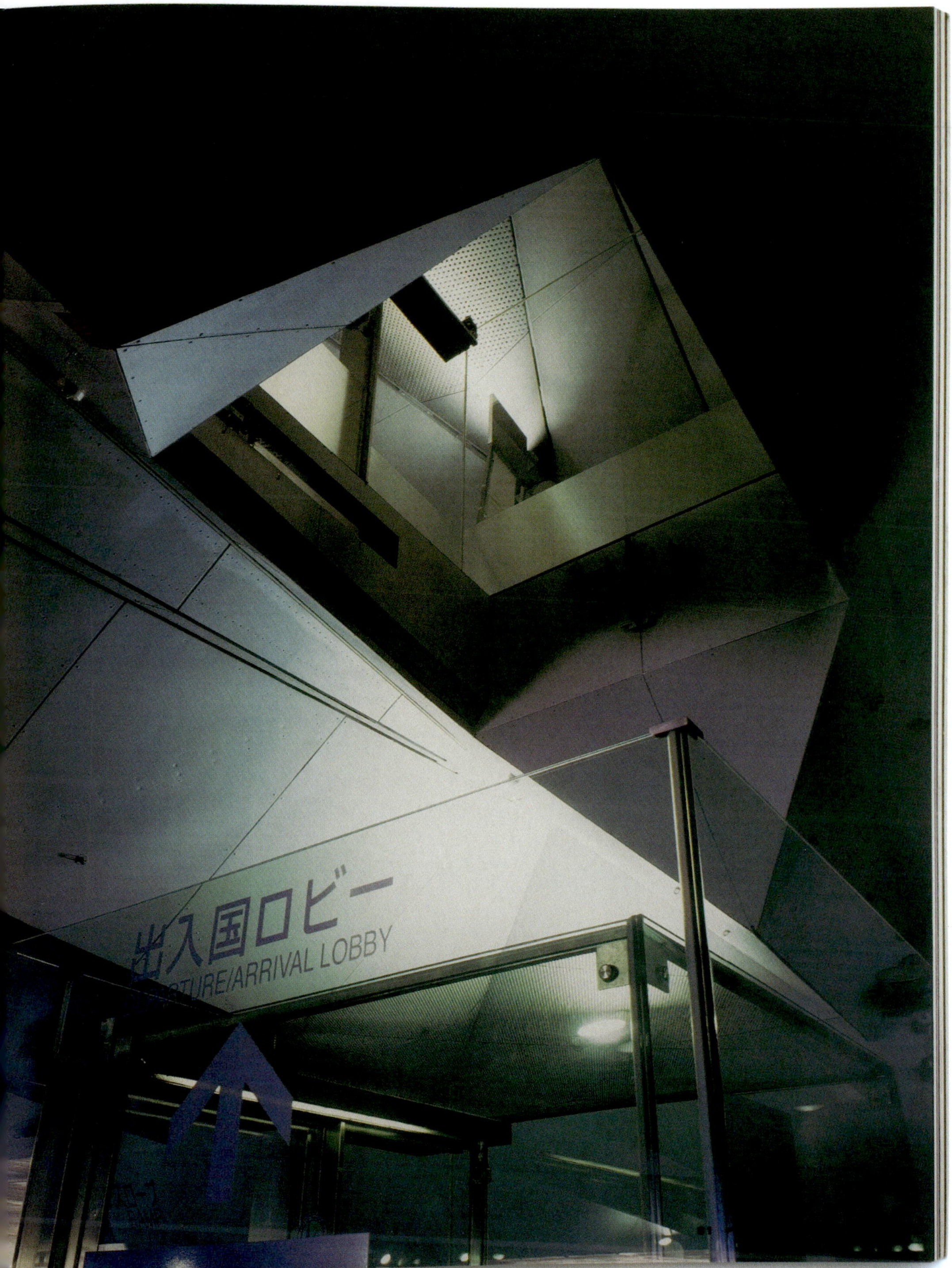

SUBJECT
Yokohama Port Terminal

LOCATION
Yokohama, Japan

DESIGNERS
Foreign Office Architects

PHOTOGRAPHER
Daichi Ano/Nacása & Partners

WRITER
Shigekazu Ohno

THIS TERMINAL IS A THEORETICAL EXPLORATION THAT GOT BUILT

palette to metal, wood and glass – conventional and unpretentious materials that help us keep our senses. In an effort to distance themselves from the preceding generation, they stress the more muted and less dramatic manner in which their work blends into its context. Point taken, but the terminal is still a prime example of the muscle-flexing architecture that defined the '90s. What's more, owing to the considerable lapse of time between design and eventual completion, the building bears the unmistakable stamp of the last decade.
And the transformation of the ground into an active surface was, theoretically speaking, yesterday's subject. This terminal, on the other hand, is a theoretical exploration that got built.

Unlike most other public facilities, the Yokohama Port Terminal is open round the clock, and locals have already embraced its roof deck as part of the public domain.
The structure might even do for Yokohama what Gehry's Guggenheim Museum has done for Bilbao, as both complexes owe much of their impact to the stark contrast with the surrounding high-rise sprawl. More than a functional work of engineering, the terminal is one of that new breed of architectural wonders, a city attraction in its own right.

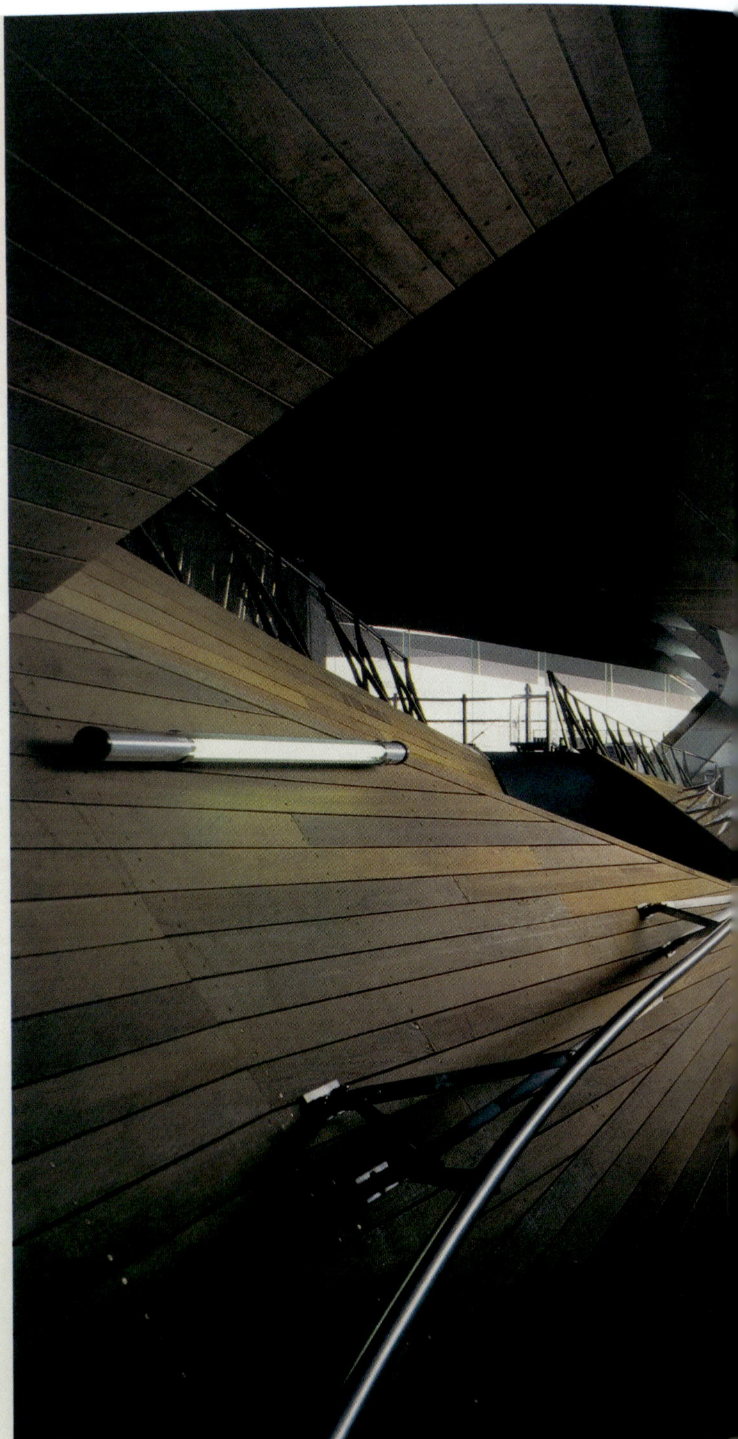

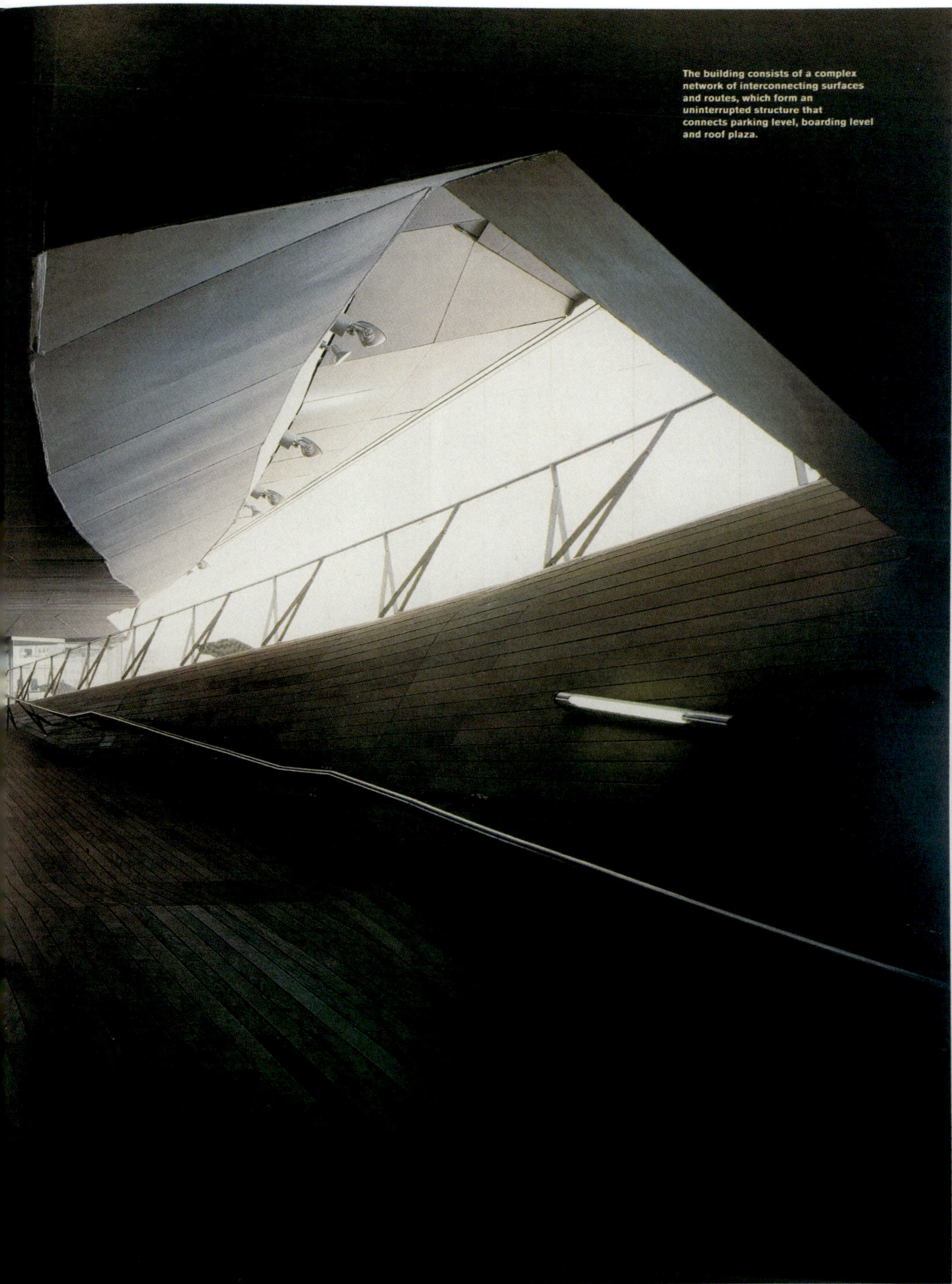

The building consists of a complex network of interconnecting surfaces and routes, which form an uninterrupted structure that connects parking level, boarding level and roof plaza.

SUBJECT
Yokohama Port Terminal

LOCATION
Yokohama, Japan

DESIGNERS
Foreign Office Architects

PHOTOGRAPHER
Daichi Ano/Nacása & Partners

WRITER
Shigekazu Ohno

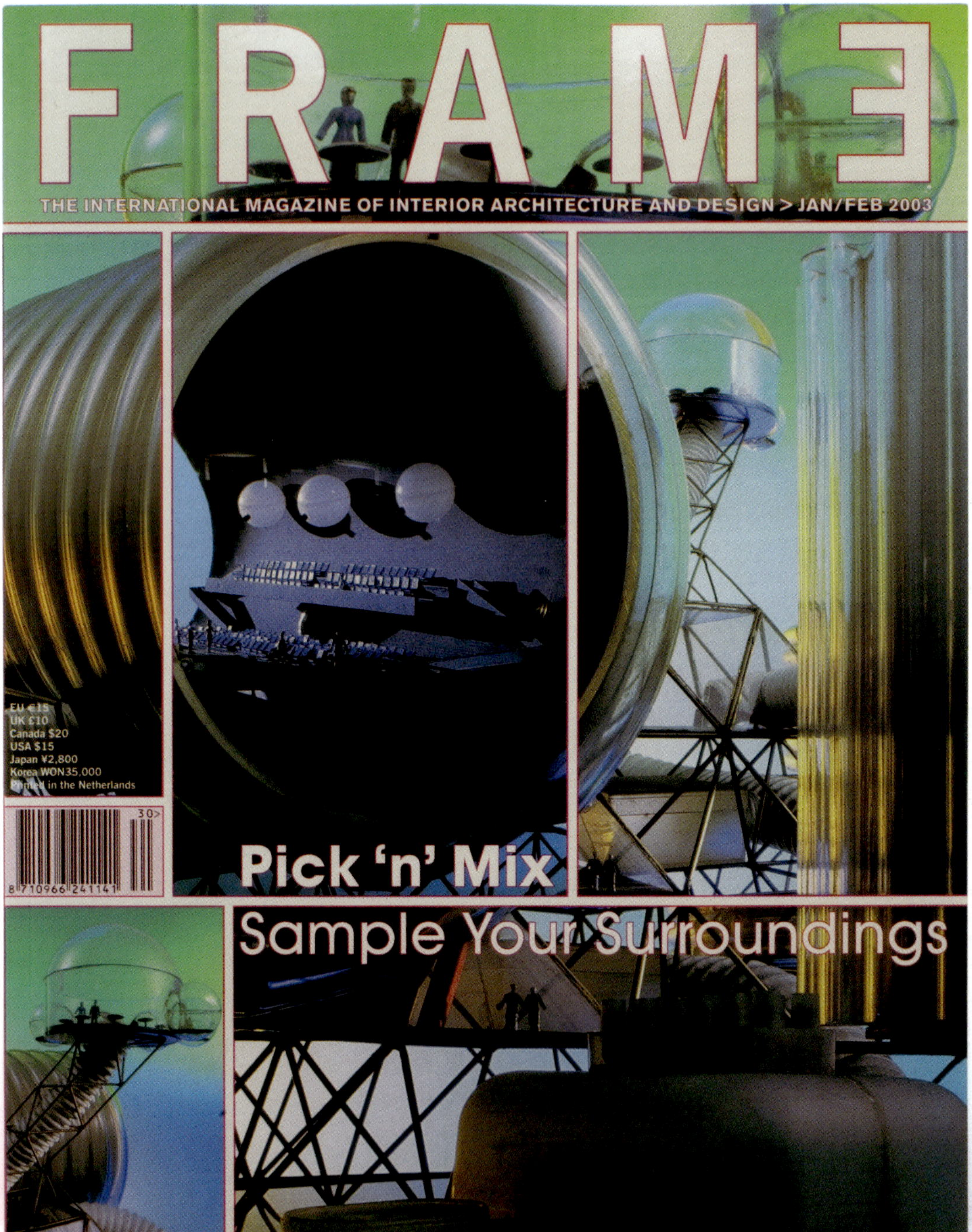

ART DIRECTORS
Cornelia Blatter & Marcel Hermans,
COMA

30

And Houses Will Grow Like Weeds

Frame editor Tessa Blokland about the project

Cornelia Blatter and Marcel Hermans, graphic designers working on their first issue of Frame at the time, suggested that we feature John Johansen. Even though the man was in his 70s, they were impressed by the contemporary look of his work. The futuristic models shown in the photographs are not only interesting, but also surprising in their use of materials. A close look at Johansen's 'buildings' reveals a fusion of ordinary items such as plastic milk jugs, vacuum hoses and pill bottles.

DESIGNER
John Johansen

SUBJECT
Portrait of
John Johansen

LOCATION
New York, USA

PHOTOGRAPHER
Michael Moran
www.moranstudio.com

WRITER
Shonquis Moreno

And Houses Will Grow Like Weeds

At 85, American architect John Johansen takes the concept of morphing to another level. Can we genetically engineer an inhabitable plant?

Text by Shonquis Moreno
Photography by Michael Moran

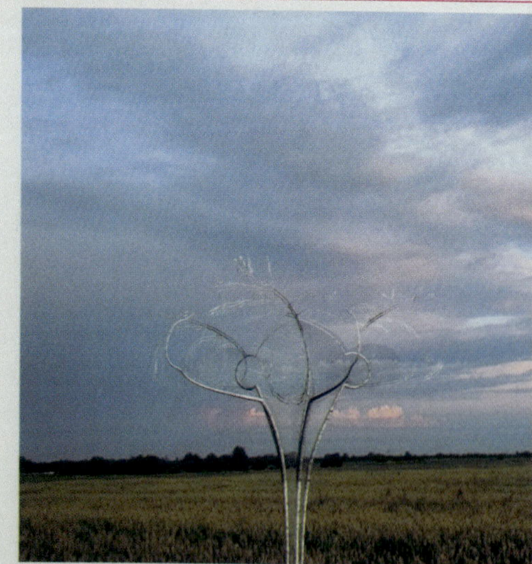

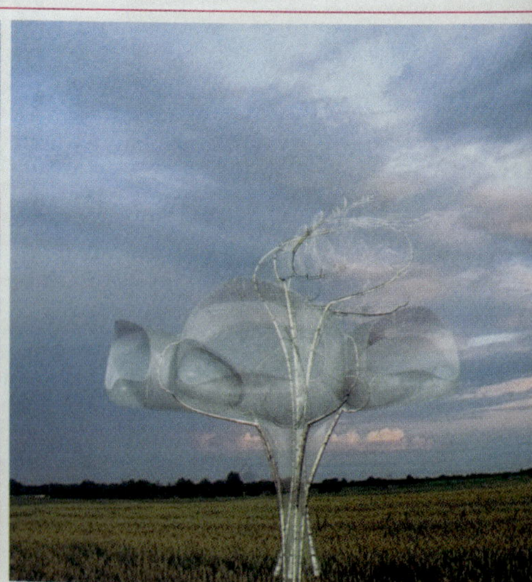

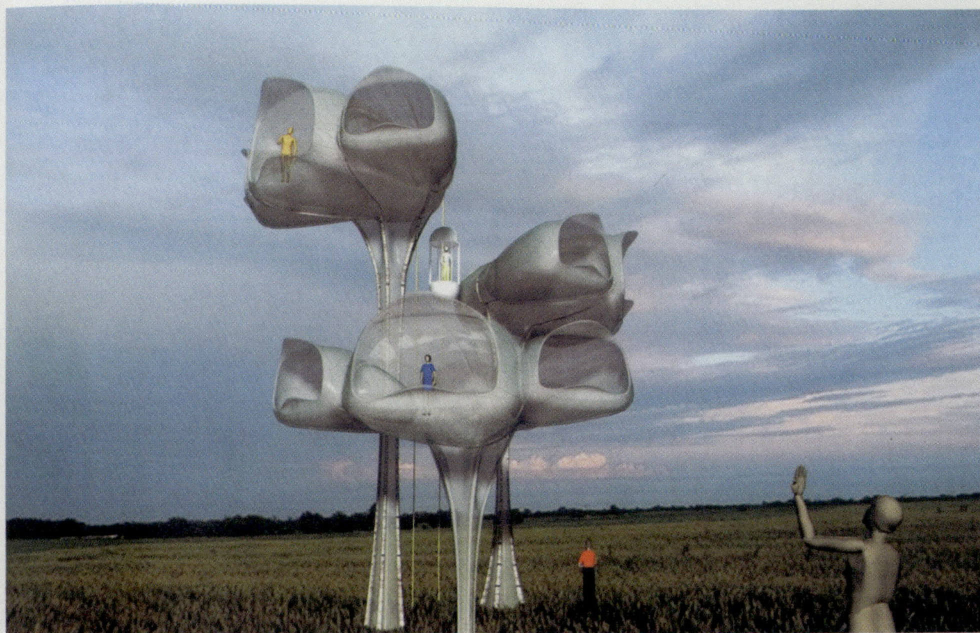

Multistory Apartment Building, 2001.
This project illustrates Johansen's thinking on the molecular engineering of complex buildings. He predicts that organisms – complete with roots, trunks, branches, platforms, membranes and pores – will one day grow from vats located at building sites. Renderings by Marcel de Winter and Dan Stoica

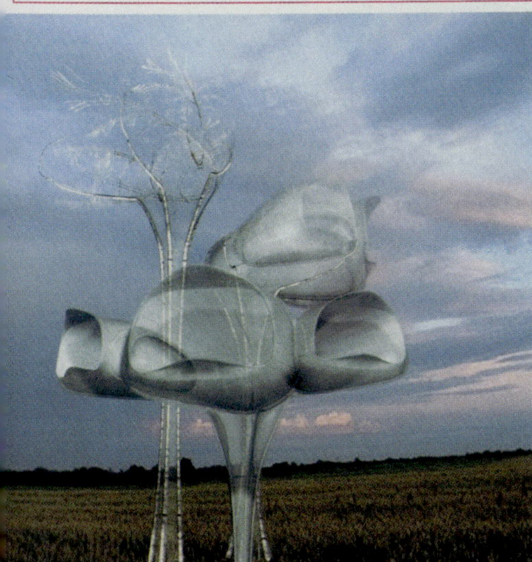

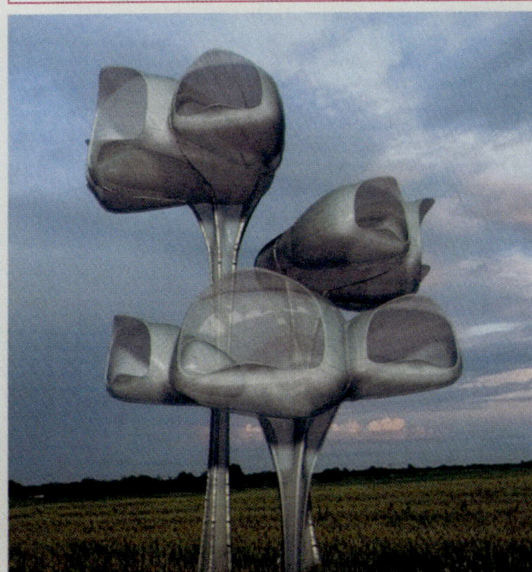

SUBJECT Portrait of John Johansen
LOCATION New York, USA
DESIGNER John Johansen
PHOTOGRAPHER Michael Moran
WRITER Shonquis Moreno

'Permutations,' says architect John Johansen slowly, and with obvious delight. 'A beautiful word. That's life, you see. Change. Unpredictability. That excites me.' What he doesn't want to hear, he continues, are sentiments like 'This is a beautiful building. Everything matches. Everything balances. That's false,' he says. 'That's what we *dream* of, not what we live.'

Johansen is a builder who yearns to balance the dream with the reality. One day, according to the physicists and engineers he's consulted, blueprints on paper will no longer exist. The alternative will take the form of code programmed into vats of liquid chemicals – intelligent cesspools from which houses will sprout like kudzu vines. In October, Princeton Architectural Press published a series of a dozen experimental projects designed by Johansen since he retired from professional activities 15 years ago. *Nanoarchitecture: A New Species of Architecture* represents Johansen's imagination unbound from practice.

Illustrating emerging building technologies from the nearly here to the speculative, Johansen's experiments exploit thin fibreglass shells currently used to build large boat hulls, kinetic structures developed by NASA for use in space, electromagnetics and, most spectacularly, molecular engineering. The architect's molecular-engineered house and a larger-scale apartment complex are the most radical and conjectural projects in the book. Their designs are based on nanotechnology, the science of building electronic circuits and devices from single atoms and molecules. It is the basis of a hypothetical manufacturing method in which objects are designed and built through the individual specification and placement of each atom. The first nanofabrication experiments took place in 1990. According to a statement made by physicist Richard Feynman 45 years earlier, however, nanotechnology has the potential to produce almost any chemically stable structure desired.

Johansen would programme the specifications for his molecular-engineered house into chemicals, place the chemicals in vats, and leave them to grow into a home. The root evolves into the foundation, membranes become walls and floors, and pores become windows. Control of lights, repairs and cleaning, though coded from the start, can be altered. The 'architect' can regulate growth to develop in stages, a few storeys at a time, or to pause and begin again later, as needed. In the apartment building, each unit can be coded and grown in a separate vat to produce flats with custom-designed interiors and cultivated-to-order layouts, lighting, textiles and furniture. It may sound like a sci-fi fantasy, but Johansen insists that the technology will be possible within a few generations. Not only possible, he argues, but inevitable.

'I'm trying to get away from the stolid, static, heavy aspect of buildings. I'm delighted that the molecular-engineered house is light. It's what I've been working toward my whole career, moving from heavy to light and from static to kinetic.' His work has also evolved from mute design to intelligent, cybernetic architecture.

'John Johansen is arguably the most experimental architect amongst his peers, at least amongst the American architects of his generation,' says Professor Haresh Lalvani, co-director of the School of Architec-

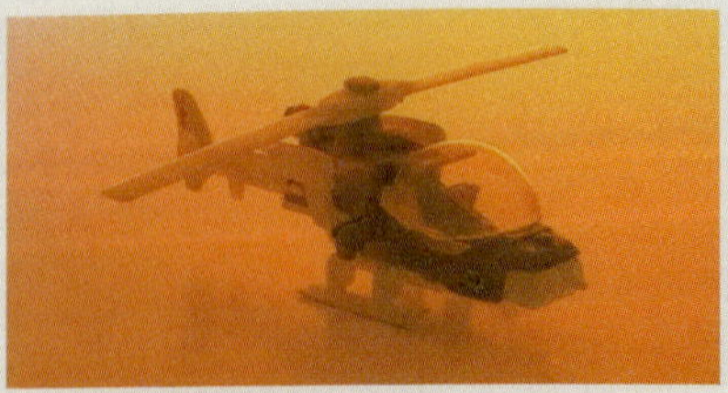

Froth of Bubbles, 1988.
Johansen's idea was to graft this conference centre onto the roof of another building. Platforms transported by helicopter are attached to a telescopic mast and enclosed within huge bubbles called 'living membranes'.

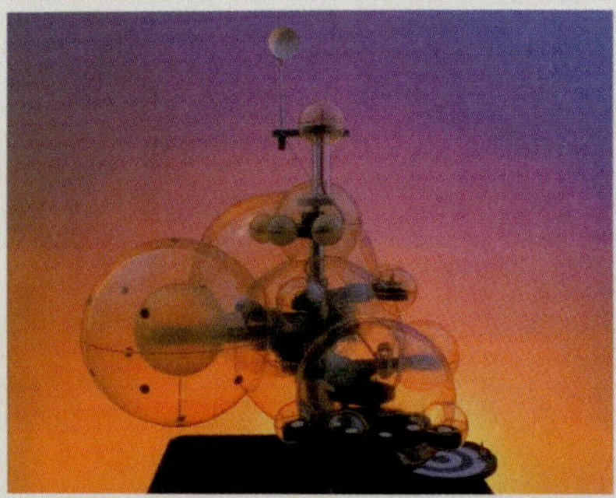

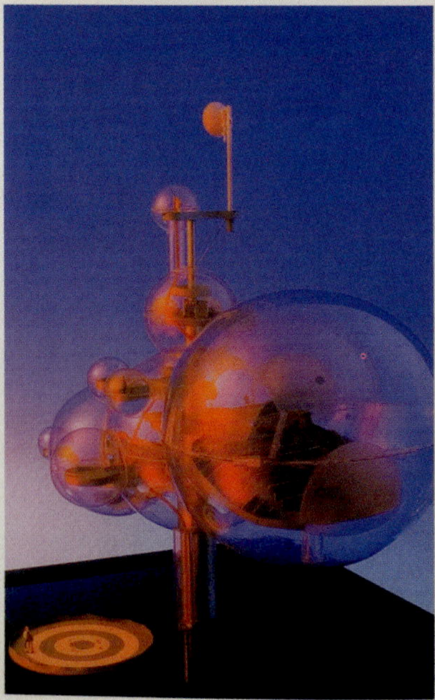

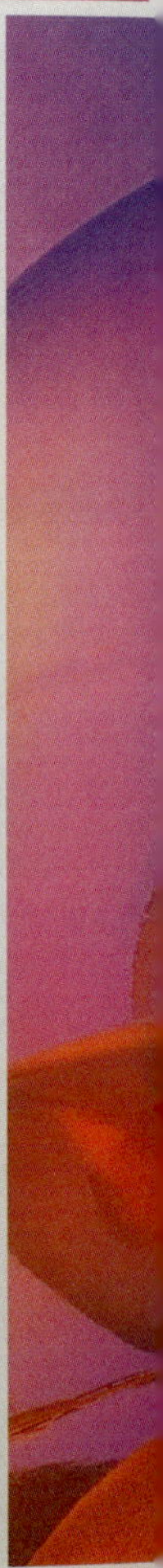

'Making things is a failure of architecture; you have to really find beautifully formed products and put them together'

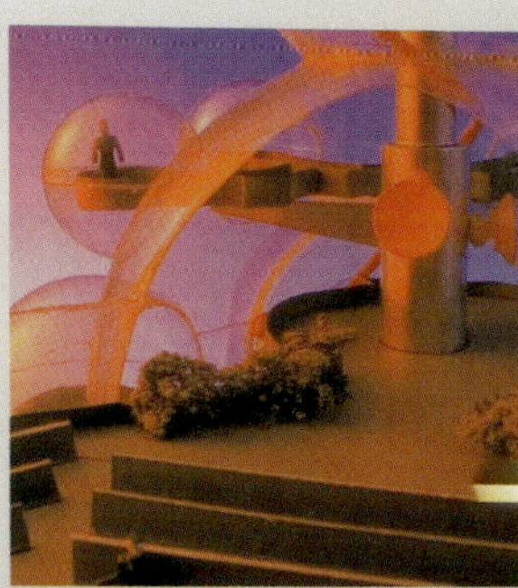

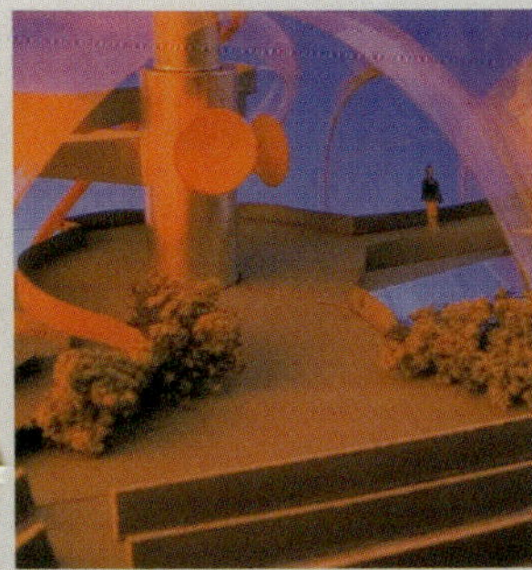

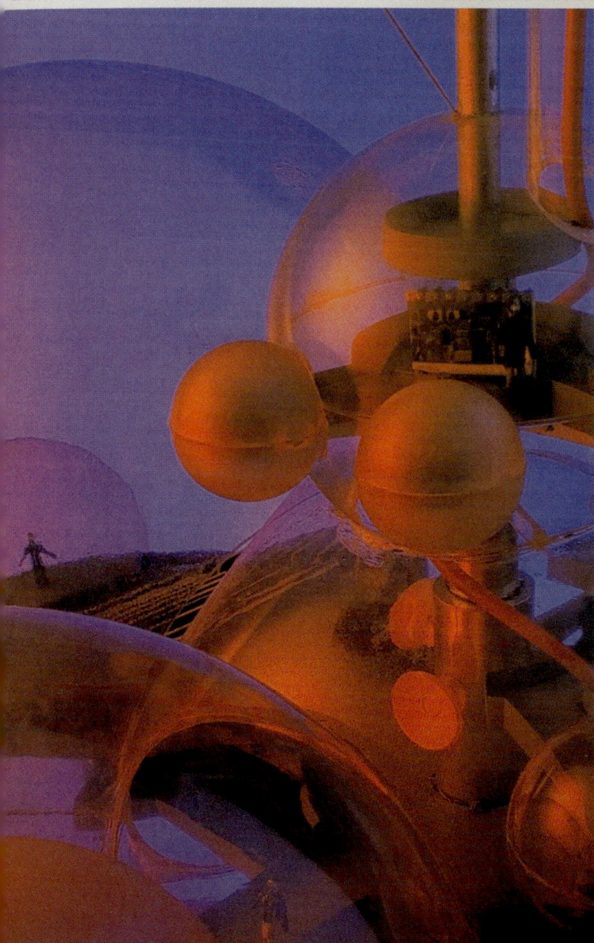

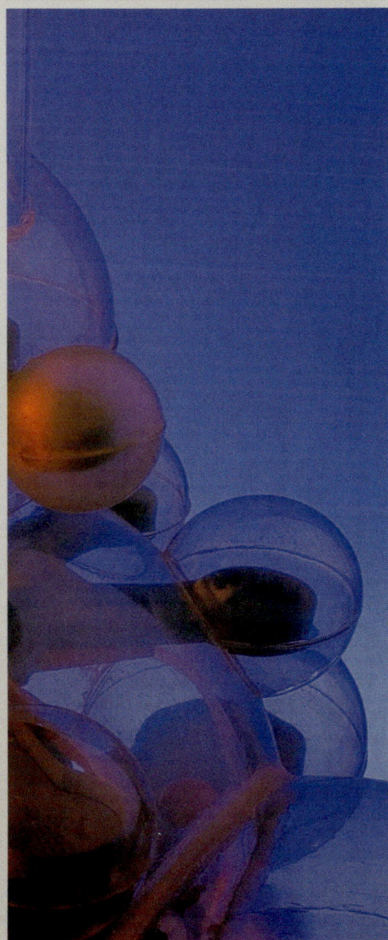

SUBJECT Portrait of John Johansen
LOCATION New York, USA
DESIGNER John Johansen
PHOTOGRAPHER Michael Moran
WRITER Shonquis Moreno

ture's Center for Experimental Structures at the Pratt Institute in New York City. Lalvani collaborated with Johansen on a project based on the geometry of a 'space labyrinth.' 'Johansen considers architecture first and foremost as "hardware",' Lalvani points out, 'and he constructs other layers of architecture around this. His architecture is shaped by building technology and by the way things are made.'

At 85, Johansen is tapering and lightening, his hair gone hoary white and his hearing gone dull. His voice, however, remains rich and firm. Having taught at Pratt for over 50 years, he speaks in the measured and thoughtful tones of a teacher. Following retirement, he says, 'I had the first opportunity to deal with pure ideas not compromised by the travails of architectural practice, lack of budget and so forth. I sat myself down. I gave myself a week. I designed 12 projects in one week. They were clumsy statements, but they dealt with distinct building technologies. Some were discarded, others added. It took me ten years to develop these more fully – approximately one per year.' Models were made from drawings done by my Pratt students.'

Johansen's Mag-Lev Theater comprises three lightweight, inflated tubes framed in steel and resting on the roof of another building. Magnetic levitation allows scenery, lighting, projections, performers and even the audience to be moved in various directions on demand. The forces of attraction and repulsion – applied through walls, floors, ceiling and the platforms themselves – enable stages or seating sections to float and move. The Metamorphic Capsule is a

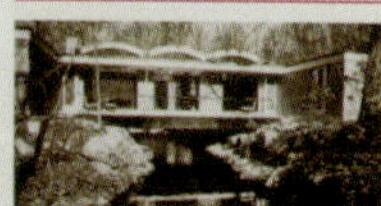

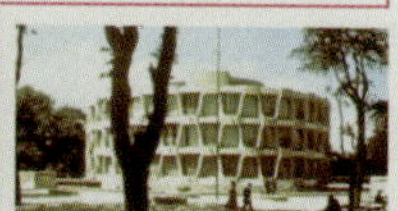

Above left: Villa Ponte (Warner House), New Canaan, Connecticut, 1957.
Above right: US Embassy, Dublin, 1963.

morphing bubble sustained by continuous internal air pressure that, using electromagnetic stimuli, can change form, opacity and colour. Power fed to various nodes alters the shape of the capsule and can even make it undulate. The whole thing can be controlled by a handheld device from within and without, or programmed remotely.

Johansen's students at Pratt built the models, under their professor's supervision, from 6-ounce plastic jugs, vacuum hoses and pill bottles. 'If you get close, you can read the Bayer aspirin label,' says Johansen. 'These materials are exactly what I want. Making things is a failure of architecture; you have to find beautifully formed products and put them together. I spend an awful lot of time on Canal Street, pawing over this stuff.' The Web project, a conference centre intended for suspension between two other buildings, is knit together from water and milk containers, the kind that come with a handle and store well for camping or emergency situations. Johansen likes to remind people that the plastic, only 1/32 of an

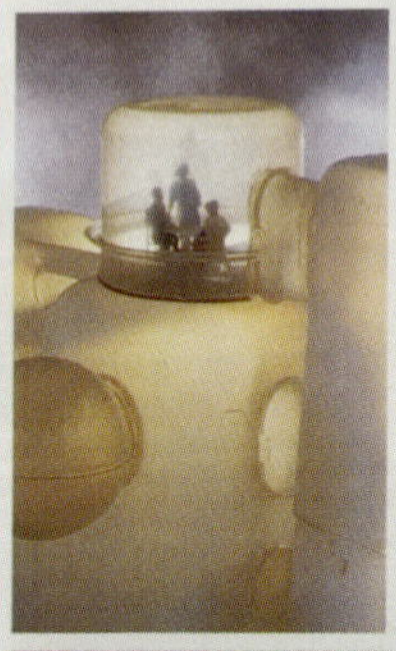

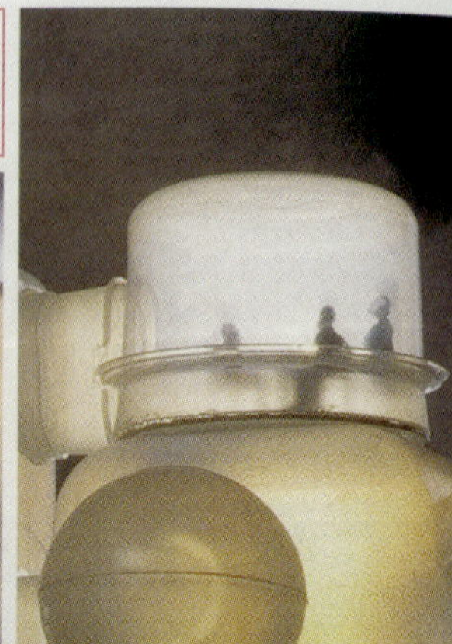

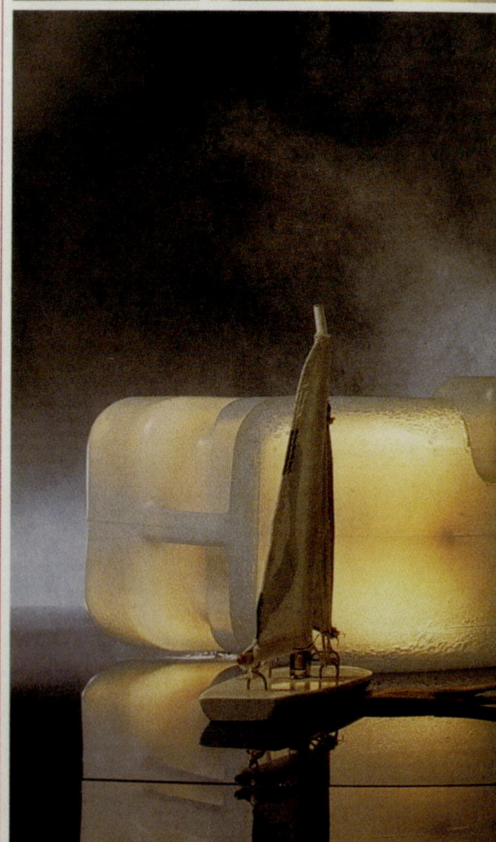

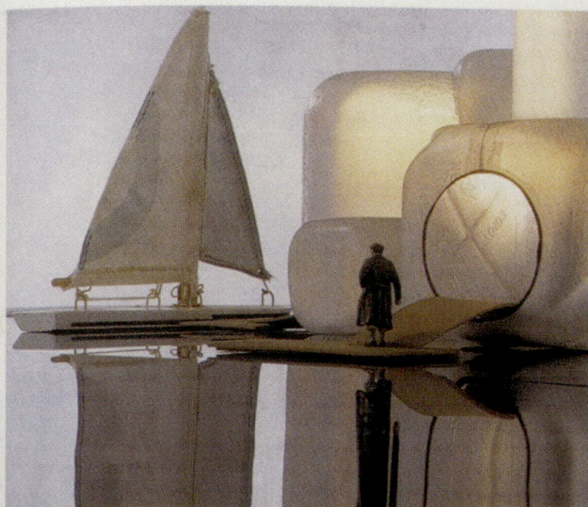

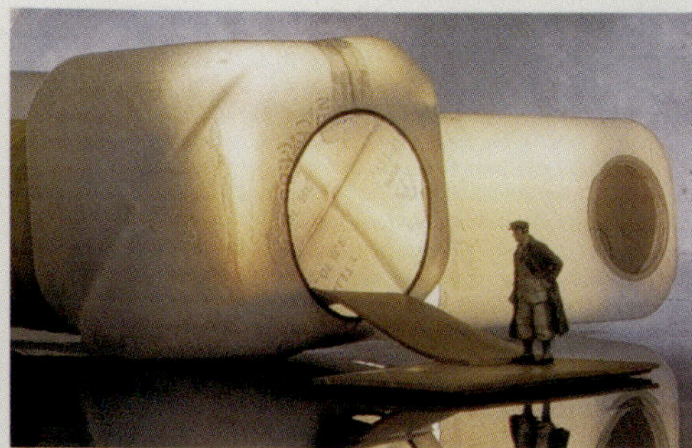

Floating Conference Center, 1997.
Here Johansen investigates the architectonic applications of fibreglass, a light but strong material capable of conveying a highly distinctive aesthetic. To make his story even more interesting, Johansen built the model in this illustration with plastic milk jugs.

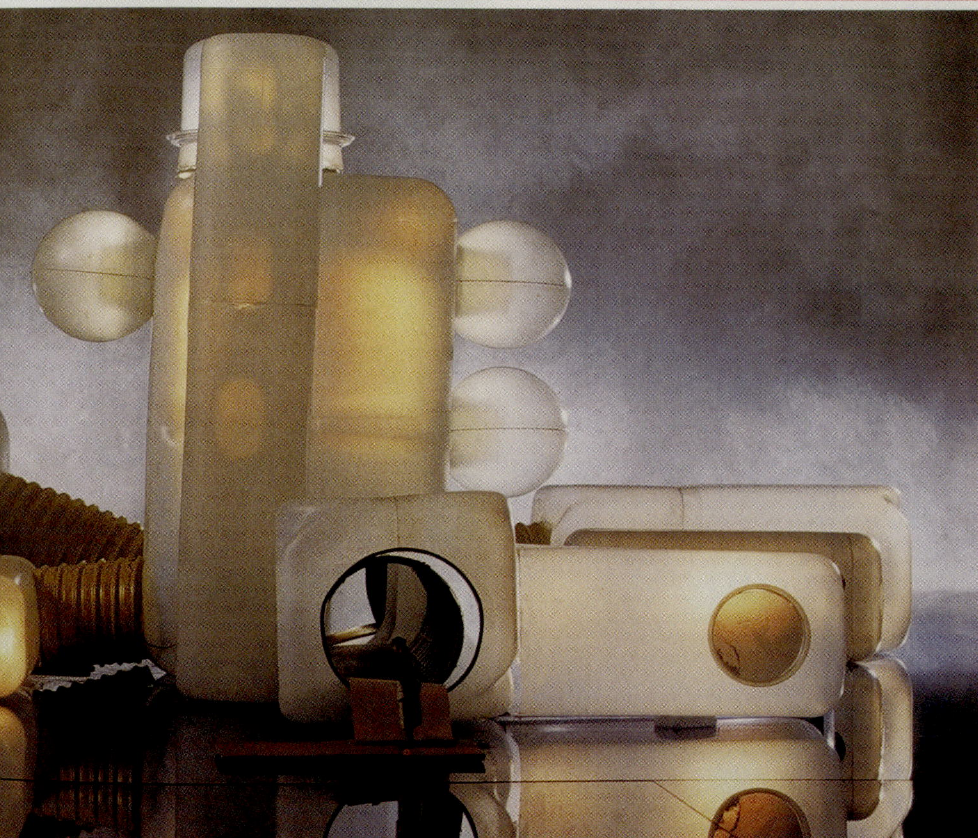

SUBJECT Portrait of John Johansen

LOCATION New York, USA

DESIGNER John Johansen

PHOTOGRAPHER Michael Moran

WRITER Shonquis Moreno

inch thick, can hold up to 50 pounds of liquid. He relishes this sort of mundane functional triumph.

The Floating House and the Floating Conference Center boast curving walls up to 175 feet high that rely on the thin fibreglass shell technology of boat construction. These slim, translucent walls encourage shifting rays of natural light to play quixotically over the interior. After dark, the house glows like a gargantuan lily from within, its petals folded for the night, its reflection on the water like a second moon. The two-storey home features a spiralling central staircase, a still pool in the public space at ground level, a roof deck, built-in balconies and built-in furniture sculpted from the same luminous, curving plastic used for the exterior. The undulant, organic and unfolding nature of Johansen's floating projects is reminiscent of his famous, unrealised, mid-1950s Spray Houses, which looked like spineless sea anemones. These were made of thin-shelled concrete sprayed onto a frame of steel rods and fastened together in one place.

Both eclecticism and experimentation have marked Johansen's career. 'I've skipped around,' he admits with some pride. 'I don't copy myself.' Born in 1916 to two New York City painters, Johansen graduated with Paul Rudolph, Philip Johnson and I.M. Pei in 1942 from Walter Gropius's Bauhaus-influenced Harvard Graduate School of Design. He spent the rest of the war years building wooden navy barracks. After World War II, Johansen worked for Marcel Breuer and, later, for Skidmore, Owings & Merrill. Even while designing and building the neoclassic Villa Ponte, which gracefully spanned a stream, and the neoclassic U.S. Embassy in Dublin, a drum embellished with twisting helixes of concrete and completed in 1962, Johansen had embarked on his sprayed concrete work. By the mid-'60s, however, he was creating brutalist buildings: heavy, modular, rational *and* chaotic. The Mummers Theater in Oklahoma was not only Johansen's favourite; it was a breakthrough and a clear influence on the Pompidou Centre built several years later in Paris. The complex in Oklahoma consists of metal-sided 'boxcars' of varying sizes, connected by tubes and vividly painted. 'This is an explosion. There's no composition here any more,' says Johansen of Mummers, as if the design surprises even him.

'What was wonderful about the Mummers building,' says Michael Webb, one of the founders of Archigram, 'was that it showed a willingness to depart from what most architects considered suitable material – concrete, glass, bricks – to use metal wall cladding and paint it bright colours. What distinguished Johansen was his willingness to adopt the ideas that were very current in the younger generation and that other practising American architects had no interest in – ideas that had to do with flexibility and expendability.'

In 1985 Johansen showed how his (unrealised) Miami Beach Hotel could be reconfigured using a crane; the method described strongly resembled the one applied to his own home, built in 1972. Probably influenced by Peter Cook's 1963 Plug-In City, both of Johansen's dramatically pyramidal structures feature 'attachment points' on a steel frame, to which rooms can be anchored and platforms added or removed.

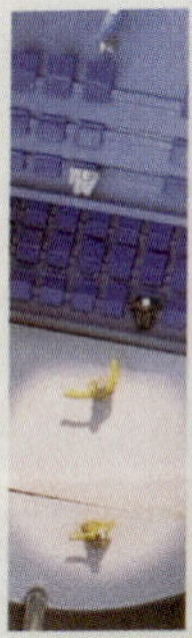

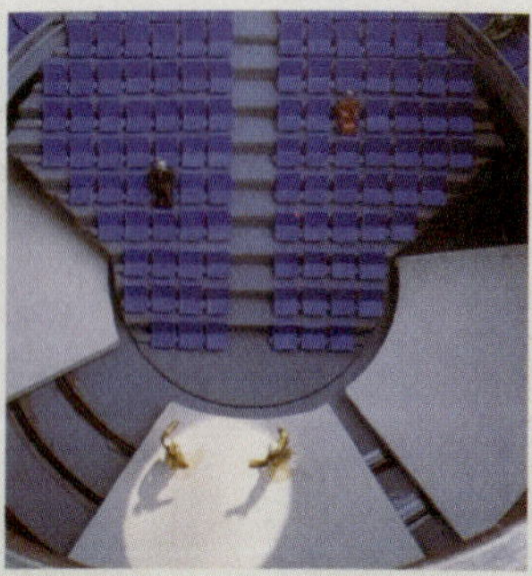

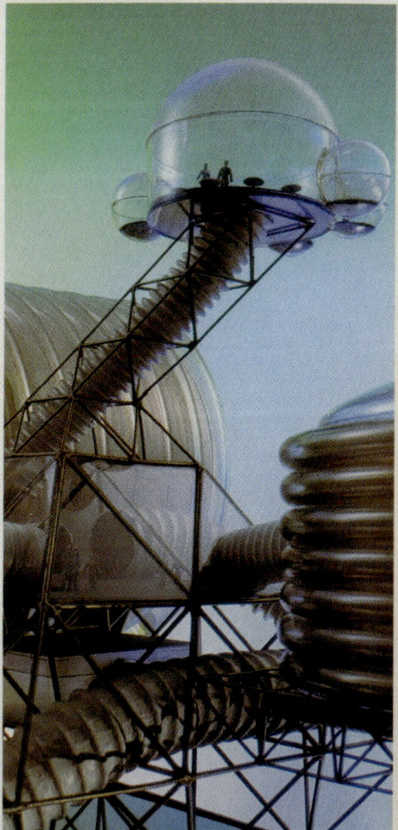

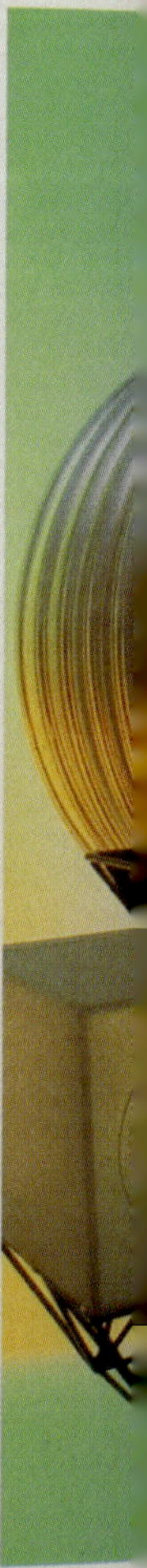

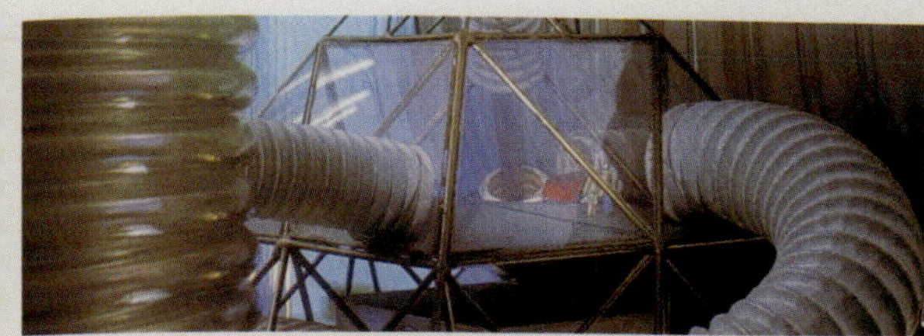

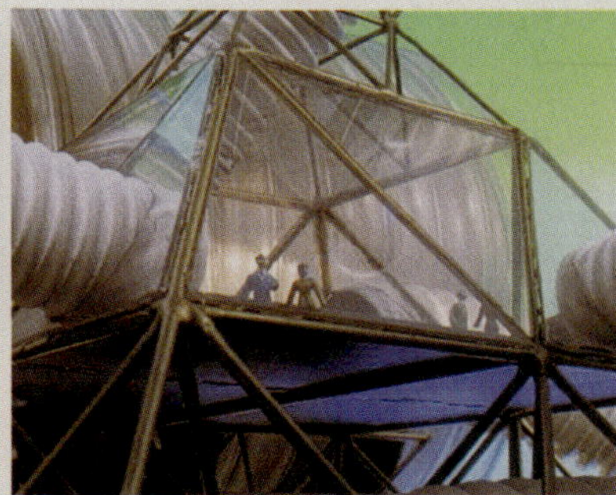

Mag-Lev Theater, 1990.
This theatre complex is based on the use of electromagnetic levitation (mag+lev). It consists of three light, inflatable structures framed in steel and resting on the roof of another building. Maglev technology can be used to lift an entire stage – or the seated audience – to give theatrical productions an unprecedented spatial dimension.

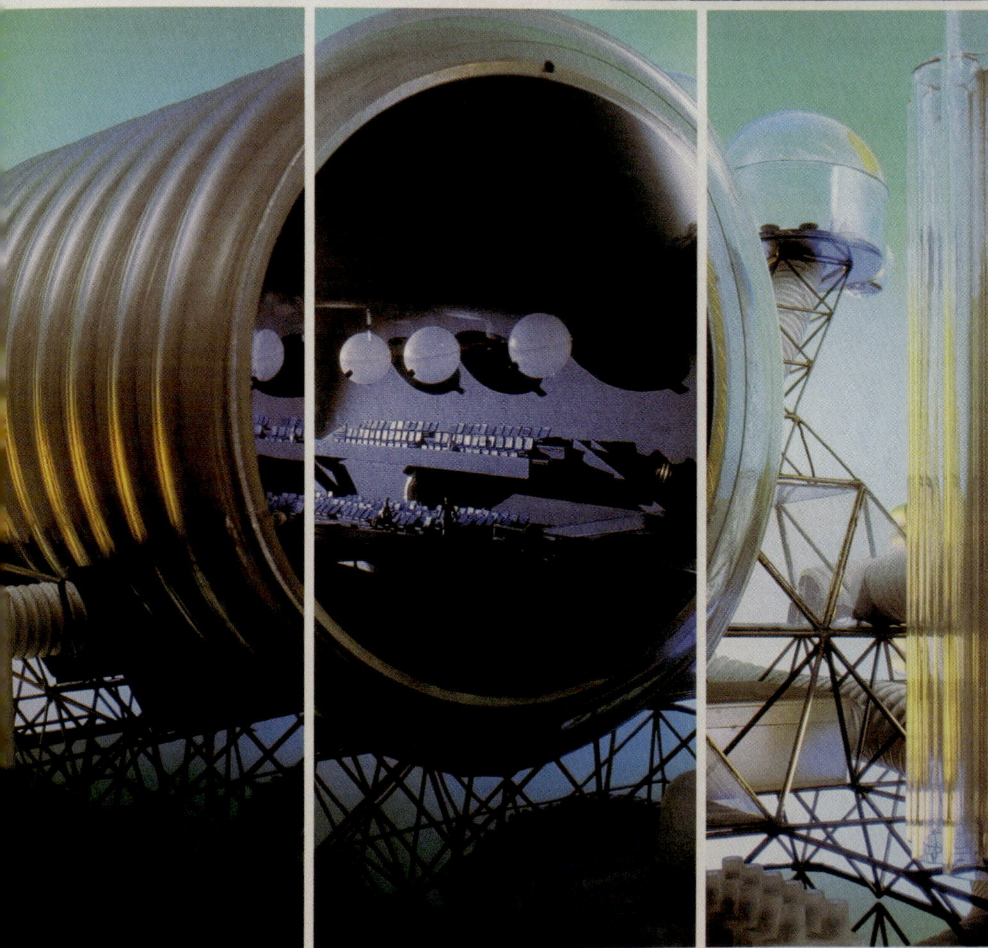

SUBJECT
Portrait of John Johansen

LOCATION
New York, USA

DESIGNER
John Johansen

PHOTOGRAPHER
Michael Moran

WRITER
Shonquis Moreno

Johansen's nanoarchitecture takes reconfiguration or 'morphability' to another level. 'Unlike the introduction of other materials, like concrete in the 1910s and 1920s, so little is known about nanotechnology,' says Webb. 'There's nothing we know about the material that would indicate an architectural application. The only limitation on what you can do with it right now is the limitation of your imagination. If such an application is attainable, the effect on architecture would be so radical that you wouldn't be able to talk about entities like apartment high-rises and bridges. How possible is this? I have no idea. It's all wonderfully crazy. The concept does resemble the growth of a plant, but can you genetically engineer a plant so that you can live in it? And how long would it take? These are questions that no one can answer at the moment. I see the book very much as the beginning of a long and comprehensive study. I'm interested in it not because it exists, but because of what it could lead to.'

Today, Johansen is attempting to increase the scale of his molecular-engineered structures by designing a cantilevered bridge, a museum and a small community on Roosevelt Island, New York. 'I have a strong belief that real changes in history occur only when building technology changes,' he says. 'Right now, I think we're waiting for a new technology, and people really don't know where to go. They're styling the whole thing – trying to style their way into the future. And that, of course, is a sort of malady.'

Johansen thoroughly enjoys technology. 'There are many reasons to resent technology and the evil aspects of it, the destructive aspects. There's always a resistance to technology and its movement. This has been true throughout history, and it has to be respected. People don't like to move fast. I do, however. I speak of "the poetics of technology" to put the subject in a different cast. Technology can be poetic. I delight in it, and I'm comfortable in it, and I'm happy with it.'

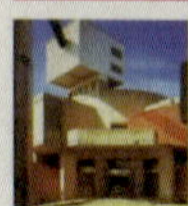
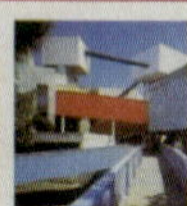
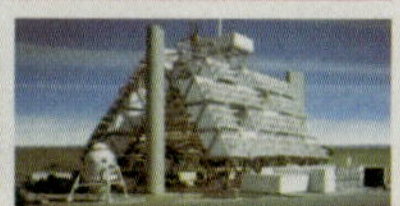

Above left and centre: Mummers Theater, Oklahoma, 1970.
Above right: Miami Beach Resort Hotel, 1985 (unrealised).

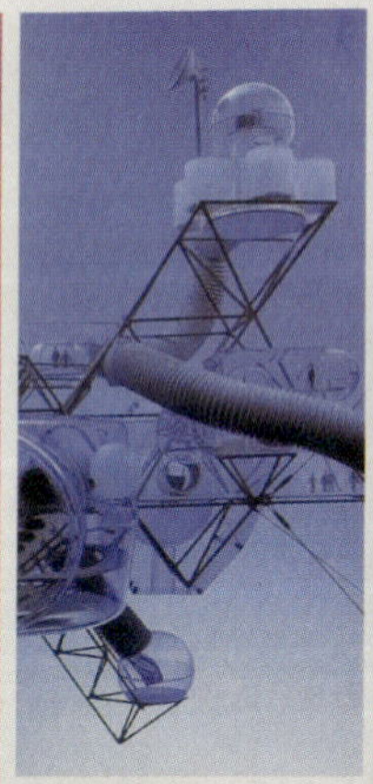

Johansen is a builder who yearns to balance the dream with the reality

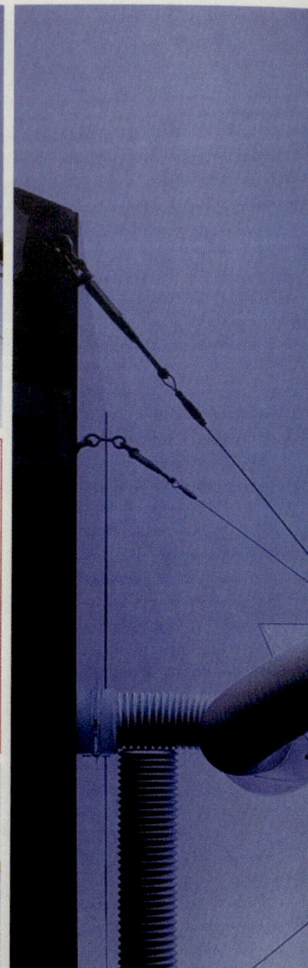
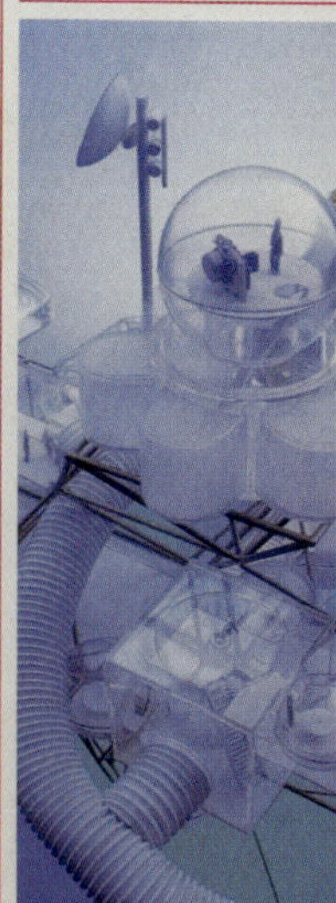
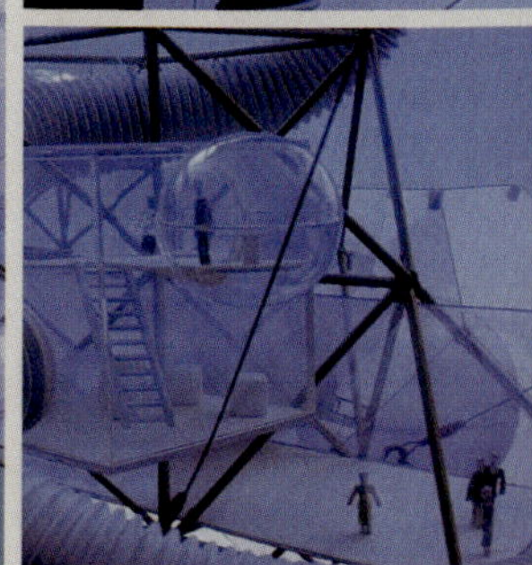

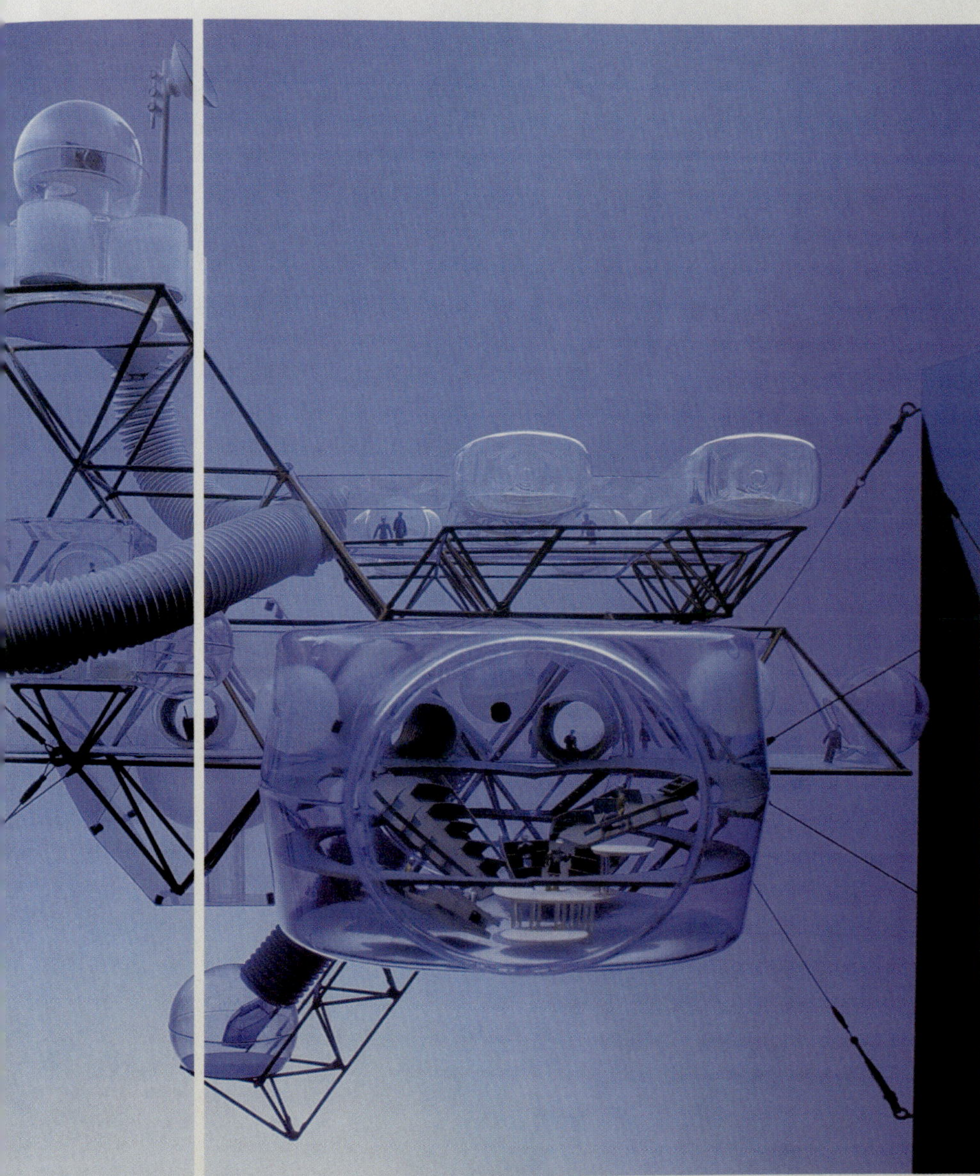

The Web, 1989.
The construction of a spider web inspired this project. Provided with a self-balancing system, the structure can be suspended between two tall buildings or under a bridge. Johansen likes the idea of the Web holding together a group of people with a common goal.

SUBJECT
Portrait of John Johansen

LOCATION
New York, USA

DESIGNER
John Johansen

PHOTOGRAPHER
Michael Moran

WRITER
Shonquis Moreno

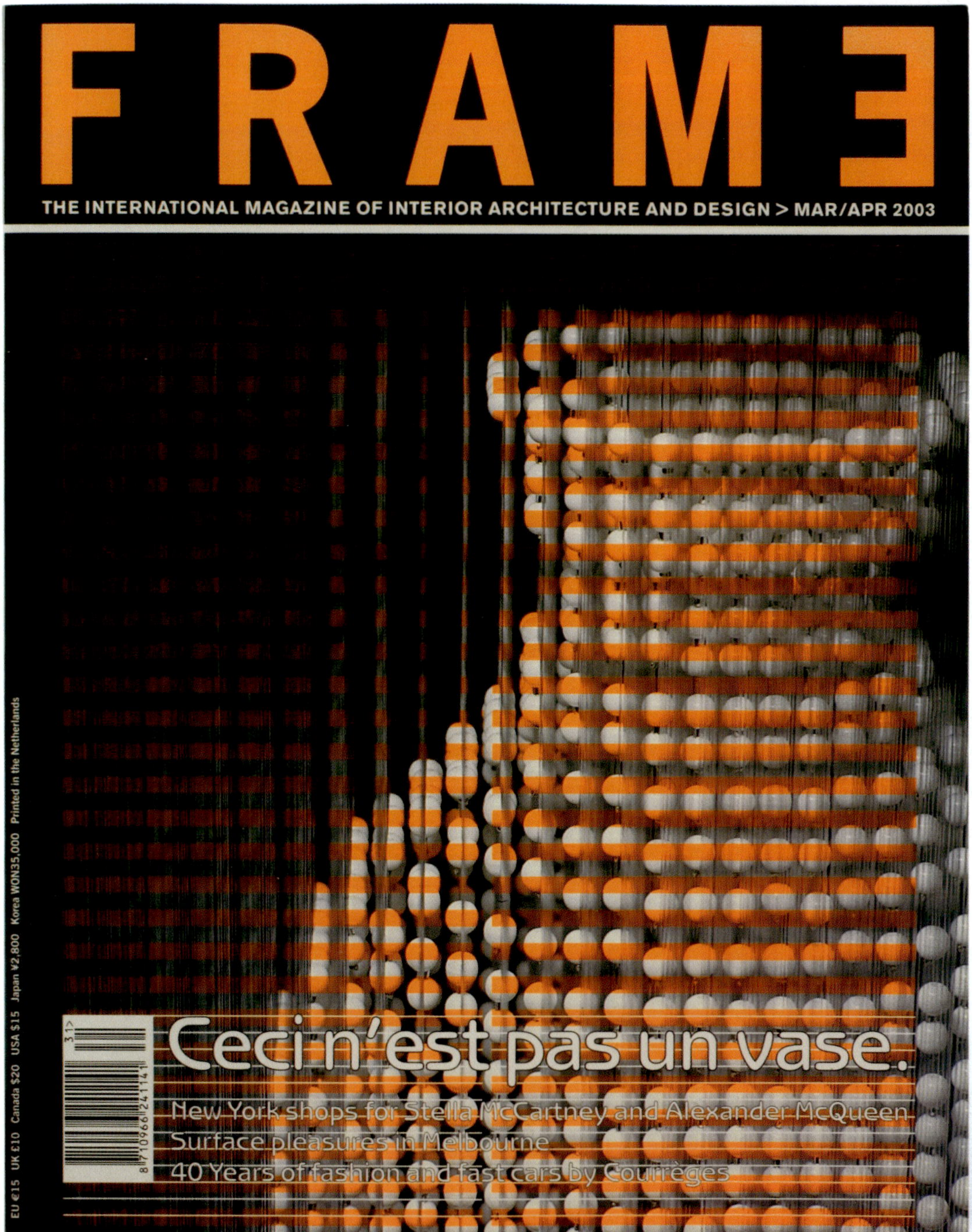

ART DIRECTORS
Cornelia Blatter & Marcel Hermans,
COMA

31

Forever Young

Designer Lucas Maassen (Untitled Nations) about the project

I think it is a clean example of an 'until recently un-explored reality'.

Comments on the published article

It explains the project in a clear and simple way.

Did the article have an effect?

The photo has been published in a series of magazines and books.

Recent news or information about the project featured in *Frame*

The original negative got lost by a London publisher, so there is only one copy left (still for sale).

Current projects

I am working on a new design platform called 'Untitled Nations' for more un-explored and un-developed realities. And I work as a freelance designer for all sorts of clients.

DESIGNER
Lucas Maassen
www.lucasmaassen.com

SUBJECT
Visual essay of the conflicts of proportion

LOCATION
Eindhoven, Netherlands

PHOTOGRAPHER
Lucas Maassen
www.lucasmaassen.com

WRITER
Louise Schouwenberg

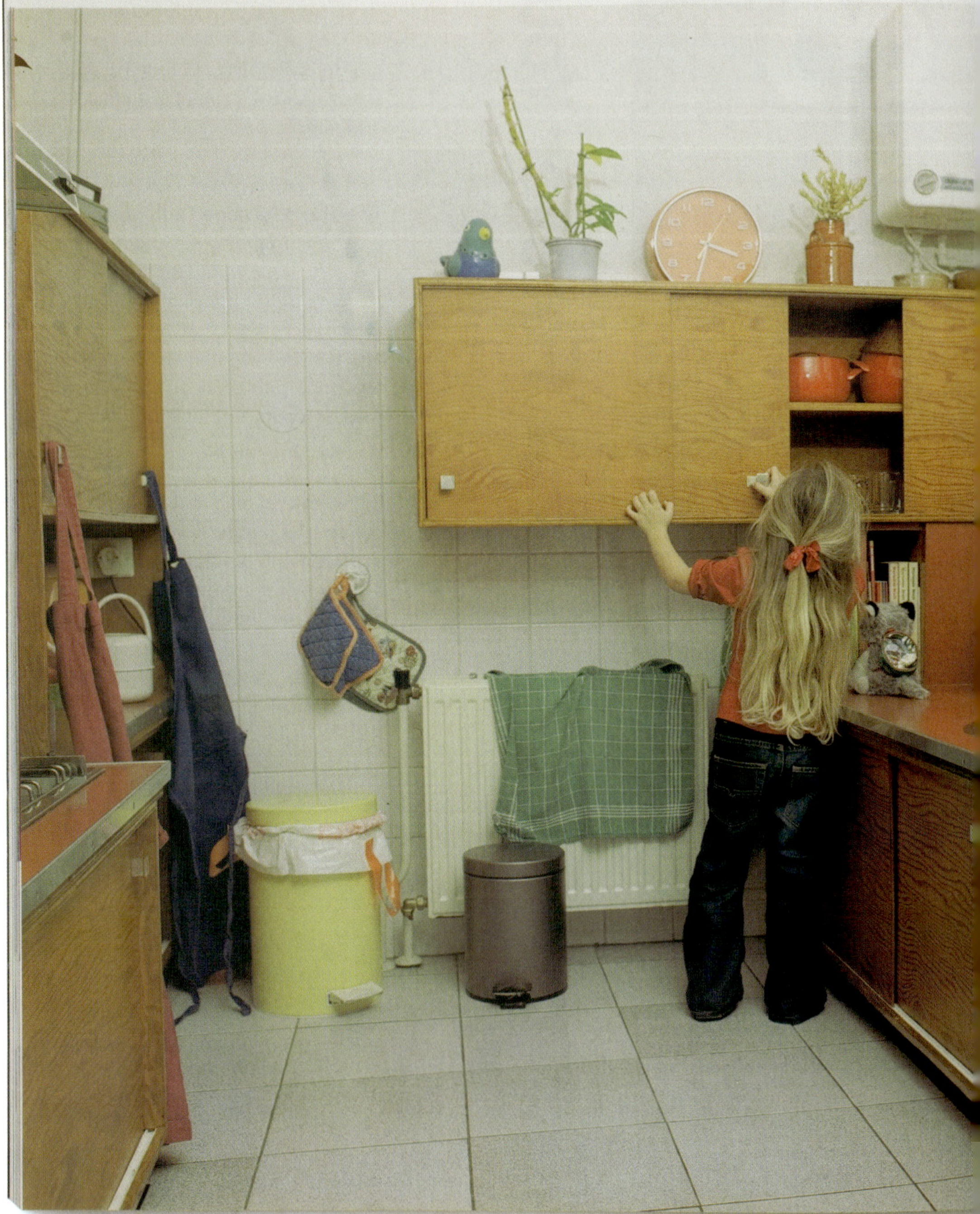

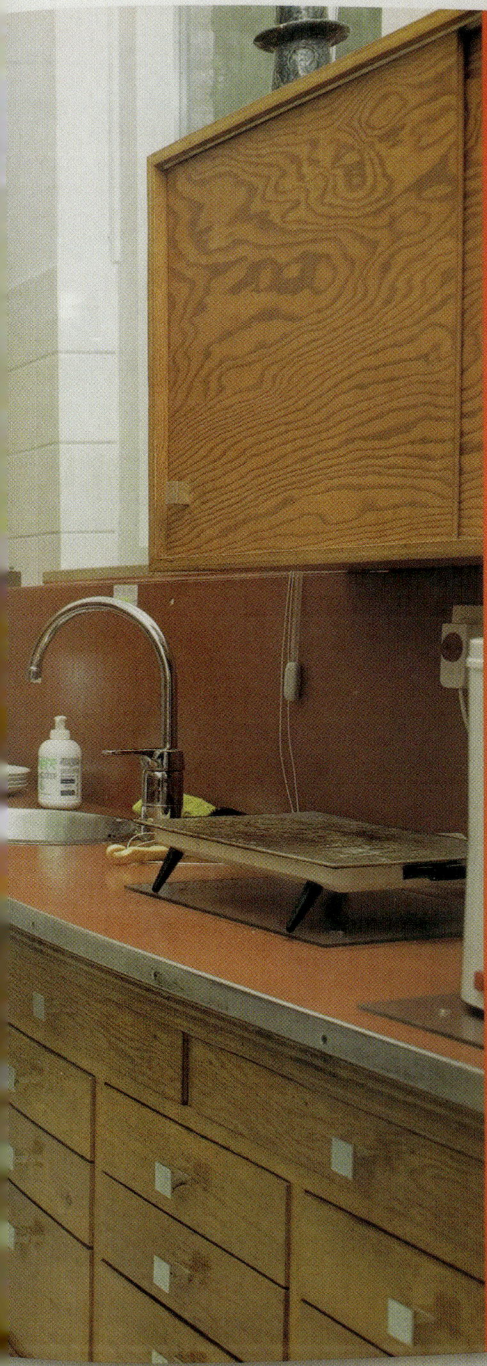

Forever Young

A girl in a kitchen. At first glance, an ordinary domestic scene. But take a closer look. Something's wrong here. The kitchen and the items in it clearly belong to the rather frowzy world of grandparents, but their proportions are incompatible – with one another and with the figure in the room.

The photograph is by Lucas Maassen, a 2001 graduate of The Design Academy in Eindhoven who presented an intriguing study of 'the conflicts of proportion' as his final academic project. In observing his surroundings, Maassen had detected the rather gruesome fact that kids and grown-ups are becoming more and more alike. The ideal age – the one that all of us apparently want to be – is around 20. Provoked by the media and masters of marketing, tiny tots walk around behaving like their parents. At the other end of the spectrum is the parade of tight-faced adults jogging by, hoping that exercise and cosmetic surgery will preserve the youthful image they're dying to perpetuate. No matter how hard they strive to turn back the clock, however, the sight of an eight-year-old jabbering into a mobile phone rubs them the wrong way.

The same disturbing feeling emanates from Lucas Maassen's photograph of an ostensibly everyday interior. Only after careful observation does it dawn on the viewer that the kitchen does not belong to a full-sized adult. The little girl has wandered into the home of a man who will always have to deal with the conflicts of proportion. By adjusting the dimensions of furniture and appliances to the size of his body, however, he manages just fine.

Louise Schouwenberg
Photography by Lucas Maassen

SUBJECT
Visual essay of the conflicts of proportion

LOCATION
Eindhoven, Netherlands

DESIGNER
Lucas Maassen

PHOTOGRAPHER
Lucas Maassen

WRITER
Louise Schouwenberg

332-333

May/June 2003

Issue 32

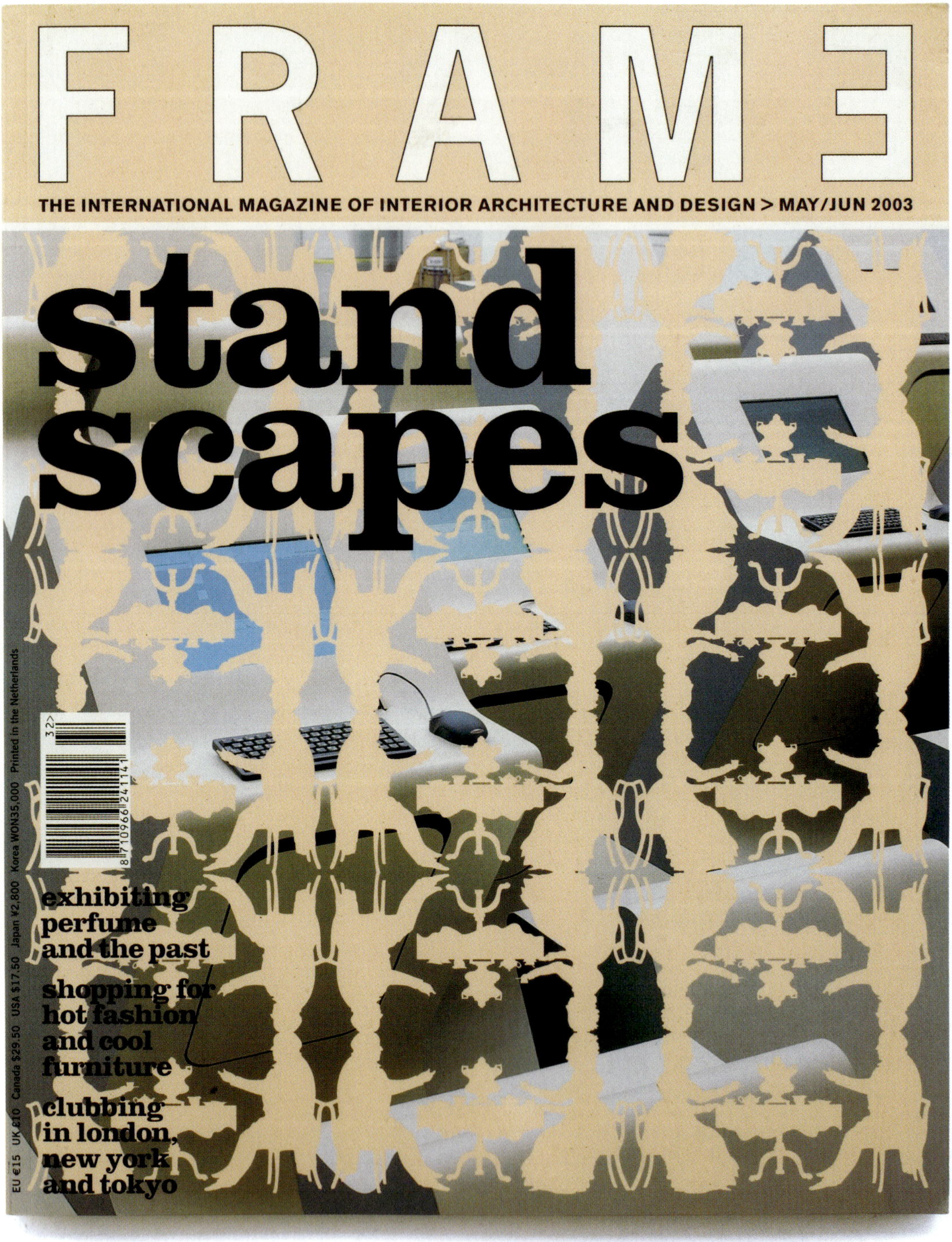

ART DIRECTORS
Cornelia Blatter & Marcel Hermans,
COMA

32

DESIGNERS
Jürgen Mayer H.
www.jmayerh.de
Konstantin Grcic
Dieter Thiel
Atelier Seitz
www.atelierseitz.de
Anne Mie Depuydt &
Erik van Daele (uapS)
www.uaps.net
Matali Crasset
www.matalicrasset.com
Fabio Novembre
Paul Wilke (La Senzo)
www.lasenzo.nl
Luc D'Hanis &
Sofie Lachaert
www.lachaert.com
Lorent & Van de Poel
www.lvdp.com

SUBJECT
Ten stand designs
on various European
trade-fairs

LOCATIONS
Cologne, Dusseldorf
and Berlin, Germany
Kortrijk, Belgium
Florence, Italy
Paris, France

PHOTOGRAPHERS
Uwe Walter
Ingmar Kurth
Constantin Meyer
Hans-Georg Esch
Olaf Schiemann
Edu van Gelder
Arrigo Coppitz
Alberto Ferrero
Reinhard Schwederski
Mark Detiffe

WRITERS
Billy Nolan
Monika von Pechmann

Stand Scapes

Designers Jürgen Mayer H. and Sebastian Finckh about the project
Innovative collaboration between Stylepark, Sikkens, GIRA, GARPA, Dombracht and J. Mayer H.

Comments on the published article
Great feature in an excellent context.

Did the article have an effect?
Big media interest.

Current projects
Metropol Parasol, Seville, Spain // Dining hall, Karlsruhe, Germany // Office building, Hamburg, Germany // Villa Max Maier, Stuttgart, Germany // Danfoss Universe, Exhibition Centre + Cafeteria, Nordborg, Denmark // Courthouse, Hasselt, Belgium.
—

Designer Paul Wilke (La Senzo), Netherlands about the project
The aim, which was to put full communication focus on the company Sedus as an innovative thinker as well as an innovative brand of office furniture, was totally fulfilled.

Comments on the published article
Yes we remember this well as a total different kind of expo concept.

Did the article have an effect?
Yes, conformation to customers and prospects that La Senzo is a competent Conceptual Design office.

Recent news or information about the project featured in *Frame*
In the Magazine's/Books; Exposant, Expovision, Messedesign Jahrbuch 2003/2004.

Stand Scapes

'Almost every man wastes part of his life attempting to display qualities which he does not possess.'

Samuel Johnson, 1752

By Billy Nolan and Monika von Pechmann

01 fold

Keyboards and screens promoting an online database are embedded in linoleum waves that invite visitors to go surfing or just lounge around looking luscious.

Stand design Jürgen Mayer H. **Client** Stylepark **Components** plywood templates, flexible ply covering, linoleum finish, keyboards, monitors **Event** XXI World Congress, Union Internationale des Architectes, Berlin, 2002 **Photography** Uwe Walter

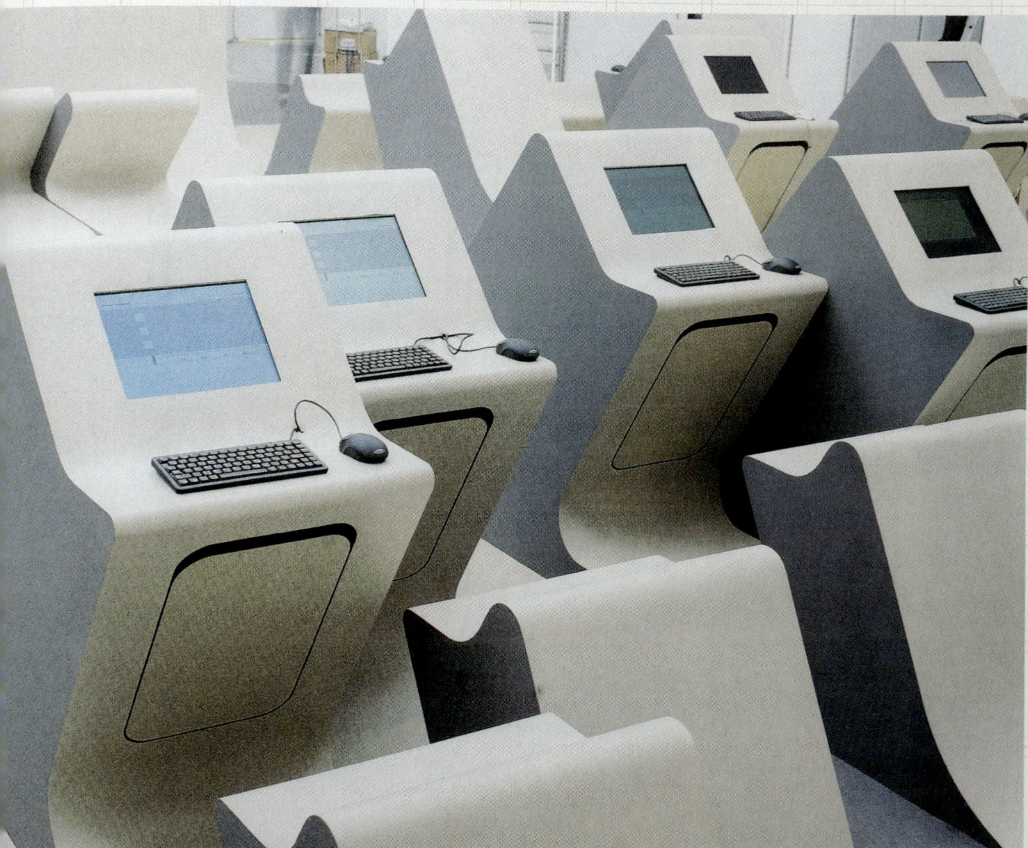

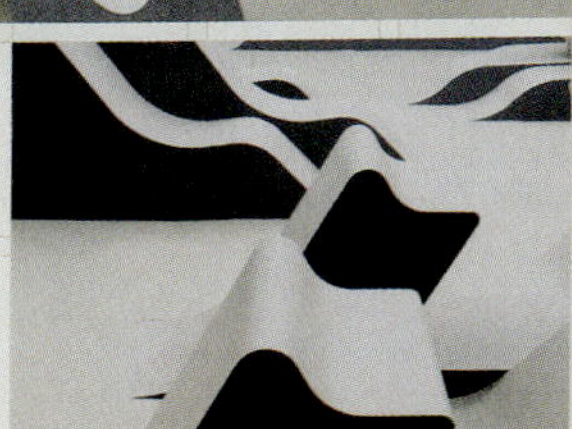

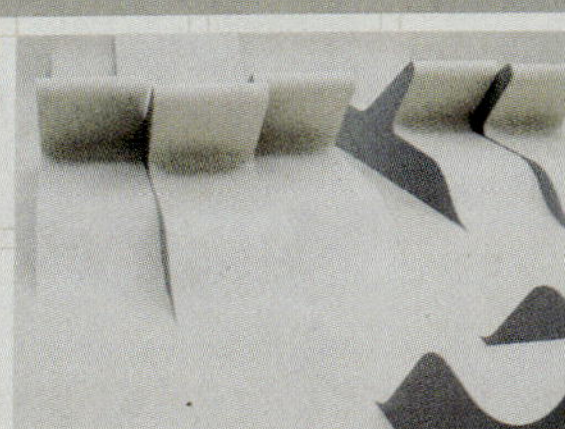

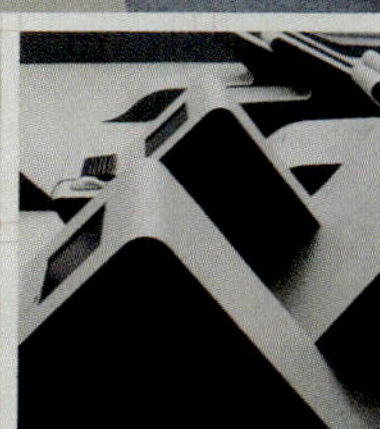

SUBJECT
Ten stand designs on various European trade fairs

LOCATIONS
Various, see page 333

DESIGNERS
Various, see page 333

PHOTOGRAPHERS
Various, see page 333

WRITERS
Billy Nolan
Monika von Pechmann

02 stack

Walk into this ideal home, an empty box, and move up, down and across the core on a platform to retrieve household objects from 17 different storage systems.

Stand design Konstantin Grcic **Client** Köln Messe **Components** particle board, metal shelving system, surrogate lift installation **Event** International Furniture Fair, Cologne, 2003 **Photography** Ingmar Kurth/Constantin Meyer

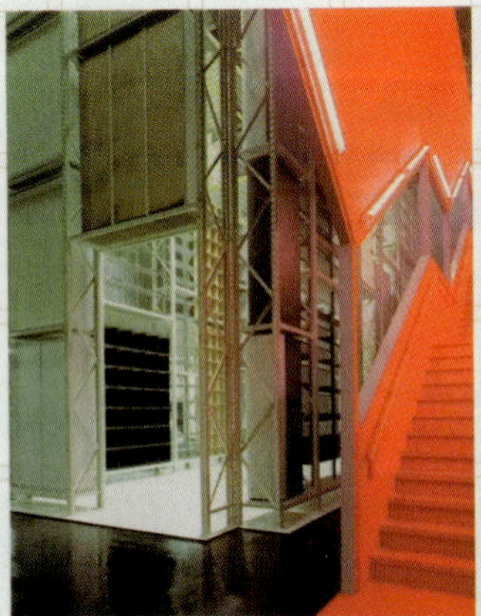

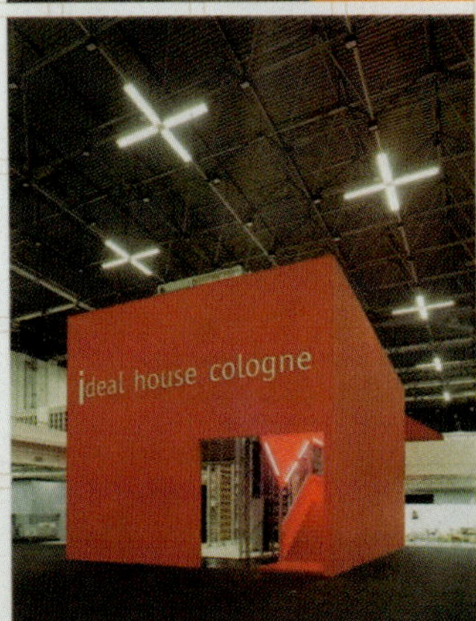

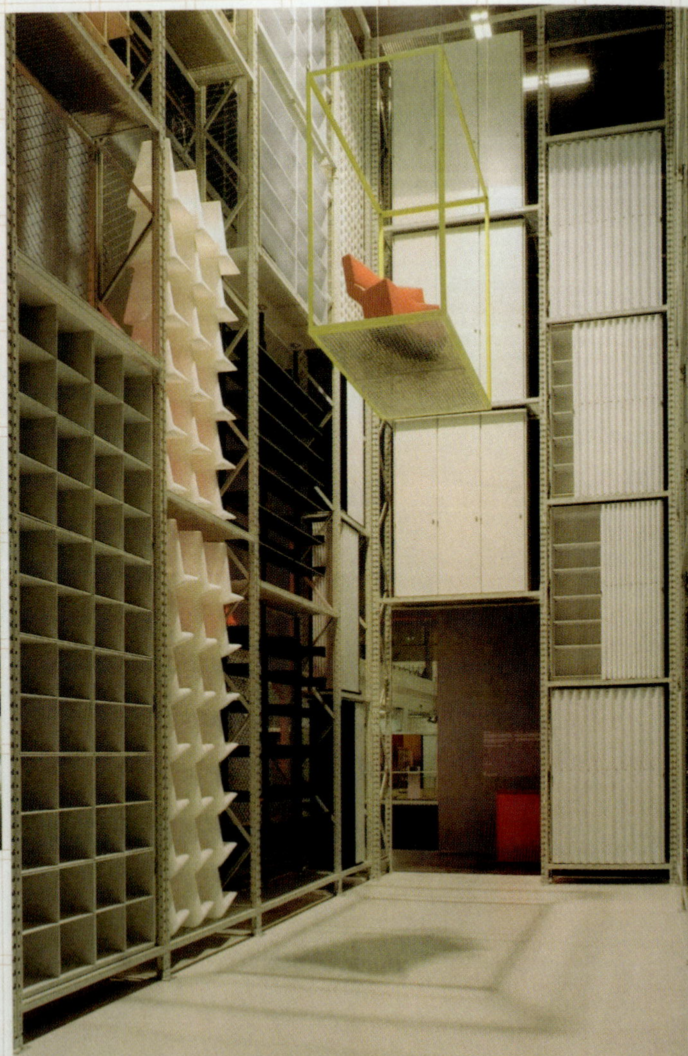

03 contrast

Ignoring the world outside, stacked cubes open up to embrace an interior arena of light. Sharp black-and-white contrasts add to the drama.

Stand design Dieter Thiel **Client** Ansorg **Components** aluminium, steel, wood **Event** EuroShop, Düsseldorf, 2002
Photography Hans-Georg Esch

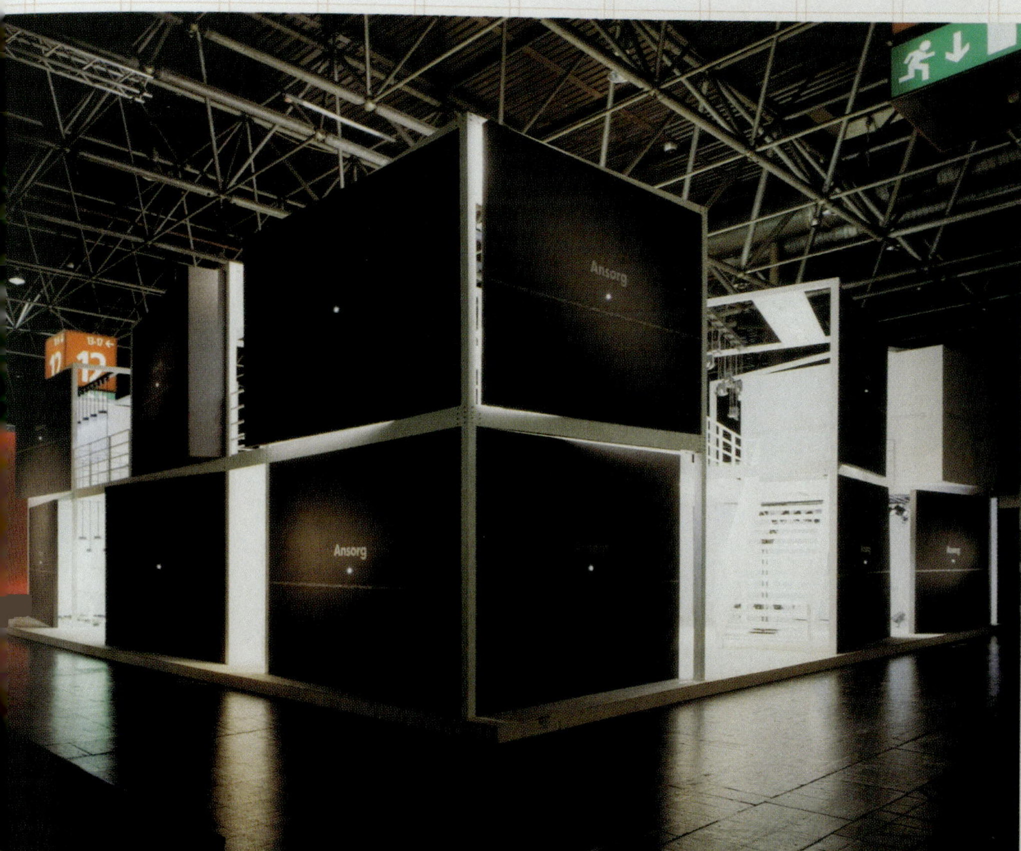

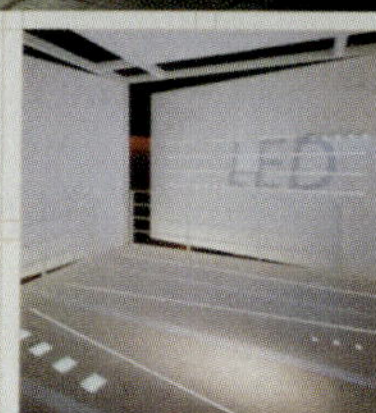

SUBJECT
Ten stand designs on various European trade fairs

LOCATIONS
Various, see page 333

DESIGNERS
Various, see page 333

PHOTOGRAPHERS
Various, see page 333

WRITERS
Billy Nolan
Monika von Pechmann

04 float

Kitchen appliances appear to hover around columns of light that gradually change colour. This expansive display of innovative cooking technology takes its cue from the powerful simplicity of the circle.

Stand design Atelier Seitz **Client** Neff **Components** MDF display units, needlefelt carpet, PVC, gauze fabric
Event HomeTech, Berlin, 2002 **Photography** Olaf Schiemann

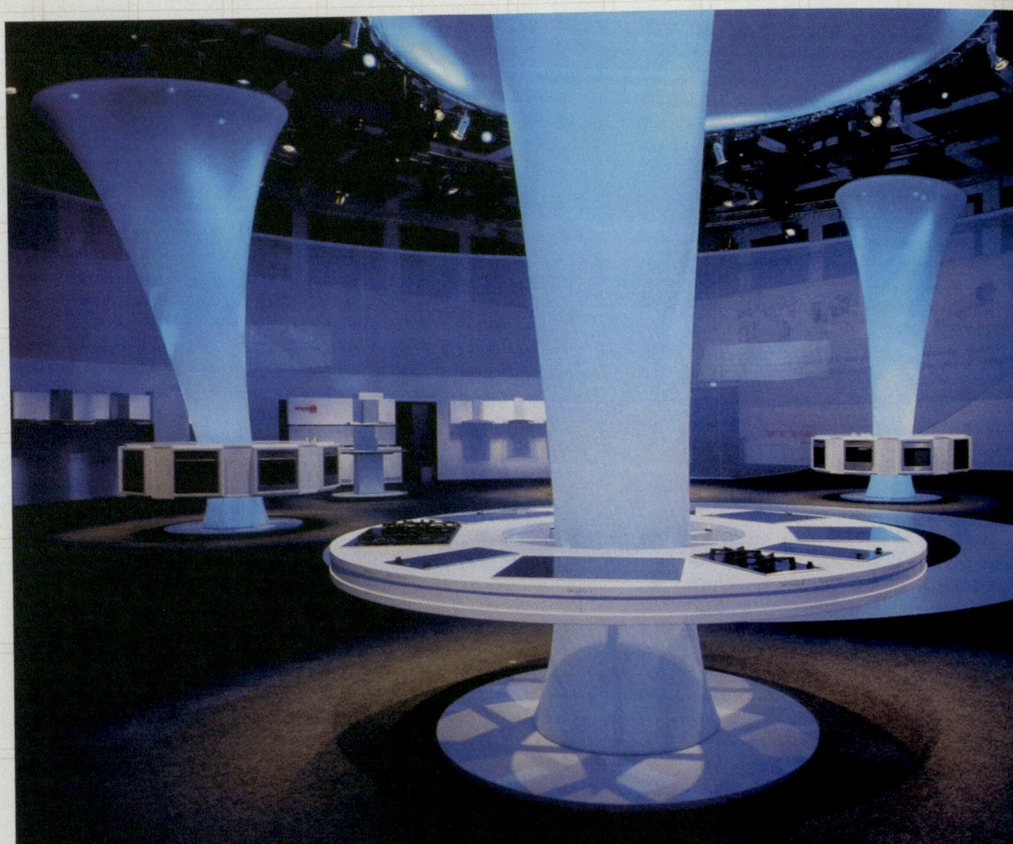

05 enchant

Stroll through a magic forest and discover 26 designer presentations displayed in a thicket of glowing stalks.

Stand design uapS, Anne Mie Depuydt & Erik van Daele **Client** Interieur Foundation **Components** 45,000 fibre-optic rods, 26 displays **Event** Interieur, Kortrijk, 2002 **Photography** Edu Van Gelder

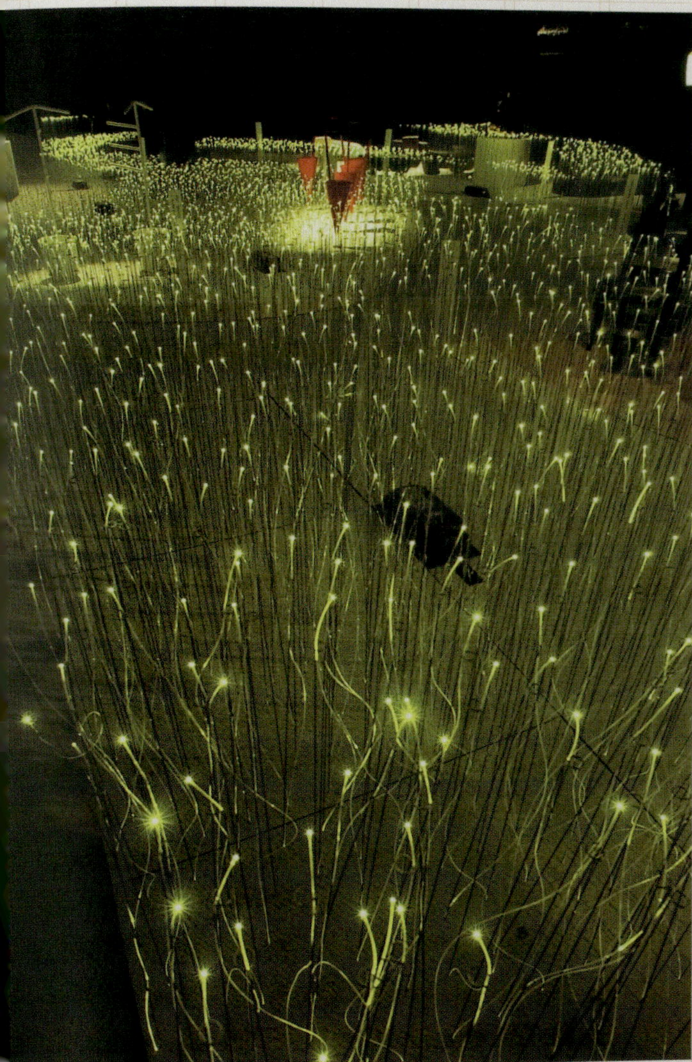

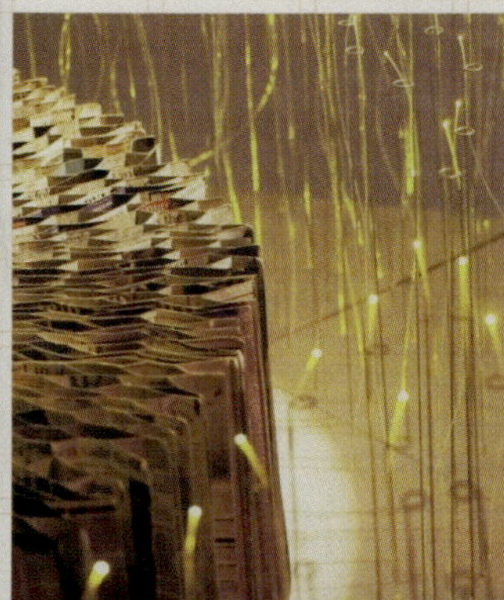

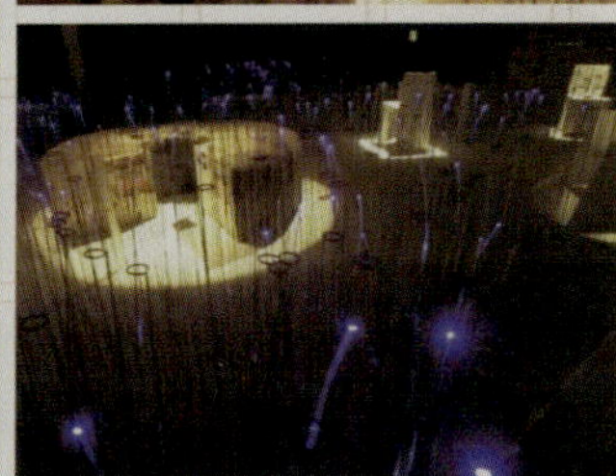

SUBJECT
Ten stand designs on various European trade fairs

LOCATIONS
Various, see page 333

DESIGNERS
Various, see page 333

PHOTOGRAPHERS
Various, see page 333

WRITERS
Billy Nolan
Monika von Pechmann

06 live

Inspired by the world of fashion, this organ-like house absorbs water, air, electricity and sound waves through a soft, sensitive filtering membrane. A home as comfortable as a second skin.

Stand design Matali Crasset **Client** Pitti Imaggine **Components** elasticated fabric, steel suspension cables
Event Pitti Imaggine Uomo, Florence, 2002 **Photography** Arrigo Coppitz

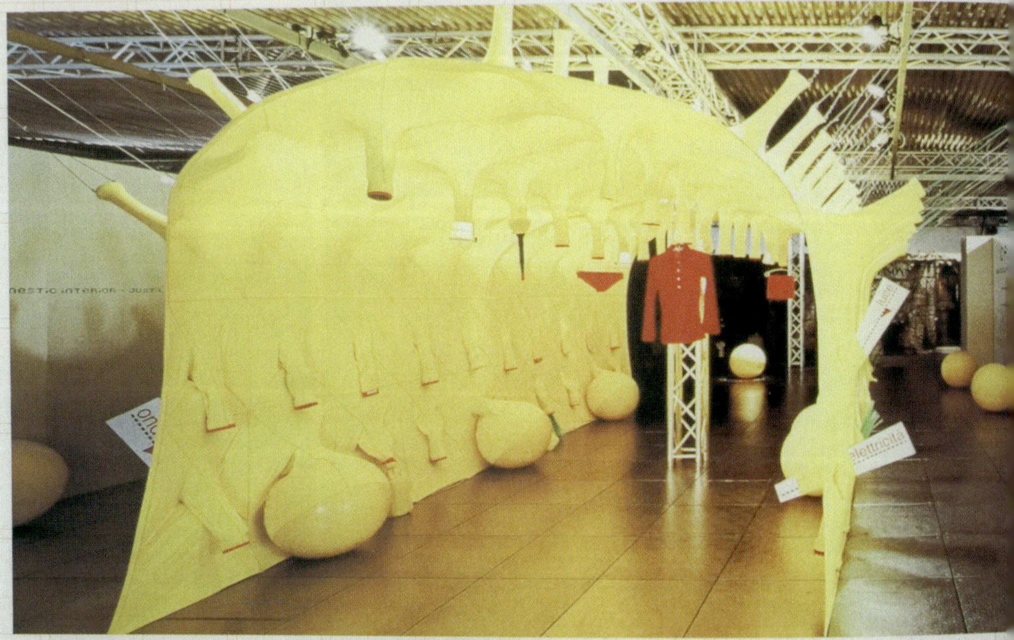

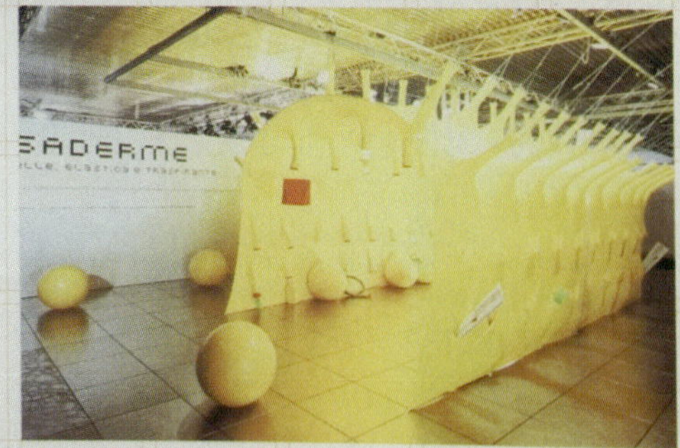

07 love

Two silhouettes covered in glass mosaic tiles – pink for her, blue for him – oppose each other and join to form a huge heart. Stand and product are one and the same.

Stand design Fabio Novembre **Client** Bisazza **Components** timber structure, glass mosaic tiles **Event** Salon du Meuble, Paris, 2003 **Photography** Alberto Ferrero

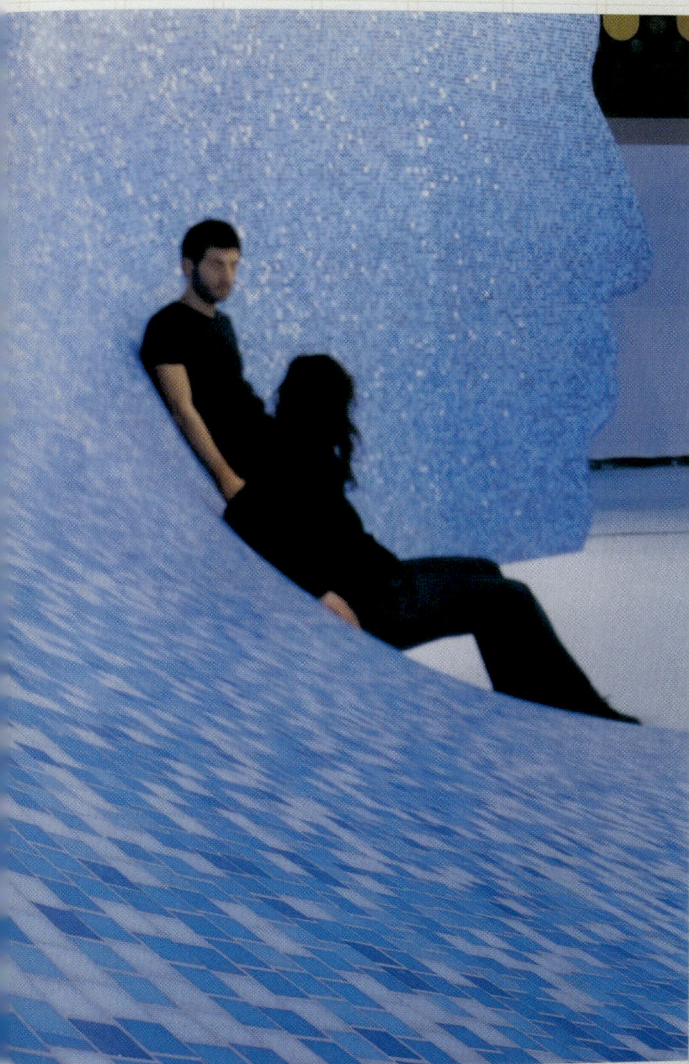

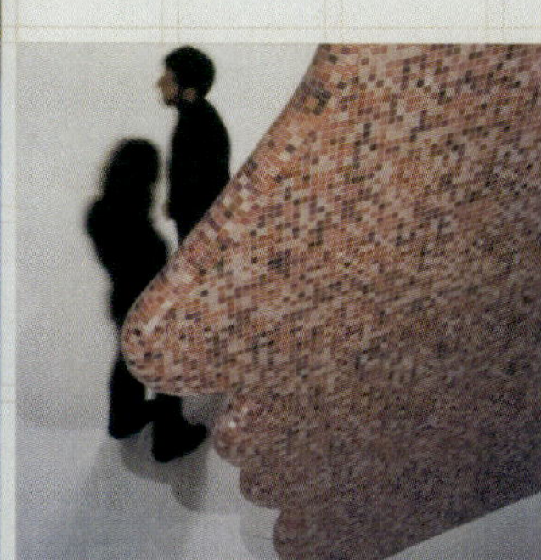

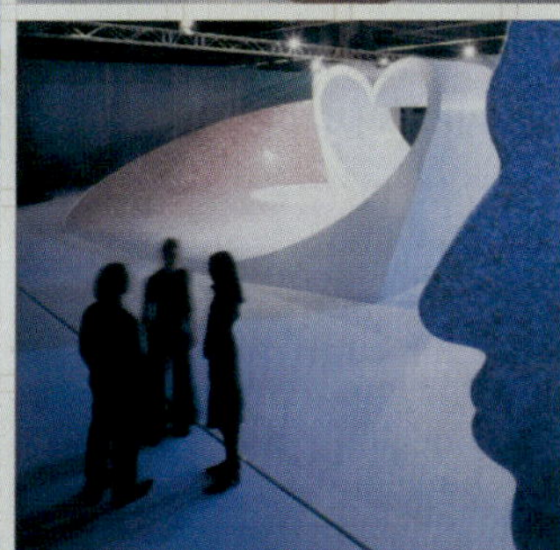

SUBJECT
Ten stand designs on various European trade fairs

LOCATIONS
Various, see page 333

DESIGNERS
Various, see page 333

PHOTOGRAPHERS
Various, see page 333

WRITERS
Billy Nolan
Monika von Pechmann

08 move

Mobility and flexibility are the thematic links that unite this line of furniture. Collapsible wigwams, small enough to fit in a sports bag, reflect this theme.

Stand design La Senzo, Paul Wilke **Client** Sedus Stoll **Components** steel and aluminium frame, elasticated fabric, high-spectrum fluorescent lighting **Event** Orgatec, Cologne, 2002 **Photography** Reinhard Schwederski

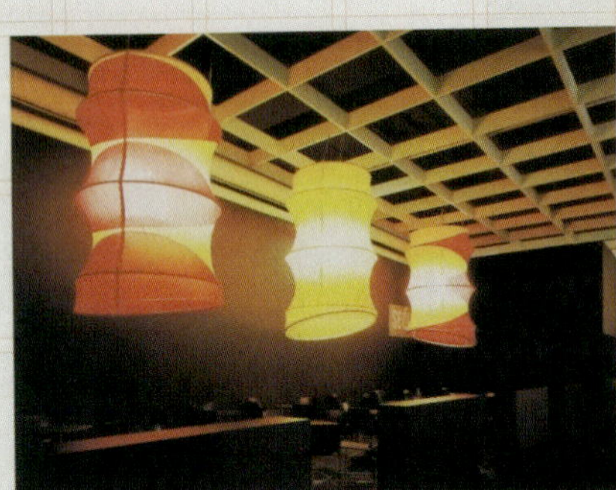

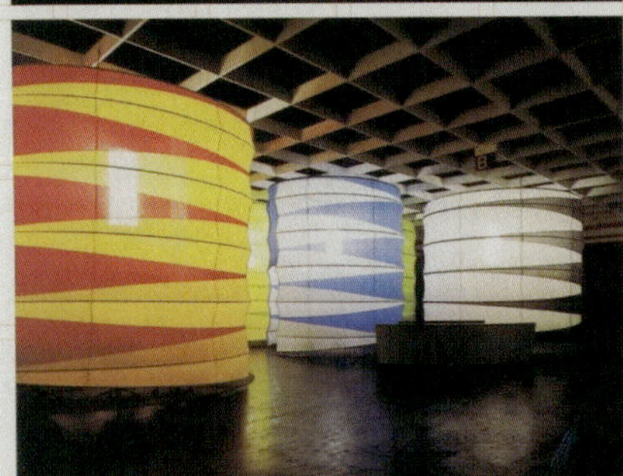

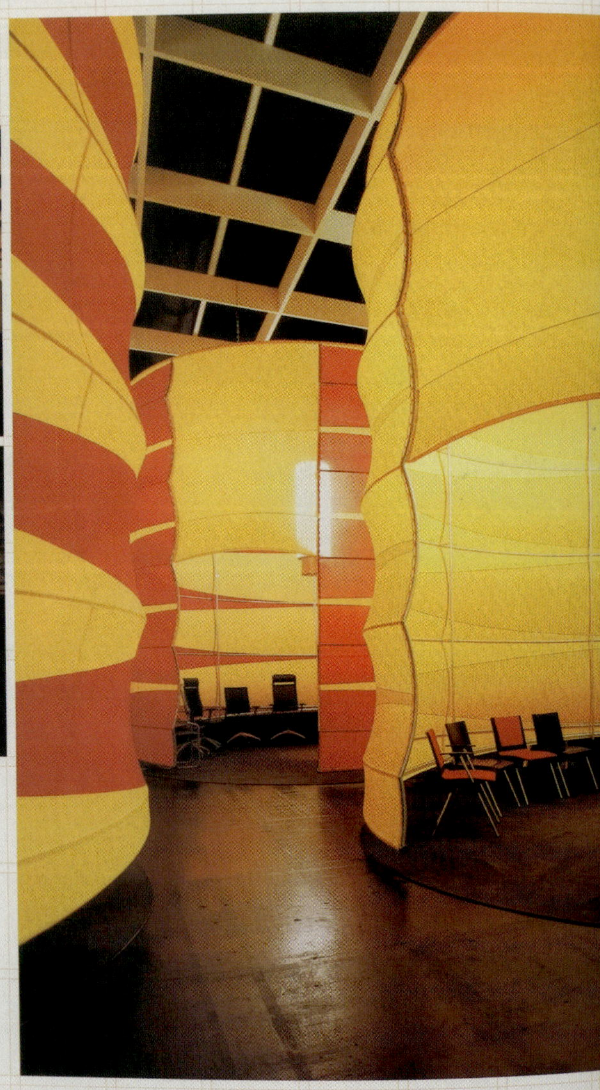

09 hide

Naked light bulbs cast a grim glow across a stage set of primitive-looking furniture – props for jewellery made of rock crystal, beads, wire and bisected coins.

Stand design, client, and photography Luc D'Hanis & Sofie Lachaert **Components** multiplex walls; multiplex floor finished in MDF; furniture in pinewood, multiplex and recycled sheet material; giant necklace of graphite spheres and white rope; 20 kg potatoes; silver french-fry fork **Event** Interieur, Kortrijk, 2002

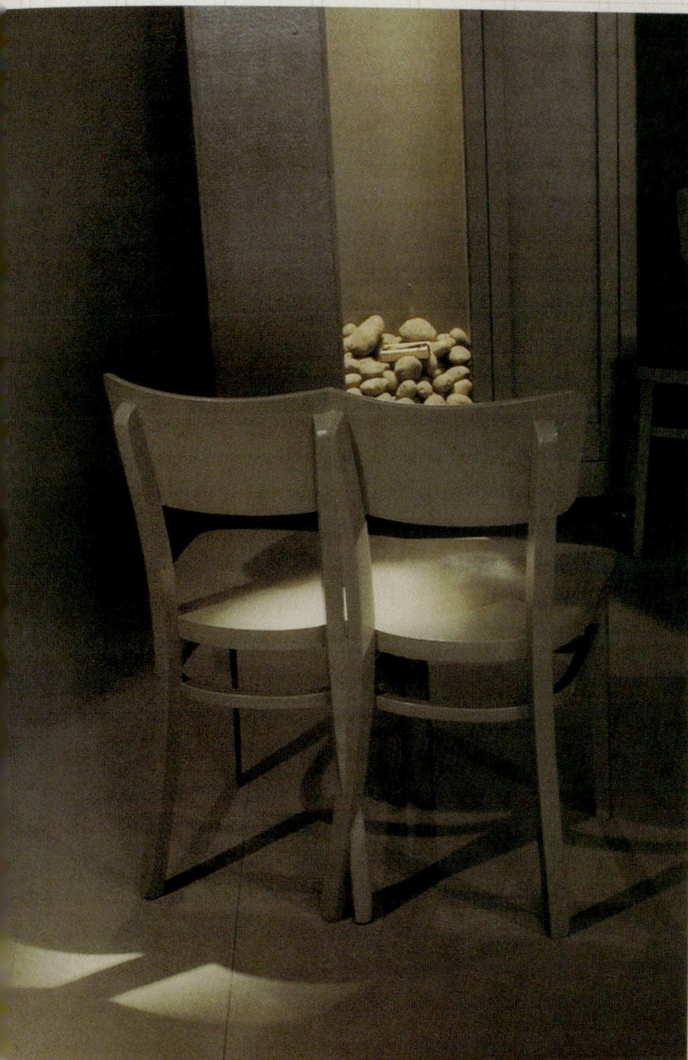

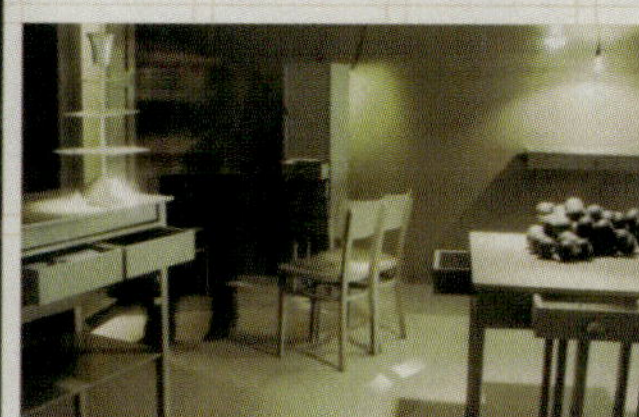

SUBJECT
Ten stand designs on various European trade fairs

LOCATIONS
Various, see page 333

DESIGNERS
Various, see page 333

PHOTOGRAPHERS
Various, see page 333

WRITERS
Billy Nolan
Monika von Pechmann

10 show

Visitors enter a luxury lab replete with soft shapes, curved lines and fresh colours. No longer is glass the cold and chilly material it once seemed.

Stand design Lorent & Van de Poel **Client** Saint-Gobain Glass **Components** wooden frame, laminate floor, plywood walls and ceiling, sandblasted glass panels **Event** Interieur, Kortrijk, 2002 **Photography** Mark Detiffe

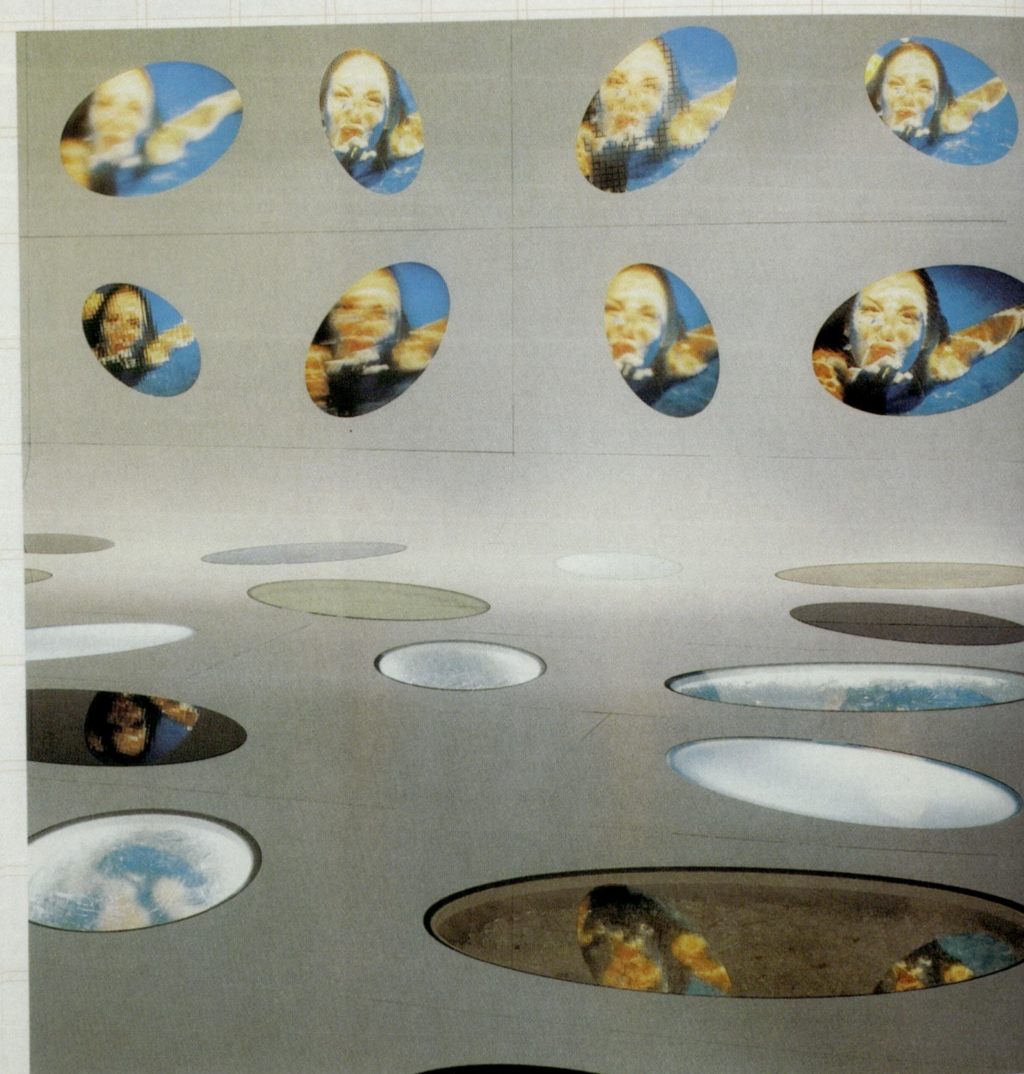

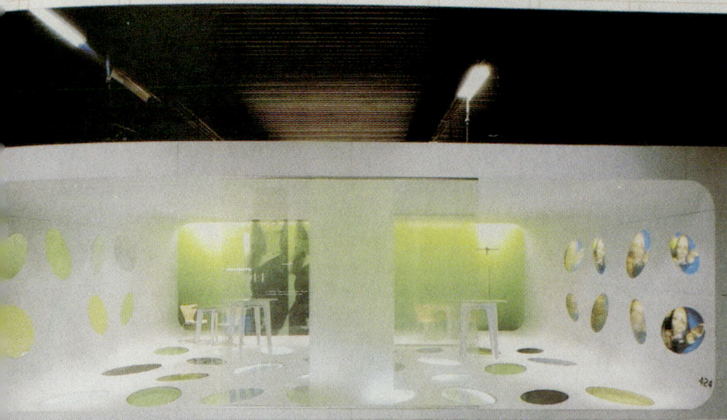

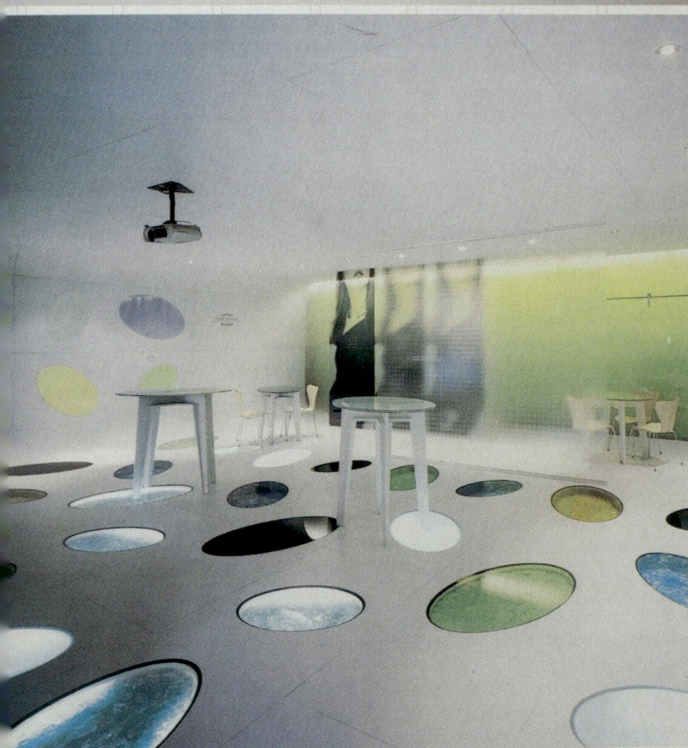

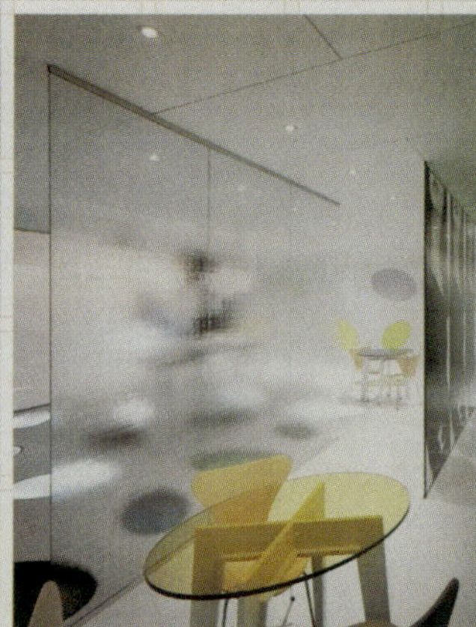

SUBJECT
Ten stand designs on various European trade fairs

LOCATIONS
Various, see page 333

DESIGNERS
Various, see page 333

PHOTOGRAPHERS
Various, see page 333

WRITERS
Billy Nolan
Monika von Pechmann

Issue 33 July/August 2003 346-347

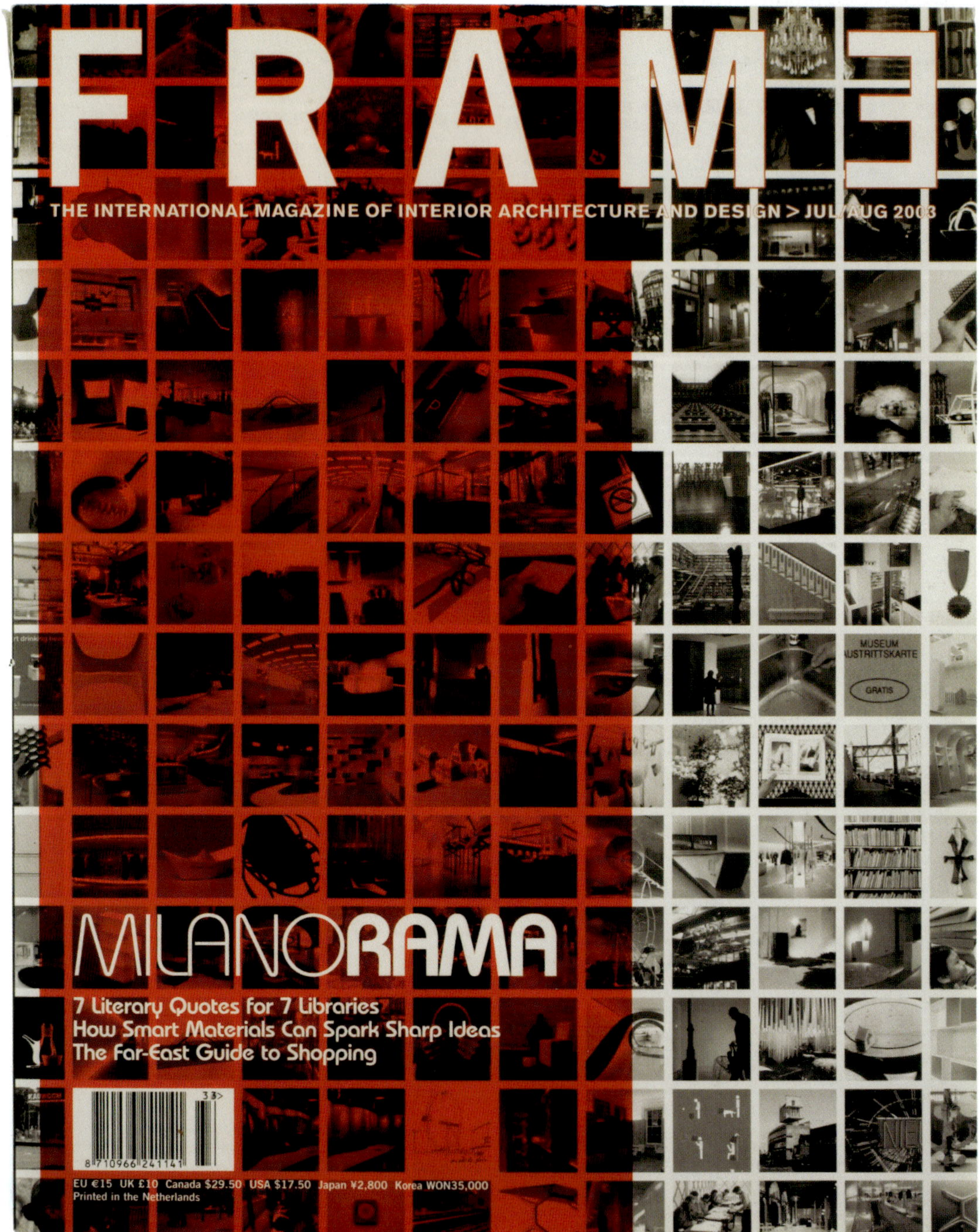

ART DIRECTORS
Cornelia Blatter & Marcel Hermans,
COMA

33

Required Reading

Photographer Corriette Schoenaerts and graphic designer Harmen Liemburg about the project

Our first collaboration and a very interesting opportunity to show a common, everyday phenomenon within a glossy-idealized-world-of-design environment.

Did the article have an effect?

Encouraging remarks everywhere we went!

Current projects

Prints for Orson + Bodil // Stamp for TPG Post // Prints supporting the architectonic design of the Rijksbelastingkantoor (Dutch Tax Building)

DESIGNERS
Italo Rota & Partners
www.studioitalorota.it
Maarten van Severen
Corriette Schoenaerts
Harmen Liemburg
Ortner+Ortner
www.ortner-ortner.de
EDIT
Andrew Miller

SUBJECT
Seven libraries

LOCATIONS
Amsterdam and
Eindhoven, Netherlands
Bologna, Italy
Dresden, Germany
Hong Kong, China
Glasgow, Scotland

PHOTOGRAPHERS
Corriette Schoenaerts
and Harmen Liemburg
www.harmenliemburg.nl
Stefano Pandini
Jannes Linders
Stefan Müller
Hoi Sit
Gary Yeung

WRITER
Frame

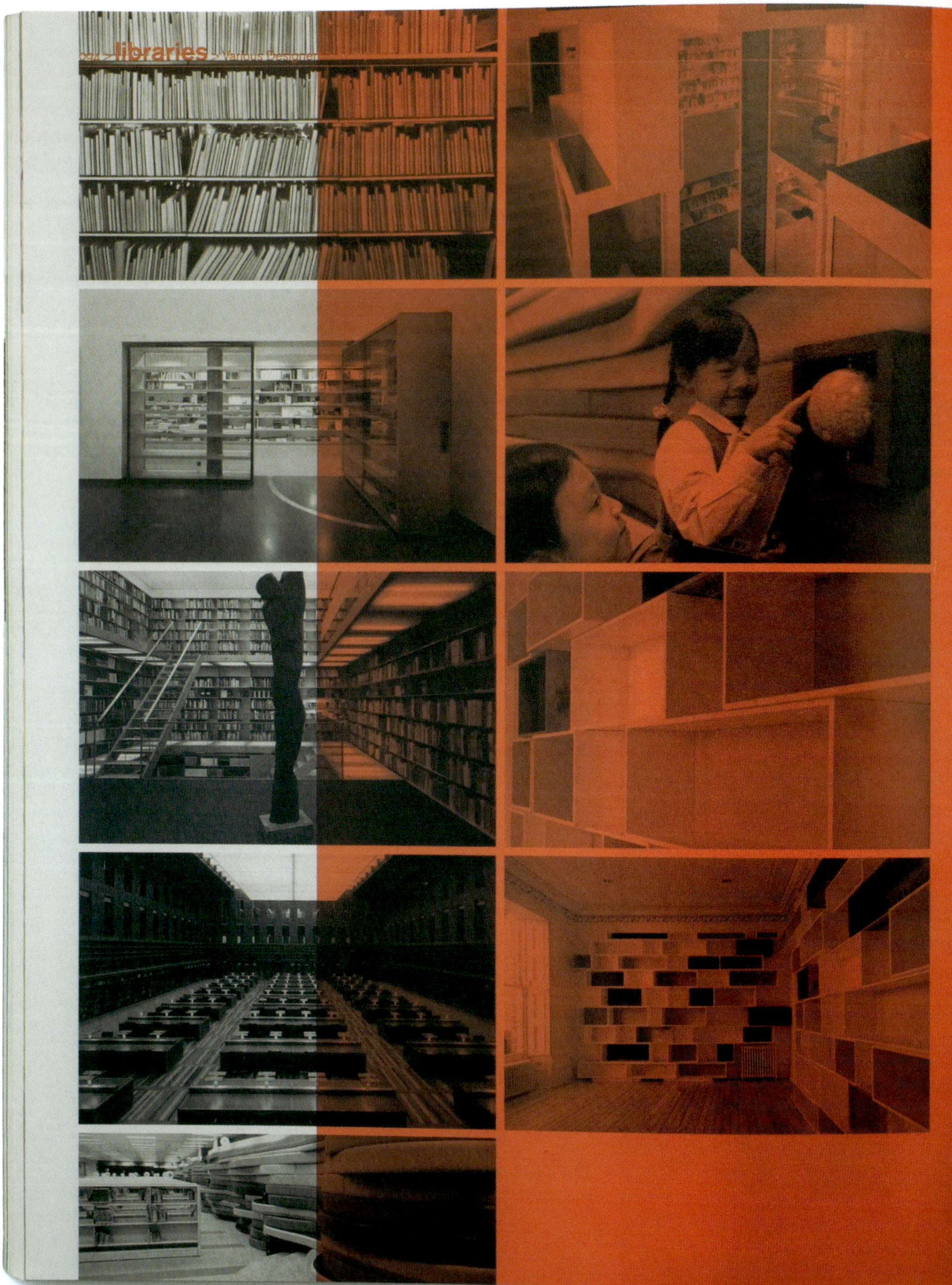
064 > libraries > Various Designer

REQUIRED**READING**

When it comes to places and spaces for books, the onward march of digital technology has not stifled innovation. If anything, library and bookshelf design is blossoming as never before. Here are seven new ways to arrange books – and the literary sources that may have inspired them.

SUBJECT
Seven libraries

LOCATIONS
Various, see page 347

DESIGNERS
Various, see page 347

PHOTOGRAPHERS
Various, see page 347

WRITER
Frame

But I think that subsequently certain libraries were created whose aim was that books should not be read, but concealed. De Bibliotheca, Umberto Eco

Amsterdam Public Library. Photography by Corriette Schoenaerts and Harmen Liemburg. Thanks to K. Hanada and H. Vogels

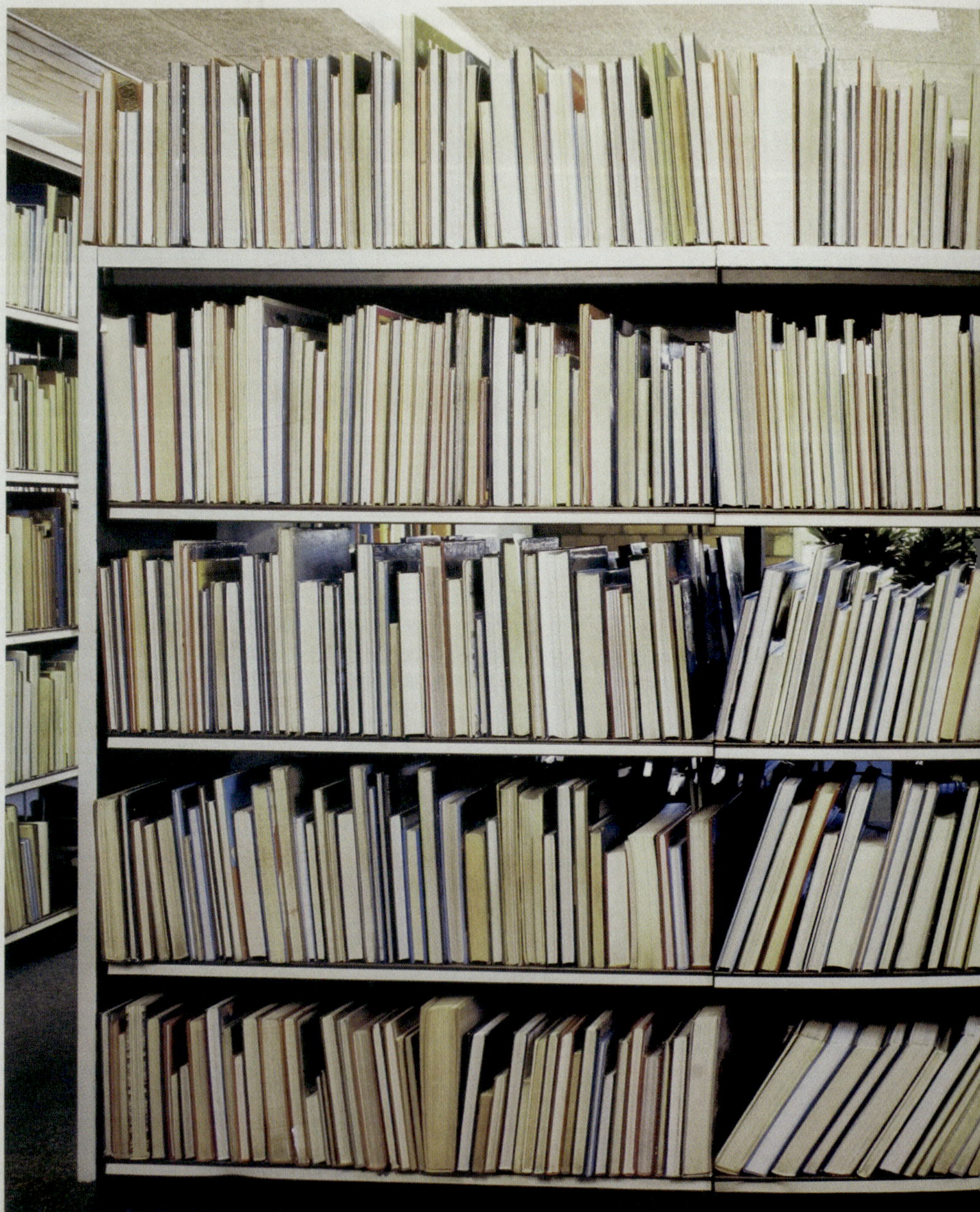

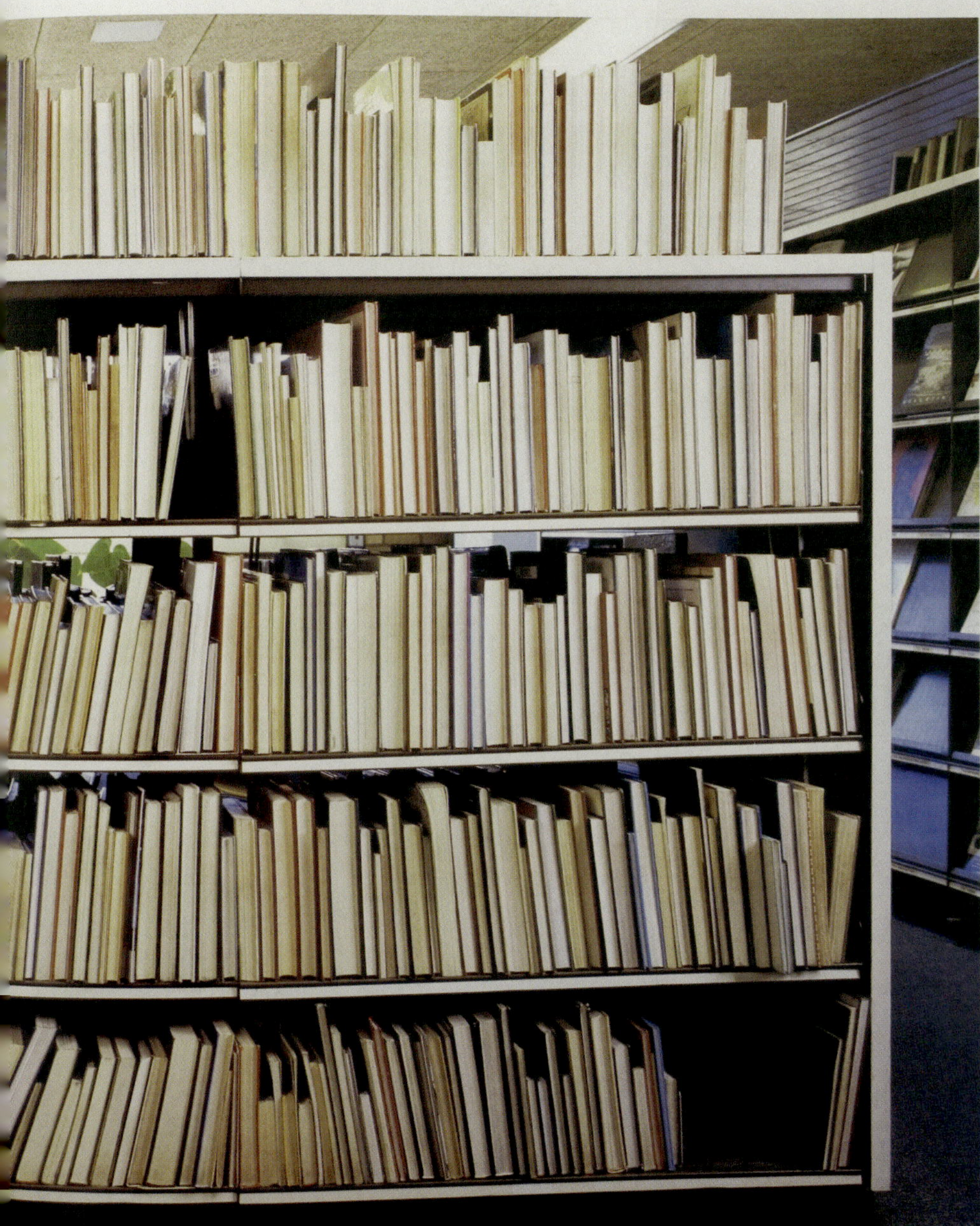

SUBJECT
Seven libraries

LOCATIONS
Various, see page 347

DESIGNERS
Various, see page 347

PHOTOGRAPHERS
Various, see page 347

WRITER
Frame

Architecture being the masterly, correct and magnificent play of masses brought together in light; the task of the architect is to vitalize the surfaces which clothe these masses, but in such a way that these surfaces do not become parasitical. Towards a New Architecture, Le Corbusier

Public Library, Bologna. Design by Italo Rota & Partners. Photography by Stefano Pandini

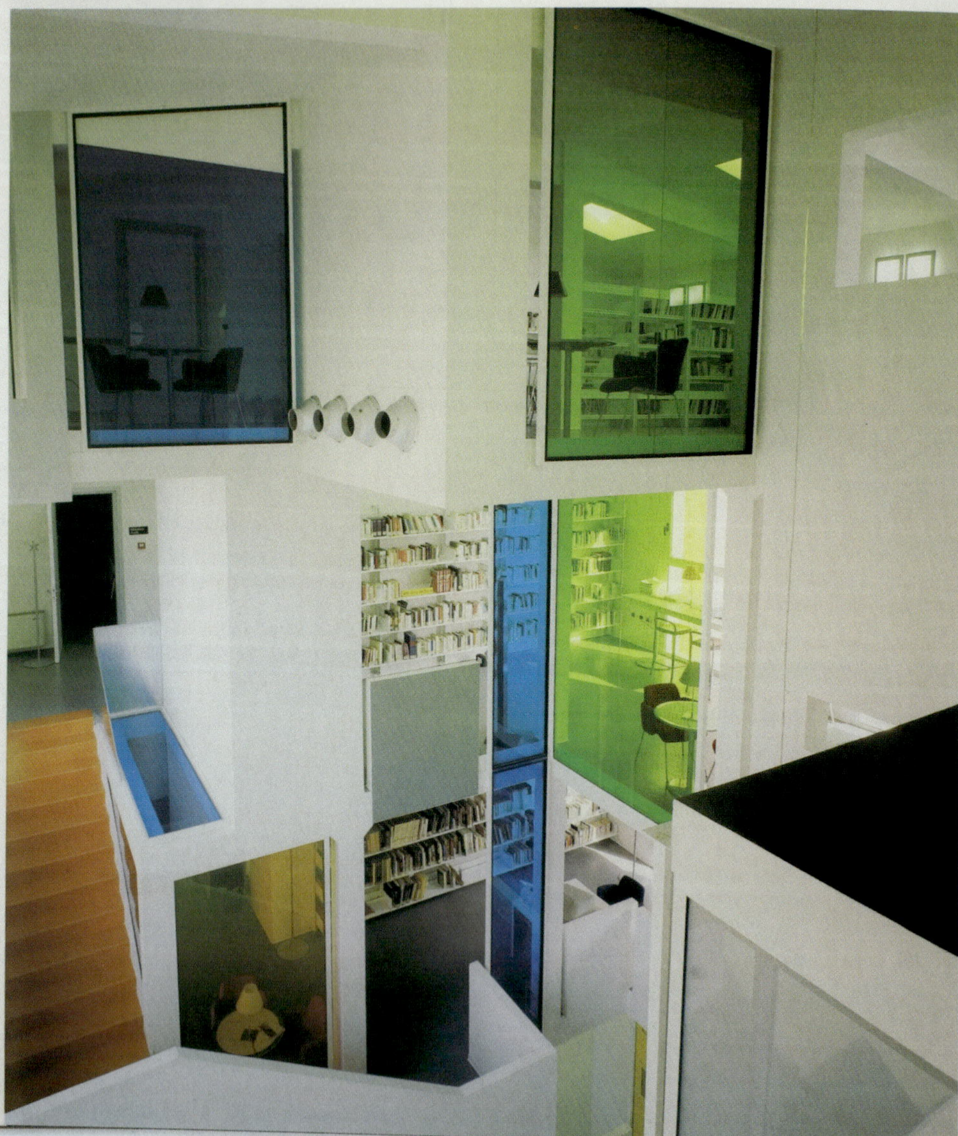

Maximum circulation = maximum sales volume The Harvard Design School Guide to Shopping / Harvard Design School Project on the City 2, Chuihua Judy Chung, Jeffrey Inaba, Rem Koolhaas, Sze Tsung Leong (ed.)

Van Abbe Museum Bookshop, Eindhoven. Design by Maarten Van Severen. Photography by Jannes Linders

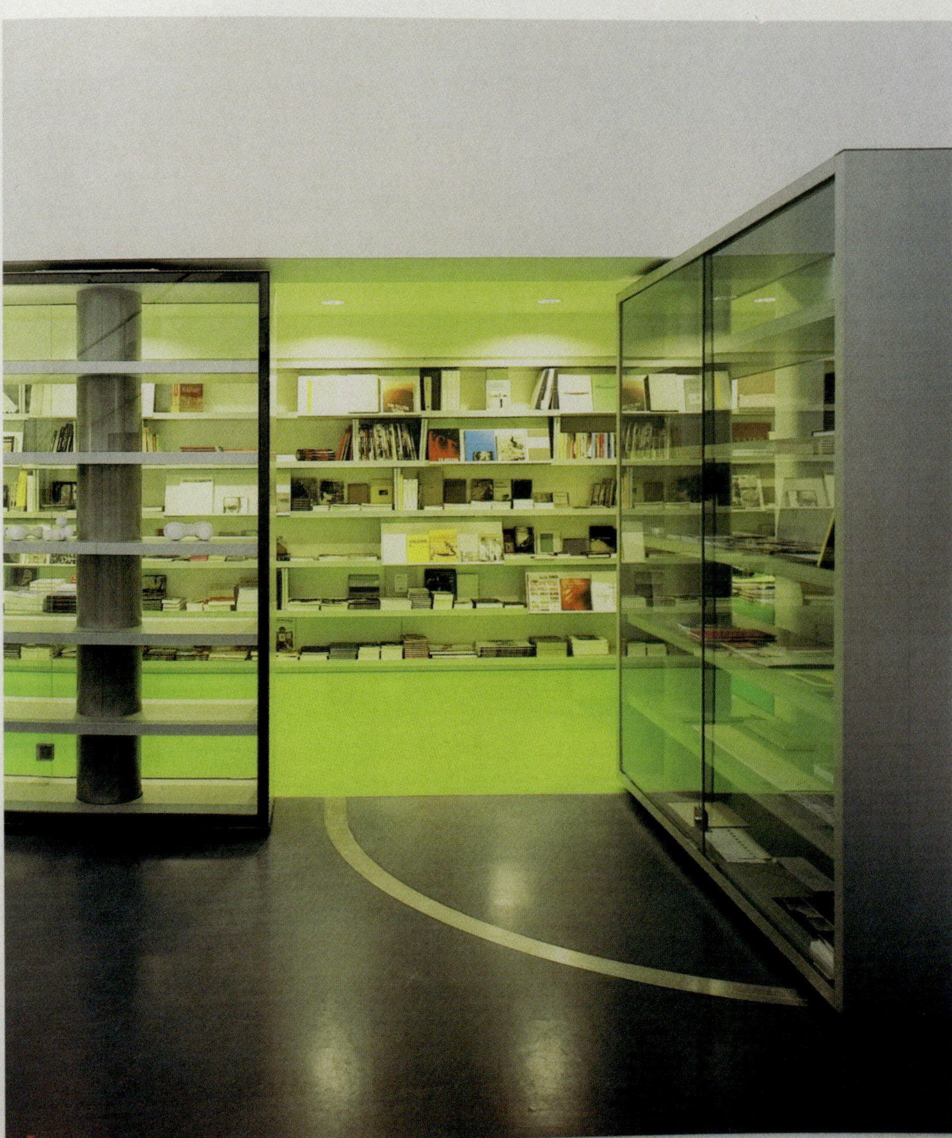

SUBJECT Seven libraries

LOCATIONS Various, see page 347

DESIGNERS Various, see page 347

PHOTOGRAPHERS Various, see page 347

WRITER Frame

Big Brother is always watching you. 1984, George Orwell

Federal Library Sachsen, State and University Library, Dresden. Design by Ortner+Ortner. Photography by Stefan Müller

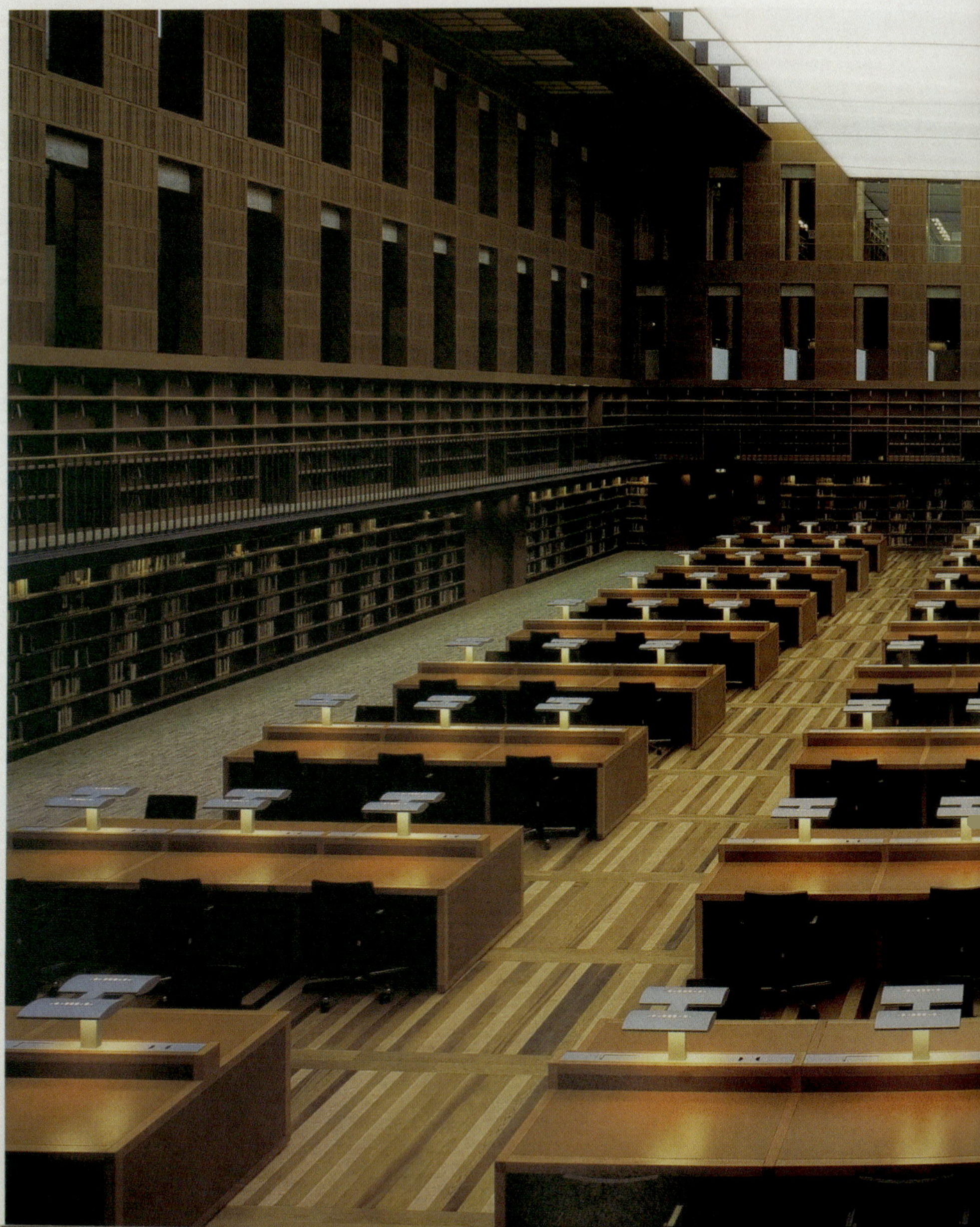

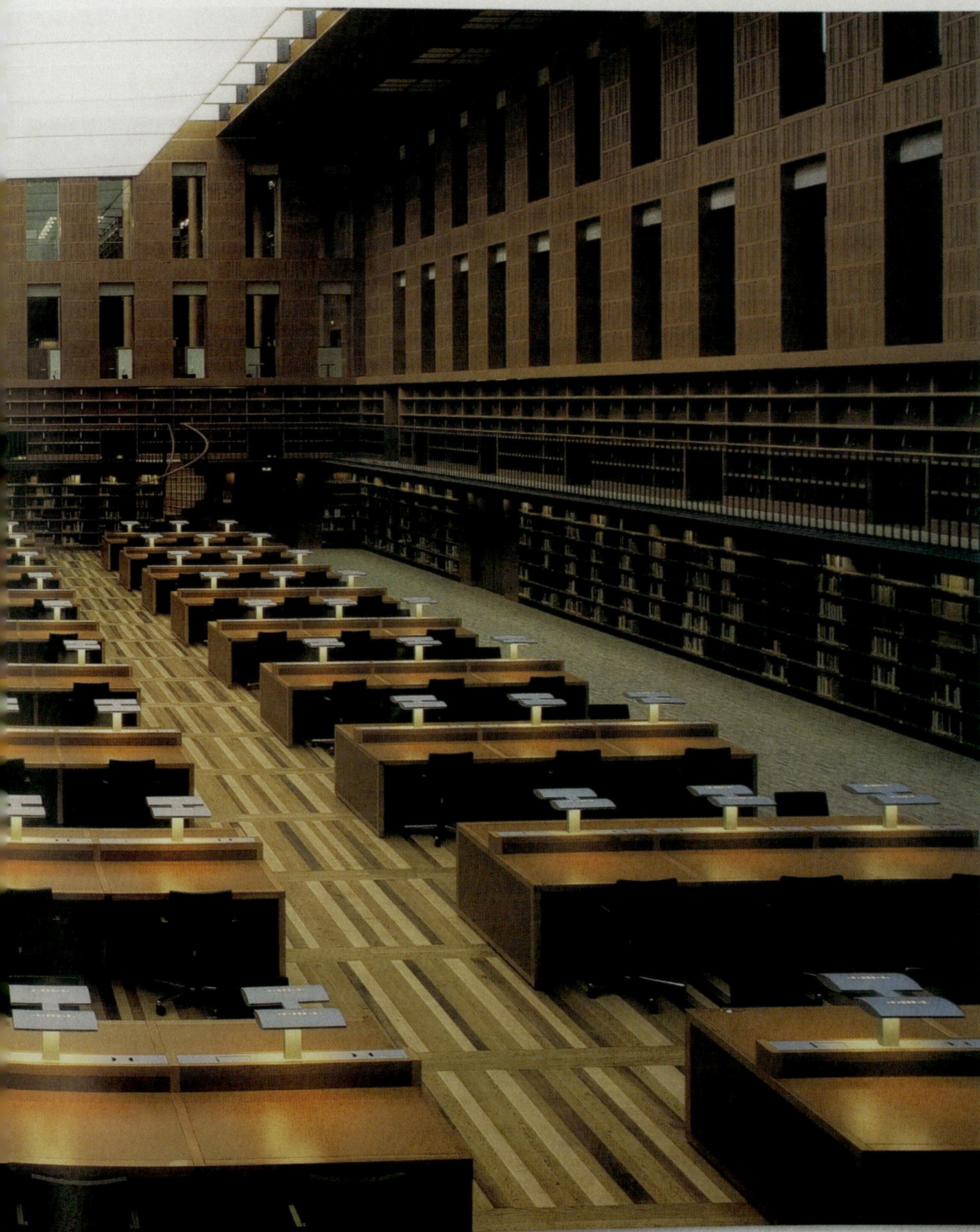

SUBJECT
Seven libraries

LOCATIONS
Various, see page 347

DESIGNERS
Various, see page 347

PHOTOGRAPHERS
Various, see page 347

WRITER
Frame

'What is the use of a book,' thought Alice, 'without pictures or conversations?' Alice's Adventures in Wonderland, Lewis Carroll

Children's Library, Fanling Public Library, Hong Kong. Design by EDIT. Photography by Hoi Sit and Gary Yeung

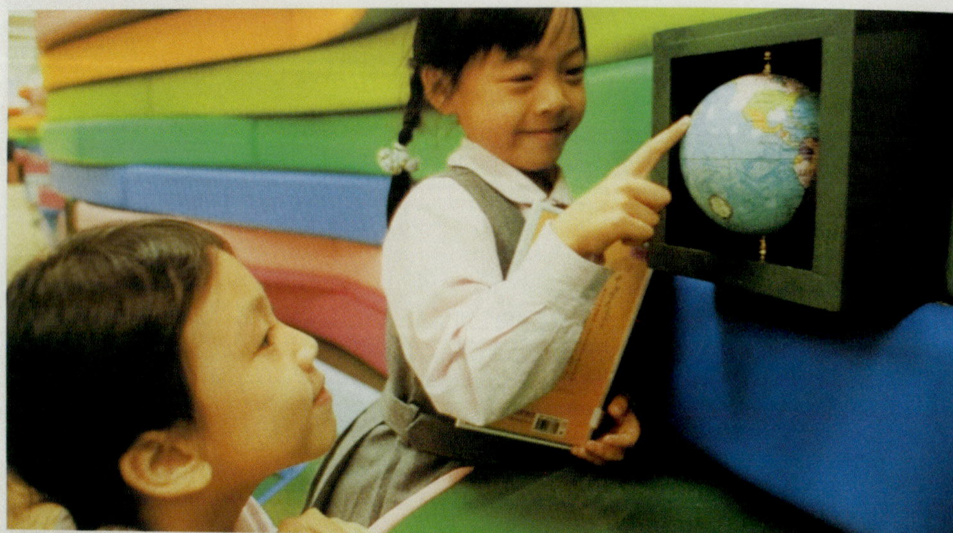

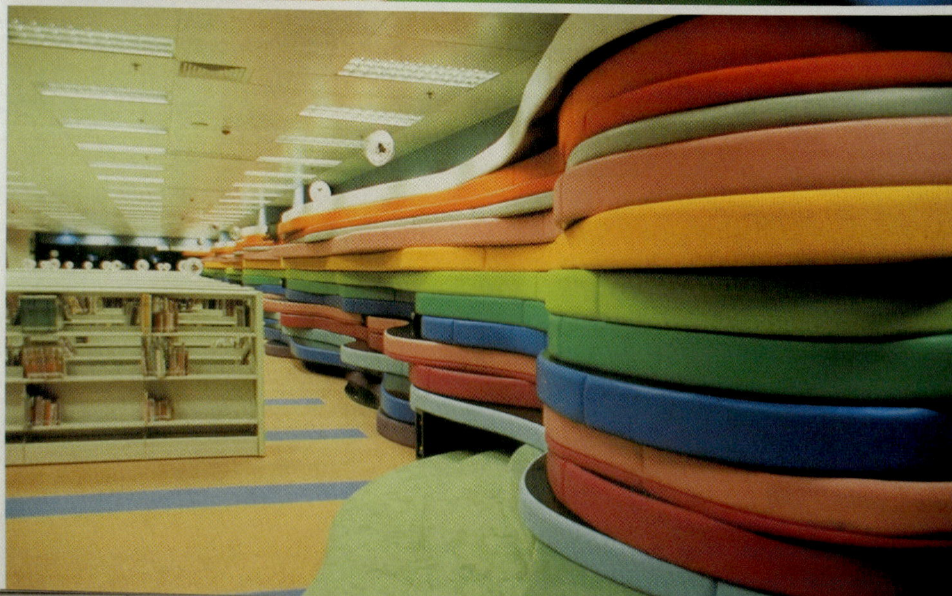

The four notebooks were identical, about eighteen inches square, with shiny covers, like the texture of a cheap watered silk. But the colours distinguished them – black, red, yellow and blue. The Golden Notebook, Doris Lessing

Private library for artist Douglas Gordon. Design by Andrew Miller. Photography by Andrew Miller

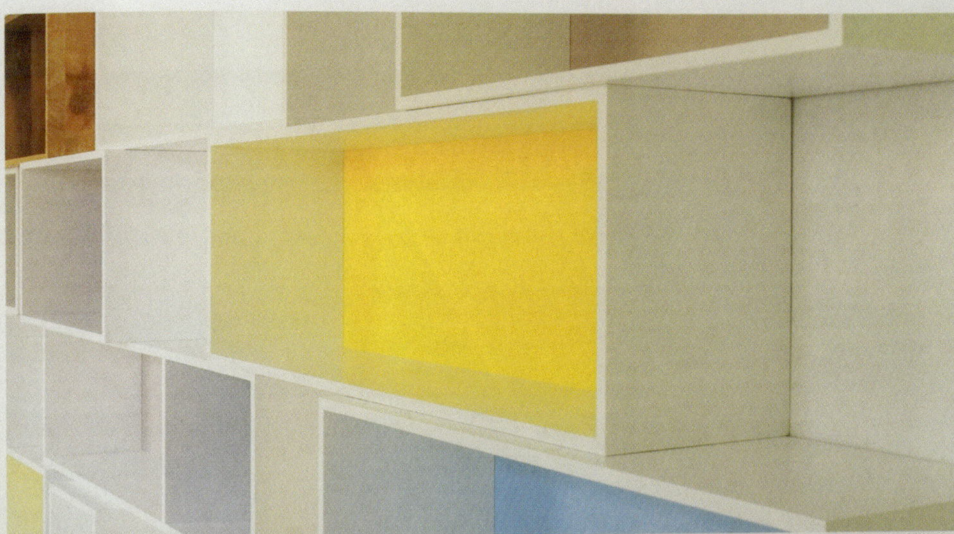

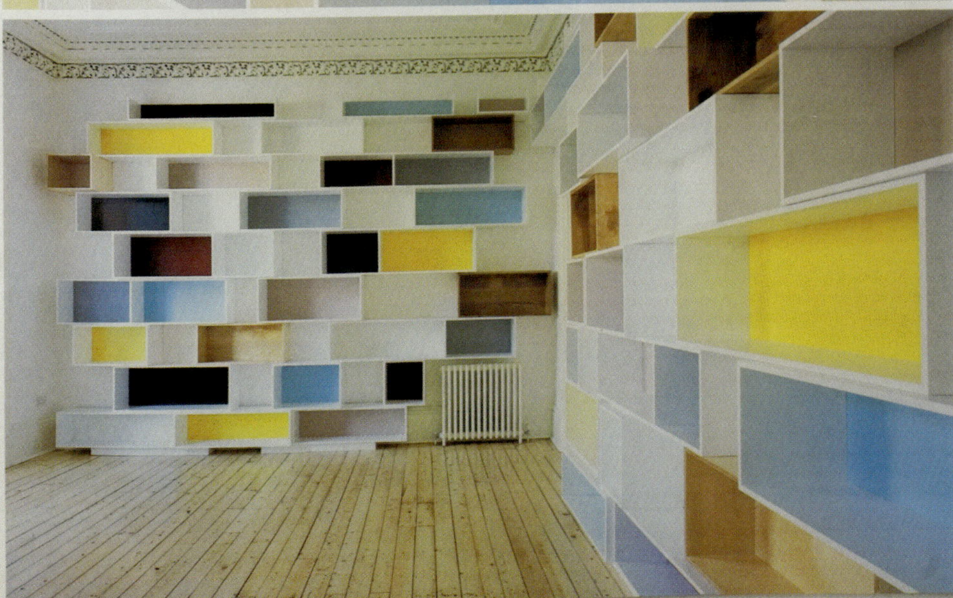

SUBJECT Seven libraries

LOCATIONS Various, see page 347

DESIGNERS Various, see page 347

PHOTOGRAPHERS Various, see page 347

WRITER Frame

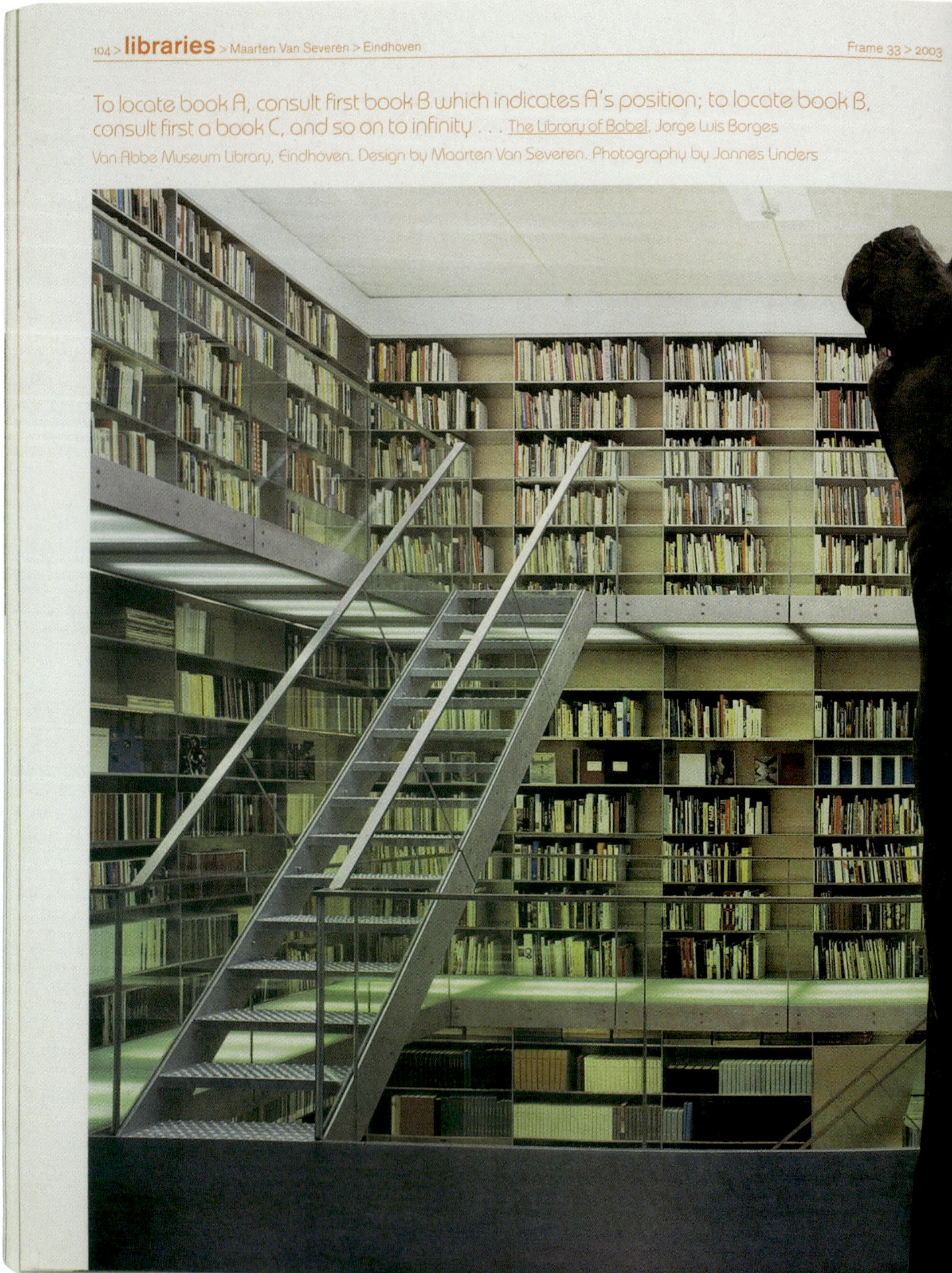

104 > **libraries** > Maarten Van Severen > Eindhoven

Frame 33 > 2003

To locate book A, consult first book B which indicates A's position; to locate book B, consult first a book C, and so on to infinity . . . The Library of Babel, Jorge Luis Borges

Van Abbe Museum Library, Eindhoven. Design by Maarten Van Severen. Photography by Jannes Linders

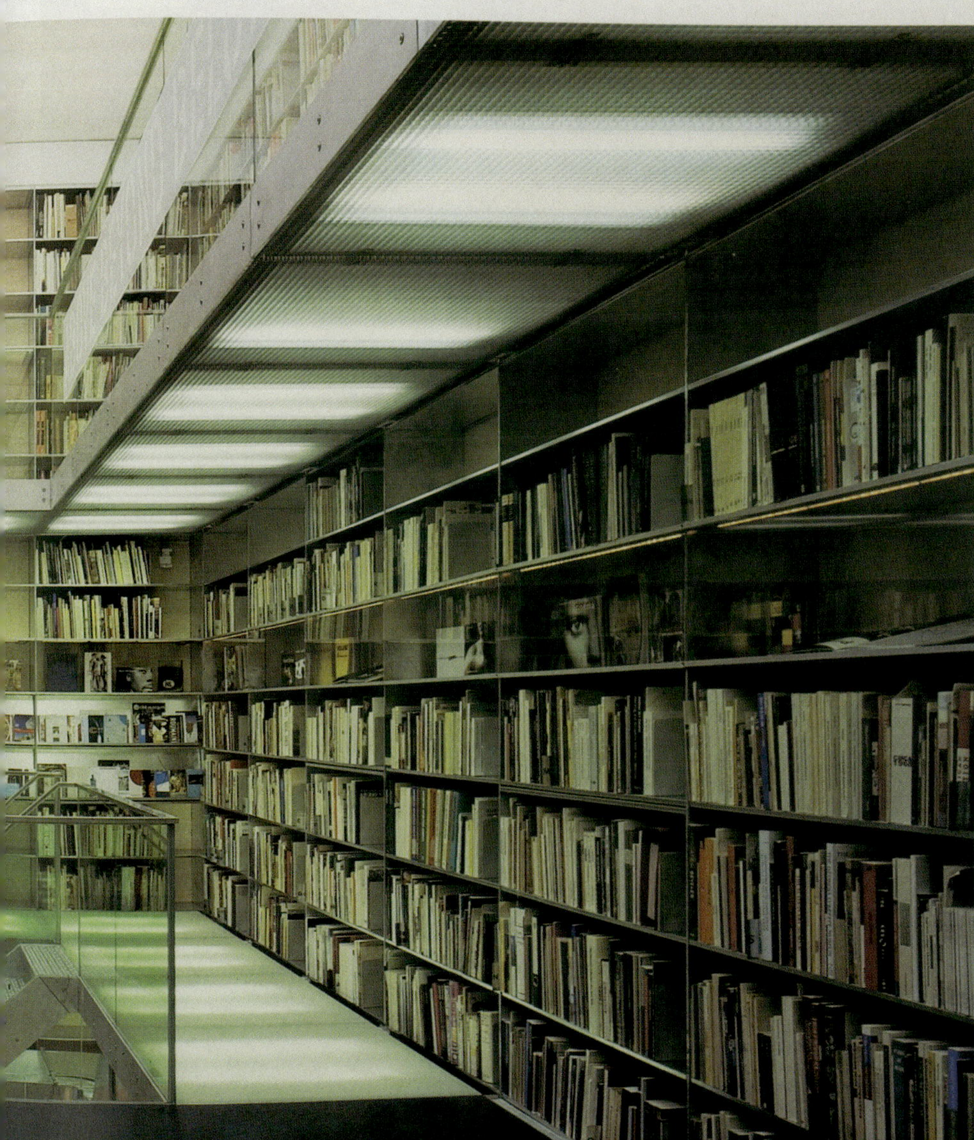

SUBJECT
Seven libraries

LOCATIONS
Various, see page 347

DESIGNERS
Various, see page 347

PHOTOGRAPHERS
Various, see page 347

WRITER
Frame

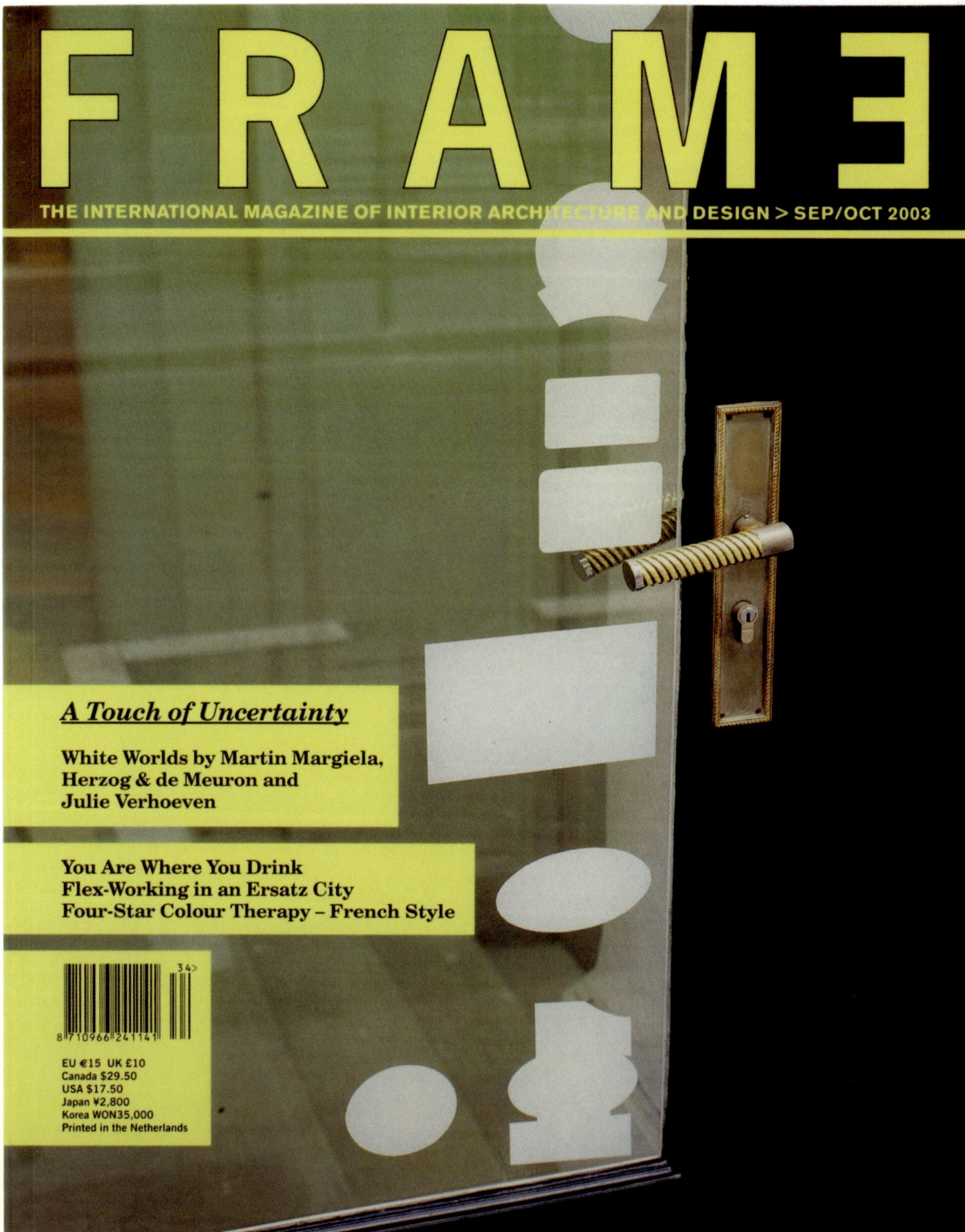

ART DIRECTORS
Cornelia Blatter & Marcel Hermans, COMA

34

Arabian Nights

Artist Hassan Hajjaj about the project

Ten years of heart, blood and soul.

Comments on the published article

Brilliant.

Did the article have an effect?

Yes. It got me more work and recognition.

Current project

I am working on a project for Tate Gallery in St Ives, to be completed in 2007.

DESIGNERS
Hassan Hajjij
Bruno Caron

SUBJECT
Tapas bar-restaurant
Andy Wahloo

LOCATION
Paris, France

PHOTOGRAPHER
Daniël Nicolas

WRITER
Chris Scott

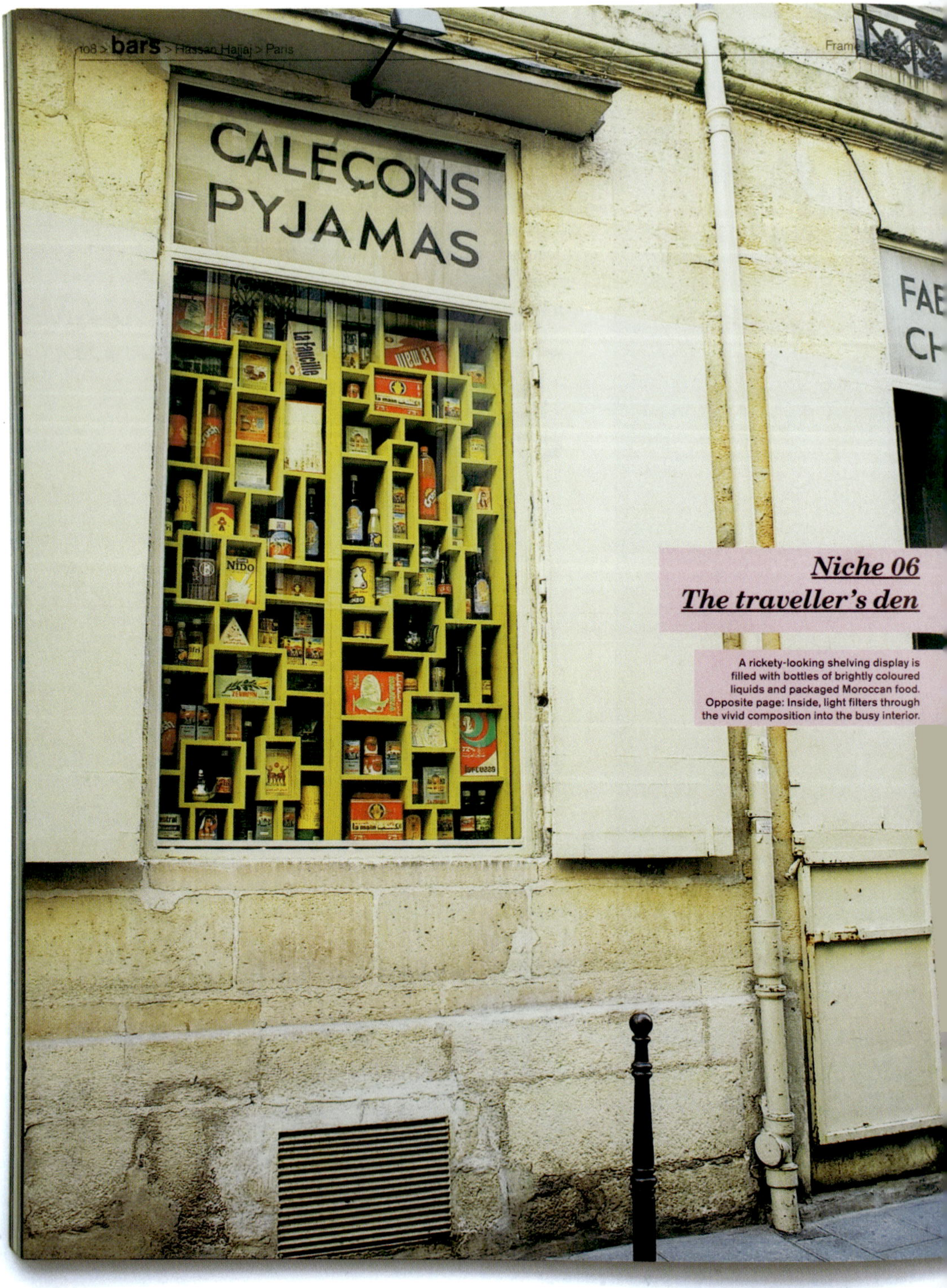

Niche 06
The traveller's den

A rickety-looking shelving display is filled with bottles of brightly coloured liquids and packaged Moroccan food. Opposite page: Inside, light filters through the vivid composition into the busy interior.

Arabian Nights

Marking a departure from big-name designers, Moroccan artist Hassan Hajjaj gives recycling an updated makeover at Parisian tapas bar-restaurant Andy Wahloo.

Chris Scott
Photography by Daniël Nicolas

Think fusion; think Andy Warhol; think pop art meets Arabia. Located on an anonymous street in Paris's third arrondissement, a recently opened tapas bar-restaurant bears a name that captures the essence of the project. Loosely translated into Arabic, Andy Wahloo means 'I have nothing', a statement that underpins the scheme.

Andy Wahloo is just one of a clutch of eateries in Paris and London owned by brothers Karim and Mourad Mazouz. Two of their more well-known London projects are Momo and the famed extravaganza known as Sketch (see *Frame* 32). In the same week that Mourad launched Sketch, Karim introduced Andy Wahloo to Paris. The interior revamp of the latter boasts none of the big names that the Mazouz brothers brought in to collaborate on Sketch. For the Paris project, they handed design responsibility over to artist Hassan Hajjaj, a man who modestly refers to himself as a 'doer'.

As great admirers of Hajjaj's work, the Mazouz brothers had earlier sponsored a Moroccan exhibition of his creative efforts. So taken were they by the setting Hajjaj selected, as well as by the way he presented his work, that inviting him to do the interior of their new Parisian project seemed to be the natural follow up.

The risk of hiring an artist to play the role of designer paid off. Hajjaj approached the task with a blend of creativity and sensitivity. In contrast to some of the Mazouz's design palaces, the Andy Wahloo project was forced to rely on a shoestring budget. As Karim points out, 'Not having the means to design a grand venue made us extra inventive. Andy Wahloo may have been done on the cheap, but it's a feel-good space nonetheless. The atmospheric vibe is all-important. Without this, you have nothing but an empty shell.'

Hajjaj celebrates the culture of his native country, Morocco, in his work. He combines childhood memories and early influences with later experiences from his urban lifestyle in London. The space is packed with an eclectic mix of interesting objects, many of which are for sale in the basement. Juxtaposed with Hajjaj's love of pop art are '70s food containers, Moroccan adverts and packaging, oriental fabrics, upturned Coca-Cola crates recycled as stools, and

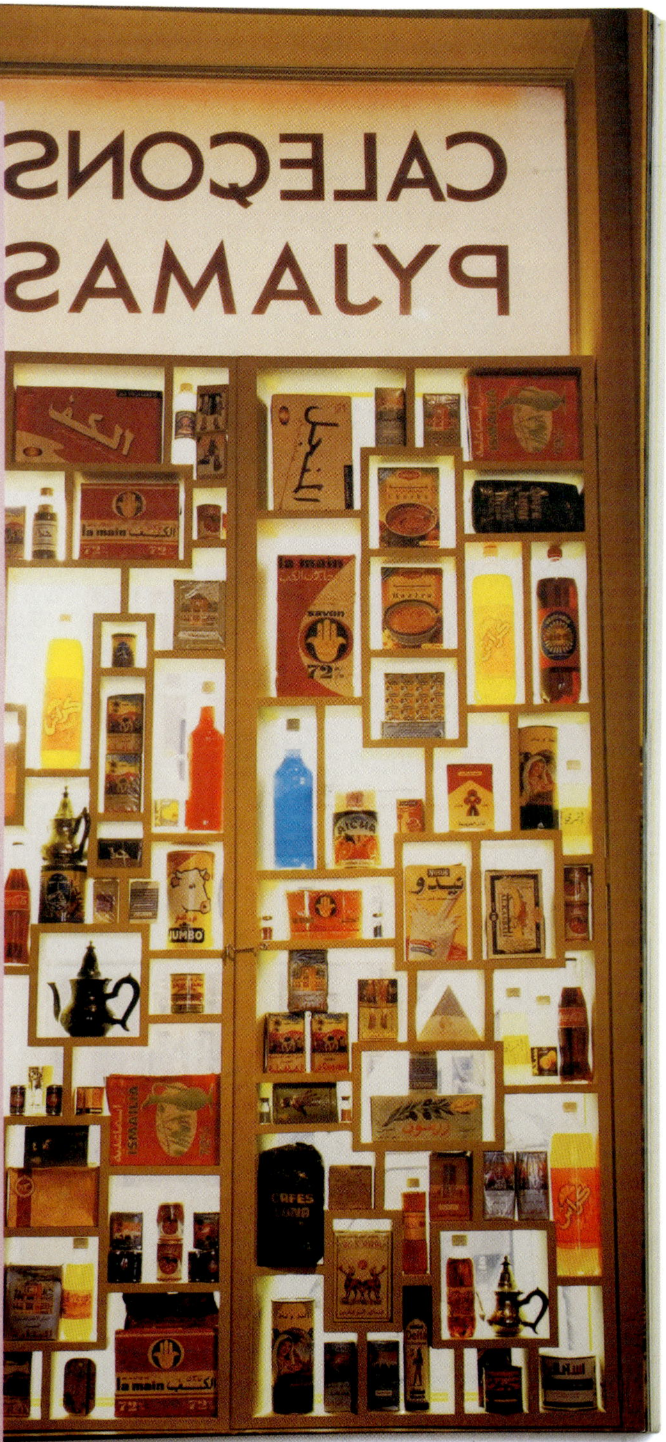

SUBJECT Tapas bar-restaurant Andy Wahloo

LOCATION Paris, France

DESIGNERS Hassan Hajjij Bruno Caron

PHOTOGRAPHER Daniël Nicolas

WRITER Chris Scott

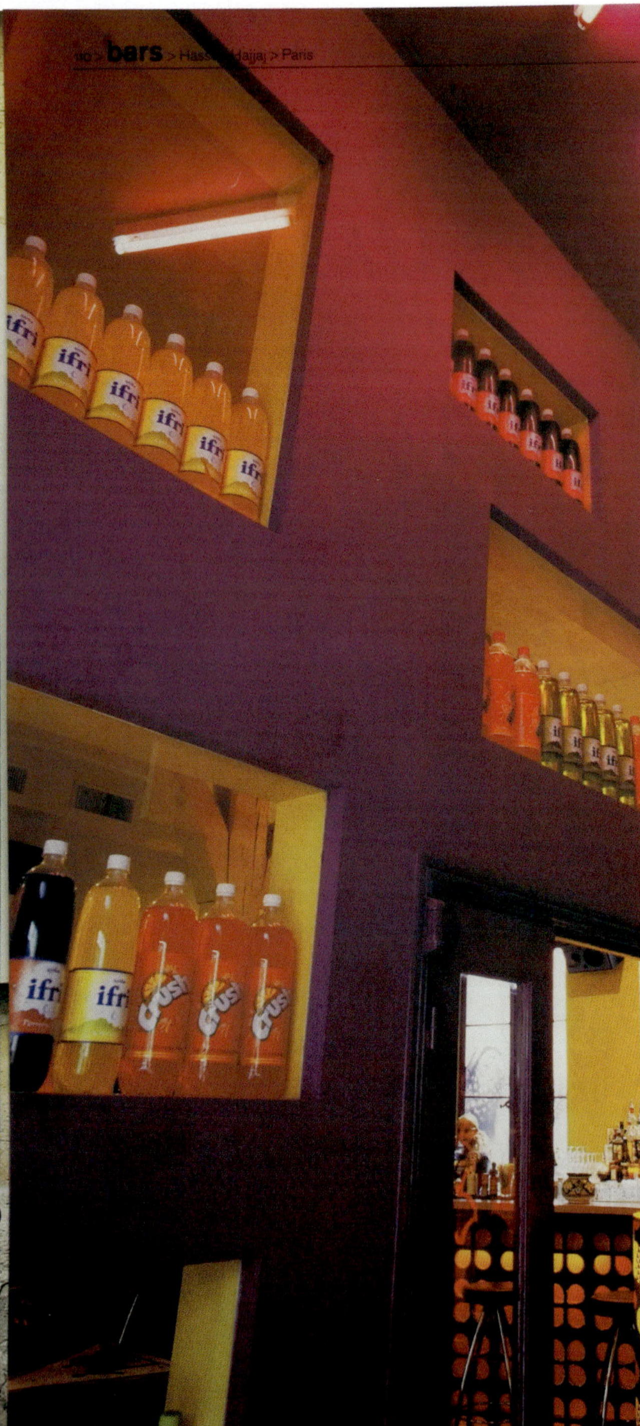

street signs that function as kitsch tabletops. The result is an excitingly animated ambience.

In keeping with the cultural values of Morocco, nothing is wasted or discarded. Design aficionados might cringe at the idea of roughly finished recycled objects used as furnishings, but Hajjaj's giant paint pots stacked with '70s patterned cushions and Coca-Cola crates piled high with kilim cushions represent an updated take on recycling.

A rickety-looking shelving display fills the high window fronting the street. Arranged in spacious splendour are bottles of brightly coloured liquids and an array of packaged Moroccan food. A dominating feature of the design, the display attracts attention from both inside and out. The intriguing Moroccan products provided the inspiration for Hajjaj's restaurant concept.

From the street, the display creates the impression that Andy Wahloo is a low-cost grocery store. Original lettering on the façade above the window gives away the former function of the space as the site of a fashion wholesaler. Inside, the display produces a beguiling effect as light filters through the vivid composition into the busy interior.

Hajjaj collaborated with architect Bruno Caron, as well as the Mazouz brothers. Caron, who advised on practical and technical aspects, also designed the bar and the central spiral light fixture that supports Hajjaj's Morocco-inspired lamps.

The deceptively casual look of Andy Wahloo is not merely a reaction to overly designed interiors. Nor is it simply the echo of a hippy hangout from the '70s, a description that underestimates the impact that the bar-restaurant has had on Parisians; local residents, as well as glitterati from the city's in-crowd, throng the place throughout the week. Developed through Hajjaj's artistic standpoint, the open brief has resulted in the type of creative freedom seldom seen in commercial interior design.

Hajjaj has crafted an appealing environment that is both comfortable and nonjudgmental. A sense of well-being permeates the space. Rather than dropping in for a been-there-done-that experience of the latest novelty, visitors quickly become regulars. Andy Wahloo invites clients to come as they are. No special treatment is reserved for celebrity guests, who have to wait their turn with the lesser mortals. Karim shrugs, 'Why give preference? We're all equal.' His is the kind of welcome that brings back devotees, time and again.

It's interesting to note that though Karim stresses how much he and others adore the ultra-stylish Sketch, given the choice, they opt for a night at Andy Wahloo. Both venues have their place, but the serious appeal of the low-budget, big-atmosphere style of Andy Wahloo takes the lounge to another level.

Artist Hassan Hajjaj juxtaposes '70s food containers, oriental fabrics and Moroccan accessories with street-sign tabletops, kilim cushions and upturned Coca-Cola crates recycled as stools. Nothing has been wasted or discarded.

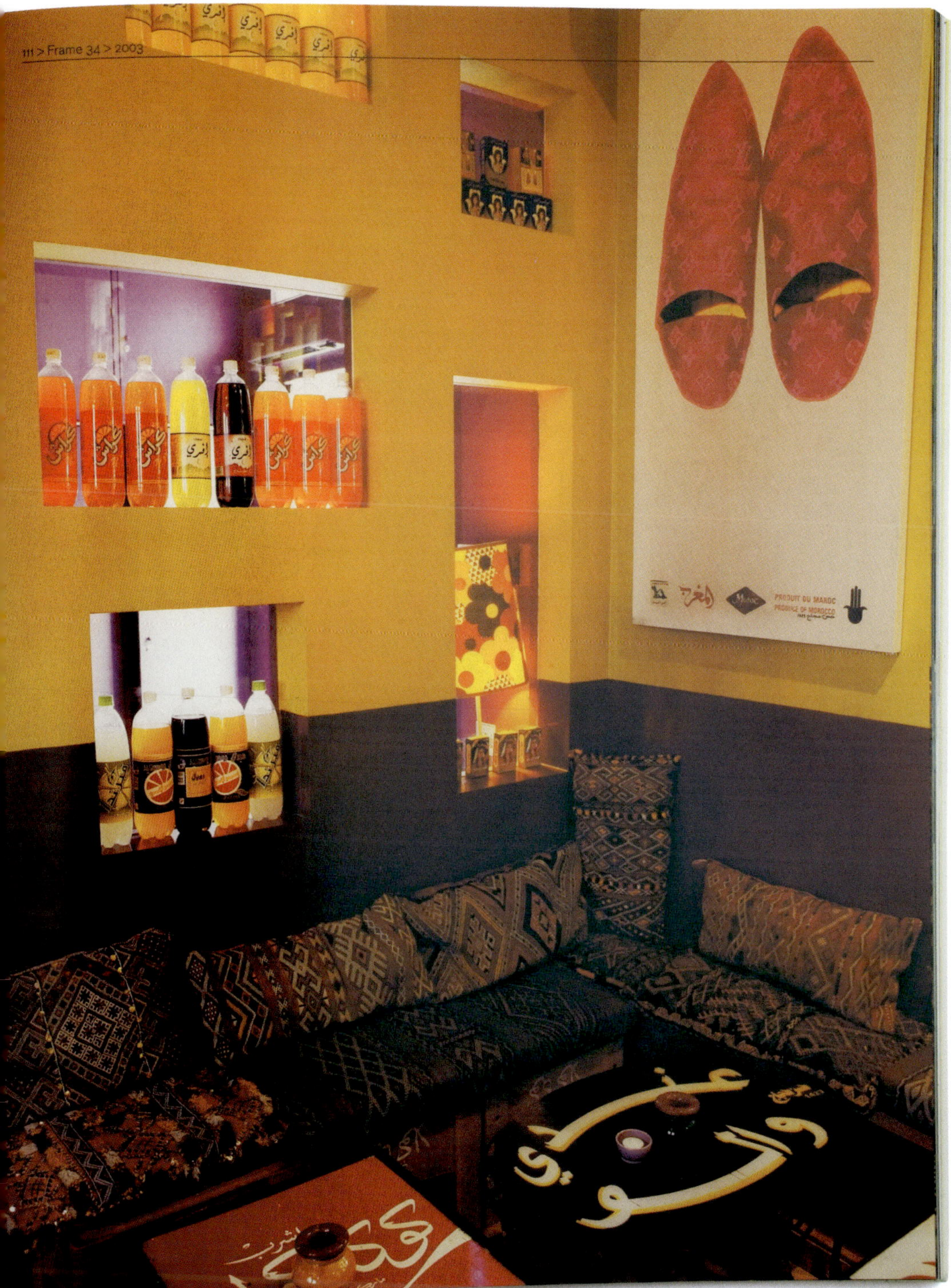

SUBJECT
Tapas bar-restaurant
Andy Wahloo

LOCATION
Paris, France

DESIGNERS
Hassan Hajjij
Bruno Caron

PHOTOGRAPHER
Daniël Nicolas

WRITER
Chris Scott

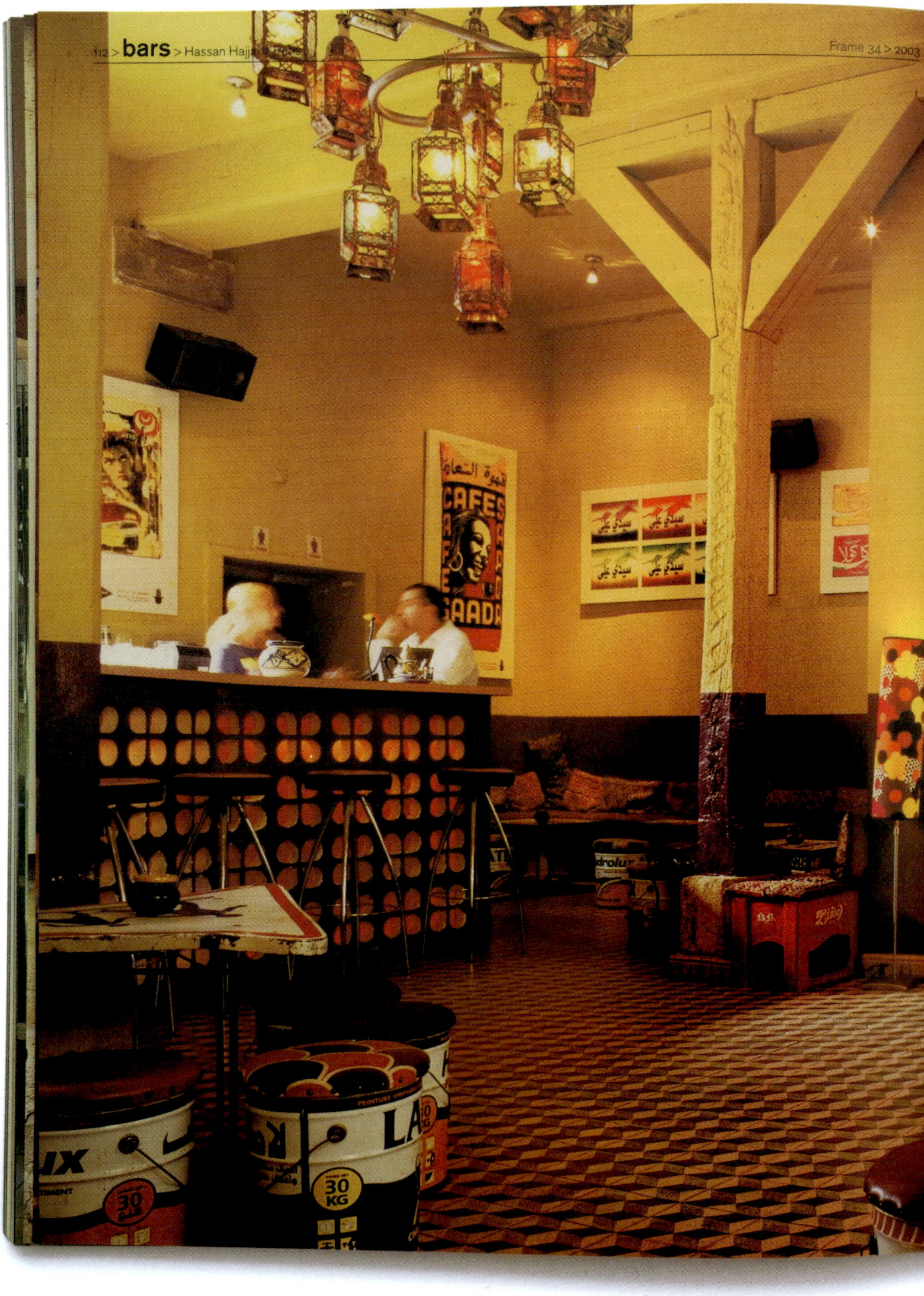

112 > bars > Hassan Hajj
Frame 34 > 2003
CAFES
SAADA
30 KG

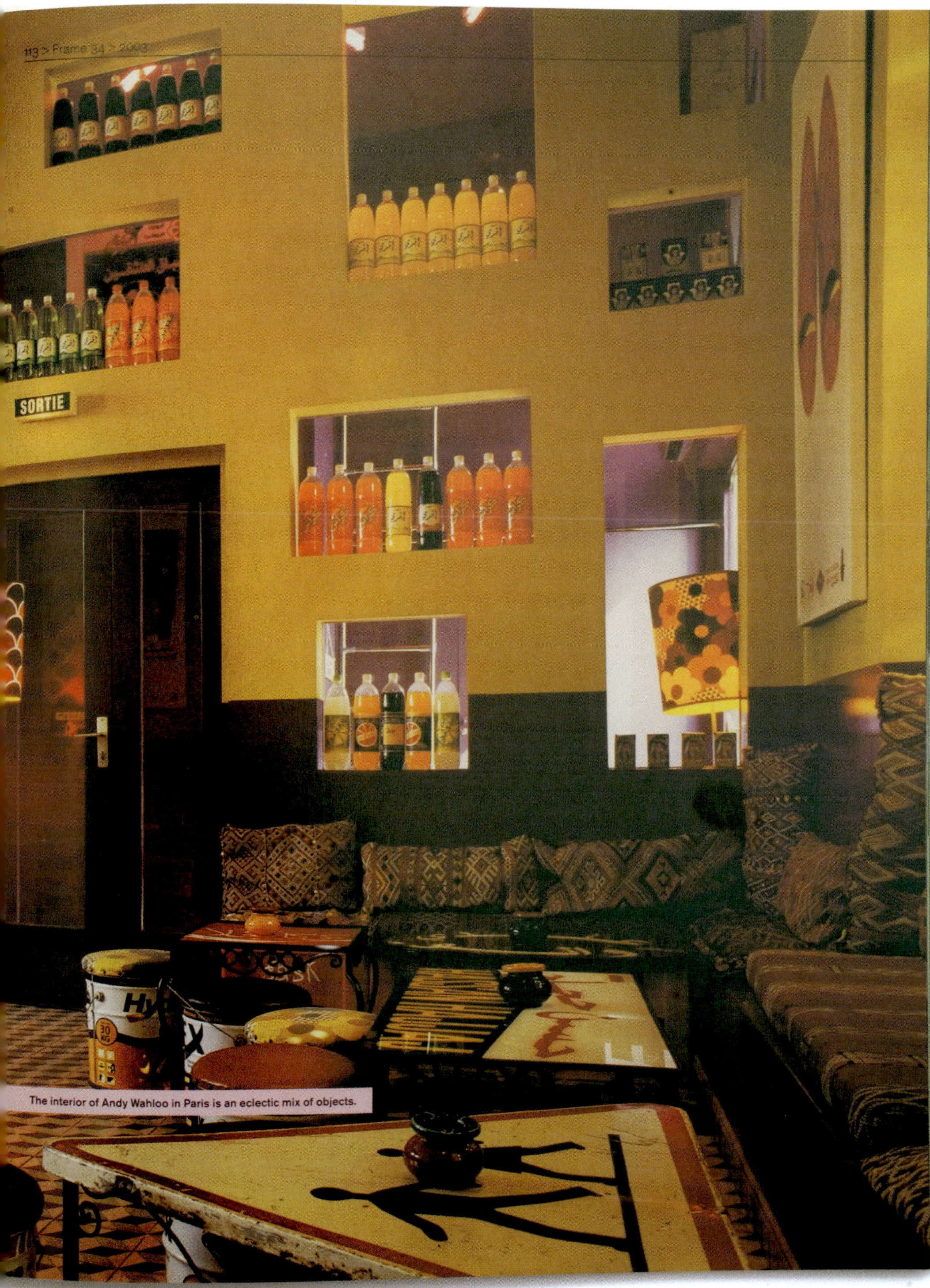

The interior of Andy Wahloo in Paris is an eclectic mix of objects.

SUBJECT
Tapas bar-restaurant
Andy Wahloo

LOCATION
Paris, France

DESIGNERS
Hassan Hajjij
Bruno Caron

PHOTOGRAPHER
Daniël Nicolas

WRITER
Chris Scott

368-369

November/December 2003

Issue 35

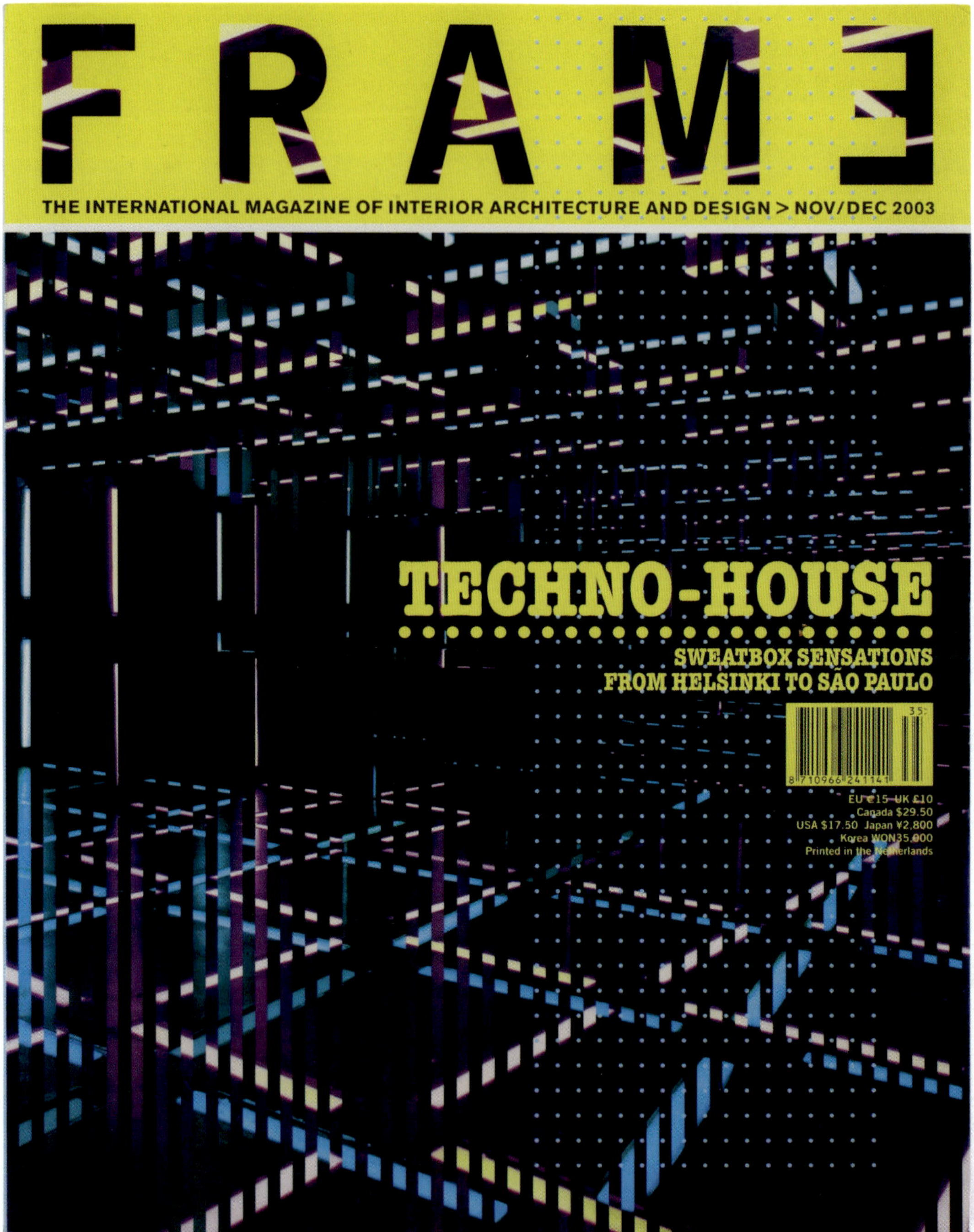

ART DIRECTORS
Cornelia Blatter & Marcel Hermans,
COMA

Cutting Edge

Designer Muti Randolph about the project

In D-edge São Paulo, my third club project, I was lucky to be able to create an environment that I dreamed about, using technology as both the source of inspiration and the means to make it real: a dance floor where people can actually see music. I am very proud and happy to realize that it turned out to be a world reference in club design.

Comments on the published article

As usual in Frame, the article is very well written, designed and printed. As well as in other stories published in Frame about my work, this article made me think about aspects of the project that I had not thought about before. It feels like taking the project to the shrink (laughs). But to be featured on the cover of the design magazine that I most respect and admire was the best part.

Did the article have an effect?

Frame is a very big reference, not only to architects and designers, but also to other publications. After being published by Frame, the project was featured in several books and magazines around the world, and became a reference and landmark in club design.

Recent news or information about the project featured in *Frame*

In its three years of life, D-edge has consolidated its position as the best underground club in São Paulo.It has been voted as such for three years in a row by the annual Cool Magazine Awards. It was elected the 4th best club in the world by a poll of djs for British clubbing magazine Mixmag last December. In February it is going through a major maintenance restoration (it is really needing it!) and the new annex will open in April.

DESIGNER
Muti Randolph
www.muti.cx

SUBJECT
Nightclub D-edge

LOCATION
São Paulo, Brazil

PHOTOGRAPHER
Rômulo Faildini

WRITER
Alberto Renault

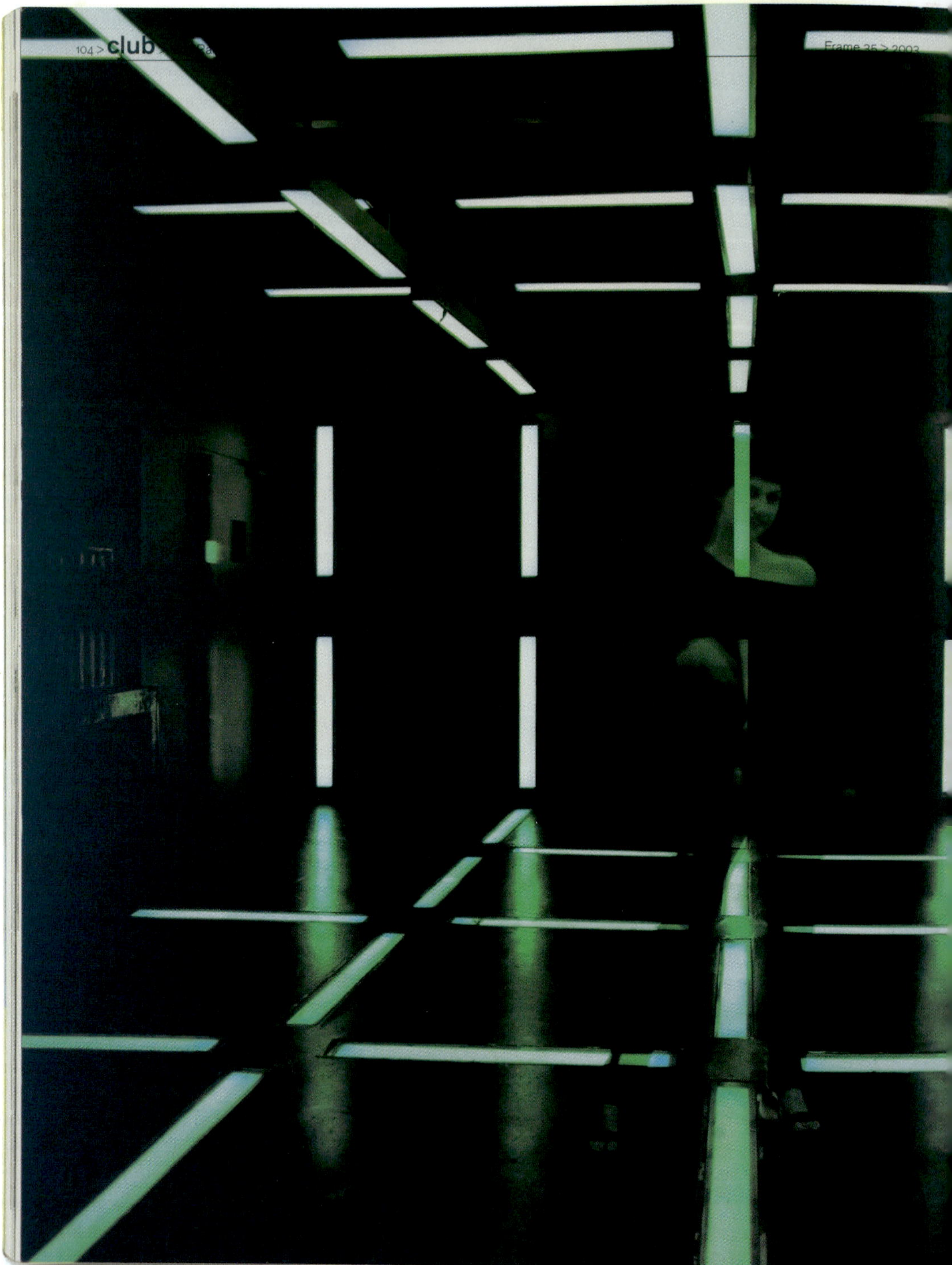
104 > club

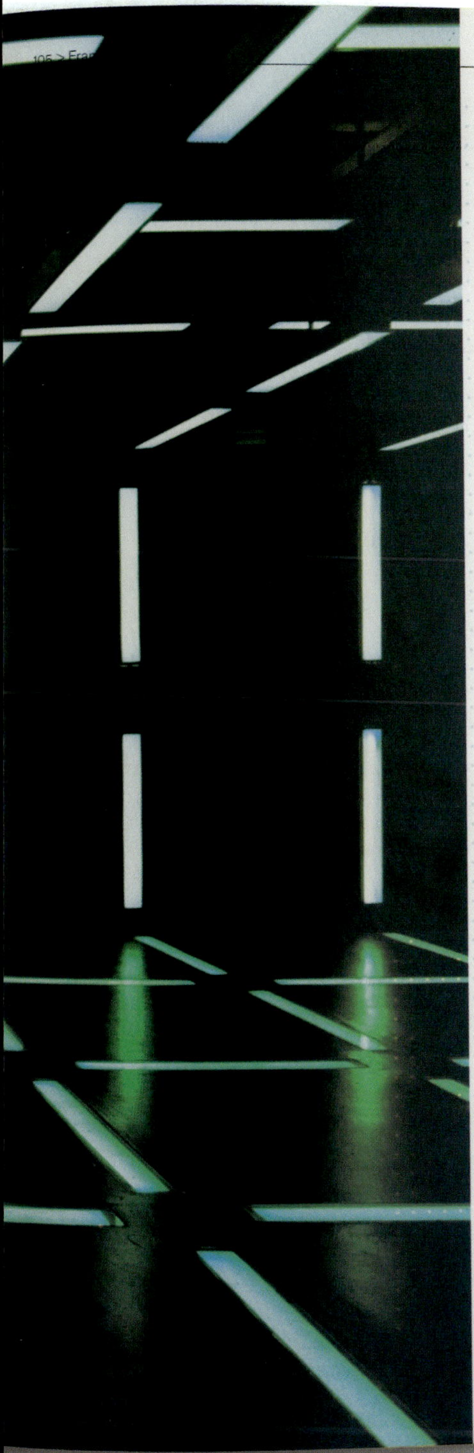

CUTTING EDGE

Muti Randolph's extraordinary lighting system defines D-Edge, a São Paulo club that's moved into the city's music scene on the merits of his computer-crazed imagination.

Text by Alberto Renault
Photography by Rômulo Fialdini

A megalopolis with an estimated population of 17 million, São Paulo is both the biggest city on the South American continent and *the* location for DJs out to fire up enthusiastic clubbers immersed in the city's rapidly expanding music scene. In one throbbing venue, young Paulistas dance as bars of light slice the air in a grid-like pattern, a battle of beams intriguingly in sync with the music. Though late-night patrons at D-Edge – a club that accommodates up to 800 – are unlikely to philosophise on the materiality of light and sound, they obviously relish the futuristic ambience. Delighted with the success of the first D-Edge in Campo Grande (see *Frame* 25), the club's DJ owner invited Brazilian Muti Randolph to participate in another stunning joint venture.

Although the local citizenry complains about traffic, street violence and pollution, São Paulo's exciting nightlife, which transforms the city into a dynamic hub of sight and sound, is not on the gripe list. D-Edge caters to the city's extraordinary range of club-bound nighthawks. And though the club shares a name with its country cousin, by all appearances the São Paulo branch of D-Edge bears little relationship to the Panton-bright space in Campo Grande. Here in the big city, a concrete floor coated in black resin meets (false) wooden walls, also black, to form a dark backdrop for a vast array of white Plexiglas lamps sunk into

A concrete floor coated in black resin meets black (false) wooden walls to form a dark backdrop for a vast array of white Plexiglas lamps sunk into walls and floor and suspended in narrow troughs from the ceiling.

WRITER Alberto Renault

PHOTOGRAPHER Rômulo Faildini

DESIGNER Muti Randolph

LOCATION São Paulo, Brazil

SUBJECT Nightclub D-edge

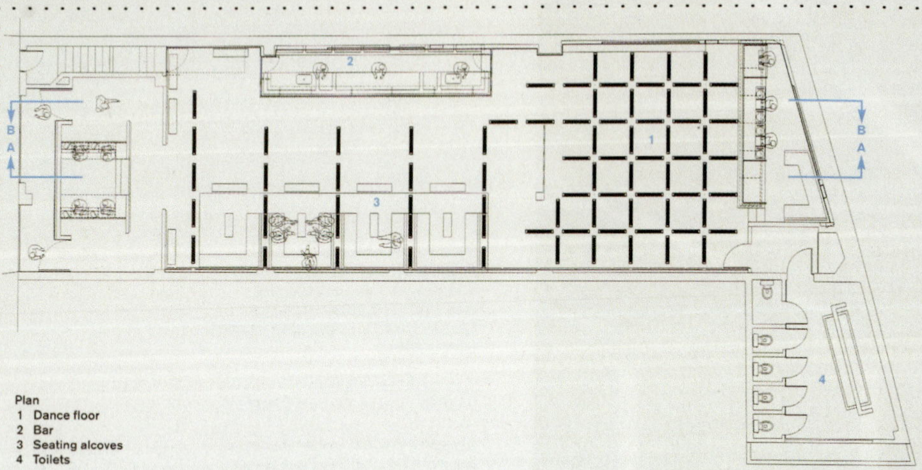

Plan
1 Dance floor
2 Bar
3 Seating alcoves
4 Toilets

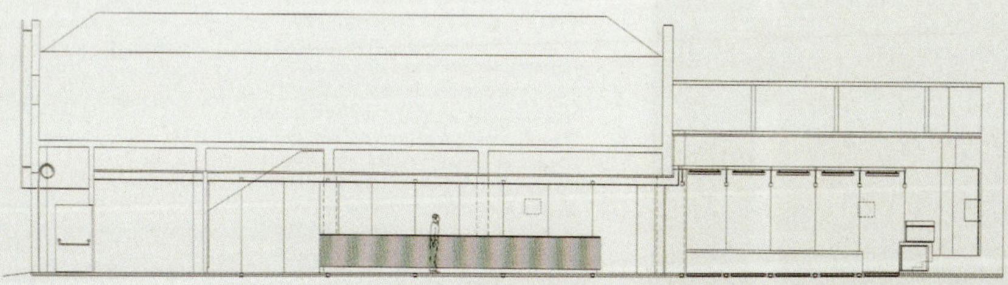

Section A-A

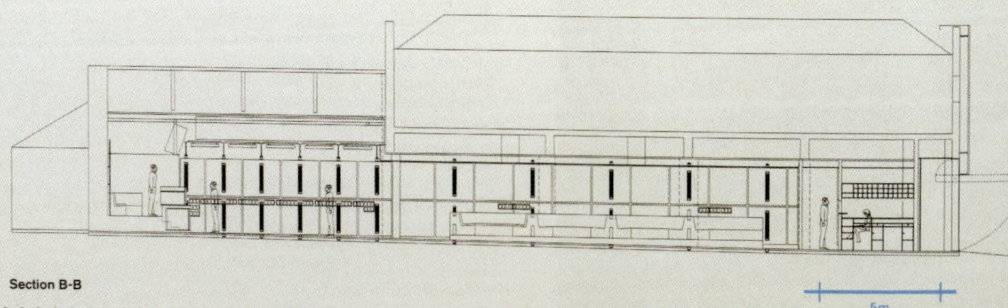

Section B-B

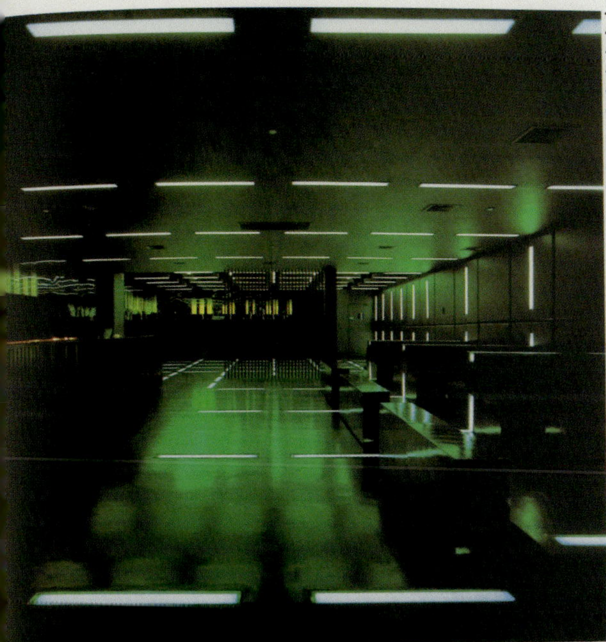

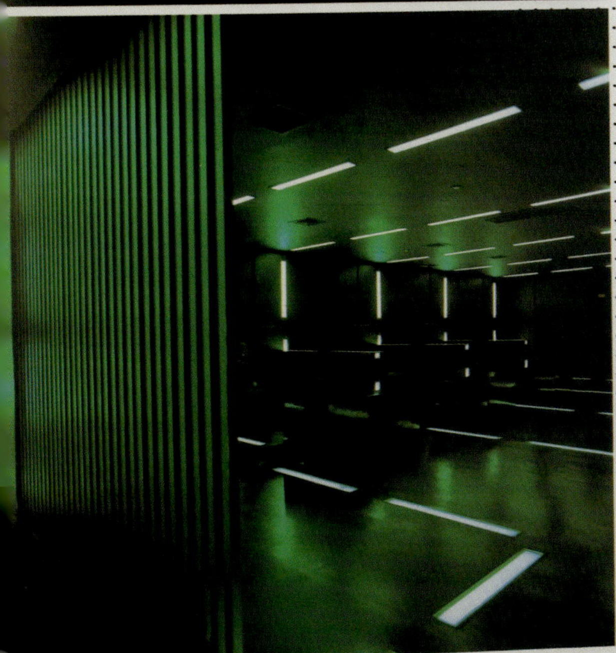

walls and floor and suspended in narrow troughs from the ceiling.

The extraordinary lighting system is geared to the music. Working in close collaboration with the DJ is a light jockey, or LJ, who operates a sophisticated program of light that moves to the music. The dramatic result of their common endeavour shapes the environment. They select one or two colours each night, depending on the party feel. For a techno-craving crowd, the club conveys a minimalist, somewhat industrial atmosphere based on a chill palette – ideal for tattooed, black-clad clubbers and after parties that go on till noon. On house or disco nights, warm hues change the space into a vibrant setting for fun-loving models and stylist wannabes who flaunt an exaggerated '80s look. Flickering frenetically on the dance floor, the lights ease into a subtler program in the bar. One interesting sequence features lights blinking in an apparently random manner, thus producing a sense of disorder that counteracts the Cartesian rationalism of the grid.

The grid of light at D-Edge is an analogue of a computer monitor: a dark foundation whose viability relies on light

'I've always been a bit geeky,' confesses Randolph, referring in particular to his lifelong addiction to video games and computers. 'I do spend too much time interacting with computers, but because of their constant presence, they end up being a source of inspiration.' Rio-based Randolph – who makes the 500-kilometre journey southwest to São Paulo for work – also recognises the downside of his passion: 'At the same time that computers give me the power and freedom to create virtual worlds, they keep me away from the real thing.' The relationship between the virtual cosmos and life beyond the monitor fascinates the graphic designer, who loves watching what he crafts on screen turn into a set for MTV, for example. Randolph compares set design to what he's done at D-Edge. Both, he explains, are projects that sweep people out of an ordinary environment and pop them into a virtual world. It's this kind of thinking that frames

Top left: The lighting system enlivens the bar area as well, albeit less dominantly. Here the light show is more easy-going than on the dance floor.
Left: Four seating alcoves line one wall of the club.

SUBJECT Nightclub D-edge
LOCATION São Paulo, Brazil
DESIGNER Muti Randolph
PHOTOGRAPHER Rômulo Fialdini
WRITER Alberto Renault

his favourite metaphor: 'The computer is to me what the mirror was to Alice.'

The grid of light at D-Edge is an analogue of a computer monitor: a dark foundation whose viability relies on light. Instead of pixels, the primary elements of this 'bitmap' are the neon light boxes that compose the grid. A computer connected to a light table that is linked, in turn, to the sound system creates an image for each beat of the music. Like pixels, each lamp emits a colour determined by the mix of red, green and blue light dictated by the software.

Muti Randolph gets clubbers moving by giving form to the sound and rhythm to the space

Club owner and DJ Renato Ratier works the decks once a week on his home turf, usually hosting house parties, which highlight his music of choice. On special occasions, well-known DJs on the national and international circuit make guest appearances for select groups of about 450 patrons with an ear for experimental and underground music.

A final comparison of the two clubs finds D-Edge Campo Grande posing as a playful set that simulates the inner workings of a computer, while the newer venue in São Paulo revolves around computerised reality. Wall art at the former mimics the circuitry of a massive machine, while hollow walls at the São Paulo venue conceal kilometres of real cables that link the main computer to the lighting system. 'The two concepts are similar,' says Randolph. 'Both are gigantic machines, but one is largely scenographic and the other actually works. What I tried to do in São Paulo was to stimulate clubbers, to really get them moving by giving form to the sound and rhythm to the space.'

D-EDG

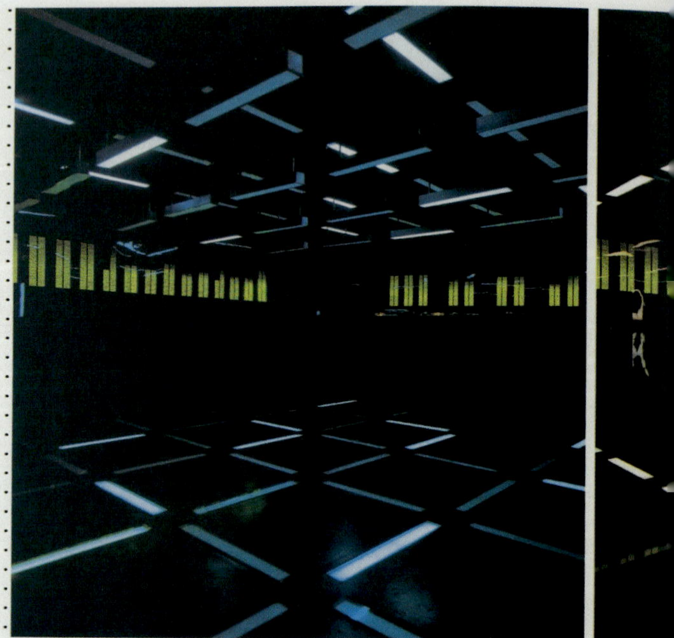

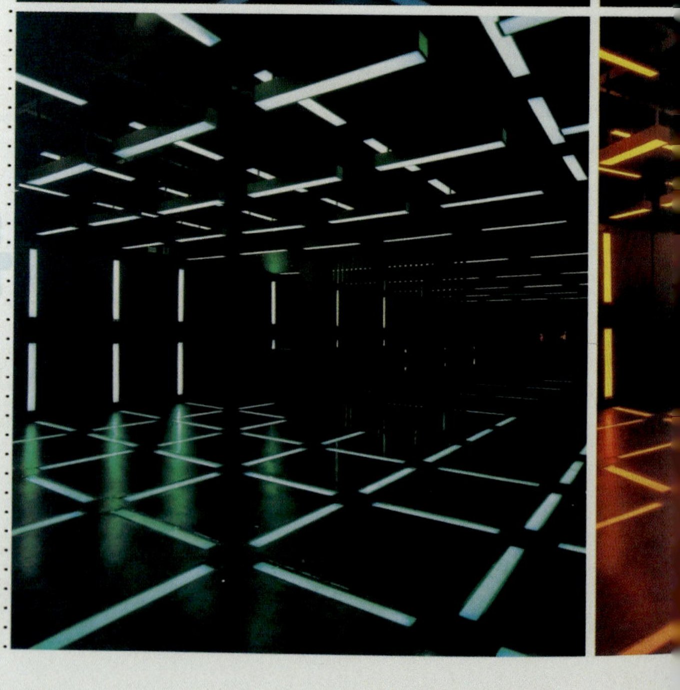

The light show is geared to the music. A DJ and an LJ (light jockey) select one or two colours each night, depending on the party feel. The giant 'equaliser' LED bars behind bar and DJ booth (top row) are the most talked-about feature of the club.

109 > Frame 35 > 2003

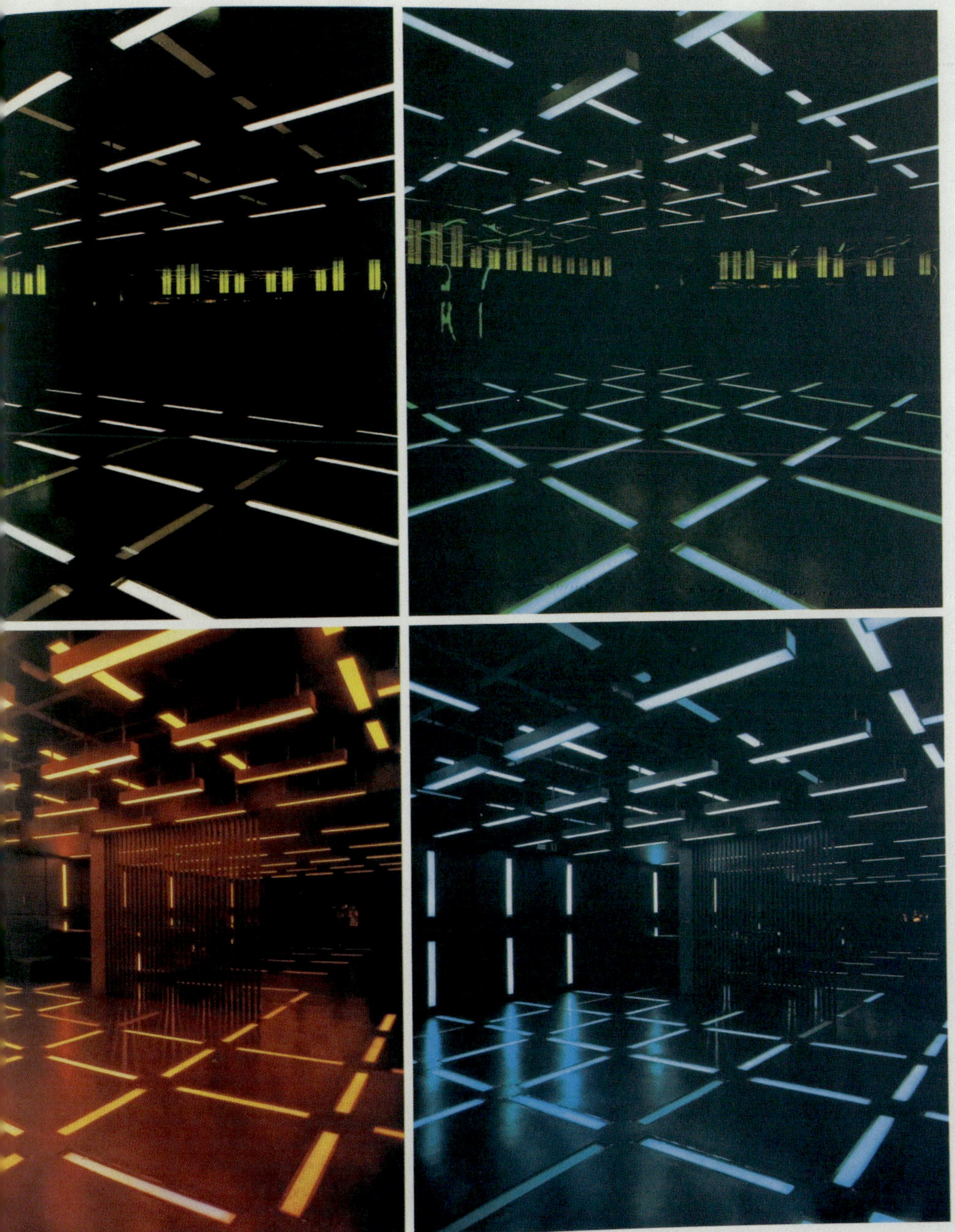

SUBJECT Nightclub D-edge

LOCATION São Paulo, Brazil

DESIGNER Muti Randolph

PHOTOGRAPHER Rômulo Faildini

WRITER Alberto Renault

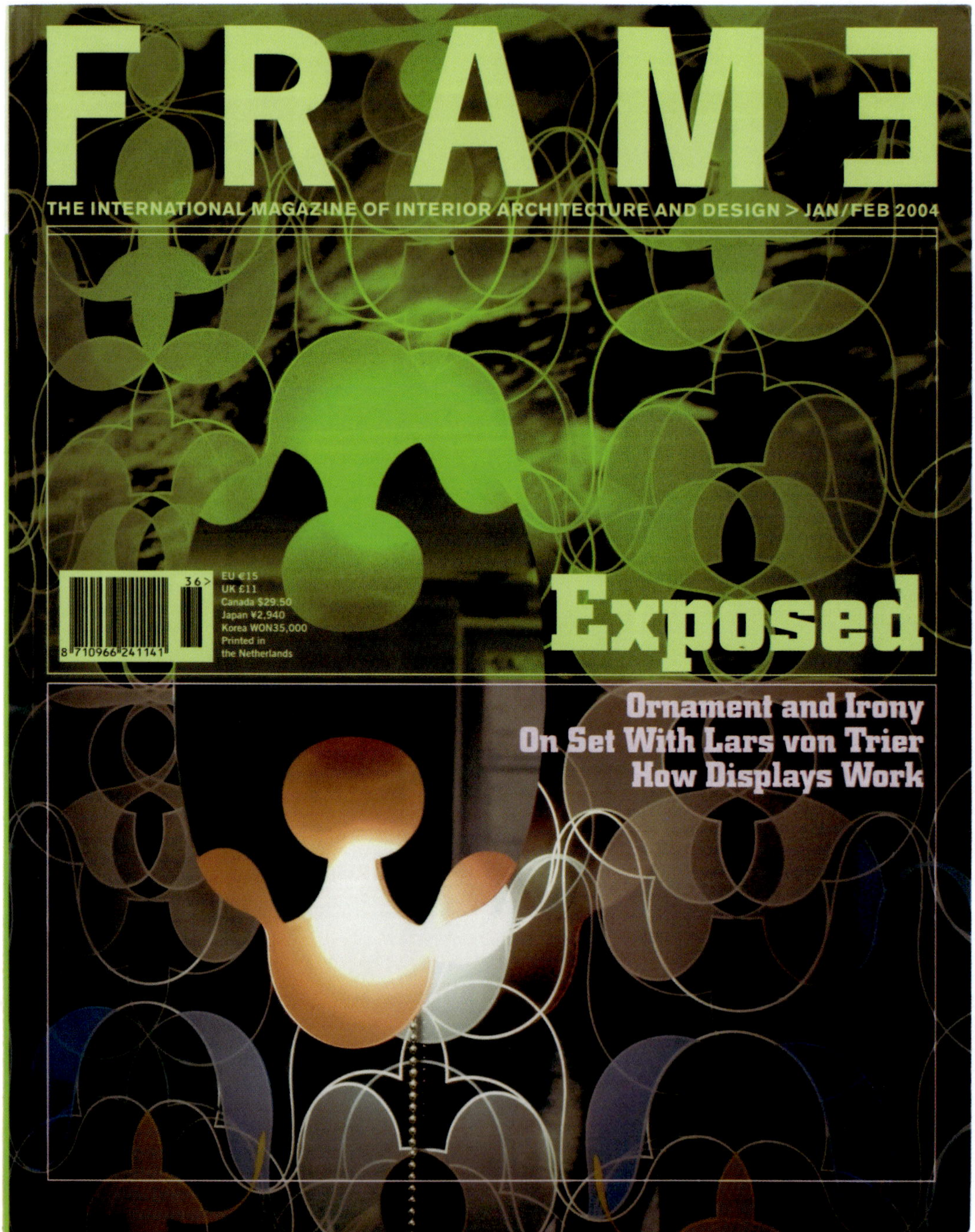

ART DIRECTORS

Cornelia Blatter & Marcel Hermans, COMA

36

Haute Cuisine

Designer Jörg Boner about the project

I still find it a very interesting project, particularly because we used product design and graphic design to achieve the final result. The project included new technologies, new materials, irony, humour, colours and ornaments.

Comments on the published article

A big pleasure! The text is a detailed description of the project.

Recent news or information about the project featured in *Frame*

It is still in use and, in Switzerland, well known.

Current projects

Product design for Nils Holger Moormann, FontanaArte and others. We will also complete another interesting interior-design project this year.

DESIGNERS
Chalet 5
www.chalet5.ch
Jörg Boner
www.joergboner.ch

SUBJECT
Cooking centre

LOCATION
Aarau, Switzerland

PHOTOGRAPHERS
Atelier Fontana

WRITER
Ralf Michel

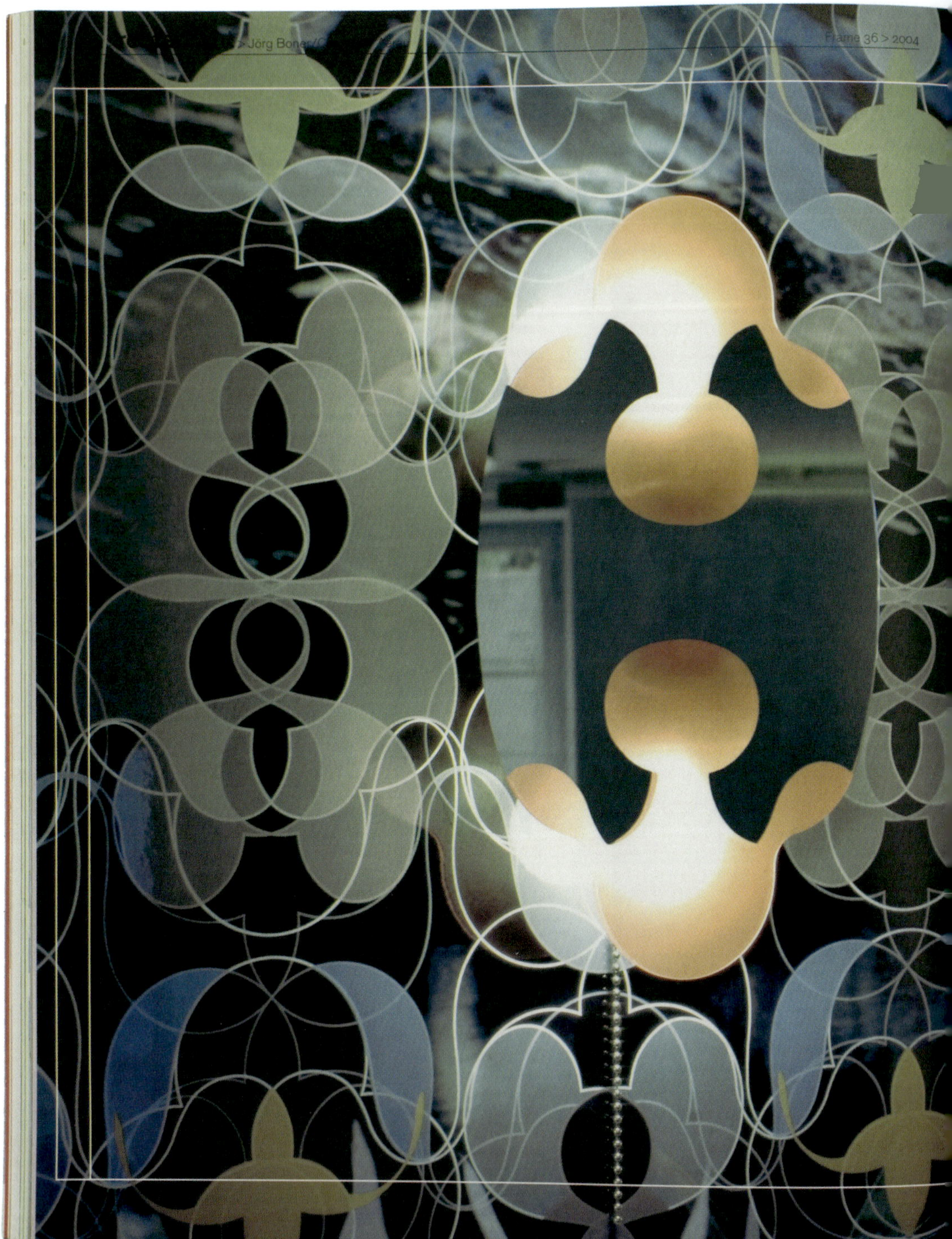
Frame 36 > 2004

Haute Cuisine

At a Swiss cooking centre, Chalet 5 and Jörg Boner raise the sampling phenomenon to a higher level.

Text by Ralf Michel. Photography by Atelier Fontana

Great chefs can take a few simple ingredients and come up with gourmet delights. To excel at their profession, they need both patience and experience, not to mention the spark of talent that tells them which flavours enhance one another and when a dish is spicy enough. Those who have truly mastered the art of cooking also possess a healthy love of experimentation.

A trip to Aarau, a town some 40 kilometres west of Zürich, confirms the mutually metaphorical nature of cooking and designing. Just beyond the heart of town, and only a five-minute walk from Herzog & De Meuron's new art centre, is a former factory complex that, until a few years ago, produced optical devices for companies such as Leica. Today the converted buildings house a number of service industries. About ten years ago chef Susanne Vögeli set up a business in the former factory canteen and called it Cookuk. Vögeli prepares meals for companies and private individuals, rents out the space for parties and other events, and conducts cookery courses: students prepare and eat their concoctions together. After a decade of growth and success, her concept *and* her industrial kitchen were showing signs of wear.

Vögeli commissioned Chalet 5 and Jörg Boner, both of Zürich, to give Cookuk a face-lift. Theirs was not an easy job. Located at basement level, the space squats beneath a ceiling only 2.35 metres high. A further consideration was Cookuk's multifunctional role as a stage for activities ranging from cookery courses and parties to lectures and workshops. The designers had to provide ample room for preparing food and serving it to crowds, but also for accommodating up to 60 people attending a lecture, for example, followed by a meal and after-dinner festivities. Fully aware of these requirements, yet restricted to a budget of €130,000, the designers knew at the outset what they were facing. They were the chefs, so to speak, who had only the bare essentials with which to prepare a delicious spread for a throng of people. What's more, the client wanted something original.

Industrial designer and interior architect Jörg Boner

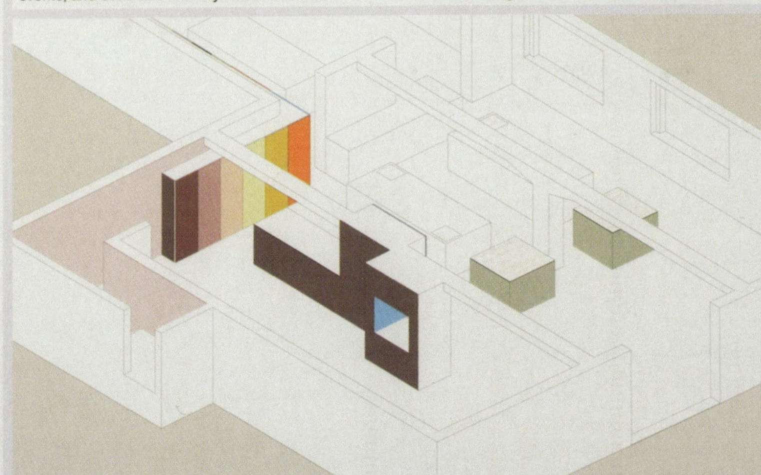

Axonometric showing aubergine entrance, brightly coloured wall cabinets and kitchen unit, including kitchen islands and a niche for audiovisual presentations.
Opposite: Wall detail with glass lamp.

WRITER Ralf Michel

PHOTOGRAPHERS Atelier Fontana

DESIGNERS Chalet 5 Jörg Boner

LOCATION Aarau, Switzerland

SUBJECT Cooking centre

'Cookuk combines a great sense of humour with the virtuoso handling of references and clichés. The refreshing result touches on trend, irony and history,' says Hannes Wettstein

approached the project with space and materiality in mind, whereas artists Karin Wälchli and Guido Reichlin of Chalet 5 set out to discover visual resources that would lend substance and depth to the interior. Together they came up with a simple recipe: expose existing features wherever possible and pack maximal originality into each new intervention.

Broadly speaking, the cooking centre consists of an industrial kitchen with two kitchen islands, a long corner banquette, easily movable tables and chairs, and a niche for audiovisual presentations. The subtlety of the design lies in the colour scheme and the details.

Before adding a thing, the designers freed the space of excess baggage associated with its past as factory and kitchen. They stripped away superfluous structures and details, levelled the floor and opened the ceiling to expose the original concrete framework. Mindful of the low ceiling, they rejected the idea of a new floor covering in favour of oiling the existing concrete floor several times, highlighting a splodgy surface that reveals the history of the building in shades of grey to olive-green. Treating the ceiling in much the same way, the designers made do with a coat of white paint.

What a difference between those minimal operations and the strategy that Boner, Reichlin and Wälchli applied to Cookuk's vertical surfaces, which they immersed in a sophisticated palette of colours, leaving nothing to chance. For starters, pale flesh tones enliven a narrow entrance devoid of daylight, welcoming visitors with a shower of warmth that raises expectations of what is to come.

Before the real fireworks go off, however, the visitor pauses briefly, confronted with a large aubergine wall that hides his view of the kitchen. Moving farther into the interior, he passes a wall of cabinets with doors of different colours, no two alike. At the midpoint of his journey he encounters the kitchen, which like all professional kitchens is a shrine to stainless steel. At one end of the room are two kitchen islands: stainless-steel-clad blocks with decorative Corian countertops.

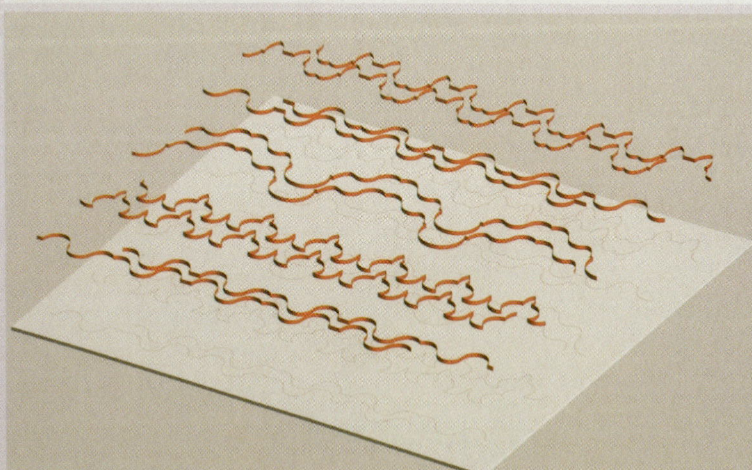

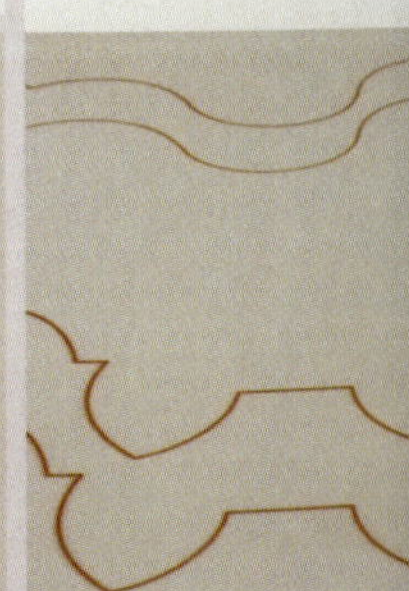

Island countertops are made of white Corian. After decorations were cut out of this surface with a computer-controlled water jet, the resulting cavities were filled with coloured liquid Corian. Opposite top: Stainless-steel kitchen with islands. Opposite bottom: View from the kitchen of the wall of coloured cabinets. One kitchen island is visible in the foreground.

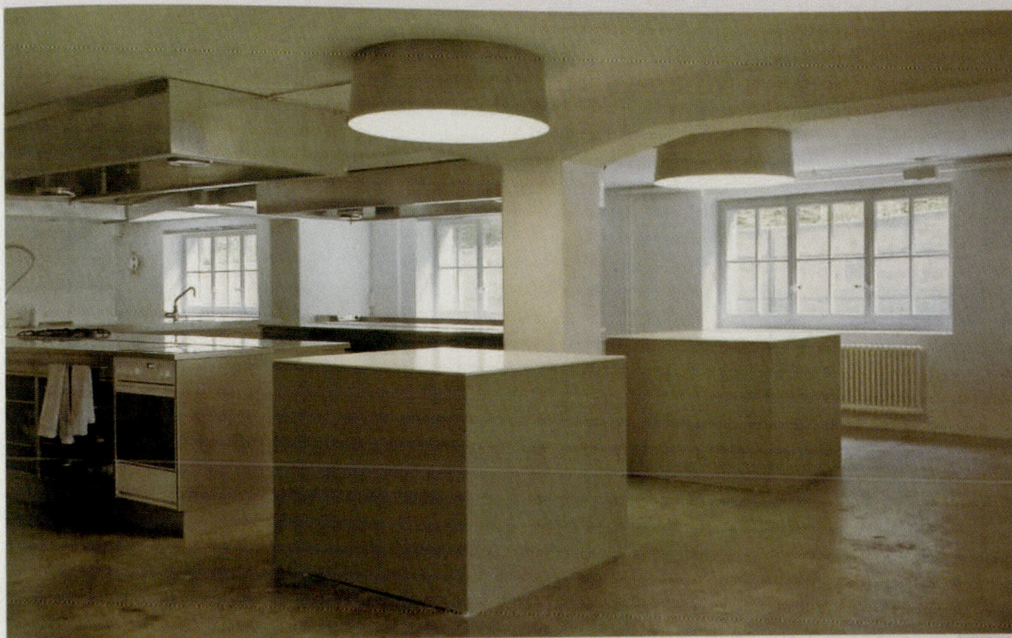

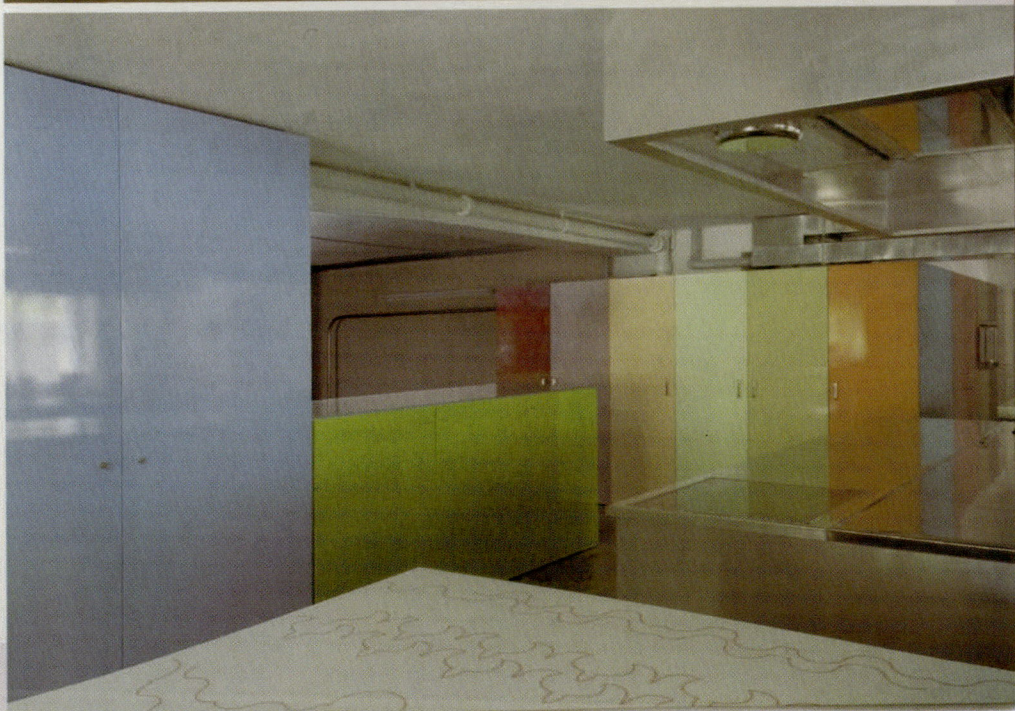

SUBJECT
Cooking centre

LOCATION
Aarau, Switzerland

DESIGNERS
Chalet 5
Jörg Boner

PHOTOGRAPHERS
Atelier Fontana

WRITER
Ralf Michel

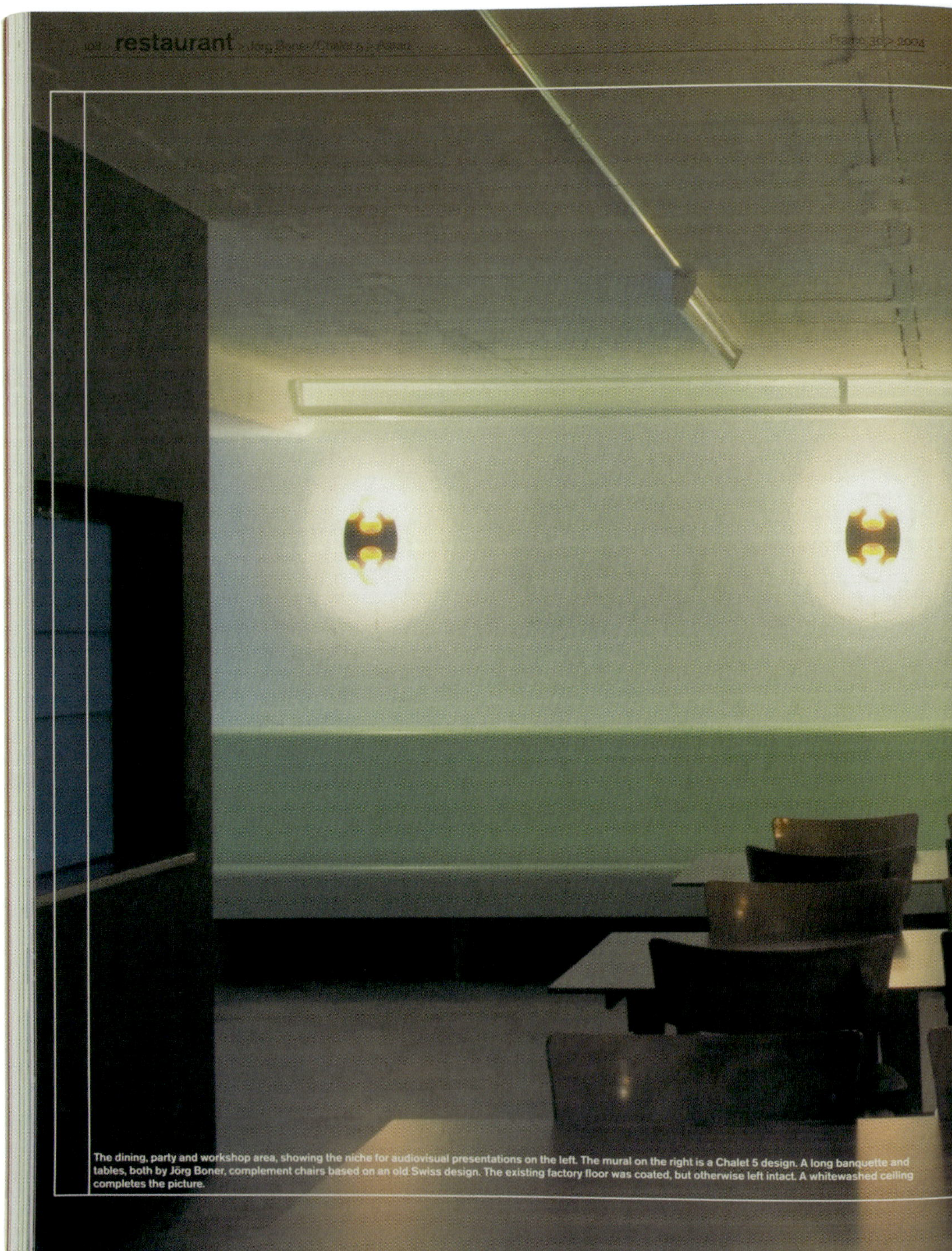

The dining, party and workshop area, showing the niche for audiovisual presentations on the left. The mural on the right is a Chalet 5 design. A long banquette and tables, both by Jörg Boner, complement chairs based on an old Swiss design. The existing factory floor was coated, but otherwise left intact. A whitewashed ceiling completes the picture.

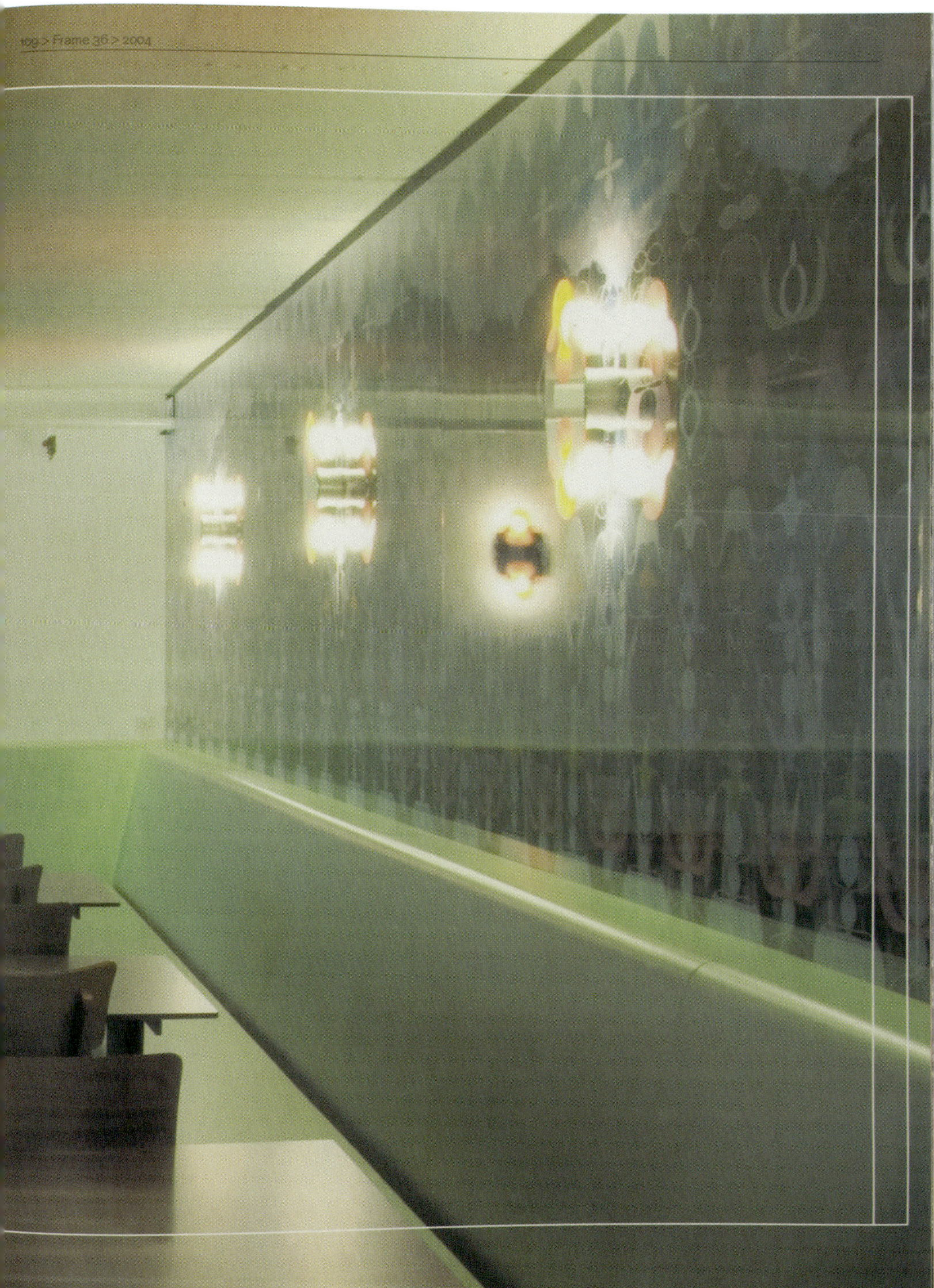

SUBJECT
Cooking centre

LOCATION
Aarau, Switzerland

DESIGNERS
Chalet 5
Jörg Boner

PHOTOGRAPHERS
Atelier Fontana

WRITER
Ralf Michel

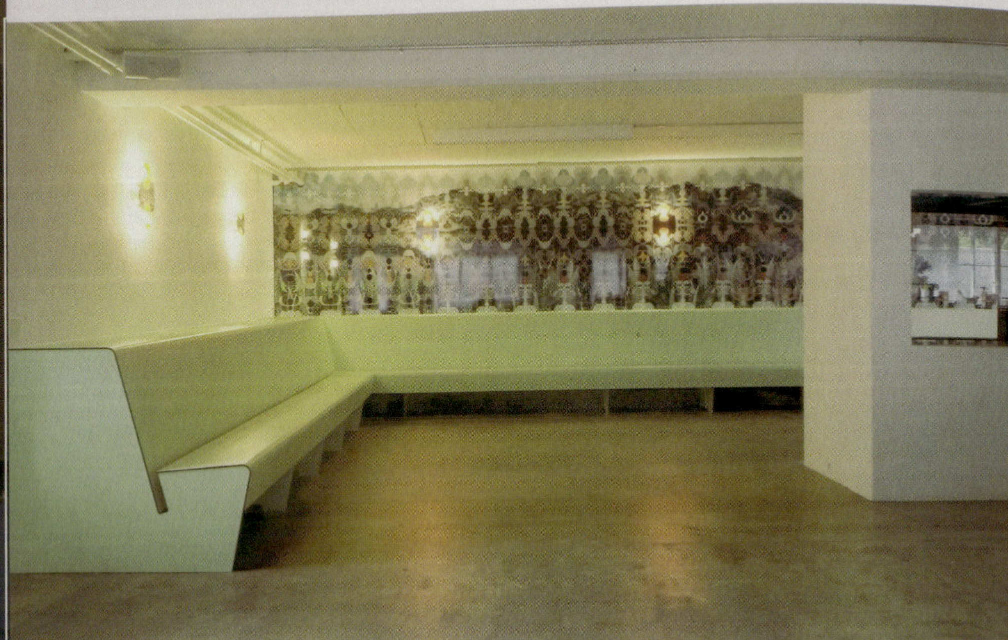

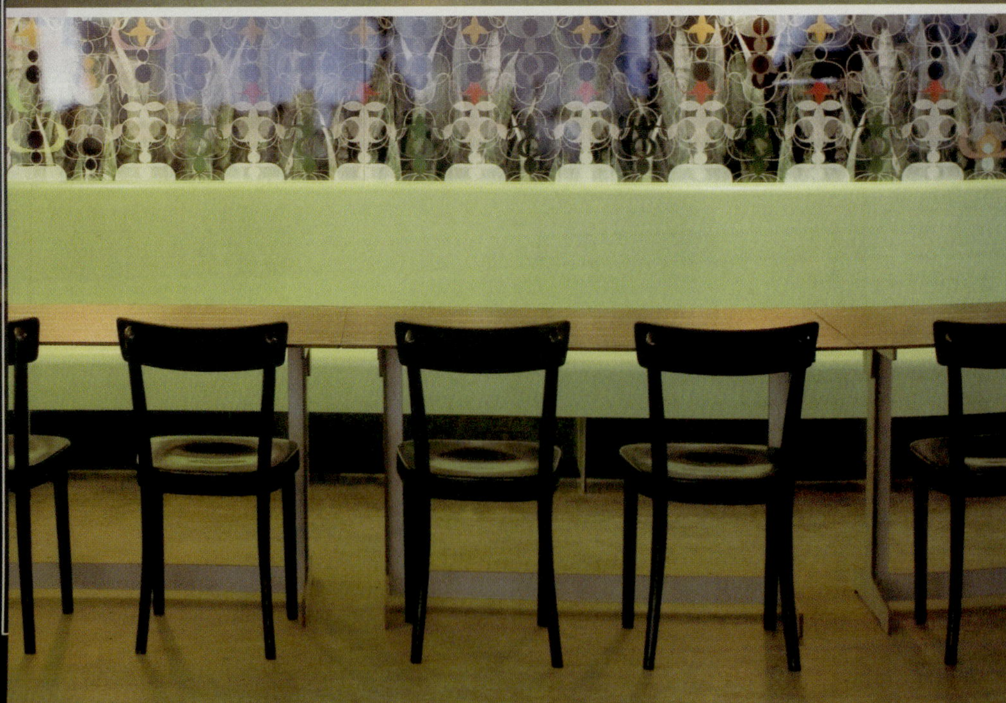

'I wanted to combine grandmother's kitchen with its industrial counterpart,' says owner Susanne Vögeli

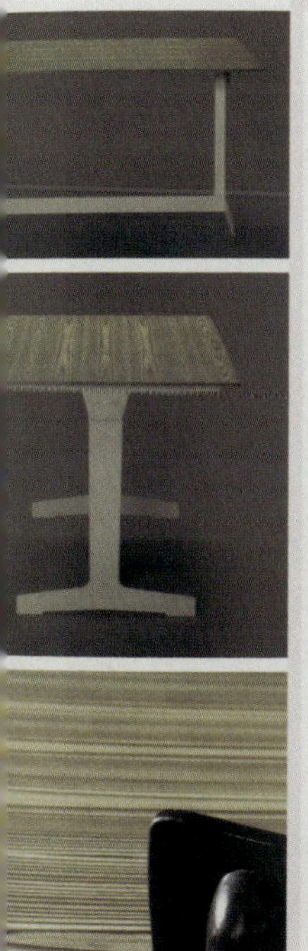

If Cookuk's heart is the kitchen, its soul is the neighbouring dining room, a multifunctional space with a long corner banquette, svelte tables and chairs, and decorative, dreamlike walls. In one sense, the shapes, materials and colours found here are in stark contrast to the high-tech aesthetic of the kitchen; at the same time, however, they complement the adjacent space and, together with it, form a harmonious entity. 'I wanted to combine grandmother's kitchen with its industrial counterpart,' says Susanne Vögeli.

Not only did the designers fulfil her wish; they did it without wallowing in the mush of cheap, kitschy nostalgia. After all, evoking memories of gran's kitchen doesn't necessarily entail a literal reproduction of the room itself. Subtle, indirect references firmly rooted in the present can be equally associative links. Although Cookuk brims with echoes of the past, the message they shout is undeniably part of the here and now.

Take the elongated banquette in the dining room. The bench harks back – at least in German-speaking Europe – to the middle-class kitchen of the past two centuries: a room where women cooked and the family gathered around a large table at mealtime. Boner's 21st-century interpretation of the concept invites diners to partake of a collective meal, but in terms of appearance and the use of materials, it has nothing in common with the historical image. The smooth, lime-green serpent nestled against the wall is composed of CNC-milled components of 10-mm HPL (high-pressure laminate) screwed together.

Tables and chairs are made in the same way, even though they look, from a distance, like cardboard furniture assembled on the spot. Unexpectedly stable, the laser-sharp tables have another surprise in store. What appears to be a simple veneered tabletop is actually a surface bearing a greatly enlarged print of a wood veneer, a photograph digitally printed on décor paper bonded to the HPL. Voilà! Tabletop doubles as ornament.

The tabletop is one of Cookuk's subtler decorative features. Far more conspicuous are kitchen islands with white Corian countertops sporting elegant red motifs. After Chalet 5's signature designs were cut out of the white surface with a computer-controlled water jet, the resulting cavities were filled with coloured liquid Corian. Not a simple procedure, as the material solidifies very quickly. The counters were then ground and polished.

A cluster of the duo's enigmatic motifs reappears at the focal point of the space: the large mural above the corner banquette. The background is a photograph of an enchanting Swiss landscape, an image repeated again and again. Above the scenic view are the hieroglyphics so typical of Chalet 5, repeated and coloured to create an abstract landscape. Dominating the foreground of the panorama, just above the banquette, are pictures of exotic plants which accentuate the cosy, bourgeois atmosphere.

This prominent mural goes to the core of Boner, Reichlin and Wälchli's interior design, in which they fuse images from the collective unconscious with deeply personal abstractions. Each component contributes to an overall design that deserves labels like 'fresh', 'contemporary' and, even more relevant, 'new'.

Swiss designer Hannes Wettstein praises the interior: 'Cookuk combines a great sense of humour with the virtuoso handling of references and clichés. The refreshing result touches on trend, irony and history.'

An equally apt description might be 'contemporary sampling at the highest level'.

Tables made of CNC-milled components of 10-mm HPL are assembled by screwing the parts together. HPL-clad tabletops bear greatly enlarged digital prints of wood veneer. Opposite: The mural mixes photographs and over a hundred abstract motifs by Chalet 5, united in one image with the aid of Photoshop. After using ink-jet printing to transfer the scene to photographic paper, the designers sealed it with a high-gloss laminate and mounted the result on four aluminium panels.

SUBJECT Cooking centre

LOCATION Aarau, Switzerland

DESIGNERS Chalet 5, Jörg Boner

PHOTOGRAPHERS Atelier Fontana

WRITER Ralf Michel

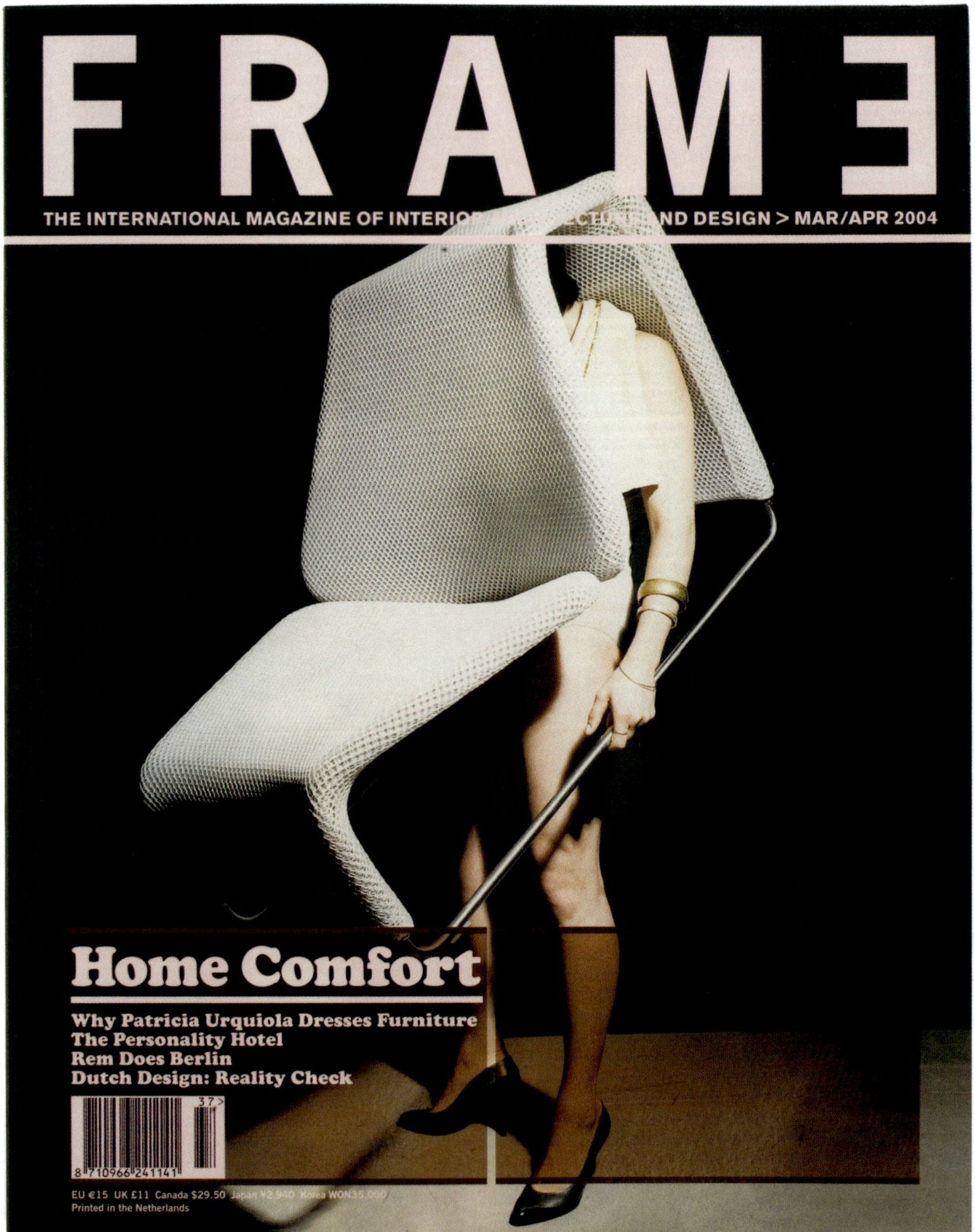

ART DIRECTORS
Cornelia Blatter & Marcel Hermans, COMA

37

Nozzles Among the Spires

<u>*Frame* editor Robert Thiemann briefed the writer</u>

'As you may know, Archigram co-founder Peter Cook has just completed the Kunsthaus (art museum) in Graz. We've just received some gorgeous pics of this building and are thinking of putting them next to an interview with Cook. Would you be interested in interviewing him? The interview should include information on Cook and Archigram, and on his small portfolio of built work. What led to the design and realization of the Kunsthaus? What does he think of the project? How do his ideas relate to the Archigram programme? What is his relationship with today's architecture? And more. We're thinking of roughly 1500 words. The next email will contain some pics of the Kunsthaus. Please let me know if you're interested.'

DESIGNERS
Peter Cook
Colin Fournier

SUBJECT
Gallery Kunsthaus Graz

LOCATION
Graz, Austria

PHOTOGRAPHER
Paul Ott
www.paul-ott.at

WRITER
David Littlefield

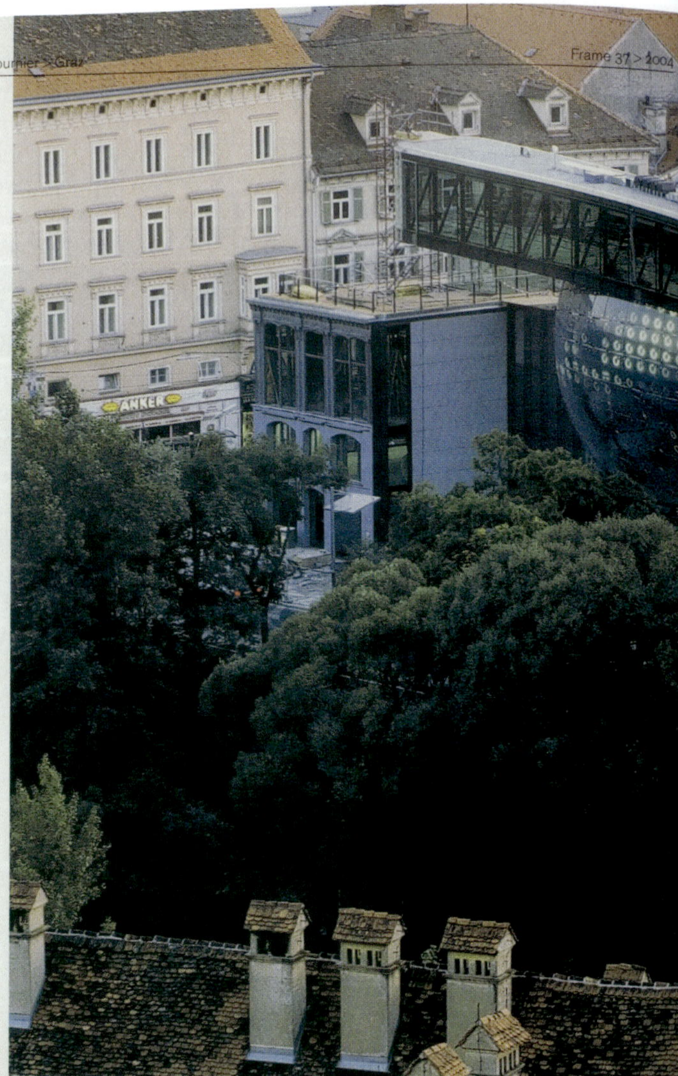

Two London academics – the architectural fantasist Peter Cook and the more practical Colin Fournier – brighten the skyline of Graz with a work of modern-day baroque.

Nozzles Among the Spires

Text by David Littlefield. Photography by Paul Ott

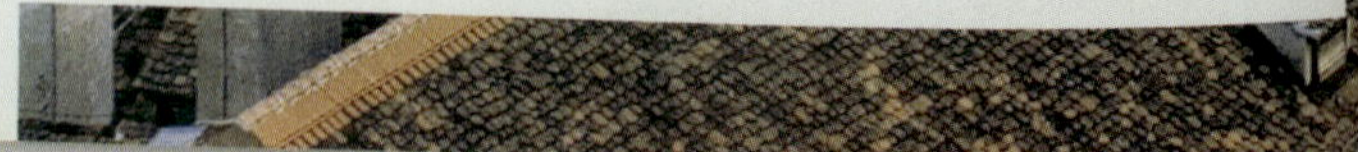

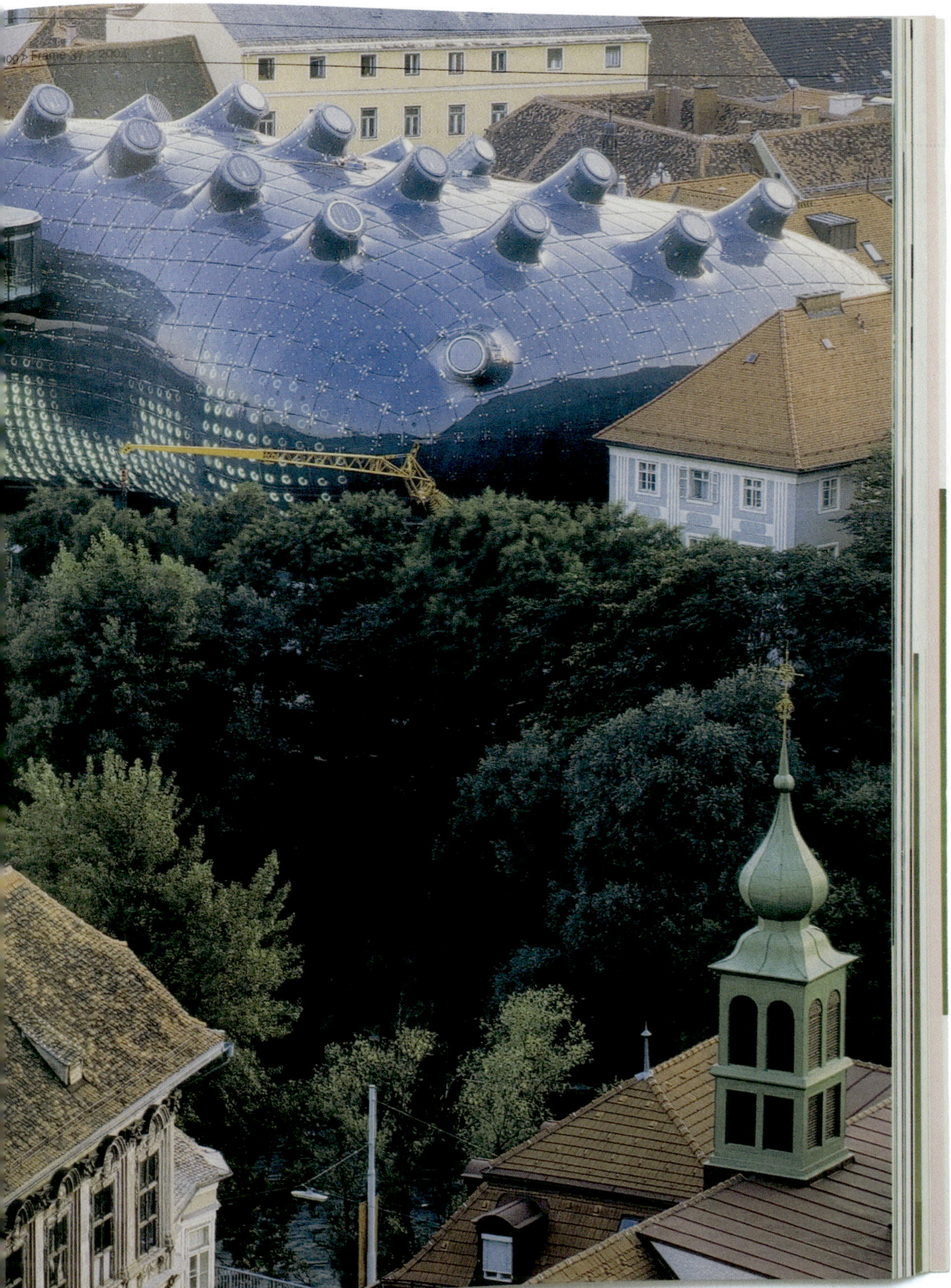

SUBJECT
Gallery Kunsthaus Graz

LOCATION
Graz, Austria

DESIGNERS
Peter Cook
Colin Fournier

PHOTOGRAPHER
Paul Ott

WRITER
David Littlefield

Peter Cook sits like he thinks – askance. Sprawled diagonally across the steel and leather of Marcel Breuer's Wassily chair, he lolls like a teenager. This is exactly how Peter Cook should loll; his sideways view of the world, his posture, speech and quickness of mind are those of someone years younger. Peter Cook is excited by newness and complains frequently of things 'boring'. Profanity slips out easily. He talks of bending and breaking rules. Little wonder he has made London's Bartlett School of Architecture one of the better institutions of its kind in Europe.

Cook talks endlessly, peeling off on one tangent after another yet somehow managing to wrap everything up and address the subject in question. He finds a vast range of topics relevant and interesting. His active refusal to be pigeonholed irritates his critics, and he, in turn, is irritated by the assumption that architecture cannot be successful unless it is a full-time occupation. The daring new Kunsthaus in Graz, Austria, designed in partnership with fellow Bartlett professor Colin Fournier, is a case in point. 'Everybody expects it to be leaking and falling down,' he says. So far it's doing neither.

'British architecture doesn't have the sharp irritation of the Swiss stuff or the barefaced zap of the Dutch stuff,' says Peter Cook

For someone who is such a pillar of the British architectural establishment – chair of the Bartlett and joint recipient, with the three remaining members of Archigram, of the 2002 RIBA Royal Gold Medal for Architecture – he is surprisingly bored by the output of his co-professionals, particularly those of the younger generation. Cook calculates that architectural generations run in 12-to-15-year cycles. He points to a clear generational difference between the Smithsons of the 1950s and Archigram, for example, and picks out Zaha Hadid and 'Willy' Alsop as designers who have put their own generation on the map of architectural process. The trouble is, Cook finds it difficult to give younger architects a positive critique.

'So, you ask, what is the next step? You look at people who are 37, 38, and they're all rather boring. Suddenly it's all become very provincial. British architecture doesn't have the sharp irritation of the Swiss stuff or the barefaced zap of the Dutch stuff. I get very despondent looking at *Building Design*,' says Cook, referring to the British weekly architecture journal. In truth, Cook doesn't lay all the blame for this malaise at the door of the younger architects. As someone who has little experience of having to make a living from getting things built, how could he? Instead, Cook lists a catalogue of ingredients that hamstring young architectural hopefuls: the sheer expense of living in London, the reluctance of clients to commission unproven architects, and the trap of securing jobs with respectable practices that drive their employees so hard they have little time or energy left for creative projects of their own. All of which leads to inoffensive 'polite modern' architecture.

'People are much more political, much more careful, much more business-like now. They're Thatcher's children. They're much more knowing,' says Cook, who would like to see more young architects operating like the theoretical, Dublin-based practice Heneghan Peng, an outfit that put together imaginative competition entries for years before finally landing the commission for a national museum overlooking the pyramids of Giza.

'I must admit that the most interesting people tend to be foreign,' says Cook, a remark that helps explain his willingness to collaborate with the Anglo-French Fournier. It also underpins the glee with which Cook talks about Kunsthaus Graz, the unlikely blue blob that has just materialized from what was, until recently, the city's red-light district. This is not, at first sight, a building that takes the slightest notice of its surroundings, but both Cook and Fournier disagree with that observation. The sweeping curves of the city's baroque church spires, for example, have long set a precedent for nonlinear design. 'That precedent was absolutely not part of any conversation during the design process, but once we'd done it, we looked at the rooftops and realized that people have been playing games for years,' says Fournier. Cook adds that even the act of challenging a site is, in a perverse sense, proof that one has at least taken notice of it.

The opening of the Kunsthaus in September 2003 was the finale to Graz's tenure as last year's European City of Culture. First planned 20 years ago for a prestigious riverside site, the Kunsthaus was blocked because of a change of local government. The scheme was revived in 1996, but the site was moved to the warren of tunnels and caves that runs through a rocky outcrop in the city centre. Cook and Fournier jointly entered the ensuing competition. The pair had first worked together in 1970 on an Archigram scheme for a cultural centre in Monte Carlo – Cook as one of the founders of the legendary architecture movement, Fournier as a talented graduate of the Architecture Association in London. This was Archigram's only chance to build something; until then, the six-man architectural think-tank had concentrated largely on generating daring ideas for lightweight, industrial, inhabitable (but generally unfeasible) structures like the Walking City and the Plug-In City. Unfortunately, the Monte Carlo project ground to a halt and the practice dissolved, leaving the spirit of Archigram to materialize in the hands of architects like Richard Rogers and Renzo Piano, who completed the Centre Pompidou in 1977.

Years later, while working together at the Bartlett, Cook and Fournier once more entertained the thought of collaboration. Graz provided the opportunity, and the pair gathered around them a group of students and recent graduates to put together a scheme. The idea was to line the caves with a complex, double-curved membrane that would protrude from the entrance like a tongue. Not only did their scheme lose; it failed to qualify for any of the seven prizes awarded, all of which went to Swiss entries. Fournier says the prizewinners were typical of Swiss design: crisp, elegant, well-handled. They had, however, 'no personality whatsoever'. Local officials agreed; in a subsequent referendum, the citizens of Graz threw out the winning scheme. Although a conservative place, Graz is an intellectual city with a long history of avant-garde thinking. Pressure was building for something remarkable to occur.

'It was quite clear the city wanted to be astonished. It has been open to new ideas for many years. People wanted some sort of provocation to introduce life to the city,' says Fournier, who was invited in 1999 to enter a second competition, this time organized by radical architect Volker Giencke. Fournier and Cook won.

The result has divided Austria and architectural opinion. According to Innsbruck-based architect Peter Lorenz, people either love it or hate it. 'It is a piece of art and definitely gets people thinking,' he says. He thinks the building conforms to the Graz tradition of commissioning daring buildings and provides a powerful statement about the city's cultural ambition. But he's not convinced that the Kunsthaus has lived up to its original promise. The competition entry showed a structure with a palpable skin of variable transparency, but the result is quite different, says Lorenz, who also mentions inadequate internal lighting.

'It's not a skin at all. The sheets of acrylic are purely decorative. You can take them all off, and there's a complete building underneath. Many architects are critical of this aspect. It's proof that a building of this sort is realizable only as a shape and not as a concept,' he says, adding that he nonetheless admires the people of Graz for commissioning a building of such architectural cheek. 'It's good for Graz. I think it's the first building of its kind, but also the last.'

After city officials had a further change of mind about location, the Kunsthaus was assigned to a tight slot in a tough part of town. The plan of the building had to comply with strict zoning laws that prohibit a new structure from coming within 2 metres of existing development. The 'funny footprint' rises two storeys to meet the

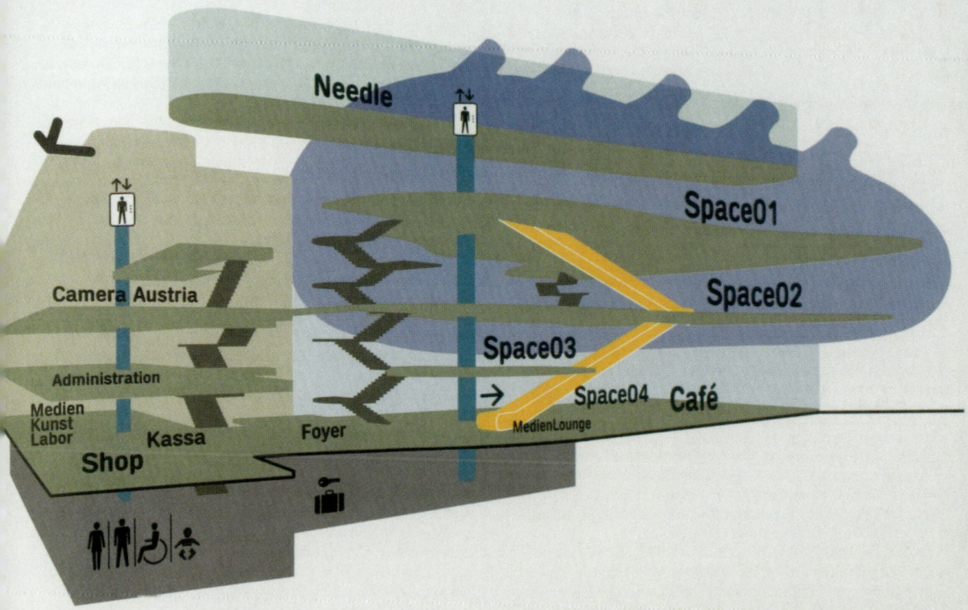

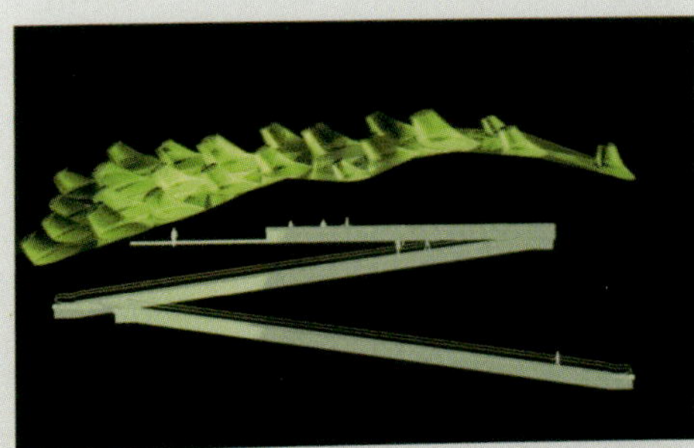

Right and above: Perspective diagrams showing the distribution of functions within the building. A travelator ascends from the ground-floor media lounge and café to upper-level galleries.
Previous spread: Kunsthaus Graz is a translucent, bubble-shaped structure that nestles among the city's pastel-coloured façades and tiled roofs.

spatial demands of the brief, and the resulting volume has been jacked up to leave the ground plane, wrapped in a double-curved membrane, free for 'street life' – one of few design elements found in the original competition entry.

Although Cook enjoys taking sides in 'the battle between the bubble people and the box people', Fournier stresses that the organic shape of the Kunsthaus emerged inevitably from the design process. 'It was not an aesthetic decision from the beginning … but we did want the building to look like an object that had fallen from the sky. At the same time, we wanted it to nest within the fabric of the city, and it does so quite comfortably,' says Fournier. Interestingly, the colour blue came about by chance. 'We were working late at night, and one of the people making the model had nothing but blue paint in his workbox.'

Cook talks about breaking the rules and working at the edge of what is permissible. If the people of Graz wanted shock, he was happy to oblige. 'I have always thought the vocabulary of architecture – and by that I mean the formal and contextual vocabulary – is overly narrow. People always try to close the rules, but what I think is more interesting is when people break those rules. The Kunsthaus backs into the existing fabric; the bum of the building sits right into it. I'm interested, quite genuinely, in how you grow out of a place rather than how you emerge on a tabula rasa.

'I'm also interested in the knock-on effect of the building that, like the Pompidou or Tate Modern, is effecting an urban shift. Graz is a big student town, with a population of around a quarter million that includes some 45,000 students – and they're finding their way to the wrong side of the river now.'

Make no mistake; finding the way to the wrong side of the river is a good thing. Whether in terms of Graz or in the wider meta-

WRITER David Littlefield
PHOTOGRAPHER Paul Ott
DESIGNERS Peter Cook, Colin Fournier
LOCATION Graz, Austria
SUBJECT Gallery Kunsthaus Graz

phorical sense, it represents a small act of rebellion, an exploration. Even the Kunsthaus itself contains moments of disobedience, or simple cheek. Cook is amused by the travelator that moves people into the heart of the building, making entry to the galleries an act of 'nonchalance' – machinery moves you to the top, and gravity brings you down. Similarly, one of the apertures in the roof, which Cook describes as 'the naughty nozzle', is deliberately misaligned. 'It's like a group of children, when one of them is looking the wrong way.'

Cook is fascinated by the act of seeing. As director of London's Institute of Contemporary Arts in the early '70s, he installed a periscope inside the gallery so that visitors could spy on passers-by. It was a roaring success. An early idea for the Kunsthaus was to use latex for the 15 nozzles, giving them the flexibility to move and track the paths of the sun, the moon and even individuals outside. Budget constraints ended this flight of fancy. For a major cultural centre, 40 million euros is not a lot.

'The organic shape was not an aesthetic decision. We wanted the building to nest within the fabric of the city; it does so quite comfortably,' says Colin Fournier

It would be easy to make the mistake of lumping the Kunsthaus together with the psychedelia of Archigram. This irritates Cook and Fournier in equal measure, although Fournier (the lesser known of the two) is further frustrated by the widespread assumption that Cook is the author of the building. The suggestion is so pervasive that it's beginning to strain their friendship.

'This is not an Archigram project,' insists Fournier, who says the two of them were co-designers, although the circulation was largely Cook's doing. 'The Archigram office closed 32 years ago and never reopened. The people working on this project are extremely sensitive to this issue. Peter is also reluctant to accept that. We've had this conversation many times.'

For his part, Cook makes one small reference to the 1960s. The 1000 bulbs buried behind the exterior panels turn the façade into a work of art; each bulb is a single 'pixel' in a huge low-res image. This gives the building something of the pizzazz of Cedric Price's 1961 Fun Palace, a scheme for an infinitely flexible and transformable entertainment centre. 'The building becomes, to quote from the Fun Palace, a "cybernetic toy",' he says, in homage to the late Price, who was closely linked with, but not part of, Archigram.

Cook has an interest in what he calls 'smart glass', which can alter its opacity and perform the dual role of window and information source. Indeed, smart glass would have been the ideal wrapping for the Kunsthaus and a solution in keeping with the original concept. 'Imagine we're sitting here looking out, not necessarily at Bloomsbury but at the cricket score or stock-market figures. Or we might want to alter the view so we can see a little of who is coming up the road. I'm interested in playing with the illusion of reality. Are we looking at a window, or are we looking at information? All of that is amazing.'

Equally amazing is the way in which Cook eases Ikea into the conversation. But what he's after is a contemporary take on Archigram's Plug-In City, which he describes as 'a romantic comment on system building' – a construction process that utilizes a finite number of standard components to create an almost infinite number of built forms. This should not be difficult, he says, because people do it already. Standard 'add-ons' like sheds, greenhouses and conservatories are being erected all the time (without the need for planning permission), with the effect that property owners are appropriating and reshaping the urban landscape, thus making it a more comfortable place to live. Cook wants to play these shed-builders at their own game and, at the same time, knock the planners (and many architects) off their perch. At a time when Britain is facing a severe housing shortage, especially in the southeast, his ideas have a considerable resonance.

'It's a pity the Ikea house is so boring. If you think Peter Cook/Ikea, you create a condition where you could build into the Thames Estuary using a combination of components. You could offer three types of toilet, 12 types of conservatory, four types of bedroom and so on. Over a scheme of 70,000 homes, you could have an amazing series of houses. I find Ikea, whenever I go there, really, really educational,' he says.

Cook is a collaborator; although he trained as an architect, he is more of a thinker, an organiser, a catalyst, and he's had to team up with people with more practical experience to get anything built. Fournier was part of Bernard Tschumi's design team for the 1982 Parc de la Villette in Paris, for example, and has planned new towns in places as diverse as California and Saudi Arabia. Cook's handful of other built work was designed in collaboration with architect and fellow academic Christine Hawley; as Cook and Hawley Architects, the pair had an association that lasted nearly 25 years (1975-98), during which time they put together a huge range of competition entries, including a scheme for Melbourne's Federation Square (for which they were short-listed). Their built work includes social housing in Berlin and Gifu City, Japan, and a steel-and-glass folly in Osaka.

If you commission Cook, don't expect him to turn up on his own; he'll invariably arrive with someone at his side, even if it's Ikea. Fournier calls him 'a man of action' little concerned with detail. Archigram's big achievement lay in successfully combining serious architectural ideas with pop imagery and science fiction, while uniting the optimism of prewar architects like Bruno Taut with the social realities of the 1960s. Cook, too, makes the most unlikely connections without embarrassment. The only thing he was embarrassed about was being caught wearing, unusually, a waistcoat and tie. He looked every inch the college professor, but that's probably what worried him.

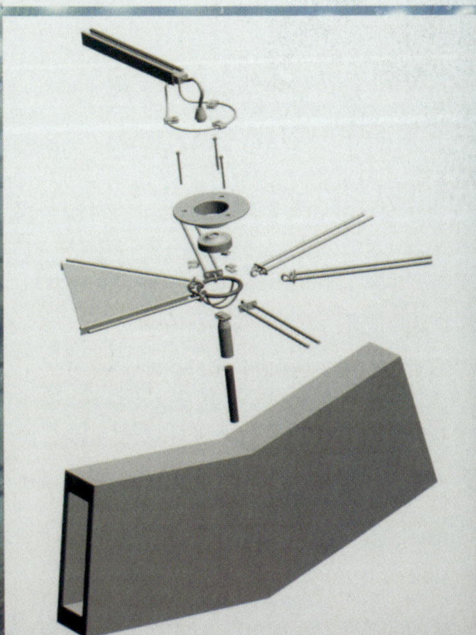

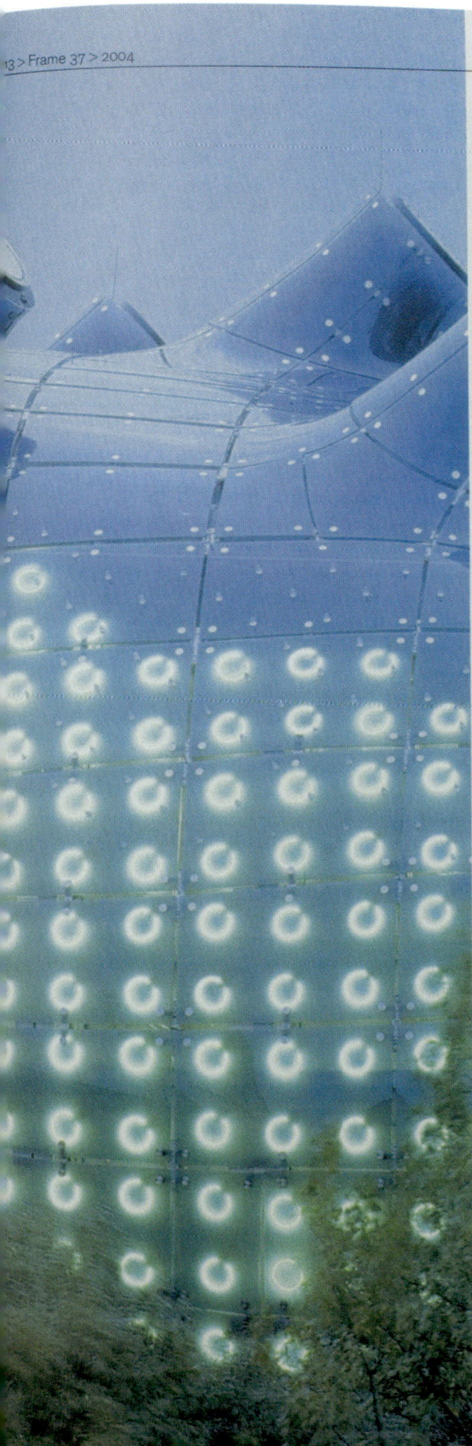

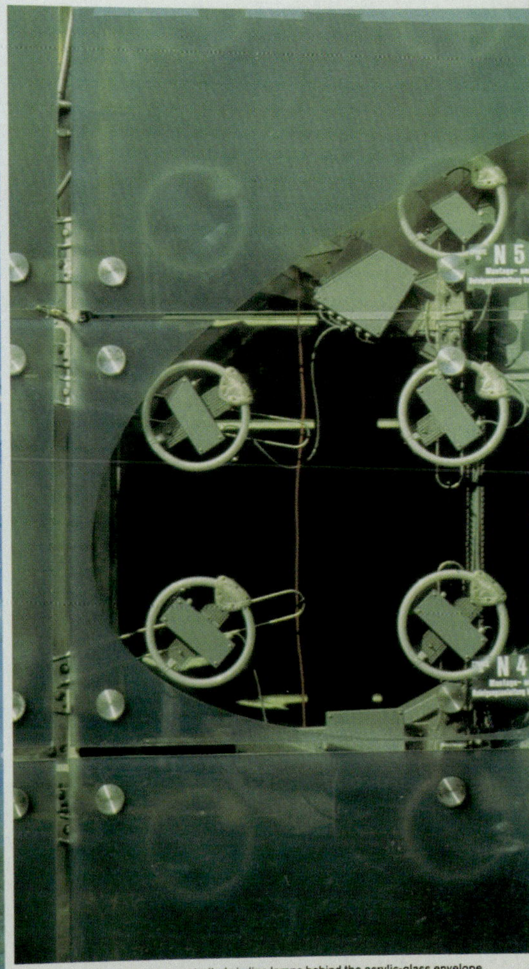

Left and above: 925 computer-controlled circline lamps behind the acrylic-glass envelope turn a section of the façade into a digital display-cum-work of art called BIX (a combination of big and pixel), designed by realities: united from Berlin. Opposite: Exploded view that shows how façade panels and light fittings are attached to roof structure.

WRITER
David Littlefield

PHOTOGRAPHER
Paul Ott

DESIGNERS
Peter Cook
Colin Fournier

LOCATION
Graz, Austria

SUBJECT
Gallery Kunsthaus Graz

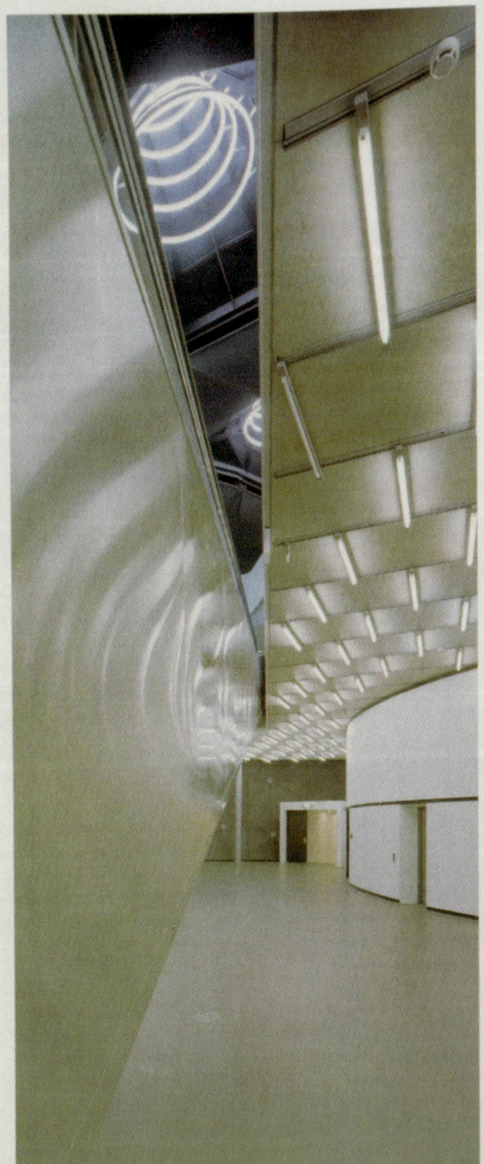

Making the most of circulation: Visitors taking the travelator from one gallery to the next observe the works on display from continuously changing angles en route.

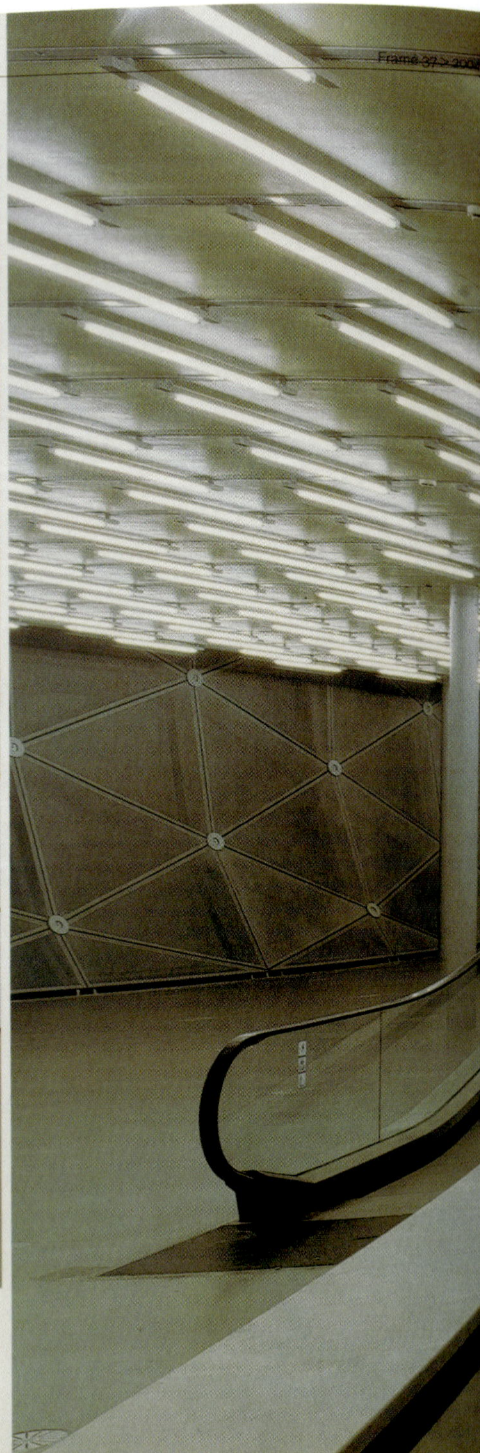

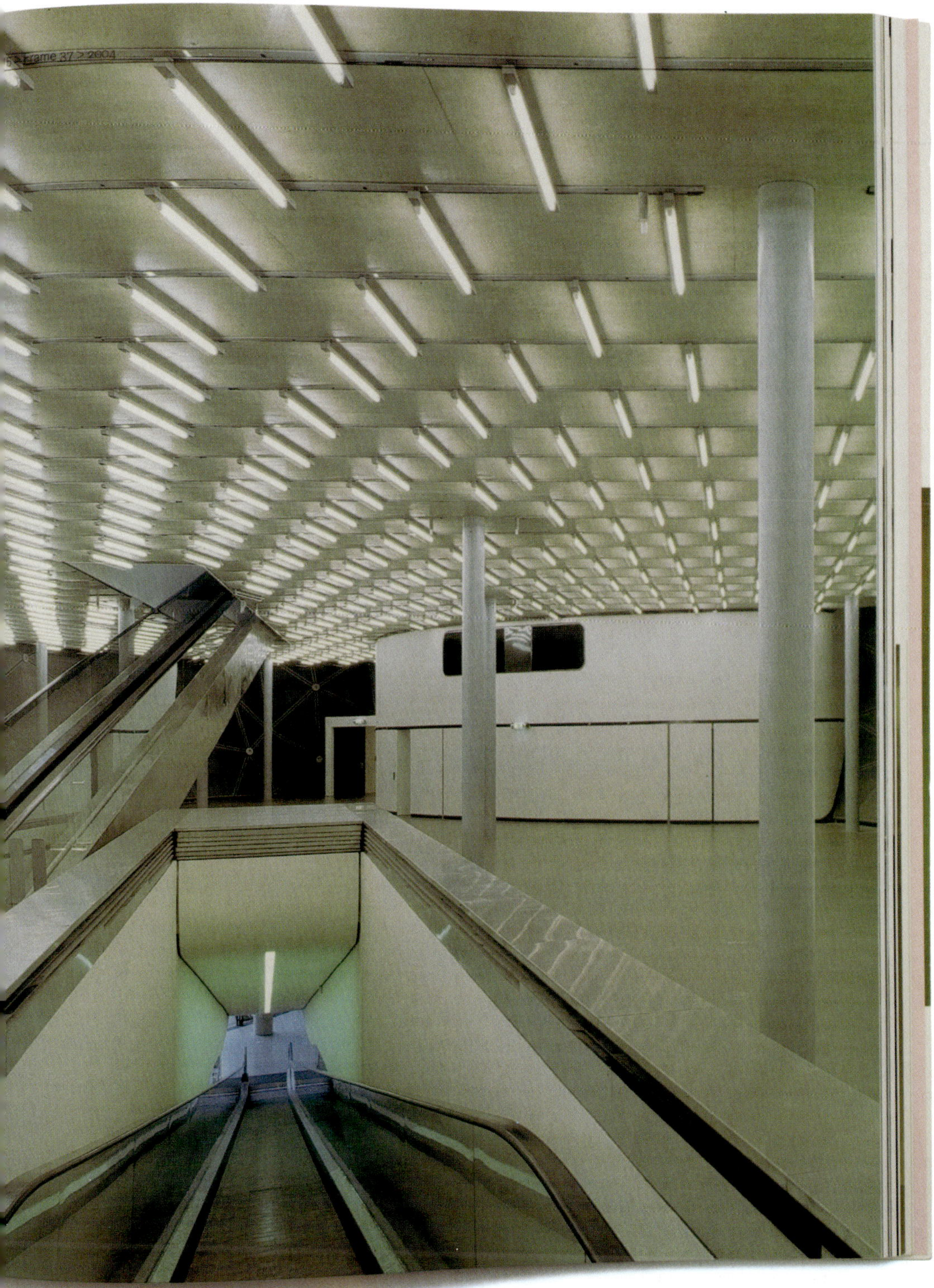

SUBJECT
Gallery Kunsthaus Graz

LOCATION
Graz, Austria

DESIGNERS
Peter Cook
Colin Fournier

PHOTOGRAPHER
Paul Ott

WRITER
David Littlefield

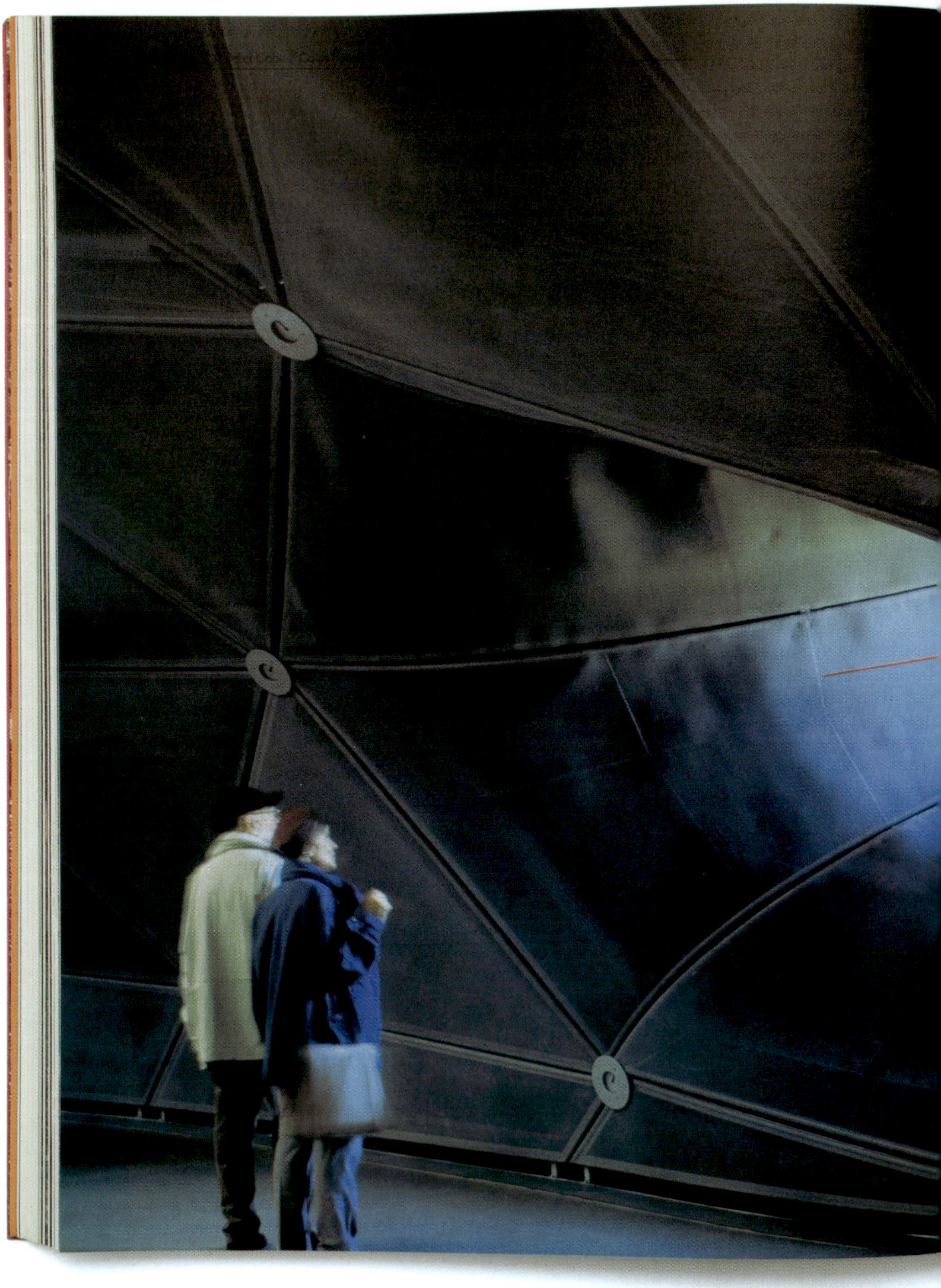

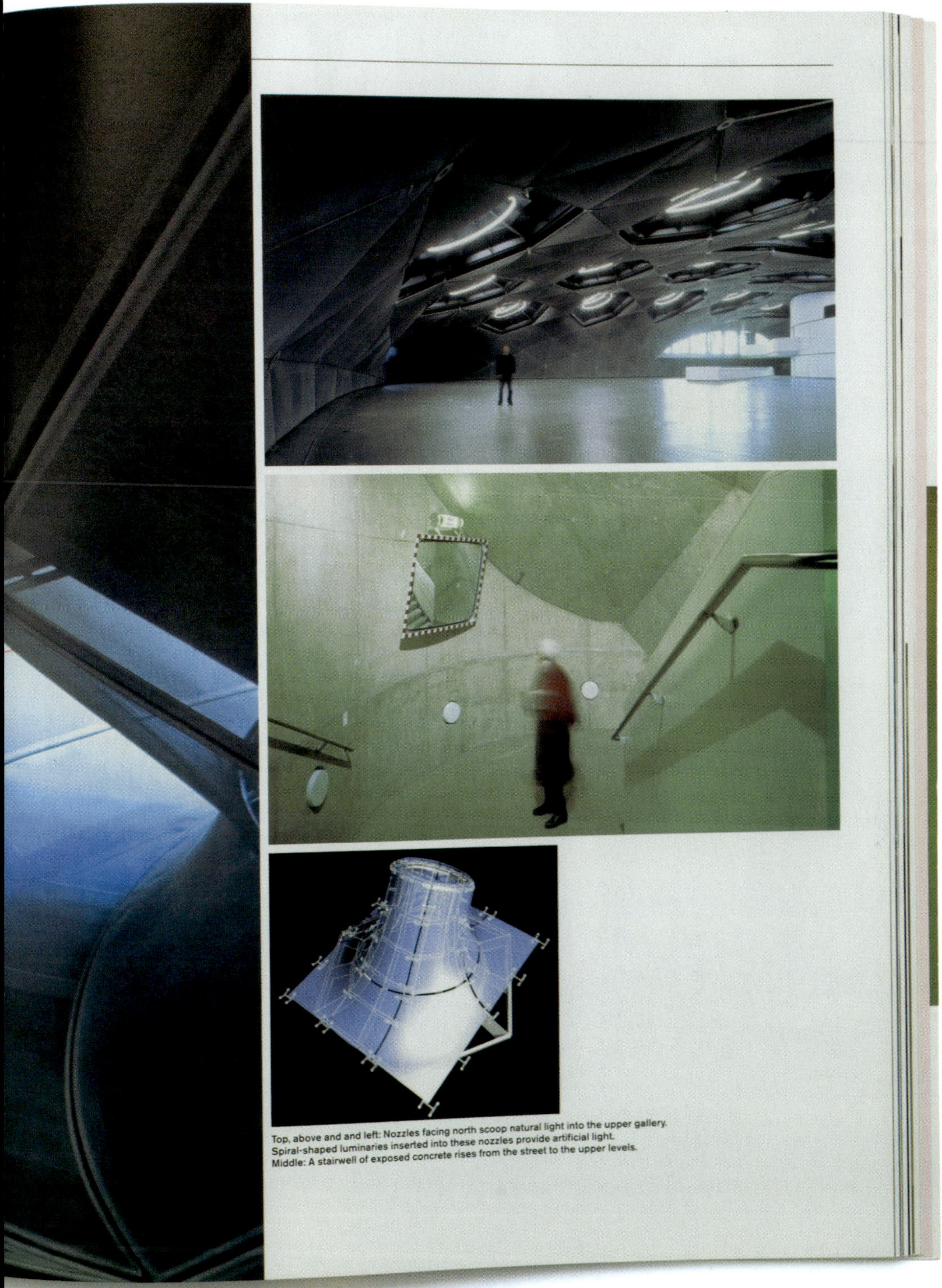

SUBJECT
Gallery Kunsthaus Graz

LOCATION
Graz, Austria

DESIGNERS
Peter Cook
Colin Fournier

PHOTOGRAPHER
Paul Ott

WRITER
David Littlefield

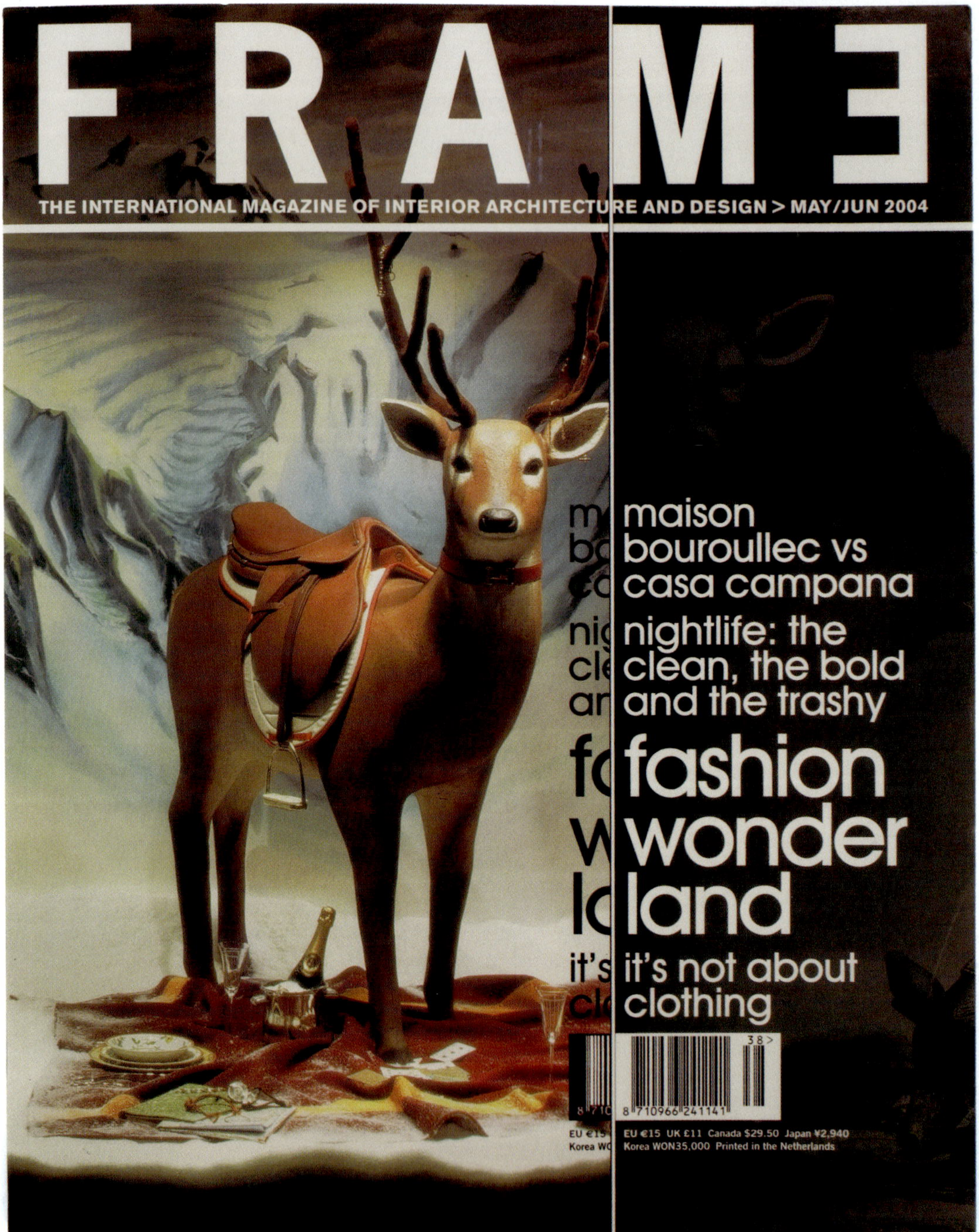

ART DIRECTORS
Cornelia Blatter & Marcel Hermans, COMA

38

Skin and Bones

Vito Acconci about the project

I like the fluidity of the store – I dislike that it only seems fluid, I wish we had made the shelves physically changeable in shape and size, either by themselves or at the will of users.

Comments on the published article

Successive pages walk you through the store, from outside to inside; photos bend, from right-hand page over to left-hand, like the PVC of the store.

Did the article have an effect?

I can't tell; maybe I don't talk to people enough.

Recent news or information about the project featured in *Frame*

It received the Gold Medal, Miami + Beach Architecture Awards for Interiors, 2005.

Current projects

Facade, West 8th Street Subway Station, Coney Island, Brooklyn, USA (recently completed) // Linear Building Up In The Trees, Anyang, Korea // Perimeter, WaterPark City Housing, Toronto, Canada // Second Facade, Imperatore development, Milan, Italy // David Wasserman's Apartment, Wasserman development, West Palm Beach, USA // Outlook Over The River, Temple Quay 2, Bristol, England // Harbor Walk & Bridge, Port de Boulogne, France.

DESIGNERS
Acconci Studio
(Vito Acconci
Peter Dorsey
Julia Loktev
Dario Nunez
Sehzat Oner
Larry Sassi
Gia Wolff)
www.acconci.com

SUBJECT
United Bamboo boutique

LOCATION
Tokyo, Japan

PHOTOGRAPHER
Nacása & Partners
www.nacasa.co.jp

WRITER
Zoë Ryan

Vito Acconci, an artist of influence in the field of architecture, wraps his latest installation around the United Bamboo boutique in Tokyo, both inside and out.

skin and bones

Text by Zoë Ryan
Photography by Nacása & Partners

Renegade artist Vito Acconci is probably best known for masturbating under a ramp at New York's Sonnabend Gallery in 1972 in the performance piece *Seedbed.* Yet since the mid-1980s Acconci has diverted his attention away from the conceptual art that first brought him notoriety, turning instead to public art and architecture as his preferred medium. His design for a pod-shaped island that floats on the River Mur in Graz and houses a performance space, café and children's playground was built as part of the city's 2003 European Capital of Culture festivities. The memorable organic form made from an interlaced steel mesh has proved so popular with locals and international visitors that a proposal to make it a permanent fixture may extend its original year-long life span.

Approaching architecture with a background in the arts, Acconci injects his designs with distinctively personal characteristics based on research he has undertaken since the '70s. Born in 1940 in New York, Acconci graduated from the Writer's Workshop at the University of Iowa in his twenties. He quickly became part of New York's emerging literary scene and assumed the name Vito Hannibal Acconci. In the early '70s he shifted his focus to performance art, film, video and photography – tools he used to conduct self-analysis and reveal latent aspects of human relationships. During the '80s Acconci's work displayed overt references to the fields of architecture and design and to issues of public and private space, themes that continue to permeate his projects.

In1988, a significant year in his career, he founded Acconci Studio in New York, an atelier for theoretical design and building that currently boasts a staff of ten architects and a steadily growing portfolio of commissions for furniture, interior design and architecture. Untrained in architecture, Acconci relies on a constant stream of collaborators who help him realize his ideas. Architects John Tagiuri and Richard Price worked with Acconci on his early projects, which were primarily public art rather than architecture. 'All we could do was to provide a marginal note to the building,' recalls Acconci. 'But we had the advantage of being outsiders – of coming in after the building had been designed. We could de-design. We could nudge into the power that the building was based on. We could infiltrate the building.'

Acconci Studio operates from an old warehouse in Dumbo, an up-and-coming area of Brooklyn. It has a relaxed working environment and is one of the few remaining places in New York where chain smokers are welcome. Vestiges of Acconci's former activities as an artist are visible in every corner, from the bank of filing cabinets overflowing with archival materials to exhibition posters on the walls and floor-to-ceiling shelving units jammed with slides that attest to his prolific career. The power of his art, unlike his design work, lives on largely in documentary form – in interviews with the artist recorded in various publications – and in Acconci's sketches, notebook entries and photographs.

To someone unfamiliar with Acconci's work, the transition from artist engaged in theoretical work to designer bent on challenging traditional notions of space might seem to be an improbable change in direction. That's not how Acconci sees it, however. 'In some ways, the emphasis of my work hasn't changed much.' He sees architecture, like much of his art, as a 'place for performance', with himself as director.

It was Acconci's reputation as a pioneering thinker that prompted Miho Aoki and Thuy Pham of New York-based fashion label United Bamboo to commission him for the new interior of their premier store in Tokyo. Located in Daikanyama, a burgeoning neighbourhood populated by low-key independent retailers and trendy bars, the store stands out as a beacon of contemporary design. Designers Aoki and Pham, natives of Japan and Vietnam respectively, founded United Bamboo in 1999. Their reinterpretation of classic American sportswear relies on innovative pattern-making techniques, unusual forms and cuts, and playful juxtapositions between materials. Envisioning a store created to reflect their distinctive design philosophy, they turned to Acconci for a design that would impart their energy for experimentation and provide an interesting counterpoint to the myriad concept stores found on street corners in cities across the globe.

It is not often that the restrictions of a project provide the inspiration, but United Bamboo's 60-square-metre floor area determined the starting point for the interior. The challenge was to introduce a concept that would not

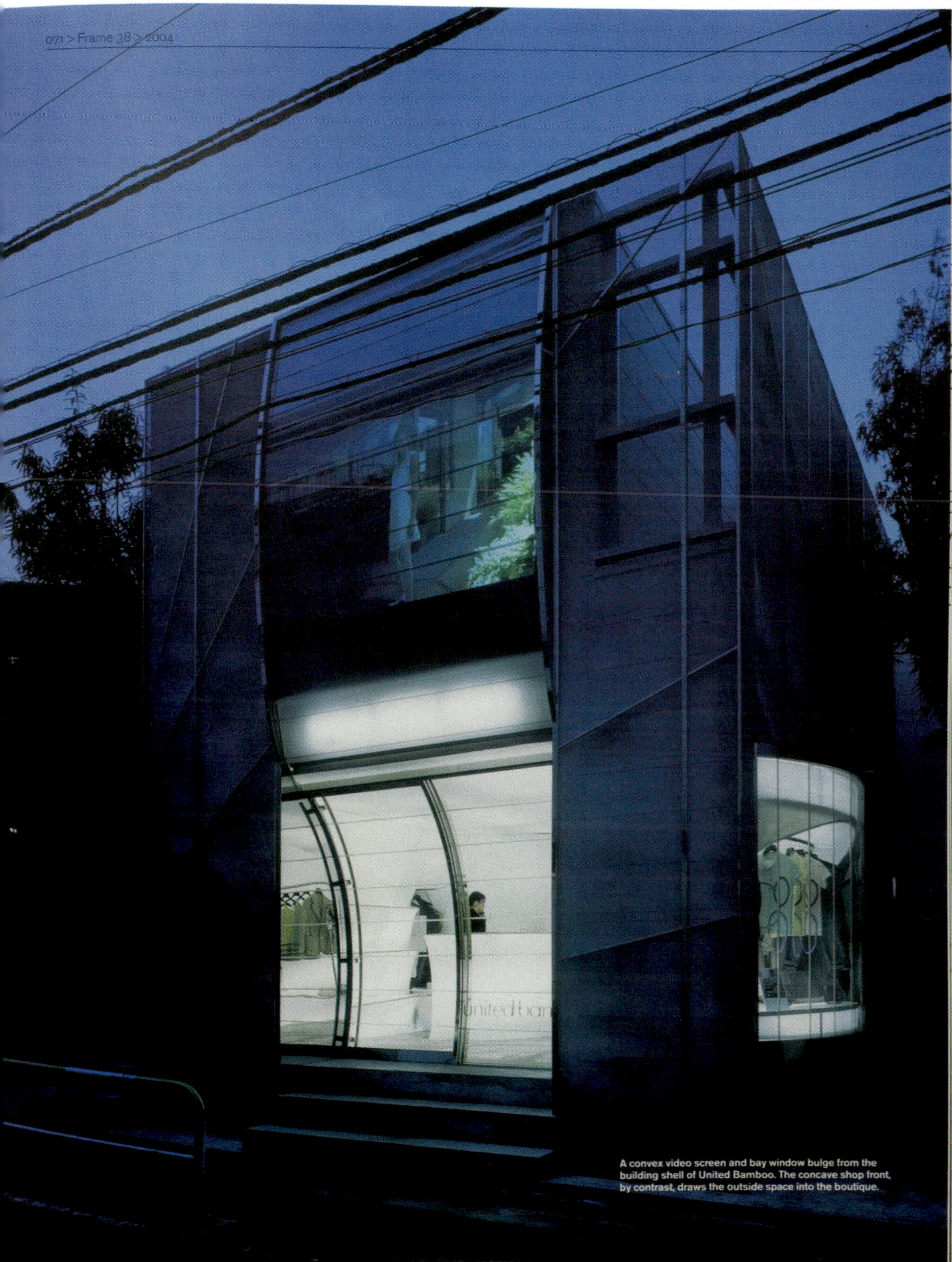

A convex video screen and bay window bulge from the building shell of United Bamboo. The concave shop front, by contrast, draws the outside space into the boutique.

SUBJECT
United Bamboo boutique

LOCATION
Tokyo, Japan

DESIGNERS
Acconci Studio

PHOTOGRAPHERS
Nacása & Partners

WRITER
Zoë Ryan

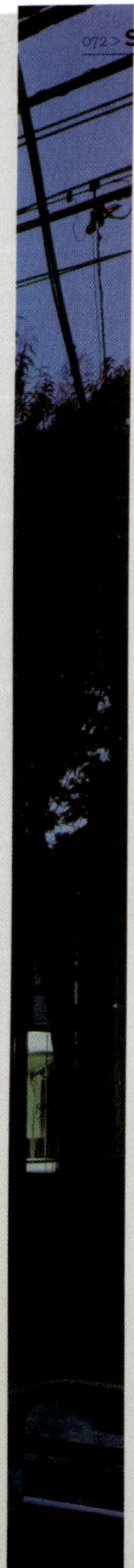

overpower the space and make it seem cramped. Keen to use light for an airy, open feel, Acconci examined whether interior surfaces could function as the source of light, thereby saving space and promising a more streamlined effect. The resulting solution is a blanket of glossy white PVC sheeting, which covers walls and ceiling and is illuminated from behind by fluorescent lights. 'I devised a push-pull method,' says Acconci. 'The ceiling comes down to meet the walls. The same surface is pulled out over a metal rod to create shelves or supports for the lights and is then pulled tight and secured at the base.' The effect is similar to that of being inside a giant tent. 'It is as if a liquid-white material has been poured into the building,' says Gwenael Nicolas of Tokyo design firm Curiosity. 'It feels more like an installation than a shop.' In addition to the shelves and supports mentioned by Acconci, the taut, well-anchored surface of PVC – tightly stretched over a metal framework – forms two simple fitting rooms that jut out from the walls and culminates in a luminous biomorphic counter at the front of the boutique. Here the PVC is secured at floor level, where it meets a channel of light skirting the circumference of the floor like way-finding lights in an aeroplane. The lights give the impression that the fabric wall units are afloat. The design is purposefully tactile. 'I wanted the store to be as soft as skin and made of fabric,' says Acconci, who aimed for 'a direct correlation between the clothes and the store'. The curvilinear shelving that juts out at irregular intersections throughout the space gently flexes when touched and glows from within.

An L-shaped tubular-steel rail protrudes into the space at eye level from each display niche. Hanging from the hook at the elbow that connects the rail to the diagonal supporting leg is an iPOD complete with earphones. Customers are encouraged to listen to the latest tunes while browsing. Pairs of floor-to-ceiling mirrors arranged in V-formations between shelving units reflect multiple views of the store, lend a sense of spaciousness to what is admittedly a tiny retail interior, and invite customers to view themselves from different angles. A similar visual device is at work in the mirrored panels that flank the doorway, making the narrow entrance seem wider.

Two bay windows draw natural light inside and create display areas that are visible from the street. Here Acconci continues his exploration of how space is experienced, both inside and out. The bulbous windows open the interior to passers-by, blurring the distinction between public and private. Tall glazed panels on the façade pursue the same theme. Topping the concave lower half, which forms a pair of sliding doors, is a convex upper section. The sculptural curve provides what Tokyo designer Akihito Fumita considers 'the building's most distinguishing feature and one that gives the store a clear presence on the street'. Illuminating the form is a row of fluorescent lights mounted behind a layer of PVC above the doorframe. 'Customers seem very comfortable in this bright space. The only other shops that are as bright are the convenience stores found on every corner in Japan', says Nicolas. 'This may be a new identity for a convenient brand.' Also lined with PVC, the glass spanning the full width of the first floor acts as a video screen visible from the street below. United Bamboo plans to project digital art and videos by emerging artists on the screen. Acconci fancies the idea of live video footage of customers shopping, thus further dismantling the boundary between activity inside the shop and life on the street.

'One difference between this store and others around it is that United Bamboo is very 'designed',' says Pham. 'Architects usually try to make a small space appear bigger or roomier by minimizing the design. I think Acconci's solution is interesting, because he has inserted a complete design into such a small space. The original structure is almost completely hidden from view.' Although a complete concept, the design is fairly low-tech. The space has a handmade feel in tune with the overriding character of the area, a neighbourhood populated with start-up, cutting-edge brands and designers' studios whose entrepreneurial nature matches that of United Bamboo. Acconci's one regret: 'I would have loved to have made a store that could grow on its own, rather than one that needs to be pushed and pulled.'

'I wanted the store to be as soft as skin and made of fabric,' says Vito Acconci, who aimed for 'a direct correlation between the clothes and the store'

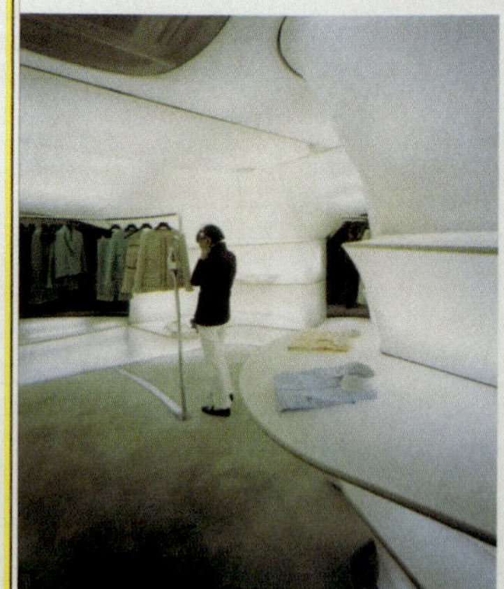

The entire interior is clothed in PVC sheeting of the kind used for projection screens. Tautly stretched over metal rods, the single surface pushes out and pulls back in high relief, creating interior fittings. Fluorescent tubes inserted behind the PVC cast a glow throughout the space.

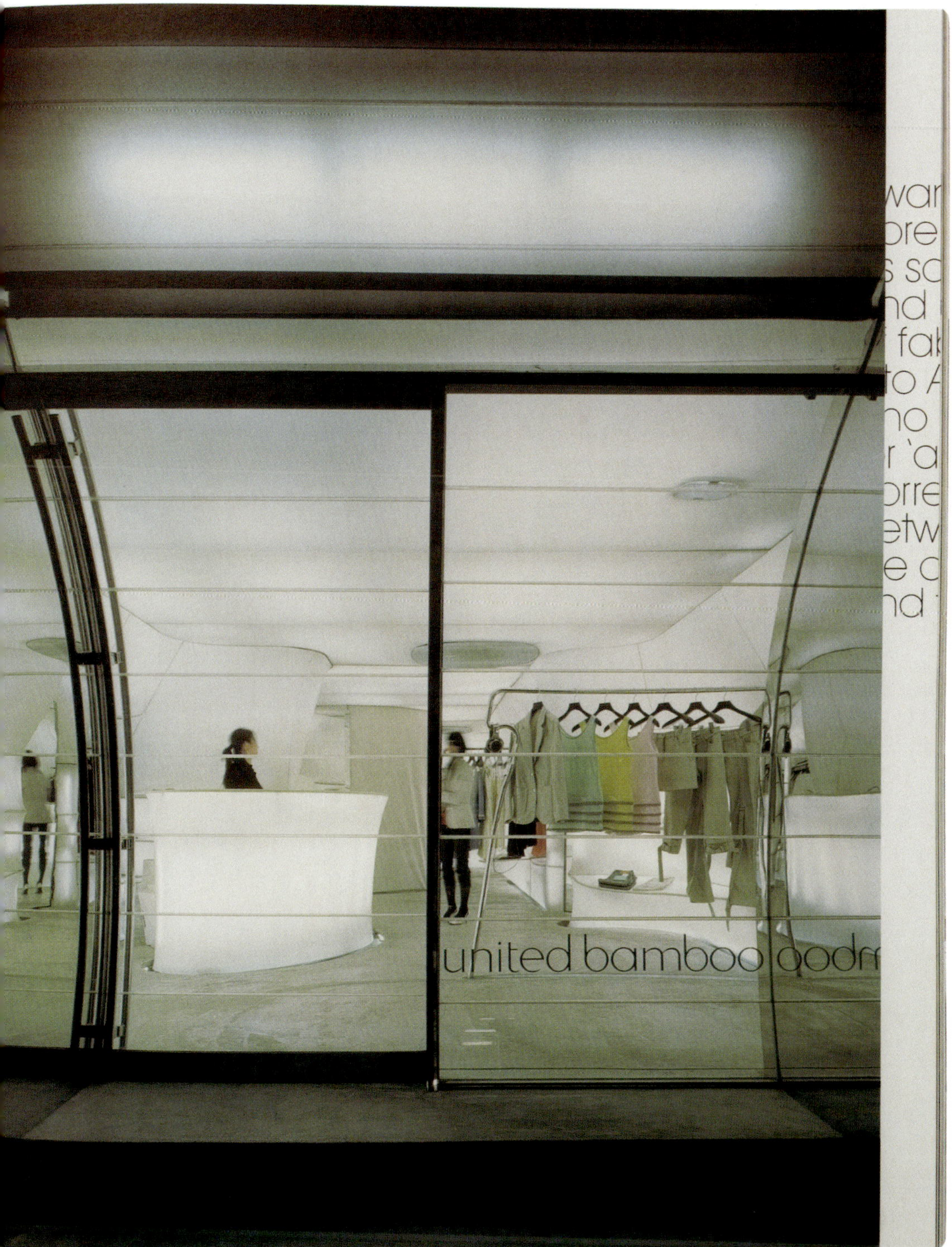

SUBJECT
United Bamboo boutique

LOCATION
Tokyo, Japan

DESIGNERS
Acconci Studio

PHOTOGRAPHERS
Nacása & Partners

WRITER
Zoë Ryan

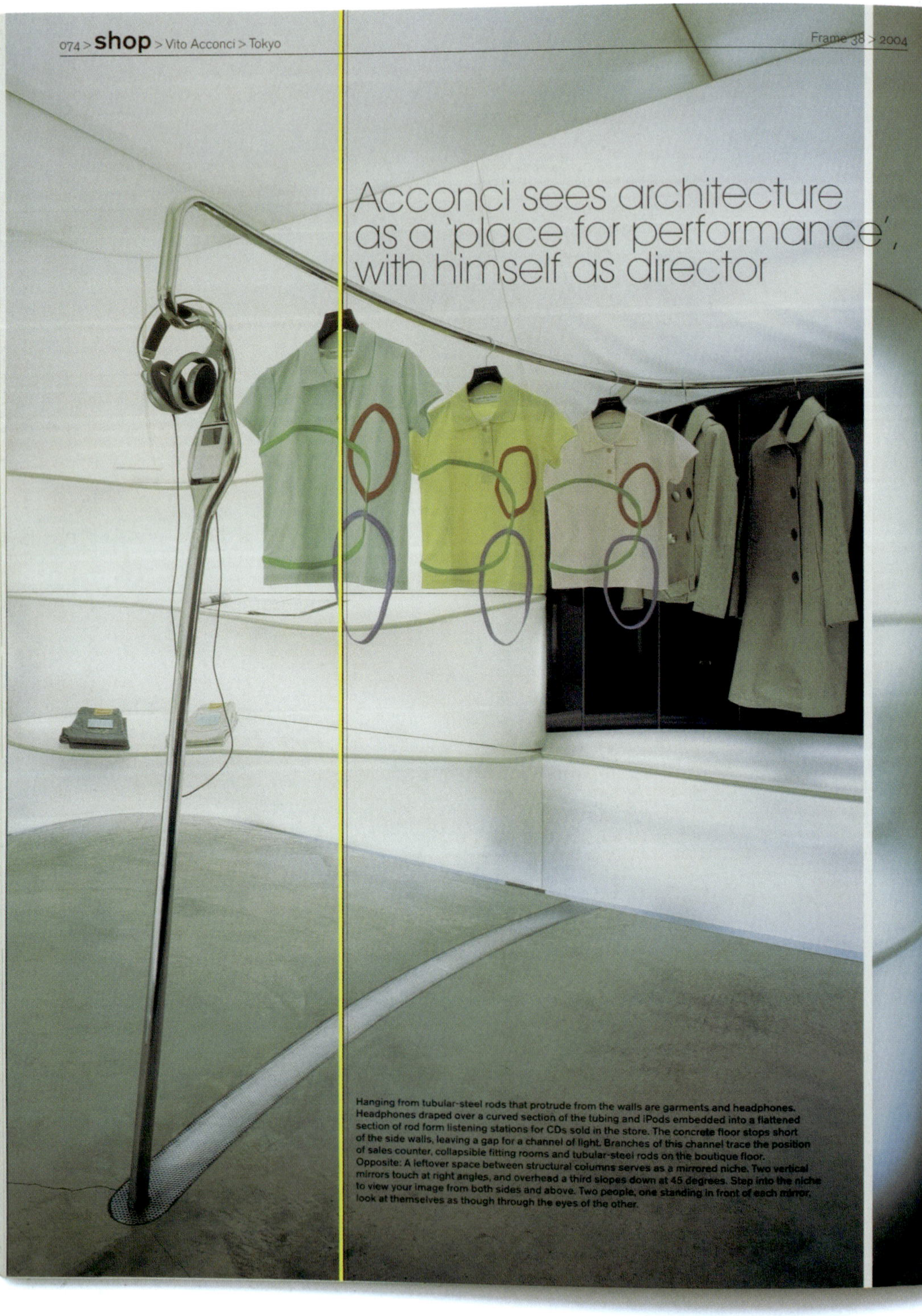

Acconci sees architecture as a 'place for performance', with himself as director

Hanging from tubular-steel rods that protrude from the walls are garments and headphones. Headphones draped over a curved section of the tubing and iPods embedded into a flattened section of rod form listening stations for CDs sold in the store. The concrete floor stops short of the side walls, leaving a gap for a channel of light. Branches of this channel trace the position of sales counter, collapsible fitting rooms and tubular-steel rods on the boutique floor.
Opposite: A leftover space between structural columns serves as a mirrored niche. Two vertical mirrors touch at right angles, and overhead a third slopes down at 45 degrees. Step into the niche to view your image from both sides and above. Two people, one standing in front of each mirror, look at themselves as though through the eyes of the other.

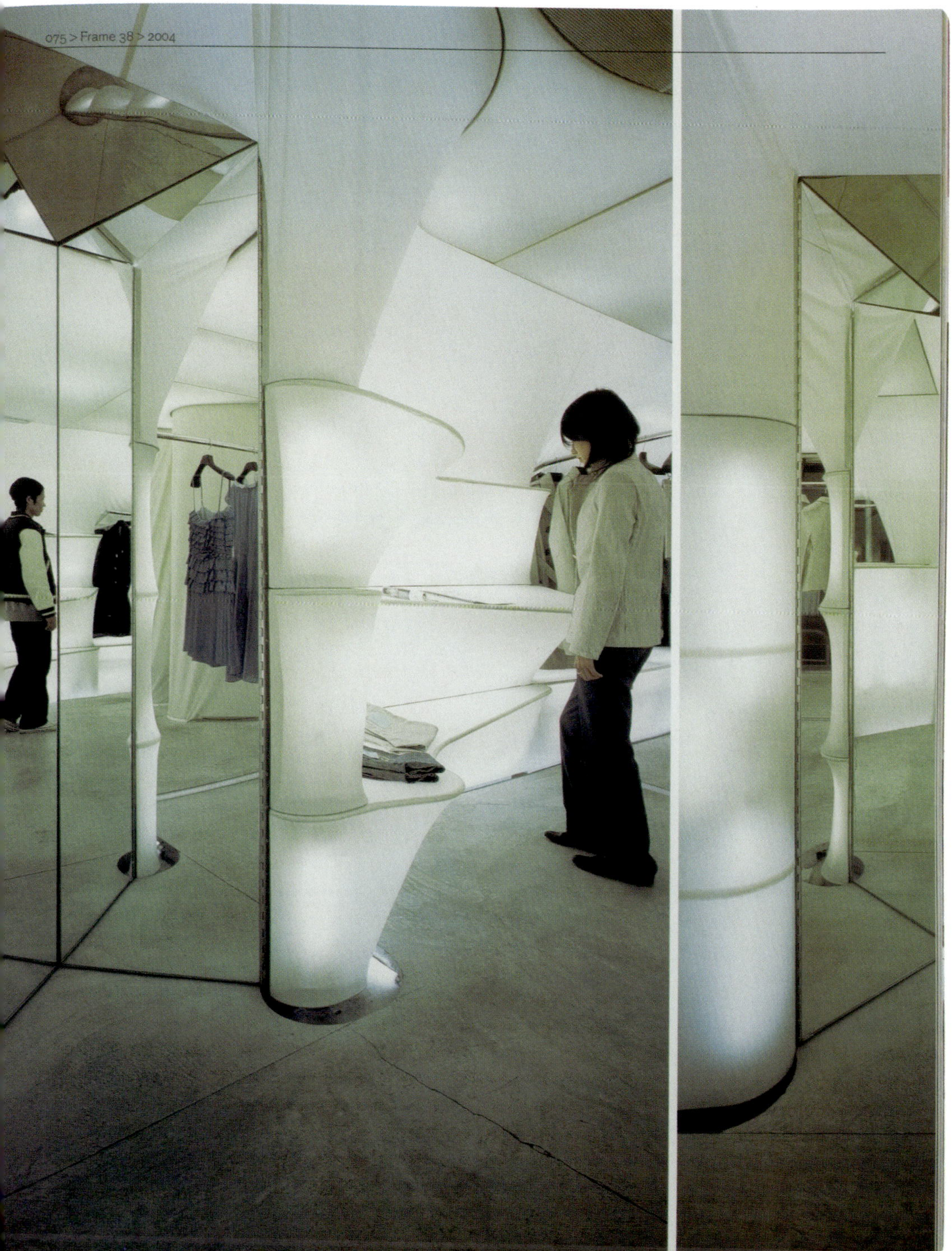

SUBJECT
United Bamboo boutique

LOCATION
Tokyo, Japan

DESIGNERS
Acconci Studio

PHOTOGRAPHERS
Nacása & Partners

WRITER
Zoë Ryan

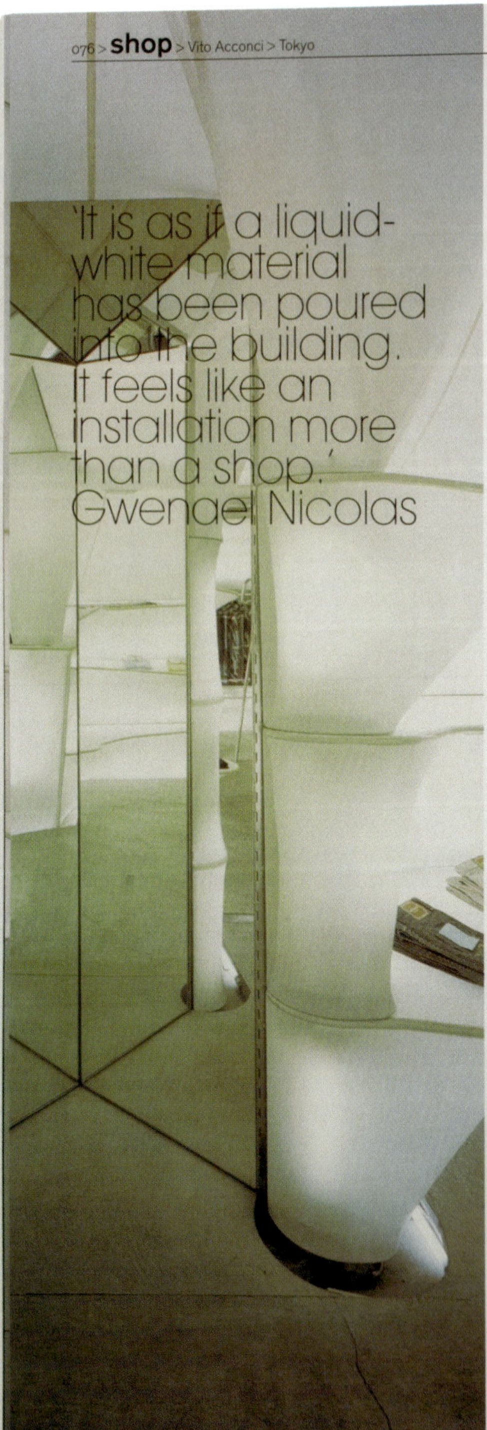

'It is as if a liquid-white material has been poured into the building. It feels like an installation more than a shop.' Gwenael Nicolas

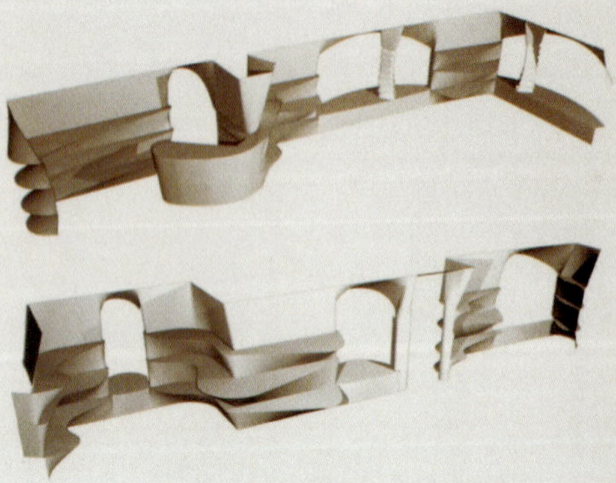

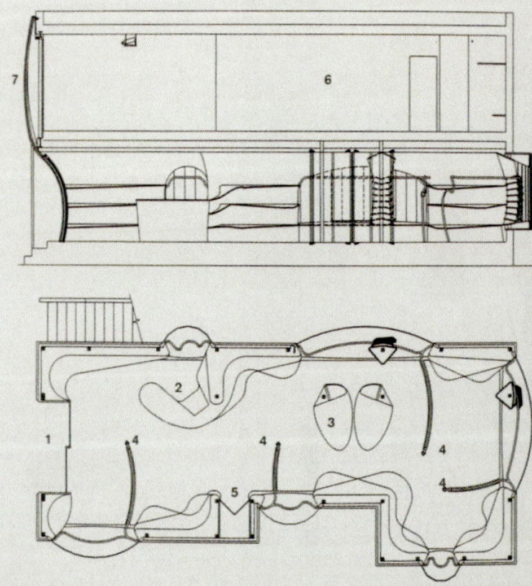

1. Entrance; 2. Cash desk; 3. Fitting rooms; 4. Clothes rail; 5. Mirrors; 6. Stockroom; 7. Video screen

Above: Axonometrics, section and plan showing PVC-clad walls articulated to form sales counter, display niches and shelving. Customers weave around protruding racks and in and out of nooks, exploring the store. Opposite top: View of the 58-square-metre interior from the entrance. The PVC skin that sheathes the interior bulges to form the sales counter. Opposite bottom: Fabric drops from the ceiling to form tent-like fitting rooms. Customers simply pull a zip to open and close the room.

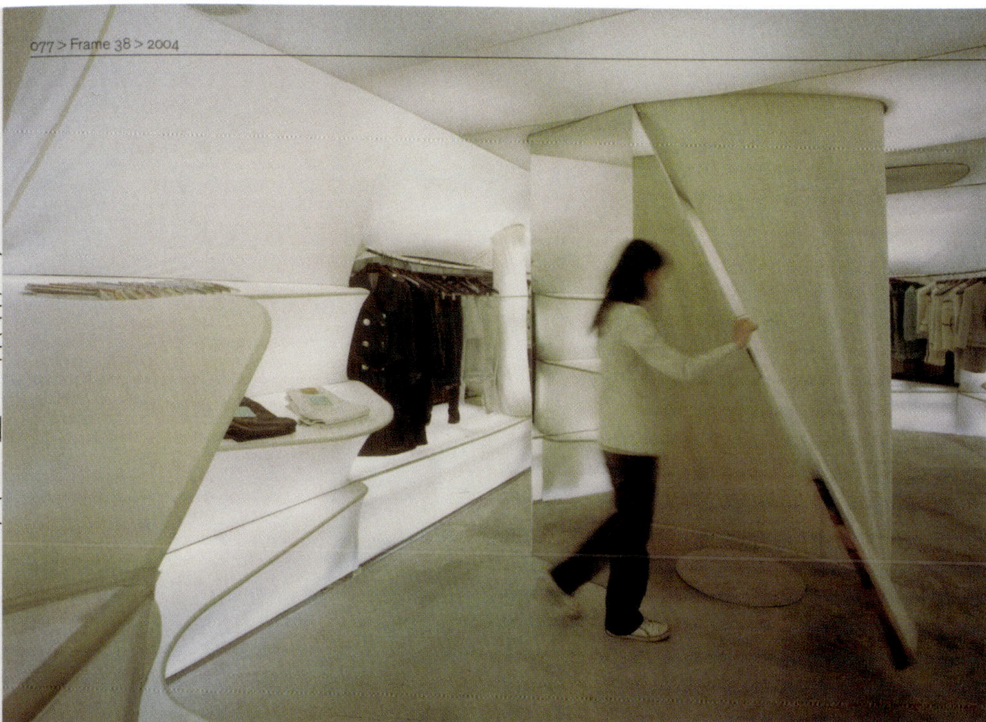

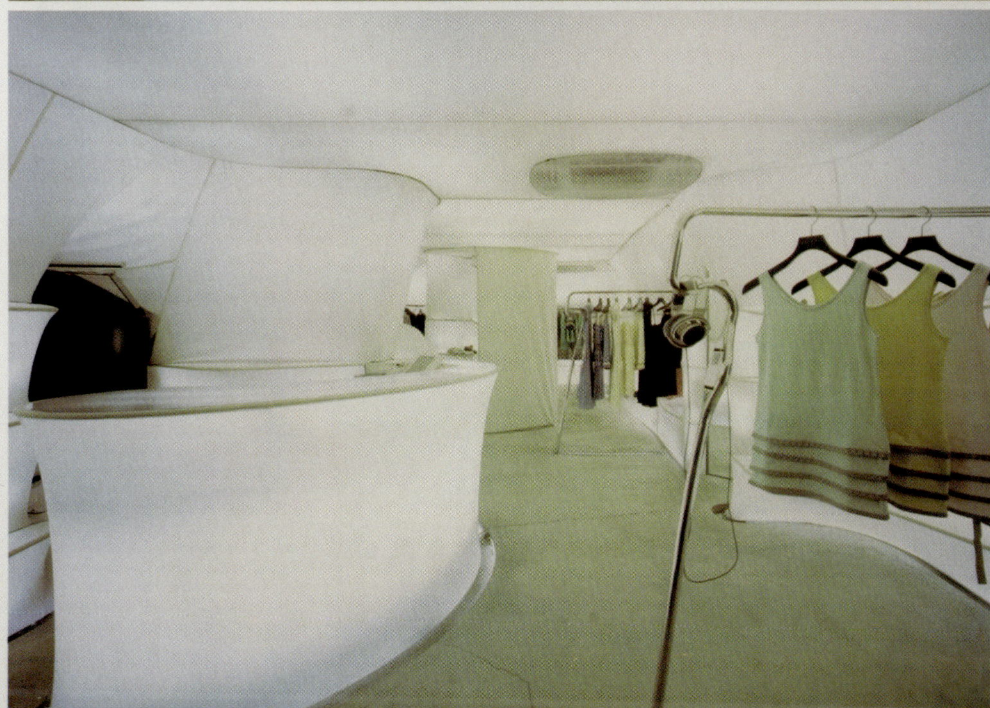

SUBJECT
United Bamboo boutique

LOCATION
Tokyo, Japan

DESIGNERS
Acconci Studio

PHOTOGRAPHERS
Nacása & Partners

WRITER
Zoë Ryan

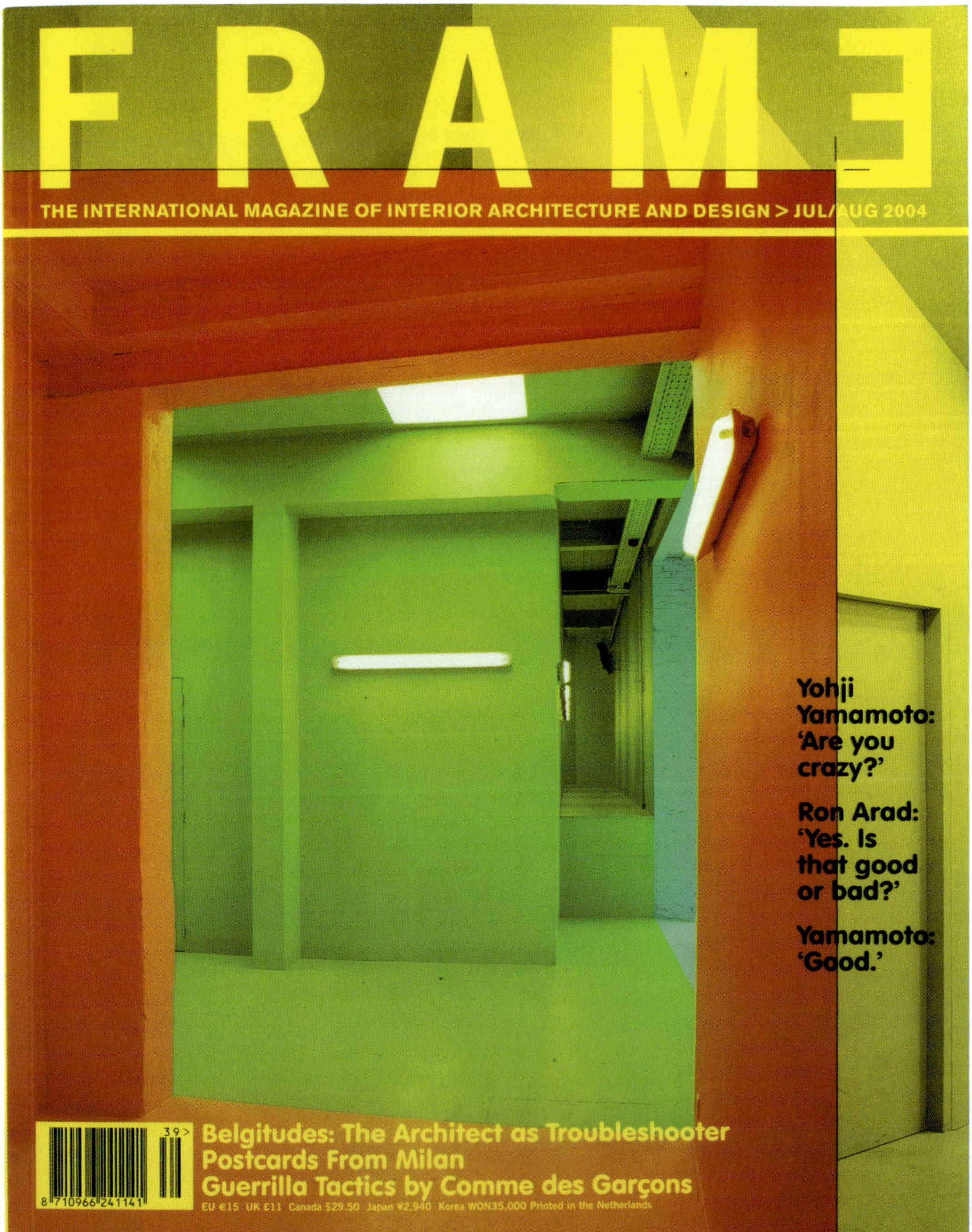

ART DIRECTORS
Cornelia Blatter & Marcel Hermans, COMA

39

Milan Open

Frame editor Tessa Blokland about the project

In 2004 Cosmit, the trade-fair organization behind the Salone del Mobile, changed its marketing strategy, and the international furniture fair became part of Milano Design Week. What had been two big events in April was about to become one even bigger event. With this in mind, the photographer Corriette Schoenaerts suggested a report that would integrate the fair into the city. Schoenaerts rented photographic equipment, and she and I walked for kilometres through the streets of Milan looking for signage: from billboards to posters. The next day she visited the fair with Robert to shoot interesting products. Back home, she spent hours at the computer 'Photoshopping' products into signage. The result was amazing; we heard from visitors to Milan who had no recollection of seeing those all those big billboards around town!

DESIGNERS
Alessi
www.alessi.it
Stephen Burks
readymadeprojects.com
Materialise
www.materialise.be
Art Academy of Latvia
www.aic.lv
Magis
www.magisdesign.com
Joohee Lee
Patricia Urquiola
Droog Design
www.droogdesign.nl
Tord Boontje
www.tordboontje.com
Ferruccio Laviani
www.laviani.com
Job Smeets (Studio Job)
www.studiojob.nl

SUBJECT
Review of Salone del Mobile/ Milano Design Week 2004

LOCATION
Milan, Italy

PHOTOGRAPHER
Corriette Schoenaerts

WRITERS
Tessa Blokland
Robert Thiemann

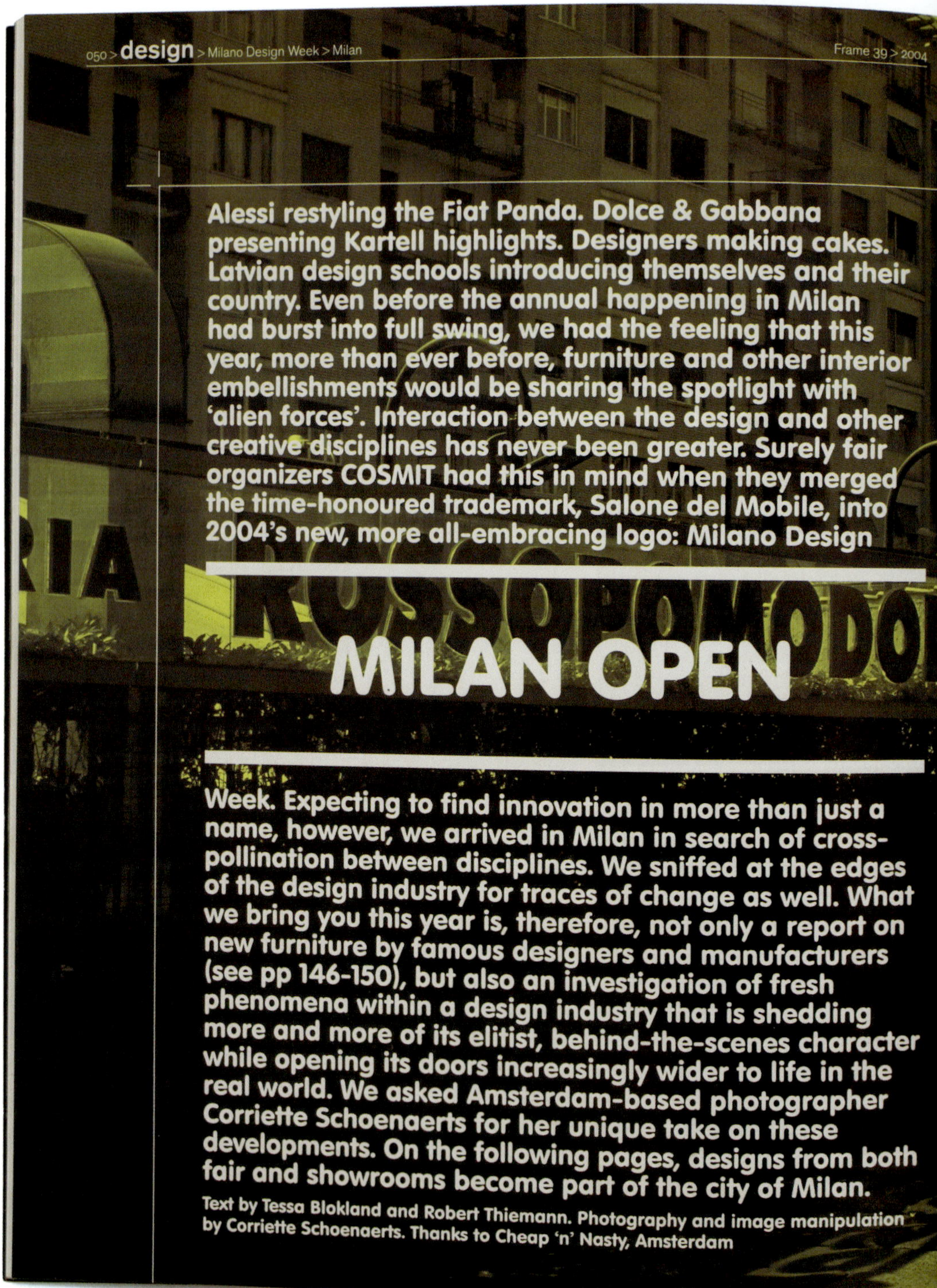

050 > **design** > Milano Design Week > Milan Frame 39 > 2004

Alessi restyling the Fiat Panda. Dolce & Gabbana presenting Kartell highlights. Designers making cakes. Latvian design schools introducing themselves and their country. Even before the annual happening in Milan had burst into full swing, we had the feeling that this year, more than ever before, furniture and other interior embellishments would be sharing the spotlight with 'alien forces'. Interaction between the design and other creative disciplines has never been greater. Surely fair organizers COSMIT had this in mind when they merged the time-honoured trademark, Salone del Mobile, into 2004's new, more all-embracing logo: Milano Design

MILAN OPEN

Week. Expecting to find innovation in more than just a name, however, we arrived in Milan in search of cross-pollination between disciplines. We sniffed at the edges of the design industry for traces of change as well. What we bring you this year is, therefore, not only a report on new furniture by famous designers and manufacturers (see pp 146-150), but also an investigation of fresh phenomena within a design industry that is shedding more and more of its elitist, behind-the-scenes character while opening its doors increasingly wider to life in the real world. We asked Amsterdam-based photographer Corriette Schoenaerts for her unique take on these developments. On the following pages, designs from both fair and showrooms become part of the city of Milan.

Text by Tessa Blokland and Robert Thiemann. Photography and image manipulation by Corriette Schoenaerts. Thanks to Cheap 'n' Nasty, Amsterdam

COSMETIC MAKEOVER

Nike and Philips team up to make portable audio equipment. Sony and Ericsson join forces to produce mobile phones. Poised to appear on the co-branding landscape later this year is the Fiat Panda Alessi. Being married to design house Alessi enhances Fiat's rather humdrum image. And the introduction of their baby at the Triennale, high temple of culture, was no accident. What has their collaboration meant to the two Italian brands? For the Panda, it's been no more than a cosmetic makeover. At the outset, Fiat's marketing department laid down strict guidelines for the design. Alessi, which asked Stefano Giovannoni to head the project, was told to focus primarily on details. Though he managed to give the body a more graphic look, the designer spent the bulk of his time on the interior, where he updated features like dashboard, gear lever and visors. Alberto Alessi, director of the company that bears his name, realizes not only that Fiat coveted the cachet attached to his popular range of household products, but also that the refurbished Panda is not a major accomplishment for Alessi. He's not left feeling used, however. 'On the contrary, I feel even stronger. Alessi stands for humour, innovation and creativity. Fiat needs me to inject these vital elements into the company.' Who knows? Perhaps an extreme makeover of the Panda is on the horizon.

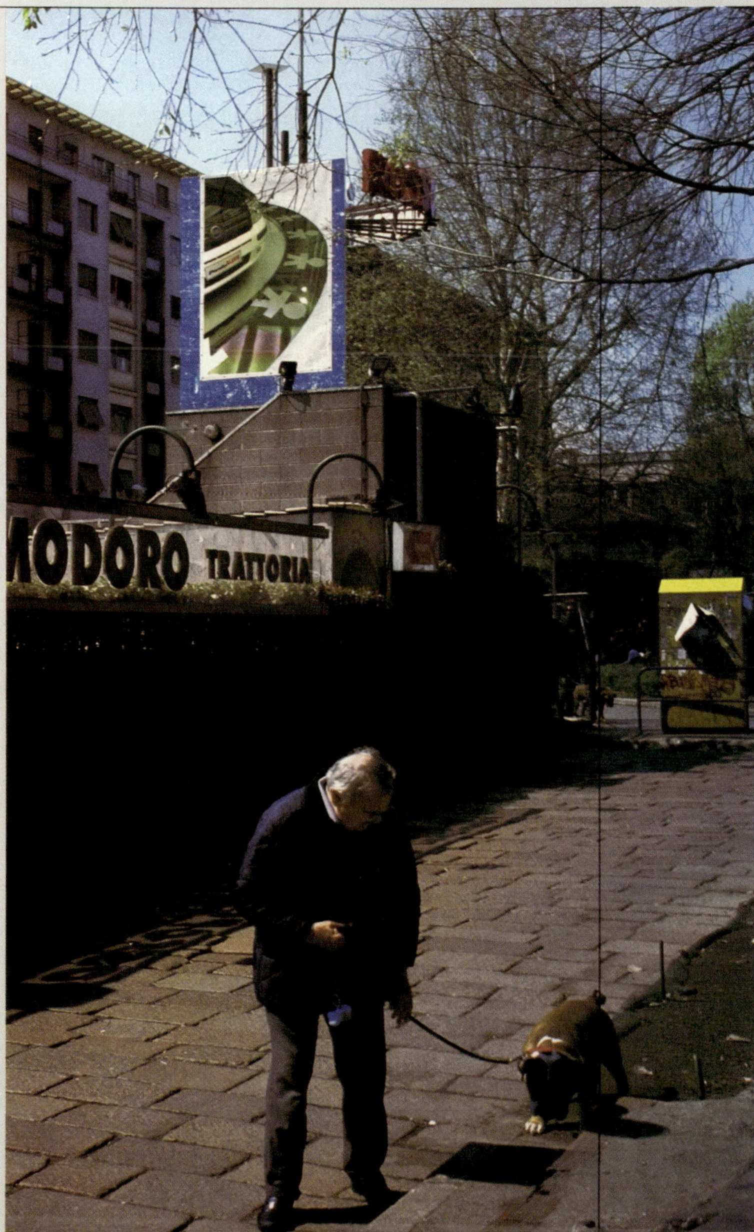

SUBJECT
Review of Salone del Mobile/
Milano Design Week 2004

LOCATION
Milan, Italy

DESIGNERS
Various, see page 409

PHOTOGRAPHER
Corriette Schoenaerts

WRITERS
Tessa Blokland
Robert Thiemann

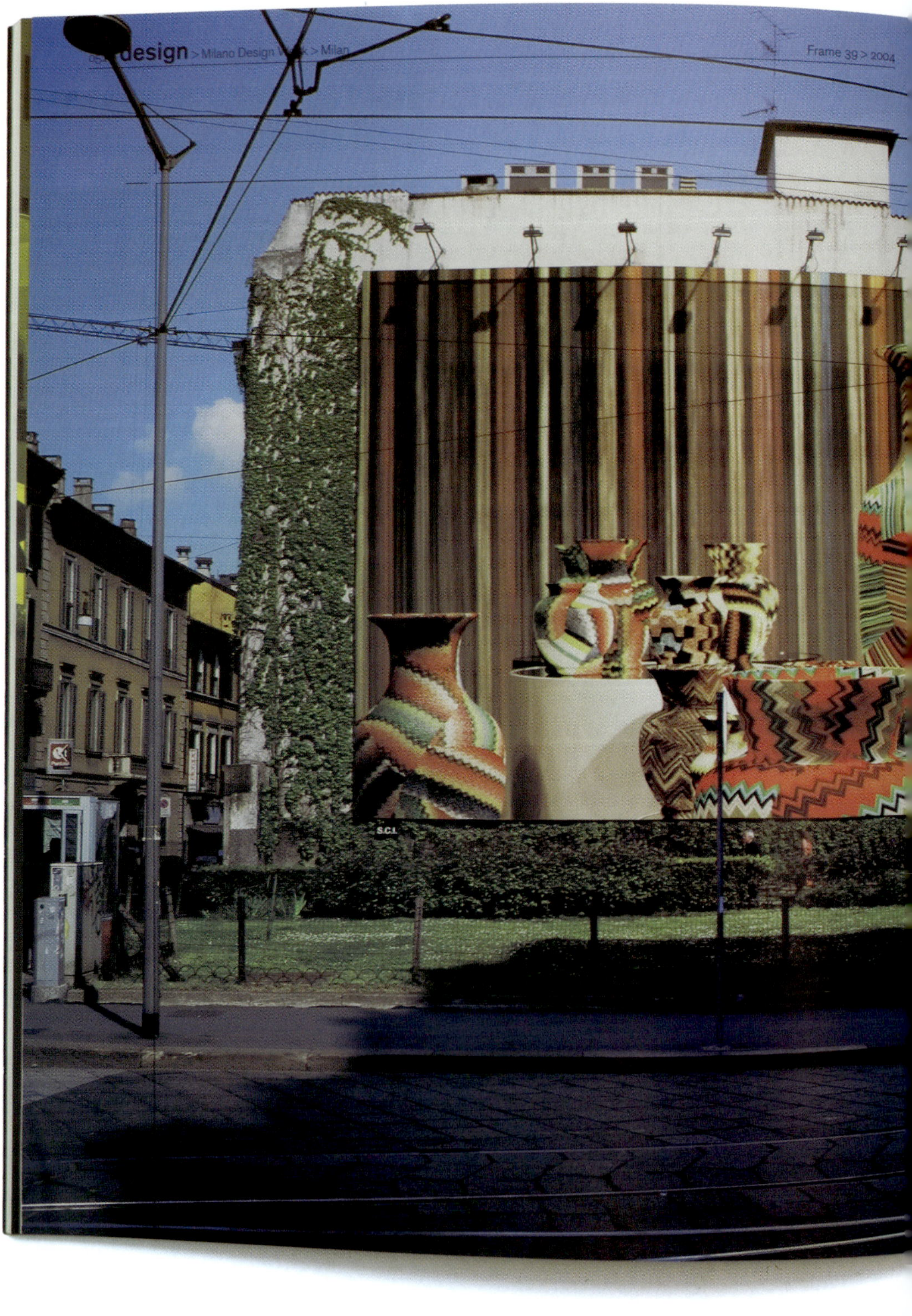
design > Milano Design
Frame 39 > 2004
S.C.I.

GO PLAY

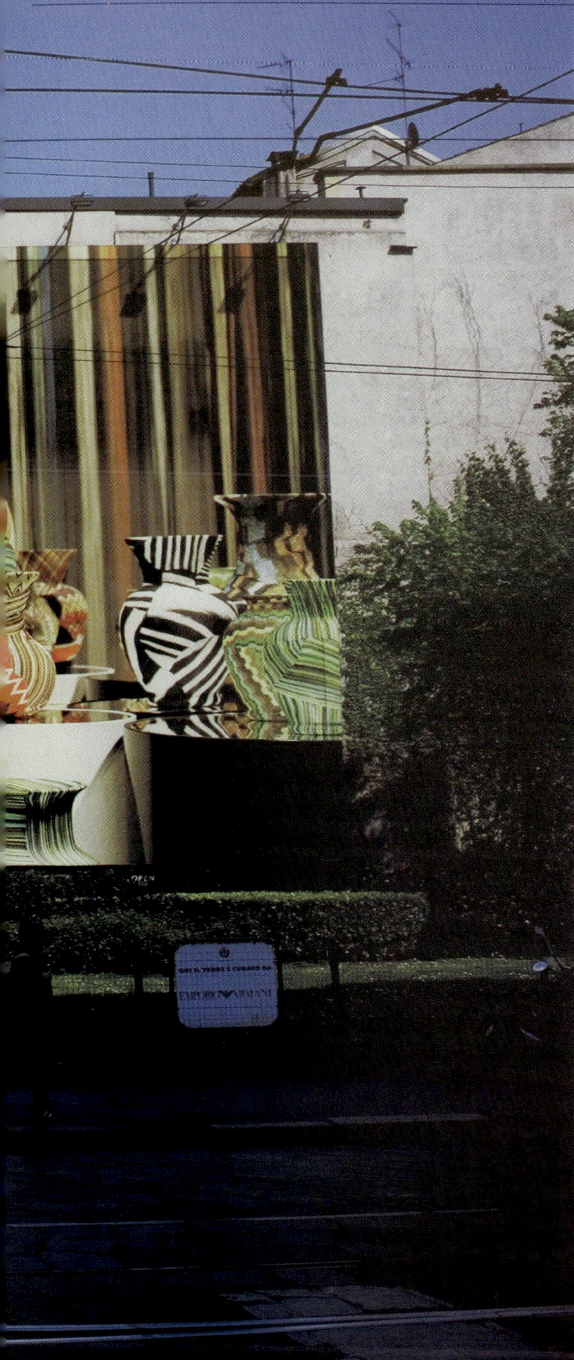

The thought currently making the rounds of the design scene is that all furniture is starting to look alike. Only a few manufacturers are willing to go out on a limb with experimental products, and even fewer are investing in research. Can the design industry look to fashion houses for help in this area? American designer Stephen Burks is delighted about his collaboration with knitwear brand Missoni. In search of an opportunity to 'do something together' at the fair, Angela and Rosita Missoni gave Burks a big bag of fabric cut-offs and sent him home with the message, 'Go play'. Fast forward to the Missoni showroom in Milan, where Burks' creations – glass vases wrapped in knitted fabric by Missoni – stole the show. As a designer for Cappellini, Moroso and Zanotta, among others, the American calls his work for the Missoni family 'an amazing experience'. He mentions 'being able to sketch ideas directly with Angela' and 'picking out fabric with Rosita' as examples of how 'incredibly accessible' these women proved to be. This type of hands-on approach to a project is seldom experienced in the furniture business. Because a number of designers are invariably involved in a furniture collection, the manufacturer has less time to invest in individual give-and-take on the long road to mass production. As a result, projects in the furniture industry are less flexible, says Burks, and less experimental. Note the following buzz words for 2005: play, hands-on, accessible.

SUBJECT
Review of Salone del Mobile/
Milano Design Week 2004

LOCATION
Milan, Italy

DESIGNERS
Various, see page 409

PHOTOGRAPHER
Corriette Schoenaerts

WRITERS
Tessa Blokland
Robert Thiemann

SLEEPING GIANT

A small presentation at the very back of Superstudio Più introduced products that could have far-reaching consequences for design as we know it. Starring in the collection that Materialise of Belgium showed were a number of exquisitely complex lamps, diminutive and fragile. Their appearance is less important, however, than the technologies used to make them: selective laser sintering and stereolithography. Selective laser sintering (SLS) is a process in which three-dimensional parts are created by fusing or sintering powdered thermoplastic materials with the heat of an infrared laser beam. The process of stereolithography (SLA), on the other hand, uses a laser beam to cure a light-sensitive liquid polymer. Both technologies are the designer's wet dream come true: Take me and shape me – I'll be whatever you desire. Never again can a manufacturer say that a design is impossible to produce. And it gets even better: SLS and SLA can also fulfil the customer's greatest wish: Every product can be made to measure. Within reach is a level of mass customization that offers each motorist a unique gear lever shaped to fit his hand. Still very expensive, these processes are currently used only for rapid prototyping. Materialise, for example, churns out about 700 prototypes a day for companies such as Alcatel, BMW, GM, Mercedes and Philips. But with the rise in orders and, consequently, an increased volume of production and noticeably decreasing cost of SLS and SLA, both technologies are beginning to attract manufacturers. When this sleeping giant finally wakes up, the design industry can prepare for a complete metamorphosis.

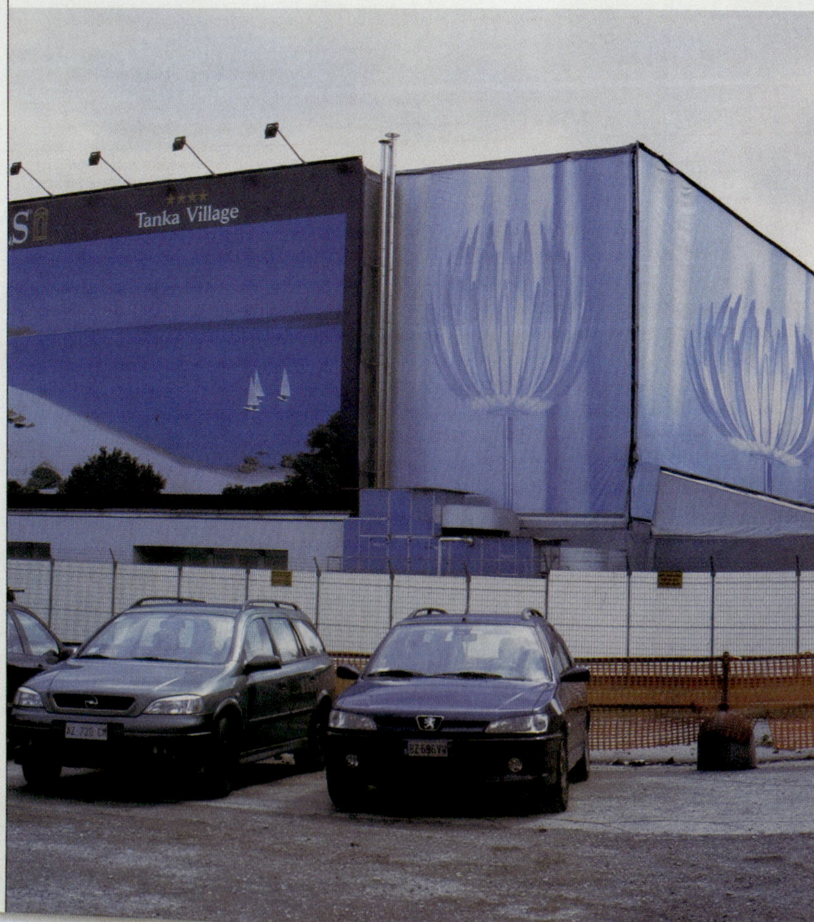

AWAKENINGS

What began years ago with the appearance of a handful of design schools in Milan has evolved into a genuine circus. No fewer than 20 schools exhibited at Salone Satellite, not to mention those that popped up at other venues throughout the city. All were there to make an impression on potential students and generous sponsors, for apparently without the business community, a design school can no longer exist. Did I say all? Correction: Two small Latvian schools arrived in Milan driven by entirely different motives. The Art Academy of Latvia and the Design and Art School of Riga wanted to show their students that design, rather than stopping at national borders, is an extremely international discipline. In addition, the Latvian schools accepted the challenge implicit in exhibiting their work alongside that of other, often bigger and richer, schools from across the globe. And finally – 'and this is no joke', Barbara Abele of the Art Academy of Latvia assured us – the schools wanted to promote their country as it prepared to enter the European Union. Their presentation had a Boontjesque touch. Branches above a surface of artificial grass held storybook cradles, symbolizing the state of design in Latvia, although Abele doubts that even a single visitor got the message. On the second day of the fair, a bus carrying 60 Latvian students pulled into Milan. That same group, says Abele, returned home highly motivated and brimming with new ideas. 'They've woken up.'

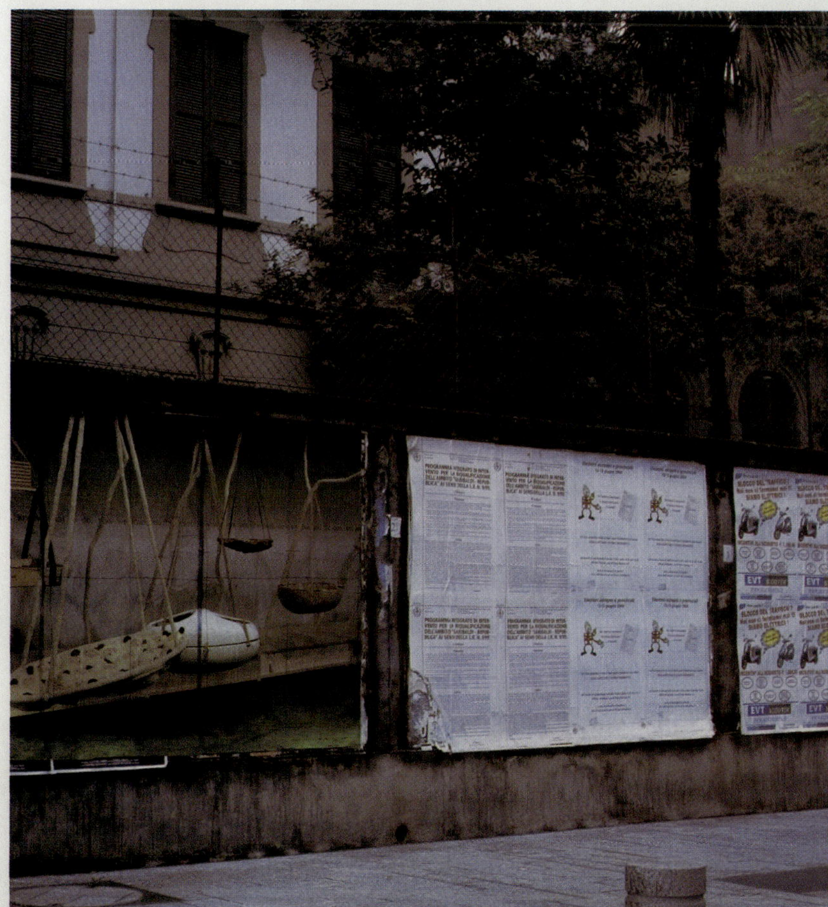

SUBJECT Review of Salone del Mobile/ Milano Design Week 2004

LOCATION Milan, Italy

DESIGNERS Various, see page 409

PHOTOGRAPHER Corriette Schoenaerts

WRITERS Tessa Blokland Robert Thiemann

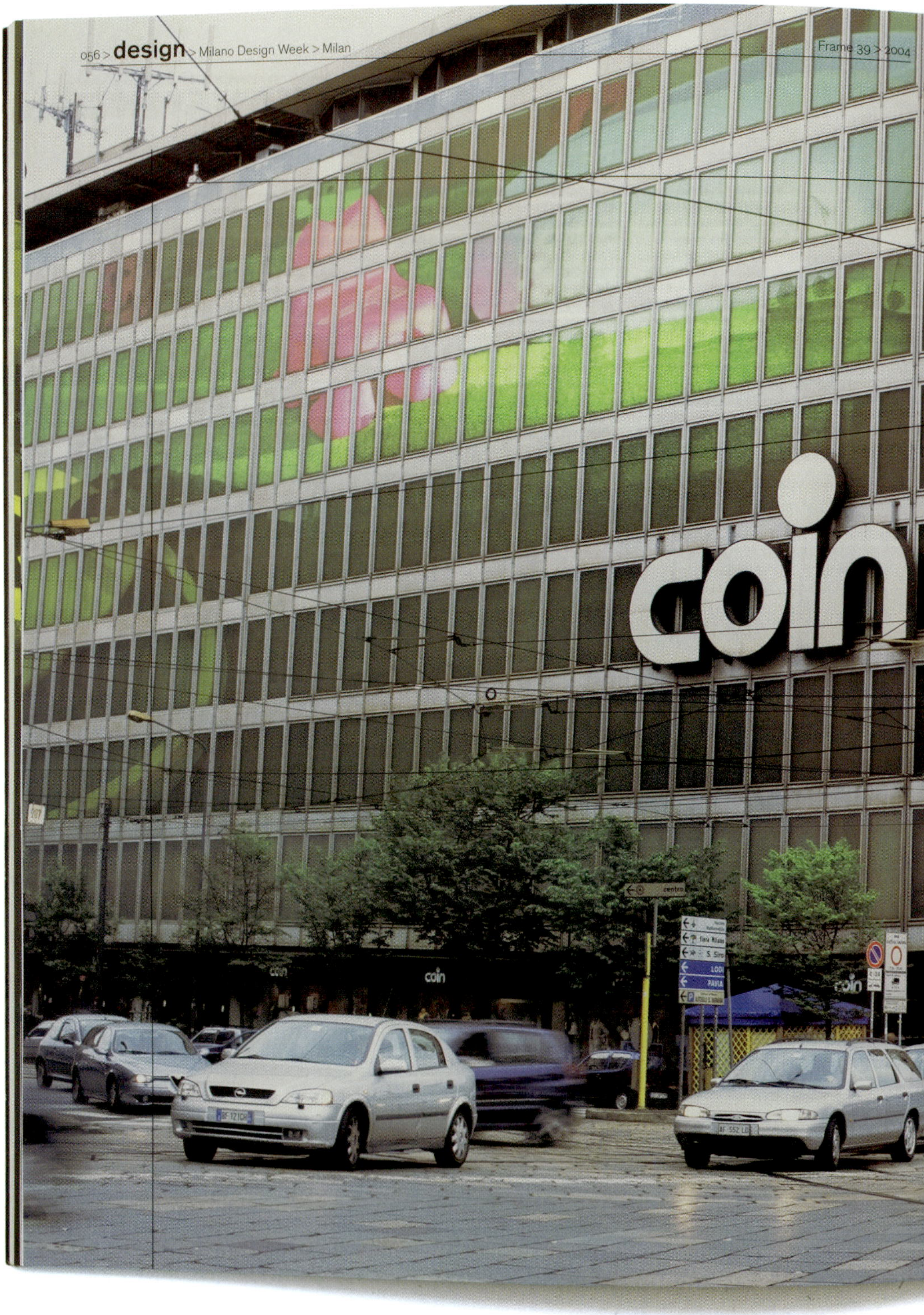
056 > design > Milano Design Week > Milan
Frame 39 > 2004
coin
centro
S. Siro
LODI
PAVIA
coin

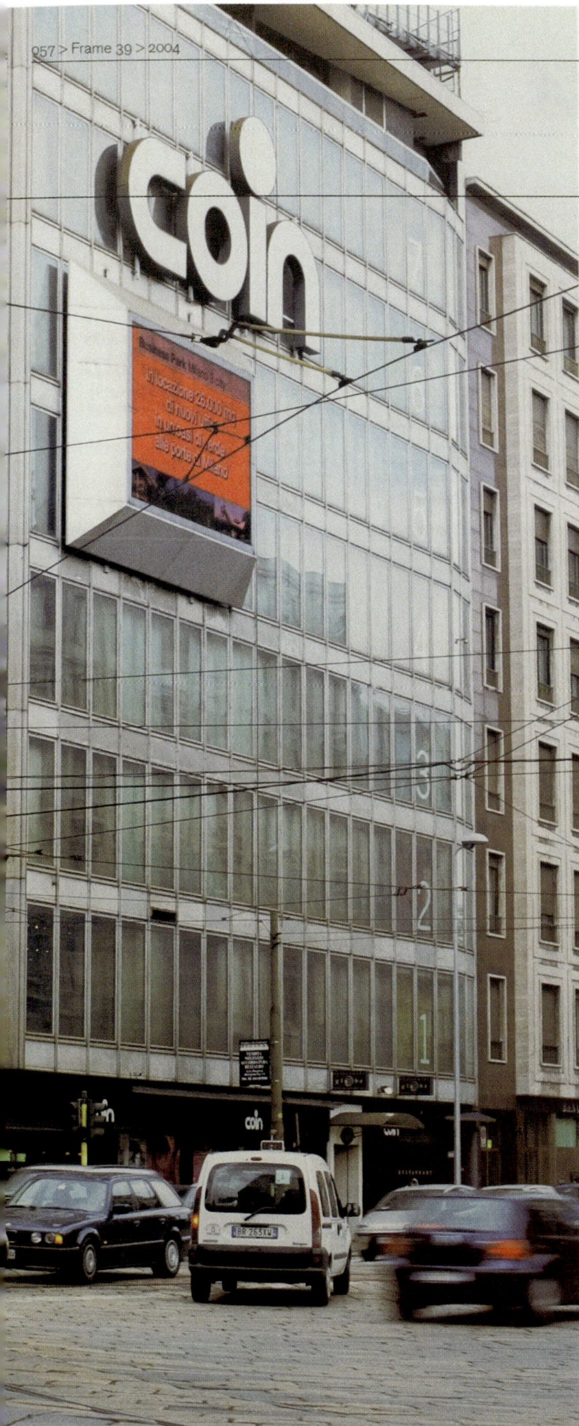

MY FIRST MAGIS

Every design-lover with children is aware of an enormous gap in the market: Reasonably priced, well-designed furniture for kids and appreciated by kids is virtually unavailable. Not counting Ikea, of course. It was only a matter of time, then, before a manufacturer found the courage to fill this gap. And Eugenio Perazza, director of Magis, was first in line when they handed out courage. Because his former boss rejected Perazza's ideas for investing more in research and design, the latter set up his own company in 1976. Asked what design means for Magis, he replies: 'Design means difference. When a product can't be distinguished from the rest, that's not design – it's styling.' And probed on the importance of marketing, he says: 'I hate marketing and managers. Design is about investment, about taking risks.' Three years ago, after searching in vain for a suitable desk for his granddaughter, Perazza set his sights on his own collection of children's furniture. The name came to him instantly: 'Me Too' is what little kids say when they want something. He approached no fewer than nine designers, briefing them in no uncertain terms on issues like manufacturing processes, recommended retail price and personal ideas on good children's furniture. For two years Magis worked on chairs, tables, bookshelves, a hand-held lamp, a puzzle carpet, even T-shirts. And throughout the entire process, it was granddaughter Anna who tested each design. Look out, Ikea. Here comes Eugenio Perazza of Magis, a guy with guts.

SUBJECT
Review of Salone del Mobile/
Milano Design Week 2004

LOCATION
Milan, Italy

DESIGNERS
Various, see page 409

PHOTOGRAPHER
Corriette Schoenaerts

WRITERS
Tessa Blokland
Robert Thiemann

JOOHEE'S CHOICE

Housed in a hall of the gigantic fairground complex, yet radiating an atmosphere all its own, is Milan's breeding ground of young talent: Salone Satellite. Rather than rows of slick stands-with-bars, visitors encounter hundreds of white cubicles, each a couple of metres square – cramped cells annually filled with young designers and small manufacturers all harbouring a single wish: to be discovered. Only a handful walk away with a story that matches the tale of Joohee Lee, a Korean-born creative employed by Samsung Design Europe (London) as an anonymous designer of mobile phones and the like. Armed with a master's degree in Design Products from London's RCA, Lee saw Milan as an opportunity to show the work he churns out in his spare time. At the Satellite show, he presented a simple, wooden 3-in-1 chair and an amusing little lamp made of electroluminescent wire and clear acrylic resin. What many a designer spends all his waking hours pursuing in vain, Lee succeeded in doing after work: grabbing the attention of an organization like Droog Design, which will be including both designs in its collection. And Droog was not alone: 'There were other, sweeter offers,' the Korean reports. With a day job that pays the rent, however, Lee passed up pounds for prestige and opted for the Dutch organization. 'I'm honoured that my first mass-produced pieces will come from Droog. It's not about money. It's about who I am.' Design in the year 2004: It's not about the product; it's about the label.

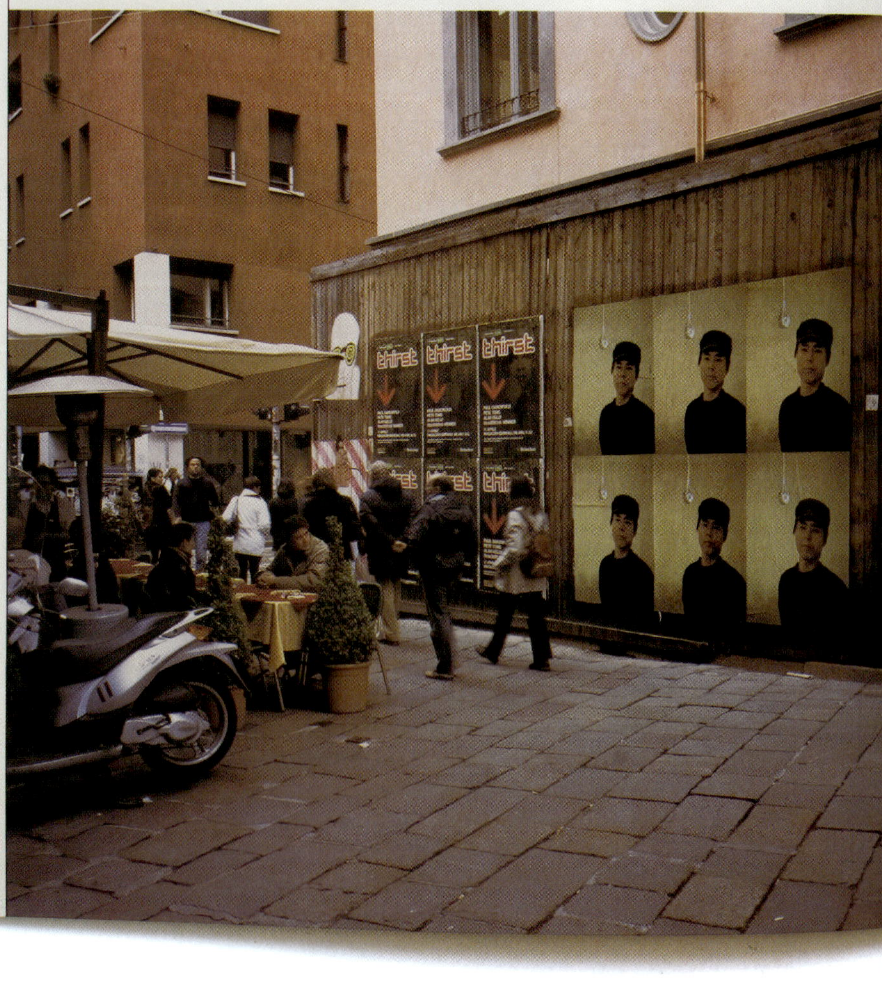

PIECE OF CAKE

Ten design schools presenting restaurant concepts and famous designers giving face-lifts to items like takeaway meals and biscuits: no one in Milan could have overlooked the discovery of food as a design topic. Just what can designers add to something as foreign to their ordinary work as edibles? The answer appeared at Sant Ambroeus, a traditional Milanese patisserie in the heart of the city. Filling five display windows of this tearoom, a space studded with Murano-glass lamps, were cakes conceived by 22 designers from around the globe. A popular Italian magazine, Case da Abitare, had asked a random sample of the world's top 50 designers to submit sketches and cake recipes. Sant Ambroeus's master pastry cook, Luciano Vismara, turned their sweet dreams into reality. Dutch designer Ineke Hans aptly describes her detour outside the boundaries of the design industry as 'delicious', albeit difficult. 'But as a designer, you can direct your thoughts to almost any subject and come up with relevant ideas.' Most of the concepts involved form. We saw a big chocolate diamond by Ora Ïto, a cake shaped like a single slice of Sachertorte by Konstantin Grcic, and a gnome with a cache of chocolate hidden in pastry tents, a Patricia Urquiola original. True, a spectacular-looking cake doesn't guarantee a sensation for the taste buds. For the crème de la crème among designers, 'form follows flavour' is not the golden rule.

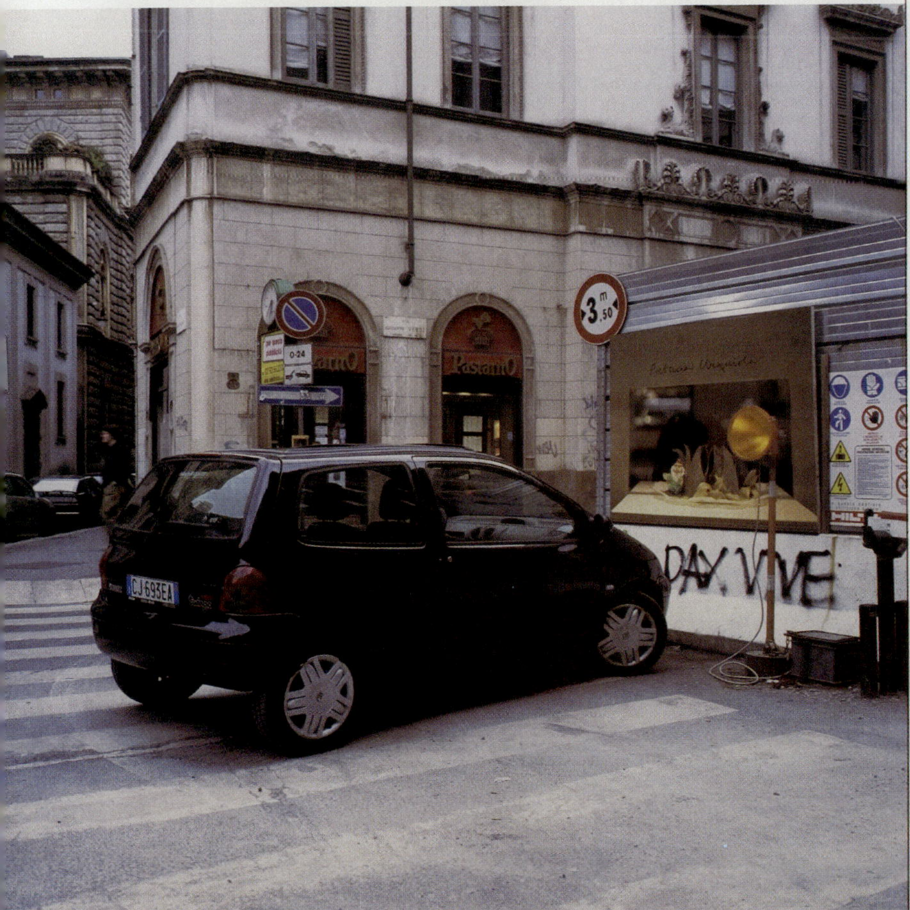

SUBJECT
Review of Salone del Mobile/
Milano Design Week 2004

LOCATION
Milan, Italy

DESIGNERS
Various, see page 409

PHOTOGRAPHER
Corriette Schoenaerts

WRITERS
Tessa Blokland
Robert Thiemann

ON A MISSION

Droog Design can never be accused of imposing a signature form on its products. For Renny Ramakers and Gijs Bakker, form is the packaging of an idea. 'If there's no message, no idea, behind a design, you can forget it. Nothing else matters,' claims Bakker. And since ideas and stories are constantly changing, and since Droog works with a varying selection of designers, the style and shape of the outfit's designs are constantly changing too. 'That's the device that makes us what we are,' says Bakker. 'We can never be pinned down. There's always room for evolution.' Once again the Dutch organization surprised us in Milan, this time with Go Slow, a project emphasizing the bright side of ageing. For 15 euros, visitors to the gallery that hosted Droog's exhibition could dine at a purpose-built restaurant run by senior citizens. The restaurant was the graduation project of a Thai student who earned her master's degree from the Design Academy Eindhoven last year. Featured in the restaurant were products that Droog had commissioned from other designers: all objects reflecting the Go Slow theme. According to Bakker, hardcore industrial designers focus primarily on the disabilities of the elderly. 'Toilets that rinse your bottom, chairs that tip forward to help the user stand up. Such cheerless items disregard the positive aspects of old age: care, attention, knowledge. Compared with the usual stuff, our Go Slow project is an eye-opener.' Bakker believes that Droog has an obligation to present a thought-provoking project every year in Milan, each time in a different way. 'If we were to stick to one style or dogma, we'd be tossed into the rubbish bin without delay. That's Droog in a nutshell.'

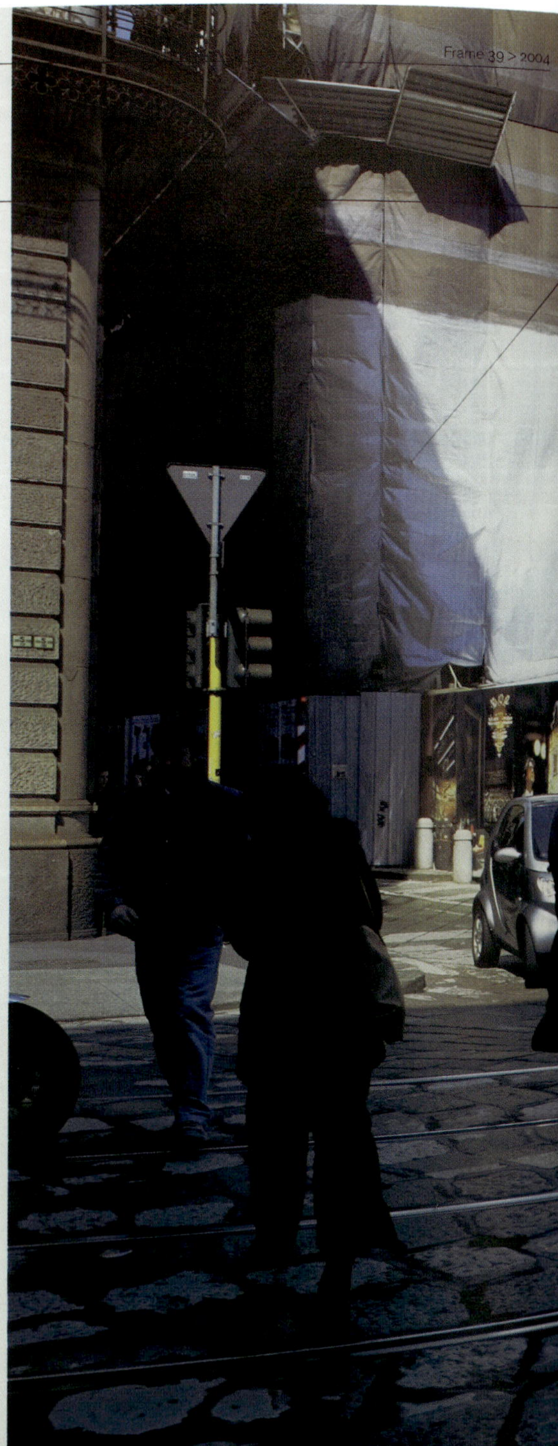

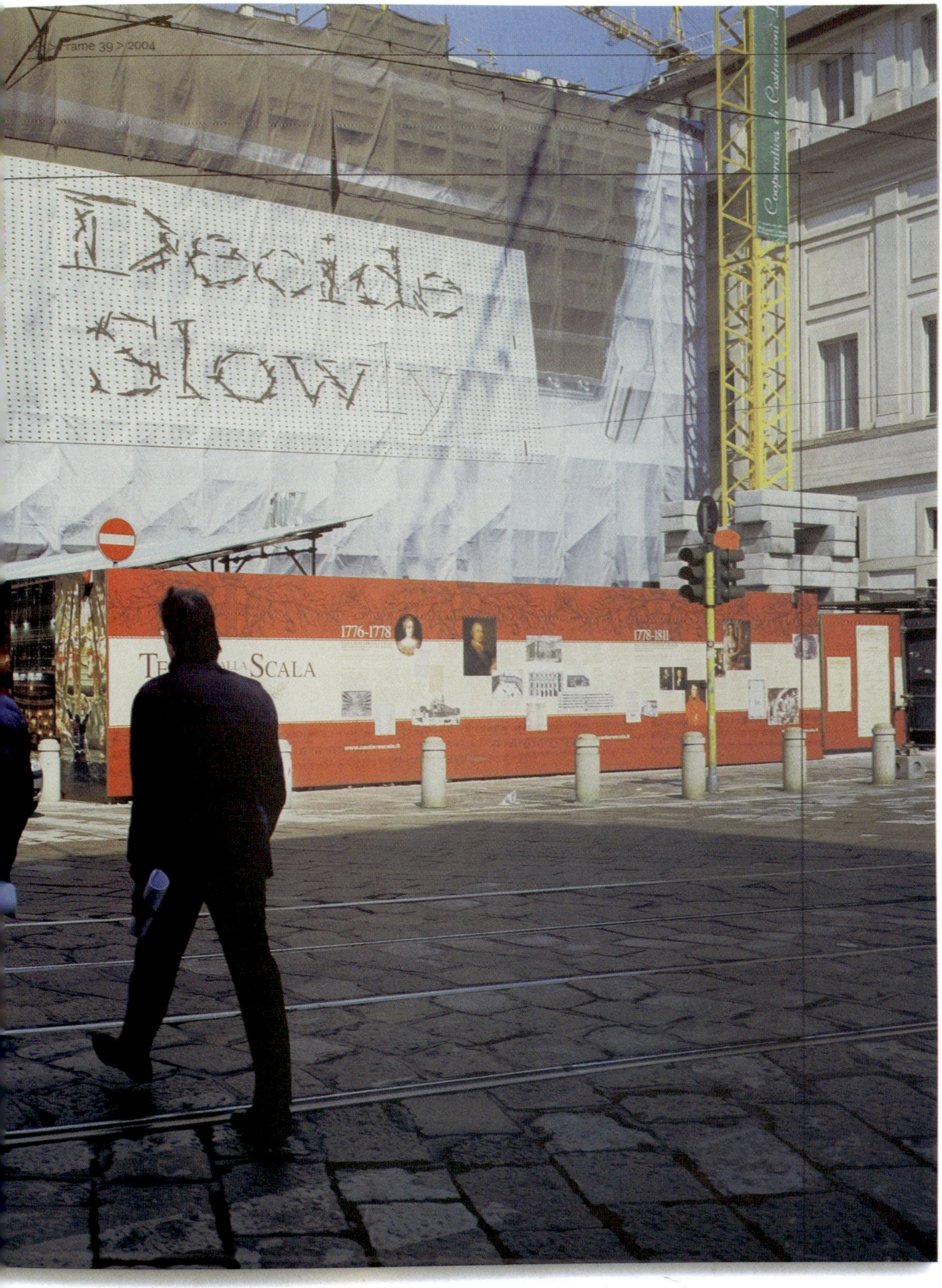

SUBJECT
Review of Salone del Mobile/
Milano Design Week 2004

LOCATION
Milan, Italy

DESIGNERS
Various, see page 409

PHOTOGRAPHER
Corriette Schoenaerts

WRITERS
Tessa Blokland
Robert Thiemann

ACT OF PASSION

By now, all major Italian design brands have showrooms in Milan. During Milano Design Week, these showrooms display more or less the same items that appear at the fair. In some cases, the showroom even replaces the furniture-fair stand. The result is a space that is shifting gradually from showroom to business zone, a space that commercializes the brand in question. Moroso's showroom policy bucks the trend by hosting a display that conveys what the company stands for. Last year Taiwanese artist Michael Lin filled the space on Via Pontaccio with his artistic decorations. Stepping into the spotlight this year was Dutchman-cum-Brit Tord Boontje, who transformed the showroom into a fairy-tale forest replete with laser-cut garlands, dreamy wall paintings and, the high point of the show, a white wedding chair. Patrizia Moroso, art director of her family's furniture company, is enchanted by Boontje's work. 'He has an outstanding ability to juxtapose old and new, naturalism and technology,' she says. 'His extraordinary approach to a décor is at once fascinating and slightly obsessive.' Moroso explains that research into fabrics and patterns has always been paramount to her company. Both Lin's and Boontje's creations reflect this research. Not tethered to a brief, the London-based Dutchman was free 'to express his romantic and highly aesthetic fantasy world'. Proving that total freedom does not indicate a lack of commitment on the part of Moroso is the company's decision to develop and manufacture even more of Boontje's work than originally anticipated.

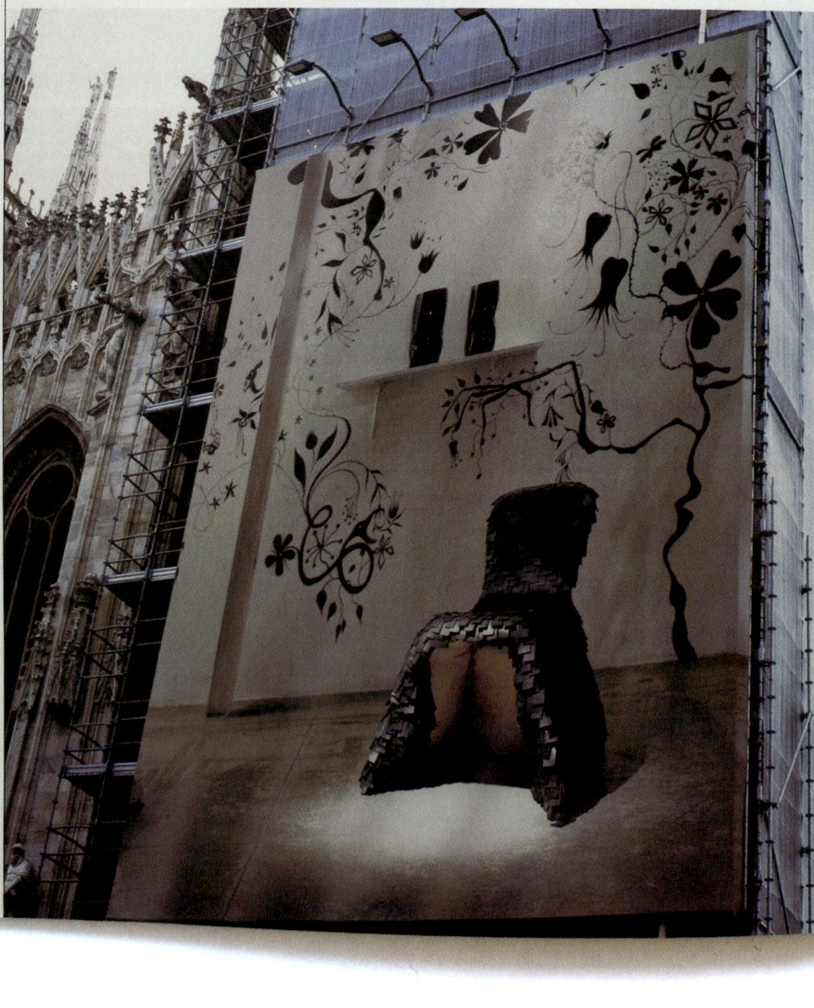

FOLLOWING FASHION

As Italy watches, the flirtation between the flashy world of fashion and the far smaller design industry moves closer to the boiling point. Fashion empires like Armani, Fendi and Versace have been dabbling in home furnishings for quite a while. It appears to be only a matter of time before major fashion labels take over the design industry. One of the few Italian fashion houses without its own line of furniture, Dolce & Gabbana, enjoyed a snog with Kartell during Milano Design Week. Domenico and Stefano showed a selection from the Kartell Museum in their flagship store on Corso Venezia. Spatial designer Ferruccio Laviani, who works for both D&G and Kartell, brought the two parties together. Claudio Luti, president of Kartell, believes the design industry can learn a lot from 'the fashion system', including 'the will needed for leadership, a precise target, brand coherence, efficient distribution, focused communication and the "cash & carry" product concept'. His attitude explains why Kartell is one of only a few furniture companies that follow a fashion-related business strategy. Having built a strong brand based on plastic furniture, Kartell is extending its range to include lighting, office furniture and upholstered furniture. With more than 50 flagship stores worldwide, Kartell is set to open another 70 stores and shops-in-shops this year. Luti is especially pleased with the impact of the collaboration on his firm: 'Kartell has gained a great return in terms of brand awareness.' What's next? A Kartell lamp designed by Domenico & Stefano or a D&G chair made by Kartell?

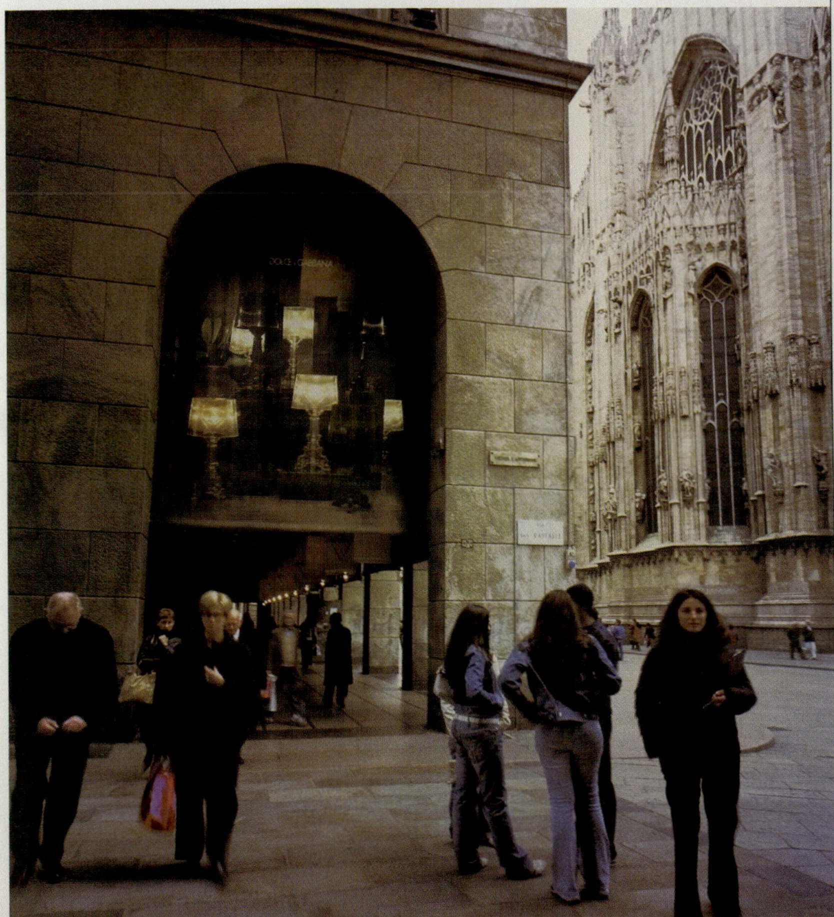

SUBJECT
Review of Salone del Mobile/
Milano Design Week 2004

LOCATION
Milan, Italy

DESIGNERS
Various, see page 409

PHOTOGRAPHER
Corriette Schoenaerts

WRITERS
Tessa Blokland
Robert Thiemann

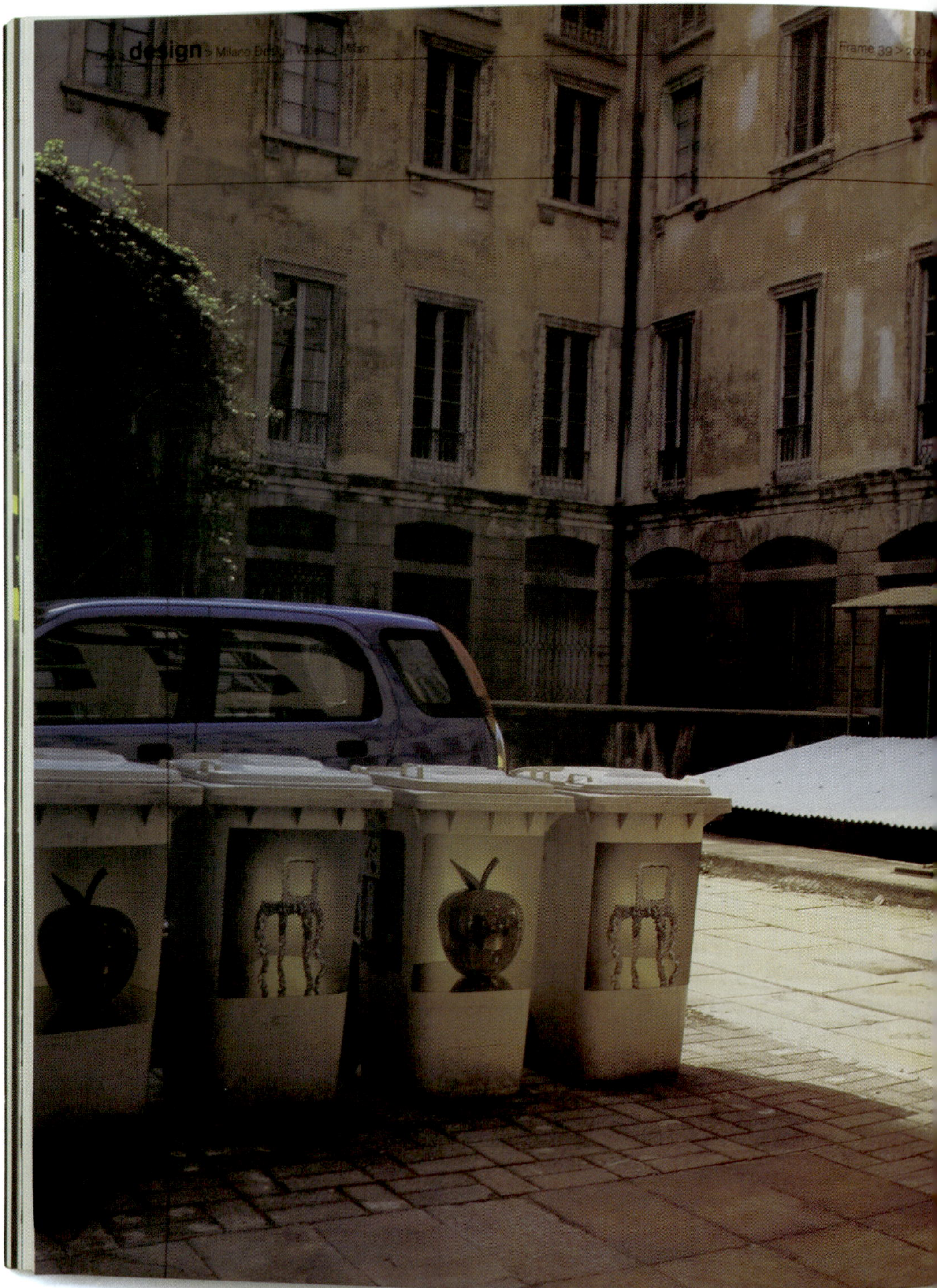
design
Frame 39 >

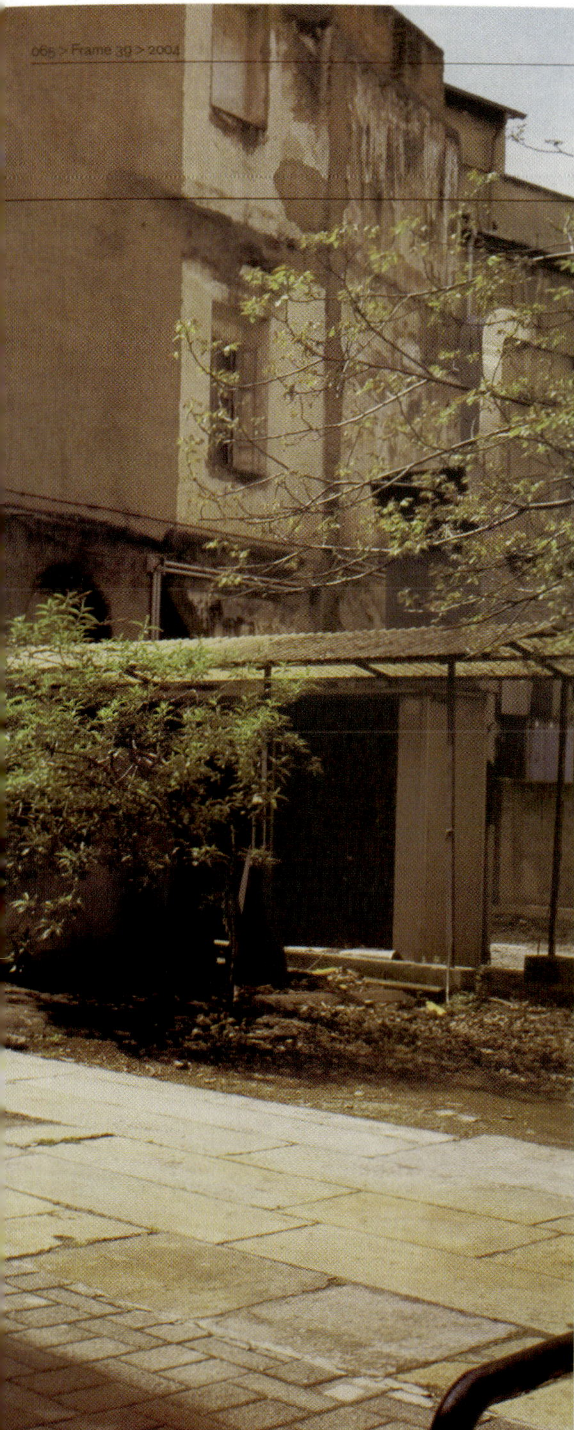

THE MYSTIC

In a time in which industrial designers are deep into personal image and style, hoping to collaborate with as many manufacturers and to produce as many lots as possible, a guy like Job Smeets is an alien presence. His work is hard to categorize, and he's not obsessed with mass production. Initially known for cartoon-like furniture, in 1999 Studio Job began drifting even further away from the path of functionality and industrial manufacturing. Emerging from the new phase were objects cast in bronze, such as a hammer and sickle, and a castle; a laser-cut aluminium necklace, which appeared in a Viktor and Rolf fashion show; and a plastic cannon. Among the designs the Dutchman presented in Milan this year were his Rock furniture, whimsical cast-aluminium chairs, and a shiny bronze apple. Smeets sees design as 'a thematic or poetic experience that expresses problems both inside and outside the discipline'. He says that aiming for the ultimate manufactured product is 'no longer the only nirvana'. If Moroso, Cappellini or Ikea were to approach him, he would 'certainly consider' working with Moroso. And although he agrees that an object should be part of its designer's identity, he finds it more interesting 'when a good designer consistently contradicts himself'. Smeets can be just as hard to comprehend as the objects he designs. Asked if perhaps he would like to design the discipline of design itself, he comes back with: 'An autonomous space is necessary to guarantee freedom, but our "voice" propagates no ultimate truth. Our objects are a possibility or, at most, a protest, but they are never dictatorial or dominant. The will to dominate is a horrible thing!'

SUBJECT
Review of Salone del Mobile/
Milano Design Week 2004

LOCATION
Milan, Italy

DESIGNERS
Various, see page 409

PHOTOGRAPHER
Corriette Schoenaerts

WRITERS
Tessa Blokland
Robert Thiemann

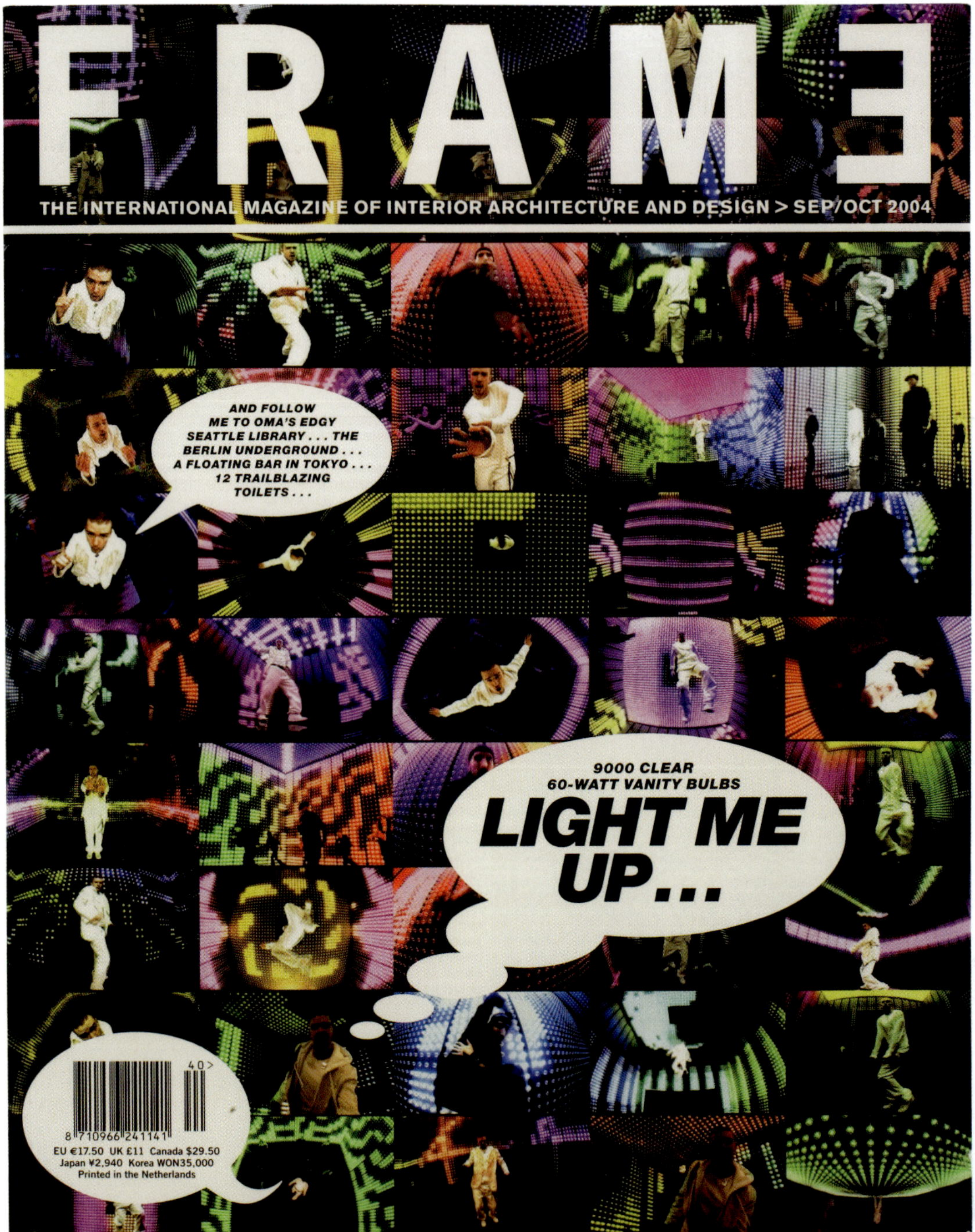

ART DIRECTORS
Cornelia Blatter & Marcel Hermans,
COMA

40

Bay Watch

Frame editor Billy Nolan about the project

A ferry boat that doubles as a floating party venue is not the kind of thing that turns heads. But this one certainly does. Why? As the work of Japanese anime artist Leiji Matsumoto, the boat looks as if it's just sailed out of one of his manga drawings. The project is an odd encounter between the worlds of interior and transport and the land peopled by cartoon characters.

DESIGNERS
Leiji Matsumoto
www.leiji-matsumoto.ne.jp
The Design Studio Takashimaya Space Creates
Yoshinori Nijou
Yoshiaki Waki

SUBJECT
Boat design

LOCATION
Tokyo, Japan

PHOTOGRAPHER
Kyosyu Mizohata

WRITERS
Paul and Ayako Greenberg

Issue 40 Bay Watch 428-429

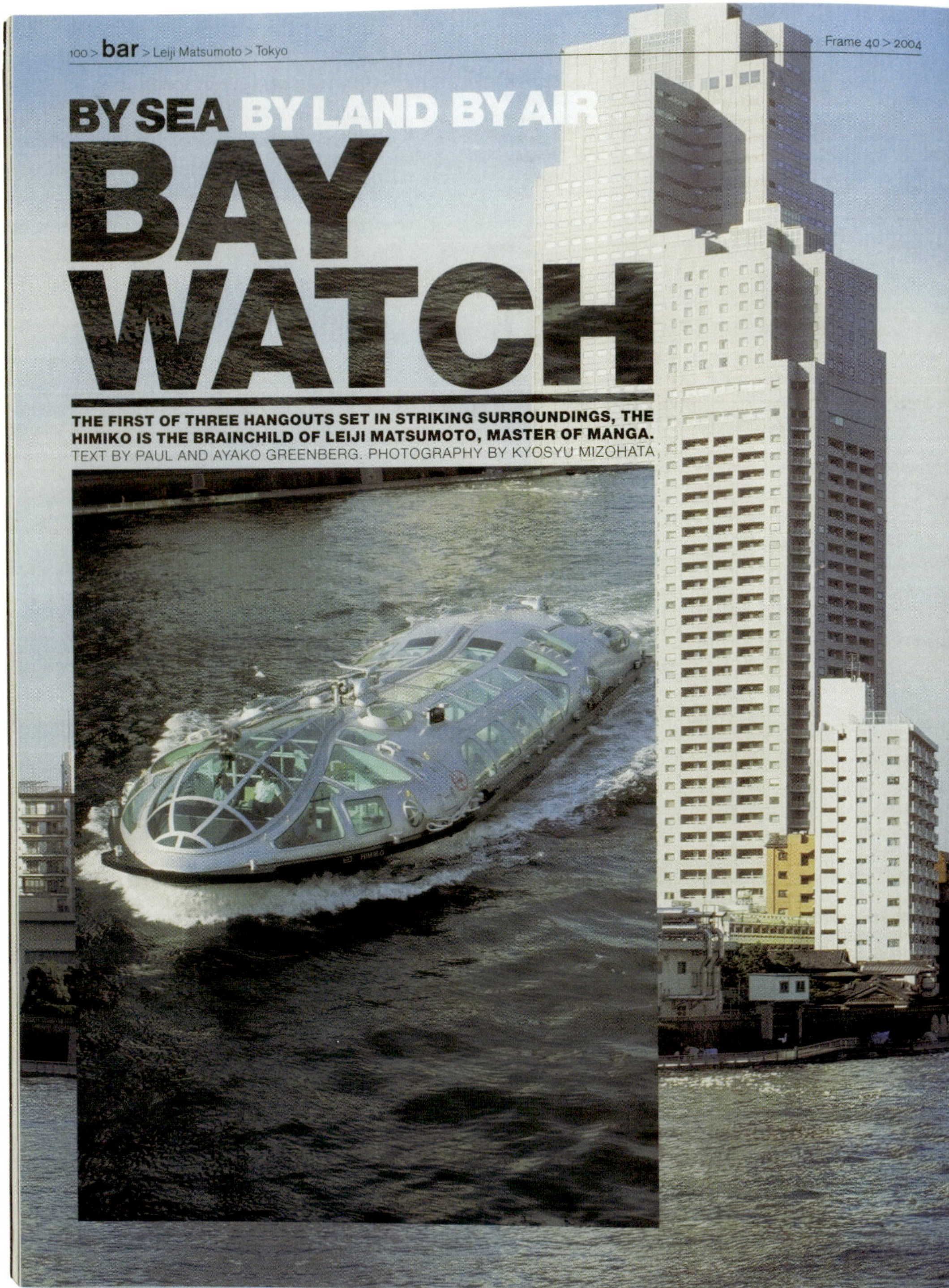

BY SEA BY LAND BY AIR

BAY WATCH

THE FIRST OF THREE HANGOUTS SET IN STRIKING SURROUNDINGS, THE HIMIKO IS THE BRAINCHILD OF LEIJI MATSUMOTO, MASTER OF MANGA.

TEXT BY PAUL AND AYAKO GREENBERG. PHOTOGRAPHY BY KYOSYU MIZOHATA

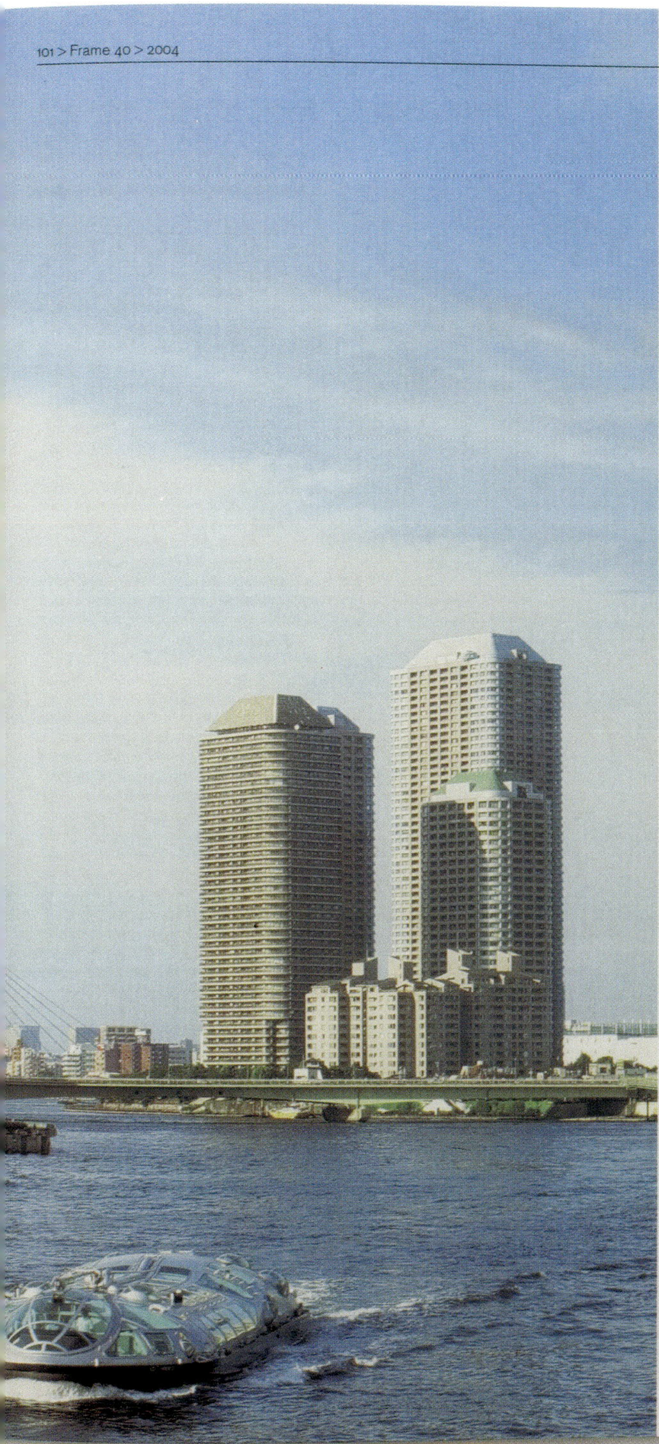

The Tokyo Cruise Ship Company (TCSC) has operated a network of river buses in Tokyo Bay and on the Sumida River since 1898. The Himiko, a vessel recently added to the fleet – she made her maiden voyage this spring – represents a new departure in boat design. The innovative look is largely down to her creator, none other than acclaimed Japanese manga and anime artist Leiji Matsumoto. Like his manga drawings, his boat embodies all the wonder and delight of youthful fantasy.

This one-off initiative represents a significant departure for TCSC, whose existing fleet of 13 vessels is pleasant though hardly awe-inspiring. 'We were looking for a boat that no-one had ever seen before,' says Mayuko Tomura of TCSC. 'A specialist shipbuilder would have placed more emphasis on functionality and efficiency and less on visual aspects. So to avoid creating the type of boat we already possess, we decided to bring in someone from a completely different discipline: anime.'

Matsumoto was chosen for three reasons. TCSC president, Mr Shinichiro Moriya, is of the post-war generation of Japanese who grew up as avid followers of the artist. Born in 1938 in Kurume, a city on the island of Kyushu, Matsumoto is renowned for creations such as Ginga Tetsudou 999 (Galaxy Express 999), Uchuu Senkan Yamato (Star Blazers) and Captain Harlock. What's more, Matsumoto's work is replete with imaginary craft of all sorts, many of which travel by water, and his animations have always displayed a clear affinity with the conceptual design of vehicles. Cross-disciplinary collaboration is not new to Matsumoto. He supervised the making of several music videos for progressive dance artists Daft Punk, as well as a full-length animated movie with a soundtrack by the group entitled 'Daft Punk & Leiji Matsumoto's Interstella 5555: The 5tory of the 5ecret 5tar 5ystem'.

The brief was simple and unconstrained: Design something new, different and exciting. Matsumoto brought together a team that included interior designer Sachiko Fukuoka of The Design Studio KK, interior contracting firm Takashimaya Space Creates KK, computer-graphics designer Yoshinori Nijou and lighting planner Yoshiaki Waki. Kanagawa Dockyard Co Ltd in Kobe built the final product.

Skimming across the surface like a light-footed crustacean, Himiko shuttles back and forth between Asakusa and Odaiba, a dockland redevelopment area full of leisure amenities.

SUBJECT **Boat design**

LOCATION **Tokyo, Japan**

DESIGNERS **Various, see page 427**

PHOTOGRAPHER **Kyosyu Mizohata**

WRITERS **Paul and Ayako Greenberg**

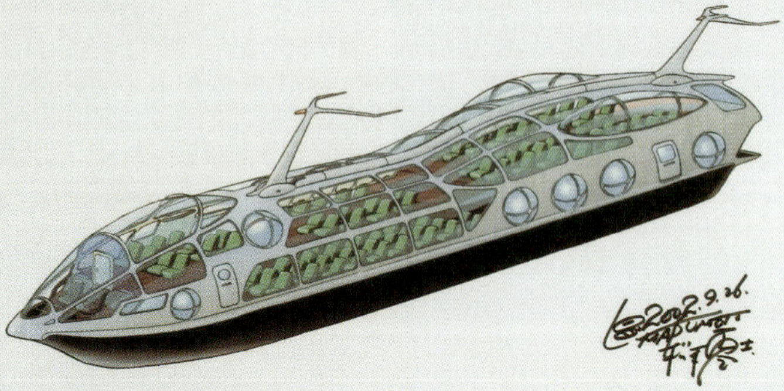

UNLIKE THEME PARKS, WHERE MAKE-BELIEVE IS SEQUESTERED FROM THE 'REAL WORLD', THE BOAT BRINGS FANTASY INTO AN EVERYDAY AND ENTIRELY FUNCTIONAL CONTEXT

Starting with a conceptual tear-drop image, Matsumoto drew inspiration from the wide array of battleships, locomotives and spacecraft that he has imagined, most notably in the Galaxy Express 999 and Star Blazers series. His first thoughts, he says, were 'to create a boat that children would spontaneously desire to board' and 'a boat that enables 360-degree viewing of the scenery'. Unlike theme parks, where make-believe is sequestered from the 'real world', the boat brings fantasy into an everyday and entirely functional context. The silver-metallic exterior, which features sections of glazing curved to match the streamlined profile, is punctuated by seven smoothly protruding oblong glass bubbles that serve as 'three-dimensional sky-viewing spots'. Eight smaller hemispheric portholes complete the picture.

The most significant technical challenge lay in realizing the unique curves of the metal frame and the glass inserts, the latter made possible thanks to technical assistance provided by the Yamaha motor company. Details such as the retractable gull-wing door and the overhead spoilers, which collapse at the press of a button as the boat passes beneath overpasses, complement the space-age image of the ship with futuristic functionality. Furthermore, this is the first small-scale ship in Japan with barrier-free universal access incorporated into its design.

The ship's interior comprises three primary spaces. The entrance opens into the hind section, where a semicircular space to one side surrounds a bar. On the other side, five steps lead down to the main passenger area, which is separated from the bulbous, glazed cockpit by a transparent divider. To offer everyone on board a relaxing voyage with unobstructed views, the number of seats has been limited to 70, although the boat can carry a maximum of 231 passengers. (The average capacity of the other vessels in the fleet is 520.) All seats line the outer walls of the central space and the rear bar area. Ultra-bright LED lamps that have been programmed to display a sequence of chang-

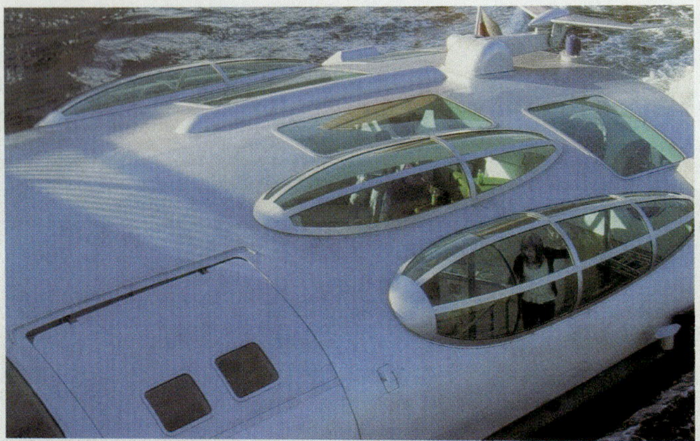

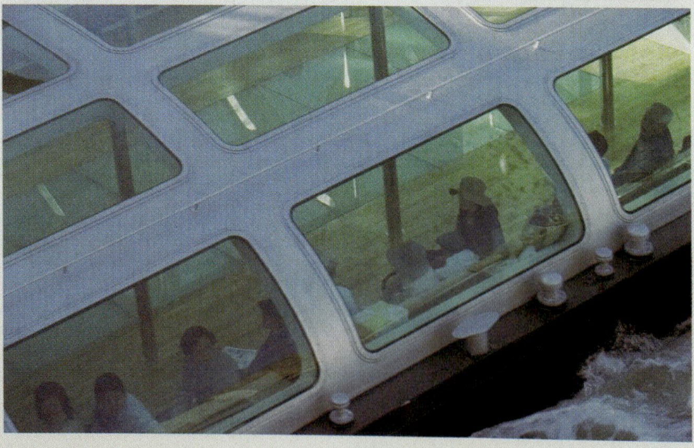

Top: Design drawing by Leiji Matsumoto, acclaimed manga artist.
Middle, bottom and opposite: Panels of curved glazing allow sightseers to enjoy unobstructed views of the shoreline. A silver-metallic shell, retractable gull-wing doors and overhead spoilers enhance the appearance of Himiko.

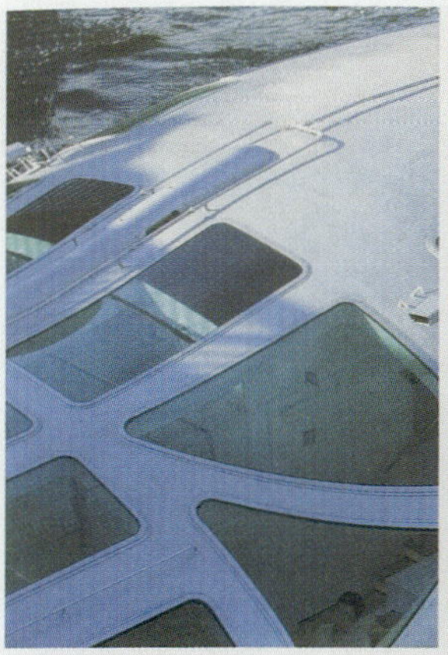
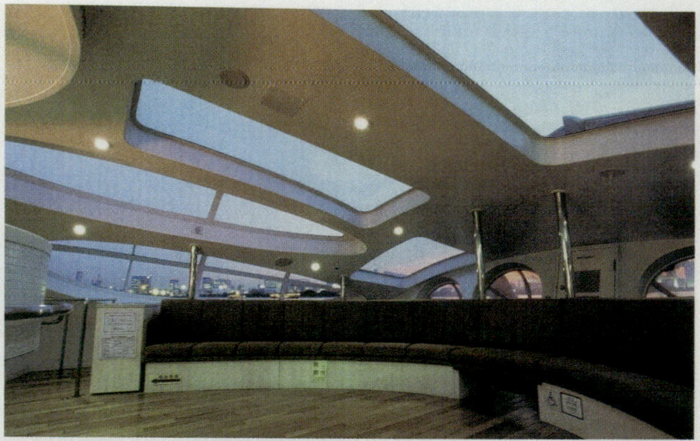
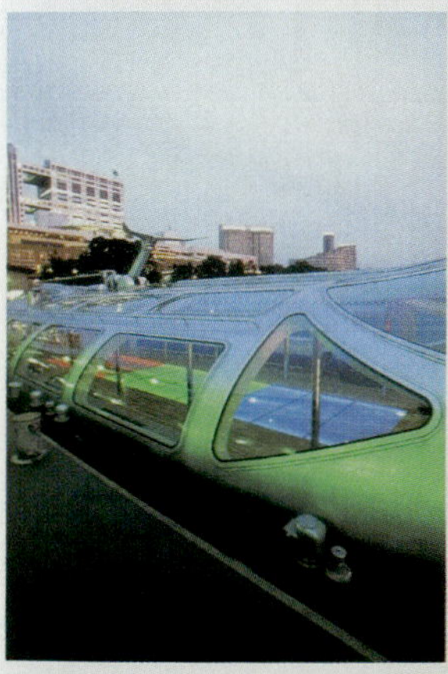
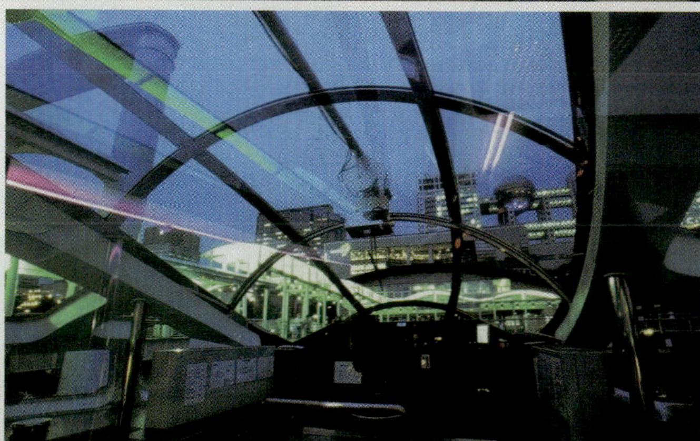
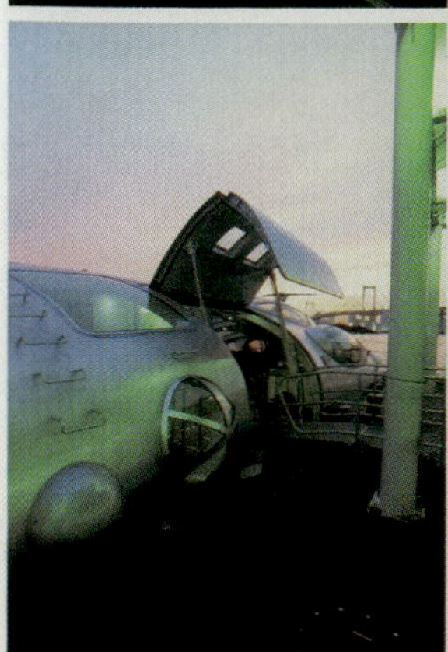
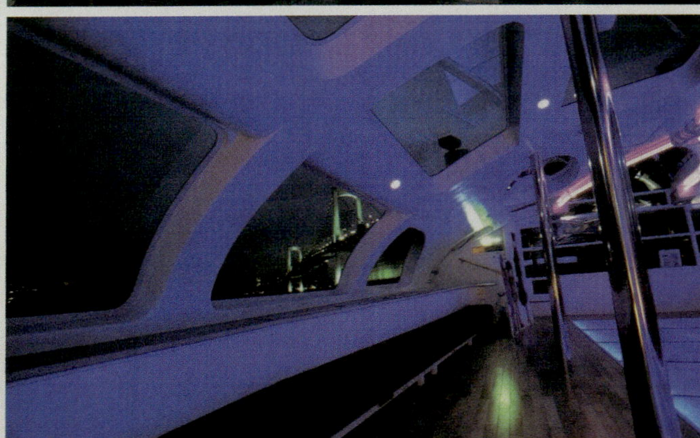

SUBJECT
Boat design

LOCATION
Tokyo, Japan

DESIGNERS
Various, see page 427

PHOTOGRAPHER
Kyosyu Mizohata

WRITERS
Paul and Ayako Greenberg

'I WANTED TO CREATE A SHIP THAT CHILDREN WOULD SPONTANEOUSLY DESIRE TO BOARD.' LEIJI MATSUMOTO

ing colours stretch down the central floor of the main passenger area. When the sun goes down, the lamps turn on, giving the ship a soft luminous glow against the darkening backdrop of the Tokyo skyline.

But the novelty of the Himiko experience isn't limited to passengers. The captain of the vessel says, 'I notice people taking pictures of us from everywhere while we are sailing. Because the cockpit is transparent, the passengers feel closer to me, and it is wonderful to sail with a 360-degree view.'

Matsumoto's animes hold a sense of nostalgia for many Japanese. Heightening the illusion of entering his phantasmal universe are on-board announcements, including instructions and commentary, that feature the voices of the three main characters from the Galaxy Express 999 series. A seamless merger of design and experience is part of the excitement that Matsumoto sought to achieve with the Himiko. 'Until I got involved with this project, I was creating imaginary sci-fi ships that could sail on the pages of comic books only,' he says 'This time, however, I was delighted to design a real ship. My wish is that the Himiko will inspire the dreams of children and carry those dreams into the world of space.'

Matsumoto's attempt to fuse fantasy with reality is not simply a flight of fancy. It is grounded in a deep desire to foster and enhance positive human interaction. His decision to name the ship Himiko is telling. 'In the third century, Yamataikoku was a collection of about 30 small communities . . . the first recognized, large-scale, organized society on the island now known as Japan,' he says. 'After an extended period of fierce fighting, Himiko emerged as the unifier, the benevolent and charismatic leader of this society. She is not only the first recorded governor of Japan, but also the first female figure in Japanese history to be recognized as powerful and influential. I wanted this boat to embody the dignity and characterize the mystery and fantasy of the ancient story of Himiko. And I want this boat to voyage across the oceans of the world to communicate dreams, gentleness, tenderness and human kindness.'

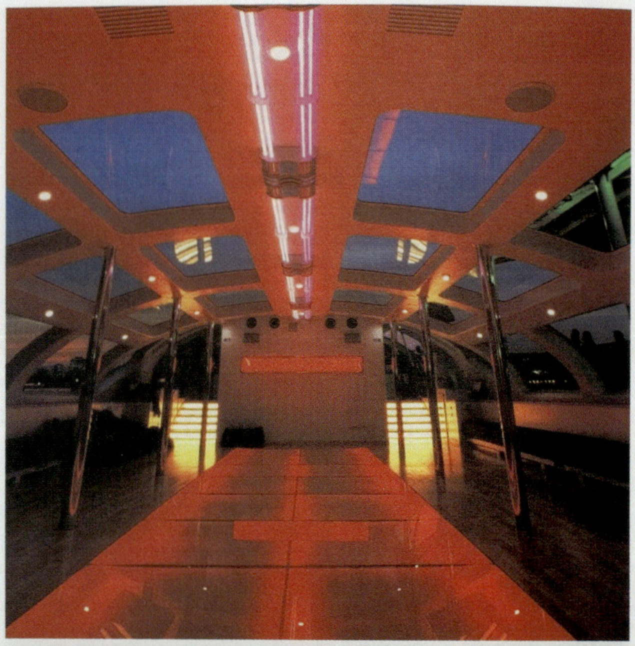

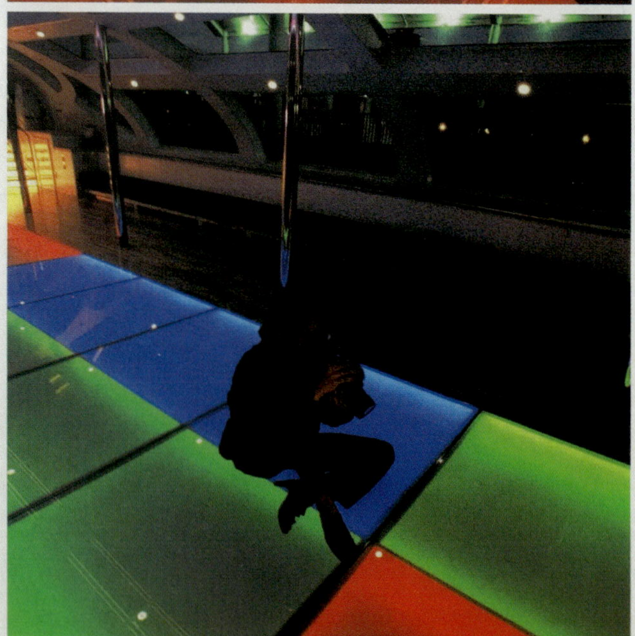

LEDs inserted into the floor of the passenger area are programmed to change colour in sequence. The 125-ton boat is 8 metres wide, 33 metres long, and has a maximum cruising speed of 12 knots.

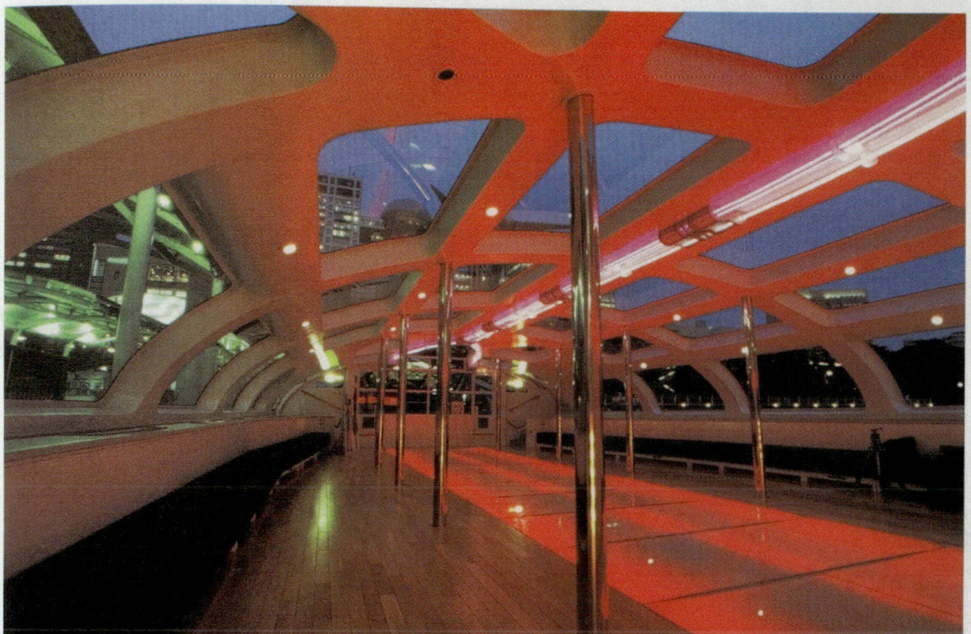

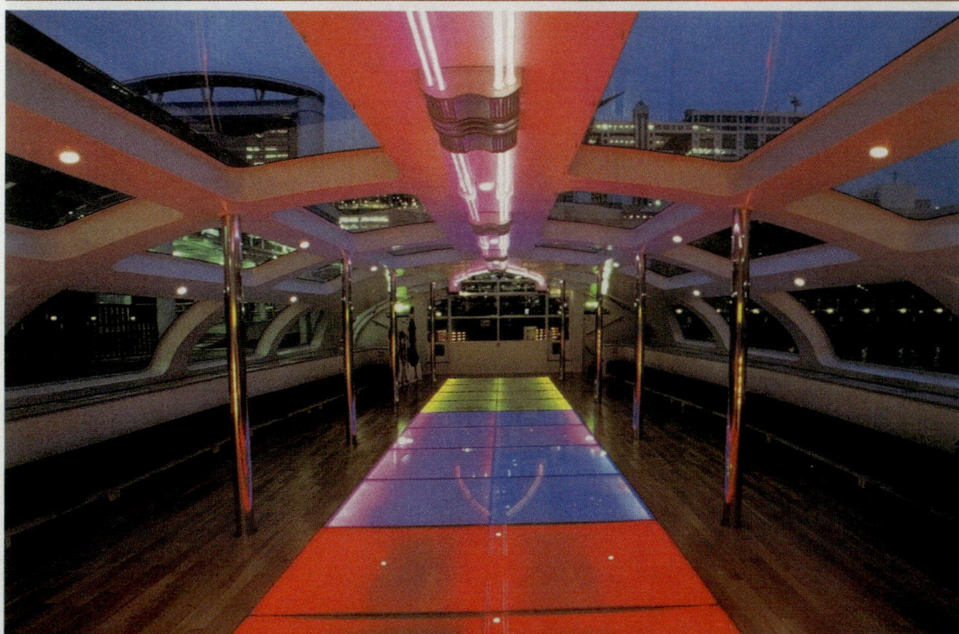

SUBJECT
Boat design

LOCATION
Tokyo, Japan

DESIGNERS
Various, see page 427

PHOTOGRAPHER
Kyosyu Mizohata

WRITERS
Paul and Ayako Greenberg

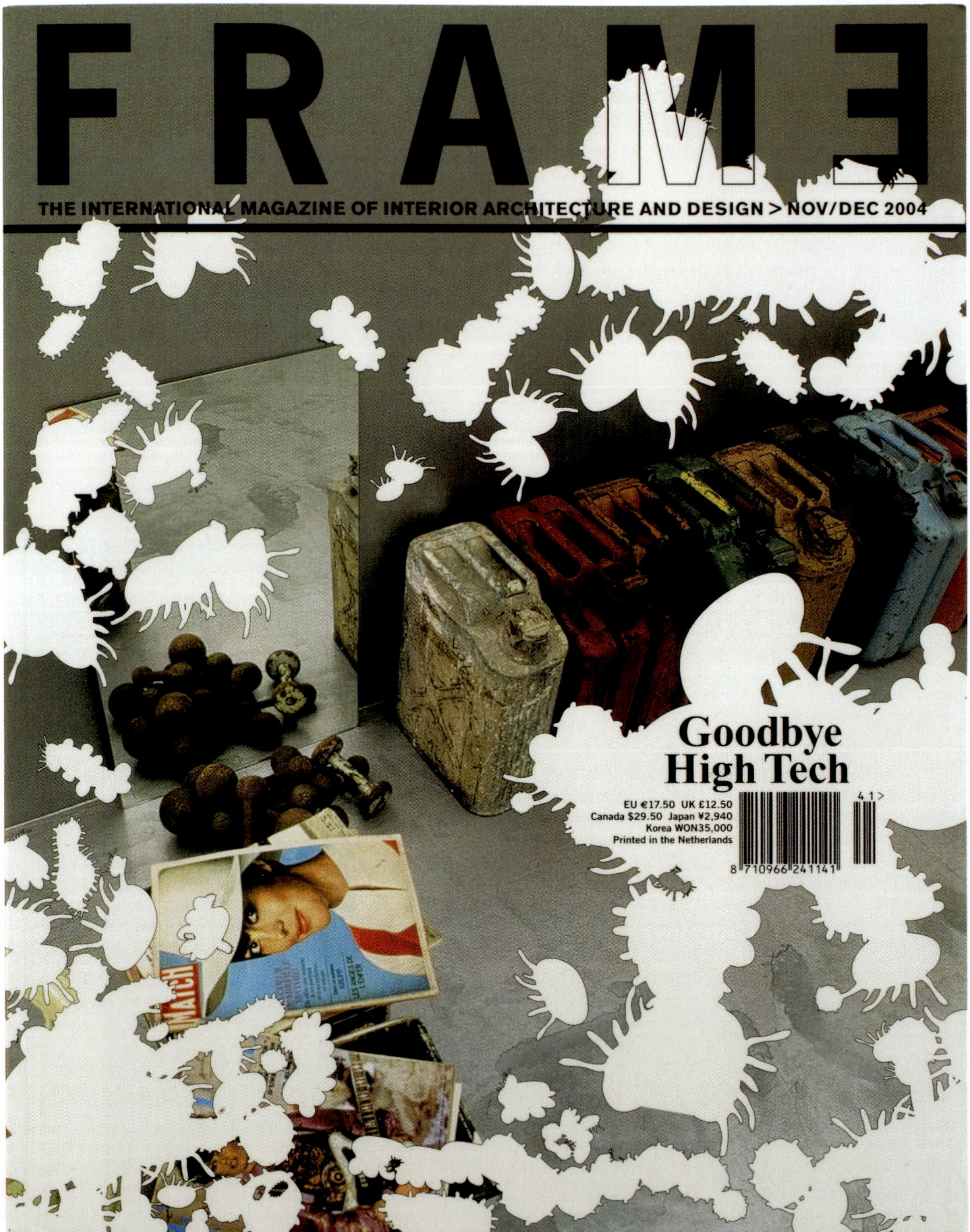

ART DIRECTORS

Cornelia Blatter & Marcel Hermans, COMA

41

Beyond Bauhaus

DESIGNER
Herman Verkerk
(EventArchitectuur)
www.eventarchitectuur.nl

SUBJECT
Orson+Bodil
fashion shop

LOCATION
Amsterdam, Netherlands

PHOTOGRAPHER
Johannes Schwartz

WRITERS
Louise Schouwenberg

Designer Herman Verkerk about the project

Shabby Glamour rules!

Comments on the published article

Both product and shop have been given an excellent context thanks to the article. Describing the integration of the development of the brand Orson + Bodil and the development of the design for the shop by EventArchitectuur makes the article a direct reflection of the way the two companies cooperate. An excellent, subdued and elegant layout with well-distributed images.

Did the article have an effect?

Recognition and understanding regarding the design principles of both Orson + Bodil and EventArchitectuur. Plus visitors to the shop/workshop from LA to Sydney, from Milan to Stockholm, etc. And a great number of applicants for design jobs at EventArchitectuur.

Recent news or information about the project featured in *Frame*

Based on the concept for the shop/workshop, the hybridization of the shop environment and Shabby Glamour, new shops are being developed for clients such as Kiki Niesten 2. We are designing an outlet for Kiki Niesten (famous shop for women's fashion) and a bookshop for Marres/Centre for Contemporary Culture – both in Maastricht. The project also helped EventArchitectuur get commissions after pitching new retail concepts to Oilily and Victoria's Secret.

Current projects

Pursuit of Happiness, art exhibition, Leidsche Rijn, Netherlands // Soft Target, art exhibition, Utrecht, Netherlands // Just In, installation for NAi exhibition on home interiors, Rotterdam, Netherlands.

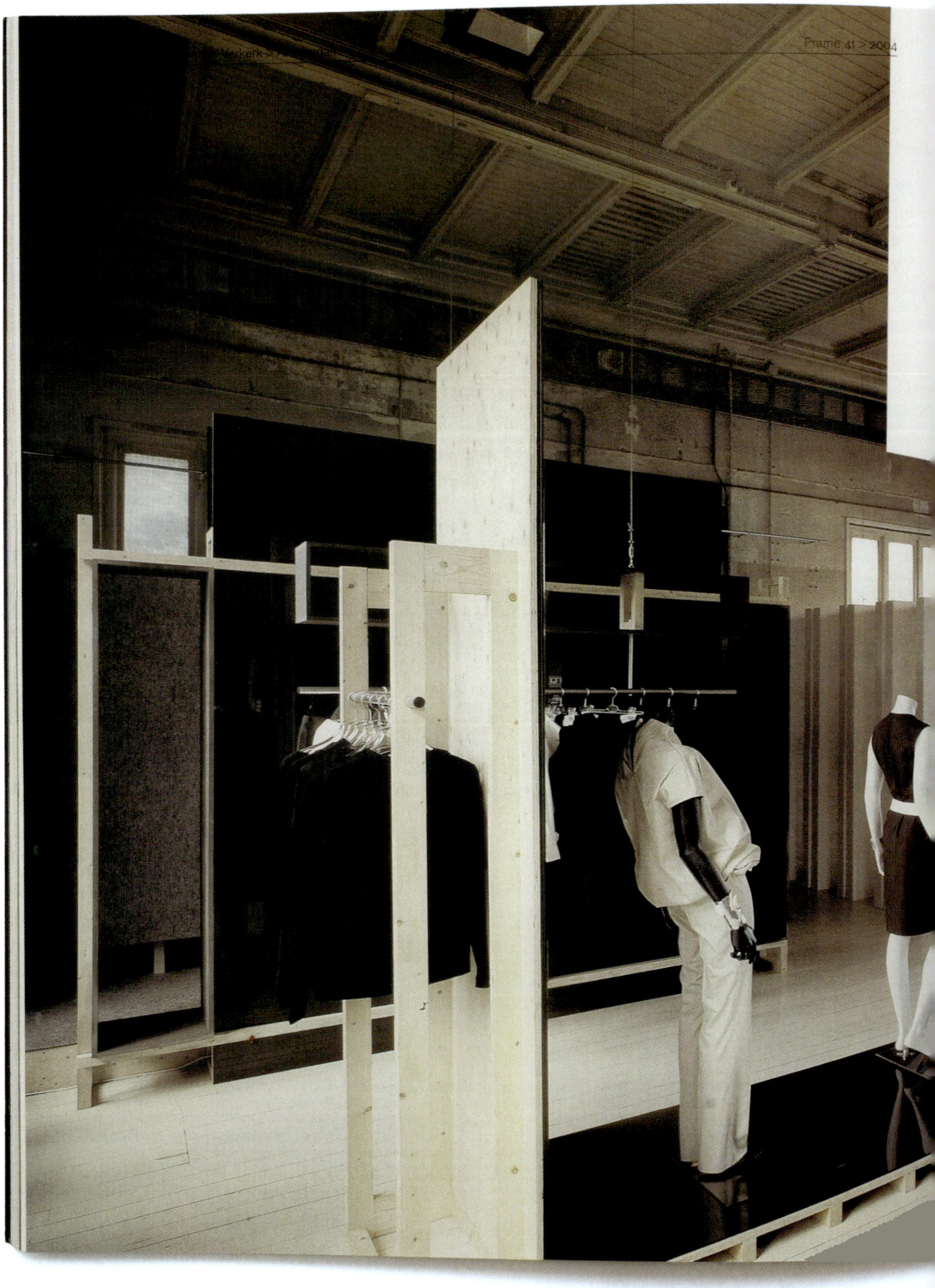
Frame 41 > 2004

Beyond Bauhaus

With its careful composition of banal materials and hidden details, Herman Verkerk's shop for Orson+Bodil dovetails seamlessly with the *prêt-à-porter* de luxe of this young Amsterdam label.

Text by Louise Schouwenberg. Photography by Johannes Schwartz

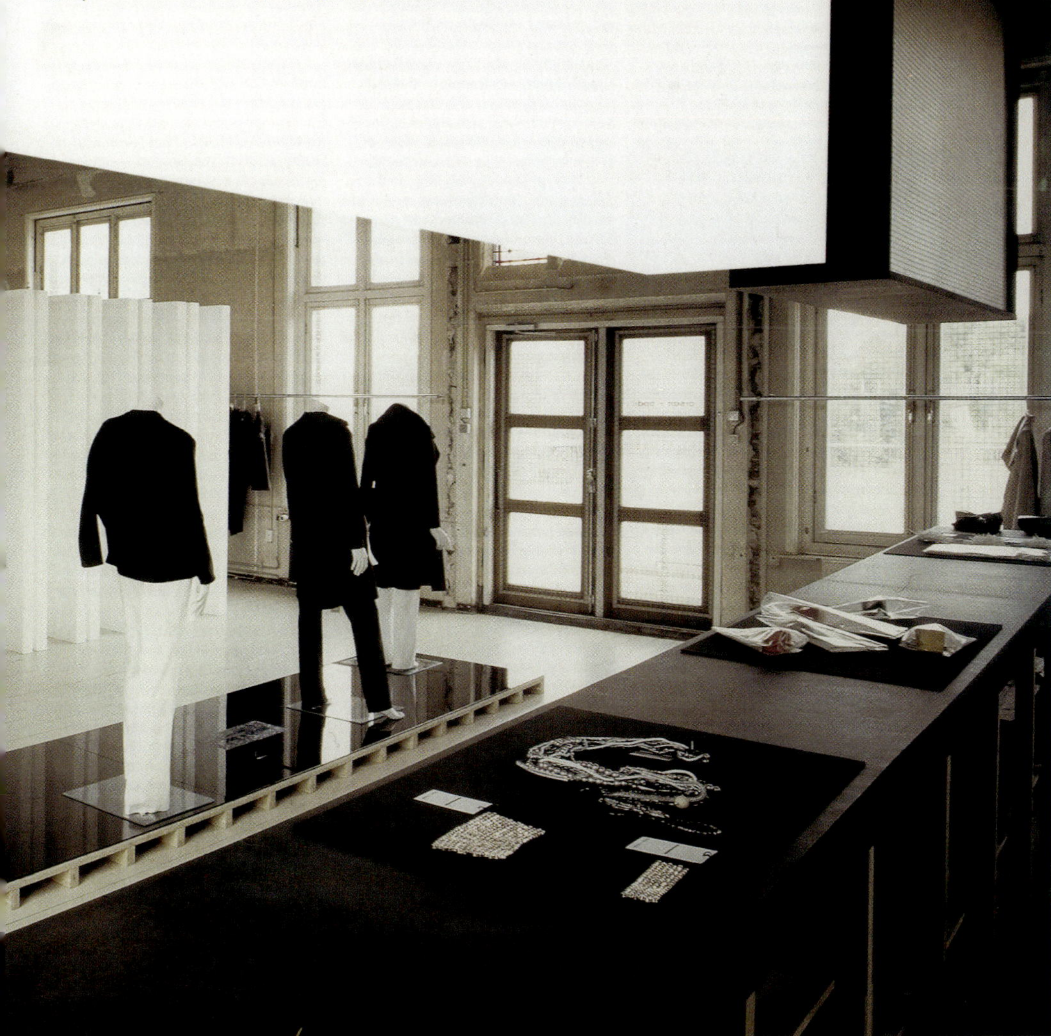

WRITER
Louise Schouwenberg

PHOTOGRAPHER
Johannes Schwartz

DESIGNER
Herman Verkerk
(EventArchitectuur)

LOCATION
Amsterdam, Netherlands

SUBJECT
Orson+Bodil fashion shop

High-flown ideals, functionalism, minimalism, traditional methods with regard for the future – it's a mix that marks a place where professional designers work side by side with students and younger colleagues. No, we're not on a nostalgic excursion to the Bauhaus in Weimar, Germany. We have arrived at the shop and workplace of Dutch fashion label Orson+Bodil. Both the interior of the buildings and the clothing designed, made and sold here reflect the modernist movement that gained popularity in the early years of the 20th century.

In 1989 Orson+Bodil came out with a line of clothing inspired by Balenciaga's sober, modernist styles of the 1950s. Ignoring postmodern designers like Jean-Paul Gaultier, who in the late '80s were taking a thematic approach to historical styles of dress, fashion designer Alexander van Slobbe (and Nannet van der Kleijn, his business partner at the time) opted for minimalism. They did strive, however, to make each collection markedly different from the last; research and experimentation were a vital part of their work.

Following a period of scepticism towards this conceptual approach to fashion design ('as dry as a stale slice of bread', wrote critic Jhim Lamoree in the Dutch magazine *HP*), the media went on to embrace the atypical clothing brand. Growing success compelled van Slobbe, now assisted by theoretician Guus Beumer, to begin mass-producing his Orson+Bodil collections, a change that would mean relinquishing 'the experiment' for a more conventional mode of operation.

Thus in 1994 the passion for experimentation was put on ice, along with the label. In its place emerged a new, industrial project: SO. Van Slobbe, as designer, and Beumer, as art director, fathered a menswear label that was to take Japan, in particular, by storm. For ten years, Beumer and van Slobbe made regular trips from their workplace in the Netherlands to Japan, the world of SO. The scale of the project – 18 SO shops and so-called 'corners', hundreds of retailers, semiannual collections comprising around 800 designs each, interim collections geared to the particular demands of the Japanese market, licences for bags, glasses, jewellery, watches and shoes – was accompanied by a set of site-specific problems. Ultimately, the distance between the design team in Amsterdam and the area of operations in Japan proved to be unbridgeable. In 2004 the duo was ready for a new round of experimentation, as well as a closer relationship with the customer. It was time to relaunch the Orson+Bodil label.

A shop-cum-workplace occupies two former transformer stations on the grounds of the Westergasfabriek (Wester Gasworks) in Amsterdam, a site that has assumed a largely cultural significance in recent years. There's not another shop in sight, which is fine with Beumer and van Slobbe, who have no desire to see Orson+Bodil evolve into a consumer Valhalla. They want the business to remain a laboratory in which, together with others, they study processes within the profession and provide opportunities for new ideas. Visitors to the complex often see van Slobbe and Beumer working together in the same space. Only 50 metres or so from the retail outlet is another workplace: CoLab. Here designers from various disciplines work on projects with and without a direct link to Orson+Bodil. It is a place where young people can carry out artistic research and develop the economic potential of their personal collections.

The collections of Orson+Bodil are a mixture of new designs and fashions from the brand's archive. The result is a clear pattern of continuity and ongoing development, which, says Beumer, is a different sort of continuity than that projected by other, more conventional labels. He says that even though fashion revolves around a desire for something new, fashion designers are thoroughly aware that today's customer also wants the assurance of a familiar shape and silhouette. Beumer is convinced that fashion is synonymous not with radical breaks in style, as in the '80s, but with continuity. Most labels conceal their 'something new' in a recontextualization of the old and in the preservation of a certain ambience and style that exemplify the brand in question, rather than giving each succeeding collection a different language of form. At Orson+Bodil, they are looking for a new approach to the recent marriage of change and continuity. Part of the plan is to provide fashions with an 'Orson+Bodil, archive' label and to mark each piece with a year: a reference to the collection that gave birth to the individual garment and a way of putting terms like 'current' into perspective. The label sees 'something new' in the step-by-step refinement of clothing as seasons pass. It's a notion that makes research and production nearly inseparable.

Shop, workplaces and company philosophy bring back memories of the Bauhaus, as illustrated by a brief history of the school. In the second half of the 19th century, objects carefully crafted by artisans gradually made way for industrial production. Not everyone saw such modern methods of manufacture as a change for the better, however. Led by Walter Gropius, several designers decided to improve the quality of functional objects by going back to traditional methods. The result was the Bauhaus. Activities in early workshops, established in Weimar (1919-1922), relied heavily on the English Arts and Crafts movement. Gropius and his followers saw the combination of arts and crafts as an effective response to the 'monstrosities of industry'. Within a guild-like structure, professional artists and designers worked together with students; experimentation and the study of materials and techniques were of primary importance. After 1922, the school experienced a shift from Arts and Crafts to functionalism and the practical nature of machines and technology. Thus the early Bauhaus rejected industry, whereas the late Bauhaus rejected traditional methods. Nonetheless, a number of concepts typify the entire Bauhaus movement (*and* modernism): idealism, a joining of creative forces, a focus on function, structure, the use of simple forms and, increasingly, an aversion to ornament.

Parallels between the Bauhaus and the philosophy of Orson+Bodil are remarkable. Ideological zeal for a Utopian future is not part of van Slobbe's mind-set, but the same cannot be said of modernist terms generally associated with architecture and design. His designs fuse function, structure, skin and silhouette in eye-pleasing garments assembled with ingenuity. Every aspect has a functional motivation. Decoration, for example, is added only in the form of detachable jewellery. Another

Previous spread: The Orson+Bodil shop occupies a former transformer station on an Amsterdam factory site. Mannequins on the mini-catwalk in the centre – an element incorporating black-painted safety glass – are wearing clothes from the Amsterdam fashion brand's 'archive' collection: a mix of new designs and fashions from 1993/94. The counter in the foreground is made of black MDF.

similarity between the two can be seen in Orson+Bodil's retail interior, which dovetails seamlessly with the clothes, creating a *Gesamtkunstwerk*. Architect Herman Verkerk stripped the existing interior of all superfluous elements to reveal the bare skin and structure of the space. With inexpensive materials like plywood and felt, and anonymous products like wooden pallets and polystyrene formwork, he created cupboards, partitions, systems for hanging garments and a black glass catwalk for lifeless mannequins. 'Shabby glamour,' is Verkerk's definition of the space. 'Frugal materials and minimal design put the glamour of the industrial product, usually aimed at perfection, into perspective. I much prefer working with industrial materials, which are generally invisible. Using them to build temporary configurations gives my work a fresh narrative quality.' The use of budget materials also corresponds to the expected life span of the shop: a mere two years. At the end of that time, Beumer and van Slobbe will examine the feasibility of their project.

Do modernist values and stylistic features have any bearing on the social developments of the 21st century? Or is 'contemporary Bauhaus' an anachronism? Verkerk, who refers to his work as 'beyond modernism', agrees that an exact duplication of such methods is not advisable. 'You can take modernist principles as a point of departure,' he says. 'After all, they have certainly proved their worth within and even beyond the field of architecture.' Verkerk's layout for the shop is based on a precise, and thus very modernist, grid. Within this grid, however, he plays with elements that veer away from modernism, such as imperfection and recycled materials. 'The idea of transience, which is linked by definition to retail environments, has never occupied a mental place or assumed a material form within modernism,' says Verkerk. 'As a style, modernism can be unmasked. As an ideology, it's unproductive. But as a body of thought, it still has an incredible amount of potential.'

Van Slobbe and Beumer also believe in a new interpretation of modernism. 'We want to enrich these premises by injecting them with sensuous and poetic elements,' says Beumer. 'Furthermore, there's no doubt that the objectives of classical modernism – abstraction, functionality, structure and the like – can be reformulated using resources available today. And certainly within the discipline of fashion, which is so caught up in its preoccupation with style.'

According to Beumer, the market for fashion has shifted from designer collections to brand-name clothes. As it eases away from traditional notions of quality, fashion is becoming largely the product of marketing strategies. The value of a garment is defined not by wearability, but by the position of its brand in relation to other brands. Consequently, the retail price is no longer determined by production costs but by overhead costs, and by what the customer is willing to pay for something that reinforces her individuality.

Another factor is the disappearance of the classic conflict between handmade and industrially manufactured products. As Beumer explains, industry has reached such a high standard that it can handle the most complex operations. 'Not even ten years ago, designers had to adapt their ideas to industry. Today, industry itself is the driving force behind the creation of ideas.' At the same time, thanks to the availability of labour from low-wage countries, handwork has become affordable *and* fast. 'One phone call to China and in two weeks we have up to 500 hand-knit sweaters. It's no longer tenable to cite traditional differences between handwork and industrial production: slow versus fast, expensive versus cheap, fine versus coarse, local versus global, one-offs versus big batches.'

Nowadays, thanks to a high level of flexibility, people producing handmade goods have joined industry to form a new partnership: sewing workshops can fill orders that run from a dozen to thousands of pieces. And freed from its negative connotations, craftsmanship can be valued for the psychological effect it exerts on the user: it not only refers to a slower pace, but also implants this deceleration, and the implied attention to detail, into the product. 'Above all else,' says Beumer, 'fashion is a visual language and has become, therefore, synonymous with visibility, with representation. What matters most is not the product, but the photograph of that product. Handicraft, which in our profession means highly skilled cutting, is refocusing attention on the three-dimensional character of a piece of clothing, rather than on the flat surface of a photo. A sleeve set in by hand, for example, not only produces a neater result than what a factory has to offer, but also highlights the materiality of the product. For us, visibility is no longer linked to the directness of a blow-up, but to the surface – the skin – of a product. Smoothness versus tactility, industrial perfection versus handmade variation, old versus new, function versus decoration: these are no longer either-or concepts. A single article of clothing may incorporate several of these seemingly opposing features.'

With all this in mind, Orson+Bodil opts for a combination of semi-manufactured products finished by hand, thus filling a niche between *prêt-à-porter* and couture. The so-called *prêt-à-porter de luxe*, says Beumer, is the final sphere in which fashion designers can operate autonomously at a time in which global brands threaten to monopolize the entire fashion market. We generally use *de luxe* to describe extravagance translatable only in terms of money. Beumer and van Slobbe, however, see luxury as a matter of refinement and attention to detail. A sleeve set in by hand, a silhouette that changes when the wearer slides her hands into trouser pockets, a skirt finished inside and out with equal care: these are details that make all the difference, without demanding attention. Herman Verkerk's retail architecture is part of this very package. He puts discarded and inconspicuous industrial materials in an unusual context: the flagship store, a symbol of luxury and extravagance. But by combining simple materials in a balanced composition, Verkerk creates an exciting field of tension. The first-time visitor experiences a subtle sense of beauty. Only later, and upon closer inspection, does she realize that her perception is based on materials that many would call worthless. A woman buying an Orson+Bodil outfit has much to discover. Her new clothes, as well as the whole shopping experience, elevate her from passive consumer to active observer.

> 'Frugal materials and minimal design put the glamour of the industrial product, usually aimed at perfection, into perspective.'
> Herman Verkerk

SUBJECT Orson+Bodil fashion shop
LOCATION Amsterdam, Netherlands
DESIGNER Herman Verkerk (EventArchitectuur)
PHOTOGRAPHER Johannes Schwartz
WRITER Louise Schouwenberg

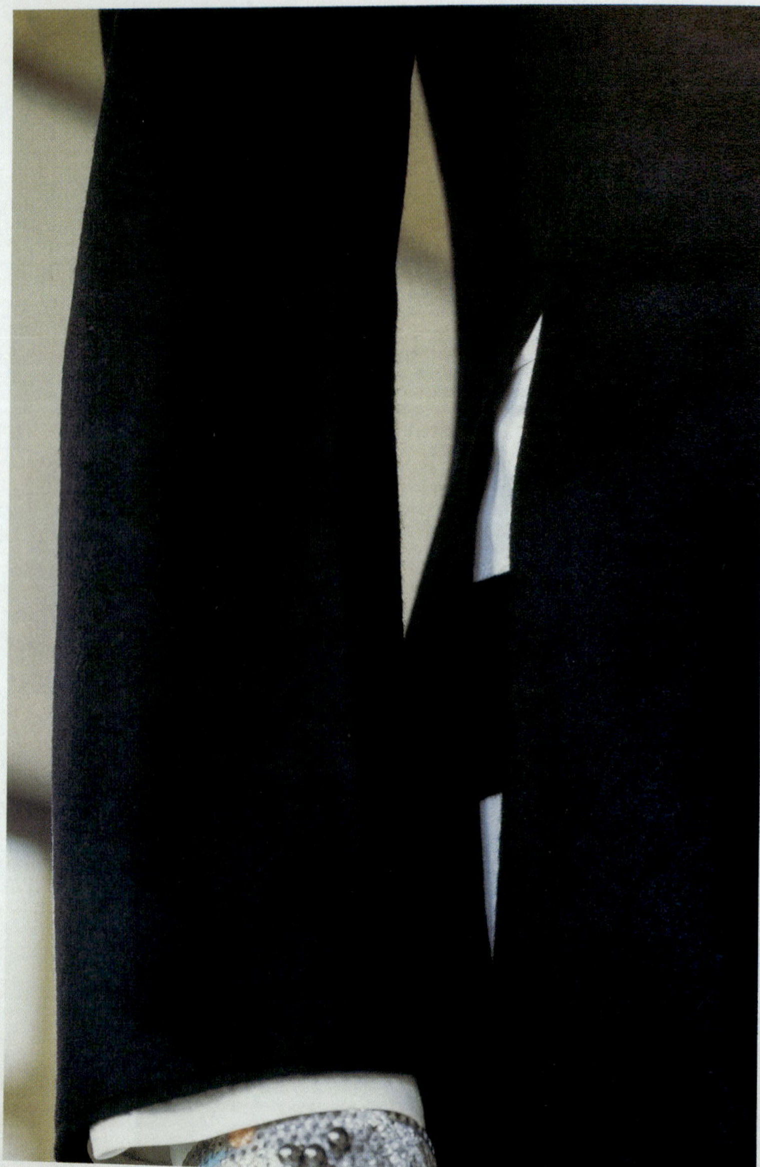

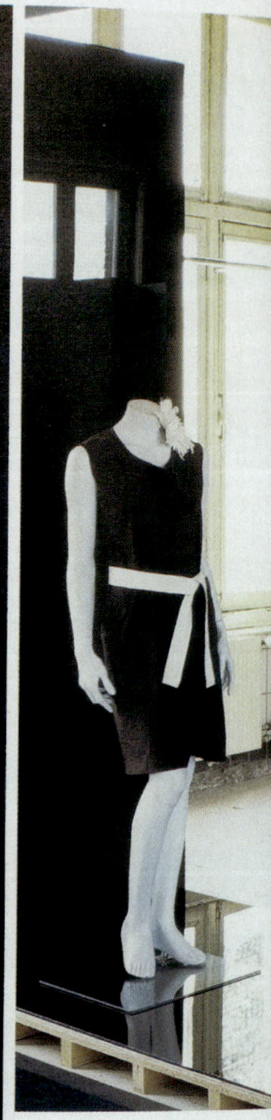

Above: Detail of a hand-finished, cashmere/merino loden coat from the 1993/94 Winter Collection. The slit in the side seam functions as both an opening (to reach the trouser pocket) and a belt loop. The embroidered bracelet features the type of aluminium mesh used for certain filters. Opposite: Apparel on the catwalk is from the 1994 Summer Collection, which highlights 'the pleat'. The L-shaped wall at the rear screens fitting rooms and serves as a display element. Like the catwalk, the wall is made of pine and safety glass, painted black. The wall's hybrid composition exhibits no clear distinction between frame and infill.

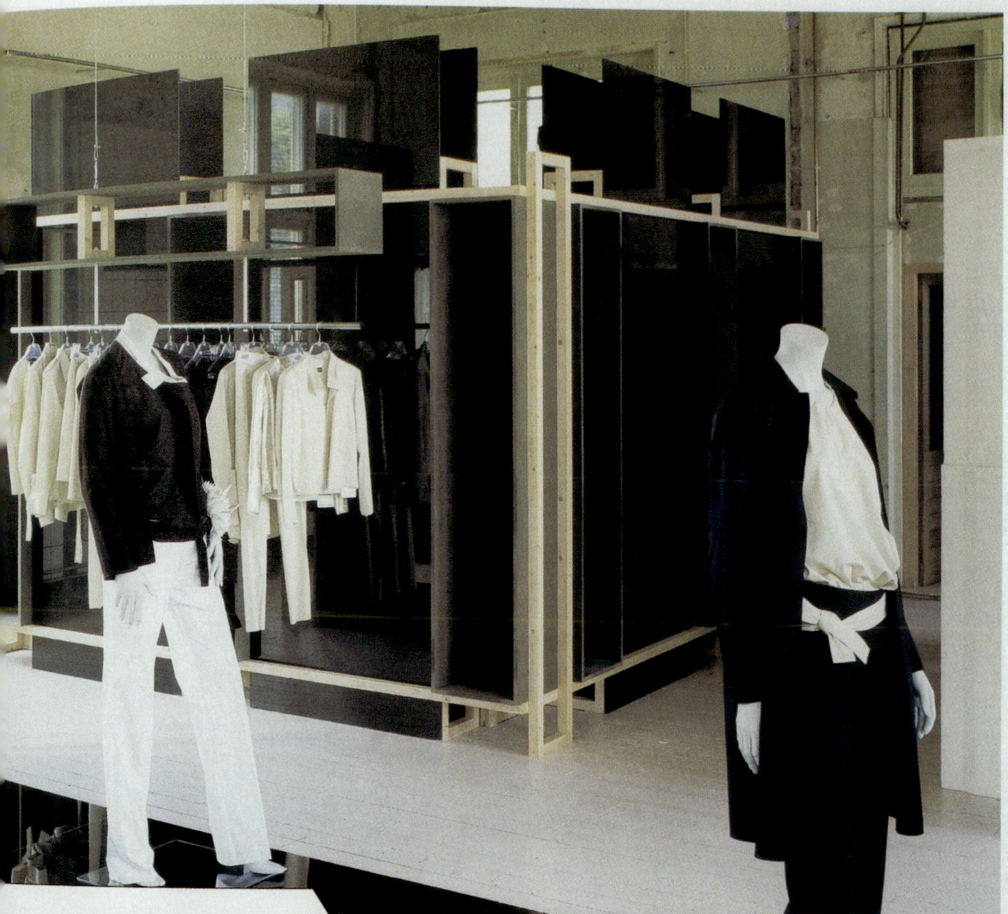

Handwork has a psychological effect on the customer: it not only refers to a slower pace, but also implants this deceleration, and the implied attention to detail, into the product

SUBJECT
Orson+Bodil fashion shop

LOCATION
Amsterdam, Netherlands

DESIGNER
Herman Verkerk
(EventArchitectuur)

PHOTOGRAPHER
Johannes Schwartz

WRITER
Louise Schouwenberg

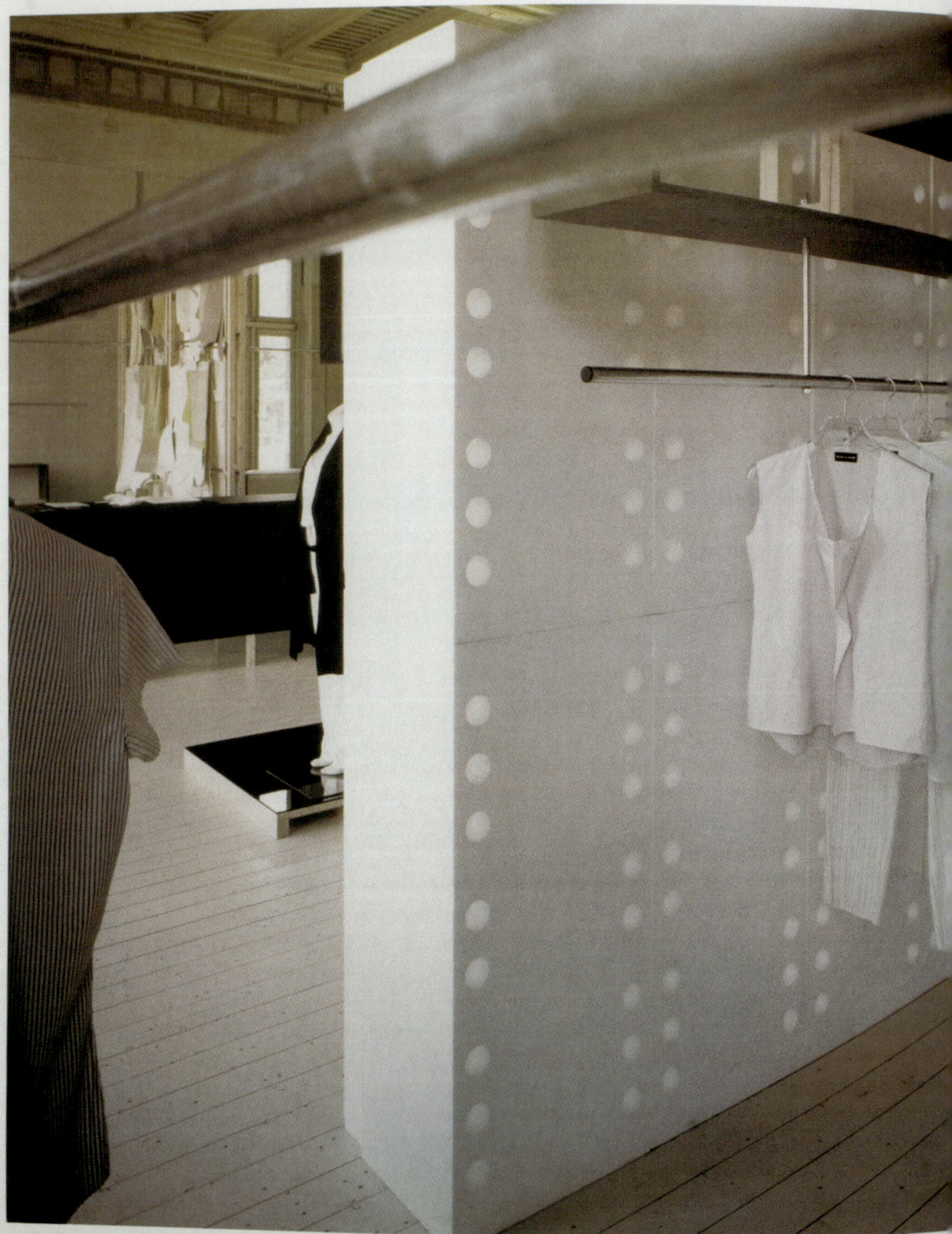

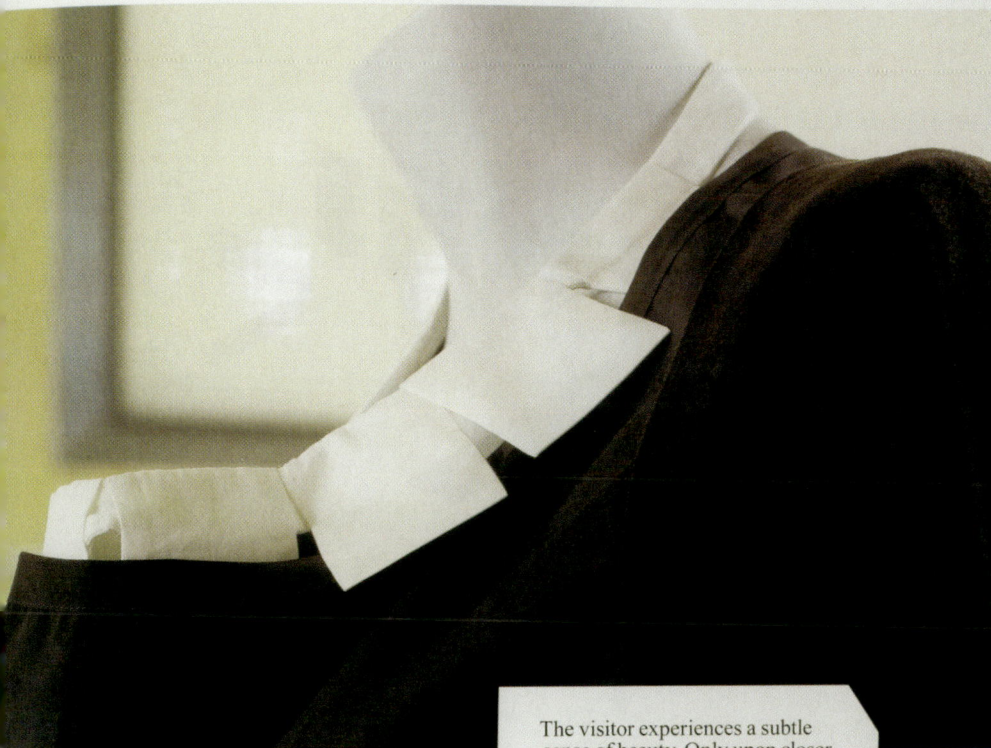

The visitor experiences a subtle sense of beauty. Only upon closer inspection does she realize that her perception is based on materials that many would call worthless

Above: Detail of pleated top and collar in 100-per-cent, satin-finish cotton from the 1994 Summer Collection. Opposite: Verkerk likes to use simple, inexpensive materials, such as ready-made polystyrene formwork for a partition wall, which takes its flexibility from the material. In the foreground is part of a galvanized-steel rack for hanging clothes.

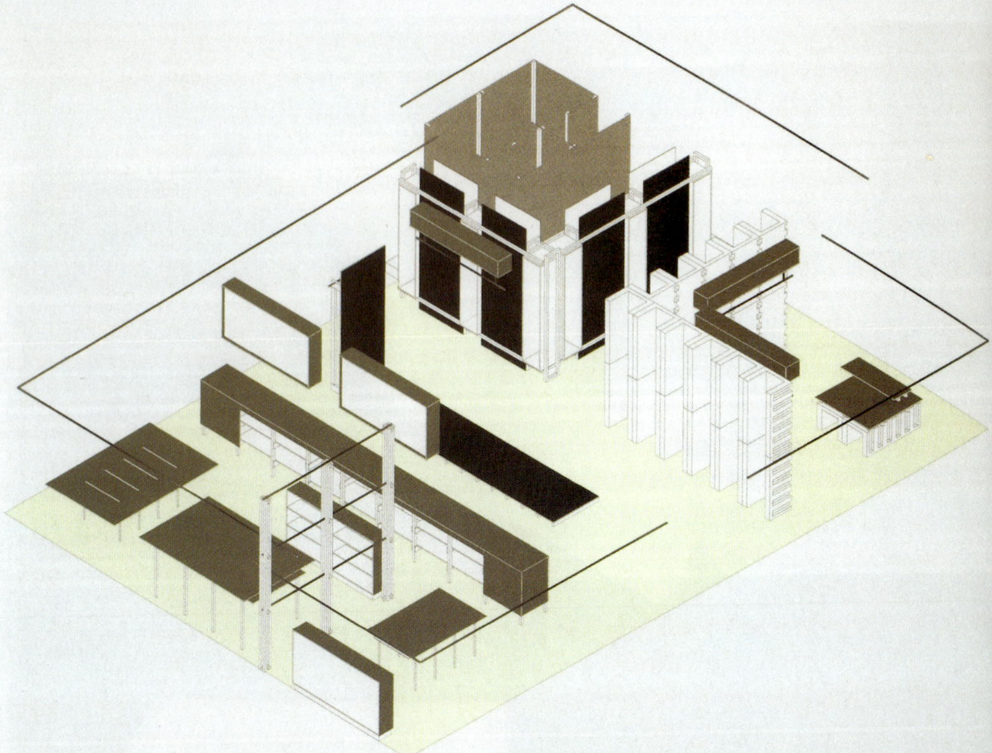

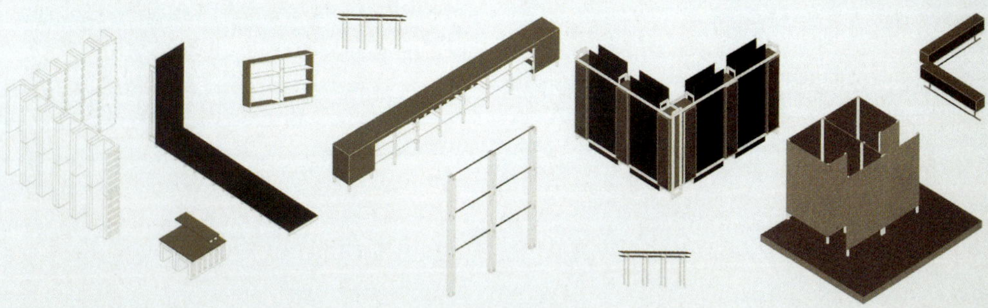

The retail interior is composed of 18 elements that combine to form an 'interior-furnishing collection'. Like the designers of the 'archive' collection, Herman Verkerk found ideas in his earlier work, which also boasts elements such as the mini-catwalk and polystyrene formwork. The axonometric shows steel tubing running at varying heights along the walls of the shop, one section of which functions as a rack for hanging clothes. Opposite: The retail space harbours confrontations between old and new, mat and shiny, hard and soft.

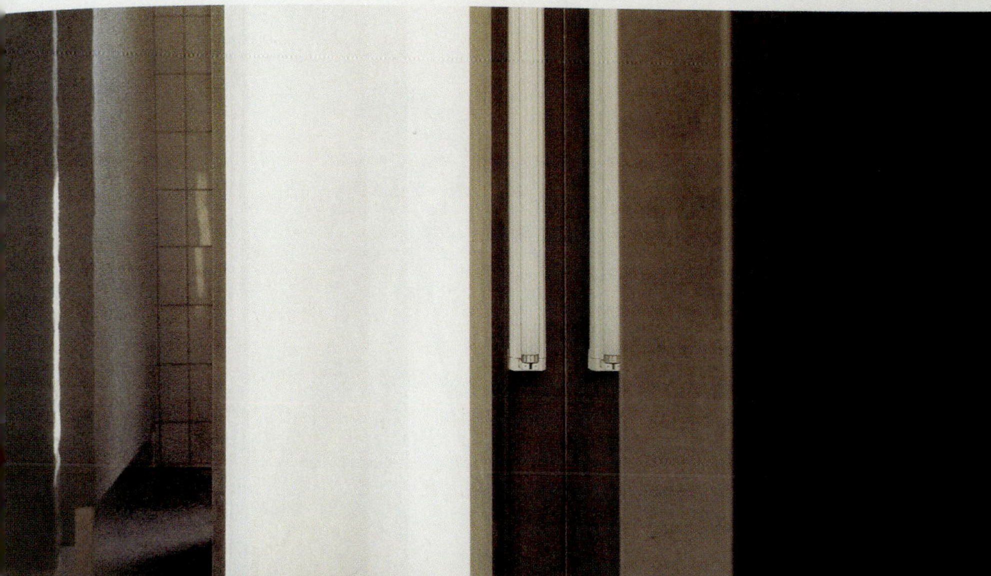

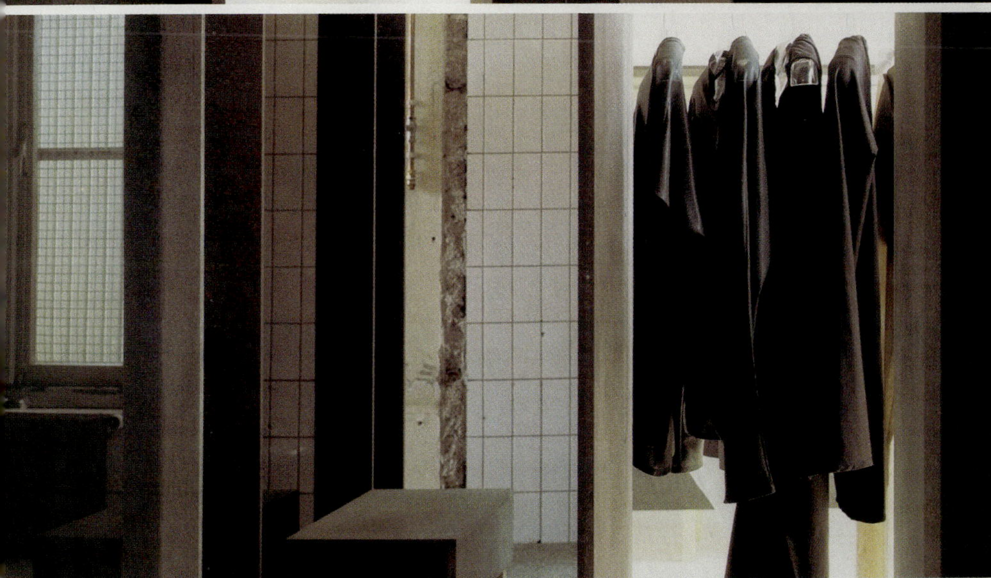

SUBJECT
Orson+Bodil fashion shop

LOCATION
Amsterdam, Netherlands

DESIGNER
Herman Verkerk
(EventArchitectuur)

PHOTOGRAPHER
Johannes Schwartz

WRITER
Louise Schouwenberg

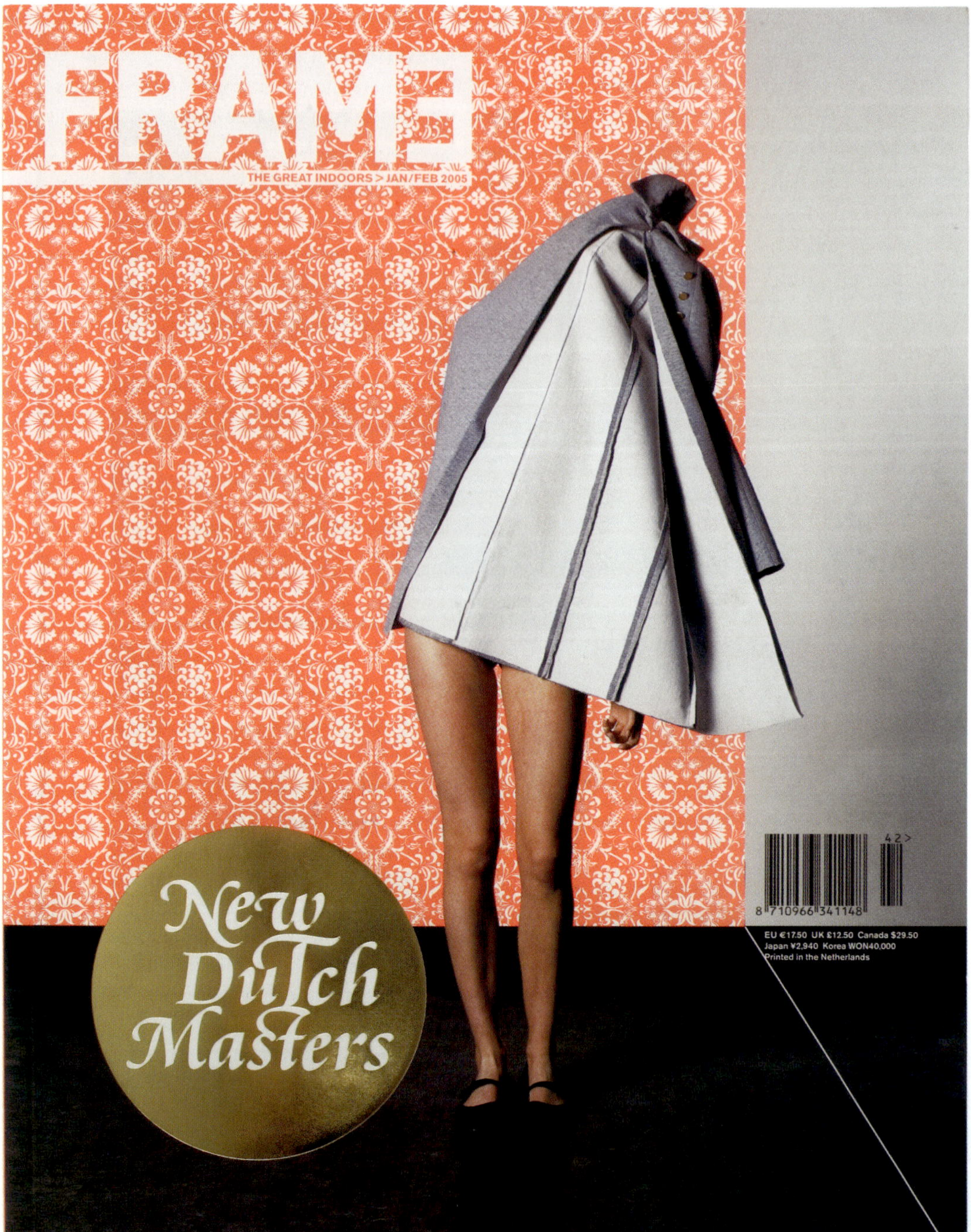

ART DIRECTORS
Cornelia Blatter & Marcel Hermans, COMA

Protect us From What Tobi Wants

DESIGNER
Tobias Wong
www.brokenoff.com

SUBJECT
Portrait of Tobias Wong

LOCATION
New York, USA

PHOTOGRAPHER
Frank Oudeman

WRITER
Shonquis Moreno

Designer Tobias Wong comments on the published article

As I wrote the editors at Frame when the article was first published: 'It's the best editorial I have received to date.' The story is well written and concise, and displays great sensitivity to the individual projects.

Did the article have an effect?

The greatest effect of the article was that it was able to capture a larger sense and approach regarding the way the work is created. I am grateful for that. It allowed the readers to see the context in which my work is produced, rather than simply showing one cool project after another.

Recent news or information about the project featured in *Frame*

Several projects featured in the article are soon to be – or are already – in full production. Others carried on as exhibition work. In addition, Carlos Salgado and I are still happily committed to each other and have just celebrated a second anniversary (with of course, an anniversary registry at mossonline.com).

Current projects

Following a solo retirement (as mentioned in the article and not so much a 'New York minute' any more), I spent 2005 successfully experimenting with collaborative projects. From 2006 onwards, I will adhere strictly to similar projects created collectively with other artists and designers alike. Will I ever return to solo work? How many times has Cher given a 'Farewell Tour'?

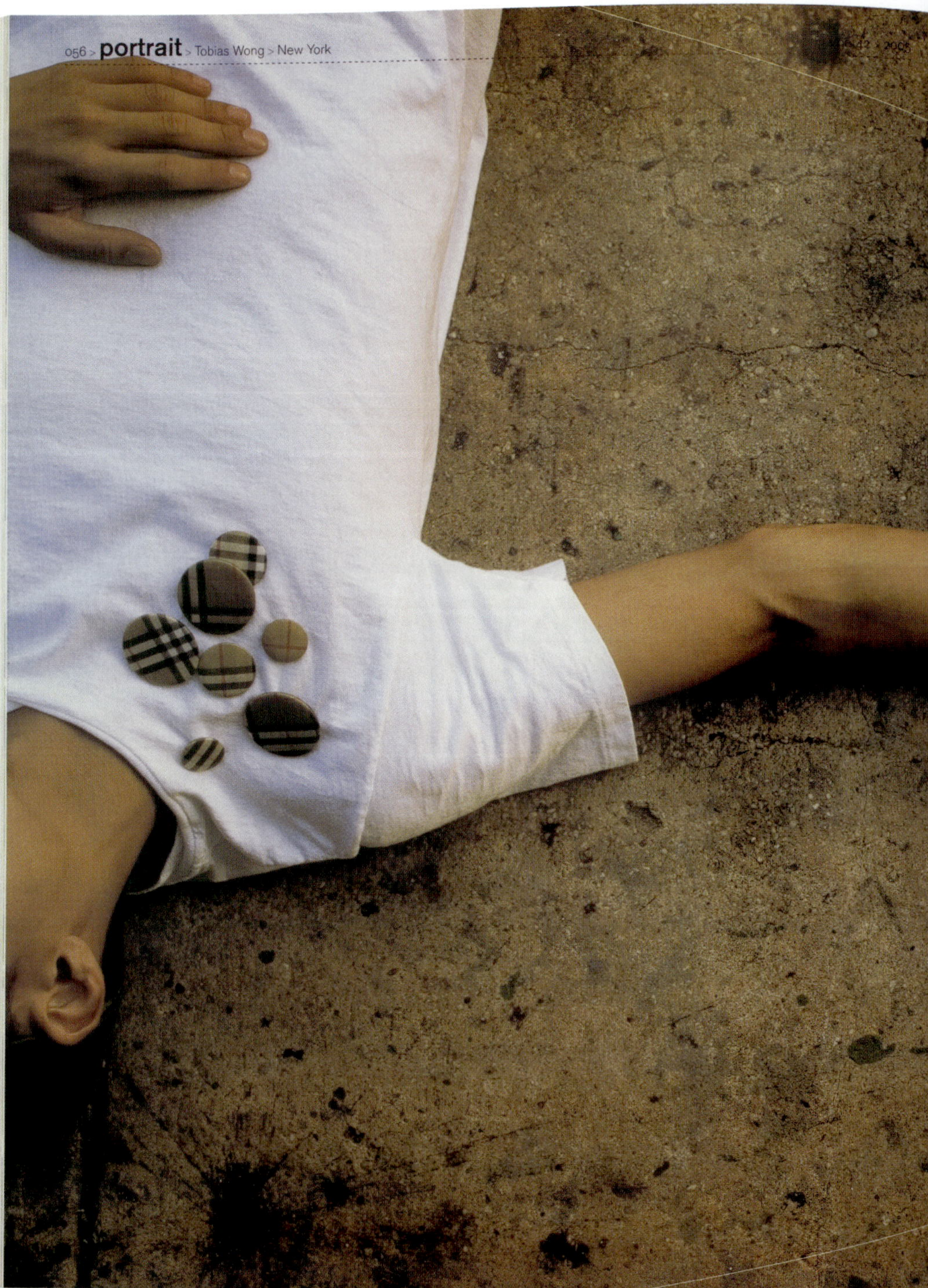

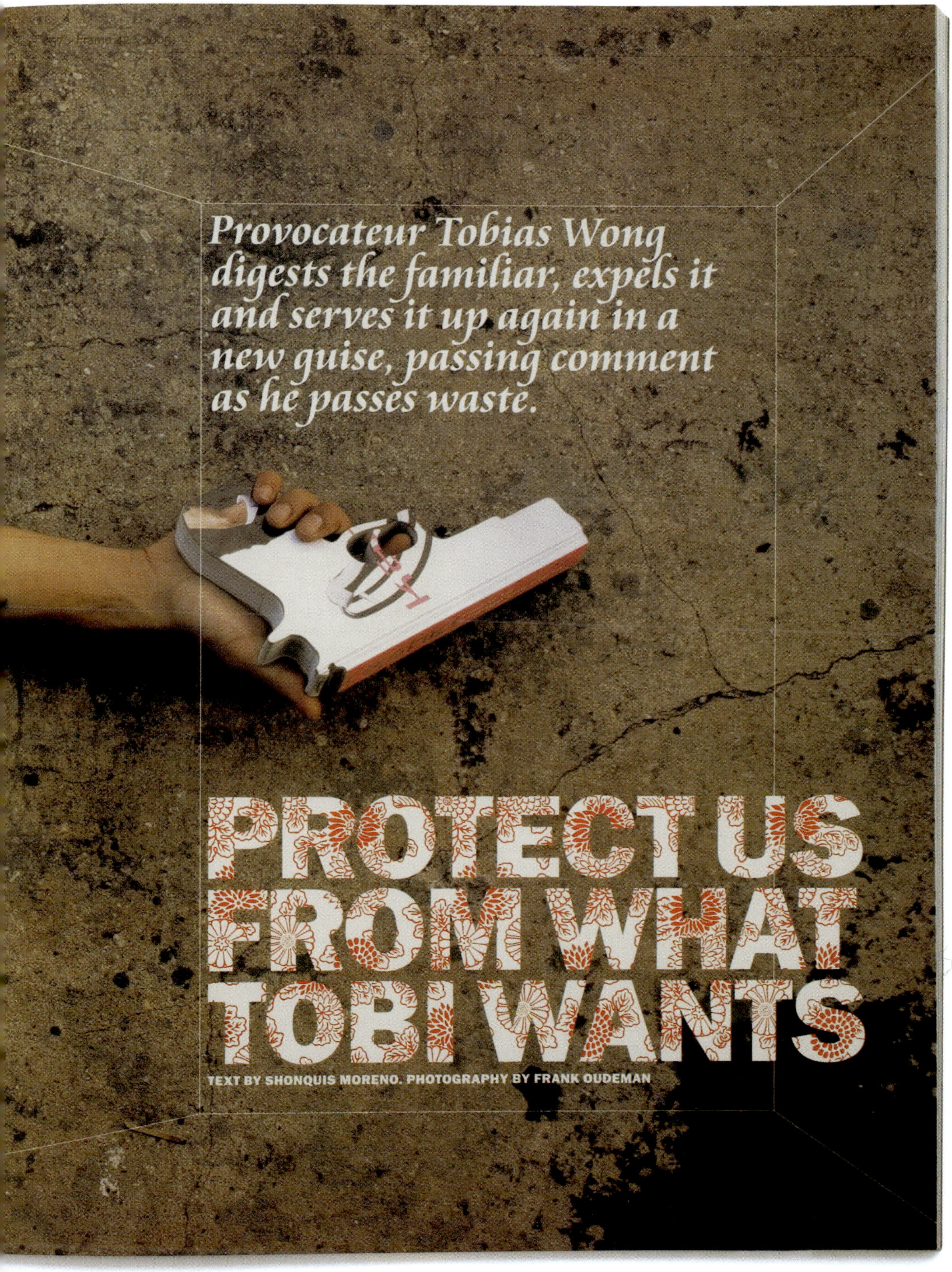

Provocateur Tobias Wong digests the familiar, expels it and serves it up again in a new guise, passing comment as he passes waste.

PROTECT US FROM WHAT TOBI WANTS

TEXT BY SHONQUIS MORENO. PHOTOGRAPHY BY FRANK OUDEMAN

SUBJECT
Portrait of Tobias Wong

LOCATION
New York, USA

DESIGNER
Tobias Wong

PHOTOGRAPHER
Frank Oudeman

WRITER
Shonquis Moreno

'Whether his pieces are artworks or design objects, they are brilliant examples of experimentation.' Constantin Boym

There are many things to talk about when you talk about Tobias Wong. There's the glory-hole ottoman he designed with a 30-centimetre shaft that is lined with pink mink and can be set to vibrate at five speeds, the lightest of which he calls 'tease' and the most vigorous of which he calls 'orgasm'. There's his ironic use of dog fur to create hats, gloves and shoes, because he saw New Yorkers treating their pets as accessories and figured he'd just eliminate the troublesome feeding and walking bits. There's his recent faux commitment to longtime (straight) design partner Carlos Salgado. And there's the now-notorious clash of Wong's romantic brand of cynicism with Karim Rashid's romantic brand of idealism in the form of a concealable gun, hand-cut from Rashid's post-9/11 monograph, *I Want to Change the World*. But at a time when America's Cooper-Hewitt National Design Museum is putting up a provocative exhibition called Design Is Not Art, Wong's work – culturally critical, occasionally manufactured, sometimes handmade, often humorous – seems to be evidence of a quiet, tectonic shift taking place between the worlds of art and design.

Wong announced his retirement from art and design in 2003, only several years into his career. He decided, for a New York minute, that he was going to drive trucks. I won't call Wong an out-and-out romantic, but only because he specifically asked me not to. What Tobi is, he likes to keep close. It's clear he's a workaholic, and he'll admit that he's both dyslexic and colourblind. Without naming the colour, he'll tell you he's only a few shades shy of a black belt in kung fu. He nibbles instead of gulping when he eats, partly because his sense of smell is extremely weak which, on the bright side, is preferable during August in New York City, where Wong has been based for the past seven years. At 29, he has the self-conscious grace and wicked sarcasm of a gay man living in a liberal pocket of a prudish country. A tiny blue square is tattooed at the centre of his chin and, on his right forearm, artist Jenny Holzer has scribbled a phrase that Tobi needled into permanence: Protect Me from What I Want. His business card reads: Protect Me from What Tobi Wants.

One of the several conventions under attack from Wong's work is the assumption that design objects perform functions that make them pedestrian, while art objects transcend function. Many people call it art precisely because it has no banal function – a simplistic, if not inaccurate, notion, since art has always been used to communicate ideas, propaganda and status; to nourish the imagination and the intellect; and, not least, to decorate. Perhaps it's more relevant to ask where the 'brief' originates. Only the naive would suggest that design is just art without ideas, or that art must be ugly to be smart. 'For the first time in a century, design is considered as cool and as interesting as art,' says Barbara Bloemink, Cooper-Hewitt co-curator of Design Is Not Art. 'Today, there is a tendency among young artists to not want to be limited by being defined as one or the other, artist or designer. And the way the public views design is not as distinct from art as it used to be. Now, some design sells for as much as contemporary art.' And some designers are earning celebrity status: Wong's renown was, perhaps, cemented by the t-shirt sold in a Brooklyn shop last year that read 'I fucked Tobias Wong'. As people like Wong begin to blend art and design, we're reminded of the subjectivity and handiwork that is sometimes the foundation of both. 'Thoreau said: "It's not what you look at, it's how you see it,"' says Bloemink. 'It really isn't whether a shovel is a functional shovel or a work of art; it's how you choose to use it. The object isn't what changes.'

This is where Tobi Wong's sparkling shit comes in. Several years ago, Wong designed capsules filled with flakes of pure silver leaf which, when consumed, make one's stool shimmer, a reference to some of the more scatological experiments made during the halcyon days of conceptual art – with a twist that is all Tobi Wong: the union of the most intimate waste matter with a material that we consider precious. Some of Wong's work is Post-It Note simple, which is an abbreviated way of saying that first you laugh at its brilliance or its audacity or both, and then you wonder why someone didn't come up with that particular idea a long time ago. Chances are that someone did, however, and Wong is the first to admit it. He has coined two terms for his work. *Paraconceptual* is about reconciling cultural commentary with aesthetics, so that even those who don't want to dissect the concept can enjoy the object – objectively. The night before Philippe Starck launched his Bubble Club chair at Kartell, for instance, Wong launched This Is a Lamp, Starck's chair stuffed with light bulbs. This is one of Wong's readydesigneds, a Duchampian readymade à la Wong: In his eyes, the success of the lamp depends not on its original form, as a chair, but on the credentials that make it a famous piece of design. 'Tobi Wong is able to look at conventional things, or at familiar design icons, and recontextualize them so they appear fresh and new. This ability is very dear to me,' says designer Constantin Boym who, like Wong, has designed products for Ray Coh's recent Conduit line. 'Whether his pieces are artworks or design objects – this issue is open to discussion and wide interpretation – they are brilliant examples of experimentation. They are about culture, about our society and its obsessions, and above all, they are often simply hilarious.'

Today, Wong says he doesn't believe in originality. He calls his work *postinteresting* because it derives from work he finds interesting. Derivative is no longer, necessarily, a dirty word. While allowing that the design and art markets are still chronically hungry for 'originality', he is happy to admit that he absorbs and digests everything around him and that his work is a byproduct of this digestive process. Having moved to New York to study sculpture at Cooper Union, Wong used to set up Crap Art stands in SoHo, selling pieces he'd made at school. At the butt-end of the internet boom, he sold 'dreams' (air-filled plastic bags) for one dollar apiece. He also made buttons from Burberry plaid and gave them away, free of charge. In using the buttons for a marketing scheme, Burberry counterfeited Tobi's counterfeit. 'The idea was that people were able to consume luxury without paying the price for it,' says Wong. And when the buttons began to appear on shopping bags, in magazines and on billboards, he saw it as his own exhibition, a way of proving that you don't necessarily need a gallery and it doesn't have to be expensive to make something interesting.

Previous spread: On the tarmac outside Eero Saarinen's Terminal 5 at New York's JFK Airport, Tobi Wong (1974) wears a tattoo handwritten by Jenny Holzer (2002) and his unauthorized Burberry Buttons (1999) later copied by … Burberry. In Tobi's hand, one of three guns (2002) hand-cut from Karim Rashid's book, *I Want to Change the World*.

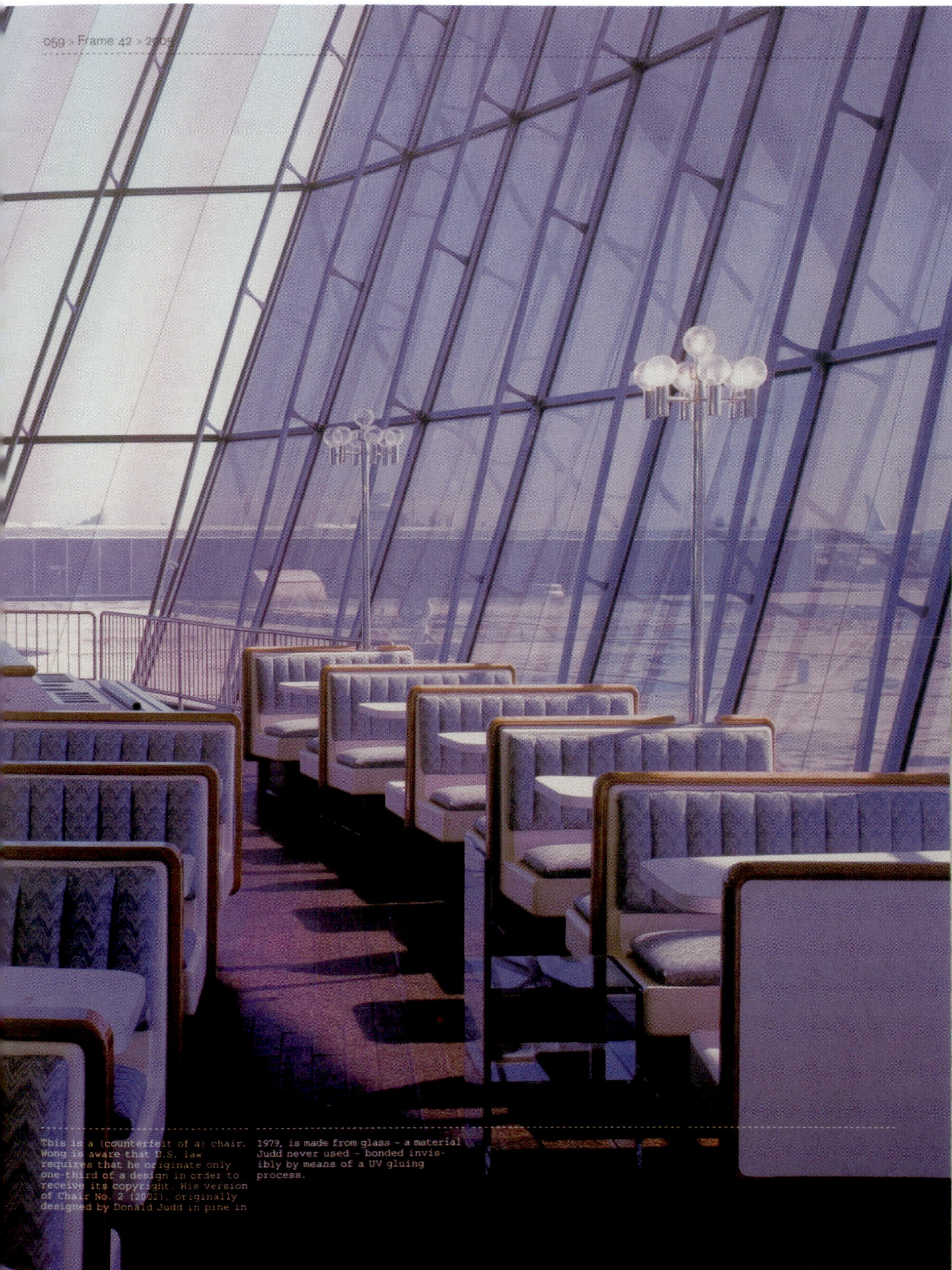

This is a (counterfeit of a) chair. Wong is aware that U.S. law requires that he originate only one-third of a design in order to receive its copyright. His version of Chair No. 2 (2002), originally designed by Donald Judd in pine in 1979, is made from glass – a material Judd never used – bonded invisibly by means of a UV gluing process.

SUBJECT Portrait of Tobias Wong

LOCATION New York, USA

DESIGNER Tobias Wong

PHOTOGRAPHER Frank Oudeman

WRITER Shonquis Moreno

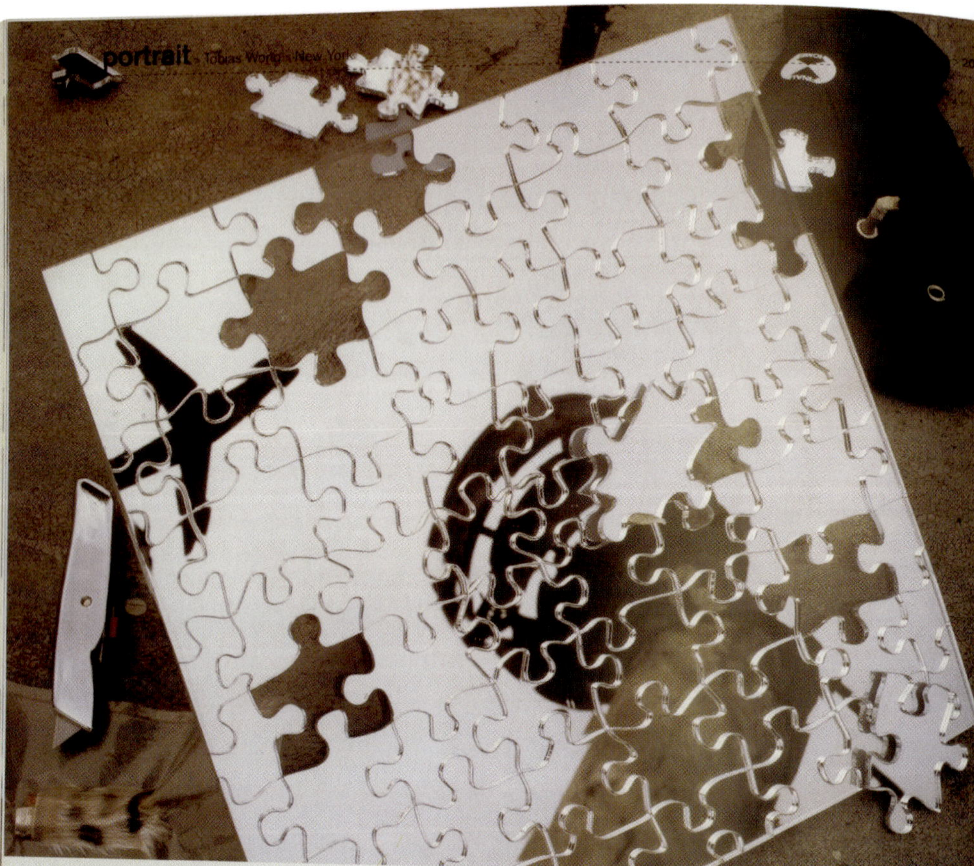

After he graduated, rather than renting a studio, Wong went to work in borrowed retail space. He camped out in Cappellini's New York showroom, stashing his Canadian passport in a pile of pillowcases and his tools in cabinets, using swinging carpet displays as partitions and storing sheets of plywood under display beds. ('It's funny where people don't look and how little they touch, even in a showroom,' he says.) He'll call it pragmatic – he funnels most of his cash into his projects – but there's a romantic notion about context that makes the circumstances surrounding his creative process an integral part of the final product.

Wong has never studied design and is not a product of the design world. His work is about how he reacts to the stuff around him. One of his strengths is an ability to absorb, and to respond pointedly to, culture and events. He addressed Mayor Bloomberg's ban on smoking in New York City bars and restaurants by designing smoking mittens, with a grommet to hold a cigarette, making the prospect of lighting up on a sidewalk in the dead of a New England winter less repellent. During a recent furore in the United States over same-sex marriage, Wong and Salgado came up with an inexpensive idea for a satellite exhibit for the International Contemporary Furniture Fair: They sent out commitment announcements – after all, they are committed (creative) partners – and entered their names in the gift registry at mossonline. The products making up their registry served as a curated online exhibition of objects that had brought the two designers together. Wong's I *Want to Change the World* gun book was a reaction to both the insecurity imposed on the West by the World Trade Center attack and the naivety and hubris of the belief that design is capable of changing anything. 'When I first saw the gun, I had a frisson of excitement mixed with anger,' says Karim Rashid. 'Art is a provocateur. Art can be slanderous, insipid, political, poetic whereas design must be more democratic, less provocative, more pragmatic, but hopefully experiential. Tobias's work borders on this line.' So who is being naive? Wong is using design as a vehicle for art and art as a material for design. Arguably, this type of work may be more capable of causing shifts than Wong is ready to admit.

Design and art may be discrete disciplines, but they overlap naturally, and the more we recognize both their similarities and differences,

Above: It's not what you look at – it's how you see it: Each piece of Wong's Mirror Puzzle (2003) has an identical surface, requiring players to match form rather than pattern. Also pictured: Box Cutter (2004), Disposable Mink Lighter (2003), and Smoking Mittens (2003) – the grommet holds a cigarette.

Opposite: Who cares if no one notices? You know it's there: Wong's Solitaire Art (2004) features diamonds that can be embedded discreetly into wall or floor.

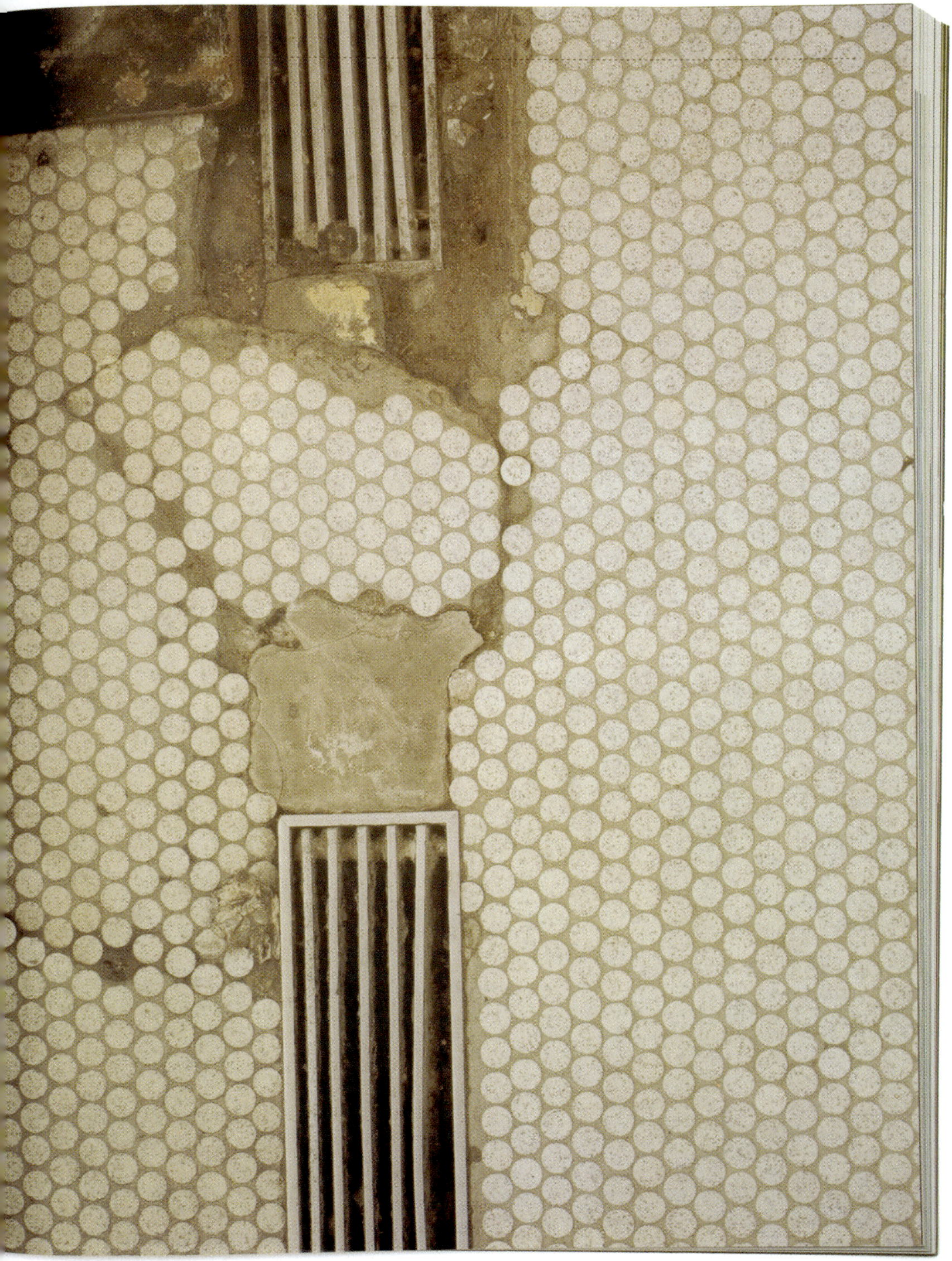

SUBJECT
Portrait of Tobias Wong

LOCATION
New York, USA

DESIGNER
Tobias Wong

PHOTOGRAPHER
Frank Oudeman

WRITER
Shonquis Moreno

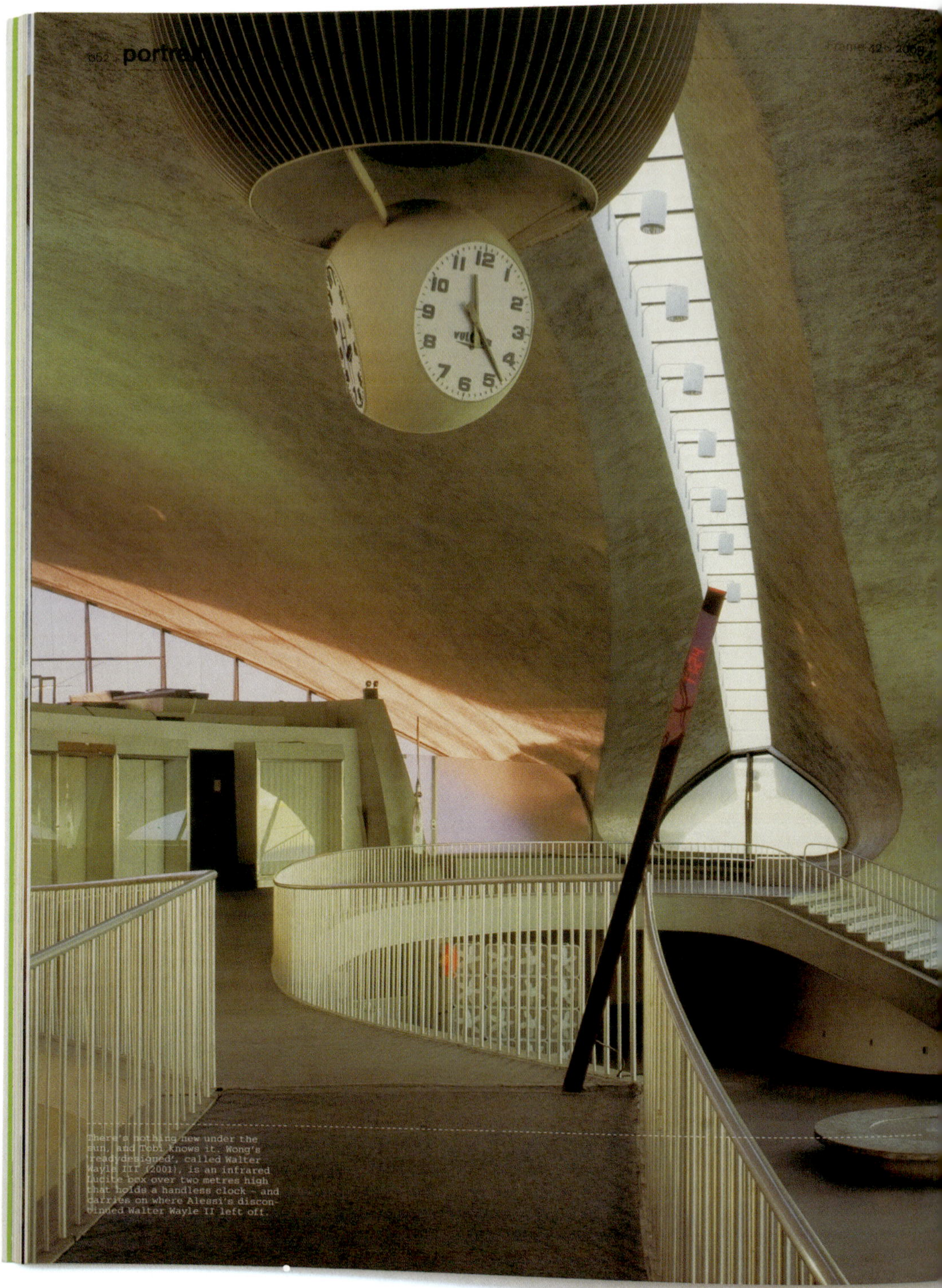

There's nothing new under the sun, and Tobi knows it. Wong's 'readydesigned', called Walter Wayle III (2001), is an infrared Lucite box over two metres high that holds a handless clock – and carries on where Alessi's discontinued Walter Wayle II left off.

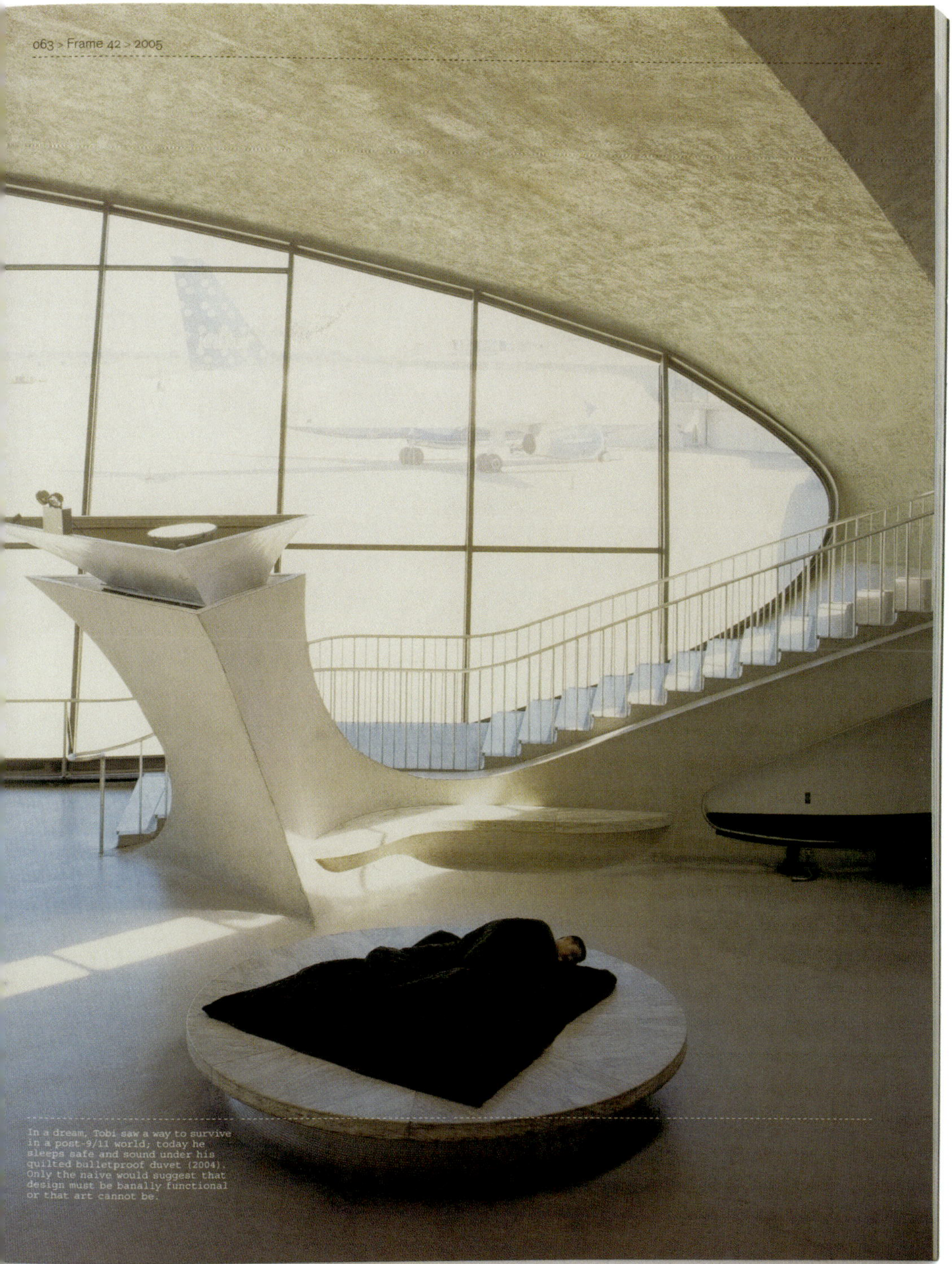

In a dream, Tobi saw a way to survive in a post-9/11 world; today he sleeps safe and sound under his quilted bulletproof duvet (2004). Only the naive would suggest that design must be banally functional or that art cannot be.

SUBJECT Portrait of Tobias Wong

LOCATION New York, USA

DESIGNER Tobias Wong

PHOTOGRAPHER Frank Oudeman

WRITER Shonquis Moreno

Issue 42 Protect us From What Tobi Wants 456-457

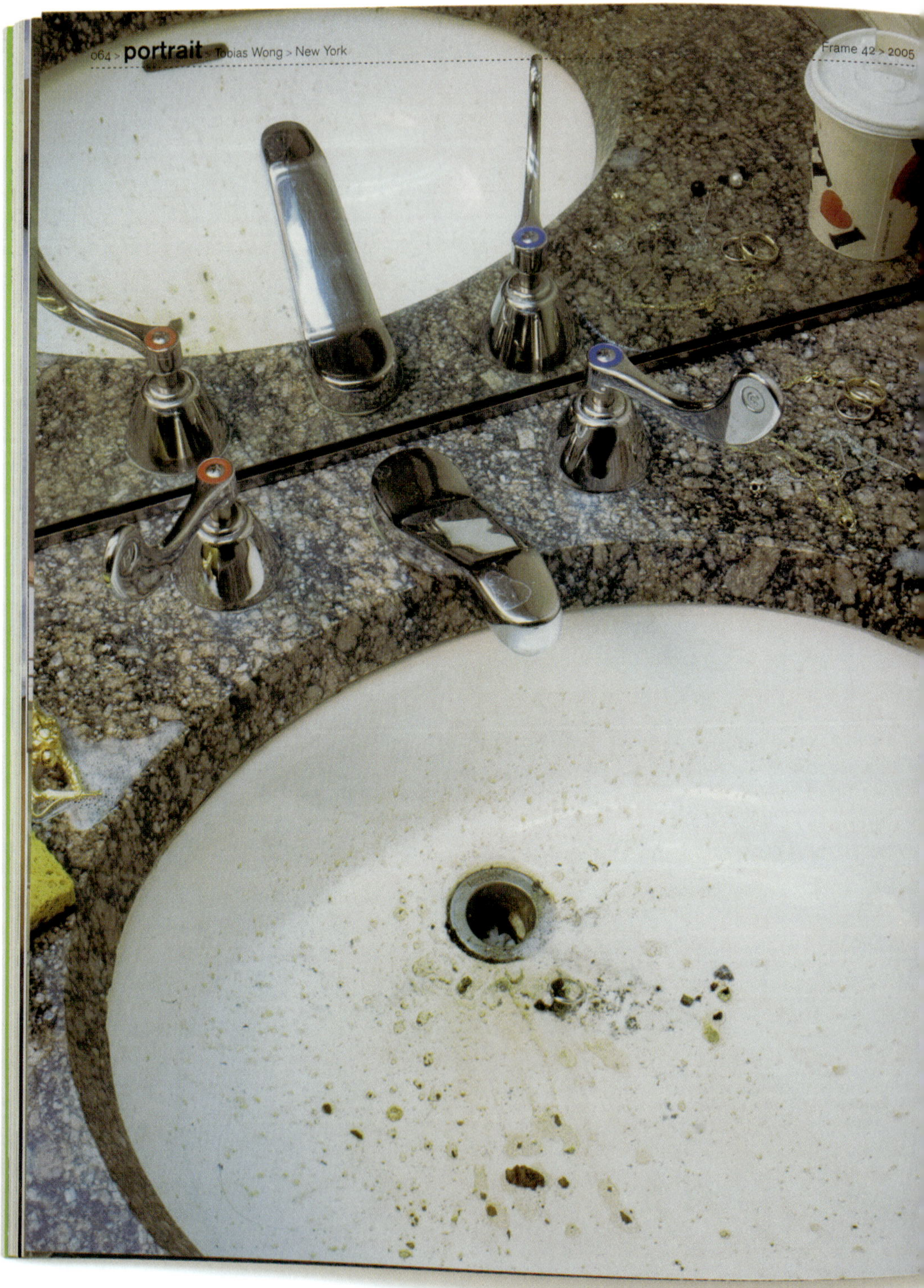

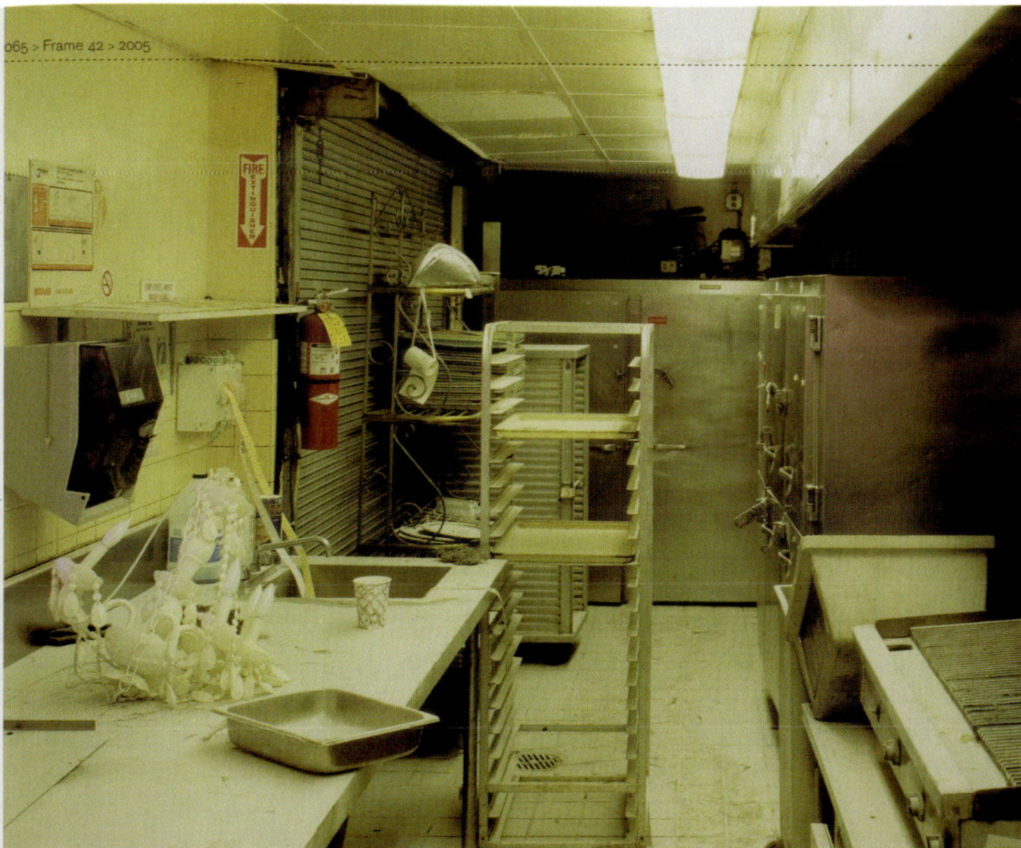

Wong insists, the easier it will be to move between them. It is partly in acknowledging, or forging an alliance between, two contrasting ideas (shit + silver) that Wong creates some of his cleverer work. He mischievously combines elegance with humour, too much information with none at all, and an intellectual or moral puzzle with the injunction to just enjoy how something looks. He'll charge $300 for a rip-away pad of 100 real dollar bills and give away other work free of charge. He'll create an object of exorbitant value and applaud when it's thrown in the dustbin. He has designed a pair of pearl earrings and a crystal chandelier dipped in industrial rubber that can be peeled away if desired. His Hidden Diamond Rings have tiny gems set inside the band where no one can see them: Who else should care? he asks us to ask ourselves.

Wong sees consumption as a vehicle for passing ideas to other people and over time, and for adding to them in the passing. For $7,500 to $25,000, you can have your purchase at the Troy store in SoHo wrapped in an original Warhol screen print and signed by Tobias Wong. Warhol, by all accounts, would have relished the idea. You can save the giftwrap if you like, but if you really don't give a shit, it goes into the bin because you've understood: It's all disposable in the end. 'I'm not giving rich people luxury,' Tobi says. 'I'm trying to share ideas. Even a poor person can go: Wow, the concept is there. It's in my head. I own it.' (That said, anyone who has trashed their Warhol-Wong giftwrap, please raise your hand.)

This winter, along with a number of international artists – including Vanessa Beecroft, Santiago Serra and Tom Sachs – Wong will transform the interior of Eero Saarinen's Terminal 5 at JFK Airport. You can ask if he's a designer or an artist until you're blue in the face, but he doesn't want to be clear on this point. 'I think I don't tend to answer that question, because I just don't know what I am,' he says, crinkling his eyes so that you can't tell if he really doesn't know, or if he just doesn't want you to know what he knows or doesn't know.

Opposite: South Sea pearl earrings (2003) dipped in industrial rubber and Hidden Diamond Rings (2004, right of basin) with gems concealed under the band so that only you know how much he cares … And if he doesn't care enough, nothing cuts deeper than the Reverse Diamond Ring (in basin). Also pictured: skull charms (with assistance of P Mohr) and the 24-carat gold toy soldier pin by DUPD, one of a number of designs 'edited' by Wong.

Above: Eye-candy, a good idea and, whatever the price, disposable: Wong's Crystal Chandelier (2003), dipped in industrial white rubber, and a paper cup illustrated with cut diamonds for Ray Coh's 2004-2005 Conduit line.

Issue 43 March/April 2005 458-459

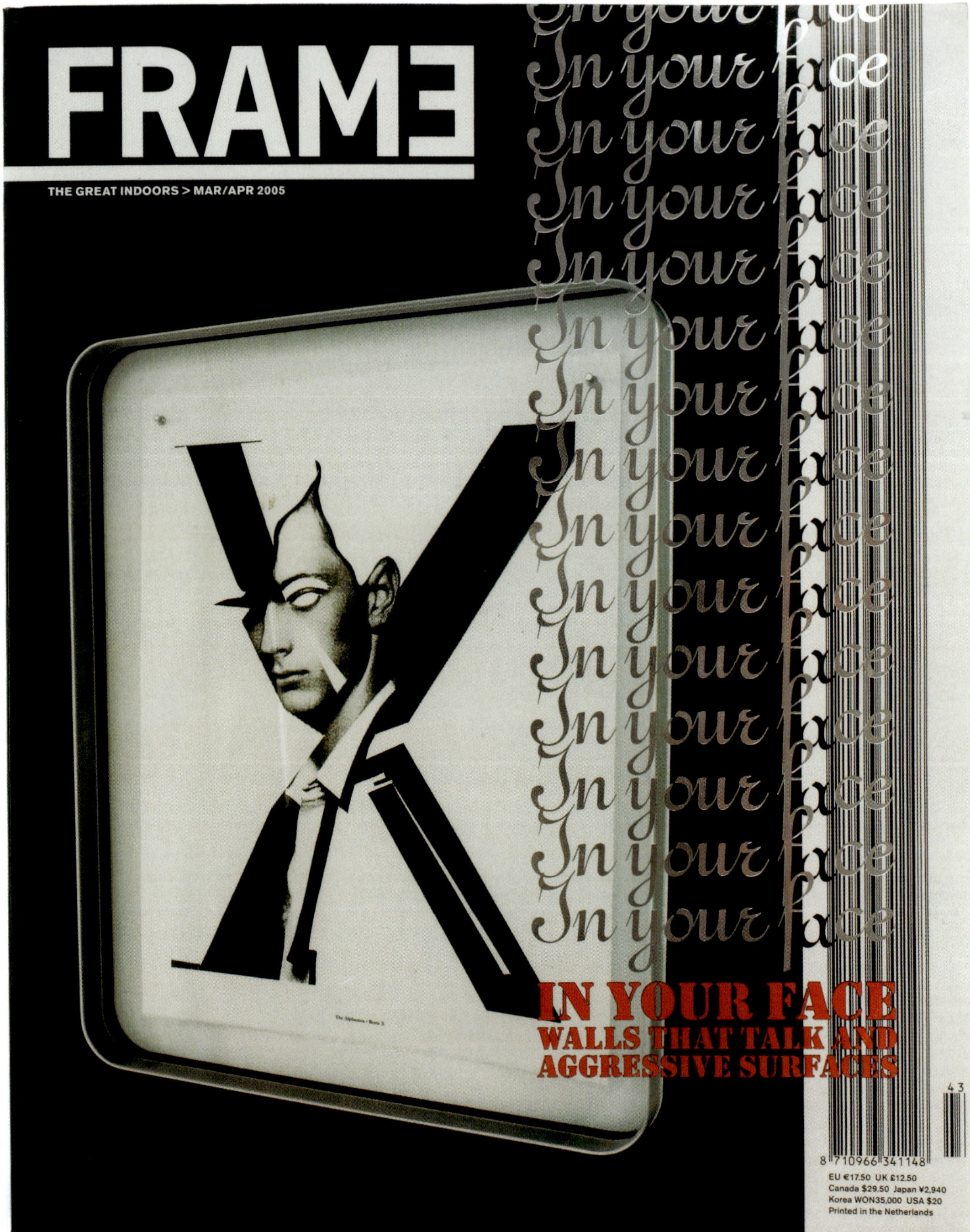

ART DIRECTORS
Cornelia Blatter & Marcel Hermans, COMA

43

California Dreamin'

Architects Clive Wilkinson, Alexis Rappaport and Hailey Soren (Clive Wilkinson Architects) about the project

We are very pleased with overall outcome of the project. The FIDM annex studio won several awards, including an award of excellence in the category of interiors (Los Angeles Business Council 2005). We were also in the finals of a competition for Best Public Space (FX International Interior Design Awards 2004).

Current projects

FIDM Library Tower (projected finish date: August 2008) // Maguire Partners // Jay Walter Thompson (projected finish date: August 2006)

DESIGNERS
Clive Wilkinson Architects
www.clivewilkinson.com

SUBJECT
LA campus of the Fashion Institute of Design and Merchandising (FIDM)

LOCATION
Los Angeles, USA

PHOTOGRAPHER
Benny Chan

WRITER
Michael Webb

CALIF

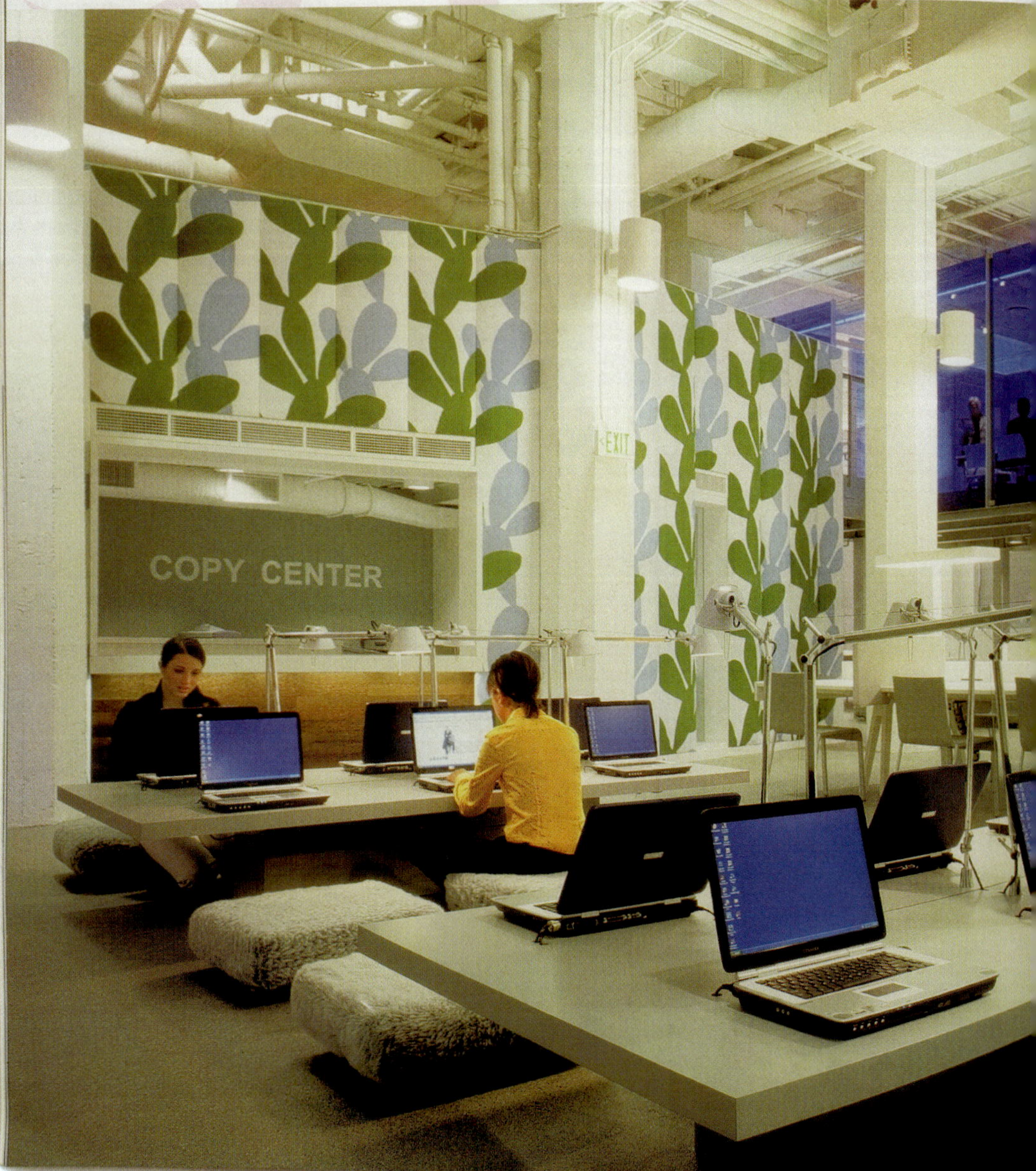

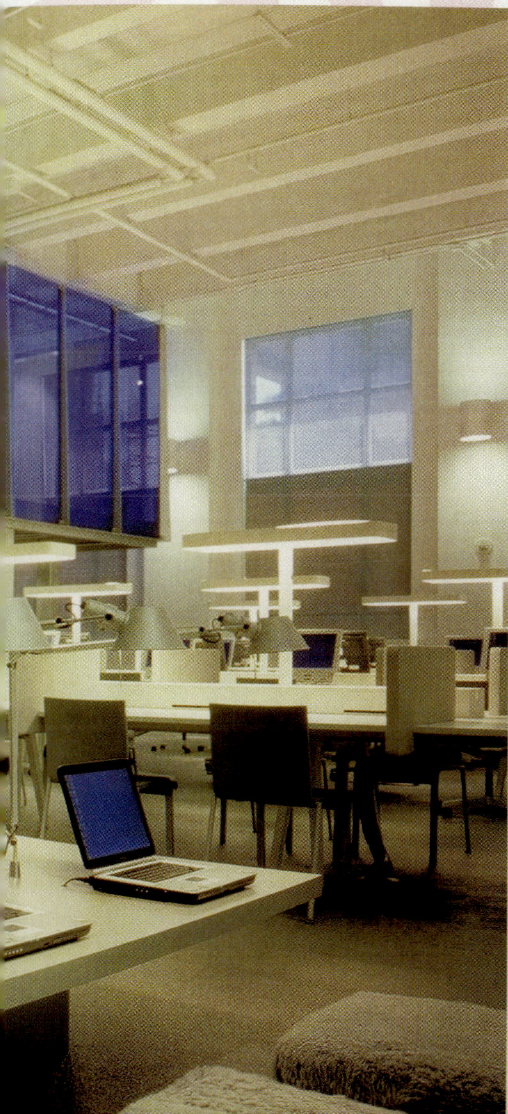

CALIFORNIA DREAMIN'

FUN FOLLOWS FUNCTION AT CLIVE WILKINSON'S STUDIO FOR STUDENTS OF FASHION AND FOLLOWERS OF THE GOOD LIFE IN SOUTHERN CALIFORNIA.

TEXT BY MICHAEL WEBB
PHOTOGRAPHY BY BENNY CHAN

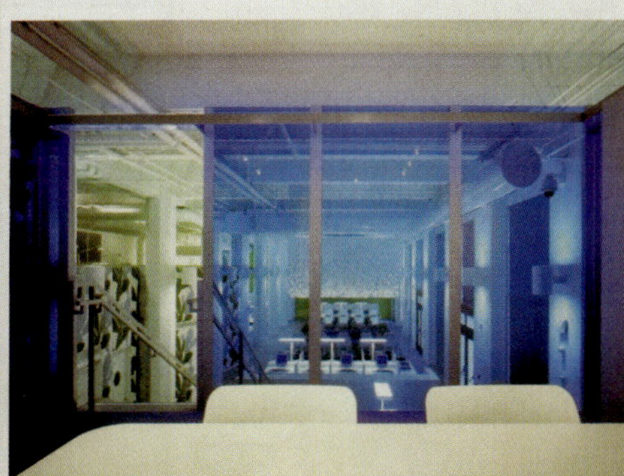

Left: study in style at the Fashion Institute of Design & Merchandising. Three-metre-valong low tables in plastic laminate and cushions covered in faux fur. Tolomeo table lamps by Michele De Lucchi and Giancarlo Fassina for Artemide.

SUBJECT LA campus of the FIDM
LOCATION Los Angeles, USA
DESIGNERS Clive Wilkinson Architects
PHOTOGRAPHER Benny Chan
WRITER Michael Webb

DREA

Students at the LA campus of the Fashion Institute of Design and Merchandising (FIDM) relish the good life of Southern California, riding a big wave, immersing themselves in a shimmering blue tank or lounging by a palm-fringed pool. Thanks to the wizardry of architect Clive Wilkinson, they can do all this in a design studio that's located in a gritty commercial neighbourhood 20 kilometres from the ocean. Lofty banking halls flanking the lobby of a 1926 beaux-arts office building have been stripped and turned into vibrant two-level study areas that are both functional and fun.

Tonian Hohberg founded FIDM in 1969 in downtown LA, conveniently close to the city's garment district, which is one of the world's larger, though perhaps least well known. As its president, Hohberg has made FIDM the pre-eminent West Coast school for fashion and related design disciplines, with an enrolment of 5000 full-time students, two-thirds of whom study in LA. The rest attend classes at satellite campuses in San Francisco, San Diego and Orange County (which Wilkinson recently re-created as a paradise of pink). When the school decided to add a design studio, Wilkinson was the logical choice to design it. Says Hohberg, 'We wanted a workplace that students would see as a home away from home – a place to inspire creativity and provoke a "wow" response.'

Breaking down the boundaries between work and play, and between solitary and group effort, is a recurring theme in Wilkinson's work for ad agencies (TBWA/Chiat/Day, Foot Cone & Belding and Mother), and the concept is equally appropriate to educational institutions. 'Every office used to be structured the same way, as a numbing repetition of closed work-spaces, meeting rooms and cubicles, regardless of what was done there,' the architect observes. 'And the same was true of educational institutions, which is why, in Hollywood high-school movies, most of the action occurs in corridors, the only place where everyone interacts.' Agencies depend on teamwork, and Wilkinson gives them a molecular structure of open-sided containers, platforms, flying staircases, juice bars, and even a place to shoot hoops: an adven-ture playground for grown-ups. Education is more about individual effort, but students socialize intensely,

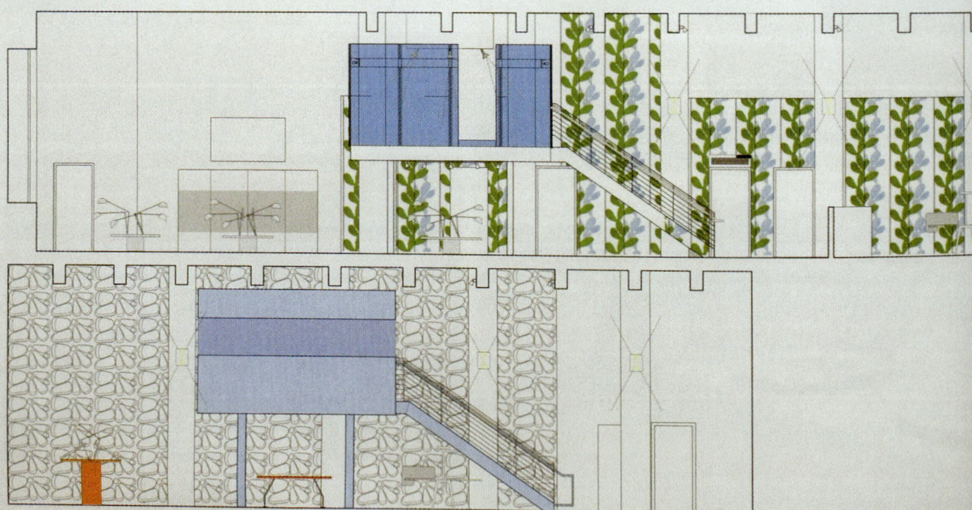

Interior elevations show how the design utilizes the full height of the refurbished banking hall to create a bright, open setting. Wilkinson inserted a series of mezzanines focused on collaborative work, while the open-plan landscape at ground level lets students choose from a variety of work settings, with or without computers.

BREAKING DOWN THE BOUNDARIES BETWEEN WORK AND PLAY, AND BETWEEN SOLITARY AND GROUP EFFORT, IS A RECURRING THEME IN WILKINSON'S WORK

and they collaborate spontaneously as often as they do while working on an assigned project.

Wilkinson likens FIDM's studio to a reading room in which laptops take the place of book stacks and indexes, and students can sit anywhere and converse freely. 'We understood that their work ranges from doodling to serious research to producing designs, and you want to provide a choice of different environments,' he says. 'For advice, we turned to the juniors in our office, and it was they who suggested that students might want to sit on the floor or find a place to flop out and snooze.' Having observed the earlier success of the design team's unconventional ideas and colours, the client embraced his recommendations.

Just as ad agencies analyse companies to develop or reinforce brand identity, Wilkinson and his staff research the history and finer points of clients to discover what they truly need and how they want to be perceived. 'We don't have a clue what we're doing until we are well into a project,' he confesses. 'It alarms clients when they ask for our vision, but we don't want to get stuck with our first idea and fail to develop a better one. Vision follows a pragmatic assembly. Even with a written programme, it takes time to arrange the parts of the body, with all the organs in place, to create a functioning mechanism.'

This determination to make things work and to accommodate the varied demands of different agencies and even different branches of the same institution, as in the case of FIDM, is the solid foundation on which Wilkinson erects colourful, inventive structures. Each of the two rooms involved in this project has a distinct identity and serves as a cockpit that brings people together in a variety of configurations. Mezzanine elements and galleries create a vertical structure within the 7-metre-high volumes, adding workspaces without crowding the main floor. 'It's important to populate a space architecturally,' says Wilkinson. 'Even when nobody's there, the room isn't forbidding, for the raised structures animate and fragment the space, giving it a welcoming scale.'

New plays off old, colour off white, and pattern off abstract forms. Wilkinson exposed the hexagonal concrete structural columns, leaving them in their original

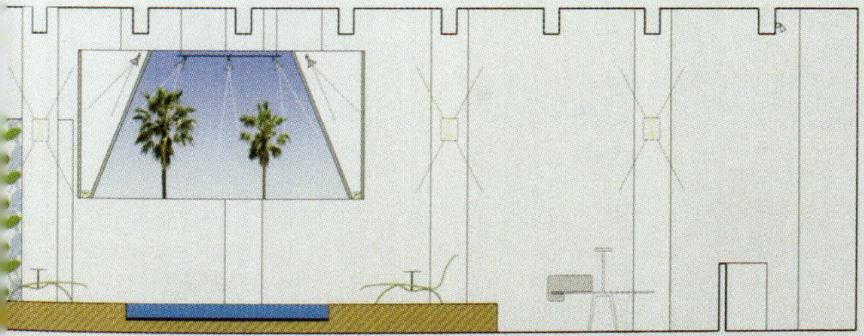

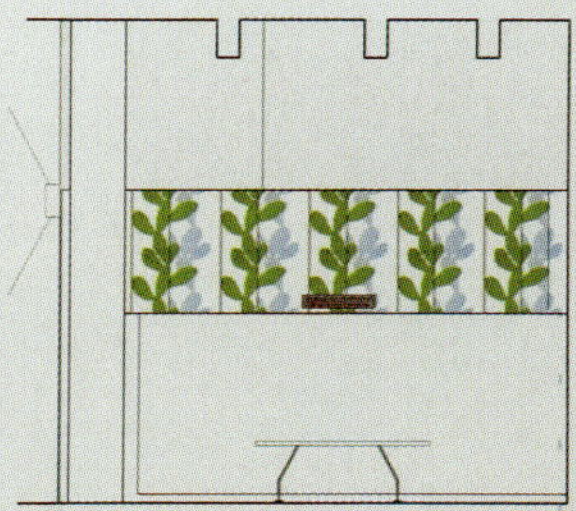

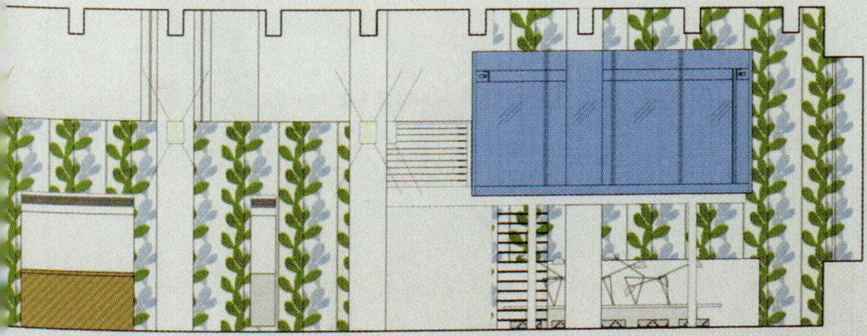

SUBJECT LA campus of the FIDM
LOCATION Los Angeles, USA
DESIGNERS Clive Wilkinson Architects
PHOTOGRAPHER Benny Chan
WRITER Michael Webb

'IT'S IMPORTANT TO POPULATE A SPACE ARCHITECTURALLY. EVEN WHEN NOBODY'S THERE, THE ROOM ISN'T FORBIDDING.' CLIVE WILKINSON

rough state and painting them white, like the walls and the raw ceiling services. The columns support light fittings, and the expanses of white diffuse natural and artificial light, making the room seem larger. They and the grey industrial carpet serve as foils to the accents of blue that evoke water, which symbolizes Southern California in the eyes of students arriving from the arid Midwest or the interior of Asia. In the East room, a blue-cushioned seating pit simulates a pool; its canopy is screen printed with palm trees, and a palm-wood deck supports custom-designed chaises with straight backs, which provide sybaritic perches for laptoppers. White canvas, laser-cut in a grid of small square openings, wraps the canopy and is backlit to suggest an office tower at night.

At the opposite end of the room, steel steps lead up to the 'tank': this meeting room, featuring whiteboards for pin-ups and projection, is enclosed in blue glass that is highlighted with a strip of blue neon. A bridge leads to a mezzanine gallery, and a low table with fur-covered stools offers an alternative to the conventional work surface with its chairs, task lights and desktop computers. Dominating the West studio is an elevated structure in the form of a cresting wave. The inner walls of both rooms are padded to absorb sound and panelled with bolts of Marimekko fabric in stylized cactus and leaf patterns. In his design of the Mother ad agency in London, Wilkinson had employed 50 of the firm's motifs, travelling to the factory in Finland to select large-scale patterns that would hold an expansive wall. Here, he chose two that reflect the ambience of an indoor garden while reminding students of Marimekko's historic contribution to design.

The 1150-square-metre design studio accommodates 150, and, though more can crowd in, there's sometimes a wait for a chaise or another prized place. FIDM leases two floors of the building, across a pocket park from their main structure, and may decide to add a third room to house the excess demand. Wilkinson is working closely with the school and expects to remodel the San Francisco campus in 2006, following the design of new headquarters for Google and J. Walter Thompson. The links between architecture, advertising and fashion are central to his work, yet each aspect retains its integrity, and all his interiors carry the signatures of both architect *and* client.

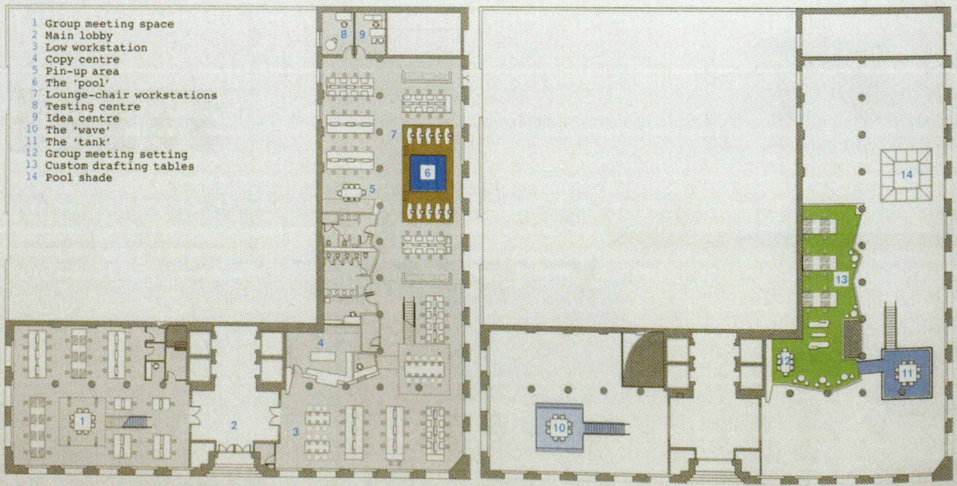

Above: plan of ground and mezzanine levels of the 1068-square-metre Fashion Institute of Design & Merchandising (FIDM).

Opposite: insulated wall panels are padded in Kaktus, a fabric by Marimekko that attenuates ambient sound. Joyn table by Ronan and Erwan Bouroullec and .04 desk chair by Maarten van Severen; both for Vitra. Perched on columns is the 'tank', a neon-illuminated, blue-glass box containing a meeting table and pin-up area.

ORNIA

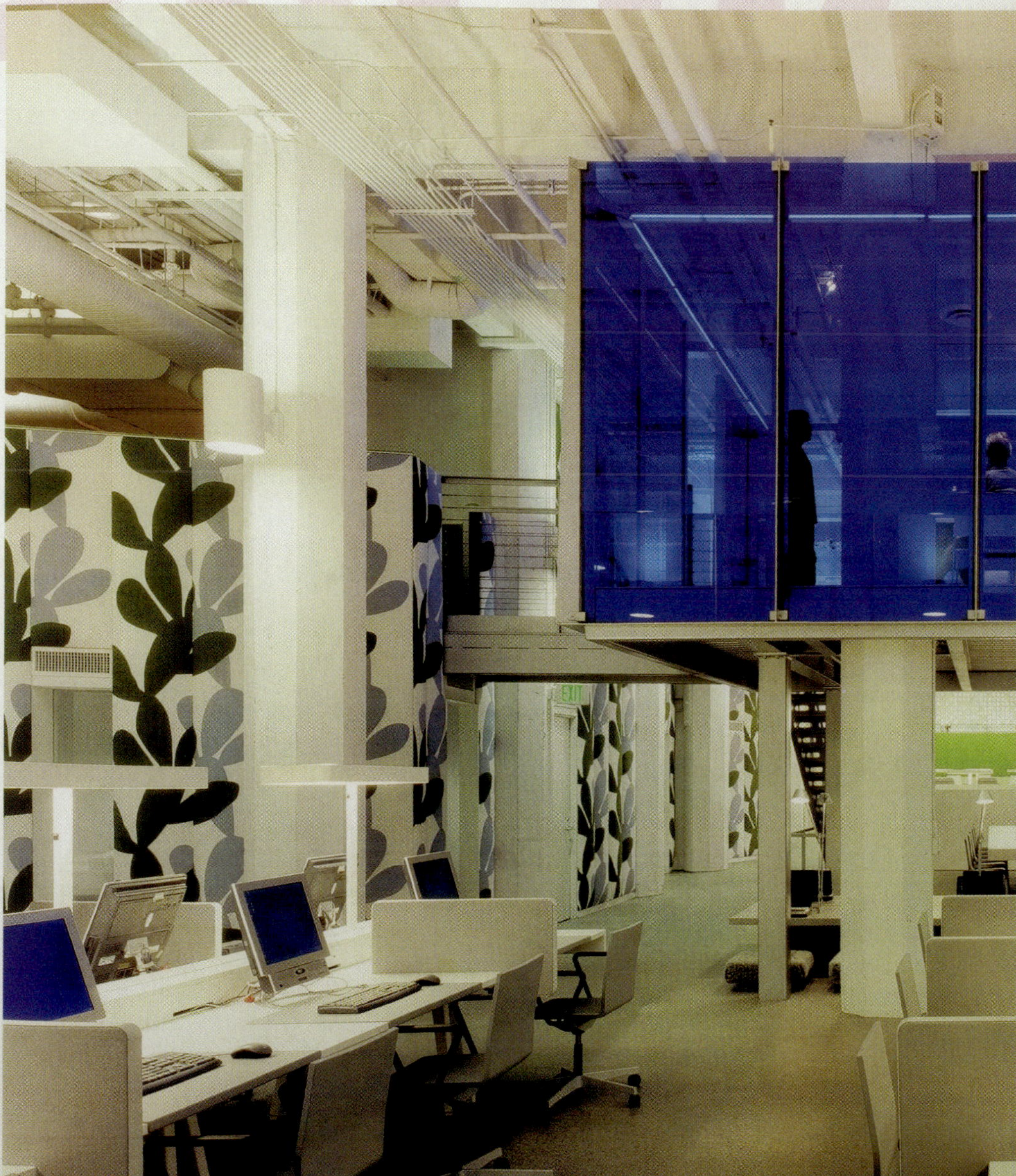

SUBJECT
LA campus of the FIDM

LOCATION
Los Angeles, USA

DESIGNERS
Clive Wilkinson Architects

PHOTOGRAPHER
Benny Chan

WRITER
Michael Webb

'THE JUNIORS IN OUR OFFICE SUGGESTED THAT STUDENTS MIGHT WANT TO SIT ON THE FLOOR OR FIND A PLACE TO FLOP OUT AND SNOOZE.' CLIVE WILKINSON

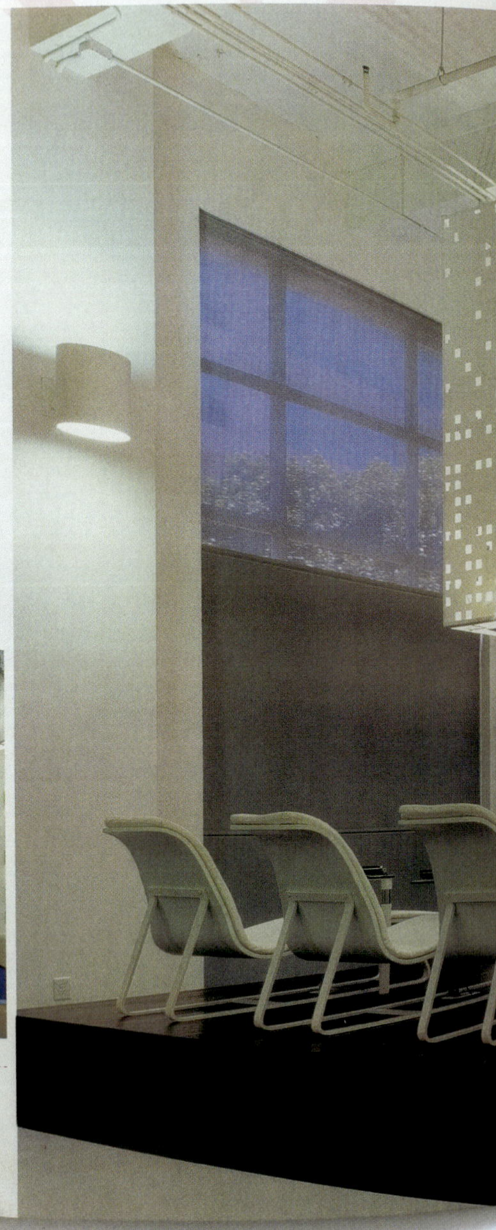

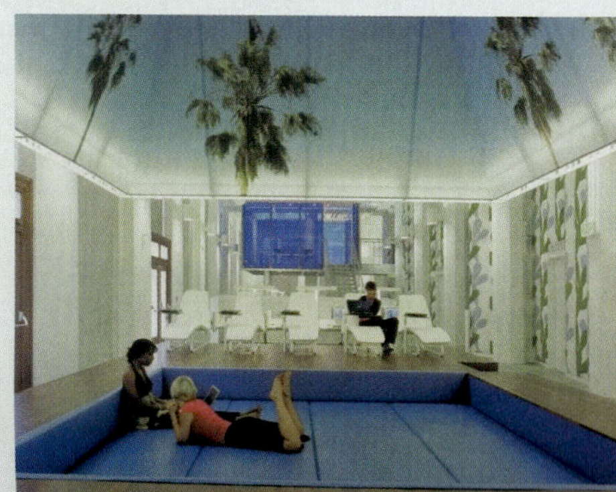

Design Studio East: to cool down, students can take a dip in the 'pool', an informal but cushy blue-upholstered work area for group or individual study surrounded by a palm-wood deck.

Those remaining poolside can enjoy the world's first chaise longues with height-adjustable laptop trays.

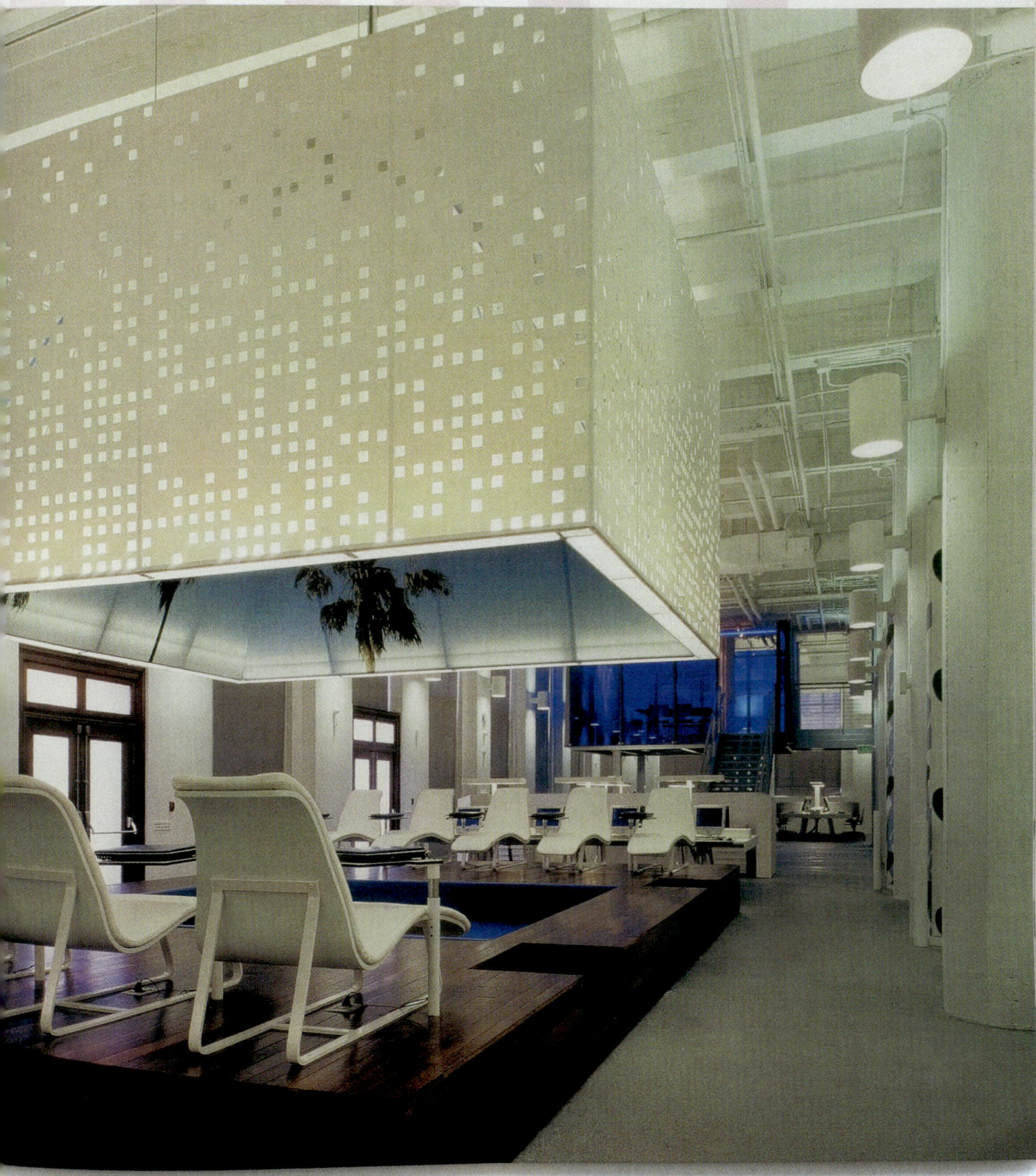

SUBJECT
LA campus of the FIDM

LOCATION
Los Angeles, USA

DESIGNERS
Clive Wilkinson Architects

PHOTOGRAPHER
Benny Chan

WRITER
Michael Webb

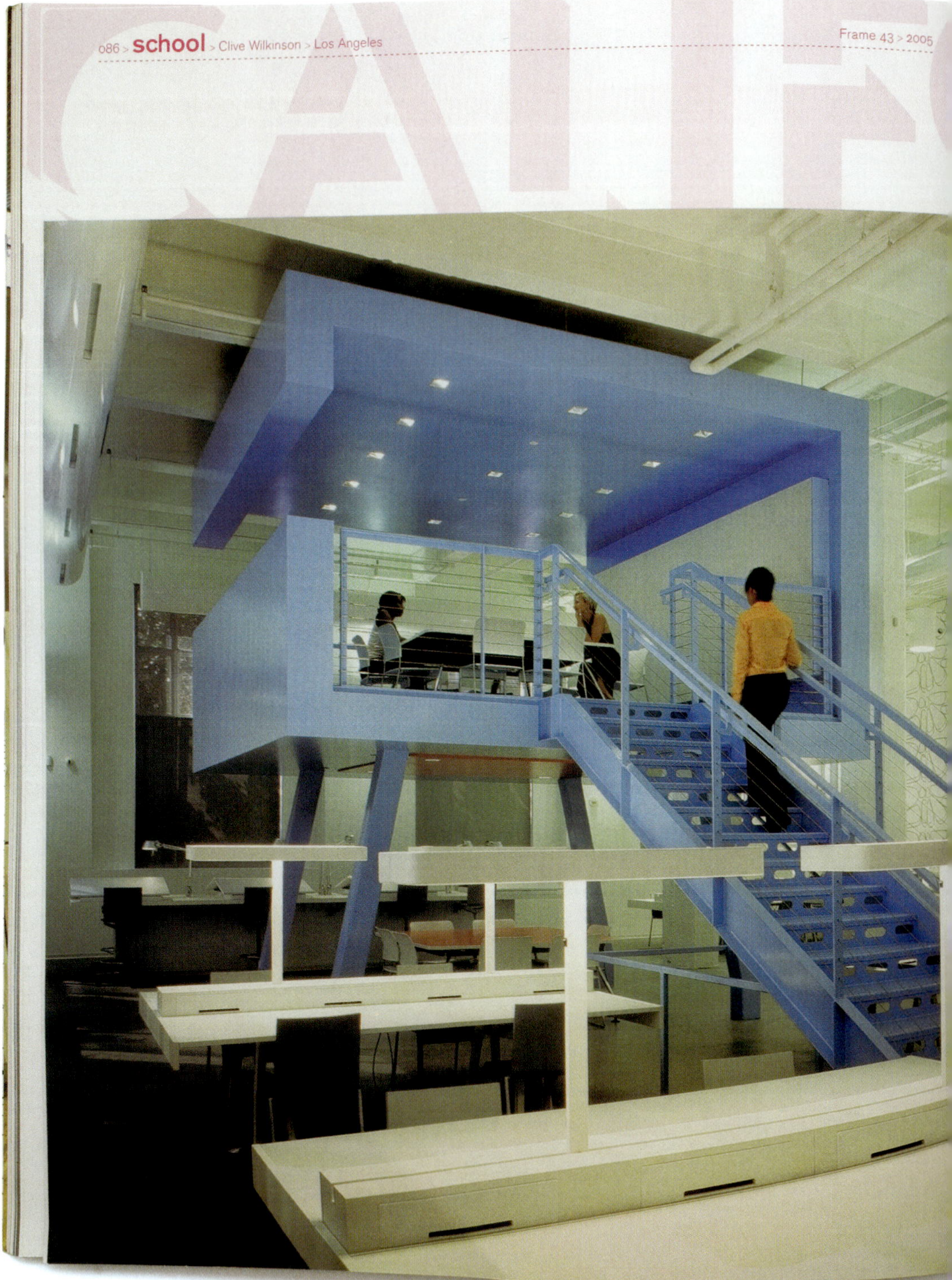
086 > school > Clive Wilkinson > Los Angeles
Frame 43 > 2005
CALIF

THE LINKS BETWEEN ARCHITECTURE, ADVERTISING AND FASHION ARE CENTRAL TO CLIVE WILKINSON'S WORK, YET EACH ASPECT RETAINS ITS INTEGRITY

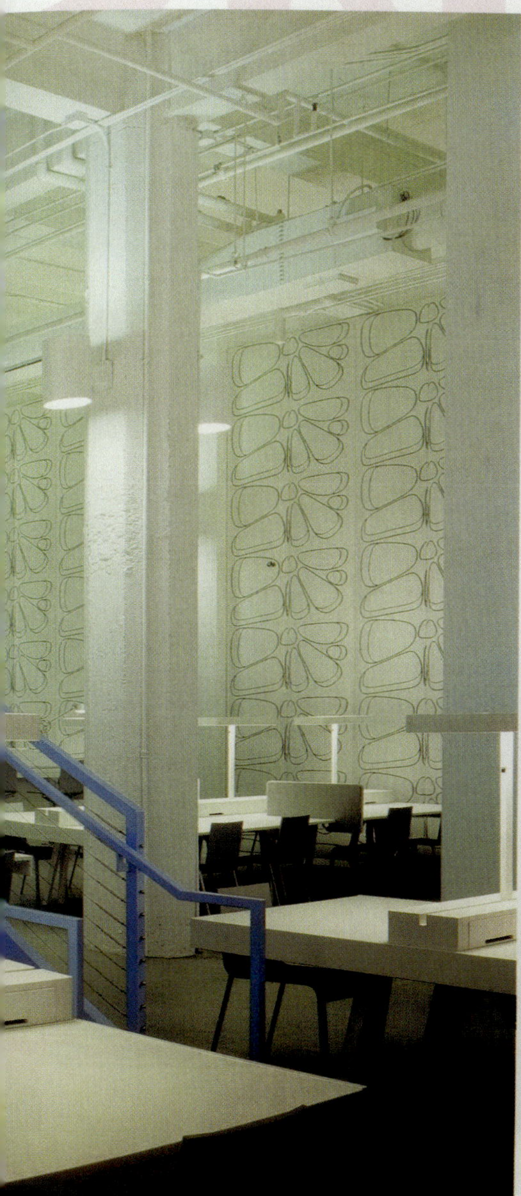

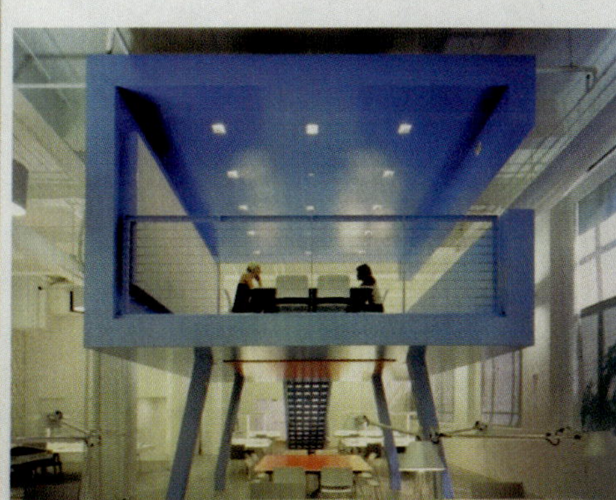

Design Studio West: the 'wave', a mezzanine sky room above the main floor, is an open collaborative area furnished with a meeting table and pin-up board for student critiques.

SUBJECT LA campus of the FIDM
LOCATION Los Angeles, USA
DESIGNERS Clive Wilkinson Architects
PHOTOGRAPHER Benny Chan
WRITER Michael Webb

Issue 44

May/June 2005

470-471

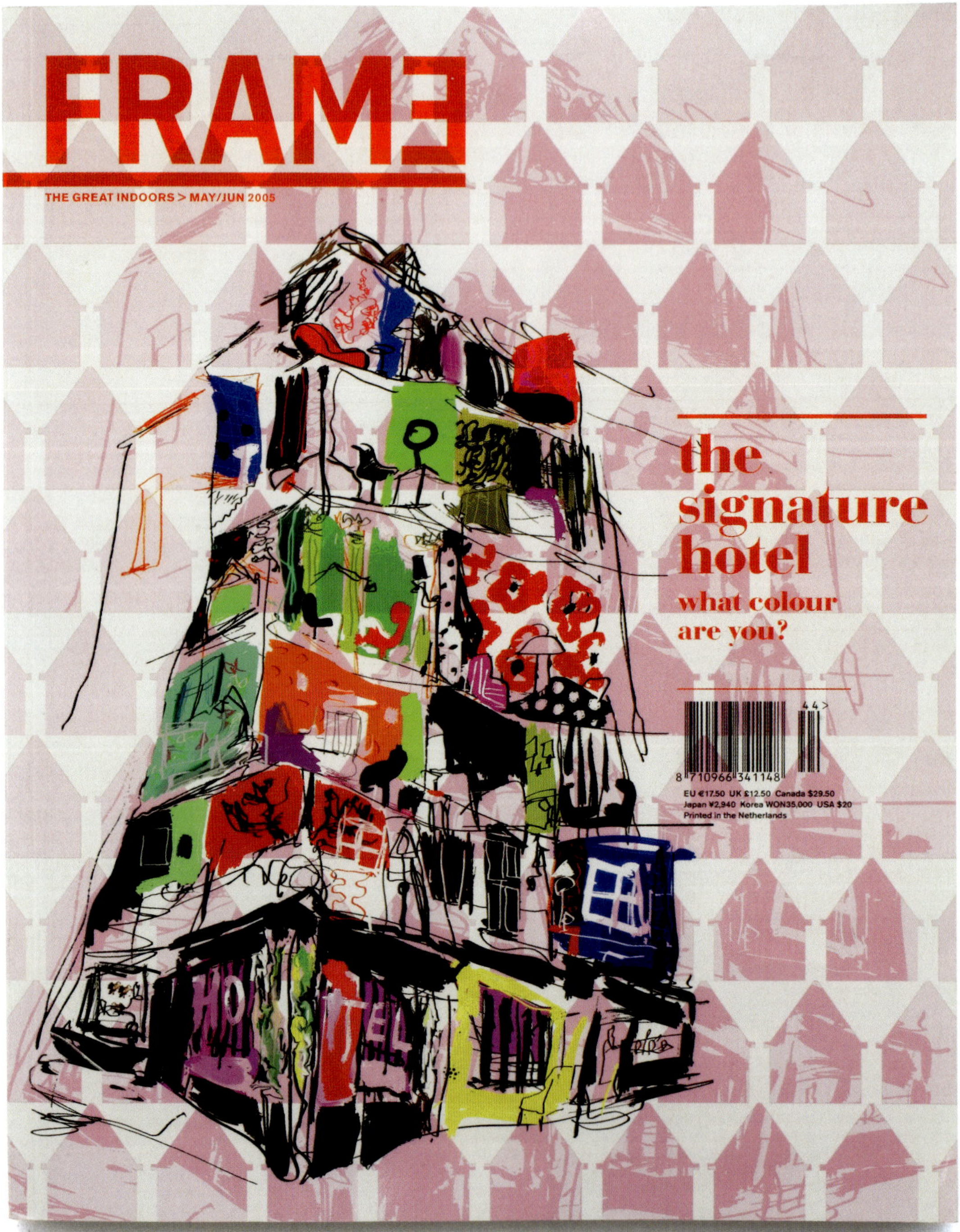

ART DIRECTORS
Cornelia Blatter & Marcel Hermans,
COMA

44

At Home With Marcel

Designer Marcel Wanders about the project

Hospitality design was completely new to me. I didn't know what was going on in the industry, what others were doing or what the latest trends were. Lute Suites was a very exciting opportunity and a project we will continue to develop. I enjoy designing for hospitality because of its size, complexity and the fact I am developing something that interacts directly with people's feelings. A hotel should offer an experience, it should be exciting and surprising with a strong visual rationale that elicits a strong emotional response. It is a great chance for me to bring together all aspects of my 3D language in one environment.

Comments on the published article

The article is really magic and presents the spirit of the place really well. What struck us were the strange props (shopping bags and cheap underwear) showing up in the photos. I remember, during the interview, having spoken about the curiosity I always have about what is behind the door of a hotel room. Having economy shopping bags and cheap underwear in the photo was a fun interpretation of this, for a high end fashionable super-chique classy design hotel. :-)

Did the article have an effect?

Lute Suites opened last January 2005 and has seen is bookings increase since, booking gets more difficult every day – it is a huge success. This is why Lute Suites (the first 'all-over-city-suites-hotel') is planning new Lute Suites to be opened in a number of places around the city at the beginning of next year in Amsterdam and probably in other European cities.

DESIGNER
Marcel Wanders
www.marcelwanders.com

SUBJECT
Hotel Lute Suites

LOCATION
Oudekerk aan de Amstel, Netherlands

PHOTOGRAPHER
Johannes Schwartz

WRITER
Brigitte van Mechelen

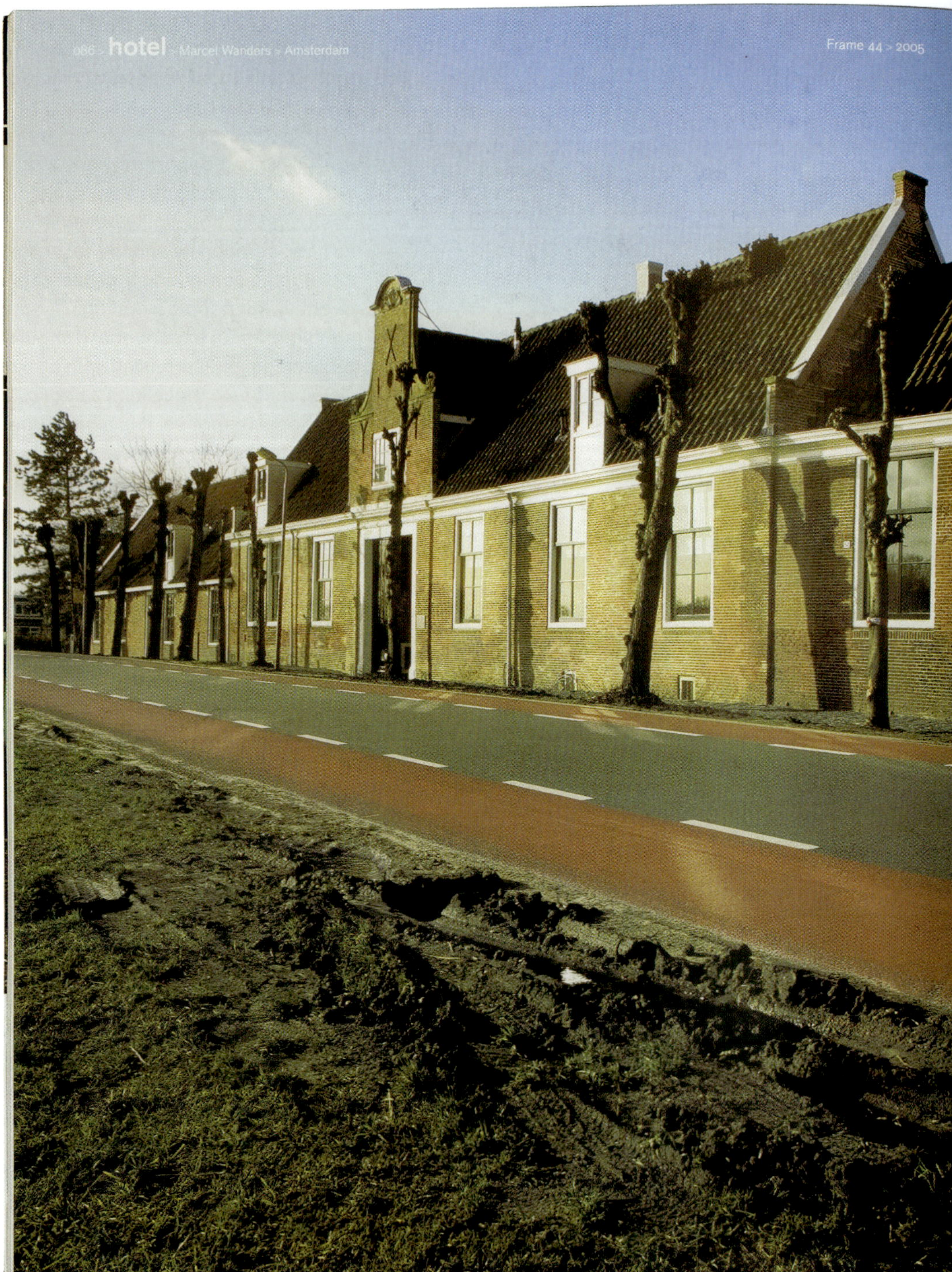

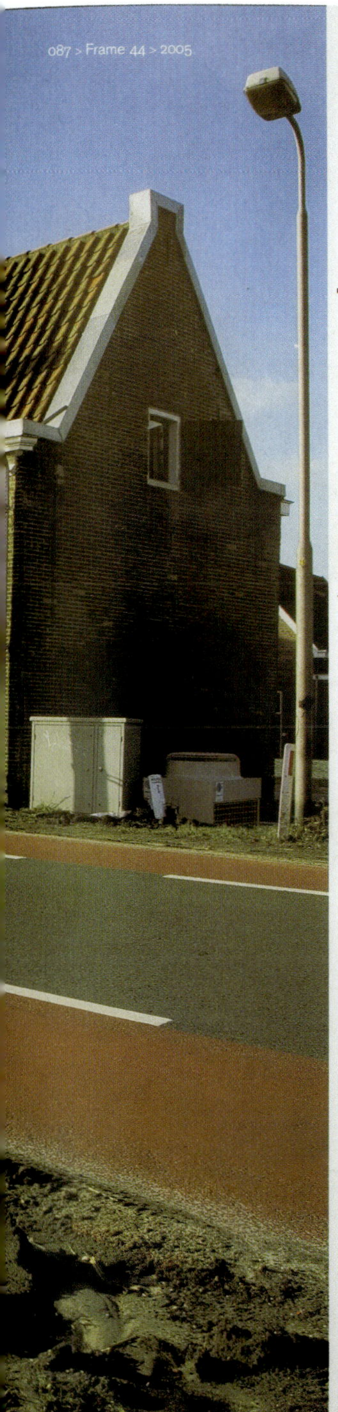

at home with marcel

On the outskirts of Amsterdam, in seven working-class houses that he's converted into luxury suites, Marcel Wanders has designed the experience of feeling at home.

Text by Brigitte van Mechelen
Photography by Johannes Schwartz

Two years after the opening of Lute, a first-rate restaurant on the site of a former gunpowder factory in Ouderkerk aan de Amstel, just outside Amsterdam, there remained on the same site a row of small, ramshackle dwellings. What a shame, thought owner and *patron-cuisinier* Peter Lute and his wife, Marieke. We've got to do something about this. But what?

What they didn't want, in any case, was a series of hotel rooms whose occupants, perched on the edges of elegant beds, are driven to raid their respective minibars out of sheer boredom. And certainly not when they've just enjoyed a banquet to which not a bite or a drop can be added. (Lute's menu lists such delicacies as oysters with *pata negra* and *grapaudine* beetroot, not to mention brill with *lardo di collonata* and salted lime.)

Enter Marcel Wanders. When Marcel dropped into Peter's restaurant for a meal, each recognized in the other a kindred spirit 'completely crazy about his profession'. The conversation that began that evening continues to this day. At least that's the story – of almost mythic simplicity – that the men tell when asked to explain the ins and outs of their joint venture. But why, I wonder, did Lute pass over a second chance to collaborate with Eline Strijkers, who designed the interior of the farm that MVRDV converted into his restaurant? His enthusiasm about her work is common knowledge. 'We're still enthusiastic,' he replies, 'but that doesn't mean we have to make the same choices all over again.'

And now there are the Lute Suites. Seven luxury apartments, each with a distinctly individual character and a range of purpose-made products, varying in size from 60 to 80 square metres and specially equipped for a prolonged stay. That means a private entrance, kitchenette, wireless internet *and* all the amenities of a hotel, including breakfast delivered to the room, if desired. Rates: €300 to €575 a night.

The Lute Suites: seven 18th-century dwellings with a view of the River Amstel; interiors renovated and designed by Marcel Wanders.

SUBJECT
Hotel Lute Suites

LOCATION
Oudekerk aan de Amstel, Netherlands

DESIGNER
Marcel Wanders

PHOTOGRAPHER
Johannes Schwartz

WRITER
Brigitte van Mechelen

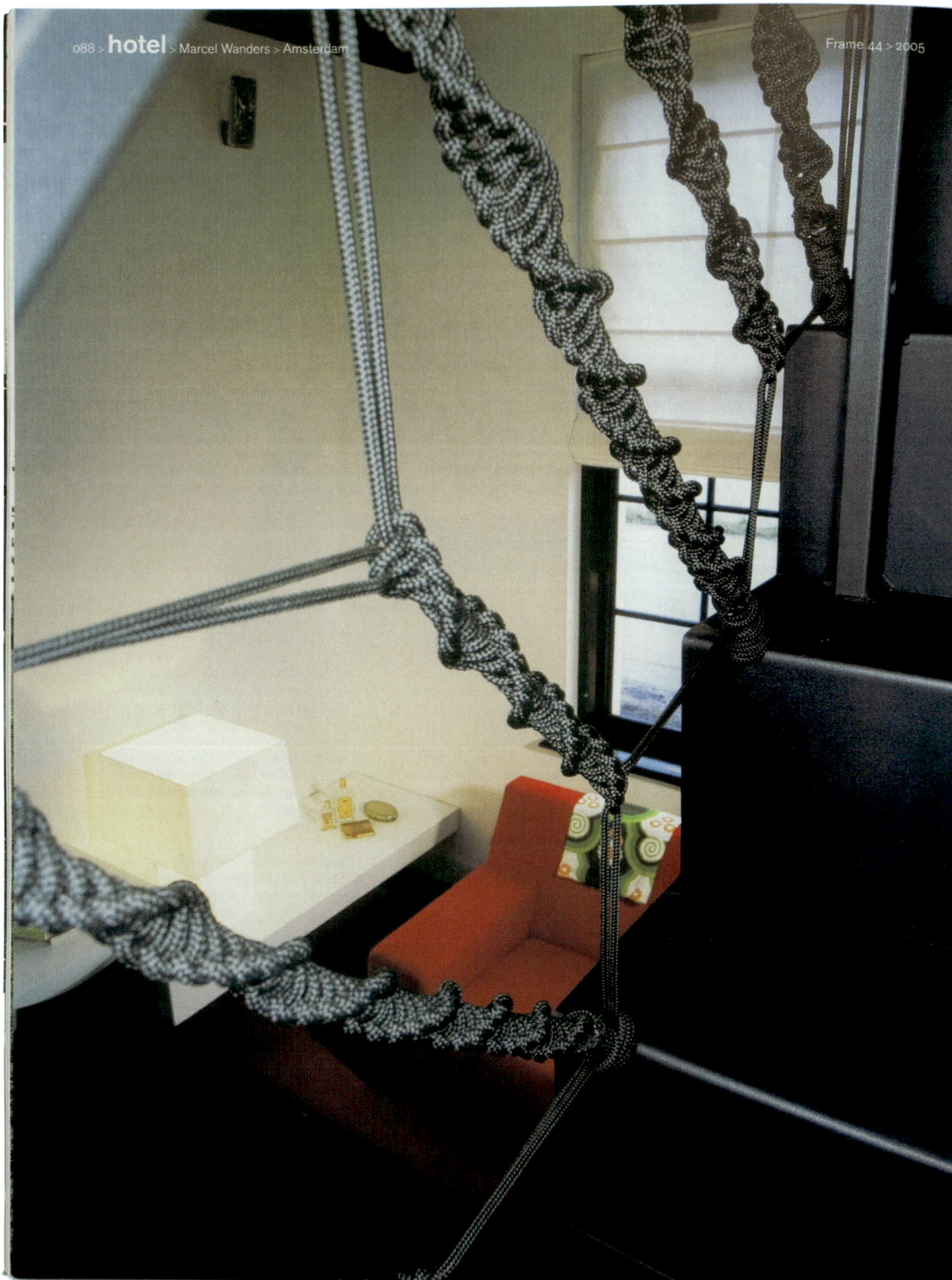
088 > hotel > Marcel Wanders > Amsterdam
Frame 44 > 2005

Wanders the interior architect?

After 14 years of product design, I've recently done some interior work. But I hadn't designed a hotel room. For Hotel on Rivington in New York, which also opened in January, I did the public areas. My first look at the inside of these houses was something of a shock. Nibble, nibble, little mouse – that says it all. We decided at once to break the place wide open, right up to the rafters. Architectonically speaking, it all went without a hitch. The rooms turned out beautifully. We had no problems in terms of products either. But everything in between – boy oh boy.

In between?

Stairs in relation to stairwells, the position of electrical outlets – we did make a few mistakes in that area. Which we eventually straightened out, of course.

Smoke chandeliers, the Bottoni sofa, VIP chairs, the one-minute-sculptures . . . did you immediately envision the project as a showroom?

It does include a lot of my things, as well as designs from Moooi. And many of the products were purpose-made for the project. Yes, by me. And by Karin Krautgartner of my studio. So, yes, I worked. Was I supposed to sit on my hands?

Angry?

Yes! Showroom . . . so whiny, such negative thinking. Everything I do, I do with all my heart.

As your slogan says . . .

Exactly. 'Marcel Wanders. Here to create an environment of love, live with passion and make our most exciting dreams come true.' For Lute, I designed the experience of 'feeling at

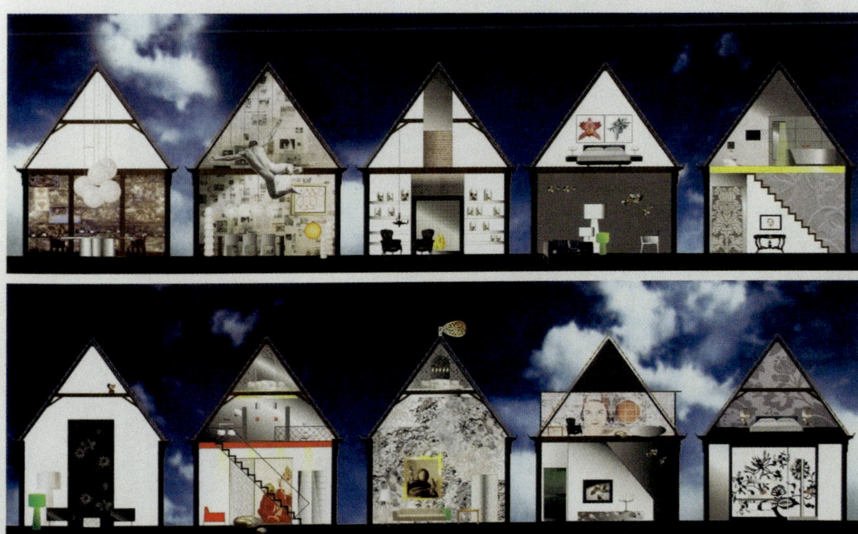

Opposite: AVL Shaker chair by Joep van Lieshout for Moooi; Double Square Light and Soft Mellow sofa, both by Marcel Wanders for Moooi; macramé balustrade by Marcel Wanders.

Above: illustration of the interiors of the seven Lute Suites by Marcel Wanders Studio.

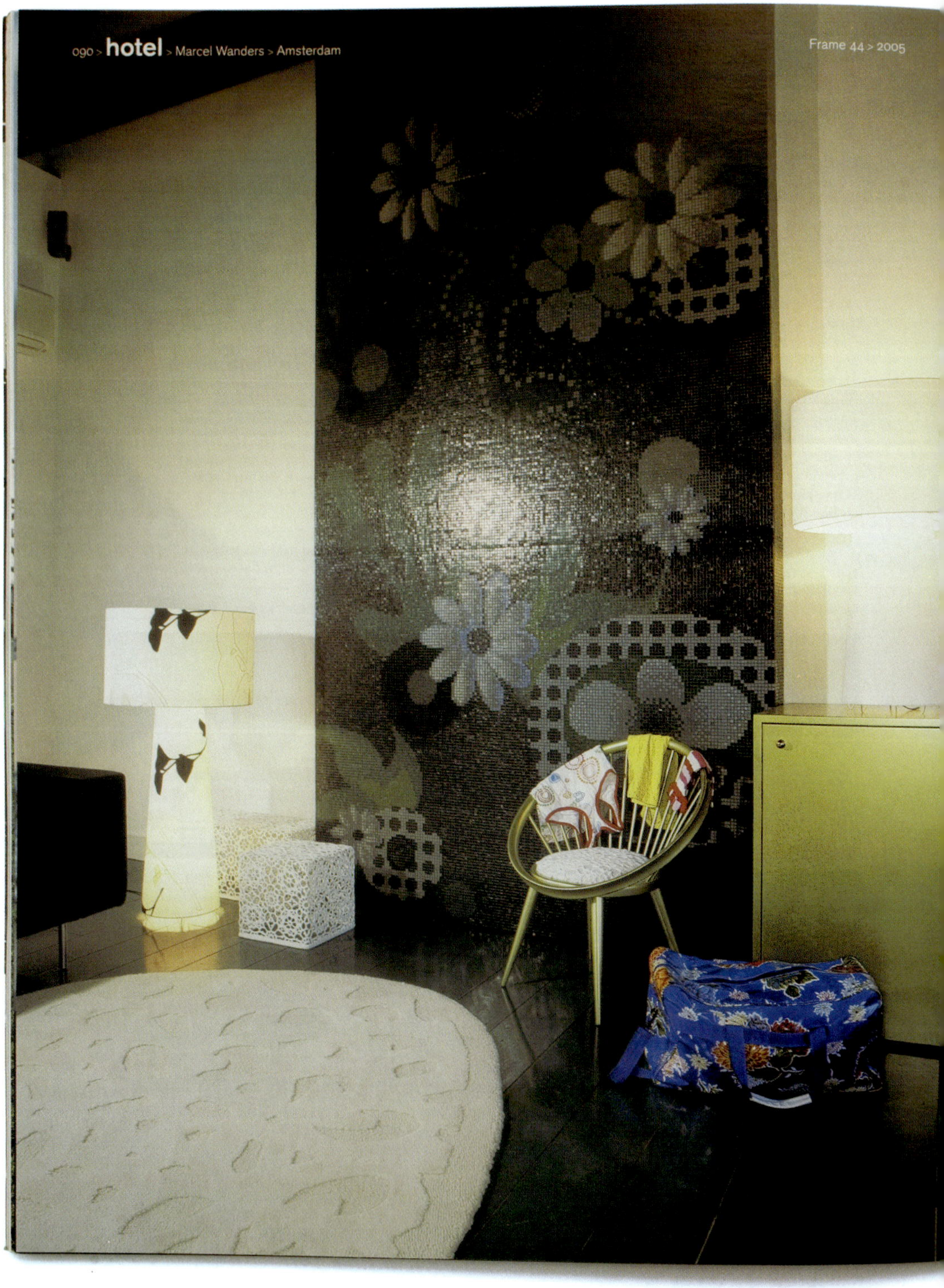

'The Suites are genuinely modern, but not in the sense of democratic and available to everyone. I'm talking about "now" design, design that's warm, that's not aloof from the general public'

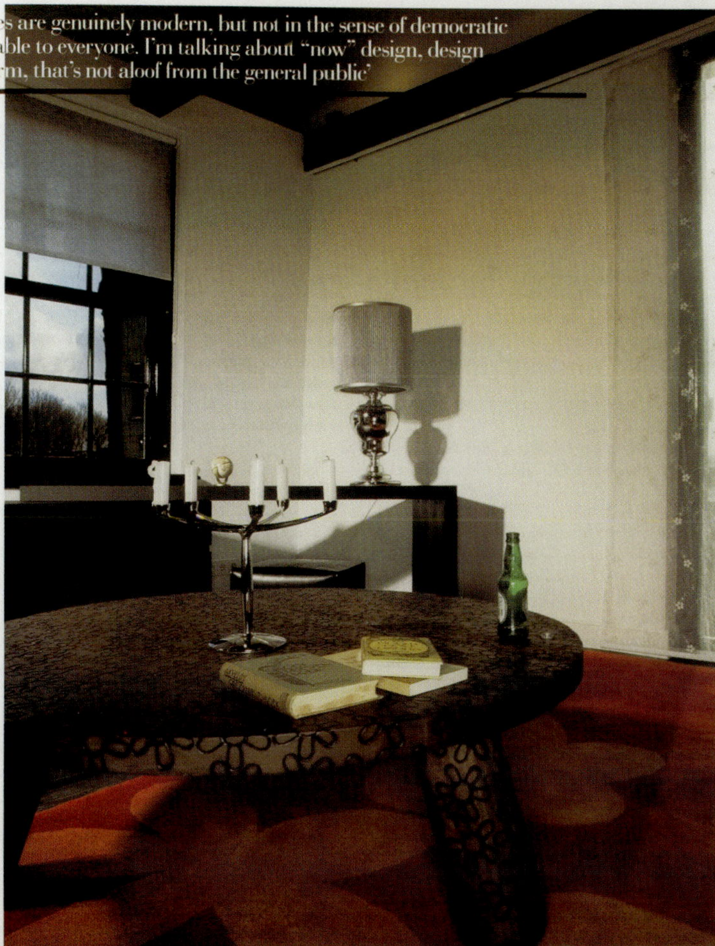

Opposite: Lute rug by Marcel Wanders Studio; Bottoni sofa and Crochet table, both by Marcel Wanders Studio for Moooi; Big Shadow lamp with special fabric cover and Little Big Shadow, both by Marcel Wanders Studio for Cappellini; mosaic from the series '...One morning they woke up' by Marcel Wanders Studio for Bisazza; second-hand chair and fridge painted in Moooi Weer Gold, by Marcel Wanders Studio for Moooi.

Above: Flower rug by Marcel Wanders Studio, Flower coffee table by Marcel Wanders for Moooi, Lucy candleholder by Marcel Wanders for Goods, Collezione Print desk and pouf by Marcel Wanders Studio for Poliform, Kaipo lamp by Edward van Vliet for Moooi.

SUBJECT
Hotel Lute Suites

LOCATION
Oudekerk aan de Amstel, Netherlands

DESIGNER
Marcel Wanders

PHOTOGRAPHER
Johannes Schwartz

WRITER
Brigitte van Mechelen

home'. Every detail is part of that experience. A nice sofa, good lighting – even clean sheets.

'Behind the door, that's where it all happens . . .'
It's a credo vital to this project. Peter's idea. It's something like 'on the plate, that's where it all happens'. Unlike those boutique hotels that have fantastic lobbies but disappointing rooms, the suites we created are good, smart, attractive, comfortable, personal and modern. Previously, there wasn't anything for people who wanted lodgings suitable for a longer stay – people looking for 'a place of their own'.

Well, visitors who want a place of their own could rent an apartment.
Umm . . . yes. But the suites offer hotel amenities. Breakfast, cleaning services. And they're genuinely modern.

As opposed to artificially modern?
I don't mean modern in the sense of democratic and available to everyone. I'm talking about 'now' design, design that's warm, that's not aloof from the general public. Design that doesn't fall prostrate on the altar of technology. Technology should enable us to make the things we think are super. The machine should know its place!

In House 1, I saw a plaster relief of a bear's head and some spoons.
Oh, that. That's a haiku – the design of the entire space is actually a haiku. When you put good things in juxtaposition, you simultaneously create a new and personal story. When I was at school, an instructor once said to me, 'Good design; you clearly gave it some thought, but why is it so ugly?' Quit bugging me, I thought. The concept – that's what it's all about. But three weeks later it occurred to me that maybe I could have it both ways. That's when I conceived a method for studying the softer side of design. Every day I arranged a collage on a grey rug. Removed a stone. Added a twig. What I learned then, I still use today.

Sounds like a very intuitive approach. Do you allow your colleagues to tinker with your arrangements?
That often makes my designs even better. But what I don't like, I throw away. After all, somebody has to make the decisions.

Here comes another question you probably won't like. Did you use your connections with manufacturers in creating the Suites?
Of course. And that's not a rotten question. What's nice about my life at the moment is that manufacturers like Boffi and Bisazza are so cooperative. Poliform, for example, sped up production of prototypes for a new bed to get them done on time. These are business connections. It's an exchange of favours.

What do you do for them?
We work with one another. Right now we're trying to put together collections of the products that we've developed for the suites. The steel floor, the wall covering, the Soap bath.

Do you see the Suites as part of a grander scheme?
It's funny, but everything that's important begins in my heart, not in my head. I've often thought about the lack of a good hotel on the water in Amsterdam – and that I'd love to make one sometime. When we started doing the Suites, I wasn't thinking about that, but once we were into it, we realized that 'behind the door, where it all happens' could occupy other sites as well, as long as we could organize the services. Why not?
So, later this year we'll be opening another seven suites in the heart of Amsterdam. Three on houseboats! It's becoming an all-over-the-city-hotel concept. Isn't that great?

Bach . . .
Ah . . . a great example of serendipity. Lute, Peter's surname, already has a strong ring to it. But when we discovered that Bach had written a series of suites for the lute, everything just fell into place.

SUBJECT
Hotel Lute Suites

LOCATION
Oudekerk aan de Amstel, Netherlands

DESIGNER
Marcel Wanders

PHOTOGRAPHER
Johannes Schwartz

WRITER
Brigitte van Mechelen

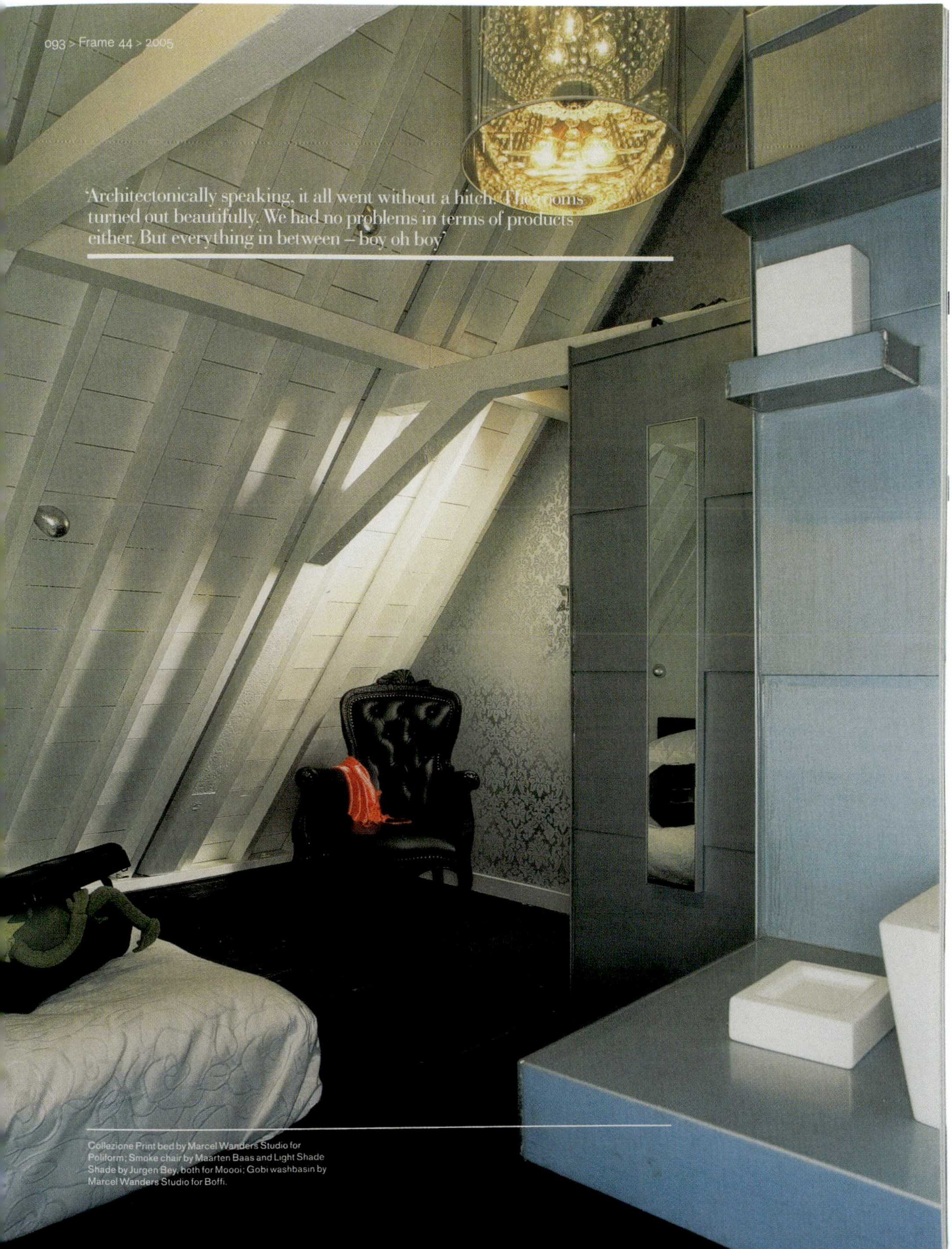

'Architectonically speaking, it all went without a hitch. The rooms turned out beautifully. We had no problems in terms of products either. But everything in between – boy oh boy'

Collezione Print bed by Marcel Wanders Studio for Poliform; Smoke chair by Maarten Baas and Light Shade Shade by Jurgen Bey, both for Moooi; Gobi washbasin by Marcel Wanders Studio for Boffi.

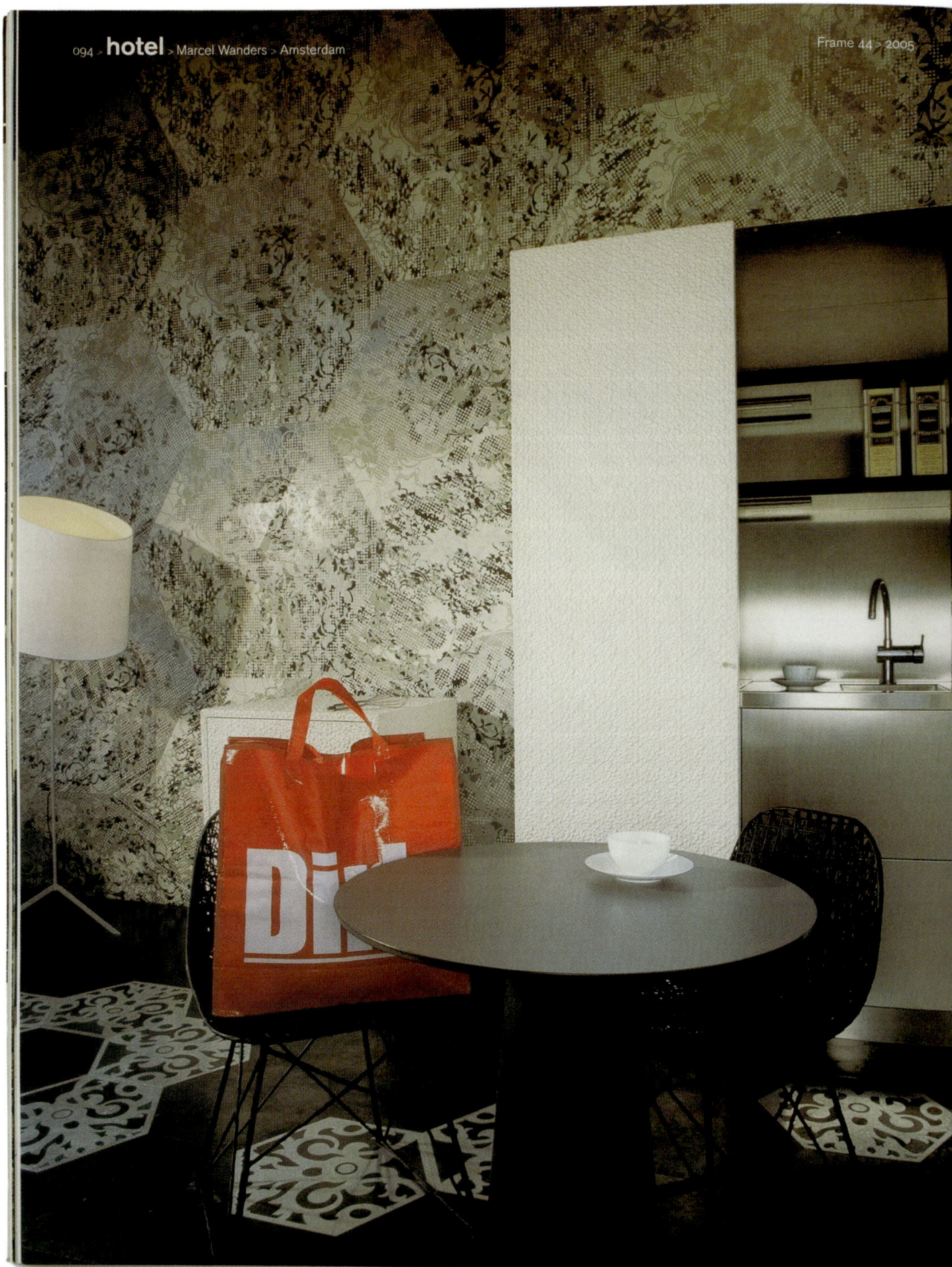
094 > hotel > Marcel Wanders > Amsterdam
Frame 44 > 2005

'Unlike those boutique hotels that have fantastic lobbies but disappointing rooms, the suites we created are good, smart, attractive, comfortable, personal and modern.' Marcel Wanders

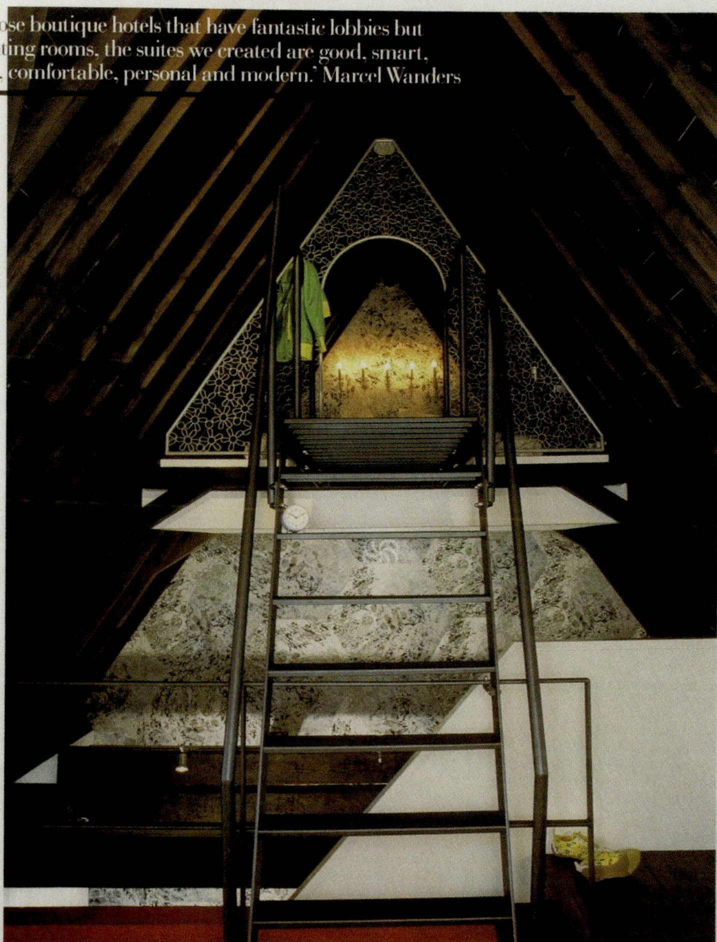

Above: Hexagon wallpaper and Flower fence by Marcel Wanders Studio, Magnum spotlight by Konstantin Grcic for Flos.

Opposite: Lamp Double Shade and Container table, both by Marcel Wanders Studio for Moooi; Carbon chair by Bertjan Pot and Marcel Wanders for Moooi; Metal tiles and Hexagon wallpaper, both by Marcel Wanders Studio; cup and saucer from J.L. Coquet Hemisphere. Kitchenette is finished with lace in epoxy.

SUBJECT
Hotel Lute Suites

LOCATION
Oudekerk aan de Amstel, Netherlands

DESIGNER
Marcel Wanders

PHOTOGRAPHER
Johannes Schwartz

WRITER
Brigitte van Mechelen

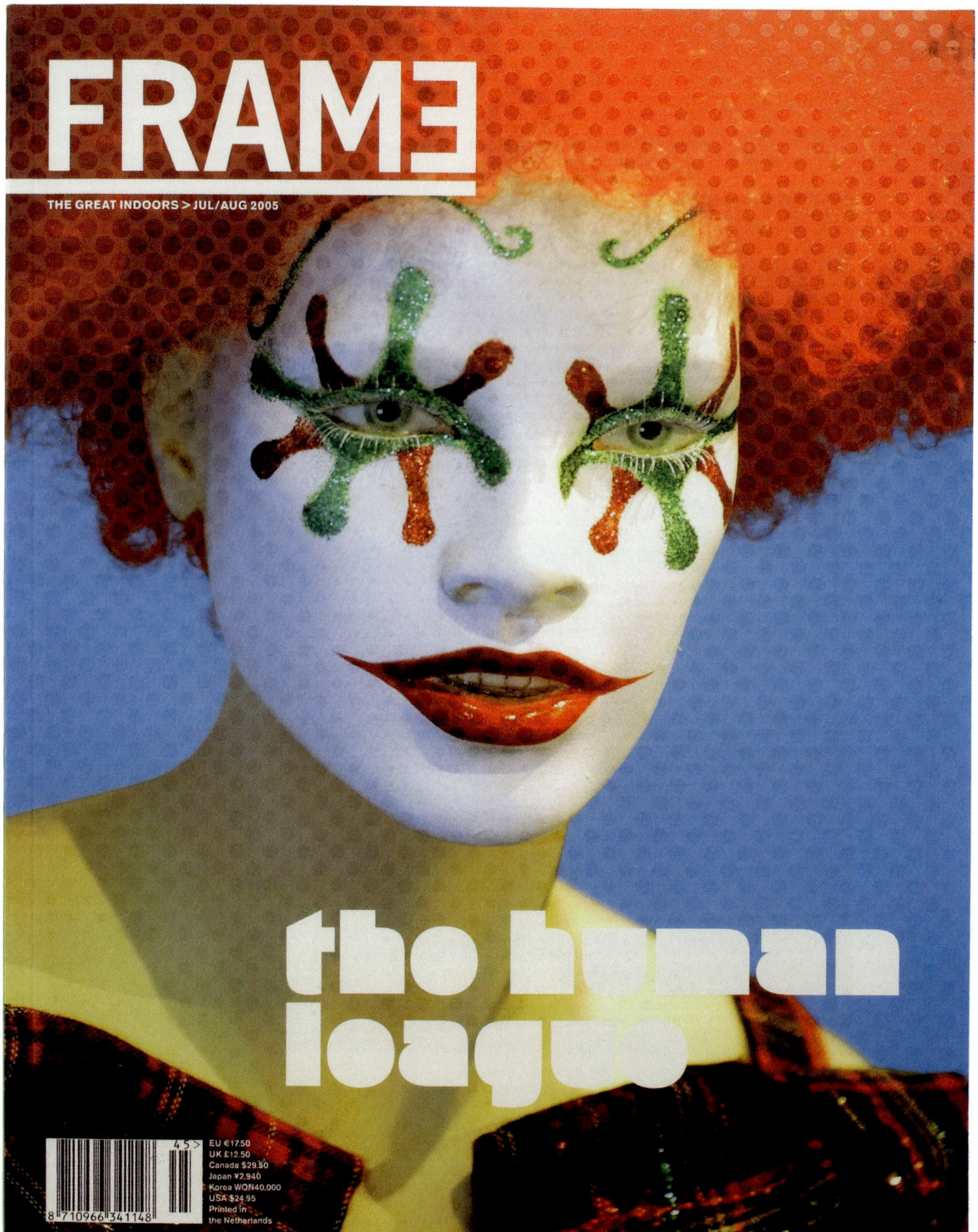

ART DIRECTORS
Cornelia Blatter & Marcel Hermans, COMA

45

All the Building Is a Stage

DESIGNERS
Michael Meredith

SUBJECT
Marionette theatre

LOCATION
Cambridge, USA

PHOTOGRAPHERS
Florian Holzherr
www.architekturfoto.net
Michael Vahrenwald

WRITER
Shonquis Moreno

Designer Michael Meredith about the project
Sometimes I love it, sometimes I'm tired of it.

Comments on the published article
Positive, encouraging. The writer (Shonquis Moreno) did an amazing job.

Did the article have an effect?
Yes, incremental, although hard to know exactly.

Current projects
More experimental projects of a similar scale along with larger projects.

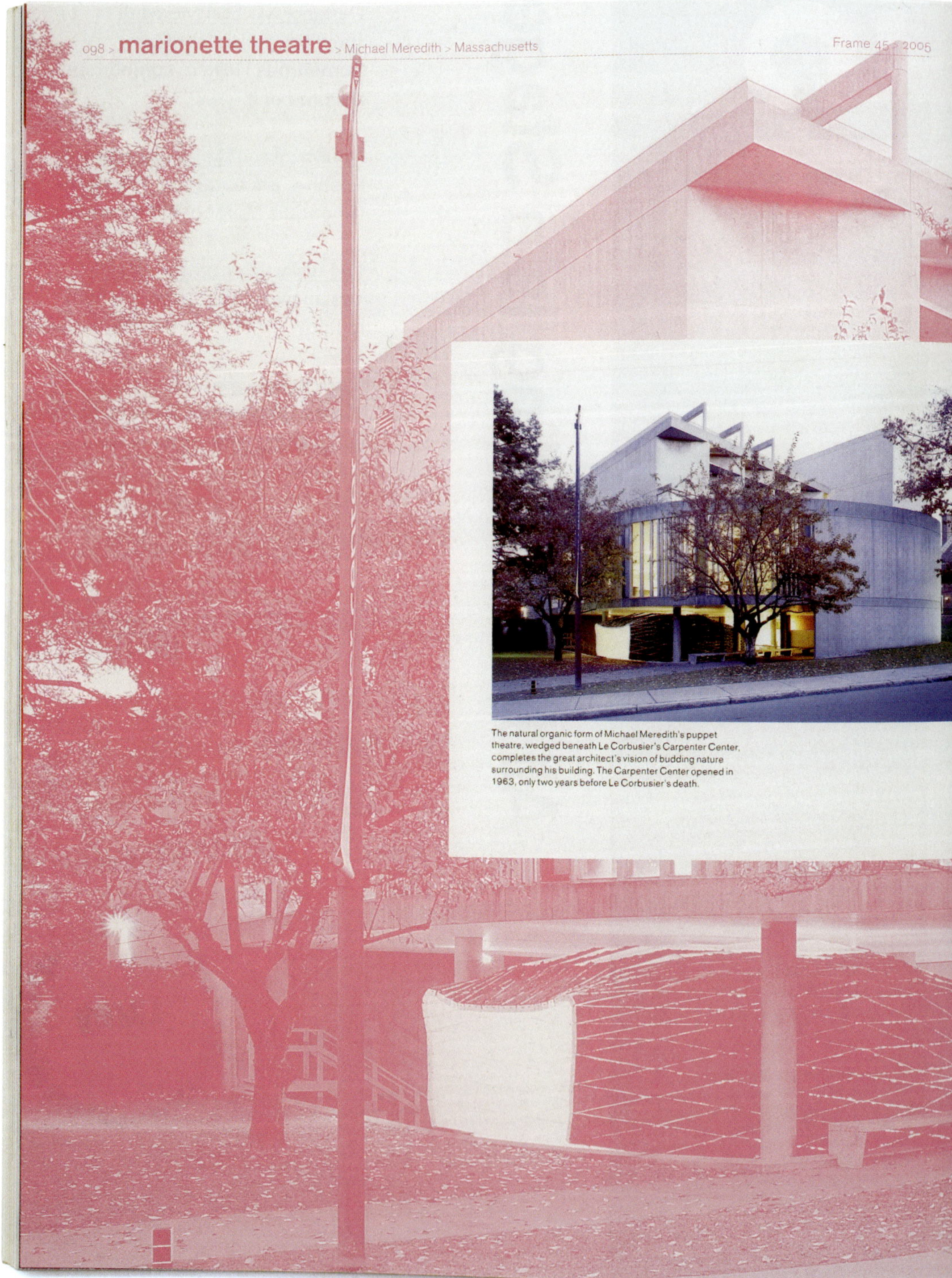

The natural organic form of Michael Meredith's puppet theatre, wedged beneath Le Corbusier's Carpenter Center, completes the great architect's vision of budding nature surrounding his building. The Carpenter Center opened in 1963, only two years before Le Corbusier's death.

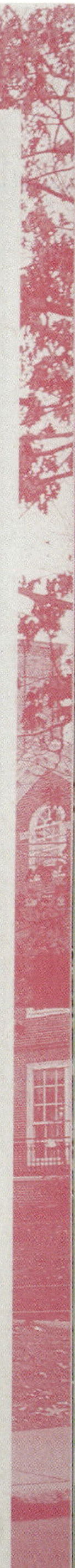

all the building is a stage

Text by Shonquis Moreno
Photography by Florian Holzherr
and Michael Vahrenwald

A tiny theatre by Harvard University professor Michael Meredith steps from beneath the shadow of its Le Corbusier-designed site to find a life of its own.

Le Corbusier designed the Carpenter Center for the Visual Arts on the Cambridge campus of Harvard University in 1963. It would prove to be the Swiss architect's only major North American building and one he hoped would serve as a summation of his design philosophy. Although the Carpenter Center did not live up to his expectations, it remains an iconic building and a reminder to the art students who toil within that envisioning something and making it are two very different things. For a short time last autumn, an amorphous theatre made of plastic and moss, a design by Harvard professor Michael Meredith, grew up beneath the broad cylinder and unadorned concrete columns of the Carpenter Center's Quincy Street courtyard. The temporarily erected, 75-square-metre pavilion became the site of six performances of a marionette 'musical' written by French conceptual artist Pierre Huyghe, and performed and filmed during a single evening late last November. The musical explored the disappointments and serendipities of the creative process, in general; and Huyghe's and Le Corbusier's disappointments and serendipities, in particular. While certainly a creature of this context, Meredith's theatre – like Pinocchio, who became a real little boy – is ambitious enough to merit a life of its own.

Huyghe made the Carpenter

WRITER
Shonquis Moreno

PHOTOGRAPHERS
Florian Holzherr
Michael Vahrenwald

DESIGNER
Michael Meredith

LOCATION
Cambridge, USA

SUBJECT
Marionette theatre

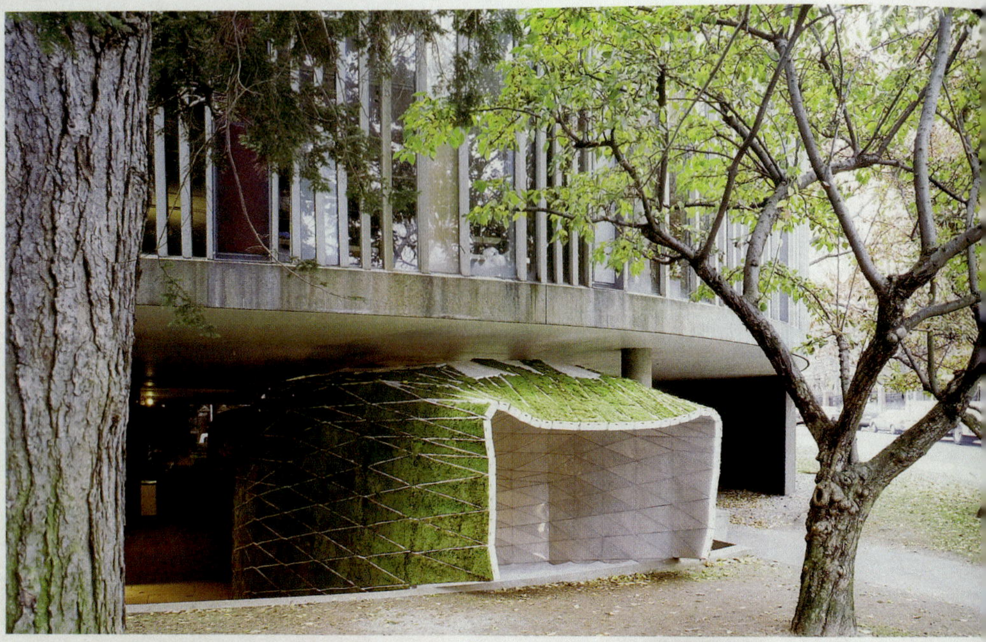

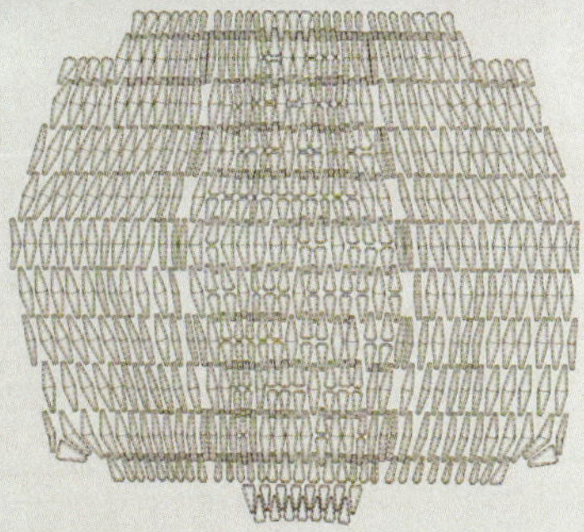

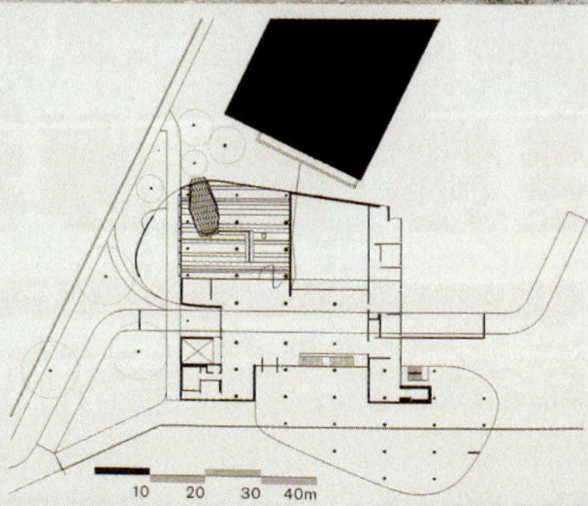

Floor plan: The Carpenter Center is raised on columns. Here, the iconic ground plan includes a small capsule-shaped structure positioned beneath the concrete structure.

Center a character – a puppet replete with moving parts – in his play. Meredith's theatre, with its flesh-and-bone (organic/synthetic) construction, also resembles a living thing. He designed it on a budget of $50,000 and assembled it with the help of students. Made from 506 diamond-shaped, heat-bent, polycarbonate panels that, bolted together, form a monocoque shell, its skin becomes its skeleton. CAD (and Rhino software, for the most part) allowed the designer to give each panel a unique shape, thus ensuring the most efficient use of the broad sheets from which the panels were cut. The network of plastic ribs formed by these panels is 7.6 centimetres deep and 4.6 metres wide (at the widest point), distributing force over the entire surface. Foam inserts stiffen each panel and provide an exterior to which living moss can adhere. From without, then, the provisional theatre resembled a growth of spongy green, a renewal of life beneath the naked concrete masonry and brute maths of Le Corbusier's building.

Whereas the exterior is nearly 'prehistoric', as Meredith puts it, the interior is 'super-modern'. The alternately reflective and light-refracting white-plastic panels form the flip side of the moss. 'The shiny geometry of the diamond panels gives an amazing scalelessness to the interior,' says Meredith. 'It looks like music to me.' Broad, shal-

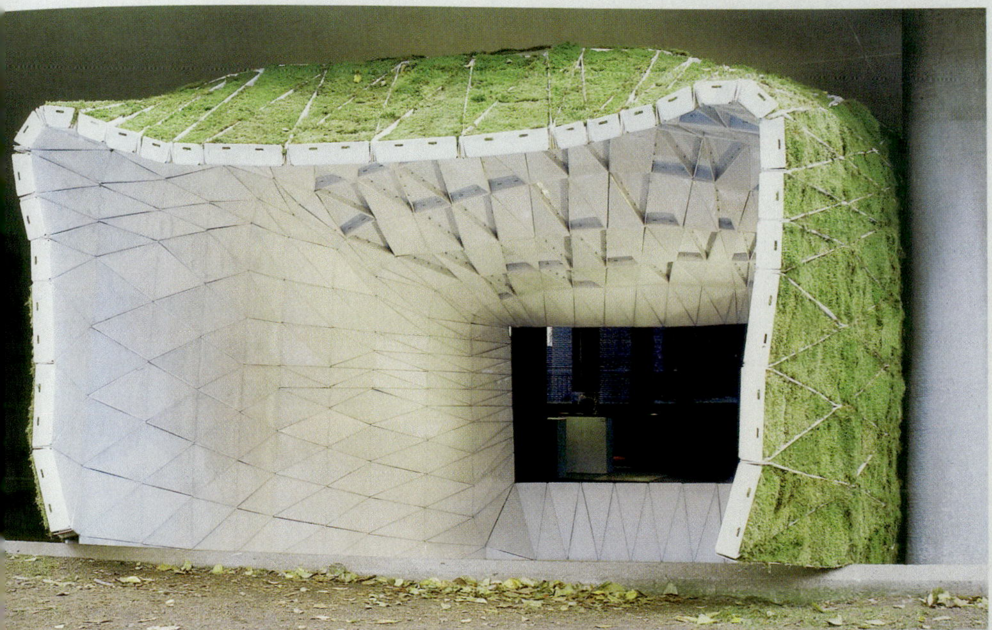

An irregular pattern of interlocking triangles and diamonds seems to have 'grown' into this corridor-cum-theatre. The moss-green colour of the exterior adds to the organic aesthetic.

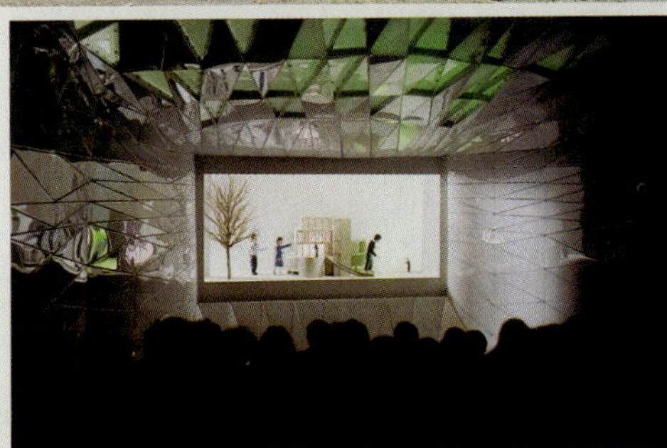

'The project is both entirely natural and entirely artificial.'
Detlef Mertins, Professor of Architecture, University of Pennsylvania.

lowly fluted tiered seating runs the width of the space. At the end that gave on to Quincy Street last autumn, the theatre's softly tapered mouth framed an existing tree, creating an almost biblical perspective. At the opposite end, visitors faced the brilliantly flat-white stage or, when the stage was removed, a view of the lobby inside the Carpenter Center (still home to Harvard's painters, filmmakers, sculptors and other artists), where the daily activities of the arts institution are played out: a performance in its own right. During the day, green peeked out from beneath the Center. At night, the plastic conducted light, disappearing completely to leave a glowing grid of green that seemed to hover there despite the weight above it. 'Many of the architects who have an interest in "organic" architecture have used digital media to explore complex shape-making,' says University of Pennsylvania architecture professor Detlef Mertins, who previously worked with Meredith at the University of Toronto. 'This work is rarely done with irony. Michael's is. His theatre seems to suggest that it's all well and fine to want to be integrated with nature and to follow nature's leads, but we need to recognize that our models of nature remain limited and artificial. The project is both entirely natural and entirely artificial.'

Meredith agrees. The interior is angular, hard and man-made;

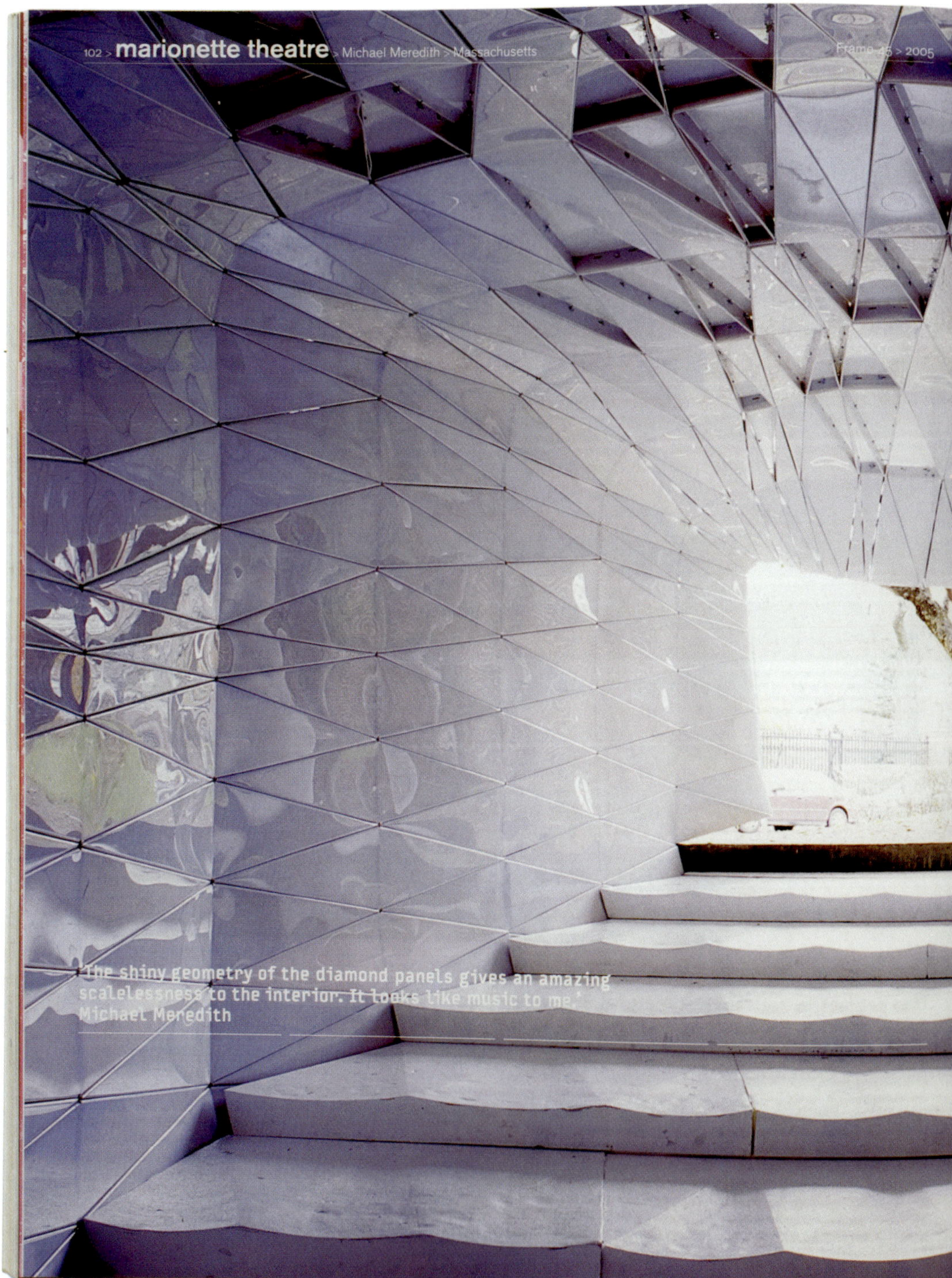
102 > marionette theatre > Michael Meredith > Massachusetts
Frame 45 > 2005
'The shiny geometry of the diamond panels gives an amazing scalelessness to the interior. It looks like music to me.'
Michael Meredith

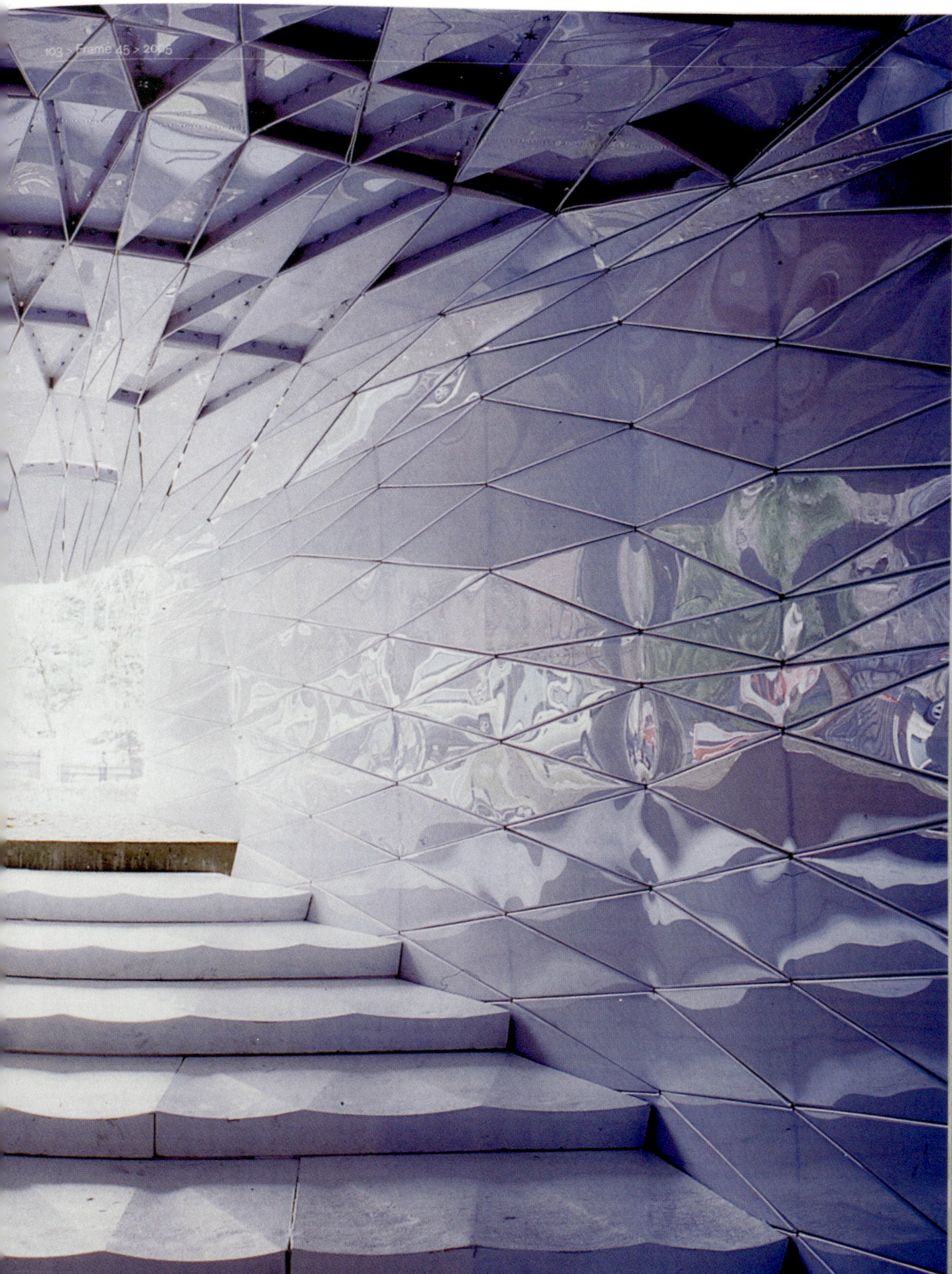

SUBJECT
Marionette theatre

LOCATION
Cambridge, USA

DESIGNER
Michael Meredith

PHOTOGRAPHERS
Florian Holzherr
Michael Vahrenwald

WRITER
Shonquis Moreno

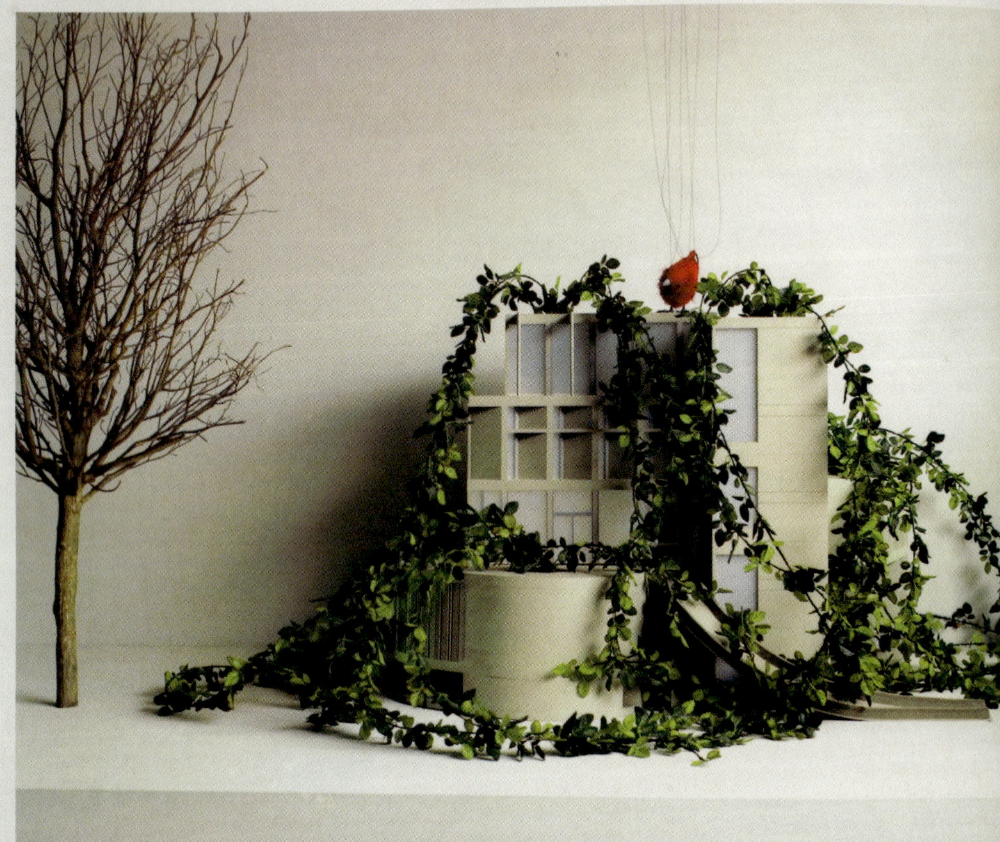

Opposite: 'The characters here include Le Corbusier, Mr. Sert, Mr. Sekler, Mr. Pierre, Ms. Linda, Mr. Scott and a Mr. or Ms. bird, a non-gendered puppet representing all those things that birds exist to represent. A black shape is Mr. Harvard. A dean of deans. This play will show the Carpenter Center in different phases along with trees, raspberries and fog and snow.' - Prologue by Liam Gillick from the puppet musical 'Working Title'.

'Scene XI: What the Birds Bring. A bird leaves a seed on the Carpenter Center's roof. Blackberry vines overwhelm the building.' - Programme notes from the puppet musical 'Working Title'

the exterior is curved, soft and natural. 'Taken together,' says Meredith, 'the two sides collapse the synthetic and organic into a single structural surface.'

This collapse of opposites both complemented Huyghe's performance and succeeded without reference to the original mandate. The extremes of the theatre's design highlighted the imposition of the natural world onto the built one and, by implication, the reverse: a phenomenon at the heart of any creative process. The play and the pavilion alluded to Le Corbusier's abandoned plans for a landscape programme that was to surround the complex with whatever greenery might grow from seeds dropped by birds. This was to become one of the architect's several failed ambitions in the course of the project, but Meredith's modernist interior and earthy exterior bring us back to the origins of the Carpenter Center as imagined by Le Corbusier. This nature-versus-nurture dialogue meshes with Huyghe's script and the building's history, but is a concept communicated by the design of the theatre in a way that is independent of both.

Theatre and play pivot on the contradictions and challenges that context, briefs and circumstances – historical or architectural – bring to bear on creativity. Squeezed beneath and connected gently to the mother building whose story it helped recount, the theatre emphasized

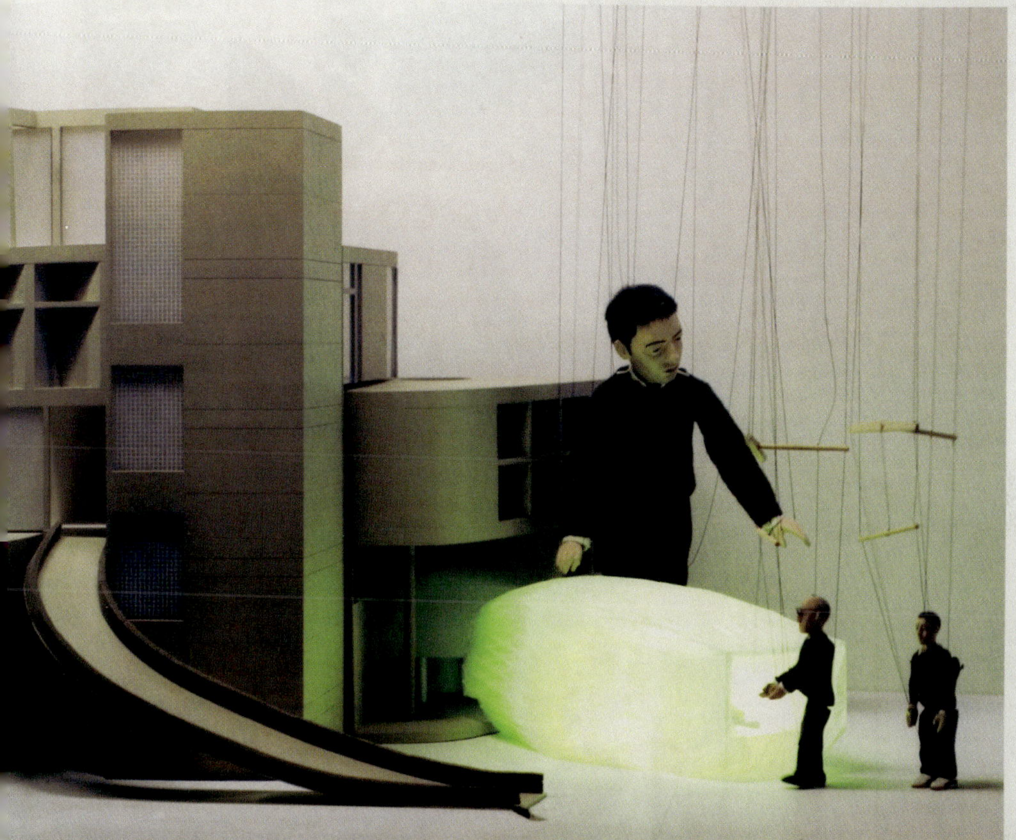

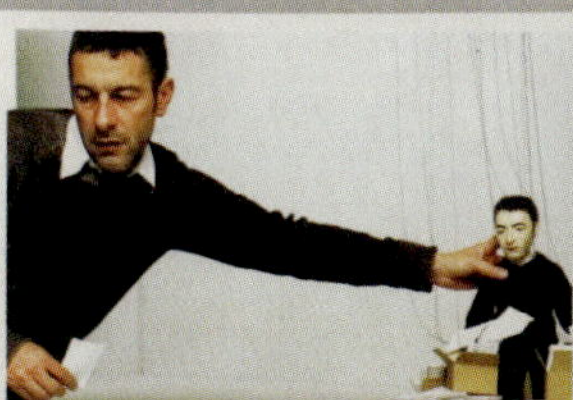

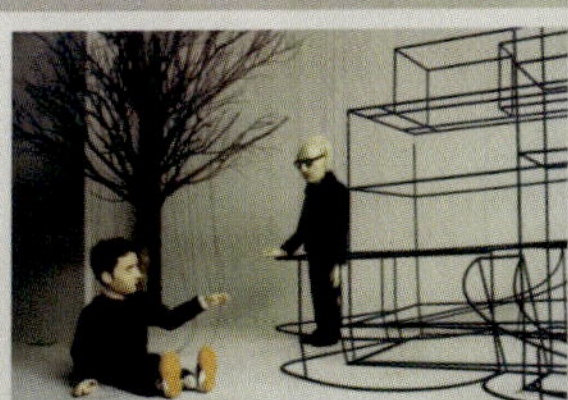

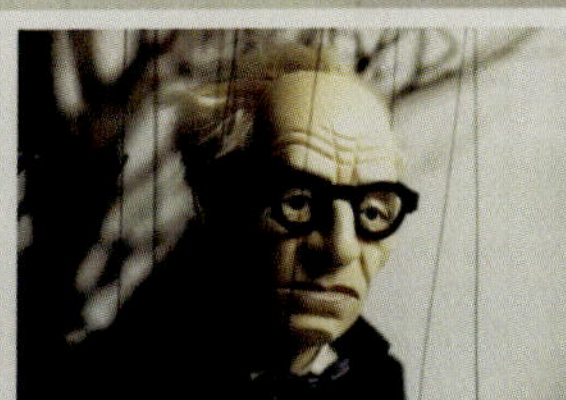

the failures and virtues of its host, such as the jettisoning of the landscape plan and the circumstances leading to the sunken courtyard. Le Corbusier had intended to build the courtyard in which the theatre was installed on grade, thus making it directly accessible from the street. Local building regulations forced him, however, to shorten the building. Instead of redesigning everything, he pushed the structure down nearly a metre, enshrining his compromise, and presumably his disappointment, in this truncation.

By virtue of its conspicuous visual dissonance, Meredith's theatre amplifies the practical challenges faced by Le Corbusier in executing his vision. Hence it is ironic that, although the theatre was designed to be assembled and disassembled for installation in various places, to date it has been merely dismantled and put into storage: a loss for the Cambridge community, at the very least. Meredith, however – perhaps bolstered by the knowledge that Le Corbusier's efforts were taken out of storage and re-examined, albeit momentarily – sounds optimistic. 'There is a chance,' he says simply, 'that the theatre will be reborn.'

SUBJECT
Marionette theatre

LOCATION
Cambridge, USA

DESIGNER
Michael Meredith

PHOTOGRAPHERS
Florian Holzherr
Michael Vahrenwald

WRITER
Shonquis Moreno

492-493

September/October 2005

Issue 46

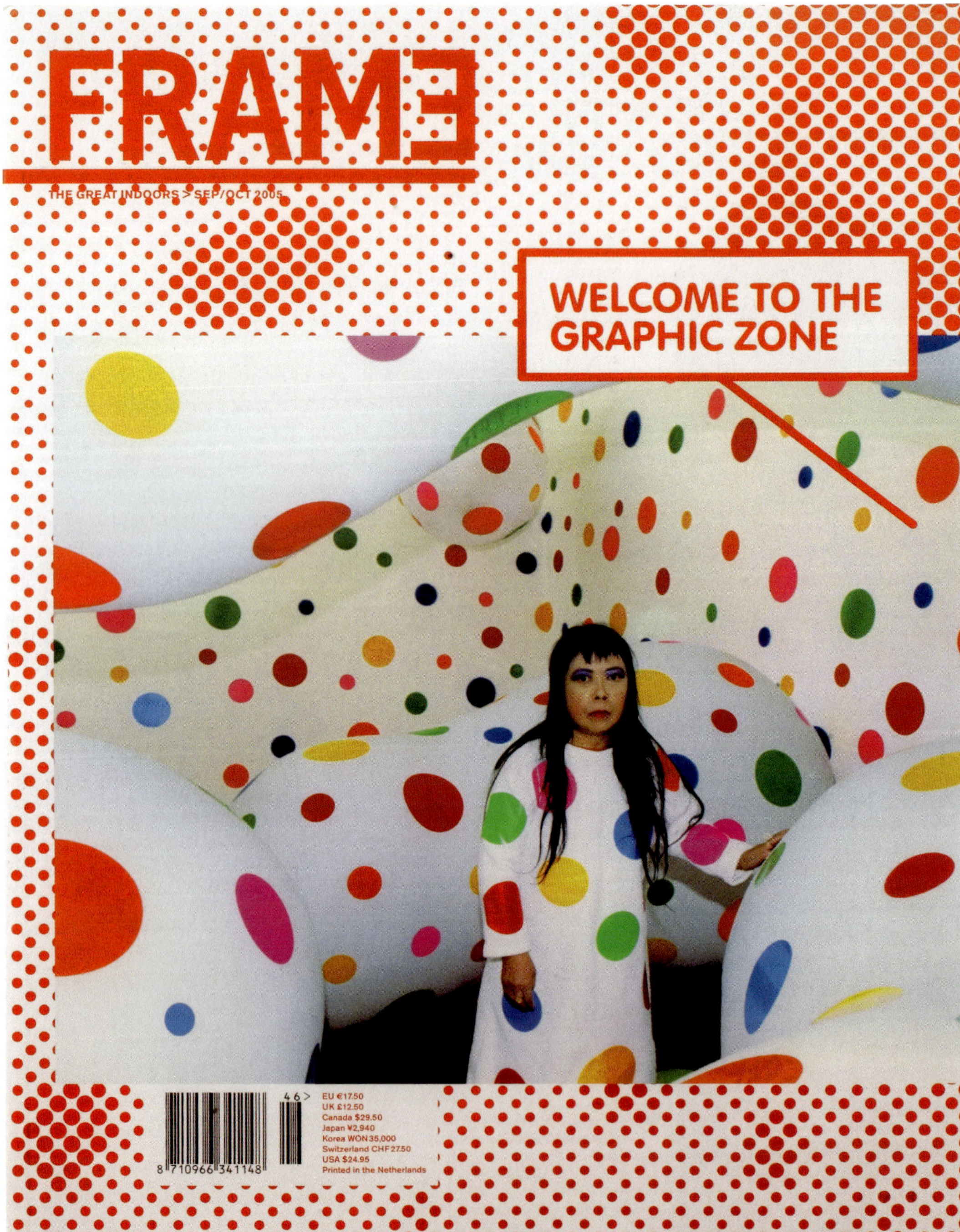

ART DIRECTORS
Cornelia Blatter & Marcel Hermans,
COMA

46

Garden of Eartly Delights

<u>Designer Eli Sudbrack (Assume Vivid Astro Focus) about the project</u>

Really nice. Into it. Honoured.

<u>Comments on the published article</u>

The best written so far on AVAF.

<u>Did the article have an effect?</u>

I don't know. I guess it did. One of our collaborators, Aleksandra Mir – she contributed a quote on us that appeared in the article – was immensely surprised at the quality of the piece. She told us she learned things about AVAF she wasn't aware of.

<u>Current projects</u>

Solo show, Massimo de Carlo, Milan, Italy // Ecstasy group show, curated by Paul Schimmel, MOCA, Los Angeles, USA // Solo show, Hiromi Yoshii Gallery, Tokyo, Japan // Infinite Painting group show, curated by Francesco Bonami, Villa Malin, Italy // Solo show, Galeria Triangulo, São Paulo, Brazil // Solo shows at John Connelly Presents and Deitch Projects, New York, USA // Launch of avaf-designed bags for Le Sport Sac.

DESIGNERS
Assume Vivid Astro Focus

SUBJECT
Portrait of Assume Vivid Astro Focus

LOCATION
New York, USA

PHOTOGRAPHERS
Mauro Restiffe
Tom Powel
Dean Sameshima
C.M. Guerio
Mariano Costa Peuser
Andy Keate

WRITER
Shonquis Moreno

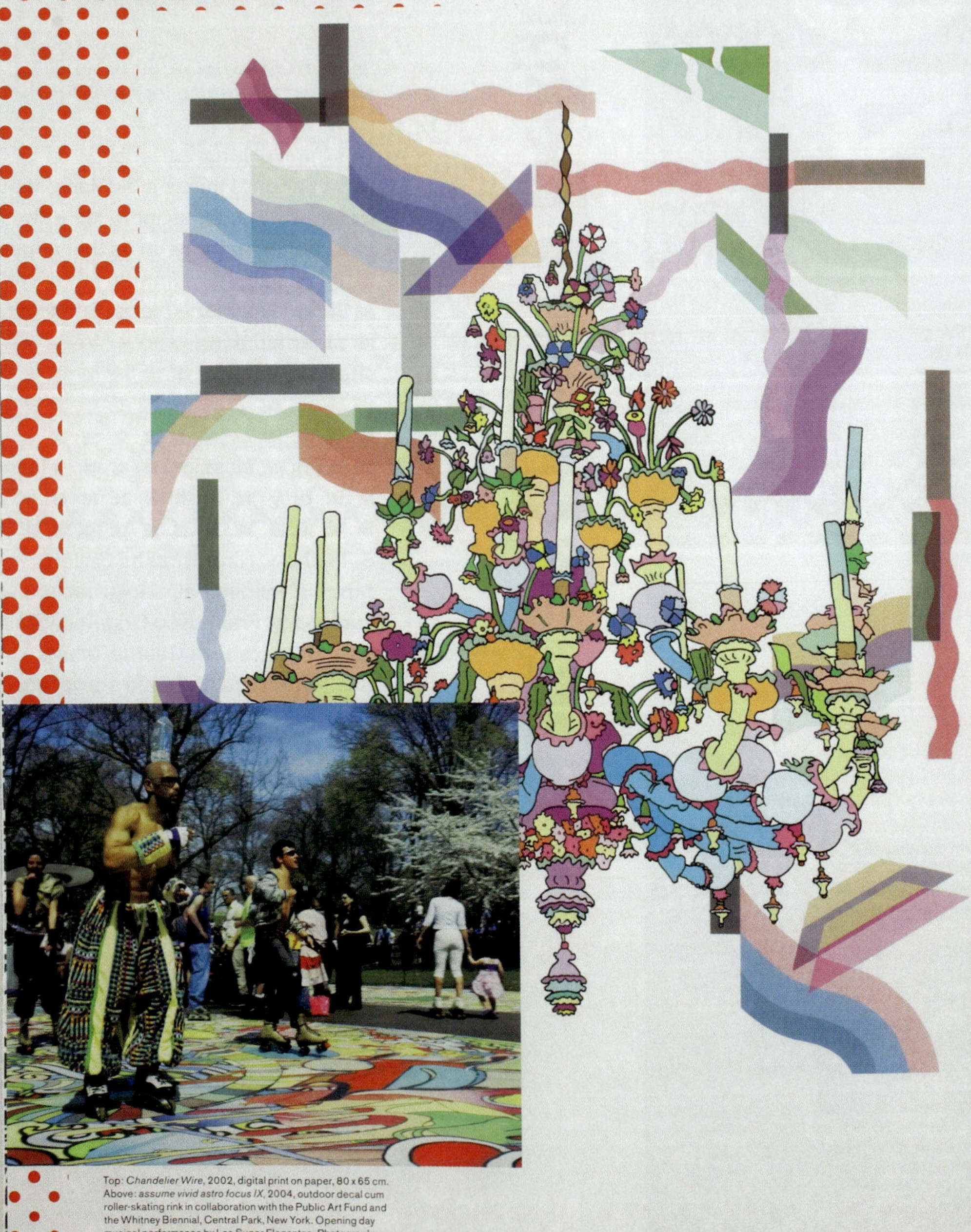

Top: *Chandelier Wire*, 2002, digital print on paper, 80 x 65 cm. Above: *assume vivid astro focus IX*, 2004, outdoor decal cum roller-skating rink in collaboration with the Public Art Fund and the Whitney Biennial, Central Park, New York. Opening day musical performance by Los Super Elegantes. Photography by Mauro Restiffe.

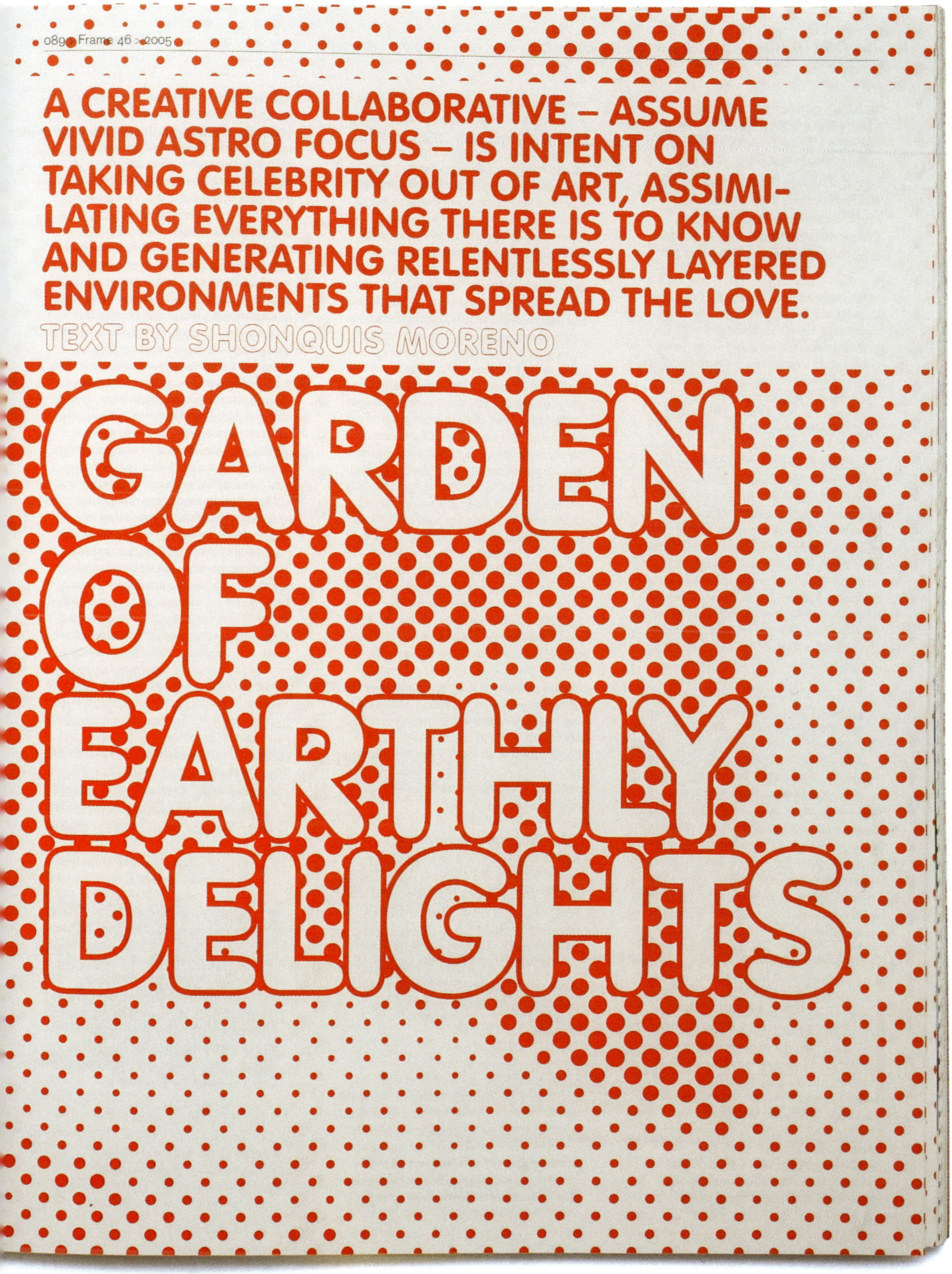

A CREATIVE COLLABORATIVE – ASSUME VIVID ASTRO FOCUS – IS INTENT ON TAKING CELEBRITY OUT OF ART, ASSIMILATING EVERYTHING THERE IS TO KNOW AND GENERATING RELENTLESSLY LAYERED ENVIRONMENTS THAT SPREAD THE LOVE.

TEXT BY SHONQUIS MORENO

GARDEN OF EARTHLY DELIGHTS

SUBJECT
Portrait of Assume Vivid Astro Focus

LOCATION
New York, USA

DESIGNERS
Assume Vivid Astro Focus

PHOTOGRAPHERS
Various, see page 493

WRITER
Shonquis Moreno

'2. Keep collecting coloring books from eBay. Look for: H.R. Pufnstuf, American Wildflowers, Space: 1999, New Kids on the Block, *Flipper, Jaws, Hart to Hart, Moonlighting, The Incredible Hulk*, American Architecture, Chips, Prismatic Designs, Bugaloos;' – From avaf To Do List III

In his caffeinated, stream-of-consciousness 'to do' lists for each project, the founder of New York-based art collaborative assume vivid astro focus (avaf) catalogues a catholic and highly eclectic range of research topics that influence the work of the group. These topics leap from *The Six Million Dollar Man*, Lazlo Maholo-Nagy, Pierre Huyghe, Felix Gonzalez-Torres, Dada and Aubrey Beardsley to Devo, U.K. artist Selfish Cunt, Memphis-movement mover and shaker Ettore Sottsass, and Land Art. One early list juxtaposes *Flipper* and *Jaws*, set apart by the mere breath of a comma. Each enumerated item is separated from its neighbour not by a full stop but by a semicolon; even when full stops do appear, there is no sense of a halt, only a reluctant pause. The teeming, Kool-Aid-coloured brain that generates these lists is that of Rio-born artist Eli Sudbrack. Sudbrack, who closes his emails with 'love, eli', demonstrates a combination of transparency and opacity that seems alternately naive and savvy. Because his intention was to make avaf a rubric for collective creativity, he will no doubt be disappointed (once again) to see his name published here. It is difficult, however, to talk about avaf without talking about Sudbrack, because he is not just one of the avaf artists; he is the group's curator and producer, handpicking his collaborators and choosing other artists' work to show within avaf pieces. Despite the reluctant centrality of the group's founder, work by avaf is increasingly the product of an ever-shifting roster of artists and a synthesis of innumerable high and lowbrow influences. Readily recognizable for its flat colour, visual overload and pop culture references, the work includes, but is by no means limited to, wall-size installations and wallpapers that do not repeat, erotic prints and prints of chandeliers, videos, stickers, decals, tattoos and entire interiors that sometimes resemble a 1960s animation brought to life. (Think montages of *Yellow Submarine* with hard edges and soft porn.)

The recipe for avaf visuals is elaborately interleaved and consists of elements that begin life as pencil sketches, silk-screen prints, tears from magazines or books, swatches of fabric, printouts from Google image searches, photographs, Xeroxes of doodles – all scanned into avaf's electronic image archive. Plucking these graphic scraps from the computer library, avaf collaborators work each image mercilessly, overdrawing by hand or in Photoshop and Illustrator. Even though a lot of figurative elements are used, avaf artists sometimes layer (or 'frankenstein,' as one artist put it) them into abstraction. It's a process in which the coillective takes diverse parts and gives them a synergistic new whole.

The same is done with people. Powerful in most of us, the compulsion to connect to others (and connect others to one another) becomes a monomania in Sudbrack. He mixes creative people – VJs, DJs, musicians, painters, sculptors, graphic designers, carpenters, tattoo artists – sometimes working with them in the studio, sometimes emailing imagery back and forth, sometimes selecting a finished piece, old or new, to pull into an avaf show. The degree of collaboration, the number of collaborators and the nature of the collaboration changes constantly. Sudbrack, who determines the starting point of all three, considers the publication of his name a failure of the work. It is tempting, despite the prettiness of avaf pieces, to suspect that this is the most intriguing question asked by the collaborations: do we need to talk about the artist to talk about the art? For Sudbrack, at least, both the starting point and authorship are irrelevant; the process of research, collaboration and accretion is everything.

'10. Wallpaper: Reference: Renaissance Unicorn Tapestry from Musée du Moyen Age, Cluny, Paris: millefleurs background on red, old and new. The red background should be a deep shade that is both attractive and difficult to look at. Make everything look really busy, tight and overwhelming.' – From avaf To Do List III

The title of avaf's *buttwallpaper* – a collection of pencil-sketched vignettes that could have come from a chinoiserie pattern if it depicted a more conventionally banal subject – speaks eloquently for itself. Even when avaf uses only a small section of a room (and eschews nudity), its wallpapers are so exuberant that they escape themselves, and the wall. Using a severe perspective or wrapping a corner of the room, they create the illusion that the wall can be walked into or that a fantasy world is bleeding through ceiling and floor. In fact, the work builds its own room within the room but without being imposed on it. The group produces wallpapers for specific interiors, their patterns determined in part by the architecture of the exhibition space. Sudbrack avoids patterns that repeat; he wants the compositions to conform to existing or imposed architectural elements. 'I've always thought of architecture and design as necessary knowledge for every person,' he says. 'These practices are ever-present in our lives and even dictate them. People should be able to model space according to their needs and not become part of a gentrified, globalized society that is always trying to flatten us towards one general view, one general behaviour, one Starbucks coffee. Architecture and design are interesting political terrains that are not explored enough; instead they are made expensive and unapproachable.' The opposite of precious, avaf prints are destroyed in the process of dissembly, since they are usually printed on self-adhesive vinyl, like a sticker, as either a photo or a grainier billboard print. For this reason, collectors buy not just a physical piece but its digital file and have the right to reprint at will. Always, the images are combined in rococo excess. 'For me, the overload represents the amount of knowledge I want to absorb,' says Sudbrack. 'It's the amount of information, or really a fraction of it, that interests or is available to me. The "to do" lists are a good example of that. In a way, they are about obsession – to not forget anything – but also about some kind of naiveté in thinking I could ever absorb it all.'

'8. Work on skylight project and graffiti topiary for MacCarren swimming pool renovation. Document most graffitis at the pool and make drawings of them. Trace water and sea motifs from Tibetan Thangka paintings and draw sea life (octopus, fish, algae, etc.). Mix all that together;' – From avaf To Do List III

For collector Rosa de la Cruz, avaf created an artificial garden of wallpapers; decals mounted on plywood, sintra board or Plexiglas; and elements that might be sculpture, seating, lighting, performance space or all four.

The project, along with two more recent shows at the Tate and the Frieze Art Fair in London, demonstrates avaf's thoughtful, interventionist take on architecture. Originally, the installation was to be a roller disco – avaf likes to create spaces that are 'activated' either by viewers or performances – but the client balked. It included a contoured grey wall on wheels called *Inverted K (William Morris)* and a white spiral staircase that had also been installed at the 2004 Whitney Biennial. At the top of the stair, visitors discovered more artwork and a fresh perspective on the room and building. Sudbrack also cut into the wall to expose three sealed windows, and then designed a wallpaper based on the incision. 'For me, many avaf pieces hint at this stupid dichotomy between art and design and furniture and architecture,' says Sudbrack. 'We would definitely like these boundaries to be blurred, of course. We don't trust the fact that all these practices are contained in their safe, separated havens. But the truth is that artwork is usually trapped in art environments that, stupidly, still demand respect and impose restrictions or censorship for fear of lawsuits and paranoid health and safety issues. At the same time, there's freedom to conceive whatever you want to conceive without needing to follow market or industrial demands. Nevertheless, you end up speaking to a restricted crowd of people. And that has been bugging me more and more.'

'16. Next music video: use bootleg version of Yoko Ono's *Walking on Thin Ice*...Work on it and extend 'AiAiAiAiAi' section almost infinitely.' – From avaf To Do List III

Set outdoors, *assume vivid astro focus IX* was a project installed in Central Park last summer; it consisted of a large sticker that formed a skating 'rink' affixed to the asphalt and a DJ booth that could accommodate skaters' equipment and fold up for nightly storage. 'Besides always making eye-popping, thoroughly researched nerd cathedrals out of cultural detritus and art-historical grabs, Eli is a great mover of people,' says New York curator Katherine Grayson who worked with avaf on a 2003 show at Deitch Gallery that included two site-specific wallpapers, a neon sign that read 'Contagious' and a tattoo parlour. 'The Central Park sticker, in particular, was so successful because he took this weird community and made them this great community-building piece that celebrated their colourful weird exuberance. It made everyone *feel* good and he *meant* it.'

Sudbrack is a sincere man and a gay man who, being both Brazilian and astute, describes America as puritanical. In 1999, several years after earning a film degree in his native country, influenced by the work of Cindy Sherman and Nan Goldin, he moved to New York to study photography at the International Center of Photography. 'For me, Eli is the total cultural experience,' says artist Aleksandra Mir, whose work was folded into the de la Cruz project. 'Always energizing and madly intense, he feeds off and gives to his friends at 10,000 miles an hour. I admire how desperate he is to reach the logical conclusion of anything he has initiated, even if it means banging his head into the wall of the typical client, who is always turned on by his energy in the beginning but eventually restricts his madness due to all sorts of bureaucracy and safety regulations. An artist of his intensity certainly needs to be tamed, but I'll trade the rhetoric of any scared client for one hour in Eli Sudbrack's maxed-out presence any day.'

The superlatives abound. Even Sudbrack will confess to having an inordinately optimistic bent that is evident in the pseudonym: 'The reason I use "assume vivid astro focus" is this Utopian idea that it's advantageous for other people to actually "assume" avaf's perspective,' he says. 'I want other people to be avaf. I want people to contaminate other people. Genius is in all of us.'

'PEOPLE SHOULD BE ABLE TO MODEL SPACE ACCORDING TO THEIR NEEDS AND NOT BECOME PART OF A GENTRIFIED, GLOBALIZED SOCIETY THAT IS ALWAYS TRYING TO FLATTEN US TOWARDS ONE GENERAL VIEW, ONE GENERAL BEHAVIOUR, ONE STARBUCKS COFFEE.' ELI SUDBRACK

SUBJECT Portrait of Assume Vivid Astro Focus

LOCATION New York, USA

DESIGNERS Assume Vivid Astro Focus

PHOTOGRAPHERS Various, see page 493

WRITER Shonquis Moreno

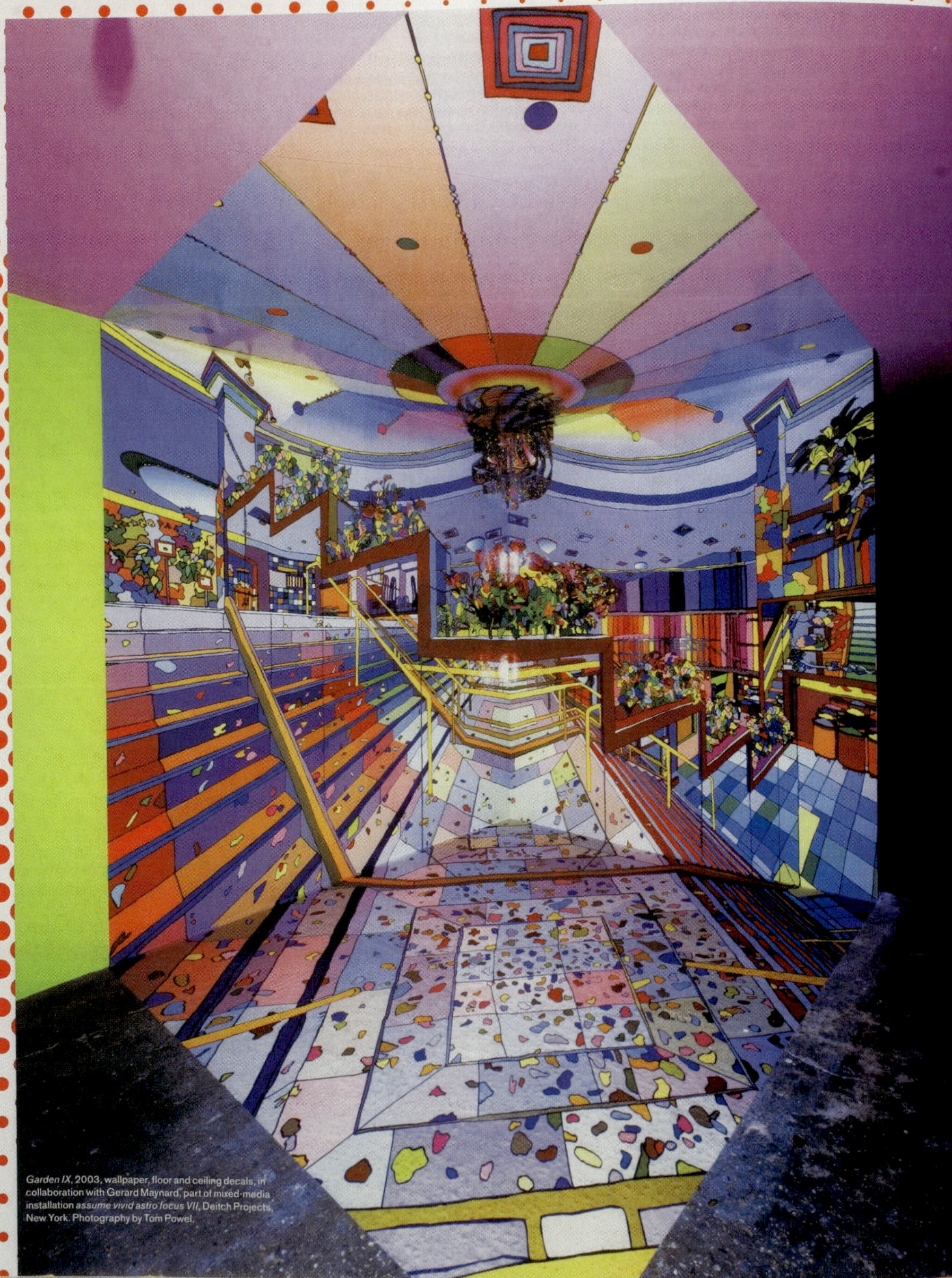

Garden IX, 2003, wallpaper, floor and ceiling decals, in collaboration with Gerard Maynard, part of mixed-media installation *assume vivid astro focus VII*, Deitch Projects, New York. Photography by Tom Powel.

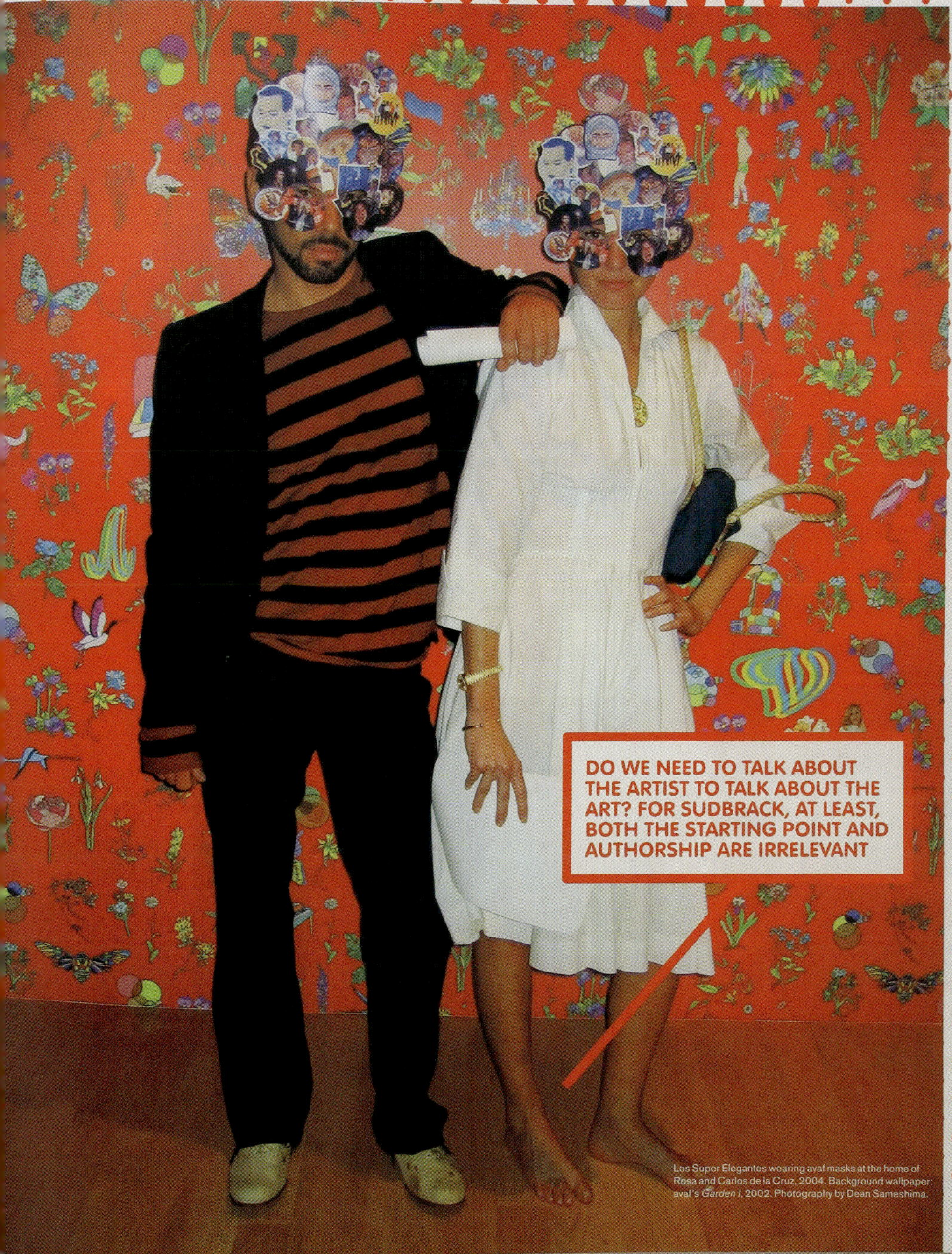

DO WE NEED TO TALK ABOUT THE ARTIST TO TALK ABOUT THE ART? FOR SUDBRACK, AT LEAST, BOTH THE STARTING POINT AND AUTHORSHIP ARE IRRELEVANT

Los Super Elegantes wearing avaf masks at the home of Rosa and Carlos de la Cruz, 2004. Background wallpaper: avaf's *Garden I*, 2002. Photography by Dean Sameshima.

SUBJECT Portrait of Assume Vivid Astro Focus

LOCATION New York, USA

DESIGNERS Assume Vivid Astro Focus

PHOTOGRAPHERS Various, see page 493

WRITER Shonquis Moreno

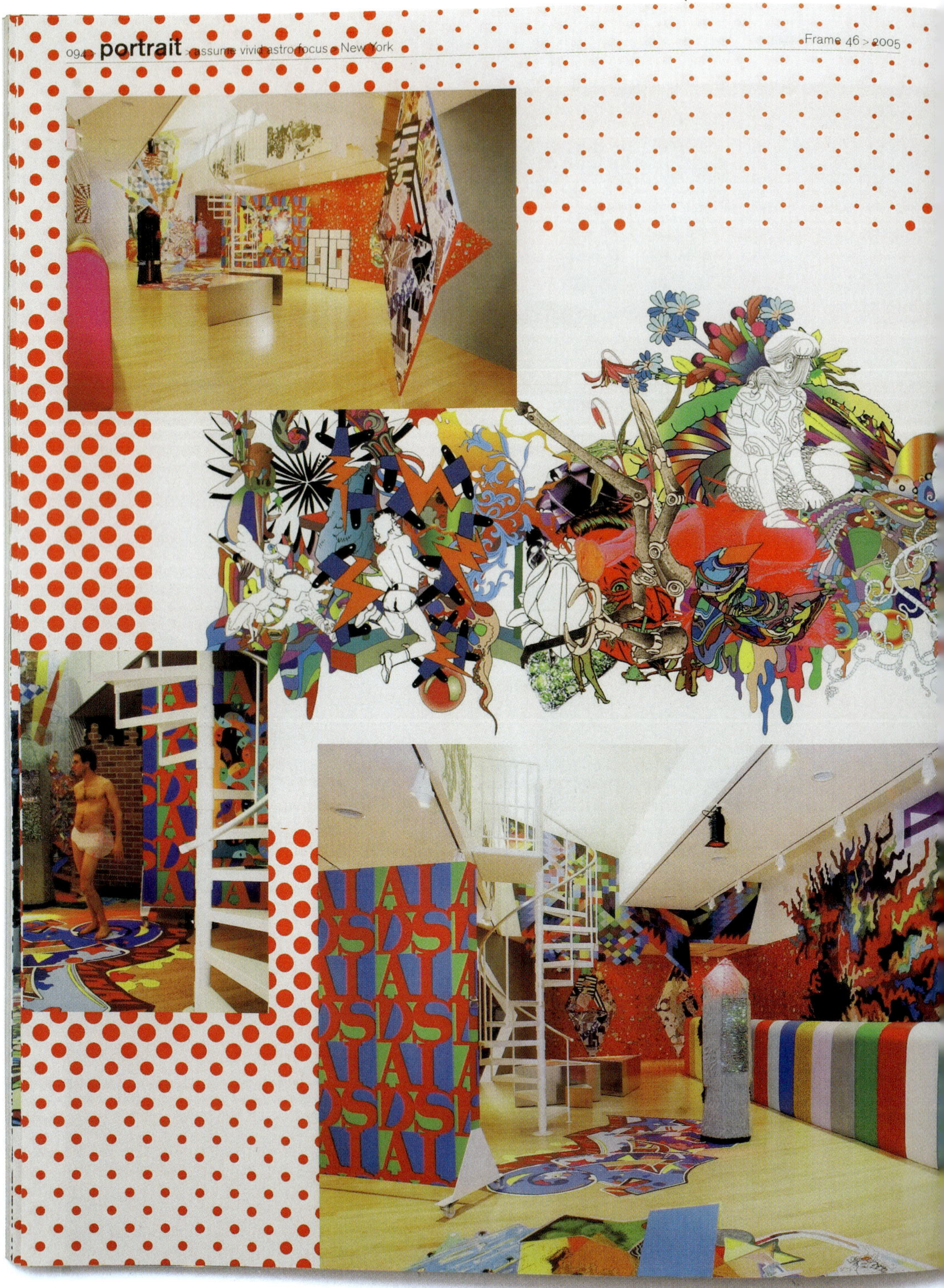
094 > portrait > assume vivid astro focus > New York
Frame 46 > 2005

PLUCKING GRAPHIC SCRAPS FROM THEIR COMPUTER LIBRARY, AVAF COLLABORATORS WORK EACH IMAGE MERCILESSLY, LAYERING (OR 'FRANKENSTEINING,' AS ONE ARTIST PUT IT) EACH FIGURATIVE ELEMENT INTO ABSTRACTION

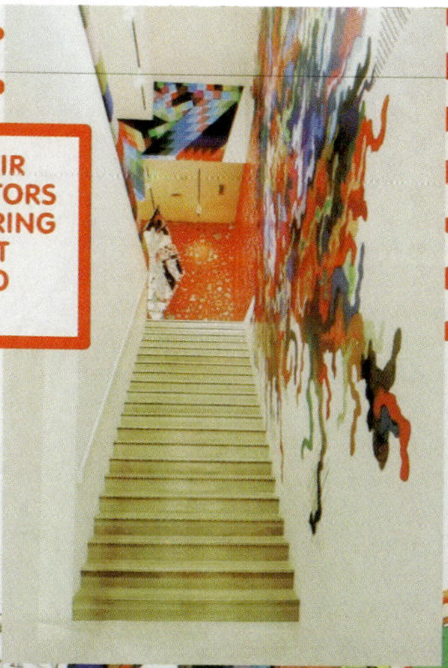

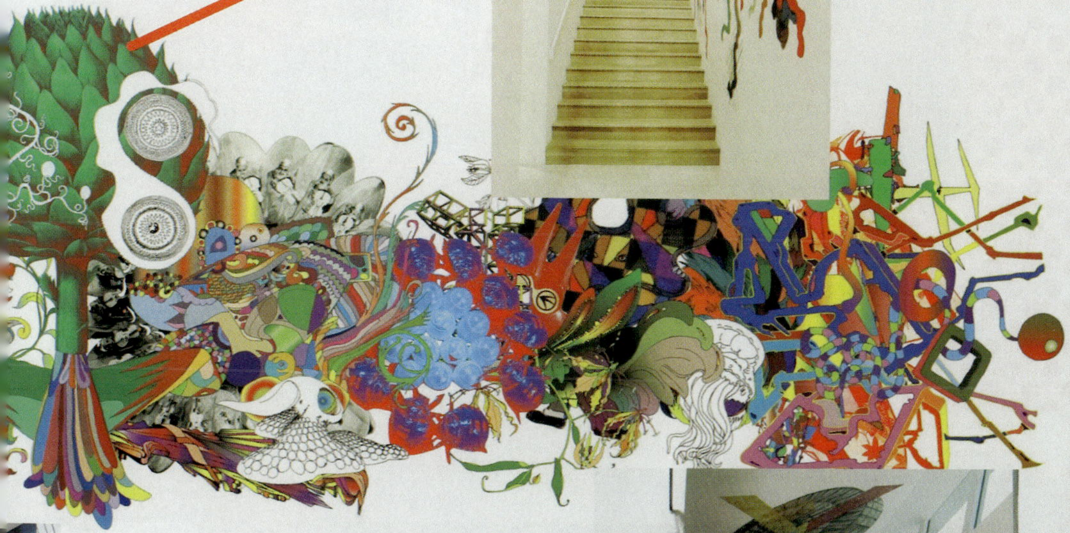

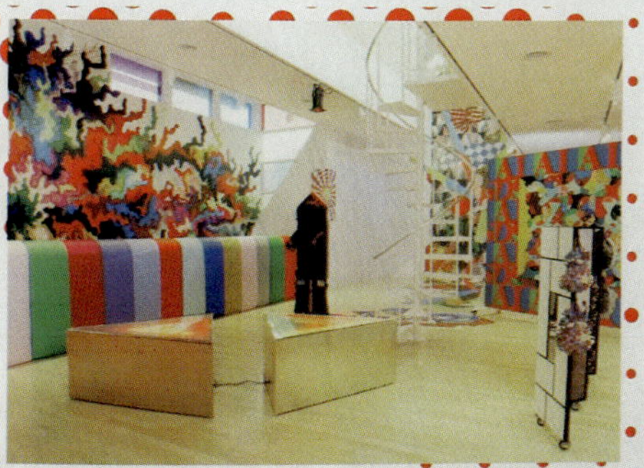

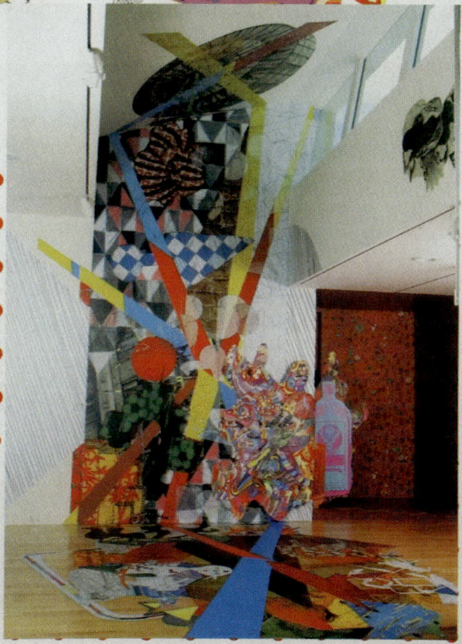

Centre: *Artichoke*, 2004, wallpaper, dimensions vary with installation.
Surrounding images: Various views of *assume vivid astro focus XI*. Photography by C.M. Guerio and (bottom right) Mariano Costa Peuser, courtesy of Rosa and Carlos de la Cruz.

SUBJECT
Portrait of Assume Vivid Astro Focus

LOCATION
New York, USA

DESIGNERS
Assume Vivid Astro Focus

PHOTOGRAPHERS
Various, see page 493

WRITER
Shonquis Moreno

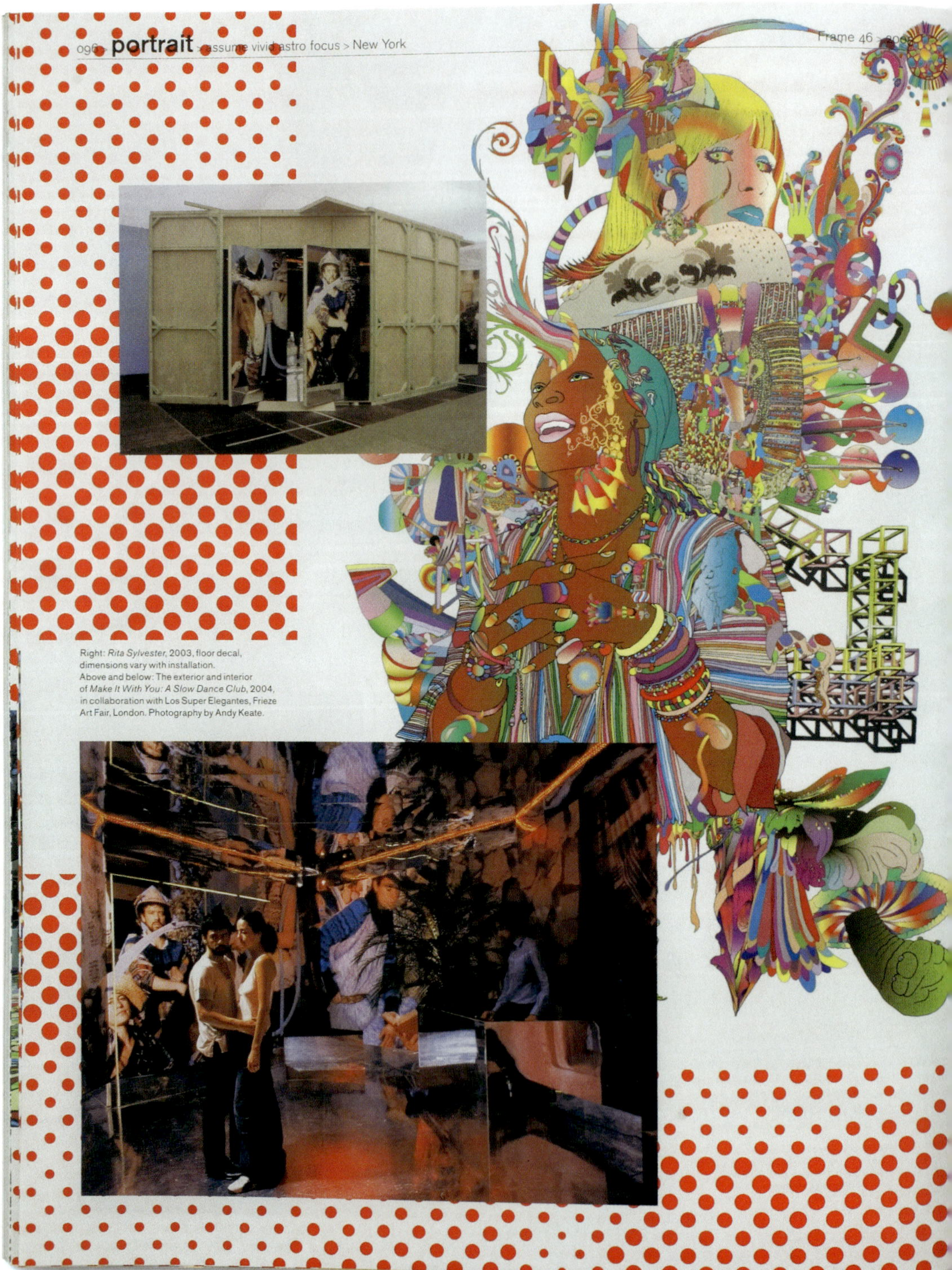

096 > **portrait** > assume vivid astro focus > New York

Frame 46 >

Right: *Rita Sylvester*, 2003, floor decal, dimensions vary with installation.
Above and below: The exterior and interior of *Make It With You: A Slow Dance Club*, 2004, in collaboration with Los Super Elegantes, Frieze Art Fair, London. Photography by Andy Keate.

'I WANT OTHER PEOPLE TO BE AVAF. I WANT PEOPLE TO CONTAMINATE OTHER PEOPLE. GENIUS IS IN ALL OF US.'
ELI SUDBRACK

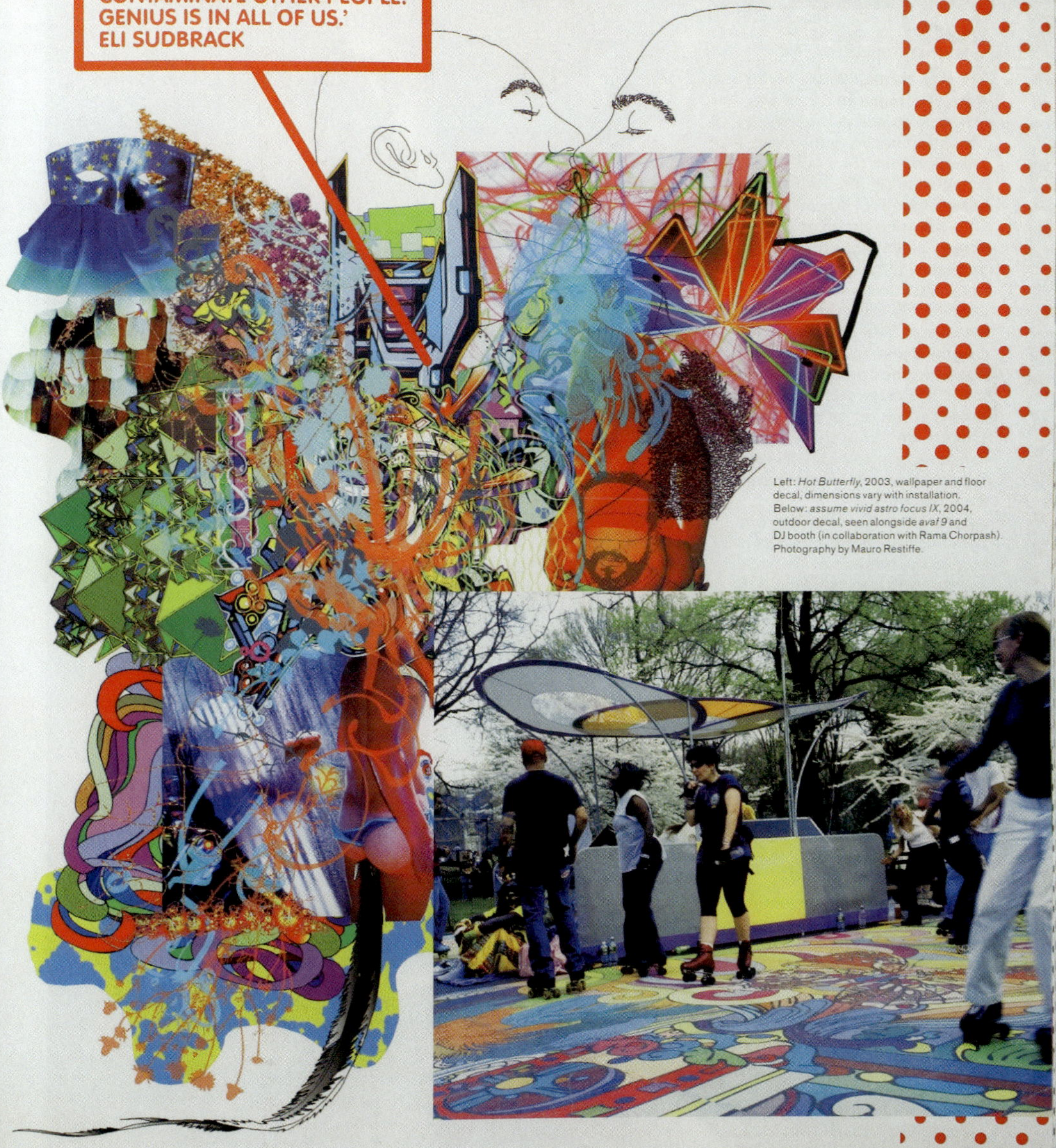

Left: *Hot Butterfly*, 2003, wallpaper and floor decal, dimensions vary with installation.
Below: *assume vivid astro focus IX*, 2004, outdoor decal, seen alongside *avaf 9* and DJ booth (in collaboration with Rama Chorpash). Photography by Mauro Restiffe.

SUBJECT
Portrait of Assume Vivid Astro Focus

LOCATION
New York, USA

DESIGNERS
Assume Vivid Astro Focus

PHOTOGRAPHERS
Various, see page 493

WRITER
Shonquis Moreno

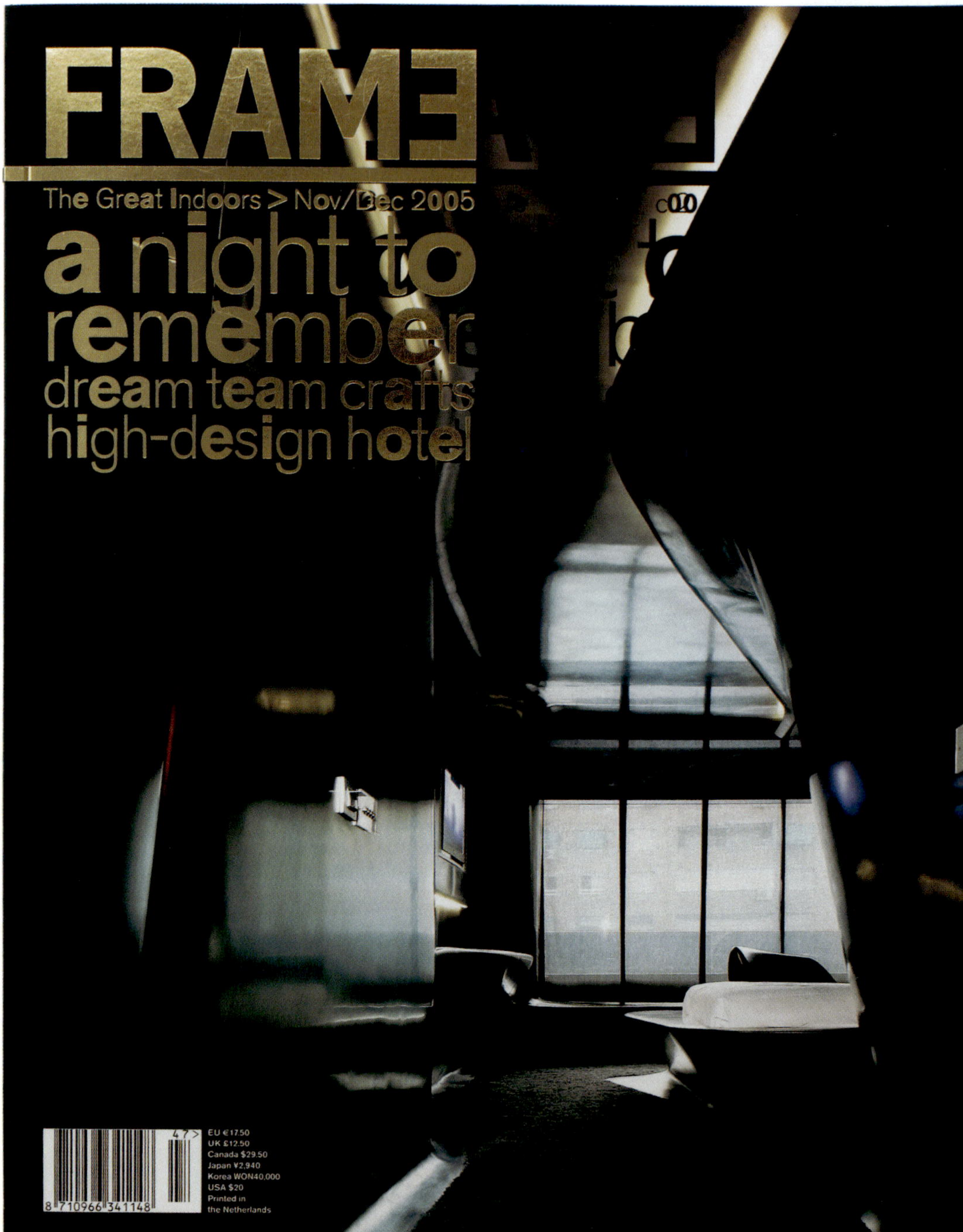

ART DIRECTORS
Cornelia Blatter & Marcel Hermans, COMA

47

Seeing Red

Frame editor Eva Schaap about the project

Tucked away between items featuring ministerial crises, murder and destruction (in a newspaper distributed free at all Dutch railway stations) was the tiny photo of a splendid red factory, spewing smoke. The picture, though vague, showed the floating stage set of Verdi's 'Il Trovatore', an opera performed on Lake Constance, near Vienna, at the Bregenzer Festspiele. We at Frame started digging for information on the factory and its 'workers'. Editor Shonquis Moreno interviewed designer Paul Steinberg and, undaunted by deadlines, wrote an article no less splendid than the pictures. A photo that could have been so easily overlooked rapidly grew into much more than a single story: soon we had a series of items on stage-set design starring people like Es Devlin and Robert Wilson. Steinberg's take on stage-set design included this remark: 'It's a unique kind of project wherethere's pressure on you to make an iconic statement, but I really wanted to include everyone in the theatre in the experience.' And so he did. Frame was delighted to help him reach an even greater audience, because even on paper the event is quite an experience.

DESIGNER
Paul Steinberg

SUBJECT
Opera set for
Verdi's 'Il Trovatore'
Vienna's Bregenz
Opera Festival

LOCATION
Bregenz, Austria

PHOTOGRAPHERS
Karl Forster
M. Tretter

WRITER
Shonquis Moreno

Seeing Red

Paul Steinberg's choleric opera set floats in the shallows of Austria's Lake Constance, reflecting the internecine affairs of doomed characters.

Text by Shonquis Moreno
Photography by Karl Forster and M. Tretter

This past summer, smokestacks belched fumes and fire into the night sky from the plein-air stage where Verdi's opera, *Il Trovatore* (The Troubadour), floated at the edge of Lake Constance at the dramatic foot of the Alps. For 60 years this lake in between Germany and Austria has been the site of the Bregenz Opera Festival, which began life aboard two gravel barges: one carried the stage and sets, the other the orchestra. Today's rather more solid mooring is anchored by an octagonal concrete core 'around which the sets are built like Venice', says set designer Paul Steinberg, on wood piles arrayed strategically in the water. Steinberg, who was responsible for the design of the fire-breathing set, is based in New York City and teaches stage design at New York University. For Verdi's 150-year-old opera, he and Bregenz's artistic directors created a formidable and wholly modern industrial set in the form of a mid-20th-century factory that will brave the Austrian winters and remain standing for two years.

The opera, set in medieval Spain, pits lovers at cross-purposes, counts against gypsies, a feudal aristocracy against the disenfranchised, rich against poor, and brother against brother. It is a tragedy that starts out badly and gets steadily, ineluctably, worse. 'The piece is a kind of nightmare that ends with the complete destruction of everyone involved,' says Steinberg. 'I almost always look for a contemporary context in which to present a theatre piece so that a contemporary audience can connect to it. They can read it not as a museum piece, but as something that speaks to the moment.'

From the beginning, director Robert Carsen envisioned something industrial, and the two settled on a timely symbol of modern capitalist society and the powerful international corporations that fuel it: a factory. Steinberg imagined it as an oil refinery – complete with serpentine duct work, ladders and smokestacks – gone wrong. A large fence keeps the gypsies outside the plant and away from the infrastructure that is the source of both power and paranoia. Sound designers rigged the set – a medley of textured plastic tubing, barrels, and perforated metal sheeting – with up-to-the-minute technology, including scores of cleverly hidden loudspeakers and microphones. At great expense (the whole set cost in the neighbourhood of €6 million), Bregenz's pyrotechnocrats made sure that the fury and yearning acted out on stage would be matched by the incendiary stacks.

And then they painted the entire thing red. 'For me,' Steinberg explains, 'colour is a kind of structural element that provides underpinning to the meaning of anything that I design. Red is about passion, danger, fire. In the environmentally correct context of Lake Constance, red is a kind of fuck-you gesture from the people in power to those who aren't. It brings attention to itself and also says stay away.'

Unlike his work in conventional theatres, Steinberg feels that constructing the *Trovatore* set was more like erecting a building than a set. Despite its complexity and scale – the entire set stretches approximately 40 metres from side to side, while the duct work and stacks provide an intricate backdrop for the exceedingly intricate plot (think of Shakespeare at his knottiest) – the designer worked hard to keep sightlines clear from every viewing angle. 'It's the unique kind of project where there's pressure on you to make an iconic statement, but we really wanted to include everyone in the theatre in the experience,' Steinberg says. 'I'm most proud of that.'

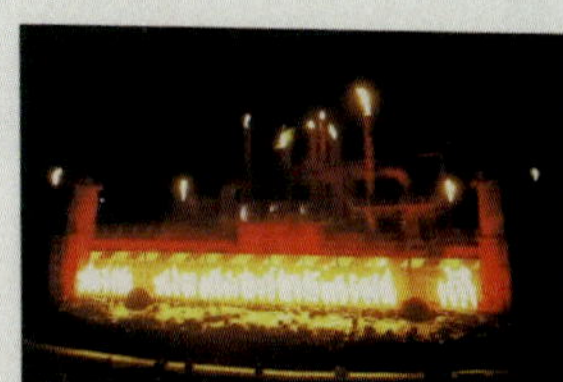

Floating on Lake Constance during Vienna's Bregenz Opera Festival, Paul Steinberg's set for Verdi's Il Trovatore – a medley of plastic tubing, barrels and perforated metal sheeting – features a blaze of fiery red smokestacks.

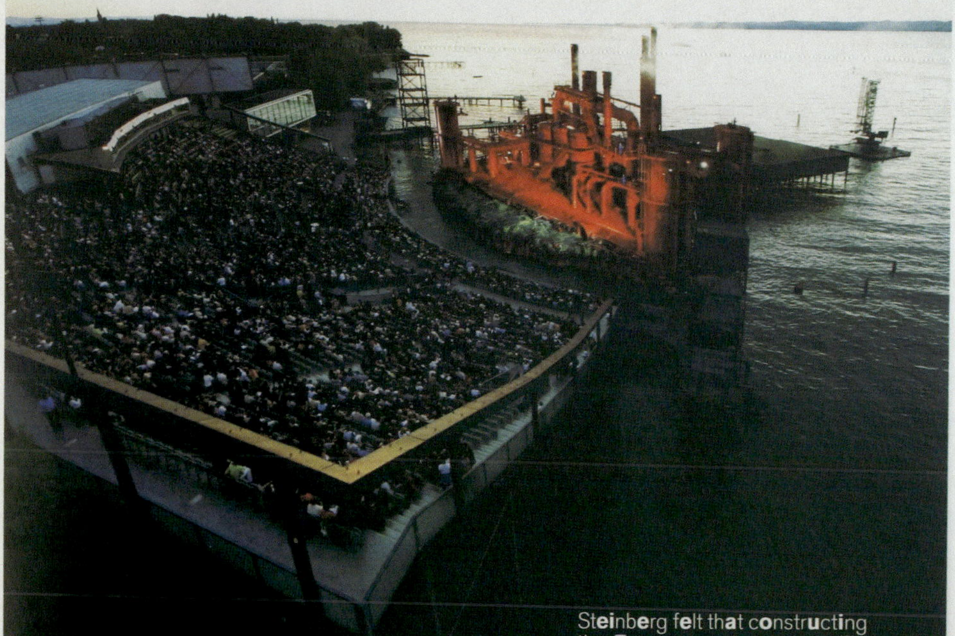

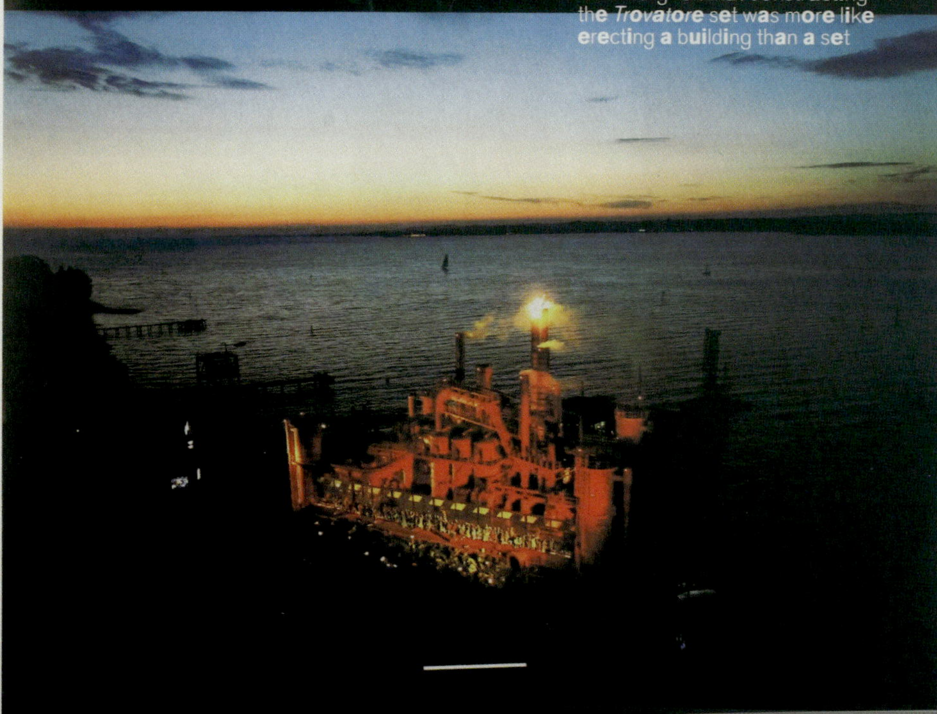

Steinberg felt that constructing the *Trovatore* set was more like erecting a building than a set

SUBJECT
Opera set for Verdi's 'Il Trovatore'

LOCATION
Bregenz, Austria

DESIGNER
Paul Steinberg

PHOTOGRAPHERS
Karl Forster
M. Tretter

WRITER
Shonquis Moreno

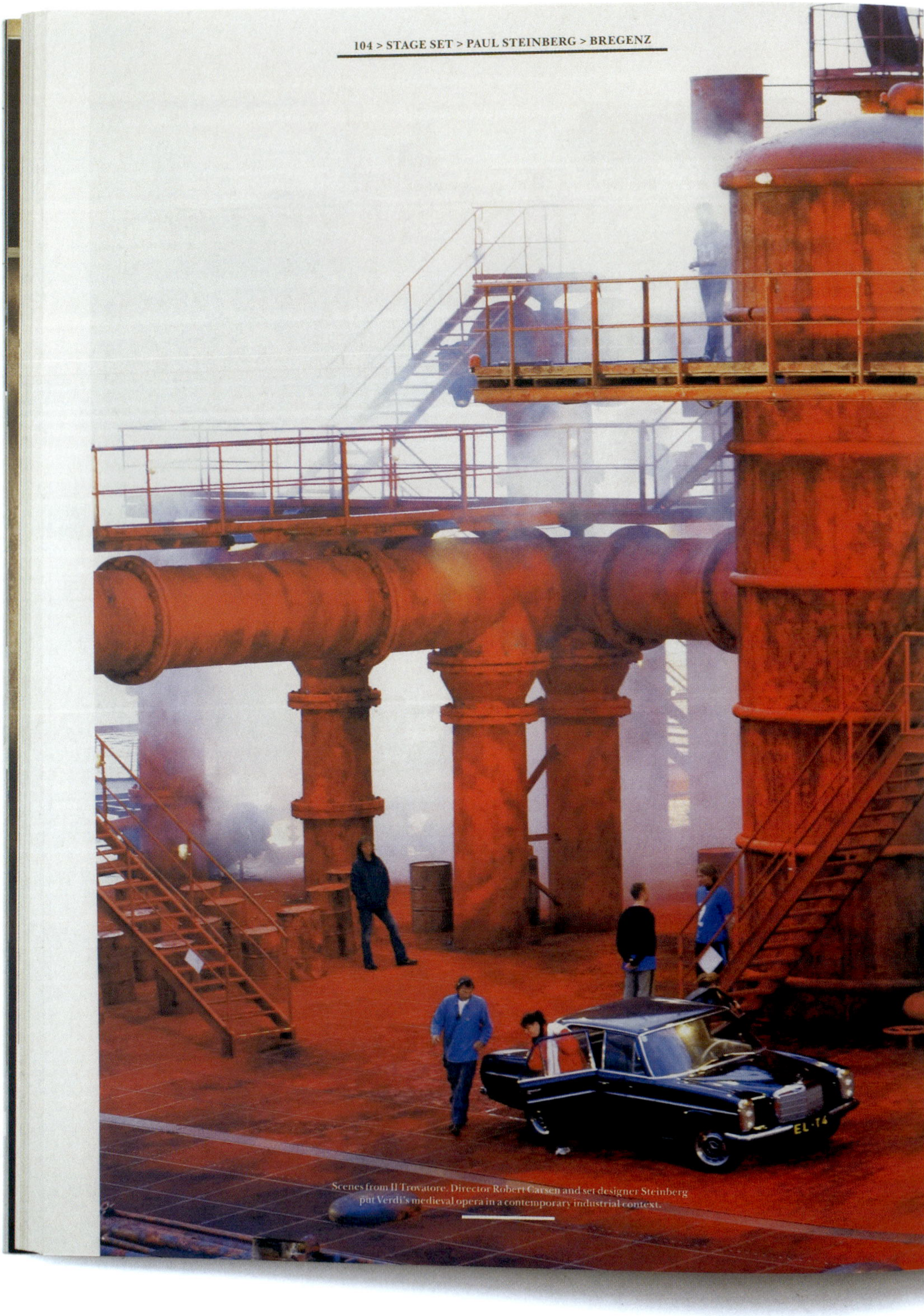

Scenes from Il Trovatore. Director Robert Carsen and set designer Steinberg put Verdi's medieval opera in a contemporary industrial context.

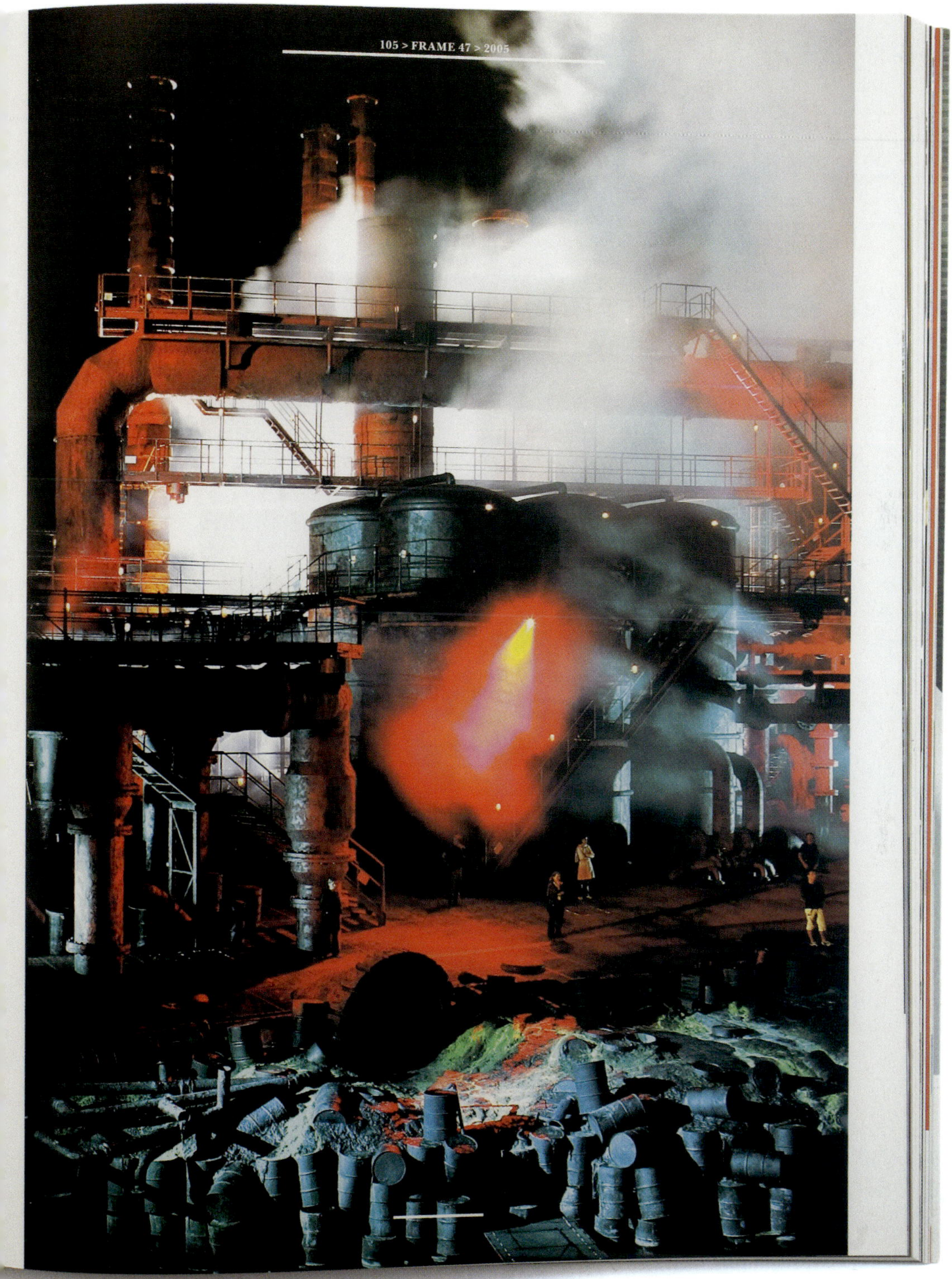

SUBJECT
Opera set for
Verdi's 'Il Trovatore'

LOCATION
Bregenz, Austria

DESIGNER
Paul Steinberg

PHOTOGRAPHERS
Karl Forster
M. Tretter

WRITER
Shonquis Moreno

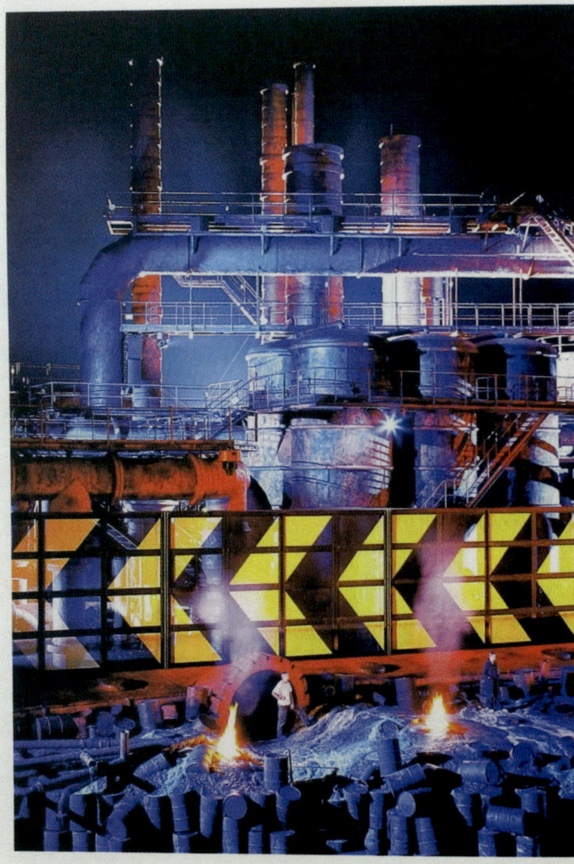

Carsen and Steinberg see the floating factory as a symbol of modern capitalist society. Sound designers rigged the set with the latest technology, including scores of hidden loudspeakers and microphones.

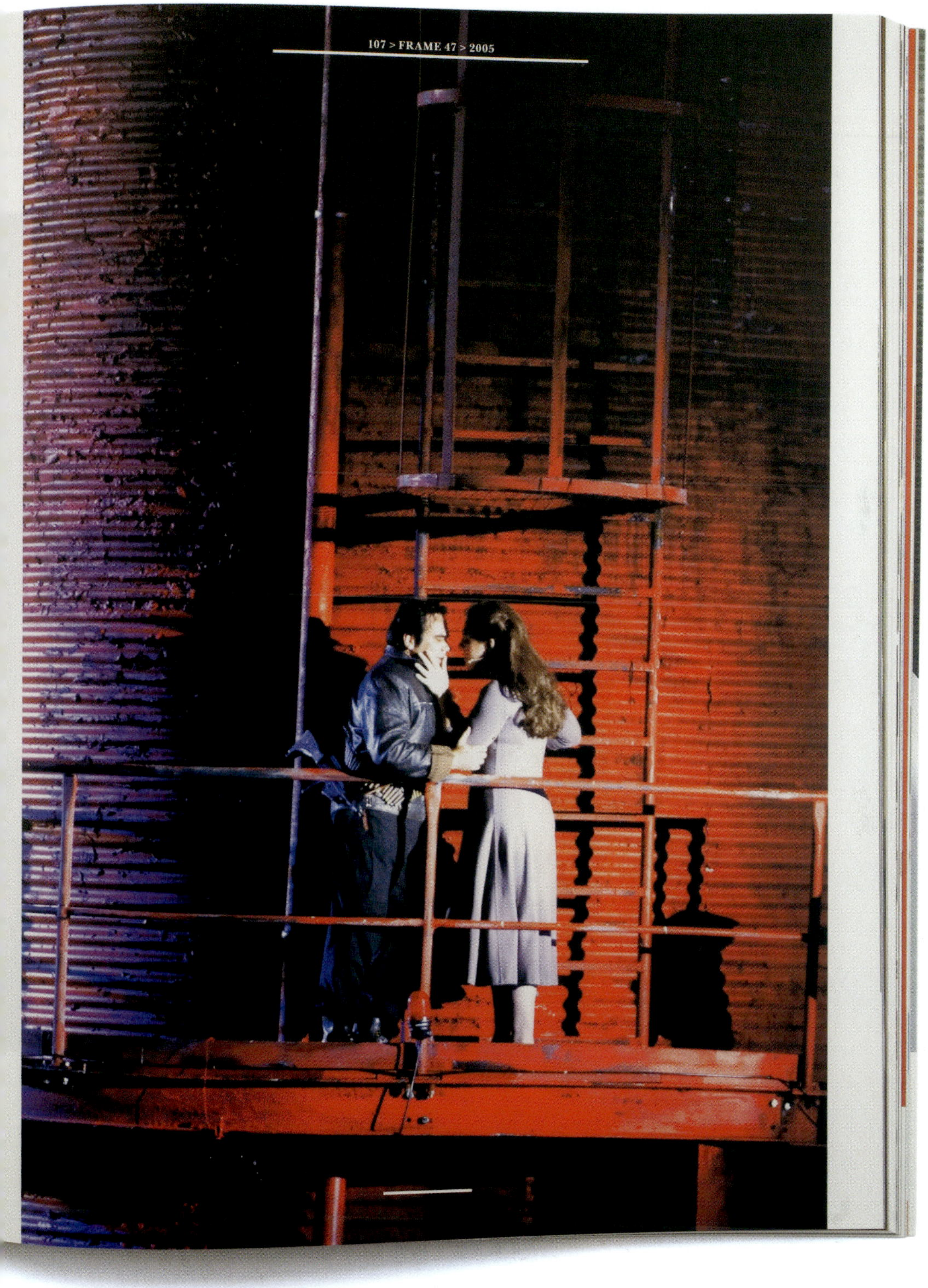

SUBJECT
Opera set for Verdi's 'Il Trovatore'

LOCATION
Bregenz, Austria

DESIGNER
Paul Steinberg

PHOTOGRAPHERS
Karl Forster
M. Tretter

WRITER
Shonquis Moreno

'I almost always look for a contemporary context in which to present a theatre piece so the audience can read it not as a museum piece, but as something that speaks to the moment'

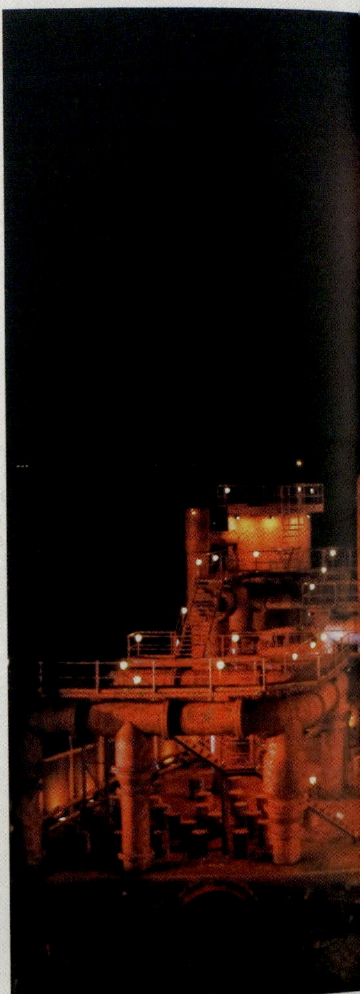

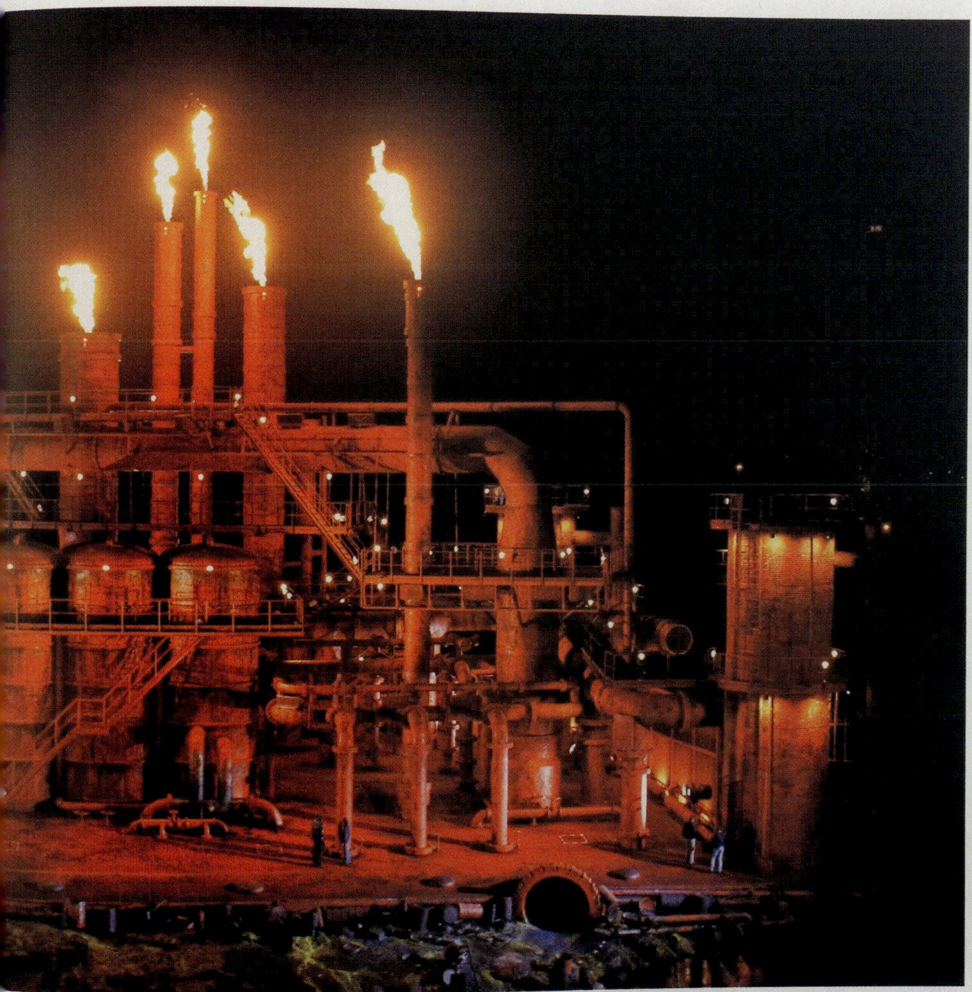

PROJECT
Opera set for
Verdi's Il Trovatore

LOCATION
Bregenz, Austria

DESIGNER
Paul Steinberg

PHOTOGRAPHERS
Karl Forster
M. Tretter

WRITER
Shonquis Moreno

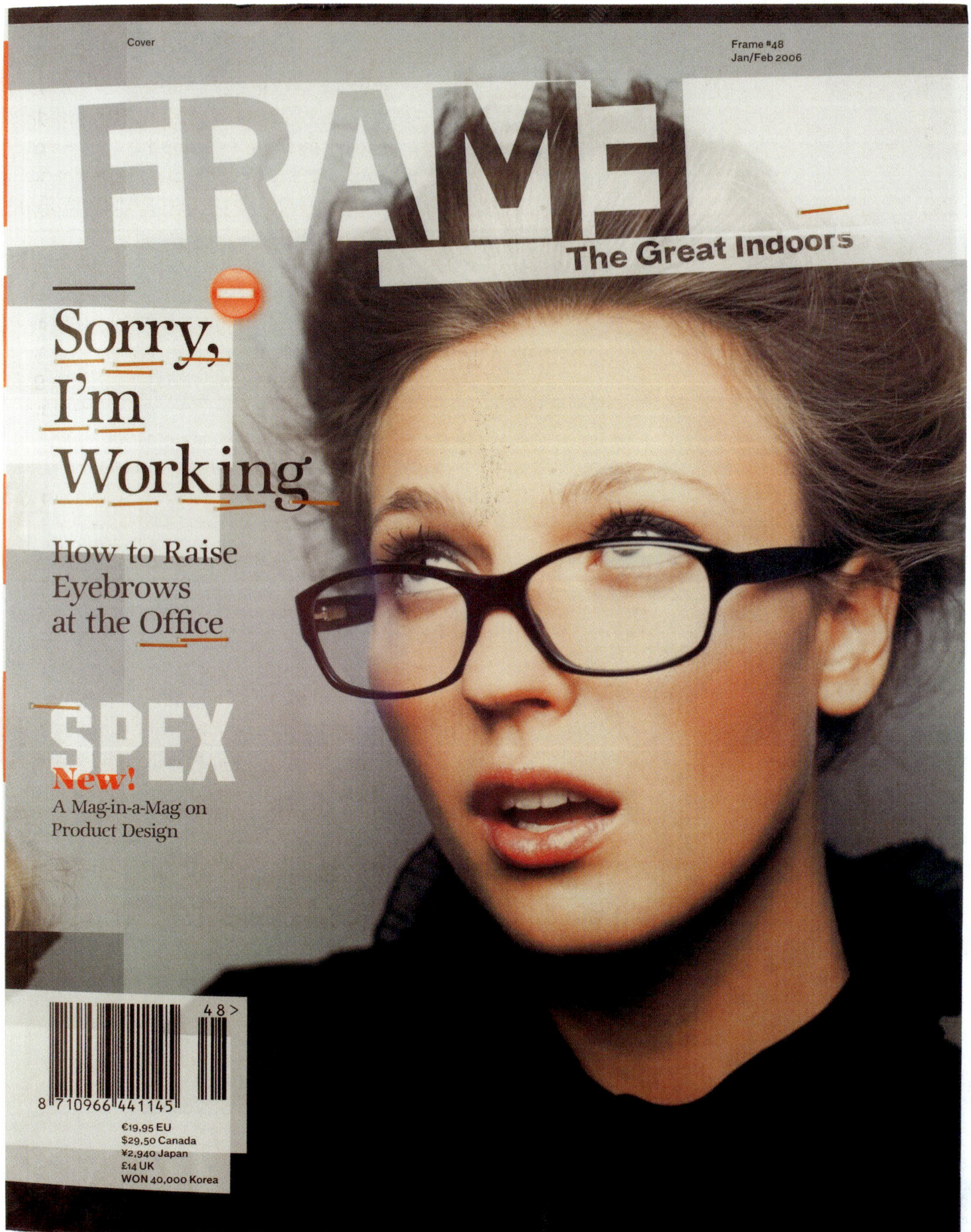

ART DIRECTORS
Jannetje in 't Veld & Toon Koehorst, Koehorst in 't Veld

Nobel Intentions

Frame editor Eva Schaap about the project

When the first pictures of the Nobel Peace Centre arrived in our inbox (among the many we receive each day), the immediate reaction was: this is something we want Frame readers to see, both in detail and from different perspectives. The building, a 19th-century railway station in Oslo, was brought back to life as the Nobel Peace Centre by London-based architect David Adjaye, who provided the venerable structure with an exciting new function. Because the building would be offering no traditional display of artefacts, Adjaye took an unconventional approach to museum design, which included the use of digital technology. The idealism exhibited at the entrance – the vision of a world with no boundaries – imbues the entire interior. Sandwiched between a series of articles on office design and a story featuring students at the Sandberg Institute in Amsterdam, the colourfully illustrated article on the Nobel Peace Centre provides readers with a perfect spot to pause in peaceful surroundings.

DESIGNERS
Adjaye Associates
www.adjaye.com

SUBJECT
Nobel Peace Centre

LOCATION
Oslo, Norway

PHOTOGRAPHER
Timothy Soar

WRITER
Beatrice Galilee

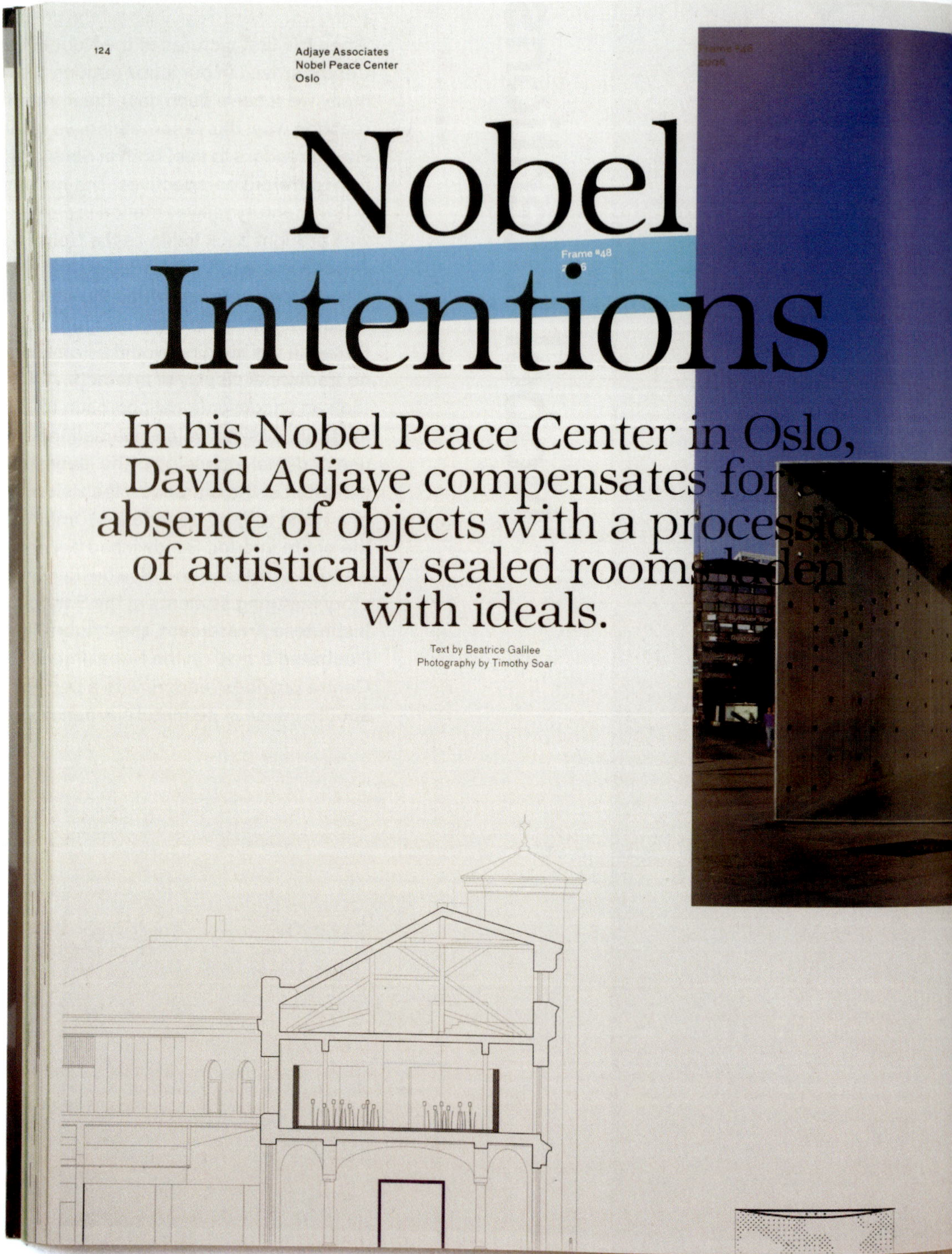

Nobel Intentions

In his Nobel Peace Center in Oslo, David Adjaye compensates for absence of objects with a procession of artistically sealed rooms laden with ideals.

Text by Beatrice Galilee
Photography by Timothy Soar

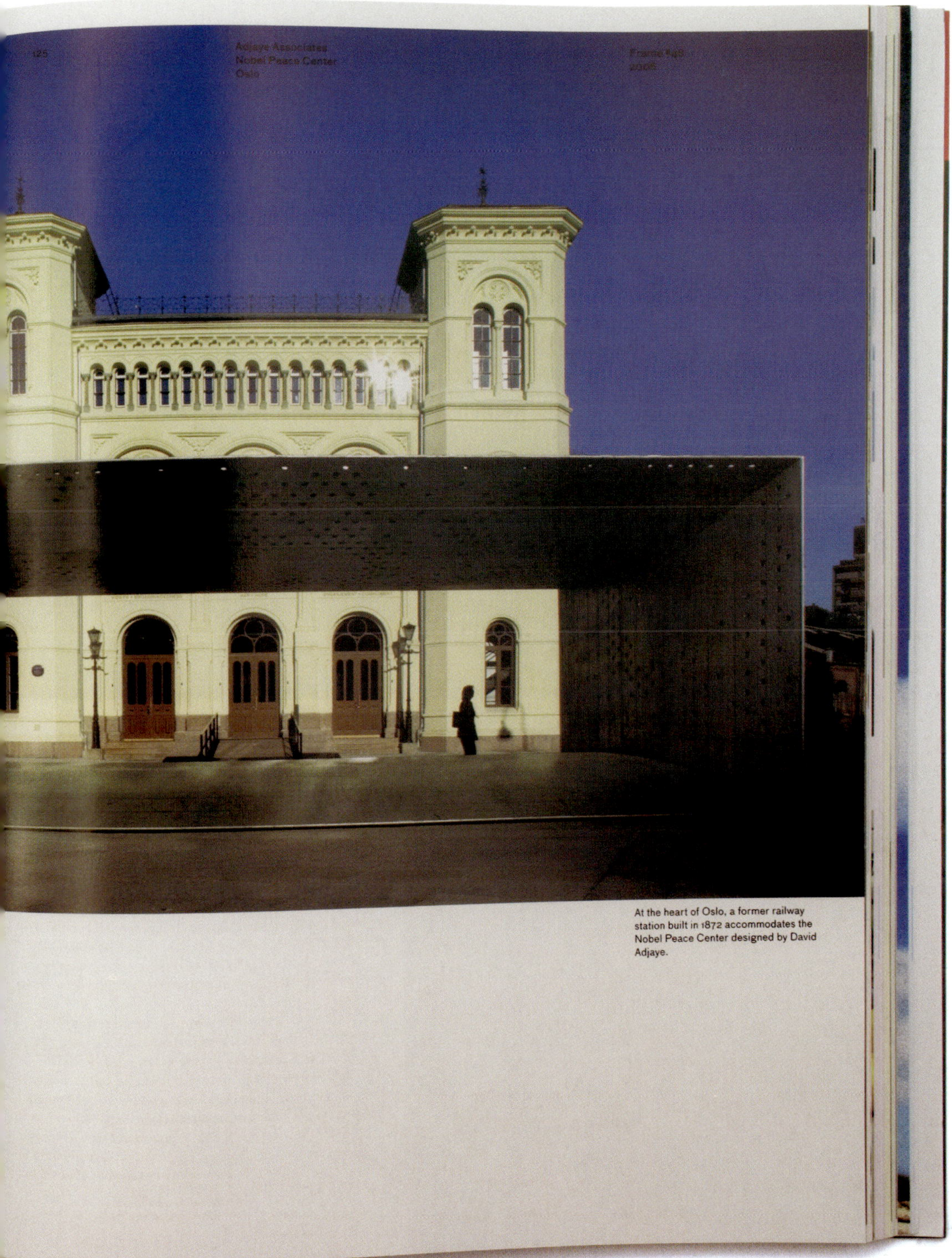

SUBJECT
Nobel Peace Centre

LOCATION
Oslo, Norway

DESIGNERS
Adjaye Associates

PHOTOGRAPHER
Timothy Soarr

WRITER
Beatrice Galilee

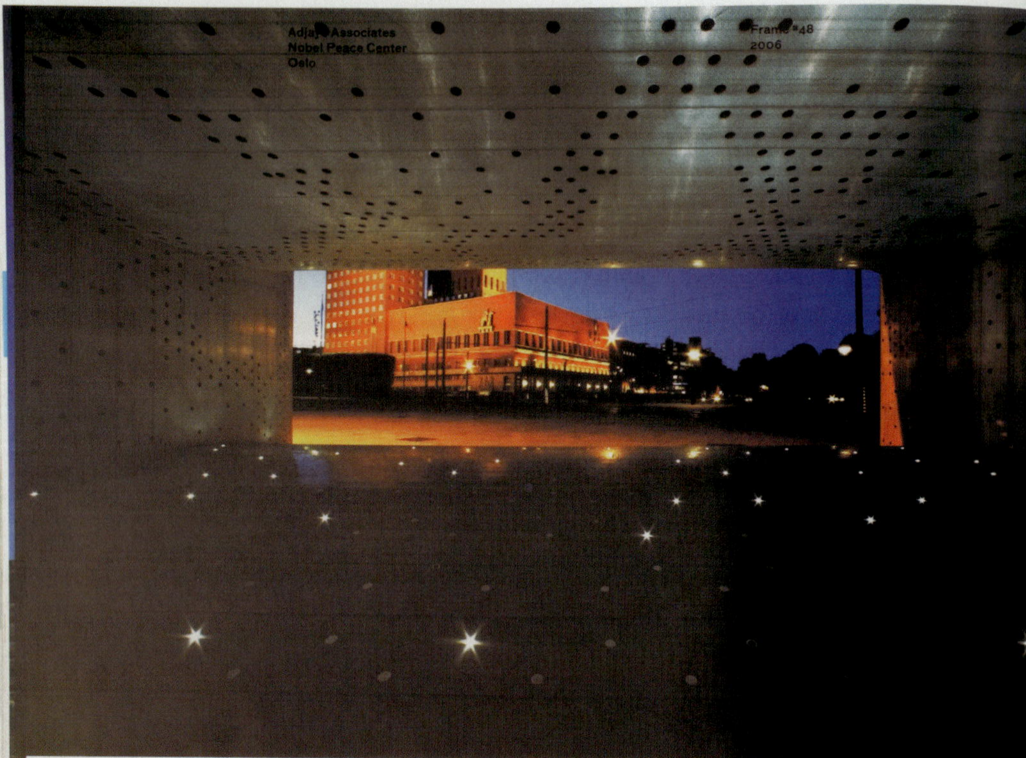

Leaving Oslo City Hall Square, the visitor walks beneath a steel canopy to reach the entrance to the 1300-sq-m Nobel Peace Center.

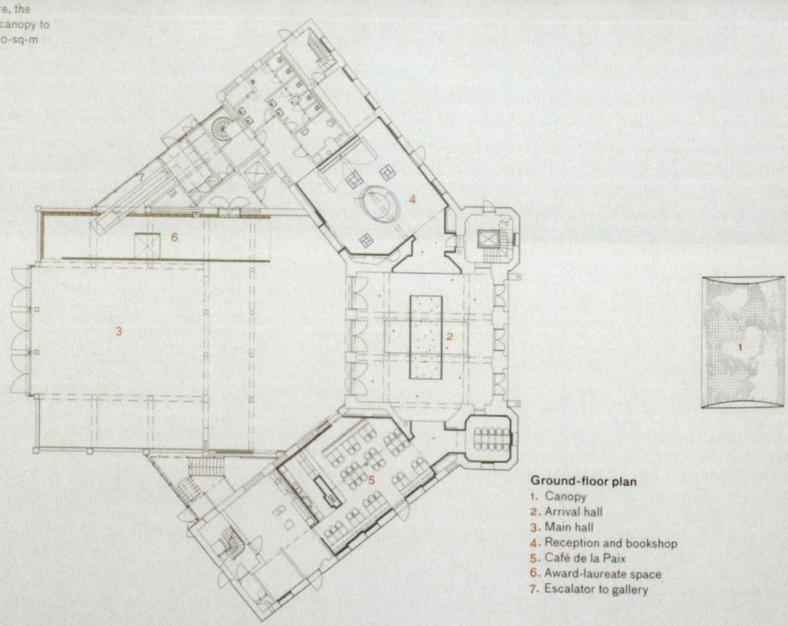

Ground-floor plan
1. Canopy
2. Arrival hall
3. Main hall
4. Reception and bookshop
5. Café de la Paix
6. Award-laureate space
7. Escalator to gallery

At the heart of Oslo, looking south over the Nordic fjords towards central Europe, a building that began as a 19th-century train station shows its new face. Fronting it is a distinctly modern steel pavilion. Peppered with seemingly random holes, the structure represents a map of the earth as a borderless global community. The pavilion welcomes visitors to the new Nobel Peace Center, a David Adjaye design. Crisply constructed by Norwegian shipbuilders, the canopy is an idealistic gesture and the only external intervention into the former station and its municipal setting. For it is within the walls of the historical transport building that Adjaye and the Nobel Committee hope to stimulate thought and reflection on war, peace and the resolution of conflict.

Five years ago the Norwegian Nobel Institute invited eight young architecture practices to design a centre that in 2005 would mark the centenary of the peaceful dissolution of the union between Norway and Sweden. The former Vestbanen railway station was chosen as the location because of its proximity to almost everything in the capital city, including City Hall, where the Nobel Prize ceremony is held each year. In its capacity as a museum, with spaces for events and exhibitions, the centre would not only record and celebrate the life of Alfred Nobel and the story of the Nobel Peace Prize, but also offer something new and inspiring to present to future generations.

Because the building would not provide a traditional display of artefacts, it quickly became apparent that a conventional approach to museum design would be inappropriate, explains project manager Grete Jarmund. 'We wanted to question ideas of peace and conflict resolution in the same way the laureates did. We were looking for somebody who could transform our written concept into visual solutions and help us tell the story.' Adjaye proposed the use of contemporary digital technology as an antithesis to a typical objects-in-glass-cases approach to museum design. His concept captured first prize and led to his first commission in Scandinavia.

The London-based architect has created a series of singular spaces, each with a task and a mood of its own. 'My architecture is a series of "host and skin" scenarios. It's placing objects within the space which have their own kind of dialogue. They become skins that completely take over, acting like second-liners in the same way that paint adds second layers to a space.' He speaks earnestly, hands clasped, but occasionally rocks backwards with a burst of unexpected laughter, as if suddenly revelling in the absurdity of it all.

There is something intrinsically artistic about the way Adjaye deals with architecture and space. Perhaps it's from his years at the RCA in London; perhaps it's simply a healthy detachment from architecture as defined by others. He's interested in conversation and in inclusion. He feels removed from an older generation that expresses ideas aesthetically and that is into 'creating light and space'. He finds this kind of thinking 'completely alienating and irrelevant for anyone not involved in architecture'.

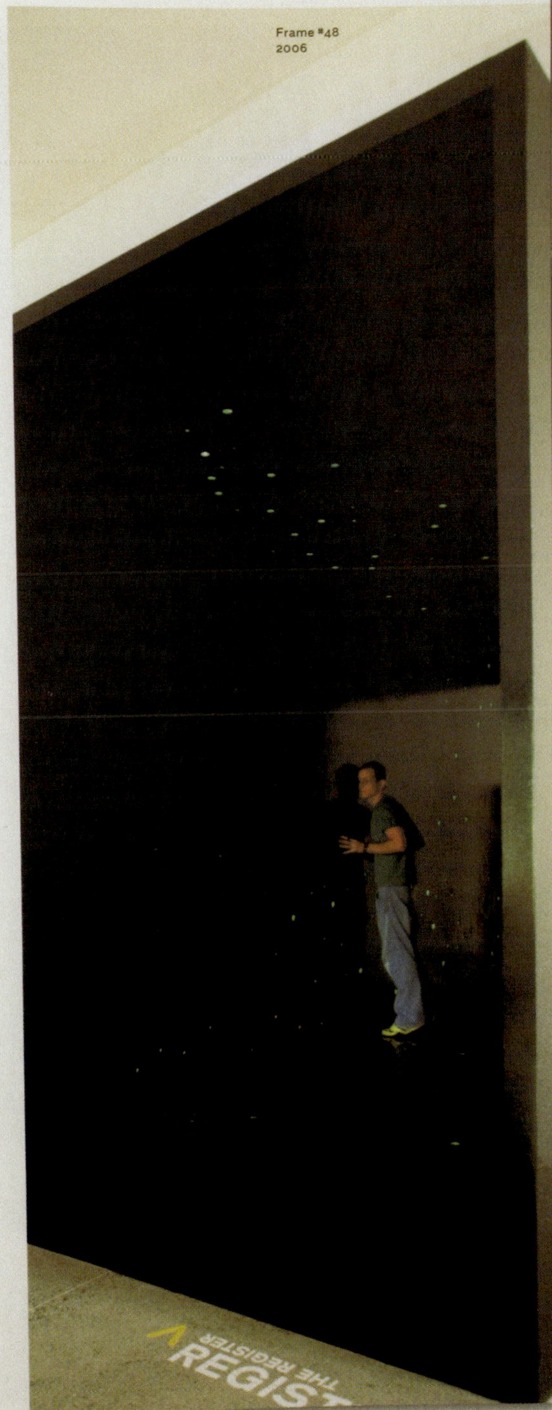

SUBJECT Nobel Peace Centre
LOCATION Oslo, Norway
DESIGNERS Adjaye Associates
PHOTOGRAPHER Timothy Soarr
WRITER Beatrice Galilee

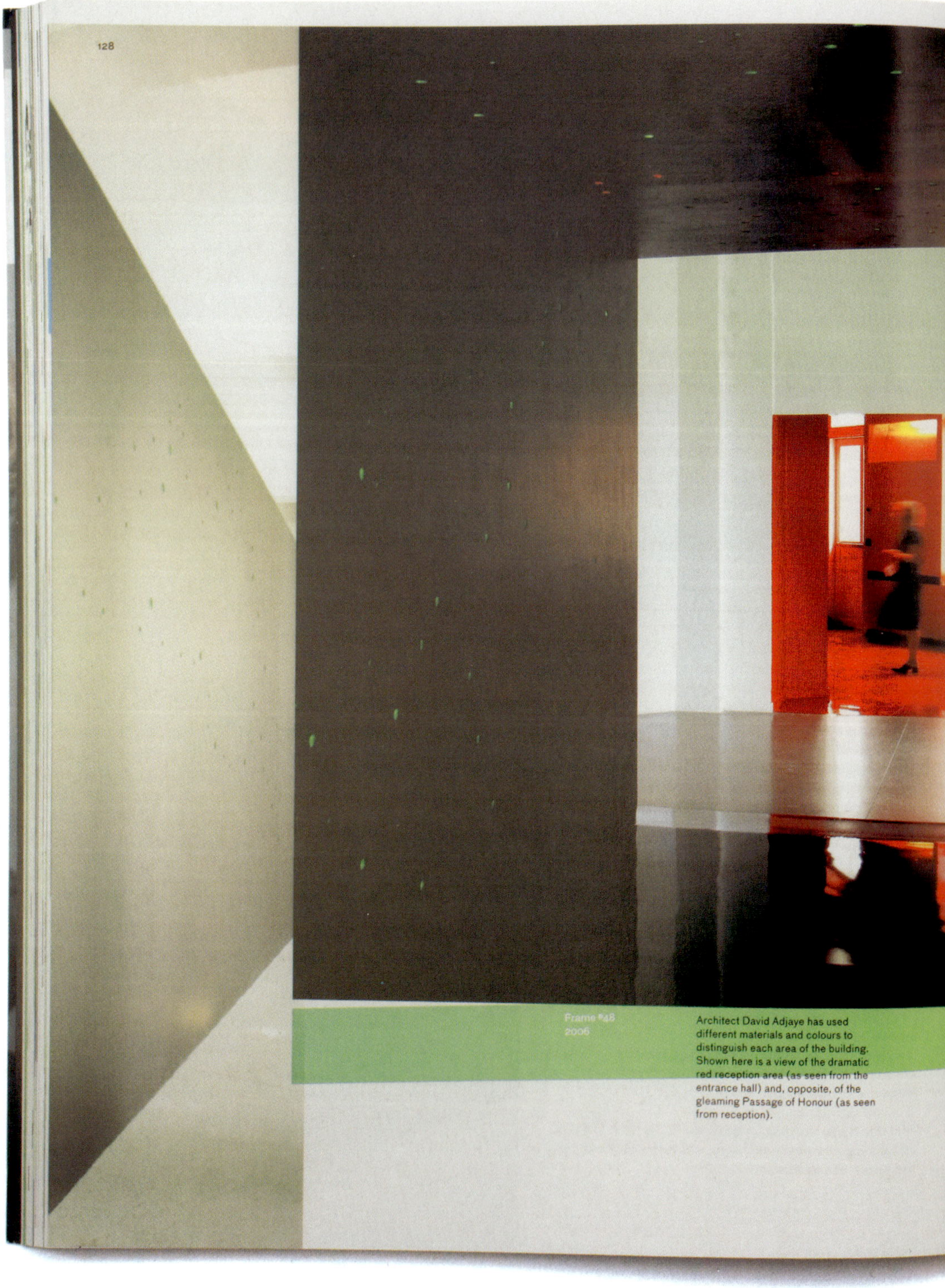

Frame #48
2006

Architect David Adjaye has used different materials and colours to distinguish each area of the building. Shown here is a view of the dramatic red reception area (as seen from the entrance hall) and, opposite, of the gleaming Passage of Honour (as seen from reception).

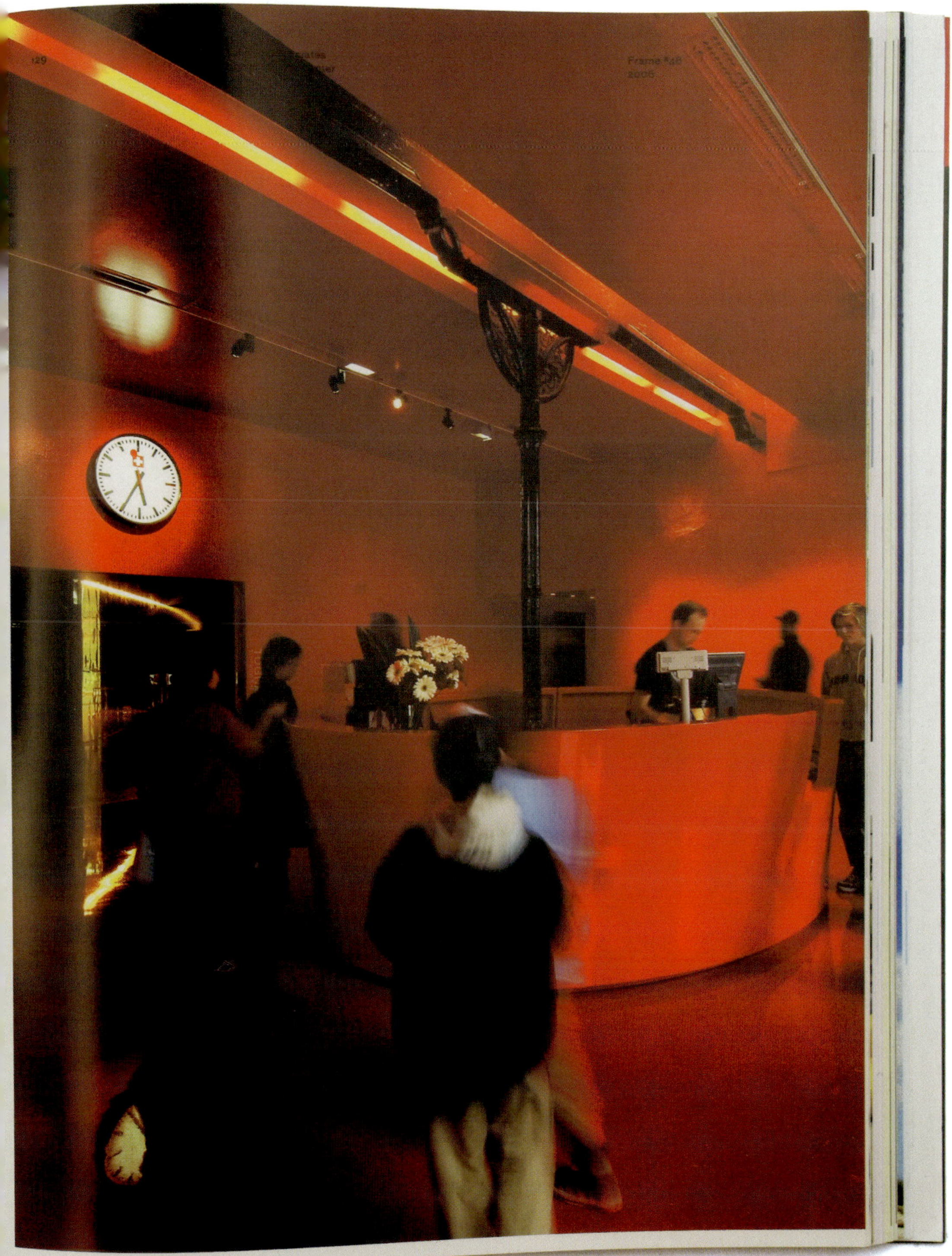

SUBJECT
Nobel Peace Centre

LOCATION
Oslo, Norway

DESIGNERS
Adjaye Associates

PHOTOGRAPHER
Timothy Soarr

WRITER
Beatrice Galilee

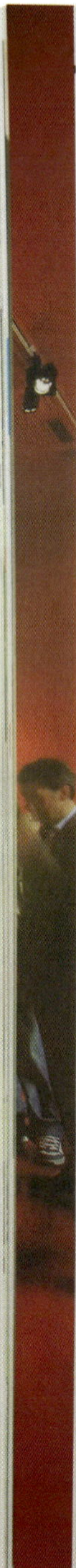

Adjaye has made his mark, skilfully sealing each room of the Vestbanen – no walls were allowed to be removed because of the building's listed status – in a manner that recalls the rich, playful simplicity of his private-house designs. Indeed, despite the change from dense urban cityscape to a less crowded Scandinavian site, Adjaye maintains that his approach remains the same. 'My work so far has been an essay in recycling and reconfiguring old buildings, and I did that in my houses in London,' he muses. 'But it's also about understanding that existing buildings hold certain memories. It's about moving past that understanding and adding another history to those buildings.'

Applying the language of a new history while telling a story of the old was not without difficulties, however. Adjaye believes that traditional museums perpetuate social distinctions. 'Public buildings are holders of a certain way of perceiving world order,' he says, 'because they talk about "us and them" in a very direct way spatially.' Resolving this dilemma meant admitting that the Peace Center was not a job for one architect alone, and in such situations Adjaye thrives by immersing himself in other disciplines and looking for helping hands. His entourage for this project included long-term collaborator and Turner-Prize winner Chris Ofili, who worked on the café; Small Design, which developed the technical infrastructure; Timon Botez, who contributed graphics; and Shin Azumi, who designed furniture for the centre.

The first example of Adjaye's sealing strategy is the black registration hall. Here, in continuation of the global theme introduced by the entrance canopy, perforations representing the world's big cities are scattered throughout the corridor, and the voices of people from hundreds of countries spill into visitors' ears from these openings. The intention, says Adjaye, was to draw attention to the multiplicity of language in the world.

An overwhelmingly red reception leads to a number of interesting features, such as 'e-houses' for exhibits, an Aalto-esque hanging space that currently houses an Amnesty International exhibition, educational facilities, and areas for discussion. Despite the dominance of information outlets, poetry prevails in the Nobel Chamber. In this cupboard-like space, visitors stand at a pulpit to watch the story of Alfred Nobel's life as shown in artist and display technologist David Small's Illuminated Manuscript, an artwork and a book like no other. Microchips embedded in each page trigger a projector above the reader's head and transform blank pages into dancing animations – an example of a tangible, human approach to digital technology that defines the project as a whole.

In keeping with the theme of Nobel prizewinners, Adjaye crafted a metaphorical 'gold medal' in the Passage of Honour, a reflective bronze-clad corridor – panelled on floor, ceiling and walls – dedicated to all who have received the distinguished award. It elicits an outpouring of seriously ego-stretching exuberance from the architect. 'Reflections above, reflections to the side, reflections everywhere,' Adjaye giggles. 'Is that narcissistic? I've really gotta sort that one out . . .'

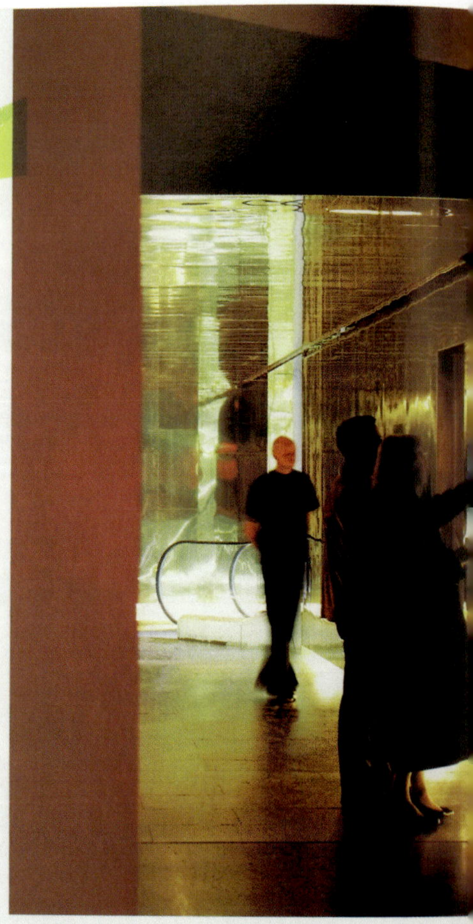

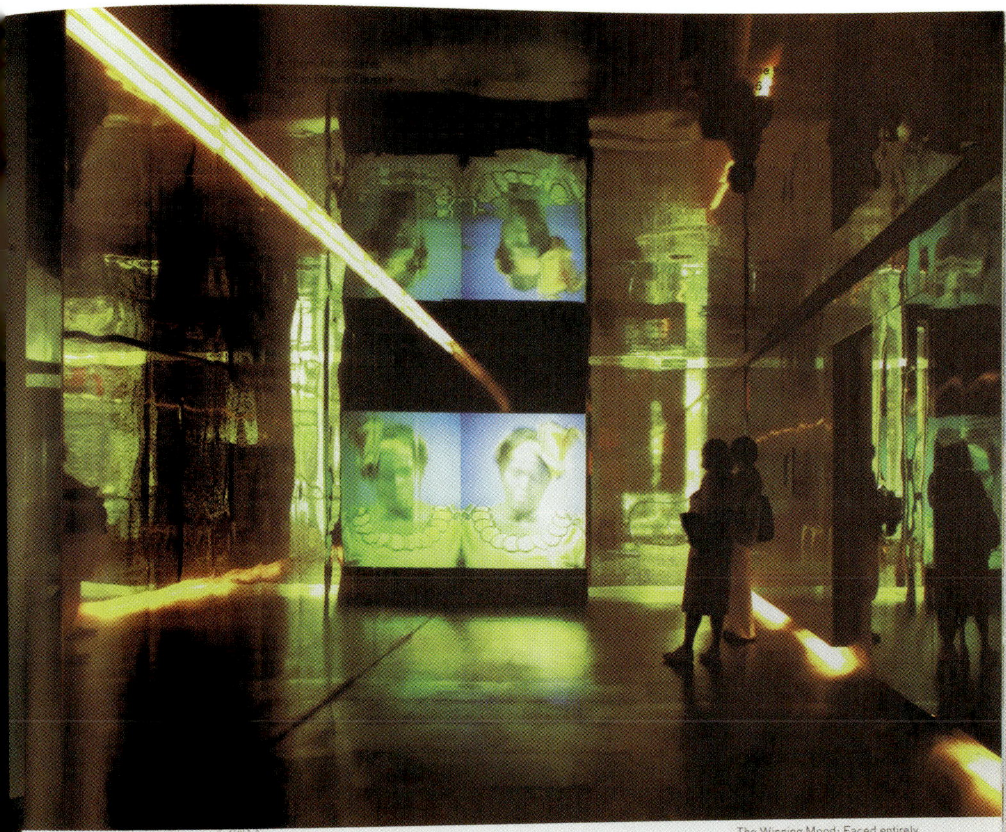

The Winning Mood: Faced entirely in bronze, the Passage of Honour celebrates the most recent recipient of the Peace Prize. The escalator at the end rises to the first-floor gallery.

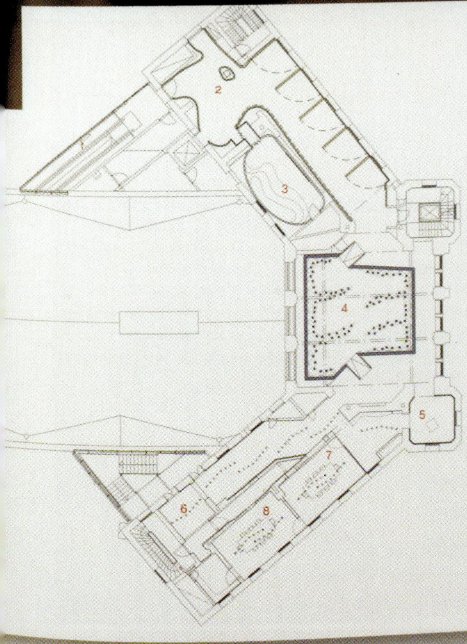

First-floor plan

1. Escalator
2. Temporary exhibition space
3. Projection room
4. Nobel Field
5. Nobel Chamber
6. E-Room
7. Education room
8. Conference room

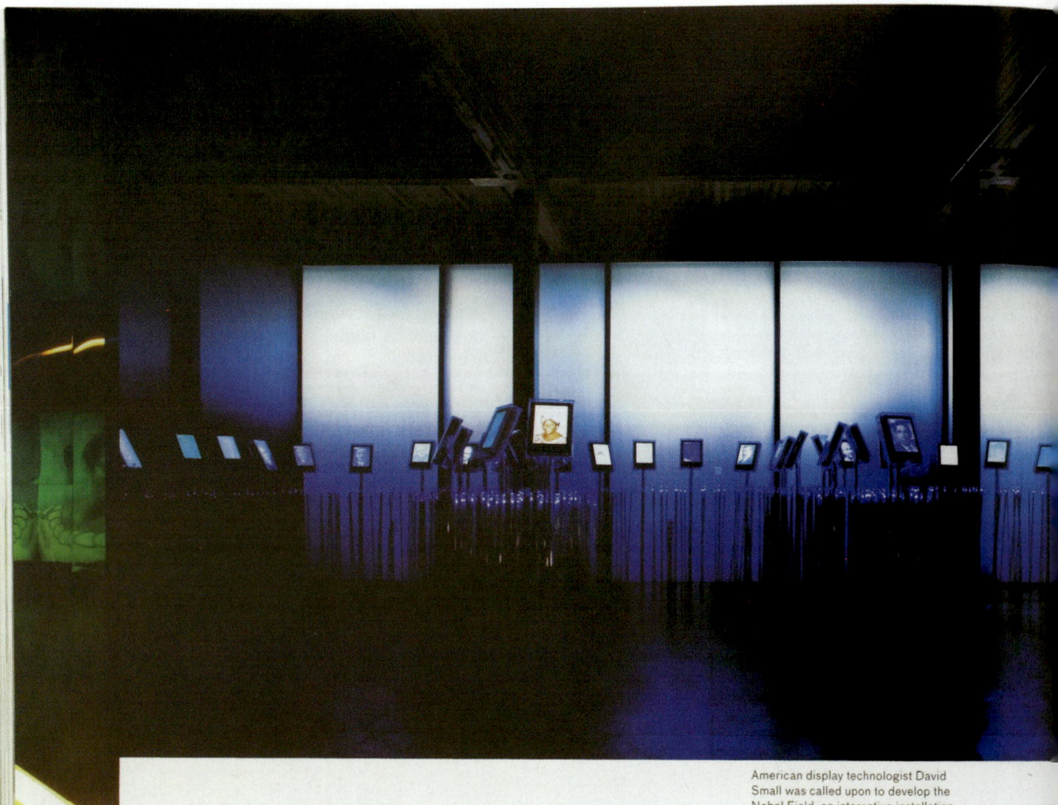

American display technologist David Small was called upon to develop the Nobel Field, an interactive installation that presents laureates on 96 screens surrounded by 1000 fibre-optic straws.

Other visual delights, and evidence of a fertile imagination, appear in the museum's Nobel Field, a space executed by David Small and his remarkable Massachusetts-based technology team. Displaying portraits of all Nobel Prize for Peace laureates, information screens rise in the 'landscape' like flowers blossoming in a digital garden. Over 50 infrared sensors monitor visitors, and as they approach a screen, its portrait dissolves and is replaced by a short feature about the laureate. The glass room pulses with a deep blue that is accompanied by ambient music written by yet another collaborator: the architect's brother, composer Peter Adjaye.

The plethora of colour ends in the final room, the Café de la Paix, where Chris Ofili's Earth Major Minor in Yellow and Green, a fractalated triangular work based on major cities, covers walls and ceiling. To the great delight of Ofili and Adjaye, because this type of work is not within the purview of Norwegian contractors, 11 painters and students from the Academy of Arts in Oslo were called in to complete the painting.

The Nobel Peace Center has an eloquence in its narrative; every room radiates a striking colour, from the dramatic red of reception, through sparkling golds and blues, to the happy ending of the green café. It's a joyful story that seems to have been conjured by a magician's fingertips. But the flair is complemented by a poignancy and an academic approach inherent in the subject matter. The idealism exhibited at the entrance – the vision of a world with no boundaries – is present throughout. 'I hope when you leave the building and go out into the world that you'll feel like a better person,' says a grinning Adjaye. 'Well, maybe you had to be a pretty good person to go there in the first place.'

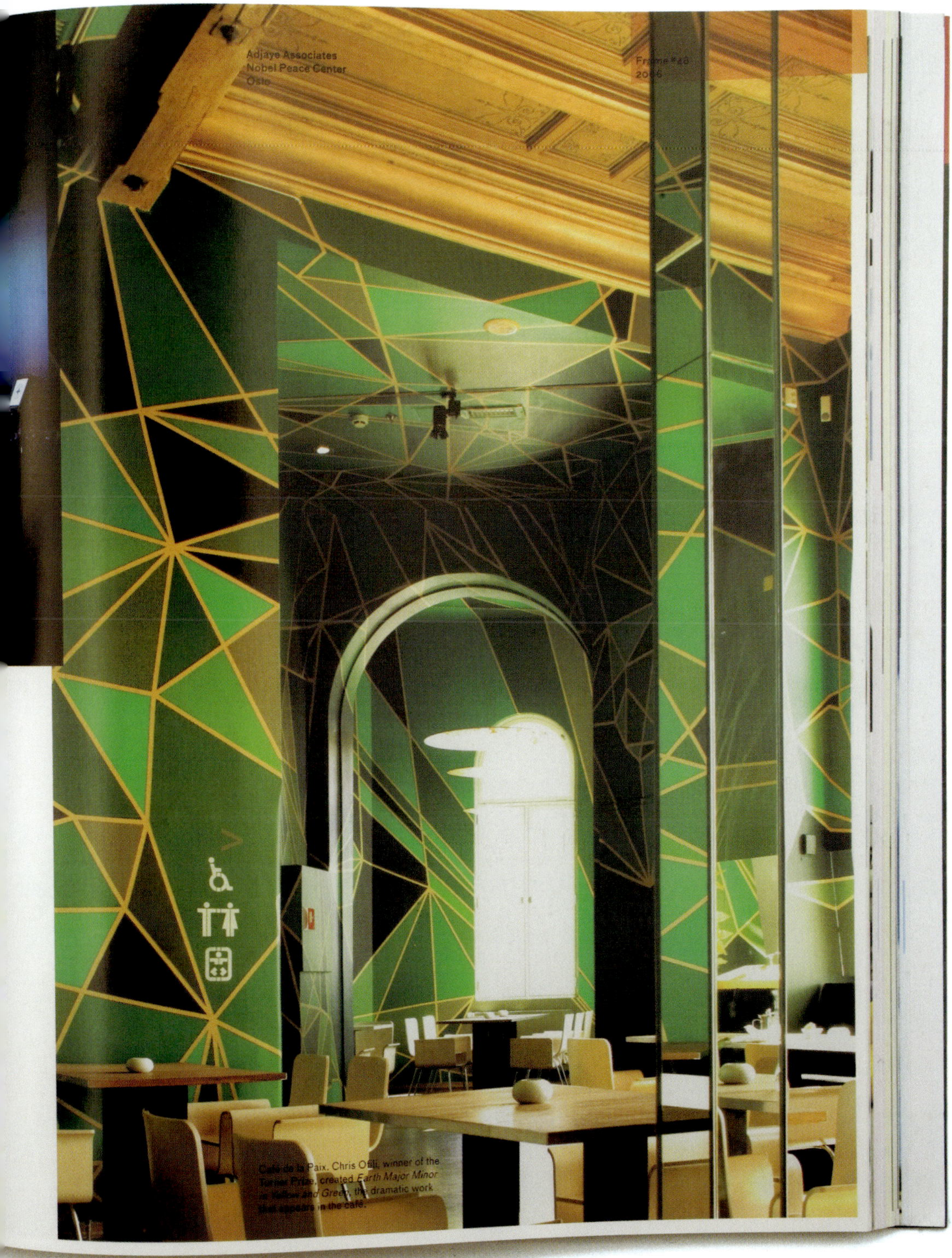

SUBJECT Nobel Peace Centre
LOCATION Oslo, Norway
DESIGNERS Adjaye Associates
PHOTOGRAPHER Timothy Soarr
WRITER Beatrice Gallilee

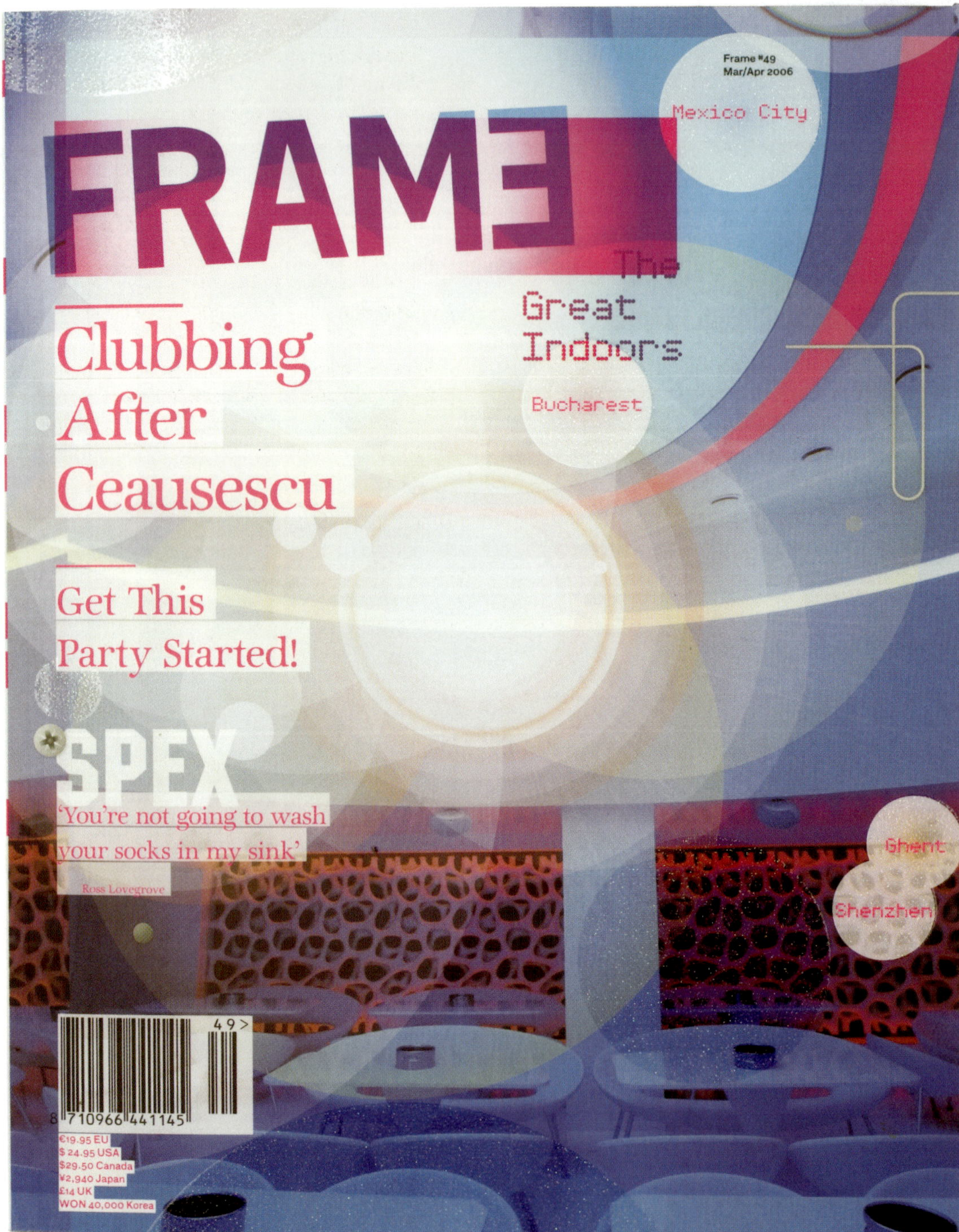

ART DIRECTORS
Jannetje in 't Veld & Toon Koehorst,
Koehorst in 't Veld

49

In Gods hands

Frame editor Billy Nolan about the project

The pitched roof and bell tower suggest the traditional church form, but the poured concrete and asymmetric entrance suggest otherwise. The heavy mood of solemnity that normally takes possession of those entering a place of prayer is lightened considerably thanks to a host of design choices: white as the dominant colour, an occasional spot of colour in the form of a chair, the slender profile of the seating, the circular seating configuration and, above all else, the spectacularly curvaceous ceiling with which the architect draws all eyes skyward, just as his Gothic predecessors did.

DESIGNERS
Ciel Rouge Création
http://cielrouge.com

SUBJECT
Church

LOCATION
Tokyo, Japan

PHOTOGRAPHER
Toshihisa Ishii

WRITER
Kanae Hasegawa

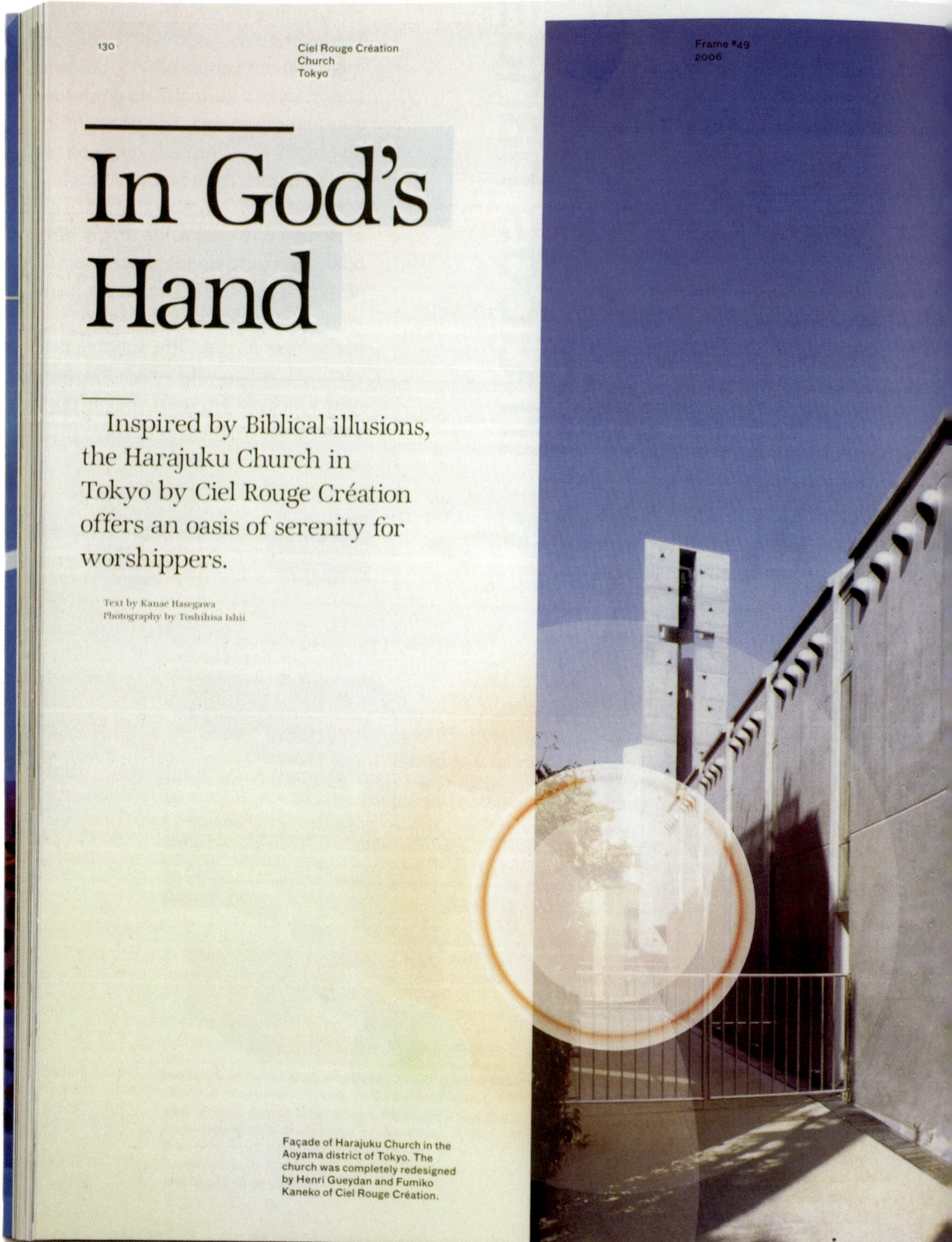

130

Ciel Rouge Création
Church
Tokyo

Frame #49
2006

In God's Hand

Inspired by Biblical illusions, the Harajuku Church in Tokyo by Ciel Rouge Création offers an oasis of serenity for worshippers.

Text by Kanae Hasegawa
Photography by Toshihisa Ishii

Façade of Harajuku Church in the Aoyama district of Tokyo. The church was completely redesigned by Henri Gueydan and Fumiko Kaneko of Ciel Rouge Création.

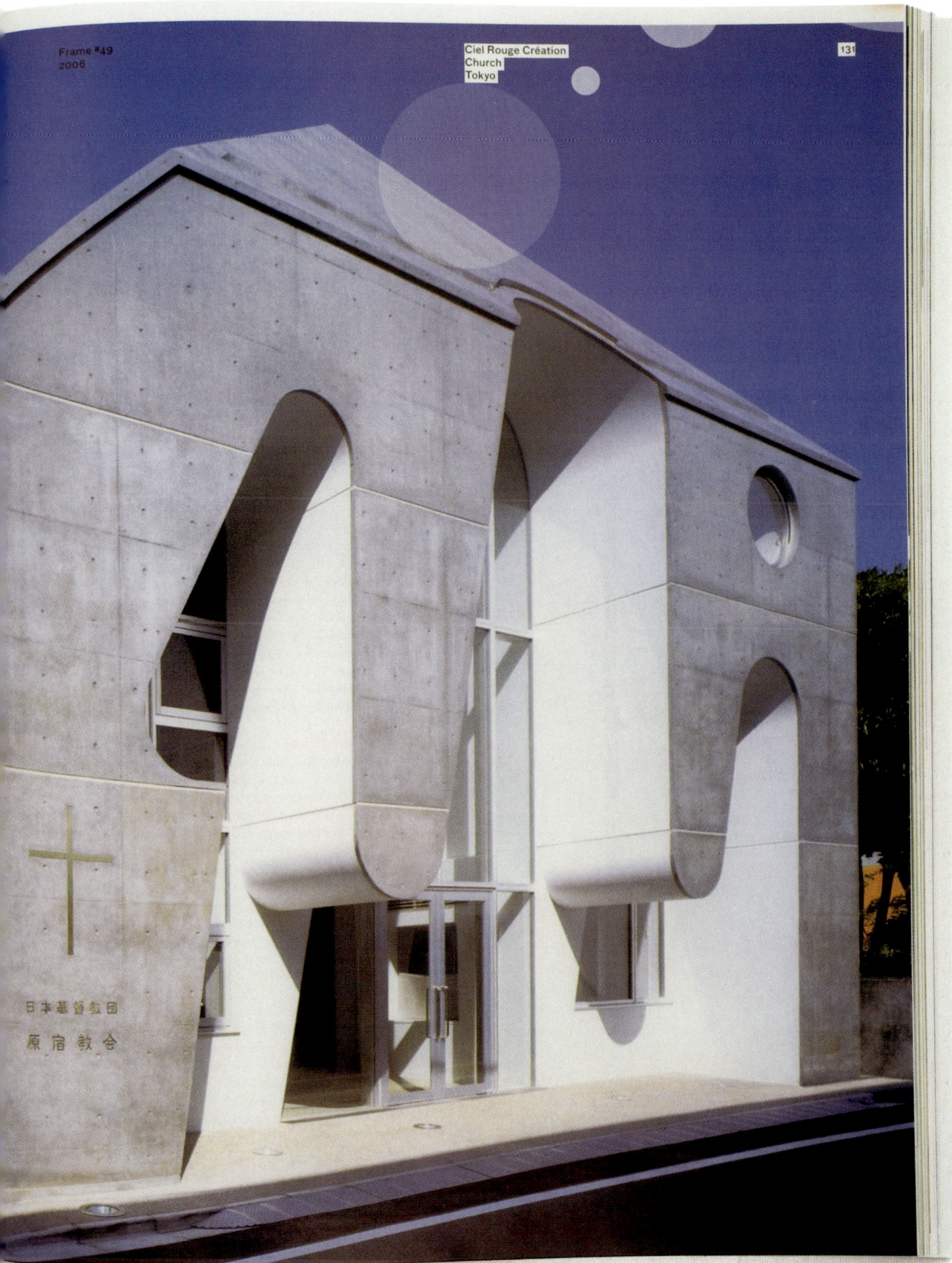

SUBJECT
Church

LOCATION
Tokyo, Japan

DESIGNERS
Ciel Rouge Création

PHOTOGRAPHER
Toshihisa Ishii

WRITER
Kanae Hasegawa

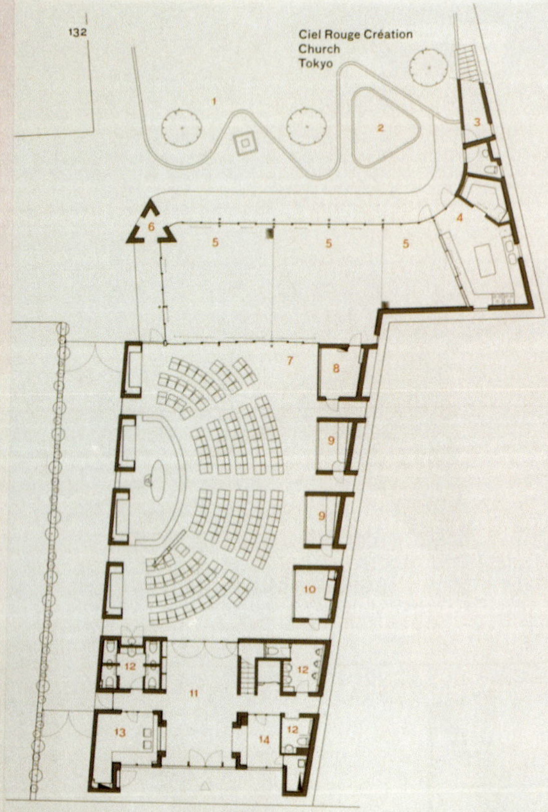

Ground floor plan

1. Garden
2. Sandbox
3. Storage
4. Kitchen
5. Hall
6. Bell tower
7. Chapel
8. Storage
9. Nursery
10. Audio room
11. Lobby
12. Toilet
13. Office
14. Cloakroom

It may sound strange to Western ears, but Japanese kindergartens often share accommodations with churches, and Christian pastors frequently serve as the headmasters of such schools. Many people in Japan believe the proximity of a religious institution is important to the education of their children. Harajuku Church, rebuilt in the autumn of 2005, is a good example of how well a church can function in a Japanese community, becoming the pride of its parishioners and a magnet for local gatherings.

> 'I imagined the chapel covered with the Holy hand, as if Christ's palm were enclosing all people within it.'
>
> Henri Gueydan of Ciel Rouge Création

Originally erected in 1904, Harajuku Church was reconstructed from the ground up following the bombardments of World War II. Even so, after 60 years the wooden framework of the church was found to be too weak to survive an earthquake. Located in Tokyo's Aoyama district and surrounded by shops such as MUJI, Starbucks and McDonald's, the church seemed to be in the wrong place to attract busy Tokyoites rushing by, intent on their destinations. Therefore, when the United Church of Christ in Japan proposed the reconstruction of the church, the pastor and his congregation had two wishes for the new building.

First and foremost, they wanted their new spiritual home to be recognizable to passers-by on the bustling streets that surround the church. Their second request was for an interior design that would show that their church is open to everyone, Christians and non-Christians alike. The selection of an architect started approximately ten years before the church finally settled on Tokyo-based architects Henri Gueydan and Fumiko Kaneko, a Japanese-French duo better known as Ciel Rouge Création. The prolonged decision-making period is attributed to Pastor Akira Tsuchihashi's wish to grant all his parishioners the opportunity to find their voices reflected in the design. 'And eventually everything has been tailored to meet the requirements of the people who visit and use the church,' he says.

Taking their cue from the medieval churches of Europe, whose stained-glass windows depict episodes from the Bible – scenes originally crafted to help the illiterate understand the life of Christ – the architects of Ciel Rouge Création designed a church whose architecture, in and

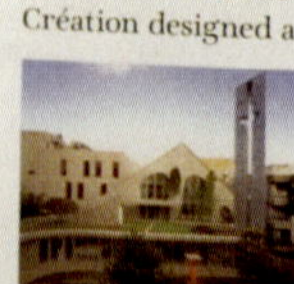

of itself, relates the story of Christ. To make the tucked-away church recognizable from the streets not only by the sound of its bells but also by its appearance, the architects constructed a 16-m-tall bell tower clearly incised with crosses down its

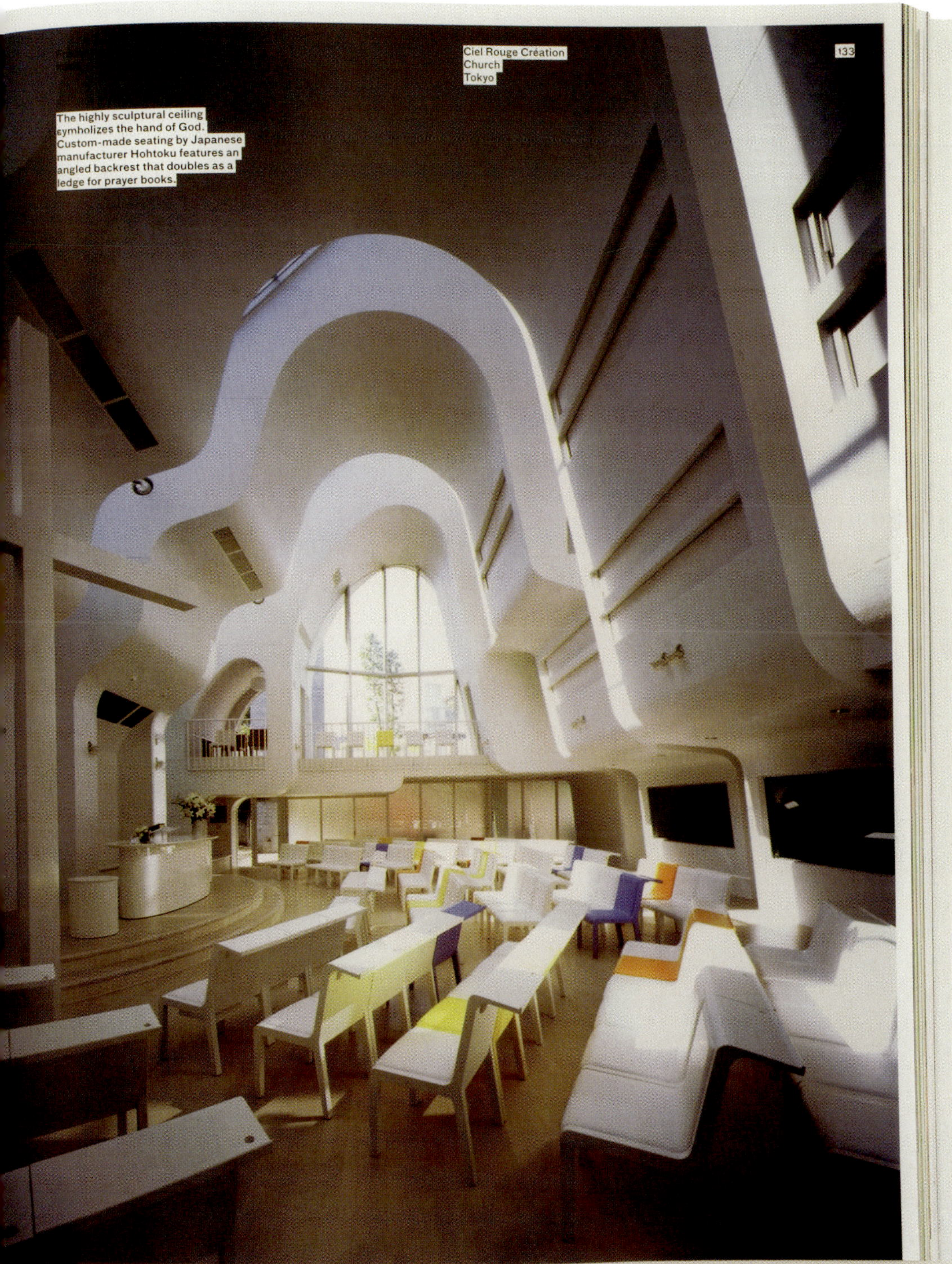
Ciel Rouge Création
Church
Tokyo

133

The highly sculptural ceiling symbolizes the hand of God. Custom-made seating by Japanese manufacturer Hohtoku features an angled backrest that doubles as a ledge for prayer books.

SUBJECT
Church

LOCATION
Tokyo, Japan

DESIGNERS
Ciel Rouge Création

PHOTOGRAPHER
Toshihisa Ishii

WRITER
Kanae Hasegawa

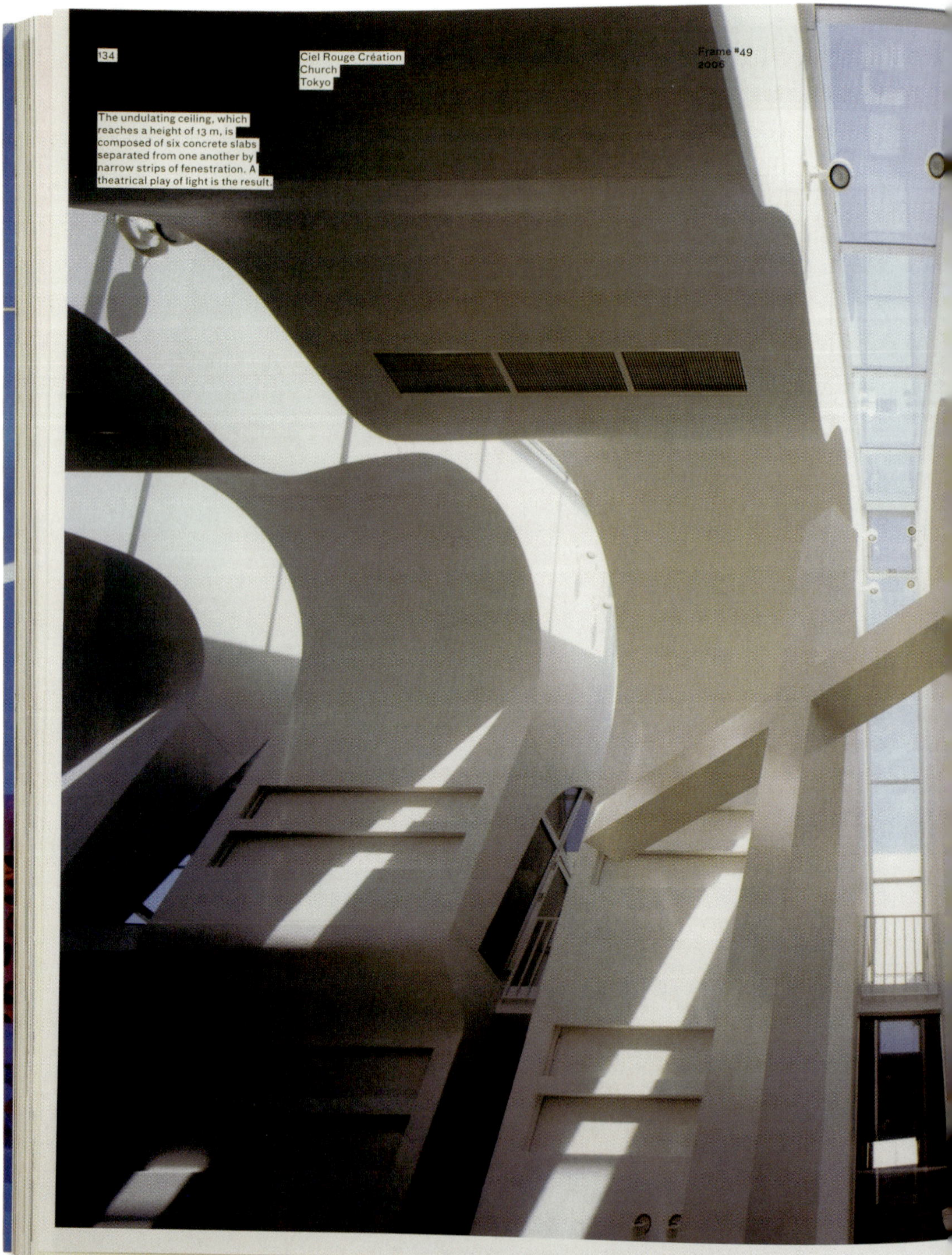
134
Ciel Rouge Création
Church
Tokyo
Frame #49
2006
The undulating ceiling, which reaches a height of 13 m, is composed of six concrete slabs separated from one another by narrow strips of fenestration. A theatrical play of light is the result.

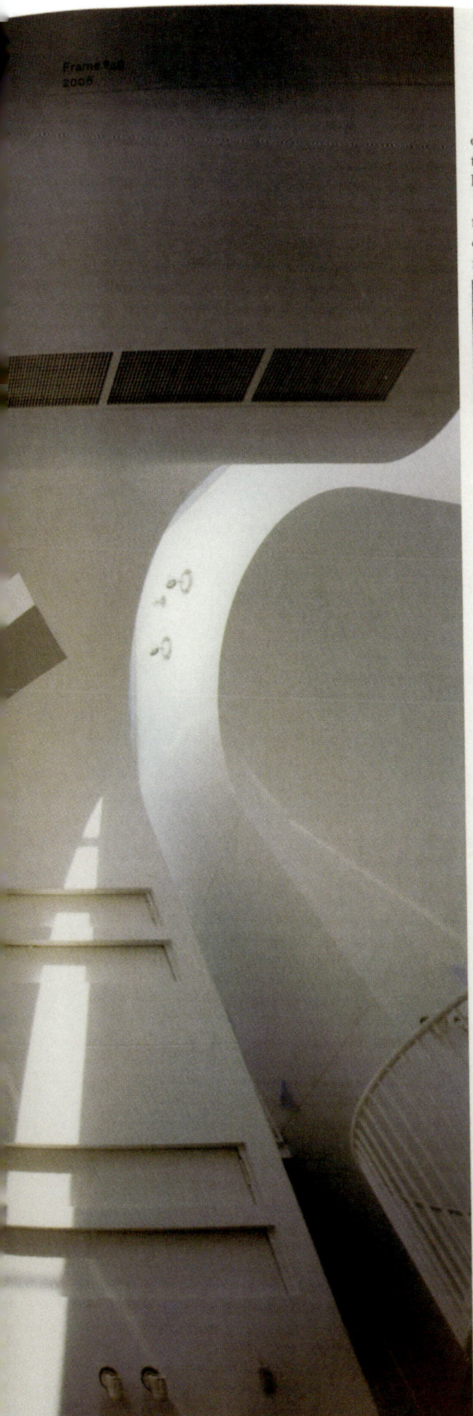

entire length. This prominent beacon sends out a message to all who pass this place: you are in the presence of the house of God.

The first encounter with the curvaceous concrete façade of what must strike many an observer as a gigantic contemporary sculpture is not particularly open and friendly, however. A trio of soaring tongues licks the

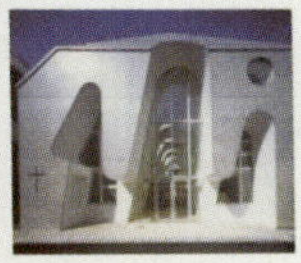

frontage, almost forcing the perils of Pentecost on unsuspecting viewers. A sharp contrast to this rather overwhelming welcome is the interior of the church, a surprisingly light and cheerful space. Visitors enter a 188-sq-m white chapel filled with daylight and capped with a gracefully curving 13-m-high ceiling sliced vertically into six structural concrete modules that are connected by fairly wide strips of fenestration. The components of this voluminously arching ceiling cast a theatrical play of light and shadow into the nave of the church. The organic forms overhead, which resemble the work of Alvar Aalto or even a blanket of clouds with light filtering through, are explained by Henri Gueydan, head of Ciel Rouge Création, who says that 'the design finds its inspiration in the Bible', emphasizing that the Good Book was the architects' sole point of departure. 'I imagined the chapel covered with the Holy hand, as if Christ's palm were enclosing all people within it.'

The cross, which stands beside the altar, can also be seen as the mast of a ship, another biblical image. Even the number of glass openings in the ceiling has an explanation: the bell tower outside and the six structural modules connected by the strips of fenestration add up to seven, a number that symbolizes the seven days of creation recorded in Genesis.

Cross Section
1. Chapel
2. Small chapel
3. Offices and communal rooms
4. Nursery

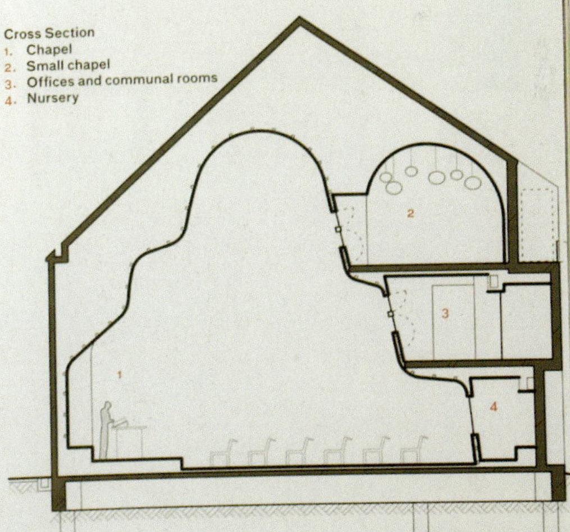

WRITER Kanae Hasegawa

PHOTOGRAPHER Toshihisa Ishii

DESIGNERS Ciel Rouge Création

LOCATION Tokyo, Japan

SUBJECT Church

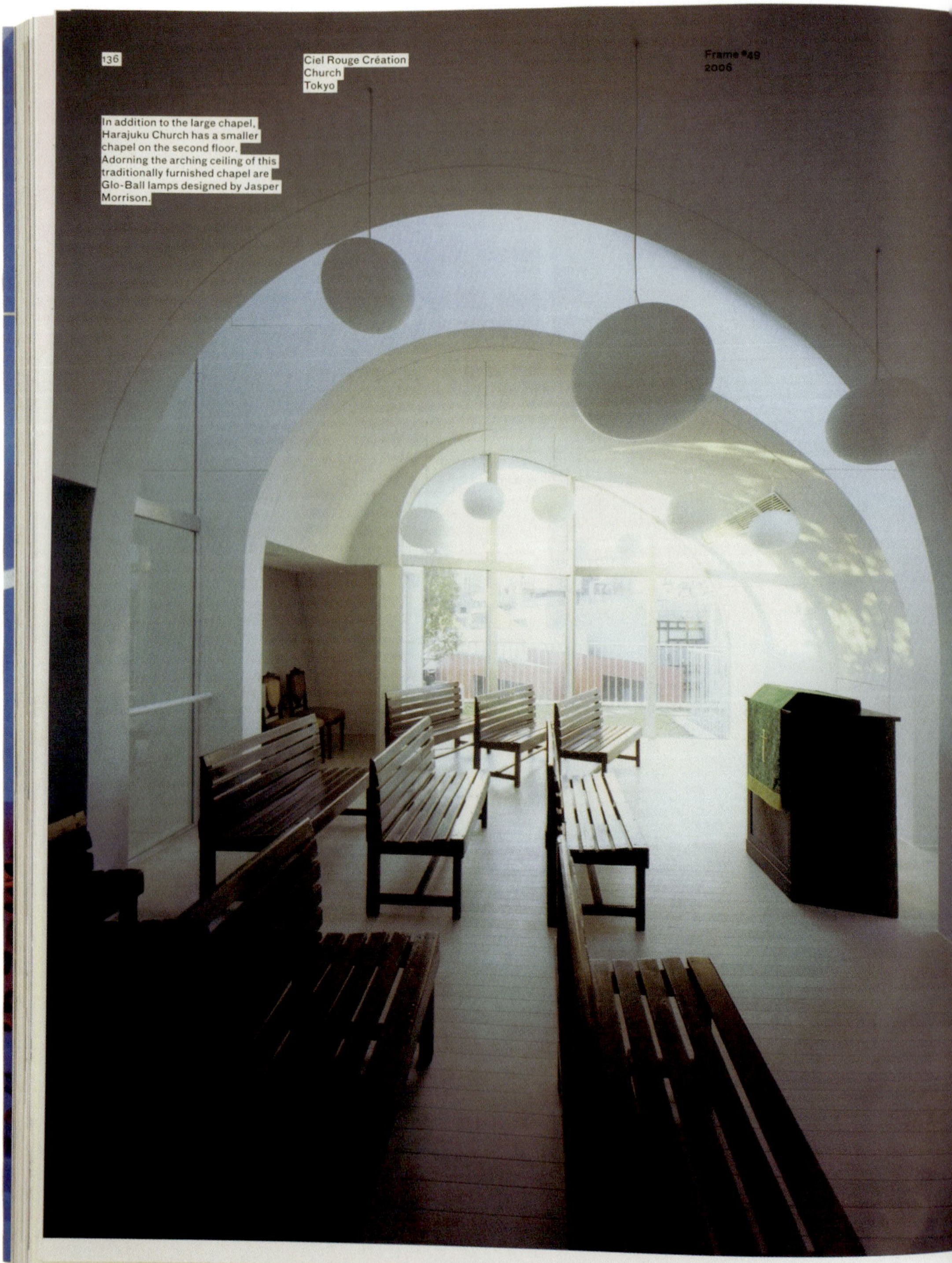

136

Ciel Rouge Création
Church
Tokyo

Frame #49
2006

In addition to the large chapel, Harajuku Church has a smaller chapel on the second floor. Adorning the arching ceiling of this traditionally furnished chapel are Glo-Ball lamps designed by Jasper Morrison.

Entering Harajuku Church takes you into a biblical world whose scenes unfold before your eyes. But 'seeing the story of Christ', says Pastor Tsuchihashi, is not the main experience offered here. 'The service is the most important thing for Christians,' he says. 'The chapel needs to function properly, and the curves of the ceiling are an essential element. It's not just a decorative ceiling. It was also designed to provide our choir with excellent acoustics.'

'I wanted to sculpt the light so that you feel the light and become the light.'

Henri Gueydan

Apart from honouring the wishes of the church to be recognizable from the street and to be visibly open to one and all, Ciel Rouge Création also thought of the children from the adjacent school, of neighbourhood residents and of visitors requiring special facilities. The side aisles and part of the first floor double as a place for after-school workshops for kindergarteners. The interior includes a kitchen where people can prepare food for church holidays and other events, as well as lifts and wheelchair-accessible doorways for mobility-impaired visitors.

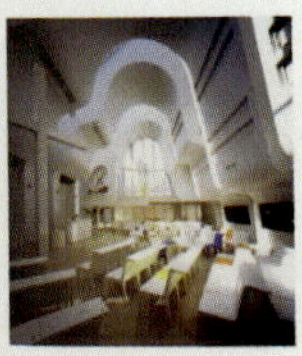

Chairs arranged in aisled, semicircular rows facing the altar, which is located at the south side of the chapel, are made by Japanese manufacturer Hohtoku. 'The service should not be about imposing dogmas as the truth,' continues Pastor Tsuchihashi. 'The church offers a time and a place for people to be with God in an open atmosphere.'

Several coloured chairs interspersed among the white ones are another suggestion that this church is not just for Christians, but open to everybody. And during the service, which begins in the early morning and lasts until noon, the sun moves slowly from one opening in the ceiling to another, beaming down on each worshipper in turn, casting a blessing on everyone in the church. Even those without knowledge of the Bible are sure to experience Harajuku Church as a work of art. In the words of Henri Gueydan: 'I wanted to sculpt the light so that you feel the light and become the light.'

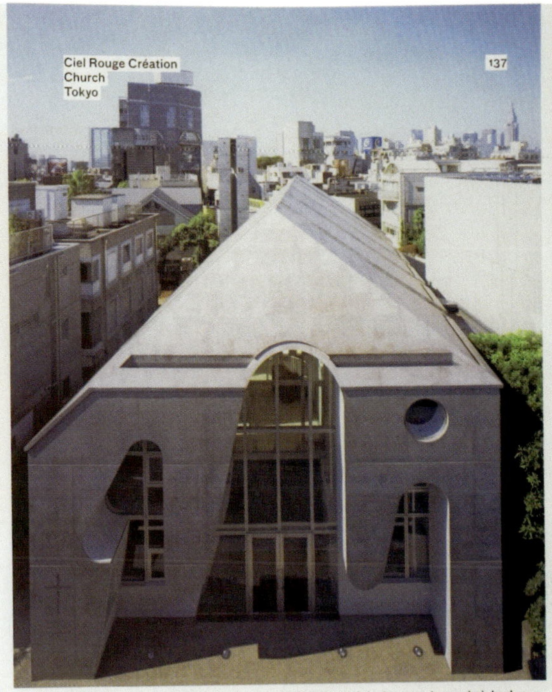

The church in Aoyama is sandwiched between office buildings and retail establishments. Thanks to the renovation, which added a striking façade and a 16-m-high bell tower, the church is now far more conspicuous.

BACK PAGES

537-592

Look Back

HISTORY & INTERVIEWS

The story behind 50 issues of Frame. How did the magazine start? Who are the founders? And: meet the art directors who gave the magazine its distinguishing looks.

FIRST ADDRESS
Nieuwe Spiegelstraat 36
1017 DG Amsterdam

TEXT
Merel Kokhuis

PHOTOGRAPHY
Krista van der Niet

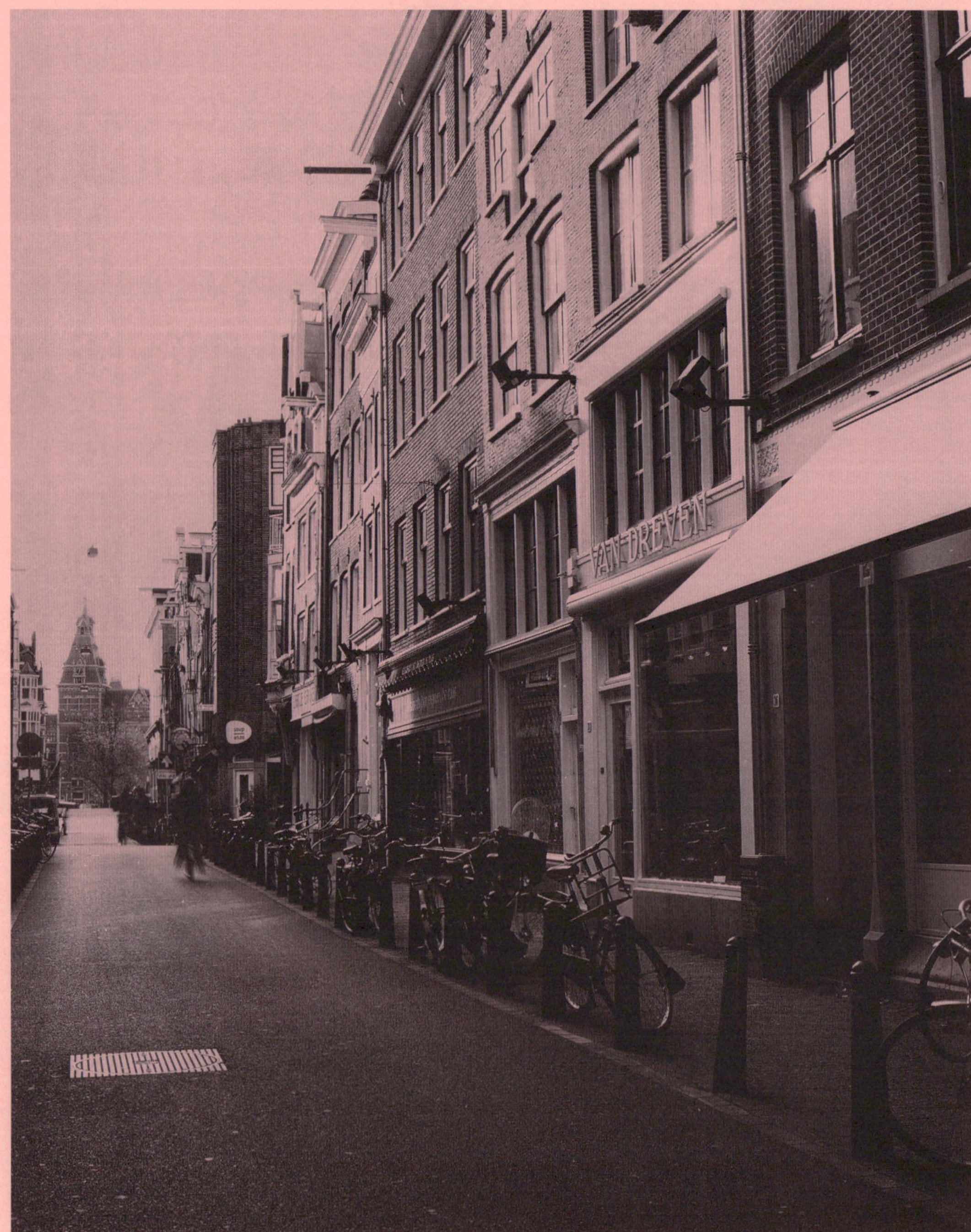

Frame and Fortune

Two magazines, eight monographs, ten books, two moves to new locations and a growing staff currently numbering sixteen. It's the dream of any dedicated entrepreneur on the brink of launching a magazine. A success story? Yes. But it wasn't always smooth sailing.

It all started in 1995, in a 'Stube' in Hanover. Peter Huiberts (now publisher of Frame) and Robert Thiemann (Frame's editor in chief) were at a trade fair in Germany representing their employer, a Dutch interior-design magazine. Two guys bursting with ideas on how to improve the magazine with a fresh approach, if only their boss would give them the go-ahead.

What can you do when you're frustrated, with no way to realize your ideas? Exactly. You consider the possibility of venturing out on your own. One drink led to another. One idea led to another. They added to the stack of note-filled beer mats between them on the table. What's the best feature of this hot magazine? What stands out in that one? What do you think of this?

Their brainstorming session soon produced a cocktail of good ideas and a concept for a magazine that would fuse British and American journalism with the aesthetic flair of an Italian publication. They speculated on the idea of Frame as a Dutch-language magazine, but in an attempt to reach as wide an audience as possible, they opted to go international. It was an ambitious undertaking. A concept they believed in without reservation.

SECOND ADDRESS
Oude Braak 16hs
1012 PS Amsterdam

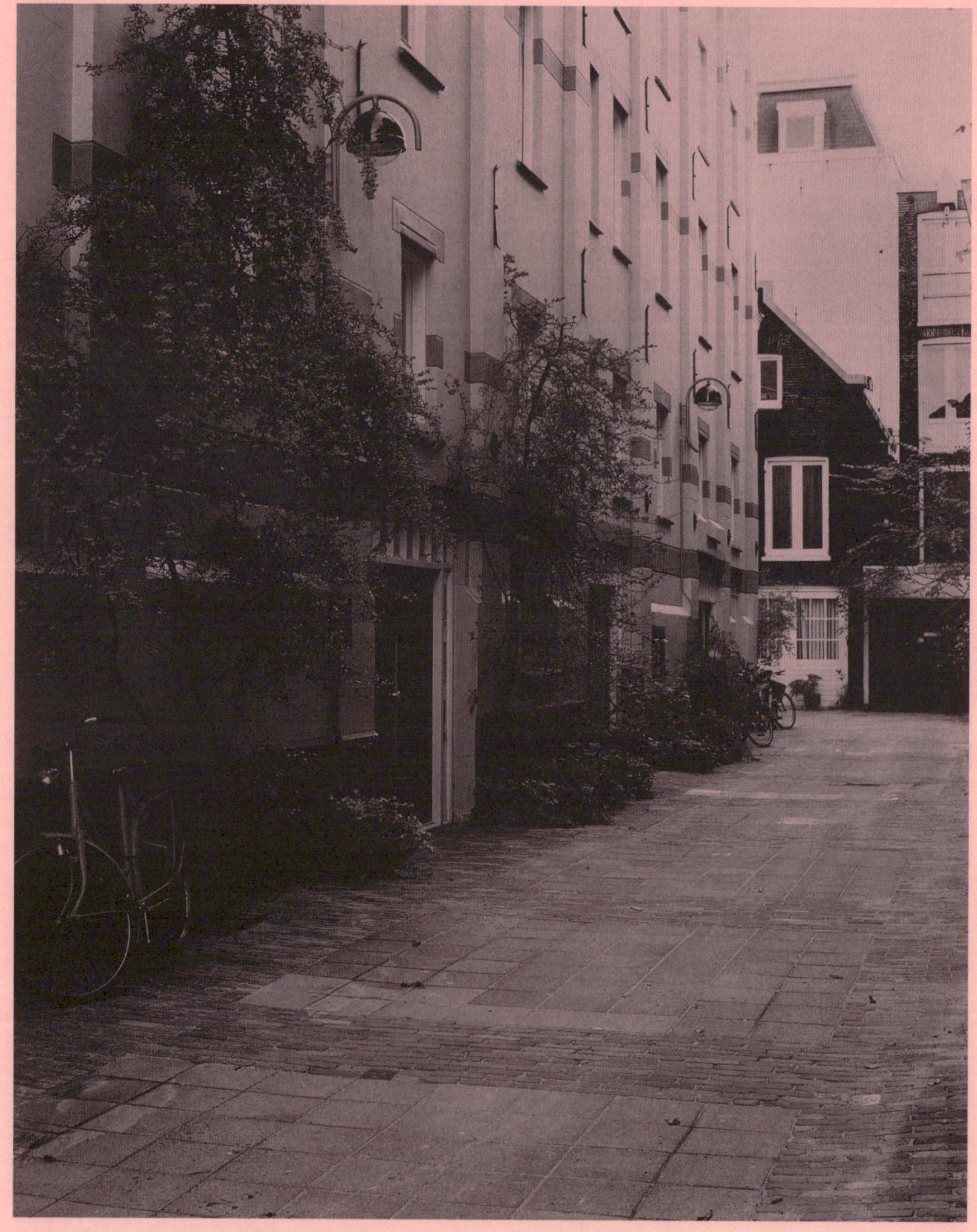

What can you do when you're frustrated with no way to realize your ideas? Exactly. You venture out on your own.

They needed a business plan. In black and white. They organized their thoughts, composed the plan, and took it to every printer and publisher around. More than a year passed before Rudolf van Wezel (then owner of BIS Publishers) saw the undeniable potential of the concept He offered Peter and Robert the attic at BIS Publishers on Nieuwe Spiegelstraat in Amsterdam for their project, a location not far from the Rijksmuseum. Badly in need of addresses they contacted designers associations in several countries. In the autumn of 1997, BIS Publishers mailed 26,000 copies of the first issue of Frame to potential subscribers, free of charge.

With bated breath, Peter and Robert spent the following days and weeks listening for the hum of the fax machine, the tread of the postman, the ring of the telephone. Waiting for subscribers. The number grew rapidly. In no time, Frame had 1000 subscribers. It's less amazing than it sounds when you know that with each subscription to Frame, the new reader received a free watch worth even more than the six magazines they were ordering.

Peter and Robert were having the time of their lives. They discovered that the best way to generate brand recognition was to show their faces at international trade fairs. They booked the cheapest hotel they could find, arrived armed with stacks of Frame magazines, and bartered to get a spot at the fair. Organizers willing to 'donate' a stand to the cause got a free advertisement in Frame in return. Their nights were filled with talk about the many enthusiastic reactions that were pouring in from people – more and more often VIPs – in the design world. Sleep was not only a waste of time but also next to impossible, thanks to the rush of adrenaline in the wake of all those positive reactions.

Peter and Robert spent the following days and weeks listening for the hum of the fax machine. Waiting for subscribers.

Two years into the venture, the the print run had decreased to mere 8000, and BIS Publishers was facing serious problems. Peter, in particular, had started to lose faith in Frame.

Although the concept clearly had considerable potential, the cost of running a magazine was more than he had anticipated. But they refused to give up. The bad patch was followed by an upswing that continues to this day. The number of subscribers is still growing steadily.

In 2001, Van Wezel sold BIS Publishers to Weekbladpers, a magazine-publishing company. The establishment of Frame Publishers, which occurred at the same time, has allowed Frame to be published independently. The magazine and its staff of four employees moved into a basement location on Oude Braak, close to Amsterdam's Central Station. What came next was a struggle to achieve widespread brand recognition, to shape the desired image for the magazine and to optimize worldwide distribution. The team succeeded in creating the perfect combination of cult and commerce. After hiring two more employees (to handle subscriptions and to reinforce the editorial staff), Peter and Robert needed more space. Their eyes fell on a corner location on Lijnbaansgracht, at the edge of Amsterdam's Jordaan district.

With the magazine moving full speed ahead, it was time to expand the firm's activities. In 2001 Frame Publishers issued its first book, a monograph of Fabio Novembre, the first of nine books devoted to a single designer or design team. Number ten was an imposing book on the interior design of remarkable workplaces: The Other Office. This volume marked the start of a series that includes shops (PowerShop and Dress Code), shop windows (Forefront), exhibition design (Grand Stand), clubs (Night Fever) and materials (Material World I and II). Two books scheduled for publication in late 2006 focus on restaurants and events.

The next addition to the company portfolio emerged in 2005, the year in which the first issue of Mark: Another Architecture rolled off the presses. A magazine with the same sense of charisma and the same area of distribution as Frame, Mark is a quarterly that features extraordinary architecture.

Rather incredibly, although it's been only eight years since the appearance of Frame 01, Frame Publishers can take credit for two magazines, eight monographs and ten books. Frame currently boasts a print run of 34,000 and more than 10,000 subscribers worldwide.

The future beckons. Seeing that the number of employees has doubled in the past year – Frame Publishers now has a total of 16 permanent staff members – the search is on for yet another, larger office space. Peter and Robert are concocting plans for a Frame Design Centre, a complex that will bring together design-related companies and people with a love of design.

Just imagine: a building that accommodates a design shop, a design hotel, showrooms displaying designer furniture and related products, and the offices of two magazines of the calibre of Frame. A dream on the distant horizon, or reality within reach? <<

CURRENT ADDRESS
Lijnbaansgracht 87hs
1015 GZ Amsterdam

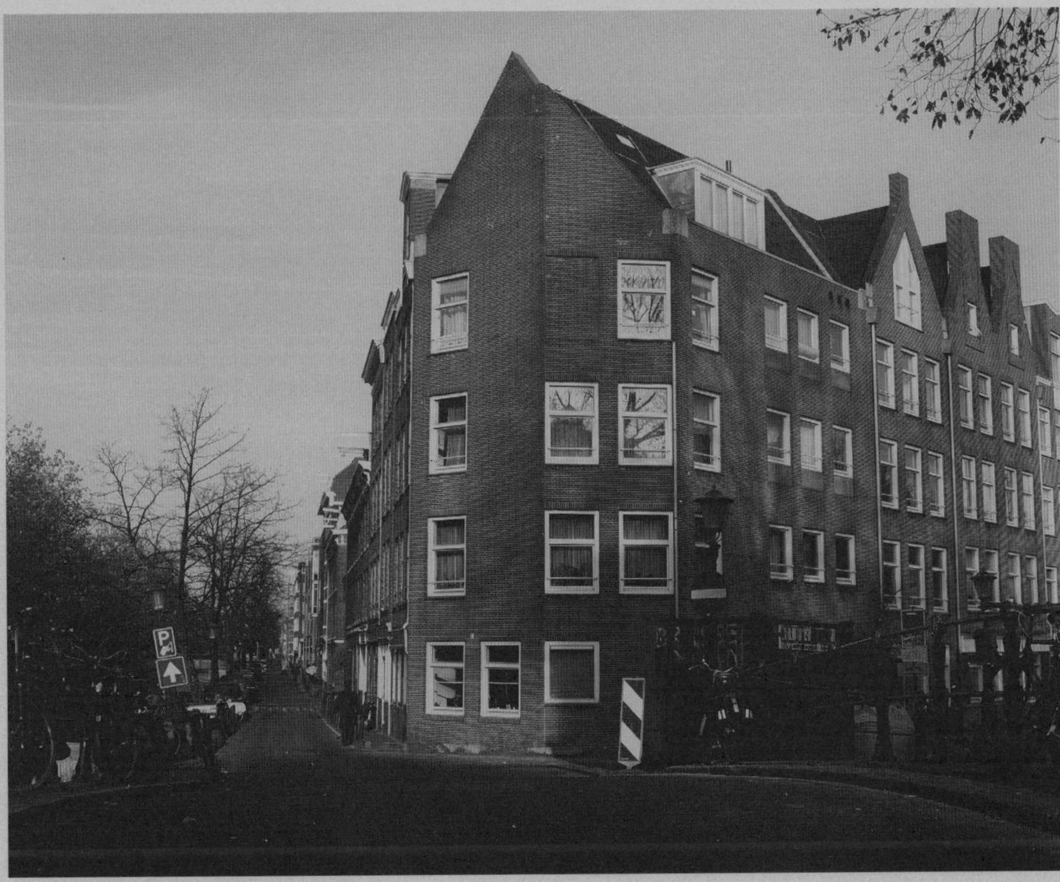

Editor Interviews Publisher

Once more in the role of a journalist, editor in chief Robert Thiemann interviews *Peter Huiberts*, the publisher of *Frame*.

Although we seem to have reached calmer waters, I must say that from the outset you and I have had some rip-roaring debates about the covers of *Frame*. What's your idea of a good cover? And do you have a favourite *Frame* cover?

A cover should be intriguing. You can get that effect by using colour or a striking image. But certain covers that I thought would have a terrific impact turned out to be flops – to the extent that we're able to judge such things, of course. Our more recent covers have struck me as being rather dull. The issues of Frame that have made the biggest positive impression on me are numbers 02, 07, 16, 25, 31, 38, 42 and 43.

Generally speaking, which other magazines – including football periodicals – have covers that grab you?

One magazine that features good covers is the Australian publication Inside. No text, no fuss, just a really great image. All the rest do more or less the same thing we do, sometimes scoring and sometimes missing the goal.

Imagine yourself as the editor in chief of *Frame* for a single issue. What would you change about the editorial formula currently in operation?

If I was sure my appearance was a one-time thing, I'd devote the entire issue to projects. Let's say 100 projects or so, each presented on one or two pages. A whole bloody barrage of projects, one after another. Then I'd sit back and wait for the reader's reaction.

What is a successful issue of *Frame* in your book? And what role does advertising play in your enjoyment of the magazine?

I have a curious preference for shops and stands. Can't figure out exactly where that comes from. And, yes, I do love great-looking ads. Bring them on! Coming across an ad that's more striking than the illustration for the adjacent article – now there's a kick.

Will you explain to me once more why advertising from the fast-moving consumer-goods sector appears in most design magazines, but not in *Frame*? Are we about to cross swords?

[He laughs.] O.K. Listen up. Unlike most design magazines, Frame has a worldwide distribution. But, to quote Johan Cruijff in reverse, every advantage has its disadvantage. The disadvantage is that because big corporations often have national budgets, they don't know how to deal with a magazine that operates internationally. This leads to a situation in which everybody wants to advertise, but nobody wants to pay. What's more, I think that a magazine gets the advertiser it deserves. Maybe we don't deserve what we don't have. Do you actually write about 'fast-moving consumer goods'? What a term. Who comes up with stuff like that?

I know that you once sold ads for some fairly obscure magazines. What's it like to work in the plush world of design and in the serious arena of architecture?

Boy, have I worked for some crazy publications. One targeted a readership made up of horse breeders. There I was, ringing the bell at a stud farm that wanted to place an ad for its prize stallion. I also recall a magazine for a hotel chain that was going downhill fast. Until it occurred to me to approach the red-light district. But I've always had a soft spot for architecture. I once started a paper for the building industry, and I was also involved with a magazine on lighting.

Is there a magazine you wouldn't work for?

Can't think of anything that I'd consider below me, as it were. Not only that, but as an ad salesman or a publisher, it's better to throw in your lot with a deplorable magazine than with one that's already a big success. If it's really bad, you can go buggering about without doing any serious damage.

You're a big football fan. Which football magazines would I find on your coffee table?

To begin with, I don't have a coffee table. And if I had one, you probably wouldn't find any football magazines on it. I buy a copy of VI [Voetbal International, a Dutch publication] once in a while, but that's more for my son Finn than for myself.

You like to compare design to football. In your opinion, which design magazines would earn a berth in the quarterfinal of the Champions League? And while you're at it, why not match each magazine with a soccer club?

Here goes. Domus, the 'grande dame' (Juve) plays Real Madrid (On Diseño). Olympique Lyon (Intramuros) goes up against Dinamo Moskou (Monitor). And AIT (Bayern Munchen) meets the surprising Swiss of Hochparterre (Grasshoppers). The fourth match is a battle between AZ (Frame) and Wallpaper (Chelsea). The home game is no problem. But away?

Now that we're on a roll, which eight manufacturers of design products will be in the quarterfinal of the next World Cup in Germany?

That's not a question I can answer. I'm biased. And you've already answered a similar question.

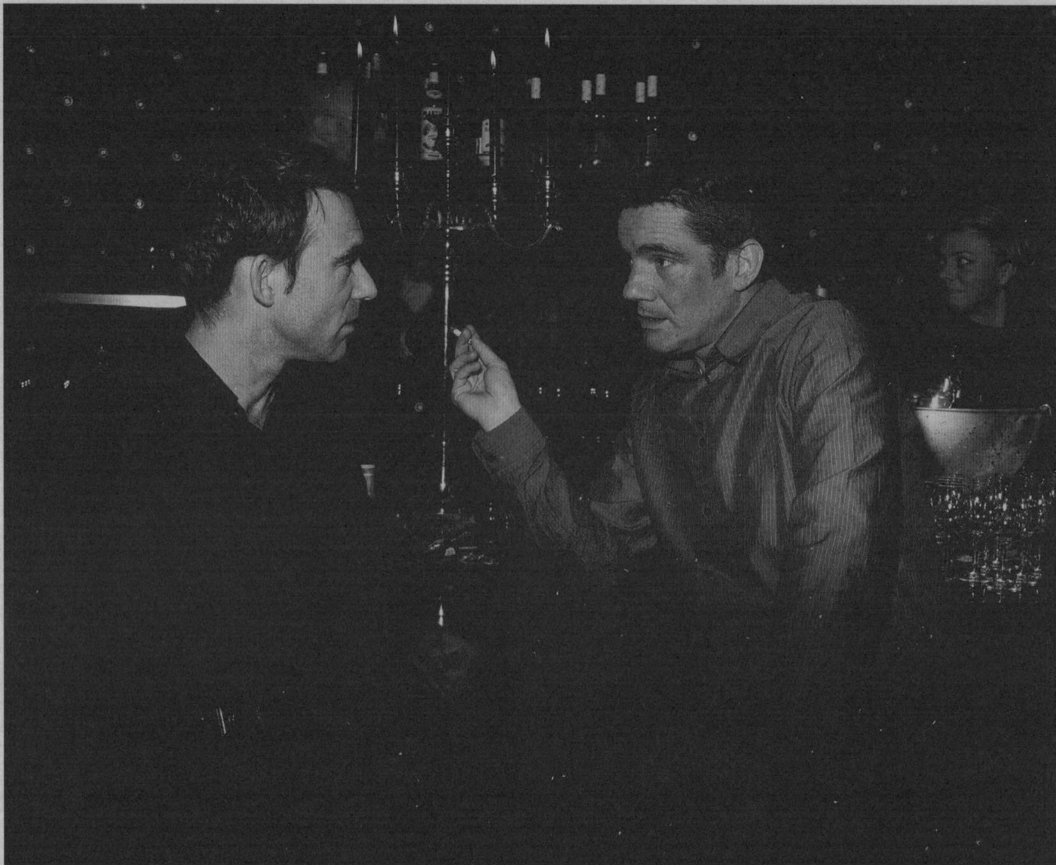

EDITOR MEETS PUBLISHER
Robert Thiemann (left) and Peter Huiberts at the launch party of Mark magazine

PHOTOGRAPHY
Krista van der Niet

You're not a guy who puts much stock in outward show. Do you have any designer furniture at home?

Not much. As a rule, designer furniture isn't very comfortable. The pieces I do have come from Hidden, Plank, Vitra and Pastoe.

When we started eight years ago, there were only two of us. Since then we've evolved into an organization with a staff of 16. How much more do you want the business to grow before you retire?

At the moment, I don't want it to grow at all. But I may not feel the same way tomorrow. My ideal is a future in which we make more books and, in time, possibly a third magazine. <<

Publisher Interviews Editor

Peter Huiberts, publisher of *Frame*, corners editor in chief *Robert Thiemann* for an exclusive one-time interview.

You get truckloads of projects from designers eager to be featured in *Frame*. How do you select the ones that make it?

We're always on the lookout for a good balance between culture and commerce, although we do occasionally lean in one direction or the other. Obviously, both interiors and products have to have an innovative aspect, and we prefer designs that are well made. A concept on its own is not enough. To be honest, most of our choices are based on a gut feeling and on the initial visual impression. If our first look at the images doesn't produce an immediately 'wow', a project has practically no chance of ending up in Frame. The toughest cases are those that push our doubt button: the ones we end up discussing, contemplating, questioning and viewing from every possible angle. Until another 'wow' project shows up, with no doubts attached, at which point the problem child often gets shifted to the bottom of the stack. We do make mistakes, of course. It's not easy to reject people, especially when you can see that they're tremendously proud of their work. And most designers fall into that category.

If you had to choose a favourite issue of *Frame*, which would it be, and why?

The issue that we're working on is always my favourite. It still holds a sense of promise. You think it will be terrific. Perfect. Finally, however, after you've worked on it for two months and it comes back from the printer's, the main feeling is usually disappointment. You spot mistakes. Things that could have been better or looked better. Ahead of you, though, is

PETER HUIBERTS:

Personally, I thought the covers of Frames 14 and 34 were disastrous. Can you name a few that you really like? And don't come right back at me with 14 and 34.

—

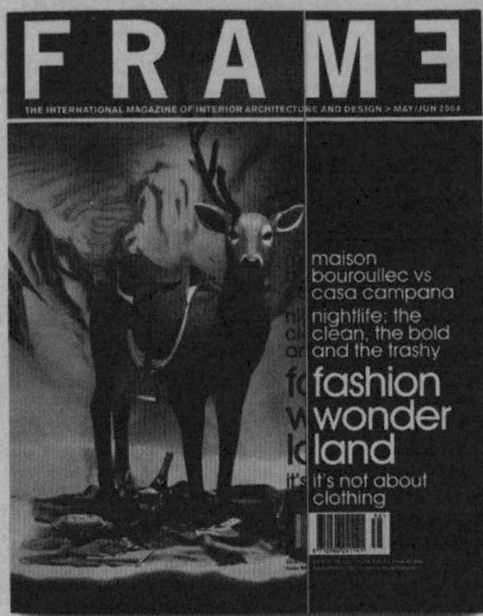

PETER'S FAVOURITE
Frame 38: One of those covers that has made the biggest impression on me

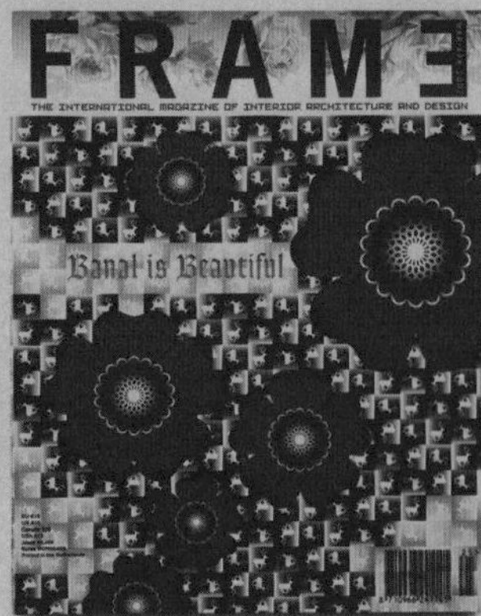

ROBERT'S FAVOURITE
Frame 25: It's attained a cult status now that not a single issue is available – other than at my house of course.

another chance to make the best issue ever. I find it extremely hard to look back at past issues and pick a favourite.

Personally, I thought the covers of *Frames* 14 and 34 were disastrous. But you didn't feel the same way about them. We're frequently at loggerheads when it comes to covers. Can you name a few that you really like? (And don't come right back at me with 14 and 34.)

I agree with you that Frame 34 wasn't one of our better covers. That was a compromise, pure and simple (no need to go into detail). The cover of Frame 14, which showed shoes behind a chalk line drawn on the floor, is still on my list of favourites, though. Very minimalist, very special and a statement on retail design. Others are the covers of Frame 25 (it's attained a cult status now that not a single issue is available – other than at my house, of course), 37 (strong, surprising image), 42 (also rather alienating, but the colour is powerful and the gold foil is a nice touch) and, yes, I like 48 (not so characteristic of Frame, maybe, but a great combination of image, typography and refinement).

A new issue of *Domus* often sends you into a blue funk. You used to have the same reaction when *Wallpaper* arrived. Do these magazines still get you down? Can you explain why?

This is becoming very therapeutic. I'm a big fan of both Domus and Wallpaper. I see Domus as an institute in the fields of design and architecture. Wallpaper had stunningly styled articles, especially in the early years, and it was always the first magazine to introduce an important new project. I also loved Wallpaper's tone of voice. And, yes, I used to be pretty miserable when one of those two magazines arrived at the office. They seemed trendier, better, more attractive and more authoritative than Frame was. And I was sure all the readers of Frame were thinking the same thing. I admit that it bordered on sheer self-destruction to study those magazines month after month. Nowadays, I rarely look at Wallpaper, and it takes me five minutes or so to page through Domus. But I still admire them very much.

Which manufacturers of design objects do you admire in a big way? Actually, it doesn't have to be 'in a big way' – or is that comment out of line?

I admire quite a few manufacturers. Examples that spring to mind, in no particular order, are Moroso, Moooi, Magis, Lapalma, Flos, Vitra, Plank, Kartell and a long list of Italian firms that make sanitaryware – oh, and I wouldn't want to forget Dornbracht, Duravit, Modular Lighting Instruments and Alu. I appreciate them for various reasons: for the way a collection is put together, for a certain way of communication, an excellent nose for talent, a desire for innovation. And a manufacturer can have a mediocre season, only to come back the next year with strong products that really impress me.

Imagine an international tournament of design – and you get to pit the best eight designers against each another in the quarterfinal.

Assuming that each designer has to represent a different country, I can picture an exciting quarterfinal with the following players: Konstantin Grcic (Germany) vs Erwan and Ronan Bouroullec (France), Hella Jongerius (Netherlands) vs Patricia Urquiola (Spain), Jasper Morrison (UK) vs Tokujin Yoshioka (Japan),Marc Newson (Australia) vs Antonio Citterio (Italy). Please don't ask me to pick the winner.

ROBERT THIEMANN:

The issue that we're working on is always my favourite. It still holds a sense of promise. You think it will be terrific.

—

Famous designers have a certain celebrity status. And some are undeniably stars with a capital S. Are you ever recognized in public?
Fortunately . . . no.

You're sometimes asked to talk about design at trade fairs or conferences. Do you believe what you're saying, or do you just do what you think is expected of you?
You know me. I'm a dead-serious guy. I arrive thoroughly prepared for every lecture, which I'm happy to say are few and far between, and do my best to give a talk tailored to the occasion. And, yes, I do believe what I say. Isn't that what's expected of me?

We're entering 2006 as we speak. Do you see a major trend in design on the horizon?
I'm gazing deep into my crystal ball. The huge boom in decoration and ornament is on the downswing. The current trend of assembling teams of big-name architects and designers to create an interior will continue for another season. Look for the first signs of ethnic influences in Western design, as well as for the emergence of wild new mixes of materials, such as furniture that features a fusion of leather and ceramics or leather and rotomoulded plastic. And be ready for extraordinary, even outlandish, colour combinations. Manufacturers of design will finally realize that haute couture is a must for enhancing a corporate image, and that there's money to be earned with diluted designs and basics. As we've already seen in the world of fashion.

What if you were to be transferred to another magazine? Give me the best scenario. And the worst.

Continuing my earlier train of thought, it would be a great honour to find myself the editor of Domus. And a magazine that would force me to kowtow to the advertising department would be out of the question, as would a magazine whose design and content meant nothing to me.

You are now responsible not only for *Frame*, but also for *Mark*, the firm's new architecture magazine.Do you have more of an affinity with architecture than with design?

No, I wouldn't say that. As for architecture, I'm still feeling my way around. Fortunately, I have some very good colleagues to go to for help.

I think of architects as a bunch of conceited complainers. At least designers seem to be up for a good time on occasion. Do you agree?

I'm starting to discover that architects are a tad more serious than designers. What we're attempting with Mark, for example, is simply not done in the realm of architectural publications: all those gridded insets and overlays, the wild fonts and cutting-edge photography, the illustrations, the graffiti. Designers are far less critical of experimentation. And I've had some really good discussions with designers who also turned out to have a great sense of humour. A few laughs. A couple of beers. A regular crowd. It will be interesting to see whether or not that type of architect crosses my path one of these days.

We set up *Frame* without being inhibited by an avalanche of personal knowledge on design. But you were amazingly fast at absorbing all the information needed for the task. Do you think you could edit a gossip magazine?

I could if it featured the private lives of designers or architects.

You studied to be a chemical engineer. How about the chemistry between Robert Thiemann and *Frame*? Is it still there?

More than ever before.

How much design can an ordinary person tolerate?

Practise moderation and enjoy.

You're probably the most fashionable guy at *Frame*. Do fashion and design always go hand in hand?

Thanks for the compliment. Yes, I think the two invariably go together. All the designers I know pay attention to how they dress, though some are more into fashion than others, of course. You see some fairly vain characters at furniture fairs. Just walk into Bar Basso some night during the Salone in Milan.

Is your house a kind of design museum, or do you buy your furniture at Ikea?

I have to admit that it's become something more of a design museum in recent years. But we also have quite a few things from Ikea, some second-hand stuff and more than a few odds and ends. My wife is no design freak, and my kids turn it all into a muddled mess anyway. It's a nice mix, and I like it.

At about the same time this book comes out, the World Cup will be under way. Who do you see in the final?

Brazil – England.

Do you ever talk to your children about design?

I'm trying not to contaminate them just yet.

Would you like to say hello to anyone?

My father and mother. <<

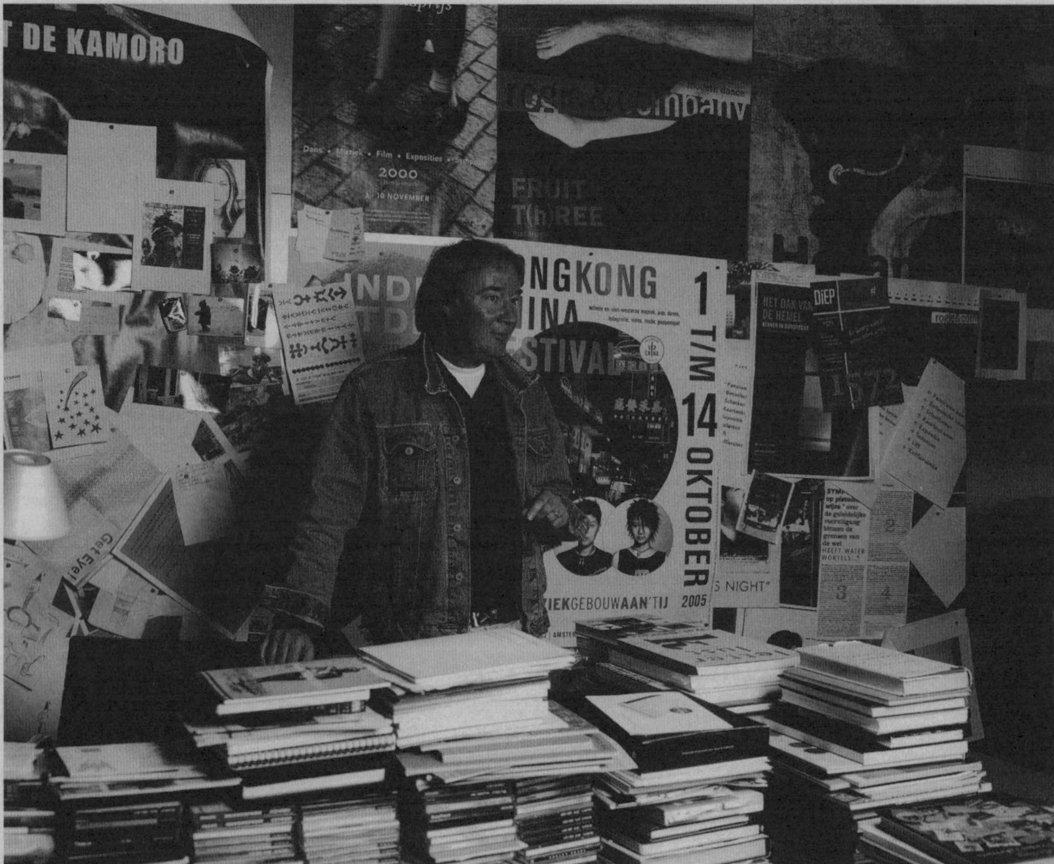

TEXT
Eva Schaap

PHOTOGRAPHY
Krista van der Niet

Eight years of *Frame: The Great Indoors* adds up to over 50 issues. Along with a number of editors, correspondents and photographers, four individuals or duos have shaped the identity and character of *Frame* in their capacity as art directors. Much of the groundbreaking work in the late 1990s was the responsibility of *Frame*'s first art director: *Ton Homburg* of *Opera Designers*.

00-17

Ton Homburg

OPERA DESIGNERS, BREDA

BACKGROUND

Ton Homburg (1952), a graduate of St Joost Academy of Fine Art in Breda, is the cofounder of Opera Designers. Homburg put his stamp on the first 17 issues of Frame, serving as the magazine's art director from December 1997 through 2000. In 1994 Opera, which had been operating in Breda since 1981, expanded to include a studio in Amsterdam. In addition to his work for permanent clients such as the National Museum of Cultural Ethnology in Leiden and Dutch magazine de Architect, Ton is currently involved in the design of books, museum interiors and shops, among which a Mexx outlet in London that opened in autumn 2005.

A stack of articles, a sheaf of photographs and some ideas on design that had been incubating in the minds of editor in chief Robert Thiemann and publisher Peter Huiberts – all aimed at a magazine they planned to call *Frame*. Doesn't sound like the easiest job for a designer.

I had known Robert before I became the art director for Frame, so I was already familiar with his thoughts on making a magazine. But you're right. It wasn't easy. At the beginning we did a lot of talking and a lot of experimenting, because in one way or another you have to make clear what you do and don't want in your magazine – what works and what doesn't. The only way to find out is to discuss what you'd like to do and then do it. After all, the content and design of Frame are what create the feeling that the magazine conveys to the reader. That feeling, however, is something you have to search for – especially at the beginning. In the three years that I was the art director at Frame, the magazine was never based on a hard and fast concept. Each new issue was the result of changes and adaptations, either because certain things were not working well or because we wanted to try something different. Art direction, which contributes to the identity and character of a magazine, is something that has to grow little by little. Ultimately, though, it comes down to common sense and a handson approach to the work.

Designing a magazine like Frame can be compared to making a painting. Articles and illustrations are colours on the palette.

So those three years with *Frame* demanded a great deal of your time and energy.

You've got to give it your all and really want to do it, or it won't work. I think the key is to believe in what you're doing. From the first moment that Robert came to me with his ideas, I believed the concept would work. At that point you go for it – in this case, making the best magazine possible. Whether or not you can actually do it is not even a point of discussion, especially at the start. The ideas and your faith in those ideas are far more important.

What was your approach to crafting a 'look' for each issue of *Frame*?

Before I even got to work on designing the upcoming issue, the editorial staff often had a general idea of what they wanted. To get the same feeling, I not only looked at the visual material, but also read several of the main articles. After that, it was a matter of ongoing discussion, trial and error, adaptation and modification. I always took the stance that basically every change or attempt was possible. Hard and fast design concepts didn't work for Frame or for me.

My contribution to the magazine was based largely on the essence of the material I got from Frame, and that goes for words as well as pictures. It was all part of a continuing conversation between them and me. There's no reason why an art director can't show his face, so to speak, but his presence shouldn't be to the detriment of text and images. The way we worked really did give Frame a chance to be the best, issue after issue. Talking, exploring, tackling something new – all that, together with the time, energy and creativity that we put into it, is what made those first years so much fun.

Are there issues that gave you more satisfaction than others? Maybe one or two that you remember with particular pride?
I recall a couple of covers that left me dissatisfied, especially in retrospect. In those days the cover was always a photograph already accompanying an article in the magazine, a system that worked better on some occasions than on others. At times it was virtually impossible to find a cover among the images used to illustrate the stories, and that made it very difficult for me. I'd come up with maybe six or seven covers, none of which was exactly what we had in mind. The cover I like best was the one we selected for Frame 14 [May/June 2000], which features three pairs of black men's shoes lined up behind a straight white chalk line drawn on a wooden floor. It shows a fragment of an Ann Demeulemeester shop in Belgium. All she did was draw the line and set the shoes next to it. Super. It's not only a great image of an interior detail but also a relevant piece of information about that particular issue: it reflects the contents of a magazine that highlighted, among other things, retail design.

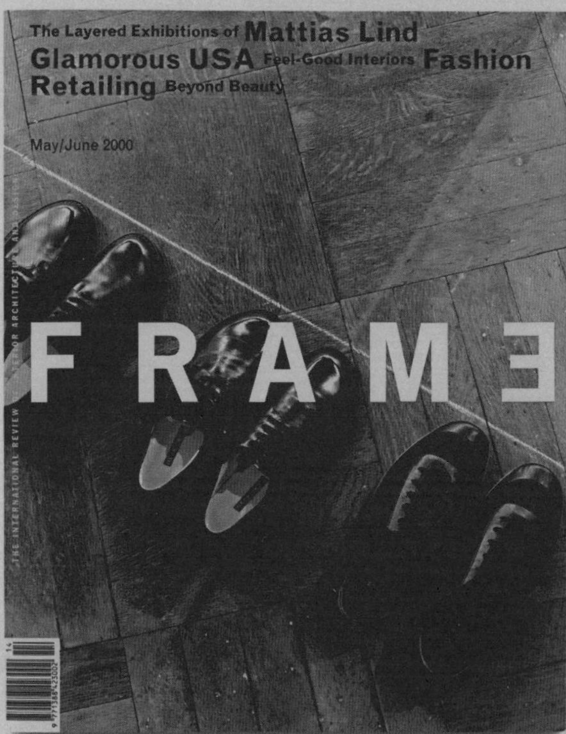

FAVOURITE COVER
Frame 14: It shows a fragment of an Ann Demeulemeester shop in Belgium. All she did was draw the line and set the shoes next to it. Super.

Everything, including the graphics, depends on the commitment of those who are making the magazine. If you don't have that, you don't have a magazine.

Are there rules for the graphic design of a magazine that you applied to *Frame*?

It's important to establish the identity of the publication. That goes for the design of every magazine. Clarity and uniformity are also important to the reader. Articles that are related to one another in terms of theme or subject, or features that are part of every issue, have to be designed as such. What's vital here is the recognition factor. But not all rules are incontrovertible. For example, I once put the Frame logo at the centre of the cover instead of at the top. Designing a magazine like Frame can be compared to making a painting. Articles and illustrations are colours on the palette. And I had to make Frame into a good painting, even knowing that a clear picture of how things should look would emerge only much later in the process. Again, it was usually discussion and experimentation that led to the obvious result. But everything, including the graphics, depends on the commitment of those who are making the magazine. If you don't have that, you don't have a magazine.

How would you describe your three years with *Frame*?

They were fantastic. But I'm not sure that I'd jump in head first all over again, or that I'd choose to go to such depths for a magazine. Frame comes out six times a year, and you work at least six months a year on those issues. At the moment I like juggling several balls at one time: books, retail interiors, exhibitions, printed matter. On the other hand, if someone were to approach me with an interesting proposal for a magazine… Who knows? I might do it all over again. During the Frame years, my business partner Marty Schoutens was incredibly important to all that was happening. I discussed everything with her and still do. She guides and motivates me. We read each other's thoughts at a glance. We agreed that after three years it was time for someone else to step into the shoes of art director – someone with different ideas and a different style. That's good for a magazine.

And now, more than eight years later, when you spot *Frame* in the shop what goes through your mind?

To begin with, I think it's just terrific that what we once created is still going strong. And I think that Frame's strength and its longevity lie in clear ideas – clear, but not stringent or obligatory – about how to make a magazine. To my knowledge, there's no better magazine in this genre.

What do you see in the future for *Frame*?

Perhaps Frame can develop activities outside the boundaries of the magazine itself. The resultant cross pollination would be smashing.

Any personal wishes for the future?

I'd like to make more books, because I love books. But I also like the great diversity of the work we're doing now. Designing is investigating and searching for forms. That's what made the creation of Frame so good, and it's that feeling that I look forward to in whatever work I do in the future.<<

TEXT
Eva Schaap

PHOTOGRAPHY
Krista van der Niet

In early 2001 Ton Homburg passed the baton to his successor, art director *Roelof Mulder*, who renewed the image of the magazine. Mulder's first project was *Frame* 18 and his last was *Frame* 29, the final issue of 2002.

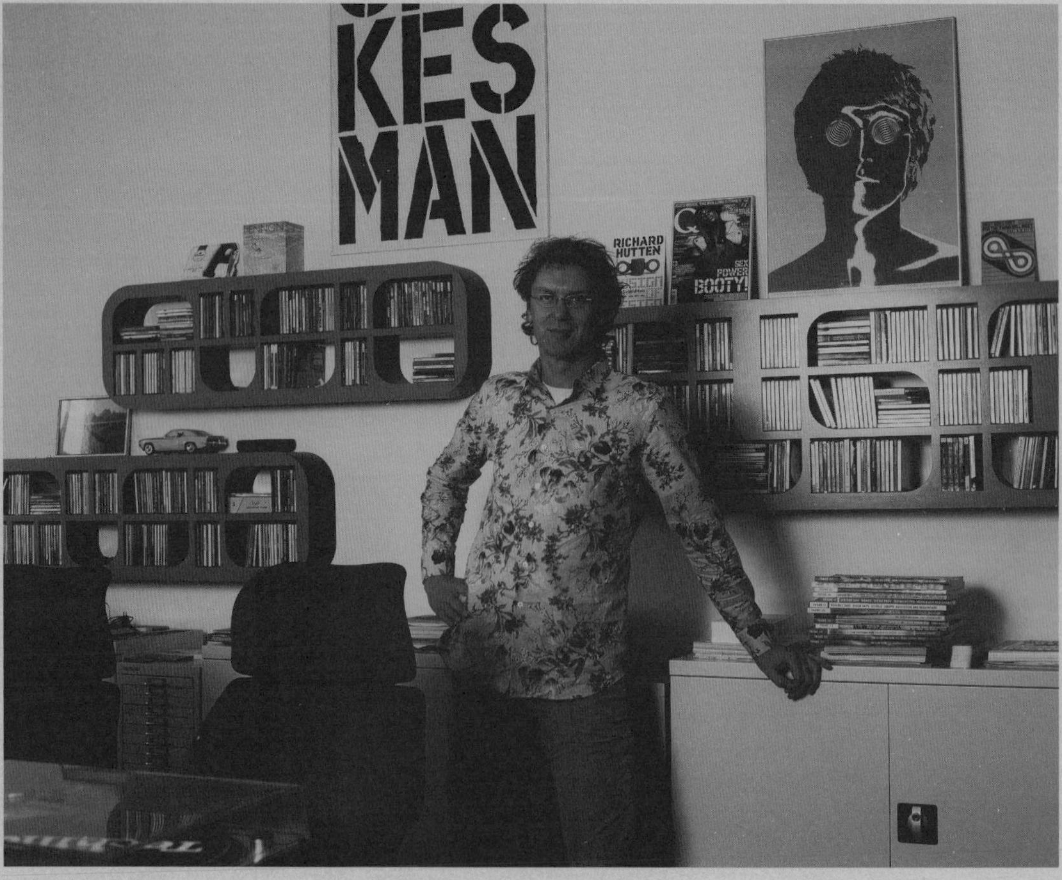

18-29

Roelof Mulder

AMSTERDAM

BACKGROUND
Roelof Mulder (1962) works as a freelance designer in Amsterdam. At the end of 2005 he completed the design of a logo and website for Platform 21, a new centre of design, fashion and creativity in Amsterdam. Mulder designed the visual material for Richard Hutten's exhibition at the Sieboldhuis in Leiden. He is also involved in the design of several books.

After Opera's Ton Homburg had determined the face of *Frame* for three years, it was your turn. Was it hard to assume the position of art director for an existing magazine?

When a magazine already exists and it's your job to take it forward, it's extremely difficult, particularly at the beginning. It takes a while before you can shake the idea of responding to the work of your predecessor or even to continue in the same direction. During my first year as art director for Frame, I felt as though it was my job to reinvent the wheel. After making things simpler for myself, that feeling subsided quite a bit during the second year. Consequently, my second year with Frame was better than the first. Furthermore, when I started at Frame the whole magazine market had collapsed, so I was hesitant to develop my ideas out of fear that Frame would suddenly stop selling or that it would be too hard for people to get used to my style of art direction. Because Frame has had three art directors to date, I can imagine that the current team [Koehorst in 't Veld] has a clearer realization that doing something entirely different is permissible. Not only that. An art director with several predecessors also understands better that the evolution of a magazine involves change and that art direction contributes to that change, and rightly so.

Cover, illustrations and the overall look of a magazine are what determine whether or not you pick up an issue and page through.

In your work for *Frame*, you introduced a specific font and design for each article, often basing your choices on the content of the piece and the accompanying photographs. Why this approach?

What I wanted, quite literally, was to add something to *Frame* that would give it a more experimental quality, something I believed went hand in hand with the content and subject matter of *Frame*. I thought that *Frame*'s experimental air should be reflected in the design of the magazine. When the experimental essence of content and design come together, the result is a unique and highly dynamic identity. Furthermore, the result of your art direction – along with *Frame*'s content and visual material – is an entity that appeals to a target group largely composed of creatives.

My approach to the design of *Frame* had never been seen within that genre of magazines before. In retrospect, the method was extremely time-consuming and labour-intensive. What's more, at that time it was not possible for the lithographer to change high-resolution images into digitally stored low-resolution material, which is easier to work with. The absence of that sort of technical tool meant that I needed a lot of time. A few issues of *Frame* clearly show just how pushed for time I was. And lack of time also had an impact on the cover, which I made only after I'd done the inside work.

Speaking of the cover, you often opted for a distinctive design, and you didn't always use a photo from one of the articles in the magazine. Why?

Book designers often begin with the cover and then come up with a concept for the contents. For *Frame*, I purposely did just the opposite. When everything inside the magazine has been completed, you know that particular issue inside out, and that's the moment to come up with a good cover design. Look at the cover of one of my favourites, *Frame* 25 [March/April 2002] and you'll see what I mean. I created an image especially for that issue, and that wasn't the only time I did that. Sometimes I used a photograph from one of the articles, but usually I conceived the cover design myself or made a compilation of various visual elements that related to one or more themes in the magazine. After all, a compiled or purpose-designed image can also be a good indicator of the contents.

What's it like to look back on your time with *Frame*?

When I look at my own work, I see that some things are good and others less good. But I'm not bothered about the mistakes I made. They're part of my attempt to try different things and to give Frame a more experimental aura. If you approach your work with a inquiring mind, I think that's always good, even if you don't reach your goal. At least you've tried. And that's still my belief. I learned a great deal in my years as art director for Frame. Even though it was labour-intensive, time-consuming work, I'd do it all over again, if only for the interaction that emerges between the various issues that you've made and the one you're working on at the moment. What you already did in a certain way you do differently in the next issue, either because it didn't work the first time or because you want to continue doing the things that turned out right. It's the sort of development that you rarely experience when you're working on individual projects.

What do you think of *Frame* now?

When I was the art director, I think it was more of a creative platform. I know that certain people really sat around waiting for the next issue, wondering what Frame would be doing this time around. In retrospect, I can say that maybe my work was a bit too experimental. But the design of a magazine should be surprising and exciting. Because cover, illustrations and the overall look of a magazine are what determine to a large extent whether or not you pick up an issue and page through it the next time you're in the shop. And I'd like to see some more of Frame's experimental attitude towards the design of the magazine. <<

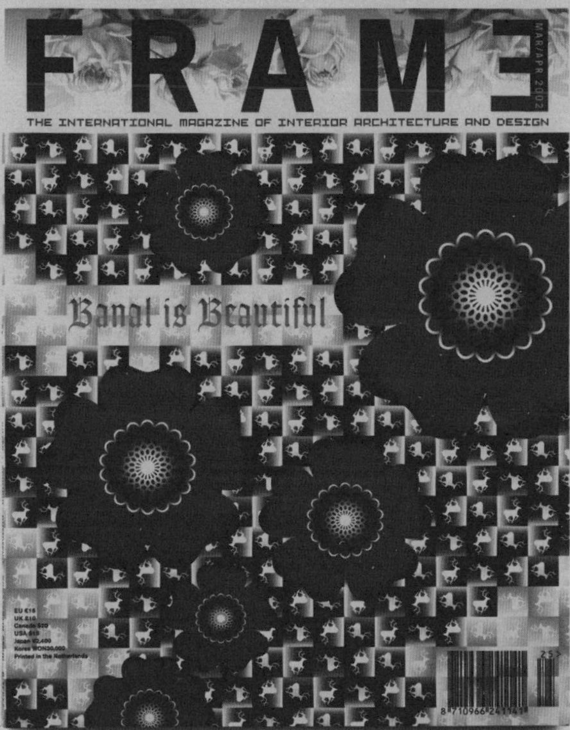

FAVOURITE COVER
Frame 25: Look at the cover of one of my favourites, and you'll see what I mean.

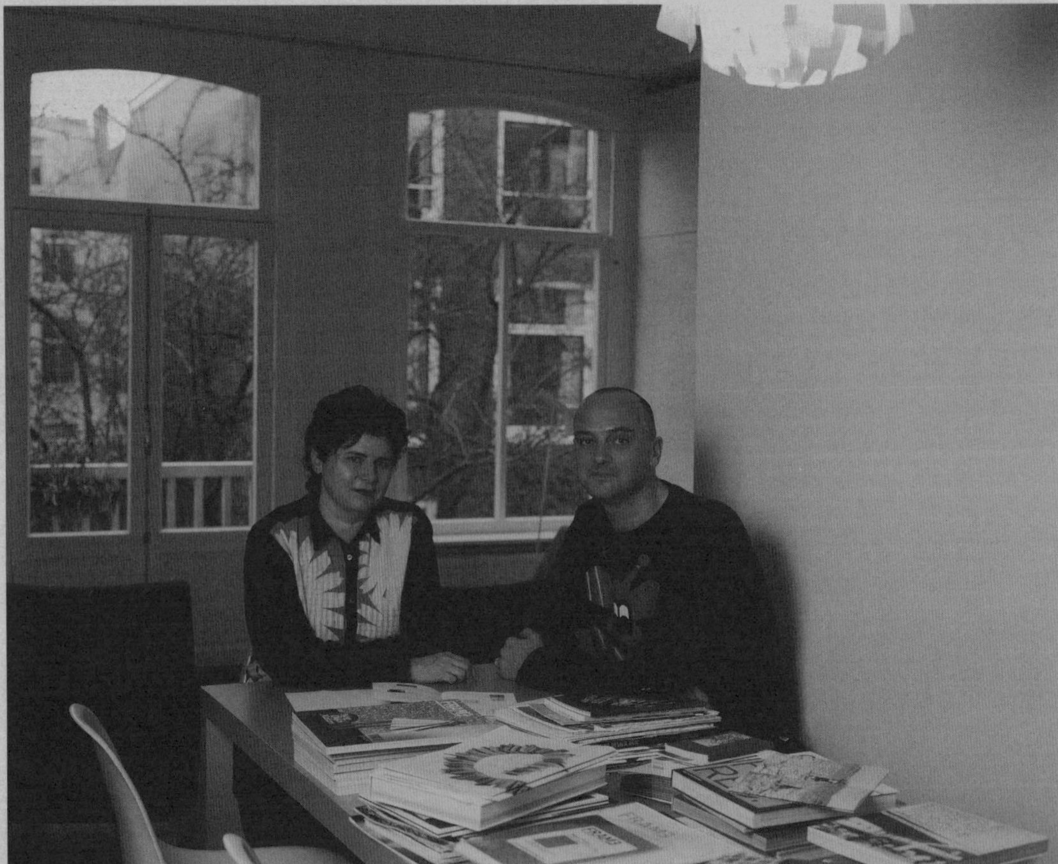

TEXT
Eva Schaap

PHOTOGRAPHY
Krista van der Niet

Frame 30, published in January 2003, marked the debut of *Cornelia Blatter & Marcel Hermans – COMA* – as the magazine's new team of art directors and designers. With the publication of *Frame* 48 in December 2005, they stepped aside as Jannetje in 't Veld and Toon Koehorst took over where COMA had left off.

30-47

Cornelia Blatter & Marcel Hermans

COMA, AMSTERDAM & NEW YORK

BACKGROUND

Established in 1996, COMA consists of Cornelia Blatter (1960) and Marcel Hermans (1961). Their portfolio includes art direction and design for magazines, books and exhibitions. Cornelia Blatter studied art at Yale University School of Art in the US, as well as at the 'Hochschule für Gestaltung und Kunst' in Switzerland. Marcel Hermans graduated from the Rietveld Academy in Amsterdam. Currently engaged in the design of several books, and the 2006 children stamp for the Dutch KPN, COMA was recently invited to participate in the National Design Triennial 2006 at the Cooper Hewitt Museum in New York. Blatter and Hermans teach at various schools.

In view of the distinctive art direction of your predecessor, Roelof Mulder, when you started working for *Frame*, did you have a strategy?

Robert and Peter didn't give us a list of requirements or a fixed concept. They simply asked us for our opinion of Frame and for our ideas on what could be changed and on how to make such changes. It was their way of getting us closely involved, right from the start, in the overall appearance and appeal of Frame. We had a picture of how Frame had evolved through the years and of what we hoped to add to that picture. We wanted to respond to Roelof Mulder's art direction with a different approach, the very approach that eventually came to characterize our design of Frame: instead of giving each article an individual treatment, we opted for a comprehensive concept that addressed the contents of the magazine as a whole. A rather unusual approach to the art direction of a periodical.

What distinguishes COMA's work for *Frame*?

Each issue of Frame has a different design. Cues for designing individual issues come from the content; as design in interior architecture changes, so does the visual logic of Frame magazine. Designing Frame is an ongoing search for the perfect shape, the ideal form. Readers and audiences witness this search, experiencing the magazine as design-in-progress, always in flux.

Did you have a particular reason for your all-in approach to the design of each issue?

One of the main tasks of the art director is to knead the richness of words and images that make up the magazine into a whole. It is also the hardest part of designing a magazine. By creating a basic look for each issue, you turn Frame into a magazine with a difference. Each issue has an 'overall design', and, as a result, all articles of the same genre begin, in terms of graphic design, in the same way. Each issue, though different from the last, has a continuity – and thus a sense of uniformity. The issues of Frame that we designed radiate serenity, without exemplifying classical art direction. The magazine's articles and illustrations offer all the diversity required.

How did you approach the design of an individual issue?

At the beginning of each new issue, we met with the editors to take a general look at the contents, to discuss the articles and to help select the images. That was often all we needed to spot a certain trend or theme that would serve as an inspiration for our basic design of the magazine. Take Frame 33, for instance, which featured several articles with a focus on transparency. A number of the projects described included coloured panels made of glass or plastic. It made us think of wearing tinted glasses, and we tried to reflect that feeling in our design. That's one example of how we translated the developments reported in Frame into the look we created for the magazine. And since Frame comes out every two months, it gives you the opportunity to experiment. Sometimes it works and sometimes it doesn't, but each new issue represents another chance. Another chance to make something that the reader thinks is terrific – or less successful. The method also guarantees that discussions with the editors are not only going to be about details like fonts and the use of colour. Every exchange of ideas focuses on the magazine as a whole. Does it say something? What does it say? Art directors and their clients should be discussing questions like these and not quibbling about which font or colour looks best.

Seen in retrospect, are there things that gave you less satisfaction than others?

We're not particularly happy about the design of Frame 30, the first issue we worked on. Which isn't so strange. That initial ho-hum issue is an inevitable part of the process. You're not going to be perfect from the outset. The next issue is a chance to correct those mistakes. Designing a magazine has its ups and downs.In the middle of working on an issue, you often have a strong feeling about the success or failure of what you're doing. Then, too, those things that go hand in hand with your personal preferences are going to turn out better. Nevertheless, you can't let personal preferences influence the design of 18 issues of a magazine. There are readers and an editorial staff to think about as well.

One of the main tasks of the art director is to knead the richness of words and images that make up the magazine into a whole.

—

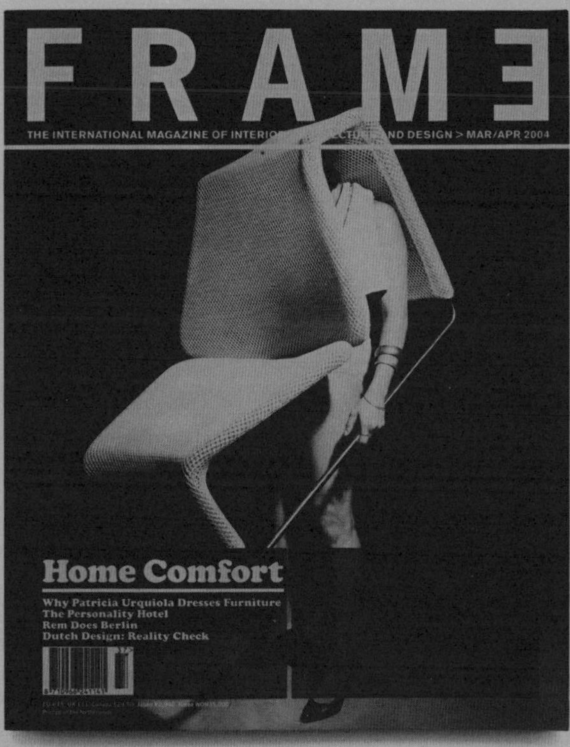

FAVOURITE COVER
Frame 37: We are definitely partial to covers with a clear, clean image.

Favourites?

Frame 43, which featured an article on the work of French designers M/M. The piece describes a project in which graphic designers and artists work together in one room. The result was a new kind of interior – a development that marked the moment and that we couldn't possibly ignore. By way of exception, we used a number of highly striking fonts and graphic layouts, giving each article a different look. The outcome was a very dense, very compact issue. As far as covers are concerned, we are definitely partial to Frame covers with a clear, clean image, such as the cover of Frame 37. And Frame 38. Though not our favourite, the cover of Frame 34 isn't so bad. It took all our powers of persuasion, but in the end we got to make it. Another cover worth mentioning is that of

In the middle of working on an issue, you often have a strong feeling about the success or failure of what you're doing.

Frame 42, an issue that is no longer available; it's a lucid image composed of various elements that don't clash but are fascinating to look at. 'Fascinating' also describes Frame 45's clown cover, which diverts attention from what Frame is all about, while making you want to know why a design magazine has a clown on its cover. Frame 46, which features artist Yayoi Kusama on the cover, is a good example of how to integrate graphics into a photograph; Kusama is the focus of the photo, but a graphic translation appears in the background and throughout the issue. Last but not least, we like the powerful styling and photography of the work of Spanish designer Patricia Urquiola, through the lens of Viviane Sassen (Frame 37) and Marcel Wanders' Lute Suites for Frame 44. After all, architecture and interiors are not sterile, slick and vacant; they exist by the grace of the user. It's true that people are not part of these photos, but the traces they leave are plain to see. Subtle yet strong. Words that apply to Frame as well.

So you wouldn't describe the art direction of *Frame* as a piece of cake?

No way. Despite all the enjoyment we got out of it, the job demanded a great deal of our time. Sometimes, when a month and a half had passed and the first articles for a new issue began to arrive, we groaned: 'Oh, no. Here we go again.' At other times our response was: 'Great! Here we go again.' It's not at all like working on a book or an exhibition, projects that are finished after a certain period of time and that's it. We developed a relationship with Frame. What was important to us and what the editors wanted to see in the magazine became more and more evident. It is a constant process of collaboration between the editorial content and it's visual interpretation.

Do you apply certain rules to the art direction of a magazine like *Frame* – or to magazine work in general?

Rules for art direction don't exist. If you base your work on rules, you exclude so many possibilities. In terms of concept, the two of us approach design in the same way, but Marcel, who studied graphic design, brings that background into our work, whereas Cornelia, who's an artist, has a more open attitude to what we do.

Would you consider doing the art direction for a magazine today?

Yes. Without a doubt. Not only did we learn a lot from the experience, but an ongoing project, coupled with worldwide exposure, proved very beneficial for COMA. We became so involved in making the magazine that it was almost as if we were part of the Frame family.

You also designed *Frame*'s stationery, business cards, posters and media kit. Do you see a relationship between these activities and the work you did for the magazine itself?

Robert mentioned the need for Frame stationery and business cards, but what would they look like? How could they reflect the idea of a magazine on interior design? We solved the problem by literally framing and wall-papering Frame's cards and stationery. Each employee's business card is distinguished by its own wallpaper pattern and colour. We designed stationery in various colours and patterns. The media kit and book ads are based on the same concept. Here, too, we put the cover of the book inside a frame. All in all, it's a matter of unity in diversity. Just like the magazine. <<

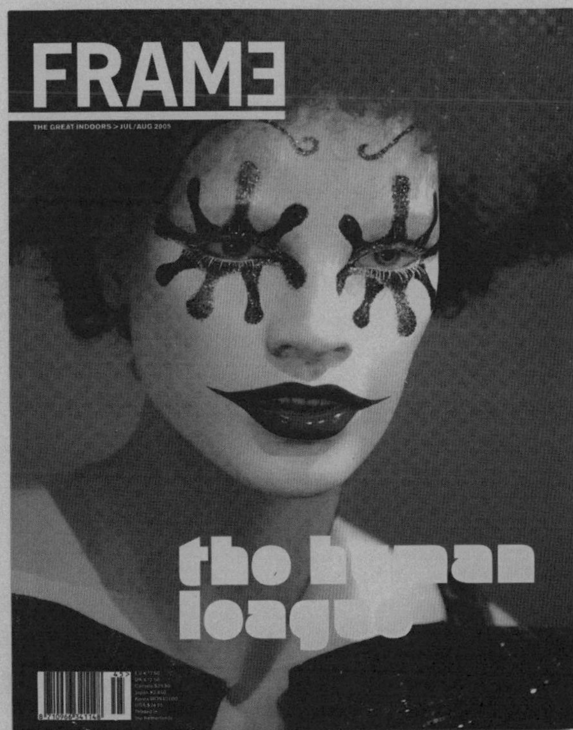

FAVOURITE COVER

Frame 45: The clown cover, which diverts attention from what Frame is all about, while making you want to know why a design magazine has a clown on its cover.

TEXT
Eva Schaap

PHOTOGRAPHY
Krista van der Niet

As we go to press, *Jannetje in 't Veld & Toon Koehorst* are *Frame*'s new art directors. In late 2005, with the publication of *Frame* 48, they took over the reins from Cornelia Blatter and Marcel Hermans of COMA, who had fulfilled the task for the three preceding years.

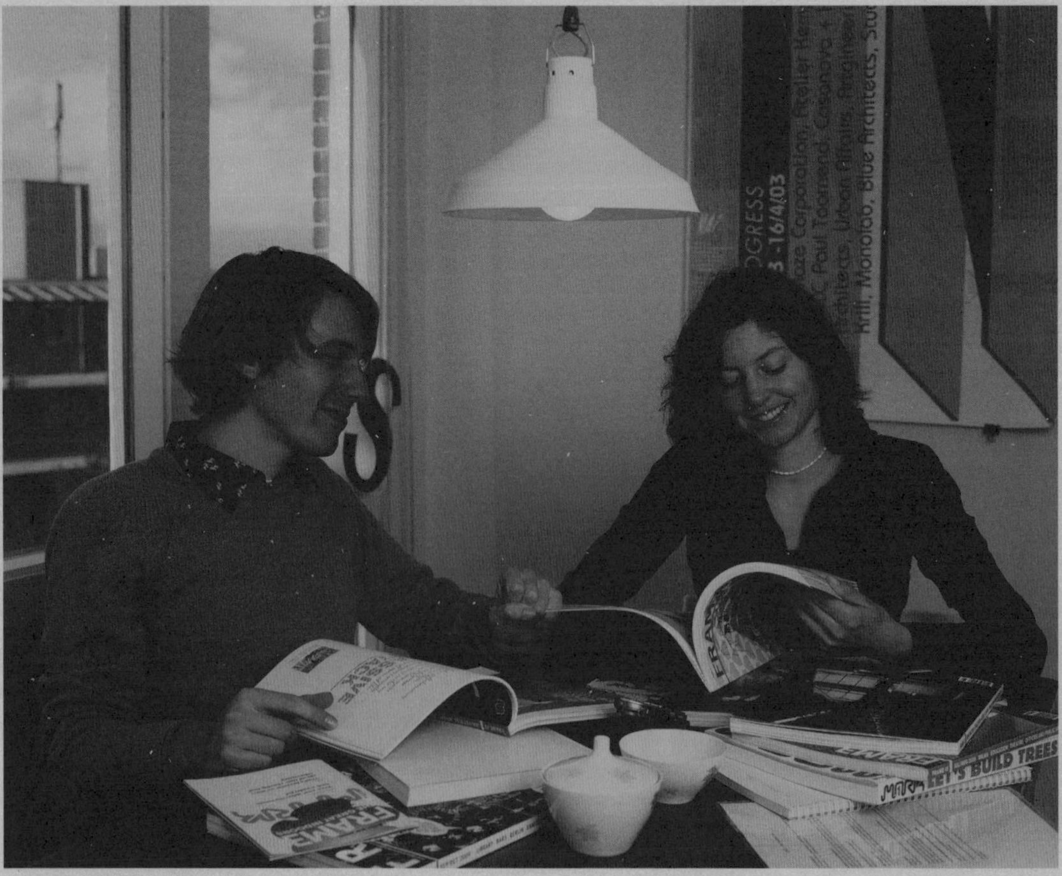

47-49

Jannetje in 't Veld & Toon Koehorst

KOEHORST IN 'T VELD, ROTTERDAM

BACKGROUND

Design duo Jannetje in 't Veld (1980) and Toon Koehorst (1981) earned their degrees in graphic design from the Academy of Art in Arnhem. They began working as freelance designers in September 2002. Prior to becoming art directors for Frame at the end of 2005, they had designed, among other things, four issues of Dutch magazine Sloom, sloom.org and Keikosato.nl, the website of Japanese artist Keiko Sato. In addition, they both teach graphic design at the Willem de Kooning Academy in Rotterdam.

Your predecessors – those whose work appeared in *Frame* prior to 2006 – were art directors who'd made a name for themselves in the profession. Stepping into their shoes must have posed quitea challenge.

TK: **Yes. Fortunately, though, we weren't completely green at the outset. First we were asked to make Frame's website, and in the process of doing that, we had an opportunity to get acquainted with Frame, the staff and the ideas that make the magazine.**

JV: **In terms of art direction, we started off slowly, looking at what our predecessors had done and determining what we thought had and hadn't worked. We made no attempt to ignore the rich history of Frame and the strong points of the magazine's earlier art direction. It would have been a pity to toss that work on the rubbish heap rather than to integrate it into plans for our own design.**

TK: **While working on the website, we were already looking into what Frame really is, what's possible with Frame and what we thought it could be. We based our approach to the website on a kind of recycling principle. And why not? Just look at all that exists already and at the wealth of information that's added every two months. Whether you're making a website or a magazine – those are not things you can disregard or pitch overboard.**

We believe Frame's strength lies in the profusion of illustrations and in their excellent quality.

What did you find in earlier issues that you were eager to bring back?

JV: Looking through old Frame magazines, we discovered elements in nearly every issue that worked well but were nowhere to be found in the following issue. That's really too bad. We'd like to prevent that from happening.

TK: We believe Frame's strength lies mainly in the profusion of illustrations and in their excellent quality. Those are the assets we want to keep and to emphasize. We've made a deliberate decision, however, to use the same font in every issue, in order to create uniformity not only within a single magazine, but also from issue to issue. If we've done that for several issues and it doesn't seem to be working, we'll make a change. No problem.

JV: It's also good to know that Frame's art directors have always worked for the magazine on a freelance basis. Just as our predecessors, we'll have the freedom to work relatively independently.

TK: What's more, it's a system that keeps both sides on their toes. For Frame, working with contracted freelance designers has been a determining factor to date. It's Frame's way of allowing art directors to give the magazine an identity and to shape its charismatic character. Other magazines don't take the same tack. In most cases, a magazine has one or more in-house art directors that are part of the editorial staff.

Do you like a magazine with a highly distinctive look?

JV & TK: Absolutely!

TK: Publishers can be somewhat hesitant when it comes to a magazine that stands out from all the rest, whereas art directors and other designers really appreciate the extraordinary. You can't be too anxious. Experimentation is good. Unlike a book, a magazine is not a one-off. It seems to me that you can go all out in creating a look for one issue and take a calmer approach the next time around. It's a good way to heighten the contrast between issues without sacrificing that 'Frame feeling' and the desired sense of uniformity. One of Frame's main characteristics – something that makes it striking and exceptional – is that every issue proves once again that you can tell stories with images and design. By maintaining that strong point and by integrating new elements into our work for Frame, we're trying to add something to the magazine.

Do you have a favourite issue of *Frame*?

JV: Frame 42 is my favourite. The design of the cover is great. There's a lot happening in that picture. And the contents also make it a very interesting issue. I like the cover of Frame 41 too. When I look at Roelof Mulder's covers as a series, I'm impressed.

TK: Roelof's covers always had a big impact at the academy. It was as though everyone who saw the latest issue had the same response: 'Oh, wow! So you can do that too.' Even though we were learning just the opposite, he showed us, for instance, that colours could not only run together but also lead to a nice result. What I like about his covers is the layered composition. And even though he used a minimum of text, the message is clear. Each cover is a design in itself. I think it's Roelof's covers that have made Frame such a big hit with designers, art directors and other creatives. From the moment he arrived, Frame counted in those circles.

Now that you're working for *Frame*, do you look at the magazine through different eyes?

JV: I do. I must admit that it hadn't been my favourite magazine. I paged through the latest issue, but I didn't make a special trip to the newsstand to buy it. Toon did, though. It was always the first magazine he grabbed off the shelf. Toon was the fan. Not me. Slowly but surely, however, Frame has become a source of inspiration. Even little things like patterns or colourways can be enough to stir my imagination.

TK: Descriptions of materials are also interesting. Especially the ones that we, as graphic designers, cannot apply to our work. While I was still at school, I was always waiting for the next issue of Frame to come out. And the best thing of all was to turn immediately to the product pages at the back, because that's where you'd find what was new or different. Sometimes I'd spot only one product that really jumped out at me because of its shape or design. That was a moment of pure pleasure. Now we're designing those pages ourselves, and that makes it all the more enjoyable. <<

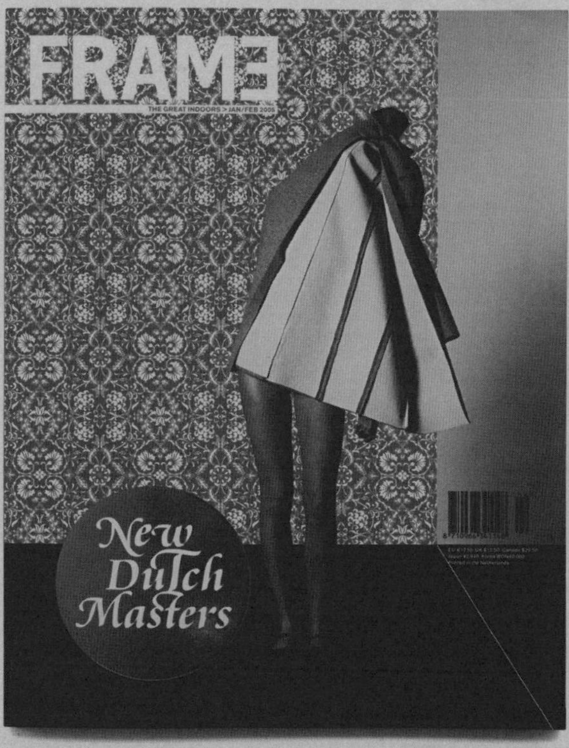

FAVOURITE COVER

Frame 42: This January/February issue is my favourite. There is a lot happening in that picture.

Feed Back

FRIENDS & ENEMIES

We delved into our archives and found your letters. You loved the magazine, you hated it, but you never stopped buying it. Thank you!

Congratulations on the new layout, which I regard as a 100-per-cent improvement over the former issues.

JEFFERY, ARCHITECT, 2002

What I like about Frame is the focus on leading-edge design ideas.

WENDY, READER, 2000

I want to commend you on your wonderful publication. For the past few years I have been reading Wallpaper. Only recently did I find your magazine, and I have to say that you blow them out of the water. No fluff or frills or unneeded pretentiousness. Just good design.

JAMIE, FRAME FAN, 2002

Frame offers a good mix: those who like looking at pictures can look at pictures, while those who want to read about the latest developments in the world of design are certainly not overlooked. Frame sees design as a big party for the entertainment of as many people as possible.

JEROEN JUNTE , JOURNALIST FOR DE VOLKSKRANT, 2000

Put me back on your list of subscribers after Frame awakens from its COMA.

ANONYMOUS, 2004

I do not want to miss an issue of Frame between subscriptions, as this would severely piss me off.

ROSS, SUBSCRIBER, 2001

Visually and design-wise, Frame is stunning and elegant. The content and editorial line-up are excellent. The articles and profiles go beyond the expected and traditional approach.

PATRICK, FRAME FAN, 2000

P.S. Change the graphic designers.

ARNE, DESIGNER, 2002

But it's not all about big names and prestigious projects. A financially stretched dotcom outfit in California that uses ocean containers to divides a huge warehouse into individual office units is equally interesting.

JEROEN JUNTE, JOURNALIST FOR DE VOLKSKRANT, *2003*

The issue of Frame 04 devoted to 'The Senses' was stunning – really good. People all over the world are calling to ask for the same publication.

LARS, DESIGNER, 1998

Those who really want to keep up to date with contemporary design are better off reading a specialist journal like Frame.

HANS DEN HARTOG JAGER, JOURNALIST FOR NRC HANDELSBLAD, *2003*

From a writer's perspective, I have found myself explaining what a good forum you provide for more than a purely descriptive style coverage on design, etc.

LAURA, WRITER, 1999

I'm desperate because I can't find your great magazine in stores since April. I can't live without it. Please help me.

PABLO, FRAME *FAN, 2000*

The positive approach (not entirely without a critical note) in a couple of the articles is uplifting. The composition is optimistic. You're making a good magazine.

ED, DESIGNER, 1999

I hereby cancel my subscription to Frame. I object to the cluttered layout of the magazine. So many different fonts and quirky tricks for presenting photography.

KLAARTJE, SUBSCRIBER, 2001

I hereby cancel my subscription to Frame. I object to the cluttered layout of the magazine. So many different fonts and quirky tricks for presenting photography.

KLAARTJE, SUBSCRIBER, 2001

Just to say . . . thank you for the relevance and the quality of the article. We are delighted.

PROTO DESIGN, DESIGN FIRM, 1999

I'm enthusiastic about the quality of the article and of the magazine itself, which represents something really new in the old world of architectural publications (and I'm not brown-nosing).

FABIO, DESIGNER, 1999

My exposure to the projects and ideas presented in your magazine has had a great impact on the progress of my own projects.

JILL, DESIGNER, 2000

Wow! What an amazing story you wrote about me. I am so honoured. I am thinking of all the things I should have said ... I could have used the opportunity to say something really deep and meaningful – to change the way the world thinks. I guess I'm just kind of simple and straight up. I feel so small and uninteresting compared with the others you interviewed. I read some those interviews today and really enjoyed them.

LINO, DESIGNER, 2001

I am not at all happy with the issue. After we had spent so much time and effort on the text, [I was disappointed] to see such a totally out-of-control layout (in my opinion), in which coloured drawings, coloured photographs and colour-applied text are combined.

LOUISA, DESIGNER, 2001

I love it. I love the look of it, the colours, the quality of both contents and pictures. A very happy and thankful devotee of Frame.

FRANCO, DESIGNER, 2002

Put me back on your list of subscribers after Frame awakens from its COMA.

ANONYMOUS, 2004

As a design teacher, I am recommending Frame as the design bible for all students here.

JORG, FRAME FAN, 1999

What on earth were you thinking of – using what looks like Monotype Century Schoolbook Bold and Bold Italic for your headings and subheadings?

ROGER, SUBSCRIBER, 2003

P.S. Change the graphic designers.

ARNE, DESIGNER, 2002

I was very sad to see, upon opening the magazine, that your graphic-design style has changed significantly from previous issues I've read. I don't know if this has happened just for this issue … I hope so, because it is nearly illegible. As a magazine for designers, the one thing I would have thought you'd want to get just right is the graphic design! Making all the vowels stand out (by making them bold or not bold) is not useful in any way; it is style over substance and, in my opinion, doesn't look very stylish either! I look forward to receiving the next issue with, I hope, a more useful graphic style.

EDWARD, SUBSCRIBER, 2005

Congratulations on the new layout, which I regard as a 100-per-cent improvement over the former issues.

JEFFERY, ARCHITECT, 2002

Frame's glossy appearance suggests the same sort of eager approach to design that design receives from the modern consumer.

JEROEN JUNTE, JOURNALIST FOR DE VOLKSKRANT, 2000

Trace Back

INDEX & PICTOGRAMS

Our favourite bars, hair salons, libraries, restaurants, workplaces and designers brought together in a handy index.

Interior Index

Interior Index

Designer Index

Designer Index

Country Index

592-000

Colophon

Frame: The Back Issue

THE ESSENTIAL GUIDE TO FRAME'S FIRST 50 ISSUES

PUBLISHERS
Frame Publishers
www.framemag.com

Birkhäuser –
Publishers for Architecture
www.birkhauser.ch

BOOK CONCEPT
Merel Kokhuis

TEXTS
Peter Huiberts, Merel Kokhuis, Eva Schaap, Robert Thiemann

ART DIRECTION & DESIGN
Robin Uleman, www.robinuleman.nl (with thanks to Harry Huybers and Sandra Rabenou)

TRANSLATION
InOtherWords
(Donna de Vries-Hermansader)

COPY EDITING
D'Laine Camp,
Donna de Vries-Hermansader

COVER & SPREAD PHOTOGRAPHY
Wouter Stelwagen

PHOTOGRAPHY
Krista van der Niet

COLOUR REPRODUCTION
Graphic Link, Nijmegen

PRINTING
D2Print, Singapore

DISTRIBUTION
ISBN-10: 90-77174-50-8
ISBN-13: 978-90-77174-50-0
Frame Publishers
Lijnbaansgracht 87
1015 GZ Amsterdam
the Netherlands
www.framemag.com

ISBN-10: 3-7643-7652-X
ISBN-13: 978-3-7643-7652-9
Birkhäuser —
Publishers for Architecture
Viaduktstrasse 42
CH-4051
Switzerland
www.birkhauser.ch

Printed on acid-free paper produced from chlorine-free pulp. TCF ∞
Printed in Singapore
987654321